Spon's
Mechanical and
Electrical Services
Price Book

2007

Spon's Mechanical and Electrical Services Price Book

Edited by

DAVIS LANGDON MOTT GREEN WALL
Engineering Services

2007

Thirty-eighth edition

Taylor & Francis
Taylor & Francis Group

LONDON AND NEW YORK

First edition 1968
Thirty-eighth edition published 2007
by Taylor & Francis
2 Park Square, Milton Park, Abingdon, Oxon OX14 4RN

Simultaneously published in the USA and Canada
by Taylor & Francis
270 Madison Avenue, New York, NY 10016

Taylor & Francis is an imprint of the Taylor & Francis Group, an informa business

Printed and bound in Great Britain by
TJ International Ltd, Padstow, Cornwall

Publisher's note
This book has been produced from camera-ready copy supplied by the authors.

British Library Cataloguing in Publication Data
A catalogue record for this book is available from the British Library

ISBN10: 0-415-39387-6
ISBN13: 978-0-415-39387-4
ISSN: 0305-4543

Contents

Electrical Installations

PART FOUR: RATES OF WAGES

Mechanical Installations

Electrical Installations

PART FIVE: DAYWORK

Preface

The Thirty Eighth edition of *SPON'S Mechanical and Electrical Services Price Book* continues to cover the widest range and depth of services, reflecting the many alternative systems and products that are commonly used in the industry as well as current industry trends.

In terms of current pricing levels, the continuing boom in the Chinese economy and it's consumption of raw materials has led to sharp increases in the price of steel and copper. Both steel and copper tube manufacturers are reporting monthly increases in prices, and whilst these have been incorporated at the time of going to press, readers are advised to check the currency of such prices before using them. Likewise, the sustained high cost of crude oil will impact on products that are oil based i.e. UPVC as well as transport costs. However, unlike last year, the picture is clearer concerning manufacturers and suppliers of plant and materials utilising steel and copper, with some increasing costs and lead times due to the increasing demand. Again, prices are current at the time of going to press, but readers are advised to check that such prices are still current before using them.

Before referring to prices or other information in the book, readers are advised to study the `Directions' which precede each section of the Materials Costs/Measured Work Prices. As before, no allowance has been made in any of the sections for Value Added Tax.

The order of the book reflects the order of the estimating process, from broad outline costs through to detailed unit rate items.

The approximate estimating section has been thoroughly reviewed to provide up to date key data in terms of square metre rates, all-in-rates for key elements and selected specialist activities and elemental analyses on a comprehensive range of building types.

The prime purpose of the Materials Costs/Measured Work Prices part is to provide industry average prices for mechanical and electrical services, giving a reasonably accurate indication of their likely cost. Supplementary information is included which will enable readers to make adjustments to suit their own requirements. It cannot be emphasised too strongly that it is not intended that these prices should be used in the preparation of an actual tender without adjustment for the circumstances of the particular project in terms of productivity, locality, project size and current market conditions. Adjustments should be made to standard rates for time, location, local conditions, site constraints and any other factor likely to affect the costs of a specific scheme. Readers are referred to the build up of the gang rates, where allowances are included for supervision, labour related insurances, and where the percentage allowances for overhead, profit and preliminaries are defined.

Readers are reminded of the service available on the Spon's website detailing significant changes to the published information. www.pricebooks.co.uk/updates

As with previous editions the Editors invite the views of readers, critical or otherwise, which might usefully be considered when preparing future editions of this work.

While every effort is made to ensure the accuracy of the information given in this publication, neither the Editors nor Publishers in any way accept liability for loss of any kind resulting from the use made by any person of such information.

In conclusion, the Editors record their appreciation of the indispensable assistance received from the many individuals and organisations in compiling this book.

DAVIS LANGDON MOTT GREEN WALL
Engineering Services
MidCity Place
71 High Holborn
London WC1V 6QS

Telephone: 0207 061 7777
Facsimile: 0207 061 7009

Copper Prices in the Services Industry

At the time of going to press, the cost of Raw Copper on the London Metals Exchange was still very volatile.

The cost of Raw Copper, increased from $3,800 per tonne in May 2005 to $4,000 in October 2005 and then over a period of time to $9,000 in May 2006. It was not until the latter part of 2Q/06 that there was any major impact on the cost of manufactured goods i.e. tube and cable. At which point, the increase was dramatic and on larger cables has been as high as 80%.

Subsequently, the Raw Material Cost has dropped to $7,475 per tonne at the time of going to Press (6 July 2006) and it now appears to be levelling out having been lower at $7,040 only 10 days prior.

However this is still double the cost of 2Q/05 and it is unlikely that there will be an immediate (if any) readjustment of manufacturing rates.

The current fall in raw materials cost is due partly to the market speculation that rising global interest rates will slow demand for the metal.

Copper prices also initially rose following concerns that supplies could be disrupted by strike action within mines in Mexico and Chile. However, long term, Investors and Traders are more concerned that the US Federal Reserve and other central banks will continue to raise rates to curb inflation and growth rather than with the prospect for a strike by workers at Chile's Escondida copper mine, the world's largest. 3.2 million Tonnes of copper production could be impacted upon by strike action over the remainder of 2006.

BHP Billiton, the world's biggest mining company and rival Producers may face disruption to as much as 18 percent of global copper supply as labour contracts come up for renewal.

Strong demand from China and India, combined with a lack of investment in new mining projects, has partly caused the surge in commodity prices. Although Customs have said that in the first five months of 2006, China imported 324,000 tonnes of refined copper, down 41.6 percent year on year, however, the building boom there is still being blamed for demand outstripping supply.

Standard Bank Plc stated in a Report on 28 June 2006 that according to charts some Traders use to predict price moves, that to "relieve the immediate downside threat" London copper must rise above $7,110 a tonne and pave the way for gains up to $7,500. As stated above, this happened in just over a week. This view of the market being in decline is an interesting and stark contrast to the Building Services industry view of the impact on finished materials.

Until stability is restored in the copper market, the impact and trend on factory production prices cannot be assured. It is the opinion of Financial Analysts that Traders are partly creating a false market, as until 2Q/06 there was no particular excessive demand or shortage. Until recently, prices could be fixed for twelve months; they are now trading on a day-to-day basis. The strength of the dollar is also impacting o such.

This edition can only reflect the market as it stands at the time of going to press. Therefore, users of the book are advised that the SPON'S M&E Price Book costs will need to be reviewed to establish movement, via the Free Quarterly updates at www.pricebooks.co.uk/updates.

Spon's International Construction Costs Handbooks

This practical series of five easy-to-use Handbooks gathers together all the essential overseas price information you need. The Hand-books provide data on a country and regional basis about economic trends and construction indicators, basic data about labour and materials' costs, unit rates (in local currency), approximate estimates for building types and plenty of contact information.

Spon's African Construction Costs Handbook 2nd Edition
Countries covered: Algeria, Cameroon, Chad, Cote d'Ivoire, Gabon, The Gambia, Ghana, Kenya, Liberia, Nigeria, Senegal, South Africa, Zambia
2005: 234x156: 368 pp Hb: 0-415-36314-4: £180.00

Spon's Latin American Construction Costs Handbook
Countries covered: Argentina, Brazil, Chile, Colombia, Ecuador, French Guiana, Guyana, Mexico, Paraguay, Peru, Suriname, Uruguay, Venezuela
2000: 234x156: 332 pp Hb: 0-415-23437-9: £180.00

Spon's Middle East Construction Costs Handbook 2nd Edition
Countries covered: Bahrain, Eqypt, Iran, Jordan, Kuwait, Lebanon, Oman, Quatar, Saudi Arabia, Syria, Turkey, UAE
2005: 234x156: 384 pp Hb: 0-415-36315-2: £180.00

Spon's European Construction Costs Handbook 3rd Edition
Countries covered: Austria, Belgium, Cyprus, Czek Republic, Finland, France, Greece, Germany, Italy, Ireland, Netherlands, Portugal, Poland, Slovak Republic, Spain, Turkey
2000: 234x156: 332 pp Hb: 0-419-25460-9: £99.00

Spon's Asla Pacific Construction Costs Handbook 3rd Edition
Countries covered: Australia, Brunei Darassalem, China, Hong Kong, India, Japan, New Zealand, Indonesia, Malaysia, Philippines, Singapore, South Korea, Sri Lanka, Taiwan, Thailand, Vietnam
2000: 234x156: 332 pp Hb: 0-419-25470-6: £180.00

To Order: Tel: +44 (0) 1264 343071 Fax: +44 (0) 1264 343005, or
Post: Taylor and Francis Customer Services, Thomson Publishing Services, Cheriton House, Andover, Hants, SP10 5BE, UK Email: book.orders@tandf.co.uk

For a complete listing of all our titles visit:
www.tandf.co.uk

Taylor & Francis
Taylor & Francis Group plc

Special
Acknowledgements

The Editors wish to record their appreciation of the special assistance given by the following organisations in the compilation of this edition.

T. Clarke plc

Electrical Engineers & Contractors
Stanhope House
116-118 Walworth Road
London SE17 1JY
Tel: 020 7358 5000
Fax: 020 7701 6265
e-mail: info@tclarke.co.uk
www.tclarke.co.uk

 HARGREAVES

DUCTWORK SPECIALISTS

Lord Street, Bury, Lancashire. BL9 0RG
Tel: 0161 764 5082 • Fax: 0161 762 2336
E-Mail: sales@senior-hargreaves.co.uk

comunica

Comunica plc
The Hallmarks
146 Field End Road
Eastcote
Pinner
Middlesex
HA5 1RJ

Tel: 020 8429 9696
Fax: 020 8429 4982
Email: enquiries@comunicaplc.co.uk
www.comunicaplc.co.uk

AXIMA

Axima Building Services
Westmead, Farnborough
Bournemouth, Hants GU14 7LP
Tel: 01252 525500
Fax: 01252 378988
www.axima.eu.com

HOTCHKISS

Hampden Park Industrial Estate
Eastbourne
East Sussex
BN22 9AX
Tel : 01323 501234
Fax : 01323 508752
E-Mail : info@Hotchkiss.co.uk
www.Hotchkiss.co.uk

Abbey

ABBEY THERMAL INSULATION LTD.
23-24, Riverside House,
Lower Southend Road, Wickford, Essex SS11 8BB
Telephone: 01268 572116- Facsimile: 01268 572117
E-mail: general@abbeythermal.com

J. & M. Insulations Limited.

For all aspects of Thermal Insulation & Fire Protection

257a Banbury Road,
Summertown, Oxford, OX2 7HN
Telephone: 01865 310220
Fax: 01865 512419
E-mail: jandminsulation@fsbdial.co.uk
www.jandminsulations.co.uk

Acknowledgements

The editors wish to record their appreciation of the assistance given by many individuals and organisations in the compilation of this edition.

Manufacturers, Distributors and Sub-Contractors who have contributed this year include:-

A C Plastics Industries Ltd
Armstrong Road
Daneshill East
Basingstoke RG24 8NU
GRP Water Storage Tanks
Tel: 01256 329334
Fax: 01256 817862
www.acplastiques.com

Alfa Laval Limited
Unit 1, 6 Wellheads Road
Farburn Industrial Estate
Dyce
Aberdeen AB21 7HG
Heat Exchangers
Tel : 01224 424300
Fax : 01224 725213
www.alfalaval.com

Aquilar Limited
Dial Post Court
Horsham Road
Rusper
West Sussex RH12 4QX
Leak Detection
Tel : 08707 940310
Fax : 08707 940320
www.aquilar.co.uk

Arrow Electronics
Edinburgh Way
Harlow
Essex CM20 2DF
Cables
Tel : 01279 441144
Fax : 01279 455704
www.arrowne.com

Axima Building Services
80 Paul Street
London EC2A 4UD
Above Ground Drainage
Tel : (020) 7729 7634
Fax : (020) 7729 9756
www.axima-uk.com

Balmoral Tanks
Balmoral Park
Loirston
Aberdeen AB12 3GY
GRP Water Storage Tanks
Tel: 01224 859000
Fax: 01224 859123
www.balmoral-group.com

Braithwaite Engineers Ltd
Neptune Works
Uskway
Newport
South Wales NP9 2UY
Sectional Steel Water Storage Tanks
Tel: 01633 262141
Fax: 01633 250631
www.braithwaite.co.uk

Brights of London Ltd
Westgate Business Park
Westgate Carr Road
Pickering
N Yorks YO18 8LX
Clock Systems
Tel: (020) 8786 8466
Fax: (020) 8786 8477
www.brightsoflondon.co.uk

Broadcrown Limited
Alliance Works
Airfield Industrial Estate
Hixon
Staffs ST18 0PF
Generators
Tel: 01889 272200
Fax: 01889 272220
www.broadcrown.co.uk

Caradon Stelrad Ideal Boilers
PO Box 103
National Avenue
Kingston-upon-Hall
North Humberside HU5 4JN
Boilers/Heating Products
Tel: 08708 400030
Fax: 08708 400059
www.rycroft.com

Carrier Air Conditioning
United Technologies House
Guildford Road
Leatherhead
Surrey KT22 9UT
Chilled Water Plant
Tel: 0870 6001100
Fax: 01372 220221
www.carrier.uk.com

Chloride Power Protection
Unit C, George Curl Way
Southampton SO18 2RY
Static UPS Systems
Tel: 023 8061 0311
Fax: 023 8061 0852
www.chloridepower.com

Communica
Chatteris Airfield
Near March
Cambridgeshire PE15 0EA
Telephone Cables
Tel : 01354 742340
www.communicaplc.co.uk

Danfoss Flowmetering Ltd
Magflo House
Ebley Road
Stonehouse
Glos GL10 2LU
Energy Meters
Tel: 01453 828891
Fax: 01453 853860
www.danfoss-randall.co.uk

Dewey Waters Limited
Cox's Green
Wrington
Bristol BS40 5QS
Tanks
Tel : 01934 862601
Fax : 01934 862604
www.deweywaters.co.uk

Diffusion
Benson Environmental Limited
47 Central Avenue
West Molesey
Surrey KT8 2QZ
Fan Coil Units
Tel: (020) 8783 0033
Fax: (020) 8783 0140
www.diffusionenv.com

Dunham-Bush Limited
8 Downley Road
Havant
Hampshire PO9 2JD
Convectors and Heaters
Tel : 02392 477700
Fax : 02392 450396
www.dunham-bush.com

E&I Engineering Ltd
14 Springtown Road
Springtown Industrial Estate
Londonderry BT48 0LY
Low Voltage Switchgear
Tel: 01504 266404
Fax: 01504 371766
www.e-i-eng.com

EMS Radio Fire & Security Systems
Limited
Technology House
Sea Street
Herne Bay
Kent CT6 8JZ
Security
Tel : 01227 369570
Fax : 01227 369679
www.emsgroup.co.uk

Engineering Appliances Ltd
Unit 11
Sunbury Cross Ind Est
Brooklands Close
Sunbury On Thames TW16 7DX
**Expansion Joints, Air and Dirt
Separators**
Tel: 01932 788888
Fax: 01932 761263
e-mail:
info@engineering-appliances.co.uk
www.engineeringappliances.com

FCS Ductwork Limited
3rd Floor
Thomas Telford House
1 Heron Quay
Canary Wharf
London E14 5JD
Fire Rated Ductwork
Tel: (020) 7987 7692
Fax: (020) 7537 5627
www.fcsgroup.co.uk

Flakt Woods Limited
Tufnell Way
Colchester CO4 5AR
Air Handling Units
Tel : 01206 544122
Fax : 01206 574434

Furse and Company Limited
Wilford Road
Nottingham NG2 1EB
Lightning Protection
Tel: 0115 863471
Fax: 0115 9860071
www.tnb.com

Hall Fire Protection Limited
186 Moorside Road
Swinton
Manchester M27 9HA
Fire Protection Equipment
Tel : 0161 793 4822
Fax : 0161 794 4950
www.hallfire.co.uk

Halton
5 Waterside Business Park
Witham
Essex CM8 3YQ
Chilled Beams
Tel : 01376 503040
Fax : 01376 503060
www.haltongroup.com

Hattersley, Newman, Hender Ltd
Burscough Road
Ormskirk
Lancashire L39 2XG
Valves
Tel: 01695 577199
Fax: 01695 578775
e-mail: uksales@hattersley-valves.co.uk
www.hattersley.com

Honeywell CS Limited
Honeywell House
Anchor Boulevard
Crossways Business Park
Dartford
Kent DA2 6QH
Control Components
Tel : 01322 484800
Fax : 01322 484898
www.honeywell.com

Hoval Limited
Northgate
Newark
Notts NG24 1JN
Boilers
Tel : 01636 672711
Fax : 01636 673532
www.hoval.co.uk

HRS Hevac Ltd
10-12 Caxton Way
Watford Business Park
Watford
Herts WD18 8JY
Heat Exchangers
Tel: 01923 232335
Fax: 01923 230266
www.hrshevac.co.uk

Hudevad
Bridge House
Bridge Street
Walton on Thames
Radiators
Tel: 01932 247835
Fax: 01932 247694
www.hudevad.co.uk

Hydrotec (UK) Limited
Hydrotec House
5 Mannor Courtyard
Hughenden Avenue
High Wycombe HP13 5RE
Chemical Treatment
Tel : 01494 796040
Fax : 01494 796049
www.hydrotec.com

IAC
IEC House
Moorside Road
Winchester
Hampshire SO23 7US
Attenuators
Tel : 01962 873000
Fax : 01962 873102
www.industrialacoustics.com

K and W Fabrications Ltd
High Street
Handcross
Haywards Heath
West Sussex RH17 6BZ
Plastic Ductwork
Tel: 01444 401144
Fax: 01444 401188

Kampmann
Benson Environmental Limited
47 Central Avenue
West Molesey
Surrey KT8 2QZ
Trench Heating
Tel: (020) 8783 0033
Fax: (020) 8783 0140
www.diffusionenv.com

Kiddie Fire Protection Services
Enterprise House
Jasmine Grove
London SE20 8JW
Fire Protection Equipment
Tel : (020) 8659 7235
Fax : (020) 8659 7237
www.kfp.co.uk

Metcraft Ltd
Harwood Industrial Estate
Littlehampton
West Sussex BN17 7BB
Oil Storage Tanks
Tel: 01903 714226
Fax: 01903 723206
www.metcraft.co.uk

Osma Underfloor Heating
18 Apple Lane
Sowton Trade City
Exeter
Devon EX2 5GL
Underfloor Heating
Tel : 01392 444122
Fax : 01392 444135
www.osmaufh.co.uk

Pullen Pumps Limited
158 Beddington Lane
Croydon CR9 4PT
Pumps, Booster Sets
Tel: (020) 8684 9521
Fax: (020) 8689 8892
www.pullenpumps.co.uk

Rycroft
Duncombe Road
Bradford BD8 9TB
Storage Cylinders
Tel : 01274 490911
Fax : 01274 498580
www.rycroft.com

SF Limited
Pottington Business Park
Barnstaple
Devon EX31 1LZ
Flues
Tel : 01271 326633
Fax : 01271 334303

Simmtronic Limited
Waterside
Charlton Mead Lane
Hoddesdon
Hertfordshire EN11 0QR
Lighting Controls
Tel : 01992 456869
Fax : 01992 445132
www.simmtronic.com

Socomec Limited
Knowl Piece
Wilbury Way
Hitchin
Hertfordshire SG4 0TY
Automatic Transfer Switches
Tel : 01462 440033
Fax : 01462 431143
www.socomec.com

Spirax-Sarco Ltd
Charlton House
Cheltenham
Gloucestershire GL53 8ER
Traps and Valves
Tel: 01242 521361
Fax: 01242 573342
www.spiraxsarco.com

Tyco Limited
Unit 6 West Point Enterprize Park
Clarence Avenue
Trafford Park
Manchester M17 1QS
Fire Protection
Tel: 0161 875 0400
Fax: 0161 875 0491
www.tyco.com

Utile Engineering Company Ltd
Irthlingborough
Northants NN9 5UG
Gas Boosters
Tel: 01933 650216
Fax: 01933 652738
www.utileengineering.com

Vokes Ltd
Henley Park
Guildford
Surrey GU3 2AF
Air Filters
Tel: 01483 569971
Fax: 01483 235384
e-mail: vokes@btvinc.com
www.vokes.com

Waterloo Air Management
Mills Road
Aylesford
Kent ME20 7NB
Grilles & Diffusers
Tel: 01622 717861
Fax: 01622 710648
www.waterloo.co.uk

Whitecroft Lighting Limited
Burlington Street
Ashton-under-Lyne
Lancashire OL7 0AX
Lighting & Luminaires
Tel: 0870 5087087
Fax: 0870 5084210
www.whitecroftlighting.com

Woods of Colchester
Tufnell Way
Colchester
Essex CO4 5AR
**Air Distribution, Fans, Anti-
vibration mountings**
Tel: 01206 544122
Fax: 01206 574434

PART ONE

Engineering Features

In this, the Thirty Eighth edition of *SPON'S Mechanical and Electrical Services Price Book,* a new section has been included for Engineering Features, dealing with current issues or technical advancements within the industry. These shall be complimented by cost models and/or itemised prices for items that form part of such.

The intention is that the book shall develop to provide more than just a schedule of prices to assist the user in the preparation and evaluation of costs.

- Revisions to Part L of the Building Regulations

- Renewable Energy Options

- Ground Water Cooling

- Fuel Cells

Revisions to Part L of the Building Regulations

The following review of the effects of the revised Part L on non-residential buildings is based on updated material originally published in a Building Cost Model on Part L, issued in *Building* Magazine on 5 August 2005. The original article was produced in collaboration with Consultants, Arup.

Introduction

Part L (2006) came into force in April 2006 at a time when awareness of the threat posed by global warming has never been higher. The 2002 revisions signaled an important shift in emphasis from energy conservation to control of emissions of carbon and other greenhouse gases. Revised Part L 2006 is building on this foundation, reducing carbon emissions from new build further, and widening the scope of work required to improve performance in existing buildings. In raising standards, the approach adopted has been to give maximum flexibility to owners and designers in selecting means of compliance.

Changes in the Approach in Part L2

Part L2 deals with non-dwellings. The major headline change is the requirement to achieve savings in carbon emissions compared to the previous 2002 standard. For new build, air-conditioned and mechanically ventilated buildings, the reduction is 28%, whereas for naturally ventilated buildings the reduction is 23.5%.

In addition to the introduction of these challenging targets for carbon emissions reduction, the Revised Part L also involves some significant changes to practice that are having an impact on the work of consultants and contractors.

- The introduction of alternative assessment methodologies for new build and existing buildings. The new build method, based on a national standard assessment methodology is concerned with comparing the carbon emissions of the proposed development with those of a notional building, which complies with the previous 2002 Regulations. The approach for existing buildings is broadly intended to encourage proportional, technically feasible and economic improvements to building performance to ensure that alterations and improvements made to an existing building contribute to a reduction in carbon emissions.

- A move away from prescriptive technical guidance. Users are required to refer to technical details taken from a wide range of sources to identify how to meet the new standards. The new Part L only sets out the required benchmarks. .

- The adoption of a single national calculation methodology. This change, driven by the EPBD, has led to the adoption of a carbon emissions method for buildings other than dwellings. As a result, the simple elemental method of calculation used under previous versions of the regulations is no longer available.

- The Regulations include a provision to include the contribution of low or zero carbon (LZC) technologies, such as solar hot water heating, in the assessment. They also provide flexibility with regard to their application whilst the technologies mature and become more widely available.

The National Calculation Methodology

The new calculation methodology required under involves inputting all aspects of the building design into an assessment tool based on an agreed methodology. The National Calculation Tool, known as iSBEM (Simple Building Evaluation Method), developed by the BRE, is currently the most common application. Others such as CECM are becoming available. The comparative emissions calculation needs to be done fairly early on in the design process and then updated once the design is fixed. This will potentially make it more difficult for the design team to handle changes, and, given the resource requirements of recalculation, this could limit design iteration, although criteria such as perimeter heat gain can be set that will allow some design iterations without recalculation. Furthermore, contractors may be less likely to propose alternative design solutions as a result of the introduction of the new methodology, if they are subsequently required to demonstrate and achieve compliance.

Impact of the Regulations on design

The revised Part L is beginning to affect all aspects of the building design. The transition period was very short, and in effect, all buildings currently being designed will have to meet the new Standard. In addition to insulation levels, proportion and type of glazing, other factors that may be improved are:

- Shading to combat perimeter heat gains, e.g. external shading, double wall facades;
- Boiler efficiencies and chiller coefficients of performance;
- Building envelope air tightness;
- Lighting efficiency, lighting control;
- Fan and pump efficiencies, air and water pressure drops;
- Energy recovery.

A key characteristic of the revisions is, although the carbon emissions target has been cut by some 28% for cooled or mechanically ventilated buildings, none of the minimum acceptable standards to which the design must comply have been changed significantly from those set in 2002. It will continue to be possible to specify, for example, highly glazed buildings, albeit that significant improvement in other aspects of building performance will be necessary. Lighting specification and lamp efficiency are examples of areas where substantial reductions in both energy consumption and cooling loads could be made at a relatively low cost.

Indicative cost implications

Davis Langdon prepared a comparison of comparative office model solutions in Summer 2005 with Consulting Engineers, Arup, in order to show how the 2006 Regulations could be met by using either 40% glazing or different combinations of 100% glazing and solar shading. The cost implications are that using a low proportion of glazing, improvements to the glass specification, services plant efficiency and the introduction of simple heat recovery measures will actually result in a small reduction in capital costs resulting from the reduction in heating and cooling loads and plant sizing.

Option	Envelope solution	Saving/Extra £/m²	Saving/Extra %
40% glazing	High performance glass	(11.82)	(0.62)
100% glazing	High performance glass, fixed external louvres	61.29	3.19
100% glazing	Clear double glazing with low-e coating, external motorised slatted blinds	96.78	5.06
100% glazing	Double wall facade with externally ventilated cavity, with motorised slatted blinds at close centres	73.12	3.82

The costs in the comparison are at 3rd Quarter 2005 prices based on a central London location. The base building costs include the category A fit-out, preliminaries and contingencies. Costs of demolitions, external works, tenant fit-out, professional fees and VAT are excluded.

Due to the complexity of the changes and the recent introduction of the Regulations, it has not been possible to assess their impact on the rates within the 'Cost Models' and the 'Approximate Estimating' sections of the SPON'S Mechanical and Electrical Price Book 2007. These sections have not been adjusted to take into account the implications of works required to comply with the revised Part L Regulations. Users should note that the items and their rates in the 'Prices for Measured Works' section are compliant.

Renewable Energy Options

RENEWABLE ENERGY, TOGETHER WITH AN ANALYSIS OF WHERE THEY MAY BE INSTALLED TO BEST ADVANTAGE.

This article focuses on building-integrated options rather than large-scale utility solutions such as wind farms which will be addressed in the 2008 Edition.

The legislative background, imperatives and incentives

In recognition of likely causes and effects of global climate change, the Kyoto protocol was signed by the UK and other nations in 1992, with a commitment to reduce the emission of greenhouse gases relative to 1990 as the base year.
The first phase of European Union Emission Trading Scheme (EU ETS) covers the power sector and high-energy users such as oil refineries, metal processing, mineral and paper pulp industries. From 1st January 2005, all such companies in all 25 EU member states must limit their CO_2 emissions to allocated levels in line with Kyoto. The EU ETS principle is that participating organisations can:

- Meet the targets by reducing their own emissions, or
- Exceed the targets and sell or bank their excess emission allowances, or
- Fail to meet the targets and buy emission allowances from other participants.

These targets can only be met by either energy efficiency measures or making using more renewable energy instead of that derived from fossil fuels.
In the UK, the Utilities Act (2000) requires power suppliers to provide some electricity from renewables, starting at 3% in 2003 and rising to15% by 2015. In a similar way to EU ETS, generating companies receive and can trade Renewables Obligation Certificates (ROCs) for the qualifying electricity that they generate. For small renewable generators, Renewable Energy Guarantee of Origin Certificates (REGOs) has been introduced, in units of 1 kWh.

The initial focus is on Carbon Dioxide (CO_2), and the goals set by the UK government are:-

- 20% emission reduction by 2010 (and 10% of UK electricity from renewable sources)
- 60% emission reduction by 2050
- Real progress towards the 60% by 2020 (and 20% of UK electricity from renewable sources)

Four years after the introduction of ROCS, it was estimated that less than 3% of UK electricity was being generated from renewable sources. A 'step change' in policy was required, and the Office of the Deputy Prime Minister (ODPM) published 'Planning Policy Statement 22 (PPS 22): Renewable Energy' in order to promote renewable energy through the UK's regional and local planning authorities.

Local Planning & Building Regulations

More than 100 local authorities have already embraced PPS 22 and its Companion Guide published in December 2004, by adopting pro-renewables planning policies. Others are expected to follow; the typical requirement is for 10% of a site's electricity or heat to be derived from renewable sources, but at least one authority has already 'raised the bar' to 15%. The London Plan states that "The Mayor will and boroughs should require major developments to show how the development would generate a proportion of the site's electricity or heat from renewables".

Energy Performance Certificates, EPC

The EU Directive on the Energy Performance of Buildings (EPBD) requires that Energy Performance Certificates must be prominently placed on all buildings open to the public and commercial buildings built, sold or let from January 2006, thus enabling prospective purchasers and tenants to be more aware of a building's energy performance.

Assessing the carbon emissions associated with the operation of buildings is now an important part of the overall early design process for planning approval. Methods are set out in:

- Part L of the UK Building Regulations: (Conservation of fuel & power)
- BREEAM - (Building Research Establishment Environmental Assessment Method)
- Standard Assessment Procedure (SAP) for Energy Rating

On-site renewable energy sources are taken into account, but developers are not allowed to rely on any 'green tariff' as part of an assessment.

Technology Options and Applications

- **Wind Generators** - In a suitable location, wind energy can be an effective source of renewable power generation. Even without grant aid, an installed cost range of £2500 to £5000 per kW of generator capacity has been established over the past few years. The most common arrangement is a machine with three blades on a horizontal axis; all mounted on a tower or, increasingly for small generators in inner city areas, on top of a building. Average site wind speeds of 4 m/s can produce useful amounts of energy from a small generator up to say 3 kW, but larger generators require at least 7 m/s. A small increase in average site wind speed will produce a large

  Third party provision through an Energy Service Company (ESCo) can be successful for larger installations co-located with or close to the host building, especially in industrial settings where there may be less aesthetic or noise issues than inner city office or residential. The ESCo provides funding, installs and operates the plant and the client signs up for the renewable electrical energy at a fixed price for a period of time.

- **Building Integrated Photovoltaics (BIPV)** - Photovoltaic materials, commonly known as solar cells, generate direct current electrical power when exposed to light. Solar cells are constructed from semi-conducting materials that absorb solar radiation; electrons are displaced within the material, thus starting a flow of current through an external connected circuit. Conversion efficiency of solar energy to electrical power is improving with advances in technology and ranges from 7% to 18% under laboratory conditions. In practice, however, allowing for typical UK weather conditions, an installation of at least $7m^2$ of the latest high-efficiency hybrid modules is needed to produce 1000 watts peak (1 kWp), yielding perhaps 800 kWh in a year. Installed costs range from £300 to £450/m^2 for roof covering, and from £850 to £1300/m^2 for laminated glass.

- **Ground source heat pumps** -The ground temperature remains substantially constant throughout the year and heat can be extracted by circulating a fluid (normally water) through a system of pipes and into a heat exchanger. An electrically driven heat pump is then used to raise the fluid temperature via the compression cycle, and hot water is delivered to the building load as if from a normal boiler (albeit at a somewhat lower temperature than a normal boiler)

  Most ground heat systems consist of a cluster of pipes inserted into vertical holes typically 50 to100 metres deep depending on space and ground type. Costs for the drilling operation vary according to location, site access and ground conditions. A geological investigation may be needed minimise the risk of failure and to improve cost certainty.

  Such systems can achieve a Coefficient of Performance (COP = heat output/electrical energy input) of between 3 and 4, achieving good savings of energy compared with conventional fossil fuels. Installed costs are in the range £800 to £1200/kW depending on system size and complexity.

- **Borehole cooling** - The constant ground temperature is well below ambient air temperature during the summer, so 'coolth' can be extracted and used to replace or, more likely commercially, to supplement conventional building cooling systems. Such borehole systems may be either 'open' – discharging ground water to river or sewer after passing it through a heat exchanger, or 'closed' – circulating a fluid (often water) through a heat exchanger and vertical pipes extending below the water table.

  Ground source heating and cooling systems are only 'partial renewable energy' because they rely on electrical power, mainly for pumping. Considerable carbon savings can be justifiably claimed, however, by avoiding the use of fossil fuel for heating and electrical power to drive conventional chillers. Indicative system costs are from £200 to £250/kW.

- **Solar Water Heating** - Simple flat-plate water-based collector panels have been used successfully on South-facing roofs over many years in the UK – especially by DIY enthusiasts prepared to devise their own simple control systems. The basic principle is to collect heat from the sun and circulate it to pre-heat space heating or domestic hot water, in either a separate tank or a twin coil hot water cylinder. Purpose-designed, evacuated tube collectors have been developed to increase performance and a typical 4 m^2 installed residential system has a cost range from £2500 to £4000 depending on pipe runs and complexity. Such a system could produce approximately 2000 kWh saving in energy use per year. Commercial systems are simply larger and slightly more complex but should achieve similar performance; low-density residential, retail and leisure developments with washrooms and showers may be suitable applications having adequate demand for hot water.

- **Biomass Boilers** - Wood chips or pellets derived from waste or farmed coppices or forests are available commercially and are considered carbon neutral, having absorbed carbon dioxide during growth. With a suitable fuel storage hopper and automatic screw drive and controls, biomass boilers can replace conventional boilers with little technical or aesthetic impact. They do, however, depend on a viable source of fuel, and there is a requirement for ash removal/disposal as well as periodic de-coking. In individual dwellings, space may be a problem because a biomass boiler does not integrate readily into a typical modern kitchen. Biomass boilers are available in a wide range of domestic and commercial sizes. For a large installation, they are more likely to form part of a modular system rather than to displace conventional boilers entirely. There is a cost premium for the biomass storage and feed system, and the cost of the fuel is currently comparable with other solid fuels. As an addition to a conventional system, installed costs could range from £200 to £250/kW.

- **Biomass Combined Heat & Power (CHP)** - Conventional CHP installations consist of either an internal combustion engine or a gas turbine driving an alternator, with maximum recovery of heat, particularly from the exhaust system. For best efficiency, there needs to be a convenient and constant requirement for the output heat energy, and the generated electricity should also be utilised locally, with any excess exported to the grid.

  Unless a source of fuel is available from landfill gas, or from a local biomass digester, then an on-site biomass to gas conversion plant would be needed to fuel the CHP engine. Considering the cost implications for biomass storage and handling as described for boilers, it appears that biomass CHP will only be viable in specific circumstances, with installed system costs in the order of £2500 to £3000/kW (electrical).

Investment 'yield' table for various renewable technologies

It can be quite difficult to compare renewable options in terms of how much energy they might save on a particular project, and how that translates into CO_2, especially if more than one option appears to be feasible. The table illustrates the potential saving per £100,000 of renewable investment i.e. £100,000 is the notional 'extra over' cost of introducing a proportion of renewable energy into the particular building service. Grant aid has been ignored in the table.

The photovoltaic options indicated include no allowance for the displacement of conventional building fabric. In practice, the 'yield per £100,000' may be higher, e.g. if PV is fully-integrated.

Renewable technology	Candidate buildings	Prerequisites	Potential barriers	Annual saving per £100,000 of capital cost	
				kWh	Kg CO_2
Tower-mounted wind generators	A	F	Environmental impact. Site space for large turbines	100,000	43,000 c.f. elec.
Building-mounted 'micro wind'	B	G	Environmental impact. Roof space for small turbines.	40,000	17,200 c.f. elec.
*Photovoltaic roof or panels	B	H	Available roof space	12,500	5,375 c.f. elec.
*Photovoltaic rain screen or glass	C	H	None	9,000	3,870 c.f. elec.
Passive solar water heating	D	J	None	50,000	9,500 c.f. gas
Ground source heat pump	B	K	Site space for pipes	40,000	7,600 c.f. gas
Borehole cooling	D	K	Site space for pipes	12,000	5,160 c.f. elec.
Biomass boilers	B	L	Environmental impact, & maintenance	100,000	19,000 c.f. gas
Biomass CHP	E	M	Environmental impact, & maintenance	28,000 + 63,000	12,000 c.f. electricity, + 12,000 c.f. gas

Key :

A - Industrial, distribution centres
B - Most types of building
C - Prestige offices or retail
D - Residential and commercial, hotels & leisure
E - Industrial, Hotel, leisure, hospital
F - Average site wind speed minimum 7 m/s

G - Average site wind speed minimum 3.5 m/s
H - Roughly south-facing, un-shaded
J - Roughly south-facing, un-shaded - for hot water
K - Feasible ground conditions
L - Space and convenient source of fuel
M - Space & convenient source of fuel - for summer heat

The following exclusions relate to all of the aforementioned indicative costs:

Inflation beyond second quarter 2005, maintenance charges, general builders work, main contractor's overheads, profit and attendance, main contract preliminaries, professional and prescribed fees, contingency and design reserve, grant aid, tax allowances, Value Added Tax. Price levels indicated are based on provincial locations.

Ground Water Cooling

THE USE OF GROUND WATER COOLING SYSTEMS, CONSIDERING THE TECHNICAL AND COST IMPLICATIONS OF THIS RENEWABLE ENERGY TECHNOLOGY.

The application of ground water cooling systems is quickly becoming an established technology in the UK with numerous installations having been completed for a wide range of building types, both new build and existing (refurbished).

Buildings in the UK are significant users of energy, accounting for 60% of UK carbon emissions in relation to their construction and occupation. The drivers for considering renewable technologies such as groundwater cooling are well documented and can briefly be summarised as follows:

- Government set targets – The Energy White Paper, published in 2003, setting a target of producing 10% of UK electricity from renewable sources by 2010 and the aspiration of doubling this by 2020.
- The proposed revision to the Building Regulations Part L 2006, in raising the overall energy efficiency of non domestic buildings, through the reduction in carbon emissions, by 27%.
- Local Government policy for sustainable development. In the case of London, major new developments (i.e. City of London schemes over 30,000m^2) are required to demonstrate how they will generate a proportion of the site's delivered energy requirements from on-site renewable sources where feasible. The GLA's expectation is that, overall, large developments will contribute 10% of their energy requirement using renewables, although the actual requirement will vary from site to site. Local authorities are also likely to set lower targets for buildings which fall below the GLA's renewables threshold.
- Company policies of building developers and end users to minimise detrimental impact to the environment.

The Ground as a Heat Source/Sink

The thermal capacity of the ground can provide an efficient means of tempering the internal climate of buildings. Whereas the annual swing in mean air temperature in the UK is around 20 K, the temperature of the ground is far more stable. At the modest depth of 2m, the swing in temperature reduces to 8 K, while at a depth of 50m the temperature of the ground is stable at 11-13°C. This stability and ambient temperature therefore makes groundwater a useful source of renewable energy for heating and cooling systems in buildings.

Furthermore, former industrial cities like Nottingham, Birmingham, Liverpool and London have a particular problem with rising ground water as they no longer need to abstract water from below ground for use in manufacturing. The use of groundwater for cooling is therefore encouraged by the Environment Agency in areas with rising groundwater as a means of combating this problem.

System Types

Ground water cooling systems may be defined as either open or closed loop.

Open loop systems

Open loop systems generally involve the direct abstraction and use of ground water, typically from aquifers (porous water bearing rock). Water is abstracted via one or more boreholes and passed through a heat exchanger and is returned via a separate borehole or boreholes, discharged to foul water drainage or released into a suitable available source such as a river. Typical ground water supply temperatures are in the range 6-10°C and typical re-injection temperatures 12-18°C (subject to the requirements of the abstraction licence).

Open loop systems fed by groundwater at 8°C, can typically cool water to 12°C on the secondary side of the heat exchanger to serve conventional cooling systems.

Open loop systems are thermally efficient but overtime can suffer from blockages caused by silt, and corrosion due to dissolved salts. As a result, additional cost may be incurred in having to provide filtration or water treatment, before the water can be used in the building.

Abstraction licence and discharge consent needs to be obtained for each installation, and this together with the maintenance and durability issues can significantly affect whole life operating costs, making this system less attractive.

Closed Loop Systems

Closed loop systems do not rely on the direct abstraction of water, but instead comprise a continuous pipework loop buried in the ground. Water circulates in the pipework and provides the means of heat transfer with the ground. Since ground water is not being directly used, closed loop systems therefore suffer fewer of the operational problems of open loop systems, being designed to be virtually maintenance free, but do not contribute to the control of groundwater levels.

There are two types of closed loop system:

Vertical Boreholes – Vertical loops are inserted as U tubes into pre-drilled boreholes, typically less than 150mm in diameter. These are backfilled with a high conductivity grout to seal the bore, prevent any cross contamination and to ensure good thermal conductivity between the pipe wall and surrounding ground. Vertical boreholes have the highest performance and means of heat rejection, but also have the highest cost due to associated drilling and excavation requirements.

Horizontal Loops – These are single (or pairs) of pipes laid in 2m deep trenches, which are backfilled with fine aggregate. These obviously require a greater physical area than vertical loops but are cheaper to install. As they are located closer to the surface where ground temperatures are less stable, efficiency is lower compared to open systems. Alternatively, coiled pipework can also be used where excavation is more straightforward and a large amount of land is available. Although performance may be reduced with this system as the pipe overlaps itself, it does represent a cost effective way of maximising the length of pipe installed and hence overall system capacity.

The Case for Heat Pumps

Instead of using the groundwater source directly in the building, referred to as passive cooling, when coupled to a reverse cycle heat pump, substantially increased cooling loads can be achieved.

Heat is extracted from the building and transferred by the heat pump into the water circulating through the loop. As it circulates, it gives up heat to the cooler earth, with the cooler water returning to the heat pump to pick up more heat. In heating mode the cycle is reversed, with the heat being extracted from the earth and being delivered to the HVAC system.

The use of heat pumps provides greater flexibility for heating and cooling applications within the building than passive systems. Ground source heat pumps are inherently more efficient than air source heat pumps, their energy requirement is therefore lower and their associated CO_2 emissions are also reduced, so they are well suited for connection to a groundwater source.

Closed loop systems can typically achieve outputs of 50W/m (of bore length), although this will vary with geology and borehole construction. When coupled to a reverse cycle heat pump, 1m of vertical borehole will typically deliver 140kWh of useful heating and 110kWh of cooling per annum, although this will depend on hours run and length of heating and cooling seasons.

Key factors Affecting Cost

- The cost is obviously dependant on the type of system used. Deciding on what system is best suited to a particular project is dependant on the peak cooling and heating loads of the building and its likely load profile. This in turn determines the performance required from the ground loop, in terms of area of coverage in the case of the horizontal looped system, and in the case of vertical boreholes, the depth and number or bores. The cost of the system is therefore a function of the building load.

- In the case of vertical boreholes, drilling costs are significant factor, as specific ground conditions can be variable, and there are potential problems in drilling through sand layers, pebble beds, gravels and clay, which may mean additional costs through having to drill additional holes or the provision of sleeving etc. The costs of excavation obviously make the vertical borehole solution significantly more expensive than the equivalent horizontal loop.

- The thermal efficiency of the building is also a factor. The higher load associated with a thermally inefficient building obviously results in the requirement for a greater number of boreholes or greater area of horizontal loop coverage, however in the case of boreholes the associated cost differential between a thermally inefficient building and a thermally efficient one is substantially greater than the equivalent increase in the cost of conventional plant. Reducing the energy consumption of the building is cheaper than producing the energy from renewables and the use of renewable energy only becomes cost effective, and indeed should only be considered, when a building is energy efficient.

- With open loop systems, the principal risk in terms of operation is that the user is not in control of the quantity or quality of the water being taken out of the ground, this being dependant on the local ground conditions. Reduced performance due to blockage (silting etc) may lead to the system not delivering the design duties whilst bacteriological contamination may lead to the expensive water treatment or the system being taken temporarily out of operation.

- In order to mitigate the above risk, it may be decided to provide additional means of heat rejection and heating by mechanical means as a back up to the borehole system, in the event of operational problems. This obviously carries a significant cost.

- If additional plant were not provided as above, then there are space savings to be had over conventional systems due to the absence of heating, heat rejection and possibly refrigeration plant.

- Open loop systems may lend themselves particularly well to certain applications increasing their cost effectiveness, i.e. in the case of a leisure centre, the removal of heat from the air conditioned parts of the centre and the supply of fresh water to the swimming pool.

- In terms of the requirements for abstraction and disposal of the water for open loop systems, there are risks associated with the future availability and cost of the necessary licenses; particularly in areas of high forecast energy consumption, such as the South East of England, which needs to be borne in mind when selecting a suitable system.

Whilst open loop systems would suit certain applications or end user clients, for commercial buildings the risks associated with this system tend to mean that closed loop applications are the system of choice. When coupled to a reversible heat pump, the borehole acts simply as a heat sink or heat source so the problems associated with open loop systems do not arise.

Typical Costs

Table 1 gives details of the typical borehole cost to an existing site in Central London, using one 140m deep borehole working on the open loop principle, providing heat rejection for the 600kW of cooling provided to the building. The borehole passes through rubble, river gravel terraces, clay and finally chalk, and is lined above the chalk level to prevent the hole collapsing. The breakdown includes all costs associated with the provision of a working borehole up to the well head, including the manhole chamber and manhole. The costs of any plant or equipment from the well head are not included.

Heat is drawn out of the cooling circuit and the water is discharged into the Thames at an elevated temperature. In this instance, although the boreholes are more expensive than the dry air cooler alternative, the operating cost is significantly reduced as the system can operate at around three times the efficiency of conventional dry air coolers, so the payback period is a reasonable one. Additionally, the borehole system does not generate any noise, does not require rooftop space and does not require as much maintenance.

Table 2 provides a comparison of the costs of a closed loop borehole system with a conventional heating and cooling system, for a new gallery in the South of England with a gross floor area of 2,400m^2. In this analysis, the pipework forming the energy loop was attached to the structural piles, so the costs of drilling and excavation are not included. Also, the ground loop does not deal with the total cooling and heating loads in the building, so additional refrigeration and heating plant is still being provided.

Ground Water Cooling

Table 1 : Breakdown of the Cost of a Typical Open Loop Borehole System

Description	Cost £

General Items

- Mobilisation, Insurances, demobilisation on Completion   20,000
- Fencing around working area for the duration of drilling and testing   2,000
- Modifications to existing LV panel and installation of new power supplies for borehole installation   14,000

Trial Hole

- Allowance for breakout access to nearest walkway   3,000
  (Existing borehole on site used for trial purposes, hence no drilling costs included)

Construct Borehole

- Drilling, using temporary casing where required, permanent casing and grouting   31,000

Borehole Cap and Chamber

- Cap borehole with PN16 flange, construct manhole chamber in roadway, rising main, header pipework, valves, flow meter   12,000
- Permanent pump   13,000

Samples

- Water Samples   300

Acidisation

Mobilisation, set up and removal of equipment for acidisation of borehole, carry out acidisation   11,500

Development and Test Pumping

- Mobilise pumping equipment and materials and remove on completion of testing   4,000
- Calibration test, pre-test monitoring, step testing   3,500
- Constant rate testing and monitoring   19,000
- Waste removal and disposal   3,000

Reinstatement

- Reinstatement and Making Good   1,500

Total   **144,300**

Table 2 : Cost Comparison of Conventional and Closed Loop Thermal Pile Cooling Installation

Description	Closed Loop Thermal Pile Cooling System		Conventional Heating and Cooling System	
	£	£/m² GIA	£	£/m² GIA
Heat Source				
• LTHW boiler – 30 kW output	6,600	2.64	-	-
• LTHW boiler – 80 kW output	-	-	17,000	6.80
• Flue system	Incl	Incl	Incl	Incl
• Gas supply	Incl	Incl	Incl	Incl
Space Heating and Air Treatment				
• LTHW pipework distribution in Plantroom, pumps, Pressurisation unit, water treatment, thermal insulation	8,500	3.40	11,000	4.40
Chilled Water Installation				
• Water cooled chiller unit (40 kW load), dry air cooler (90kW load), pipework distribution in Plantroom (excluding condenser water), pumps, pressurisation unit, water treatment ,thermal insulation	25,000	10.00	-	-
• Water cooled chiller unit (90 kW load), dry air cooler • (180kW load), pipework distribution in Plantroom (excluding condenser water), pumps, pressurisation unit, water treatment ,thermal insulation	-	-	38,000	15.20
• Condenser water pipework, valves, pumps	18,000	7.20	25,500	10.20
Energy Piles				
• Pipework to 50 nr piles – total 50 kW	45,000	18.00	-	-
• Header and pipework connections	7,000	2.80	-	-
• Connection to heat pump from header	9,000	3.60		
Heat Pump Package Unit				
• Water to heat pump unit incl all associated Controls, 63 kW cooling load, 53 kW heating load	38,000	15.20	-	-
Total	**157,100**	**62.84**	**91,500**	**36.60**

Costs exclude – Secondary heating and cooling circuits within the building, Main Contractor's overheads, preliminaries, Profit and attendances, professional and prescribed fees, contingency and design reserve, VAT

Fuel Cells

Fuel cells are electrochemical devices that convert the chemical energy in fuel into electrical energy directly, without combustion, with high electrical efficiency and low pollutant emissions. They represent a new type of power generation technology that offers modularity, efficient operation across a wide range of load conditions, and opportunities for integration into co-generation systems. With the publication of the energy white paper in February this year, the Government confirmed it's commitment to the development of fuel cells as a key technology in the UK's future energy system, as the move is made away from a carbon based economy.

There are currently very few fuel cells available commercially, and those that are available are not financially viable. Demand has therefore been limited to niche applications, where the end user is willing to pay the premium for what they consider to be the associated key benefits. Indeed, the UK currently has only one fuel cell in regular commercial operation. However, fuel cell technology has made significant progress in recent years, with prices predicted to approach those of the principal competition in the near future.

Fuel Cell Technology

A fuel cell is composed of an anode (a negative electrode that repels electrons), an electrolyte membrane in the centre, and a cathode (a positive electrode that attracts electrons). As hydrogen flows into the cell on the anode side, a platinum coating on the anode facilitates the separation of the hydrogen gas into electrons and protons. The electrolyte membrane only allows the protons to pass through to the cathode side of the fuel cell. The electrons cannot pass through this membrane and flow through an external circuit to form an electric current.

As oxygen flows into the fuel cell cathode, another platinum coating helps the oxygen, protons, and electrons combine to produce pure water and heat.

The voltage from a single cell is about 0.7 volts, just enough for a light bulb. However by stacking the cells, higher outputs are achieved, with the number of cells in the stack determining the total voltage, and the surface area of each cell determining the total current. Multiplying the two together yields the total electrical power generated.

In a fuel cell the conversion process from chemical energy to electricity is direct. In contrast, conventional energy conversion processes first transform chemical energy to heat through combustion and then convert heat to electricity through some form of power cycle (e.g. gas turbine or internal combustion engine) together with a generator.
The fuel cell is therefore not limited by the Carnot efficiency limits of an internal combustion engine in converting fuel to power, resulting in efficiencies 2 to3 times greater.

Fuel Cell Systems

In addition to the fuel cell itself, the system comprises the following sub-systems:

- A fuel processor – This allows the cell to operate with available hydrocarbon fuels, by cleaning the fuel and converting (or reforming) it as required.
- A power conditioner – This regulates the dc electricity output of the cell to meet the application, and to power the fuel cell auxiliary systems.
- An air management system – This delivers air at the required temperature, pressure and humidity to the fuel stack and fuel processor.
- A thermal management system – This heats or cools the various process streams entering and leaving the fuel cell and fuel processor, as required.
- A water management system – Pure water is required for fuel processing in all fuel cell systems, and for dehumidification in the PEMFC.

The overall electrical conversion efficiency of a fuel cell system (defined as the electrical power out divided by the chemical energy into the system, taking into account the individual efficiencies of the sub-systems) ranges from 35-55%. Taking into account the thermal energy available from the system, the overall or cogeneration efficiency is 75-90%.

Also, unlike most conventional generating systems (which operate most efficiently near full load, and then suffer declining efficiency as load decreases), fuel cell systems can maintain high efficiency at loads as low as 20% of full load.

Fuel cell systems also offer the following potential benefits:

- At operating temperature, they respond quickly to load changes, the limiting factor usually being the response time of the auxiliary systems.

- They are modular and can be built in a wide range of outputs. This also allows them to be located close to the point of electricity use, facilitating cogeneration systems.
- Noise levels are comparable with residential or light commercial air conditioning systems.
- Commercially available systems are designed to operate unattended and manufactured as packaged units.
- Since the fuel cell stack has no moving parts, other than the replacement of the stack at 3-5 year intervals there is little on-site maintenance. The maintenance requirements are well established for the auxiliary system plant.
- Fuel cell stacks fuelled by hydrogen produce only water, therefore the fuel processor is the primary source of emissions, and these are significantly lower than emissions from conventional combustion systems.
- Since fuel cell technology generates 50% more electricity than the conventional equivalent without directly burning any fuel, CO_2 emissions are significantly reduced in the production of the source fuel.
- Potentially zero carbon emissions when using hydrogen produced from renewable energy sources.
- The facilitation of embedded generation, where electricity is generated close to the point of use, minimising transmission losses.
- The fast response times of fuel cells offer potential for use in UPS systems, replacing batteries and standby generators.

Types of Fuel Cell

There are four main types of fuel cell technology that are applicable for building systems, classed in terms of the electrolyte they use. The chemical reactions involved in each cell are very different.

Phosphoric Acid Fuel Cells (PAFCs) are the dominant current technology for large stationary applications and have been available commercially for some time. The only working fuel cell installation in the UK, in Woking, uses a PAFC, rated at 200kW. There is less potential for PAFC unit cost reduction than for some other fuel cell systems, and this technology may be superseded in time by the other technologies.

The Solid Oxide Fuel Cell (SOFC) offers significant flexibility due to its large power range and wide fuel compatibility. SOFCs represent one of the most promising technologies for stationary applications. There are difficulties when operating at high temperatures with the stability of the materials, however, significant further development and cost reduction is anticipated with this type.

The relative complexity of Molten Carbonate Fuel Cells (MCFCs) has tended to limit developments to large scale stationary applications, although the technology is still very much in the development stages.

The quick start-up times and size range make Proton Exchange Membrane Fuel Cells (PEMFCs) suitable for small to medium sized stationary applications. They have a high power density and can vary output quickly, making them well suited for transport applications as well as UPS systems. The development efforts in the transport sector suggest there will continue to be substantial cost reductions over both the short and long term.

All four technologies remain the subject of extensive research and development programmes to reduce initial costs and improve reliability through improvements in materials, optimisation of operating conditions and advances in manufacturing. It is expected that all types will be commercially available in limited markets by 2006 and with mass market availability by 2010.

The market for fuel cells

The stationary applications market for fuel cells can be sectaries as follows:

- Distributed generation/CHP – For large scale applications, there are no drivers specifically advantageous to fuel cells, with economics (and specifically initial cost) therefore being the main consideration. So, until cost competitive and thoroughly proven and reliable fuel cells are available, their use is likely to be limited to niche applications such as environmentally sensitive areas from 2005. Wider commercialisation is likely closer to 2010, with high temperature cells (MCFC's and SOFC's) being most suitable, although PEMFC's may preferable in specific areas, i.e. where hydrogen is available.

- Domestic and small scale CHP – The drivers for the use of fuel cells in this emerging market are better value for customers than separate gas and electricity purchase, reduction in domestic CO_2 emissions, and potential reduction in electricity transmission costs. However, the barriers of resistance to distributed generation, high capital costs and competition from Stirling engines needs to be overcome. Commercialisation depends on cost reduction, and successful demonstration which is expected to begin in the next 2-3 years, leading to wider commercialisation before 2010. Systems based on SOFC's and PEMFC's are being developed for this application.

- Small generator sets and remote power – The drivers for the use of fuel cells are high reliability, low noise and low refuelling frequencies, which cannot be met by existing technologies. Since cost is often not the primary consideration, fuel cells will find early markets in this sector. Existing PEMFC systems are close to meeting the requirements in terms of cost, size and performance. Small SOFC's have potential in this market, but require further development.

Cost comparison

Table 1 provides an indication of the capital and operating costs of the different fuel cell types, together with comparative figures for the existing technologies. The projected figures for the fuel cell technologies are based on economies typically achieved through mass manufacture.

The projected costs for 2007 show the fuel cell technologies still being significantly more expensive than the existing technologies. To extend fuel cell application beyond niche markets, their cost needs to reduce significantly. The successful and wide-spread commercial application of fuel cells is dependant on the projected cost reductions indicated, with electricity generated from fuel cells being competitive with current centralised and distributed power generation.

It has been estimated that if the cost reductions are met, fuel cells could achieve up to 50% penetration of the global distributed energy market by 2020.

Typical Current Project Costs

Table 2 gives an indication of the typical cost breakdown to be expected for the installation of a fuel cell system in the UK. This is based on information from the manufacturer and from economic evaluation/feasibility studies, since with only one working system installed to date in the UK there is no available accurate cost data.

The unit is a standard commercially available PAFC, complete with fuel processor, fuel stack and power conditioning system. The parameters are as follows:

- Rating – 200kW/235kVA, 400V, 3ph
- Power generating efficiency – 40%
- Heat output – 204kW, 60°C hot water
- Fuel, consumption – Natural gas, 54m^3/hr
- External location
- Size – 5.5m x 3m x 3m (h)
- Weight – 20 tons
- Noise level at full load – 62dBA at 10m

The above illustrates the fact that the costs to supply, install and set to work a modestly sized fuel cell unit are prohibitively high, compared with incumbent generator technologies. Despite being the only commercially available unit, only 220 units have been sold worldwide, and so the full benefits of volume manufacture have not been realised, and a proportion of the costs associated with developing the unit is included within the unit cost.

Conclusions

Despite significant growth in recent years, fuel cells are still at a relatively early stage of commercial development, with prohibitively high capital costs preventing them from competing with the incumbent technology in the market place. However, costs are forecast to reduce significantly over the next five years as the technology moves from niche applications, and into mass production.

However, in order for these projected cost reductions to be achieved, customers need to be convinced that the end product is not only cost competitive but also thoroughly proven, and Government support represents a key part in achieving this.

The Governments of Canada, USA, Japan and Germany have all been active in supporting development of the fuel cell sector through integrated strategies, however the UK has been slow in this respect, and support has to date been small in comparison. It is clear that without Government intervention, fuel cell applications may struggle to reach the cost and performance requirements of the emerging fuel cell market.

Free Updates

with three easy steps...

1. Register today on
   www.pricebooks.co.uk/updates

2. We'll alert you by email when new
   updates are posted on our website

3. Then go to
   www.pricebooks.co.uk/updates
   and download.

All four Spon Price Books – *Architects' and Builders'*, *Civil Engineering and Highway Works*, *External Works and Landscape* and *Mechanical and Electrical Services* – are supported by an updating service. Three updates are loaded on our website during the year, in November, February and May. Each gives details of changes in prices of materials, wage rates and other significant items, with regional price level adjustments for Northern Ireland, Scotland and Wales and regions of England. The updates terminate with the publication of the next annual edition.

As a purchaser of a Spon Price Book you are entitled to this updating service for the 2007 edition – free of charge. Simply register via the website www.pricebooks.co.uk/updates and we will send you an email when each update becomes available.

If you haven't got internet access we can supply the updates by an alternative means. Please write to us for details: Spon Price Book Updates, Spon Press, 2 Park Square, Milton Park, Abingdon, Oxfordshire, OX14 4RN.

Find out more about spon books
Visit www.sponpress.com for more details.

New books from Spon

Table 1 – Cost Comparison for Stationary Generation Equipment

Period	Fuel Cell Type			
	PEMFC	**PAFC**	**MCFC**	**SOFC**
Capital Costs (£/kW)	£	£	£	£
2004	2,600-6,500	2,000-3,400	2,000-5,000	5,000-10,000
2006	1.500	1,900	1,850	2,300
2010 (Projected)	700	1,500	950	1,000
2015 (Projected)	500	1,300	700	750
2020 (Projected)	300	1,100	550	550
Operating Costs (£/kW/h)				
2004	0.04-0.06	0.08	0.04-0.08	0.05
Maintenance Costs (£/kW/h)				
2006	0.003-0.01	0.003-0.01	0.003-0.01	0.003-0.01

Period	Conventional Systems		
	Internal Combustion - Generator	**Micro Turbine - Generator**	**Gas Turbine**
Capital Costs (£/kW)	£	£	£
2004	-	-	-
2006	200-820	450-820	450-570
2010 (Projected)	-	-	-
2015 (Projected)	-	-	-
2020 (Projected)	-	-	-
Operating Costs (£/kW/h)			
2004	-	-	-
Maintenance Costs (£/kW/h)			
2006	0.005-0.01	0.002-0.01	0.002-0.06

Table 2 – Breakdown of Typical Project Costs

Item	£
Fuel Cell Stack	225,000
Fuel Processor	85,000
Plant	113,600
Labour	113,400
Total	**567,000**
Delivery to Site (Outside USA)	20,000

Installation Costs (Site and Application dependant)

Standard (Generation Only)	50,000
Non Standard	80,000
CHP	100,000-200,000
Spares	10% of Capital Cost
Maintenance	16,000-26,000 pa
Total (£/kW)	**3,549-4,415**

Fuel Stack

Fuel Stack Replacement	150,000 every 4 years
Fuel Stack Refurbishment	120,000 every 4 years

Note – Excludes Incoming Gas Supply, BWIC, Main Contractor's Overheads ad Profit, attendance, Preliminaries and Professional Fees etc, VAT

Typical
Engineering Details

In addition to the Engineering Features, Typical Engineering Details are included. These are indicative schematics to assist in the compilation of costing exercises. The user should note that these are only examples and cannot be construed to reflect the design for each and every situation. They are merely provided to assist the user with gaining an understanding of the Engineering concepts and elements making up such.

ELECTRICAL

- Typical Simple 11kV Network Connection For LV Intakes Up To 1000kVA

- Typical 11kV Network Connections For HV Intakes 1000kVA To 6000kVA

- Static UPS System - Simplified Single Line Schematic For a Single Module

- Typical Data Transmission (Structured Cabling)

- Typical Networked Lighting Control System

- Typical Standby Power System, Single Line Schematic

- Typical Fire Detection and Alarm Schematic

- Typical Block Diagram - Access Control System (ACS)

- Typical Block Diagram - Intruder Detection System (IDS)

- Typical Block Diagram - Digital CCTV

MECHANICAL

- BMS Controls For Low Pressure Hot Water (LPHW)

- BMS Controls For Primary Chillers and Chilled Water

- Fan Coil Unit System

- Displacement System

- Chilled Ceiling System (Passive System)

- Chilled Beam System (Passive System)

- Variable Air Volume (VAV)

- Variable Refrigerant Volume System (VRV)

- Alternative All Air System (FGU)

- Reverse cycle heat pump

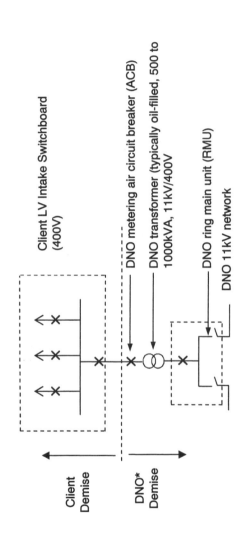

Client LV Intake Switchboard
(400V)

DNO metering air circuit breaker (ACB)

DNO transformer (typically oil-filled, 500 to
1000kVA, 11kV/400V)

DNO ring main unit (RMU)

DNO 11kV network

Client
Demise

DNO*
Demise

Note : * DNO - Distribution Network Operator

Typical Simple 11kV Network Connection For LV Intakes Up To 1000kVA

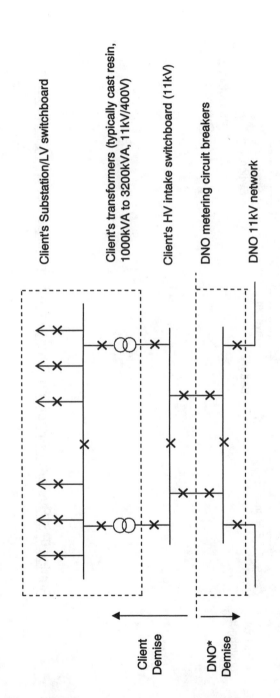

Client's Substation/LV switchboard

Client's transformers (typically cast resin, 1000kVA to 3200kVA, 11kV/400V)

Client's HV intake switchboard (11kV)

DNO metering circuit breakers

DNO 11kV network

Client Demise

DNO* Demise

Note : * DNO - Distribution Network Operator

Typical 11kV Network Connections For HV Intakes 1000kVA To 6000kVA

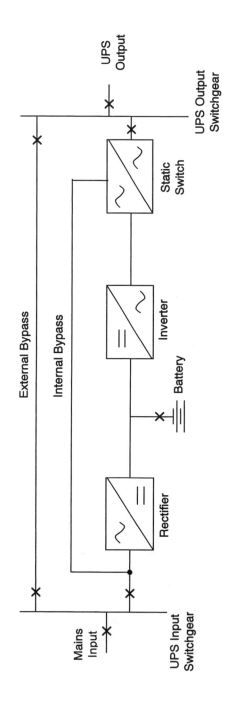

Static UPS System - Simplified Single Line Schematic For a Single Module

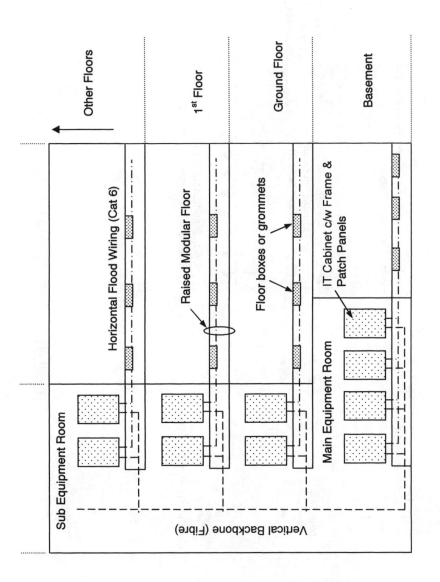

Typical Data Transmission (Structured Cabling)

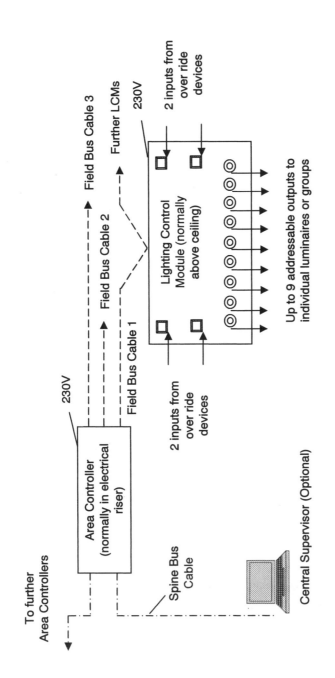

To further
Area Controllers

230V

Area Controller
(normally in electrical
riser)

Field Bus Cable 3

Field Bus Cable 2

Further LCMs

230V

Field Bus Cable 1

2 inputs from
over ride
devices

2 inputs from
over ride
devices

Lighting Control
Module (normally
above ceiling)

Up to 9 addressable outputs to
individual luminaires or groups

Spine Bus
Cable

Central Supervisor (Optional)

Typical Networked Lighting Control System

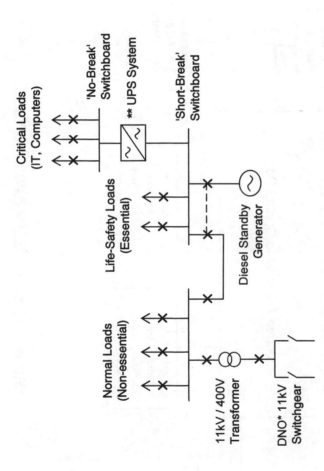

Critical Loads
(IT, Computers)

'No-Break'
Switchboard

** UPS System

'Short-Break'
Switchboard

Life-Safety Loads
(Essential)

Diesel Standby
Generator

Normal Loads
(Non-essential)

11kV / 400V
Transformer

DNO* 11kV
Switchgear

Note : * DNO - Distribution Network Operator
** UPS - Uninterruptible Power Supply

Typical Standby Power System, Single Line Schematic

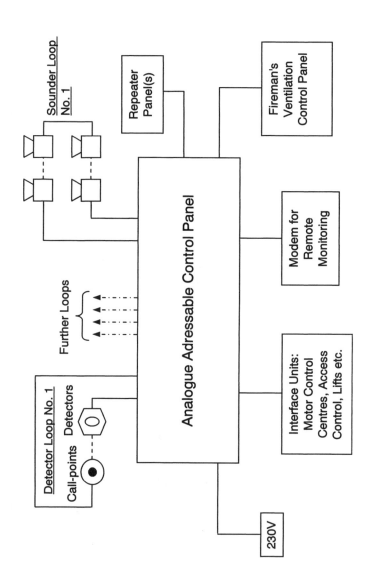

Typical Fire Detection and Alarm Schematic

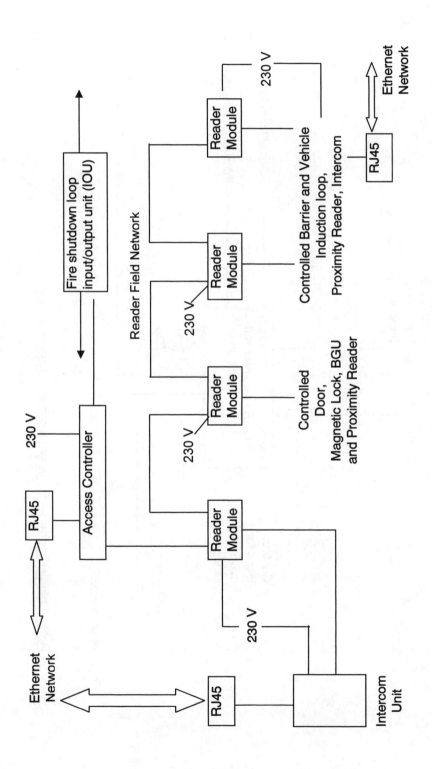

Typical Block Diagram - Access Control System (ACS)

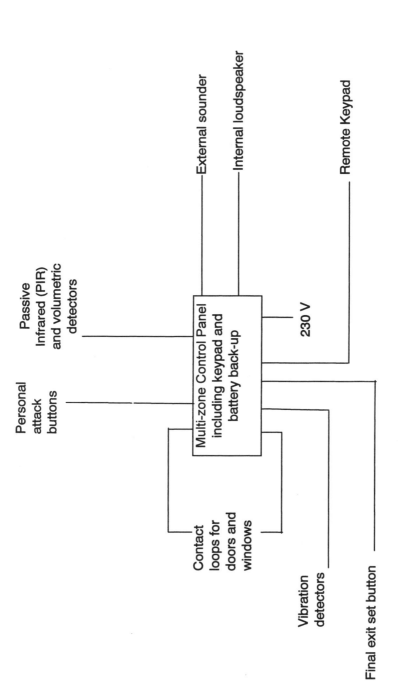

Personal
attack
buttons

Passive
Infrared (PIR)
and volumetric
detectors

External sounder

Internal loudspeaker

Multi-zone Control Panel
including keypad and
battery back-up

230 V

Remote Keypad

Contact
loops for
doors and
windows

Vibration
detectors

Final exit set button

Typical block diagram - Intruder Detection System (IDS)

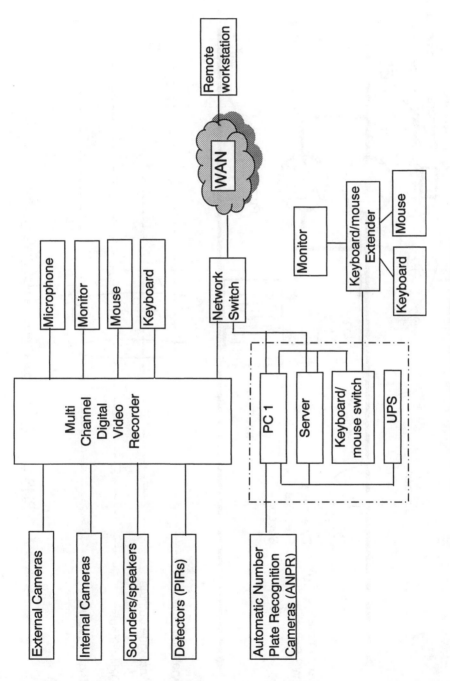

Typical Block Diagram - Digital CCTV

External Cameras

Internal Cameras

Sounders/speakers

Detectors (PIRs)

Automatic Number Plate Recognition Cameras (ANPR)

Multi Channel Digital Video Recorder

Microphone

Monitor

Mouse

Keyboard

Network Switch

PC 1

Server

Keyboard/ mouse switch

UPS

Monitor

Keyboard/mouse Extender

Mouse

Keyboard

WAN

Remote workstation

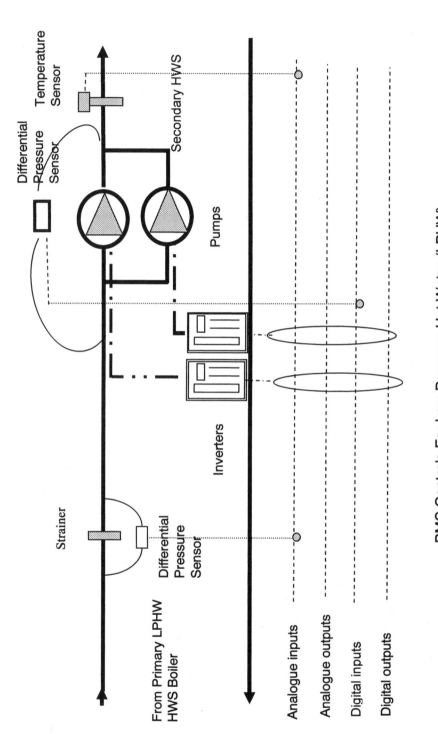

BMS Controls For Low Pressure Hot Water (LPHW)

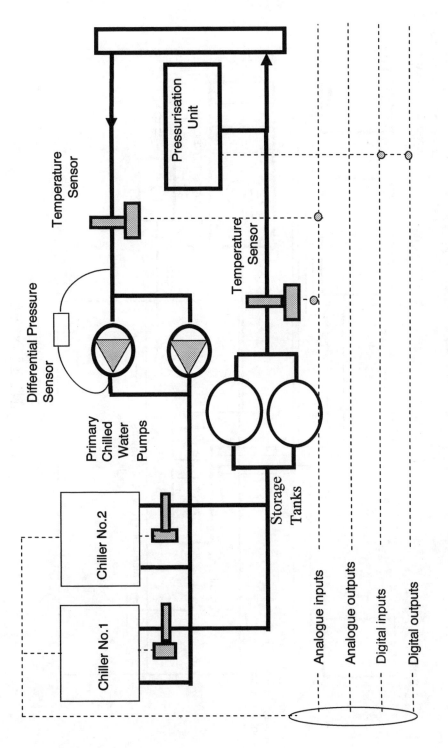

Temperature Sensor

Differential Pressure Sensor

Primary Chilled Water Pumps

Pressurisation Unit

Temperature Sensor

Chiller No.2

Chiller No.1

Storage Tanks

Analogue inputs ------------

Analogue outputs ------------

Digital inputs ------------

Digital outputs ------------

BMS Controls For Primary Chillers and Chilled Water

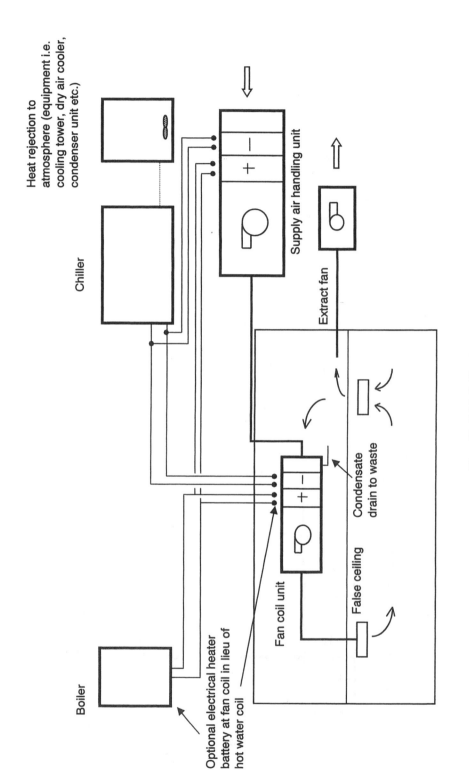

Heat rejection to atmosphere (equipment i.e. cooling tower, dry air cooler, condenser unit etc.)

Chiller

Boiler

Supply air handling unit

Extract fan

Condensate drain to waste

Fan coil unit

False ceiling

Optional electrical heater battery at fan coil in lieu of hot water coil

Fan Coil Unit System

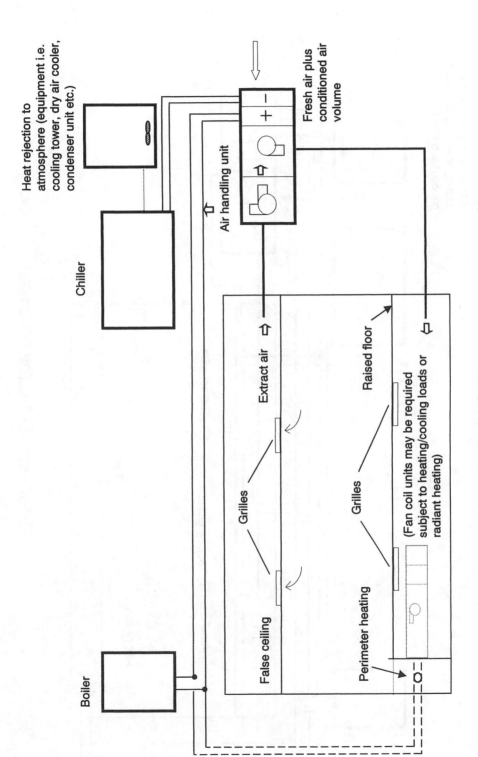

Heat rejection to atmosphere (equipment i.e. cooling tower, dry air cooler, condenser unit etc.)

Chiller

Boiler

Air handling unit

Fresh air plus conditioned air volume

Extract air

Raised floor

Grilles

Grilles

False ceiling

Perimeter heating

(Fan coil units may be required subject to heating/cooling loads or radiant heating)

Displacement System

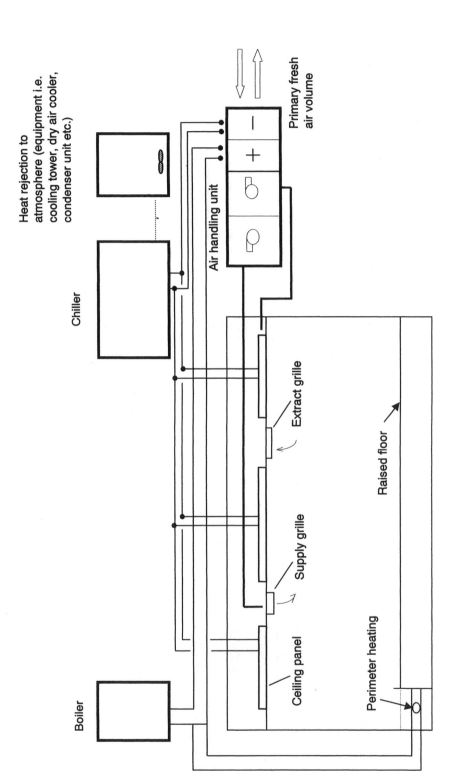

Heat rejection to atmosphere (equipment i.e. cooling tower, dry air cooler, condenser unit etc.)

Chiller

Boiler

Air handling unit

Primary fresh air volume

Extract grille

Supply grille

Ceiling panel

Perimeter heating

Raised floor

Chilled Ceiling System (Passive System)

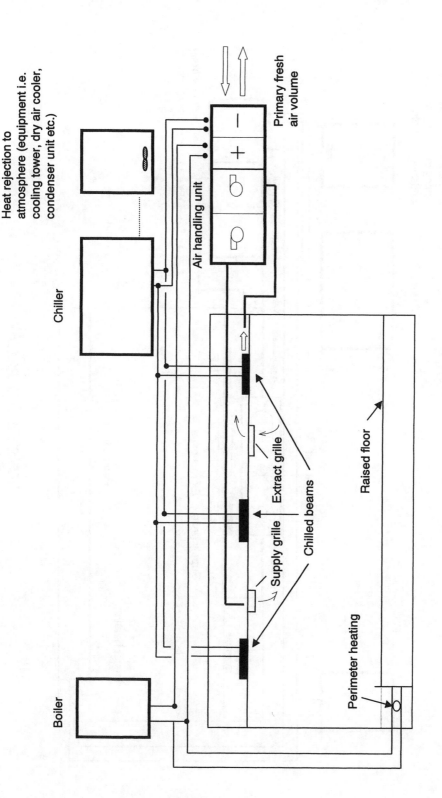

Heat rejection to atmosphere (equipment i.e. cooling tower, dry air cooler, condenser unit etc.)

Chiller

Boiler

Air handling unit

Primary fresh air volume

Extract grille

Supply grille

Chilled beams

Raised floor

Perimeter heating

Chilled Beam System (Passive System)
Active Option Connects Air Supply Duct To Chilled Beams & Deletes Supply Grilles

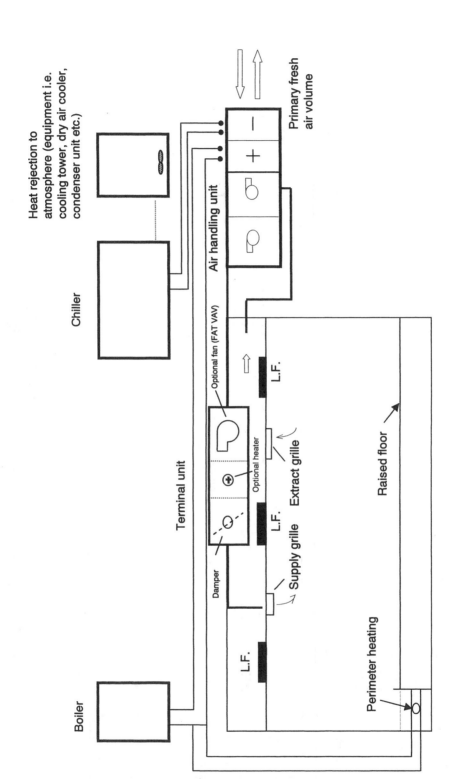

Variable Air Volume (VAV)

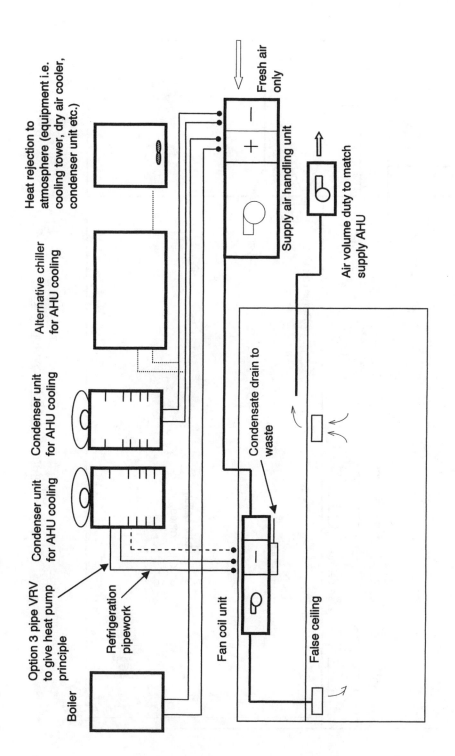

Heat rejection to atmosphere (equipment i.e. cooling tower, dry air cooler, condenser unit etc.)

Alternative chiller for AHU cooling

Condenser unit for AHU cooling

Option 3 pipe VRV to give heat pump principle

Condenser unit for AHU cooling

Boiler

Refrigeration pipework

Fan coil unit

False ceiling

Condensate drain to waste

Fresh air only

Supply air handling unit

Air volume duty to match supply AHU

Variable Refrigerant Volume System (VRV)

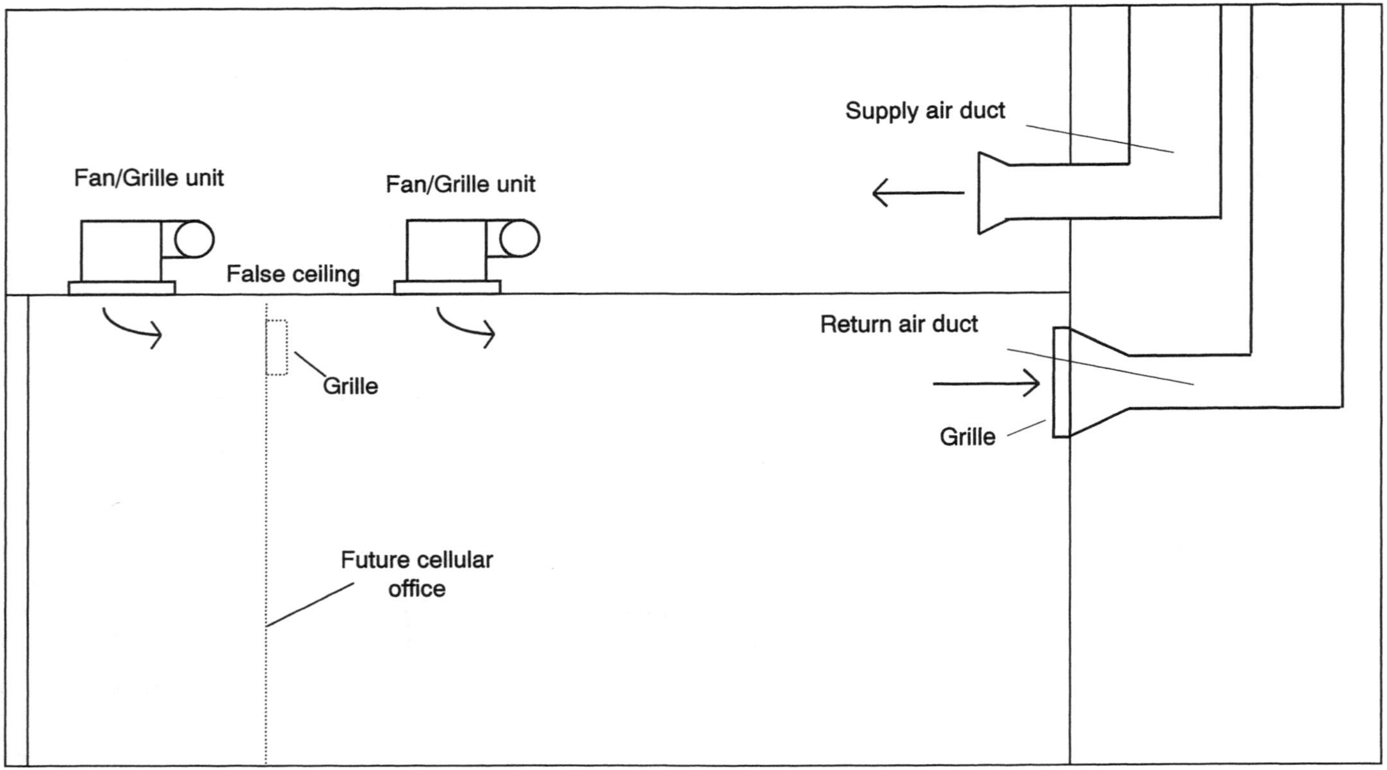

Supply air duct

Return air duct

Grille

Fan/Grille unit

Fan/Grille unit

False ceiling

Grille

Future cellular office

Alternative All Air System (FGU)

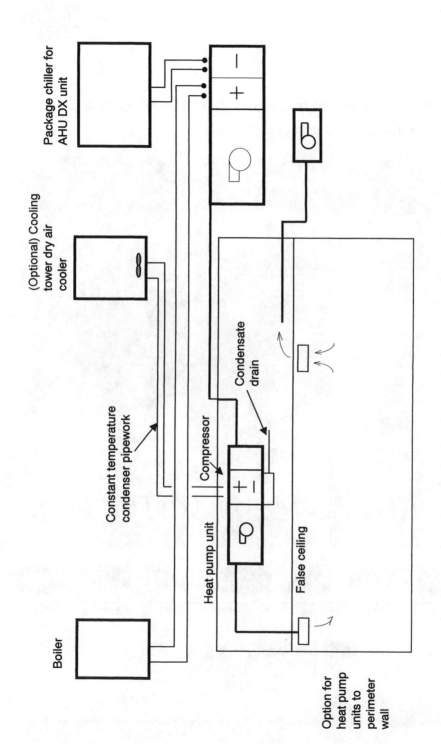

Package chiller for AHU DX unit

(Optional) Cooling tower dry air cooler

Constant temperature condenser pipework

Compressor

Heat pump unit

Condensate drain

False ceiling

Boiler

Option for heat pump units to perimeter wall

Reverse Cycle Heat Pump

DAVIS LANGDON

We maximise value and reduce risk for clients investing in
infrastructure, construction and property

Project Management | Cost Management | Management Consulting | Legal Support | Specification Consulting | Engineering Services | Property Tax & Finance

www.davislangdon.com

DAVIS LANGDON

EUROPE & MIDDLE EAST
office locations

EUROPE & MIDDLE EAST

ENGLAND

DAVIS LANGDON
LONDON
MidCity Place
71 High Holborn
London WC1V 6QS
Tel: (020) 7061 7000
Fax: (020) 7061 7061
Email: simon.johnson@davislangdon.com

BIRMINGHAM
75-77 Colmore Row
Birmingham B3 2HD
Tel: (0121) 710 1100
Fax: (0121) 710 1399
Email: david.daly@davislangdon.com

BRISTOL
St Lawrence House
29/31 Broad Street
Bristol BS1 2HF
Tel: (0117) 927 7832
Fax: (0117) 925 1350
Email: alan.francis@davislangdon.com

CAMBRIDGE
36 Storey's Way
Cambridge CB3 0DT
Tel: (01223) 351 258
Fax: (01223) 321 002
Email: laurence.brett@davislangdon.com

LEEDS
No 4 The Embankment
Victoria Wharf
Sovereign Street
Leeds LS1 4BA
Tel: (0113) 243 2481
Fax: (0113) 242 4601
Email: duncan.sissons@davislangdon.com

LIVERPOOL
Cunard Building
Water Street
Liverpool
L3 1JR
Tel: (0151) 236 1992
Fax: (0151) 227 5401
Email: andrew.stevenson@davislangdon.com

MAIDSTONE
11 Tower View
Kings Hill
West Malling
Kent ME19 4UY
Tel: (01732) 840 429
Fax: (01732) 842 305
Email: nick.leggett@davislangdon.com

MANCHESTER
Cloister House
Riverside
New Bailey Street
Manchester
M3 5AG
Tel: (0161) 819 7600
Fax: (0161) 819 1818
Email: paul.stanion@davislangdon.com

MILTON KEYNES
Everest House
Rockingham Drive
Linford Wood
Milton Keynes
MK14 6LY
Tel: (01908) 304 700
Fax: (01908) 660 059
Email: kevin.sima@davislangdon.com

NORWICH
63 Thorpe Road
Norwich NR1 1UD
Tel: (01603) 628 194
Fax: (01603) 615 928
Email: michael.ladbrook@davislangdon.com

OXFORD
Avalon House
Marcham Road
Abingdon
Oxford OX14 1TZ
Tel: (01235) 555 025
Fax: (01235) 554 909
Email: paul.coomber@davislangdon.com

PETERBOROUGH
Clarence House
Minerva Business Park
Lynchwood
Peterborough PE2 6FT
Tel: (01733) 362 000
Fax: (01733) 230 875
Email: stuart.bremner@davislangdon.com

PLYMOUTH
1 Ensign House
Parkway Court
Longbridge Road
Plymouth PL6 8LR
Tel: (01752) 827 444
Fax: (01752) 221 219
Email: gareth.steventon@davislangdon.com

SOUTHAMPTON
Brunswick House
Brunswick Place
Southampton SO15 2AP
Tel: (023) 8033 3438
Fax: (023) 8022 6099
Email: chris.tremellen@davislangdon.com

LEGAL SUPPORT
LONDON
MidCity Place
71 High Holborn
London WC1V 6QS
Tel: (020) 7061 7000
Fax: (020) 7061 7061
Email: mark.hackett@davislangdon.com

MANAGEMENT CONSULTING
LONDON
MidCity Place
71 High Holborn
London WC1V 6QS
Tel: (020) 7061 7000
Fax: (020) 7061 7005
Email: john.connaughton@davislangdon.com

SPECIFICATION CONSULTING
Davis Langdon Schumann Smith
STEVENAGE
Southgate House
Southgate
Stevenage
SG1 1HG
Tel: (01438) 742 642
Fax: (01438) 742 632
Email: nick.schumann@schumannsmith.com

MANCHESTER
Cloister House
Riverside, New Bailey Street
Manchester M3 5AG
Tel: (0161) 819 7600
Fax: (0161) 819 1818
Email: richard.jackson@davislangdon.com

ENGINEERING SERVICES
Davis Langdon Mott Green Wall
MidCity Place
71 High Holborn
London WC1V 6QS
Tel: (020) 7061 7777
Fax: (020) 7061 7009
Email: barry.nugent@mottgreenwall.co.uk

PROPERTY TAX & FINANCE
LONDON
Davis Langdon Crosher & James
MidCity Place
71 High Holborn
London WC1V 6QS
Tel: (020) 7061 7077
Fax: (020) 7061 7078
Email: tony.llewellyn@crosherjames.com

BIRMINGHAM
102 New Street
Birmingham
B2 4HQ
Tel: (0121) 632 3600
Fax: (0121) 632 3601
Email: clive.searle@crosherjames.com

SOUTHAMPTON
Brunswick House
Brunswick Place
Southampton SO15 2AP
Tel: (023) 8068 2800
Fax: (0870) 048 8141
Email: david.rees@crosherjames.com

SCOTLAND

DAVIS LANGDON
EDINBURGH
39 Melville Street
Edinburgh EH3 7JF
Tel: (0131) 240 1350
Fax: (0131) 240 1399
Email: sam.mackenziel@davislangdon.com

GLASGOW
Monteith House
11 George Square
Glasgow G2 1DY
Tel: (0141) 248 0300
Fax: (0141) 248 0303
Email: sam.mackenzie@davislangdon.com

PROPERTY TAX & FINANCE
Davis Langdon Crosher & James
EDINBURGH
39 Melville Street
Edinburgh EH3 7JF
Tel: (0131) 220 4225
Fax: (0131) 220 4226
Email: ian.mcfarlane@crosherjames.com

GLASGOW
Monteith House
11 George Square
Glasgow G2 1DY
Tel: (0141) 248 0333
Fax: (0141) 248 0313
Email: ken.fraser@crosherjames.com

WALES

DAVIS LANGDON
CARDIFF
4 Pierhead Street
Capital Waterside
Cardiff CF10 4QP
Tel: (029) 2049 7497
Fax: (029) 2049 7111
Email: paul.edwards@davislangdon.com

PROPERTY TAX & FINANCE
Davis Langdon Crosher & James
CARDIFF
4 Pierhead Street
Capital Waterside
Cardiff CF10 4QP
Tel: (029) 2049 7497
Fax: (029) 2049 7111
Email: michael.murray@crosherjames.com

IRELAND

DAVIS LANGDON PKS
CORK
Hibernian House
80A South Mall
Cork. Ireland
Tel: (00 353 21) 4222 800
Fax: (00 353 21) 4222 801
Email: alangmaid@dlpks.ie

DUBLIN
24 Lower Hatch Street
Dublin 2, Ireland
Tel: (00 353 1) 676 3671
Fax: (00 353 1) 676 3672
Email: mwebb@dlpks.ie

GALWAY
Heritage Hall
Kirwan's Lane
Galway
Ireland
Tel: (00 353 91) 530 199
Fax: (00 353 91) 530 198
Email: joregan@dlpks.ie

LIMERICK
8 The Crescent
Limerick, Ireland
Tel: (00 353 61) 318 870
Fax: (00 353 61) 318 871
Email: cbarry@dlpks.ie

SPECIFICATION CONSULTING
DUBLIN
24 Lower Hatch Street
Dublin 2, Ireland
Tel: (00 353 1) 676 3671
Fax: (00 353 1) 676 3672
Email: jhartnett@dlpks.ie

SPAIN

DAVIS LANGDON EDETCO
BARCELONA
C/Muntaner, 479, 1-2
Barcelona 08021
Spain
Tel: (00 34 93) 418 6899
Fax: (00 34 93) 211 0003
Contact: Francesc Monells
Email: barcelona@edetco.com

GIRONA
C/Salt 10 2on
Girona 17005
Spain
Tel: (00 34 97) 223 8000
Fax: (00 34 97) 224 2661
Contact: Francesc Monells
Email: girona@edetco.com

RUSSIA

RUPERTI PROJECT SERVICES
INTERNATIONAL
MOSCOW
Office 5, 15 Myasnitskaya ul
Moscow 101000
Russia
Tel: (00 7 495) 933 7810
Fax: (00 7 495) 933 7811
Email: anthony.ruperti@davislangdon.co

LEBANON

DAVIS LANGDON
BEIRUT
PO Box 13-5422-Shouran
Beirut
Lebanon
Tel: (00 9611) 780 111
Fax: (00 9611) 809 045
Contact: Muhyidden Itani
Email: DLL.MI@cyberia.net.lb

BAHRAIN

DAVIS LANGDON
MANAMA
3rd Floor Building 256
Road No 3605
Area No 336
PO Box 640
Manama
Bahrain
Tel: (00 973) 1782 7567
Fax: (00 973) 1772 7210
Email: steven.coates@davislangdon-bahrain.com

UNITED ARAB EMIRATES

DAVIS LANGDON
DUBAI
PO Box 7856
No. 410
Oud Metha Office Building
UAE
Tel: (00 9714) 32 42 919
Fax: (00 9714) 32 42 838
Email: neil.taylor@davislangdon-dubai.com

QATAR

DAVIS LANGDON
DOHA
PO Box 3206
Doha
State of Qatar
Tel: (00 974) 4580 150
Fax: (00 974) 4697 905
Email:steven.humphrey@davislangdon-qatar.com

EGYPT

DAVIS LANGDON
CAIRO
35 Misr Helwan Road
Maadi 11431
Cairo
Egypt
Tel: (00 20 2) 526 2319
Fax: (00 20 2) 527 1338
Email: bob.ames@dlegypt.com

Specialist Service Lines
Project Management | Cost Management | Management Consulting | Legal Support | Specification Consulting | Engineering Services | Property Tax & Finance

Specialist Sectors
Arts | Commercial Offices | Distribution | Education | Food Processing | Health | Heritage | Hotels & Leisure | Industrial | Infrastructure | Public Buildings | Regeneration | Residential | Retail | Sports | Transportation

Davis Langdon LLP is a member firm of Davis Langdon & Seah International, with offices throughout Europe and the Middle East, Asia, Australasia, Africa and the USA

2nd Edition
Spon's Irish Construction Price Book

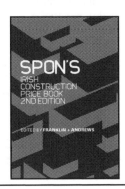

Franklin + Andrews

This new edition of *Spon's Irish Construction Price Book*, edited by Franklin + Andrews, is the only complete and up-to-date source of cost data for this important market.

• All the materials costs, labour rates, labour constants and cost per square metre are based on current conditions in Ireland

• Structured according to the new Agreed Rules of Measurement (second edition)

• 30 pages of Approximate Estimating Rates for quick pricing

This price book is an essential aid to profitable contracting for all those operating in Ireland's booming construction industry.

Franklin + Andrews, Construction Economists, have offices in 100 countries and in-depth experience and expertise in all sectors of the construction industry.

April 2004: 246x174 mm: 448 pages
HB: 0-415-34409-3: £130.00

To Order: Tel: +44 (0) 1264 343071 Fax: +44 (0) 1264 343005, or
Post: Taylor and Francis Customer Services, Thomson Publishing Services, Cheriton House, Andover, Hants, SP10 5BE, UK Email: book.orders@tandf.co.uk

For a complete listing of all our titles visit:
www.tandf.co.uk

Taylor & Francis
Taylor & Francis Group plc

Approximate Estimating

Spon's Estimating Cost Guide to Minor Works, Alterations and Repairs to Fire, Flood, Gale and Theft Damage

Bryan Spain

All the cost data you need to keep your estimating accurate, competitive and profitable.

Do you work on jobs between £50 and £50,000? Then this book is for you.

Specially written for contractors, quantity surveyors and clients carrying out small works, this book contains accurate information on thousands of rates each broken down to labour, material overheads and profit.More than just a price book, it gives easy-to-read professional advice on setting up and running a business.

December 2005: 216x138 mm: 320 pages
PB: 0-415-38213-0: £27.99

To Order: Tel: +44 (0) 1264 343071 Fax: +44 (0) 1264 343005, or
Post: Taylor and Francis Customer Services, Thomson Publishing Services, Cheriton House, Andover, Hants, SP10 5BE, UK Email: book.orders@tandf.co.uk

For a complete listing of all our titles visit:
www.tandf.co.uk

Taylor & Francis
Taylor & Francis Group plc

DIRECTIONS

The prices shown in this section of the book are average prices on a fixed price basis for typical buildings tendered during the second quarter of 2006. Unless otherwise noted, they exclude external services and professional fees.

The information in this section has been arranged to follow more closely the order in which estimates may be developed, in accordance with the RIBA stages of work;

a) Cost Indices and Regional Variations – These provide information regarding the adjustments to be made to estimates taking into account current pricing levels for different locations in the UK.

b) Feasibility Costs – These provide a range of data (based on a rate per square metre) for all-in engineering costs, excluding lifts, associated with a wide variety of building types. These would typically be used at work stage A/B (feasibility) of a project.

c) Elemental Rates – The outline costs for offices have been developed further to provide rates for the alternative solutions for each of the services elements. These would typically be used at work stage C, outline proposal.

Where applicable, costs have been identified as Shell and Core and Fit Out to reflect projects where the choice of procurement has dictated that the project is divided into two distinctive contractual parts.

Such detail would typically be required at work stage D, detailed proposals.

d) All-in-Rates – These are provided for a number of items and complete parts of a system i.e. boiler plant, ductwork, pipework, electrical switchgear and small power distribution, together with lifts and escalators. Refer to the relevant section for further guidance notes.

e) Elemental Costs – These are provided for a diverse range of building types; offices, laboratory, shopping mall, airport terminal building, supermarket, performing arts centre, sports hall, luxury hotel, hospital and secondary school. Also included is a separate analysis of a building management system for an office block. In each case, a full analysis of engineering services costs is given to show the division between all elements and their relative costs to the total building area. A regional variation factor has been applied to bring these analyses to a common London base.

Prices should be applied to the total floor area of all storeys of the building under consideration. The area should be measured between the external walls without deduction for internal walls and staircases/lift shafts i.e. G.I.A (Gross Internal Area).

Although prices are reviewed in the light of recent tenders it has only been possible to provide a range of prices for each building type. This should serve to emphasise that these can only be average prices for typical requirements and that such prices can vary widely depending on variations in size, location, phasing, specification, site conditions, procurement route, programme, market conditions and net to gross area efficiencies. Rates per square metre should not therefore be used indiscriminately and each case needs to be assessed on its own merits.

The prices do not include for incidental builder's work nor for profit and attendance by a Main Contractor where the work is executed as a sub-contract: they do however include for preliminaries, profit and overheads for the services contractor. Capital contributions to statutory authorities and public undertakings and the cost of work carried out by them have been excluded.

Where services works are procured indirectly, i.e. ductwork via a mechanical Sub Contractor, the reader should make due allowance for the addition of a further level of profit etc.

COST INDICES

The following tables reflect the major changes in cost to contractors but do not necessarily reflect changes in tender levels. In addition to changes in labour and materials costs, tenders are affected by other factors such as the degree of competition in the particular industry, the area where the work is to be carried out, the availability of labour and the prevailing economic conditions. This has meant in recent years that, when there has been an abundance of work, tender levels have tended to increase at a greater rate than can be accounted for solely by increases in basic labour and material costs and, conversely, when there is a shortage of work this has tended to result in keener tenders. Allowances for these factors are impossible to assess on a general basis and can only be based on experience and knowledge of the particular circumstances.

In compiling the tables the cost of labour has been calculated on the basis of a notional gang as set out elsewhere in the book. The proportion of labour to materials has been assumed as follows:
Mechanical Services - 30:70, Electrical Services - 50:50, (1976 = 100)

Mechanical Services

Year	First Quarter	Second Quarter	Third Quarter	Fourth Quarter
1997	361	356	358	363
1998	365	363	368	373
1999	368	363	370	384
2000	384	386	388	400
2001	401	401	405	411
2002	411	411	410	442
2003	443	446	447	456
2004	458	464	467	482
2005	486	487	492	508
2006	513	515	517	533P
2007	535P	537F	539F	555F

Electrical Services

Year	First Quarter	Second Quarter	Third Quarter	Fourth Quarter
1997	411	411	411	411
1998	410	423	422	422
1999	433	432	431	446
2000	458	464	465	468
2001	485	484	484	487
2002	508	508	508	513
2003	530	533	533	541
2004	571	574	576	589
2005	607	608	607	615
2006	630	631	632	642P
2007	660P	660F	661F	672F

(P = Provisional)
(F = Forecast)

COST INDICES

Regional Variations

Prices throughout this Book apply to work in the London area (see Directions at the beginning of the Mechanical Installations and Electrical Installations sections). However, prices for mechanical and electrical services installations will of course vary from region to region, largely as a result of differing labour costs but also depending on the degree of accessibility, urbanisation and local market conditions.

The following table of regional factors is intended to provide readers with indicative adjustments that may be made to the prices in the Book for locations outside of London. The figures are of necessity averages for regions and further adjustments should be considered for city centre or very isolated locations, or other known local factors.

Greater London	1.00	Yorkshire & Humberside	0.97
South East	1.00	North West	0.97
South West	0.99	North East	0.97
East Midlands	0.97	Scotland	0.98
West Midlands	0.96	Wales	0.99
East Anglia	0.99	Northern Ireland	0.96

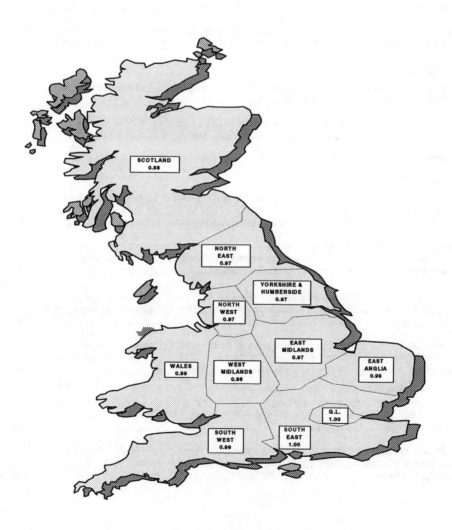

RIBA STAGE A FEASIBILITY COSTS

TYPICAL SQUARE METRE RATES FOR ENGINEERING SERVICES

The following examples indicate the range of rates within each building type for engineering services, excluding lifts etc, utilities services and professional fees. Based on Gross Internal Area (GIA).

Industrial Buildings   £/m² GIA

Factories

Owner occupation: Includes for rainwater, soil/waste, LTHW heating via HL radiant heaters, BMS, LV installations, lighting, fire alarms, security, earthing... 120

Owner occupation: Includes for rainwater, soil/waste, sprinklers, LTHW heating via HL radiant heaters, local air conditioning, BMS, HV/LV installations, lighting, fire alarms, security, earthing..................... 180

Warehouses

High bay for owner occupation: Includes for rainwater, soil/waste, LTHW heating via HL gas fired heaters, BMS, HV/LV installations, lighting, fire alarms, security, earthing... 95

High bay for owner occupation: Includes for rainwater, soil/waste, sprinklers, LTHW heating via HL radiant heaters, local air conditioning, BMS, HV/LV installations, lighting, fire alarms, security, earthing
.... 195

Distribution Centres

High bay for letting: Includes for rainwater, soil/waste, LTHW heating via HL gas fired heaters, BMS, HV/LV installations, lighting, fire alarms, security, earthing... 120

High bay for owner occupation: Includes for rainwater, soil/waste, sprinklers, LTHW heating via HL radiant heaters, local air conditioning, BMS, HV/LV installations, lighting, fire alarms, security...... 225

Office Buildings 5,000m² to 15,000m²

Offices for letting

Shell & Core and Cat A non air conditioned; Includes for rainwater, soil/waste, cold water, hot water via local electrical heaters, LTHW heating via radiator heaters, toilet extract, controls, LV installations, lighting, small power (landlords), fire alarms, earthing, security & IT wireways............................. 250

Shell & Core and Cat A non air conditioned; Includes for rainwater, soil/waste, cold water, hot water, LTHW heating via perimeter heaters, toilet extract, controls, LV installations, lighting, small power (landlords), fire alarms, earthing, security wireways, IT wireways.. 260

Shell & Core and Cat A air conditioned; Includes for rainwater, soil/waste, VRV 3 pipe heat pumps , toilet extract, BMS, LV installations, lighting, small power (landlords), fire alarms, earthing, security & IT wireways, ... 395

Shell & Core and Cat A air conditioned; Includes for rainwater, soil/waste, cold water, hot water via local electrical heaters, LTHW heating via perimeter heaters, 2 pipe , toilet extract, BMS, LV installations, lighting, small power (landlords), fire alarms, earthing, security & IT wireways............... 410

Offices for owner occupation

Non air conditioned; Includes for rainwater, soil/waste, cold water, hot water via local electrical heaters, dry risers, LTHW heating via radiator heaters, toilet extract, controls, LV installations, lighting, small power (landlords), fire alarms, earthing, security, IT wireways... 280

Non air conditioned; Includes for rainwater, soil/waste, cold water, hot water, dry risers, LTHW heating via perimeter heaters, toilet extract, BMS, LV installations, lighting, small power (landlords), fire alarms, earthing, security, IT wireways 290

RIBA STAGE A FEASIBILITY COSTS

TYPICAL SQUARE METRE RATES FOR ENGINEERING SERVICES *continued*

Office Buildings *continued*

Offices for owner occupation *continued* £/m² GIA

Air conditioned; Includes for rainwater, soil/waste, cold water, hot water via local electrical heaters, dry risers, LTHW heating via perimeter heating, 4 pipe air conditioning, toilet extract, kitchen extract, sprinkler protection, BMS, LV installations, life safety standby generators, lighting, small power, fire alarms, earthing, security, IT wireways small power, fire alarms L1/P1, earthing, security, and IT wireways   480

Health and Welfare Facilities

District general hospitals
Natural ventilation Includes for rainwater, soil/waste, cold water, hot water, dry risers, medical gases, LTHW heating via perimeter heating, toilet extract, kitchen extract, BMS, LV installations, standby generation, lighting, small power, fire alarms, earthing/lightning protection, nurse call systems, security, IT wireways...   425

Natural ventilation Includes for rainwater, soil/waste, cold water, hot water, dry risers, LTHW heating via perimeter heaters, localised VAV air conditioning, kitchen/toilet extract, BMS, LV installations, standby generation, lighting, small power, fire alarms, earthing/lightning protection, nurse call systems, security, IT wireways   600

Private hospitals
Air conditioned; Includes for rainwater, soil/waste, cold water, hot water, dry risers, medical gases, LTHW heating, 2 pipe air conditioning, toilet extract, kitchen extract, BMS, LV installations, standby generation, lighting, small power, fire alarms, earthing, nurse call systems, nurse call system, security, IT wireways   630

Air conditioned; Includes for rainwater, soil/waste, cold water, hot water, dry risers, medical gases, LTHW heating , 4 pipe air conditioning, kitchen/toilet extract, BMS, LV installations, standby generation, lighting, small power, fire alarms, earthing, nurse call system, security, IT wireways...   650

Day care unit
Natural ventilation; Includes for rainwater, soil/waste, cold water, hot water via local electrical heaters, medical gases, LTHW heating, toilet extract, kitchen extract, BMS, LV installations, lighting, small power, fire alarms , earthing, security, IT wireways...   450

Comfort cooled; Includes for rainwater, soil/waste, cold water, hot water, medical gases, LTHW heating, DX air conditioning, kitchen/toilet extract, BMS, LV installations, lighting, small power, fire alarms, earthing, security, IT wireways   470

Entertainment and Recreation Buildings

Non Performing
Natural Ventilation; Includes for rainwater, soil/waste, cold water, central hot water, dry risers, LTHW heating, toilet extract, kitchen extract, controls, LV installations, lighting, small power, fire alarms, earthing, security, IT wireways...   330

Comfort cooled; Includes for rainwater, soil/waste, cold water, hot water via local electrical heaters sprinklers/dry risers, LTHW heating , DX air conditioning, kitchen/toilet extract, BMS, LV installations, lighting, small power, fire alarms, earthing, security, IT wireways   525

RIBA STAGE A FEASIBILITY COSTS

TYPICAL SQUARE METRE RATES FOR ENGINEERING SERVICES *continued*

Entertainment and Recreation Buildings *continued*   £/m² GIA

Performing Arts (With Theatre)
Natural Ventilation; Includes for rainwater, soil/waste, cold water, central hot water, sprinklers/dry risers, LTHW heating, toilet extract, kitchen extract, controls, LV installations, lighting, small power, fire alarms, earthing, security, IT wireways   575

Comfort Cooled; Includes for rainwater, soil/waste, cold water, central hot water, sprinklers/dry risers, LTHW heating, DX air conditioning, kitchen/toilet extract, BMS, LV installations, lighting including enhanced dimming/scene setting, small power, fire alarms, earthing, security, IT wireways including for production, audio and video recording, EPOS system, ...   700

Sports Halls
Natural ventilation; Includes for rainwater, soil/waste, cold water, hot water gas fired heaters, LTHW heating, toilet extract, BMS, LV installations, lighting, small power, fire alarms, earthing, security   170

Comfort Cooled; Includes for rainwater, soil/waste, cold water, hot water via LTHW heat exchangers, LTHW heating, air conditioning via AHU's, toilet extract, BMS, LV installations, lighting, small power, fire alarms, earthing, security.   295

Multi Purpose Leisure Centre
Natural ventilation; Includes for rainwater, soil/waste, cold water, hot water gas fired heaters, LTHW heating, toilet extract, BMS, LV installations, lighting, small power, fire alarms, earthing, security, IT wireways   295

Comfort cooled; Includes for rainwater, soil/waste, cold water, hot water LTHW heat exchangers, LTHW heating, air conditioning via AHU, pool hall supply/extract, kitchen/toilet extract, BMS, LV installations, lighting, small power, fire alarms, earthing, security, IT wireways   400

Retail Buildings

Open Arcade
Natural ventilation; Includes for rainwater, soil/waste, cold water, hot water, sprinklers/dry risers, LTHW heating, toilet extract, smoke extract, BMS, LV installations, life safety standby generators, lighting, small power, fire alarms, public address, earthing/lightning protection, security, IT wireways   290

Enclosed Shopping Mall
Air conditioned; Includes for rainwater, soil/waste, cold water, hot water, sprinklers/dry risers, LTHW heating, air conditioning via AHU's, toilet extract, smoke extract, BMS, LV installations, life safety standby generators, lighting, small power, fire alarms, public address, earthing/lightning protection, CCTV/security, IT wireways people counting systems   510

Department Stores
Air conditioned; Includes for rainwater, soil/waste, cold water, hot water, sprinklers/dry risers, LTHW heating, air conditioning via AHU's, toilet extract, smoke extract, BMS, LV installations, life safety standby generators, lighting, small power, fire alarms, public address, earthing, lightning protection, CCTV/security, IT wireways, people counting systems (footfall).   370

RIBA STAGE A FEASIBILITY COSTS

TYPICAL SQUARE METRE RATES FOR ENGINEERING SERVICES *continued*

Retail Buildings *continued*   £/m² GIA

Supermarkets
Comfort cooled; Includes for rainwater, soil/waste, cold water, hot water, sprinklers, LTHW heating, 2 pipe air conditioning, toilet extract, BMS, LV installations, lighting, small power, fire alarms, earthing, lightning protection, security, IT wireways, refrigeration...   390

Air conditioned; Includes for rainwater, soil/waste, cold water, hot water, sprinklers, LTHW heating, air conditioning via AHU's, toilet extract, BMS, LV installations, lighting, small power, fire alarms, earthing, lightning protection, security, IT wireways, refrigeration...   590

Educational Buildings

Secondary Schools (Academy)
Natural Ventilation; Includes for rainwater, soil/waste, cold water, central hot water, LTHW heating, toilet extract, BMS, LV installations, lighting, small power, fire alarms, earthing, security, IT wireways   290

Natural vent with comfort cooling to selected areas (BB93 compliant); Includes for rainwater, soil/waste, cold water, central hot water, LTHW heating, DX air conditioning, general supply/extract, toilet extract, BMS, LV installations, lighting, small power, fire alarms, earthing, security, IT wireways   355

Scientific Buildings

Educational Research
Comfort cooled; Includes for rainwater, soil/waste, cold water, central hot water, dry risers, compressed air, medical gases, LTHW heating, 4 pipe air conditioning, toilet extract, fume, BMS, LV installations, lighting, small power, fire alarms, earthing, lightning protection, security, IT wireways   730

Air conditioned; Includes for rainwater, soil/waste, laboratory waste, cold water, central hot water, specialist water, dry risers, compressed air, medical gases, steam, LTHW heating, 4 pipe air conditioning, Comm's room cooling, toilet extract, fume extract, BMS, LV installations, UPS, standby generators, lighting, small power, fire alarms, earthing, lightning protection, security, IT wireways   1255

Commercial Research
Air conditioned; Includes for rainwater, soil/waste, laboratory waste, cold water, central hot water, specialist water, dry risers, compressed air, medical gases, steam, LTHW heating, 4 pipe air conditioning, Comm room cooling, toilet extract, fume extract, BMS, LV installations, UPS, standby generators, lighting, small power, fire alarms, earthing, lightning protection, security, IT wireways   1255

Air conditioned; Includes for rainwater, soil/waste, laboratory waste, cold water, central hot water, specialist water, dry risers, compressed air, medical gases, steam, LTHW heating, VAV air conditioning, toilet extract, fume extract, cold rooms, BMS, LV installations, UPS...   1880

Comfort cooled throughout (sealed façade) to comply with BB93 including rainwater, soil/waste, cold water, central hot water, dry risers, compressed air, medical gases, LTHW heating, 4 pipe air conditioning, toilet extract, fume, BMS, LV installations, lighting, small power, fire alarms, earthing, lightning protection, security IT wireways...   1440

RIBA STAGE A FEASIBILITY COSTS

TYPICAL SQUARE METRE RATES FOR ENGINEERING SERVICES *continued*

£/m² GIA

Hotels

1 to 3 Star; Includes for rainwater, soil/waste, cold water, hot water, dry risers, LTHW heating via radiators, toilet/bathroom extract, kitchen extract, BMS, LV installations, lighting, small power, fire alarms, earthing, lightning protection security, IT wireways 350 to 550
....

4 to 5 Star; Includes for rainwater, soil/waste, cold water, hot water, sprinklers, dry risers, 4 pipe air conditioning, kitchen extract, toilet/bathroom extract, BMS, LV installations, life safety standby generators, lighting, small power, fire alarms, earthing, security, IT wireways........................... 750

Airport Buildings

Air conditioned; Includes for rainwater, soil/waste, cold water, hot water via local electrical heaters, dry risers, LTHW heating via perimeter heating, 2 pipe air conditioning, toilet extract, kitchen extract, BMS, LV installations, lighting, small power, fire alarms, earthing, lightning protection, security, IT wireways
.... 617

Air conditioned; Includes for rainwater, soil/waste, cold water, hot water via local electrical heaters sprinklers, dry risers, LTHW heating via perimeter heaters, VAV air conditioning, Comms room cooling, kitchen extract, toilet extract, car park extract, BMS, LV installations, lighting, small power, fire alarms, earthing, lightning protection, security, IT wireways 790

RIBA STAGE C ELEMENTAL RATES

ELEMENTAL RATES FOR ALTERNATIVE ENGINEERING SERVICES SOLUTIONS

The following examples of building types indicate the range of rates for alternative design solutions for each of the engineering services elements based on Gross Internal Area for the Shell and Core and Net Internal Area for the Fit Out. Fit Out is assumed to be to Cat A standard.

Consideration should be made for the size of the building, which may affect the economies of scale for rates i.e. the larger the building the lower the rates

OFFICES MECHANICAL SERVICES	Shell & Core £/m² GIA.	Fit Out £/m² NIA.
Sanitaryware		
Building up to 3,000m²	7 to 10	-
Building over 3,000m² to 15,000m² (low rise)	5 to 8	-
Disposal installation		
Building up to 3,000m²	15 to 25	-
Building over 3,000m² to 15,000m²	10 to 15	-
Water installation		
Building up to 3,000m²	5 to 10	-
Building over 3,000m² to 15,000m²	10 to 15	-
LPHW Heating Installation; including gas installations		
Building up to 3,000m²	30 to 40	40 to 50
Building over 3,000m² to 15,000m²	25 to 30	30 to 40
Air Conditioning; including ventilation		
Comfort Cooling:		
2 pipe fan coil for building up to 3,000m²	50 to 65	90 to 100
2 pipe fan coil for building over 3,000m² to 15,000m²	45 to 60	80 to 90
2 pipe variable refrigerant volume (VRV) for building up to 3,000m²	35 to 45	60 to 70
Full air conditioning:		
4 Pipe fan coil for building up to 3,000m²	75 to 90	130 to 150
4 Pipe fan coil for building over 3,000m² to 15,000m²	65 to 80	110 to 130
3 pipe variable refrigerant volume for building up to 3,000m²	60 to 70	110 to 120
Ventilated (active) chilled beams for building over 3,000m² to 15,000m²	65 to 80	130 to 140
Chilled beam exposed services for building over 3,000m² to 15,000m²	65 to 80	220 to 240
Concealed passive chilled beams for building over 3,000m to 15,000m²	65 to 80	120 to 130
Chilled ceiling for building over 3,000m² to 15,000m²	65 to 80	210 to 220
Chilled ceiling/perimeter beams for building over 3,000m² to 15,000m²	65 to 80	230 to 250
Displacement for building over 3,000m² to 15,000m²	75 to 90	70 to 90

RIBA STAGE C ELEMENTAL RATES

ELEMENTAL RATES FOR ALTERNATIVE ENGINEERING SERVICES SOLUTIONS *continued*

OFFICES **MECHANICAL SERVICES** *continued*	Shell & Core £/m² GIA.	Fit Out £/m² NIA.
Ventilation systems		
Building up to 3,000m²	25 to 30	-
Building over 3,000m² to 15,000m²...	20 to 25	-
Fire Protection over 3,000 m² to 15,000m²		
Dry risers	3 to 5	-
Sprinkler installation	15 to 25	20 to 25
BMS Controls: including MCC panels and control cabling		
Full air conditioning, fan coil/chilled ceiling...	15 to 20	15 to 20
Full air conditioning, chilled beam...	15 to 20	15 to 20

RIBA STAGE C ELEMENTAL RATES

ELEMENTAL RATES FOR ALTERNATIVE ENGINEERING SERVICES SOLUTIONS *continued*

OFFICES ELECTRICAL SERVICES	Shell & Core £/m² GIA.	Fit Out £/m² NIA.
LV Installations		
Standby generators (life safety only)		
Buildings up to 3,000m²...	16 to 25	-
Buildings over 3,000m² to 15,000m²...	10 to 15	-
LV distribution		
Buildings up to 3,000m²...	15 to 20	-
Buildings over 3,000m² to 15,000m²...	20 to 25	-
Lighting Installations (including lighting controls and luminaries)		
Buildings up to 3,000m²...	10 to 15	40 to 60
Buildings over 3,000m² to 15,000m²...	15 to 20	40 to 60
Small Power		
Buildings up to 3,000m²...	4 to 8	-
Buildings over 3,000m² to 15,000m²...	2 to 5	-
Protective Installations		
Earthing		
Buildings up to 3,000m²...	2 to 3	1 to 2
Buildings over 3,000m² to 15,000m²...	2 to 3	1 to 2
Lightning Protection		
Buildings up to 3,000m²...	3 to 4	-
Buildings over 3,000m² to 15,000m²...	2 to 3	-
Communication Installations		
Fire Alarms (single stage)		
Buildings up to 3,000m²...	6 to 10	8 to 10
Buildings over 3,000m² to 15,000m²...	6 to 8	8 to 10
Fire Alarms (phased evacuation)		
Buildings over 3,000m² to 15,000m²...	10 to 12	10 to 15
IT (Wireways only)		
Buildings up to 3,000m²...	2 to 3	-
Buildings over 3,000m² to 15,000m²...	2 to 3	-

RIBA STAGE C ELEMENTAL RATES

ELEMENTAL RATES FOR ALTERNATIVE ENGINEERING SERVICES SOLUTIONS *continued*

OFFICES **ELECTRICAL SERVICES** *continued*	Shell & Core £/m² G.I.A.	Fit Out £/m² N.I.A.
Security		
Buildings up to 3,000m²..	10 to 12	-
Buildings over 3,000m² to 15,000 m².......................................	8 to 10	-
Electrical Installations for Mechanical Plant		
Buildings up to 3,000m²..	5 to 8	-
Buildings over 3,000m² to 15,000 m².......................................	5 to 8	-

RIBA STAGE C ELEMENTAL RATES

ELEMENTAL RATES FOR ALTERNATIVE ENGINEERING SERVICES SOLUTIONS *continued*

HOTELS
MECHANICAL SERVICES   £/m² GIA.

Sanitaryware and above ground disposal installation

2 to 3 Star...	20 to 30
4 to 5 Star...	20 to 30

Water installation

2 to 3 Star...	28 to 38
4 to 5 Star...	40 to 50

LPHW Heating Installation; including gas installations

2 to 3 Star...	30 to 45
4 to 5 Star...	30 to 45

Air Conditioning; including ventilation

2 to 3 Star – 4 pipe Fan coil..	200 to 240
4 to 5 Star – 4 pipe Fan coil..	200 to 240
2 to 3 Star - 3 pipe variable refrigerant volume	130 to 150
4 to 5 Star - 3 pipe variable refrigerant volume	130 to 150

Fire Protection

2 to 3 Star - Dry risers..	8 to 12
4 to 5 Star - Dry risers..	8 to 12
2 to 3 Star - Sprinkler installation....................................	5 to 10
4 to 5 Star - Sprinkler installation....................................	5 to 10

BMS Controls; including MCC panels and control cabling

2 to 3 Star...	10 to 12
4 to 5 Star...	20 to 30

HOTELS
ELECTRICAL SERVICES

LV Installations

Standby generators (life safety only)

2 to 3 Star...	10 to 20
4 to 5 Star...	10 to 20

LV distribution

2 to 3 Star...	25 to 35
4 to 5 Star...	35 to 45

RIBA STAGE C ELEMENTAL RATES

ELEMENTAL RATES FOR ALTERNATIVE ENGINEERING SERVICES SOLUTIONS *continued*

HOTELS
ELECTRICAL SERVICES *continued*

	£/m² GIA.
Lighting Installations	
2 to 3 Star	15 to 25
4 to 5 Star	15 to 25
Small Power	
2 to 3 Star	5 to 10
4 to 5 Star	10 to 15
Protective Installations	
Earthing	
2 to 3 Star	1 to 2
4 to 5 Star	1 to 2
Lightning Protection	
2 to 3 Star	1 to 2
4 to 5 Star	1 to 2
Communication Installations	
Fire Alarms	
2 to 3 Star	10 to 15
4 to 5 Star	10 to 15
IT	
2 to 3 Star	10 to 20
4 to 5 Star	10 to 20
Security	
2 to 3 Star	15 to25
4 to 5 Star	15 to25
Electrical Installations for Mechanical Plant	
2 to 3 Star	5 to 8
4 to 5 Star	5 to 8

RIBA STAGE C ELEMENTAL RATES

ELEMENTAL RATES FOR ALTERNATIVE ENGINEERING SERVICES SOLUTIONS *continued*

RESIDENTIAL MECHANICAL & ELECTRICAL SERVICES	Shell & Core £/m² GIA.	Fit Out £/m² NIA.
Sanitaryware and above ground disposal installation		
Affordable	0 to 5	25 to 30
Private	0 to 5	50 to 75
Disposal installation		
Affordable	10 to 15	5 to 15
Private	15 to 20	9 to 16
Water installation		
Affordable	25 to 30	30 to 65
Private	27 to 32	35 to 65
Heat Source		
Affordable	7 to 10	0 to 15
Private	9 to 12	0 to 20
Space Heating & Air Treatment		
Affordable	5 to 10	20 to 45
Private	50 to 65	100 to 230
Ventilation (to façade)		
Affordable	5 to 10	20 to 25
Private	5 to 15	20 to 40
Electrical Installations		
Affordable	25 to 30	35 to 55
Private	35 to 40	65 to 85
Gas Installations		
Affordable	1 to 2	0 to 8
Private	2 to 3	4 to 8
Protective Installations		
Affordable	4 to 6	-
Private	6 to 8	-
Communication Installations		
Affordable	20 to 30	15 to 30
Private	25 to 35	25 to 55
Special Installations		
Affordable	10 to 15	2 to 5
Private	10 to 20	2 to 5

Note: The range in cost differs due to the vast diversity in services strategies available. The lower end of the scale reflects all electric schemes (not always be possible due to Part L requirements) or local plant within apartment schemes, such as combi boilers, local ventilation etc. The high end of the scale is based on good quality apartments, which includes comfort cooling, home network installations, video entry, higher quality of sanitaryware and lighting.

ALL-IN-RATES

ALL-IN-RATES FOR PRICING MECHANICAL APPROXIMATE QUANTITIES

	Cost per Point £
ABOVE GROUND DRAINAGE	
Soil and Waste	350 – 400
WATER INSTALLATIONS	
Cold Water	350 - 400
Hot Water	400 - 450

	Cost Per kW £
HEAT SOURCE	
Gas fired boilers including gas train and controls	25 - 30
Gas fired boilers including gas train, controls, flue, Plantroom pipework, valves and insulation, pumps and pressurisation unit	65 – 120

SPACE HEATING AND AIR TREATMENT

	Cost per kW £
CHILLED WATER	
Air cooled R134a refrigerant chiller including control panel, anti vibration mountings	120 - 140
Air cooled R134a refrigerant chiller including control panel, anti-vibration mountings, plantroom pipework, valves, insulation, pumps and pressurisation units	160 - 220
Water cooled R134a refrigerant chiller including control panel, anti vibration mountings	60 – 80
Water cooled R134a refrigerant chiller including control panel, anti-vibration mountings, plantroom pipework, valves, insulation, pumps and pressurisation units	110 - 190
Absorption steam medium chiller including control panel, anti-vibration mountings, plantroom pipework, valves, insulation, pumps and pressurisation units	170 - 270

	Cost per kW (heat rejection) £
HEAT REJECTION	
Open circuit, forced draft cooling tower	50 – 60
Closed circuit, forced draft cooling tower	65 - 75
Dry Air	50 – 60

ALL-IN-RATES

ALL-IN-RATES FOR PRICING MECHANICAL APPROXIMATE QUANTITIES *continued*

SPACE HEATING AND AIR TREATMENT *continued*

	Cost per kPa £
PUMPS	
Pumps including flexible connections, anti-vibration mountings	12 - 60
Pumps including flexible connections, anti-vibration mountings, plantroom pipework, valves, insulation and accessories	40 - 130

DUCTWORK

The rates below allow for ductwork and for all other labour and material in fabrication, fittings, supports and jointing to equipment, stop and capped ends, elbows, bends, diminishing and transition pieces, regular and reducing couplings, volume control dampers, branch diffuser and 'snap on' grille connections, ties, 'Ys', crossover spigots, etc., turning vanes, regulating dampers, access doors and openings, hand-holes, test holes and covers, blanking plates, flanges, stiffeners, tie rods and all supports and brackets fixed to structure.

	Per m^2 of duct £
Rectangular galvanised mild steel ductwork as HVCA DW 144 up to 1000mm longest side	40 – 45
Rectangular galvanised mild steel ductwork as HVCA DW 144 up to 2500mm longest side	45 – 55
Rectangular galvanised mild steel ductwork as HVCA DW 144 3000mm longest side and above	60 – 65
Circular galvanised mild steel ductwork as HVCA DW 144	45 – 55
Flat oval galvanised mild steel ductwork as HVCA DW 144 up to 545mm wide	45 – 50
Flat oval galvanised mild steel ductwork as HVCA DW 144 up to 880mm wide	50 – 55
Flat oval galvanised mild steel ductwork as HVCA DW 144 up to 1785mm wide	60 – 65

	Cost per m^3/s £
PACKAGED AIR HANDLING UNITS	
Air handling unit including LPHW pre-heater coil, pre-filter panel, LPHW heater coils, chilled water coil, filter panels, inverter drive, motorised volume control dampers, sound attenuation, flexible connections to ductwork and all anti-vibration mountings.	4,500 – 6,000

EXTRACT FANS

Extract fan including inverter drive, sound attenuation, flexible connections to ductwork and all anti-vibration mountings	1,000 – 2,000

ALL-IN-RATES

ALL-IN-RATES FOR PRICING MECHANICAL APPROXIMATE QUANTITIES *continued*

PROTECTIVE INSTALLATIONS

SPRINKLER INSTALLATION   £

Recommended maximum area coverage per sprinkler head:
Extra light hazard, 21 m² of floor area
Ordinary hazard, 12 m² of floor area
Extra high hazard, 9 m² of floor area

Sprinkler equipment installation, pipework, valve sets, booster pumps and water storage...   50,000 - 75,000

Price per sprinkler head; including pipework, valves and supports.......................   175

PROTECTIVE INSTALLATIONS

HOSE REELS AND DRY RISERS

Wall mounted concealed hose reel with 36 metre hose including approximately 15 metres of pipework and isolating valve:

Price per hose reel..   1,500

100mm dry riser main including 2 way breeching valve and box,, 65mm landing valve, complete with padlock and leather strap and automatic air vent and drain valve.

Price per landing..   1,400

COMMUNICATIONS INSTALLATIONS

SECURITY

ACCESS CONTROL SYSTEMS

Door Mounted access control unit inclusive of door furniture, lock plus software. Including up to 50 meters of cable and termination. Including documentation testing and commissioning

Internal single leaf door...	1,100
Internal double door	1,250
External single leaf door	1,200
External Double leaf door...	1,350
Management control PC with printer software and commissioning up to 1000 users...	9,250

CCTV INSTALLATIONS

CCTV Equipment inclusive of 50 m of cable including testing and commissioning

Internal camera with Bracket	900
Internal camera with Housing..	950
Internal PTZ camera with Bracket..	1,400
External fixed camera with housing..	950
External PTZ camera dome...	2,100
External PTZ camera dome with power...	2,600

ALL-IN-RATES

ALL-IN-RATES FOR PRICING MECHANICAL APPROXIMATE QUANTITIES *continued*

IT INSTALLATIONS

DATA CABLING

Complete channel link including, patch leads, cable, panels, testing and documentation (excludes cabinets and/or frames as well as containment).

	Cost per point £
Cat 5e up to 5,000 outlets..........	60.00
Cat 5e 5,000 to 15,000 outlets.........	55.00
Cat 6 up to 5,000 outlets....	67.00
Cat 6 5,000 to 15,000 outlets...	65.00

PIPEWORK

HOT AND COLD WATER *excludes insulation, valves and ancillaries etc*

Light gauge copper tube to EN1057 R250 (TX) formerly BS 2871 part 1 table X with joints as described including allowance for waste, fittings and supports assuming average runs with capillary joints up to 54mm and bronze welded thereafter

	Cost per metre £
Horizontal High Level Distribution	
15mm	24.82
22mm	26.88
28mm	31.09
35mm	39.87
42mm	47.66
54mm	59.89
67mm	81.89
76mm	95.81
108mm	125.68
Risers	
15mm	16.57
22mm	19.52
28mm	25.68
35mm	32.91
42mm	39.14
54mm	49.86
67mm	76.94
76mm	90.62
108mm	119.46
Toilet Areas etc at Low Level	
15mm	44.58
22mm	52.31
28mm	65.45

ALL-IN-RATES

ALL-IN-RATES FOR PRICING MECHANICAL APPROXIMATE QUANTITIES *continued*

PIPEWORK *continued*

LTHW AND CHILLED WATER *excludes insulation, valves and ancillaries etc*

Black heavy weight mild steel tube to BS1387 with joints in the running length, allowance for waste, fittings and supports assuming average runs

	Cost per metre	
	LTHW £	*Chilled Water* £
Horizontal Distribution – Basements etc		
15mm	36.60	37.78
20mm	37.67	40.73
25mm	45.08	45.85
32mm	52.65	53.43
40mm	59.91	60.62
50mm	72.65	73.77
65mm	81.43	82.20
80mm	107.69	108.37
100mm	135.84	137.00
125mm	173.19	173.83
150mm	200.77	201.60
200mm	290.80	291.27
250mm	384.17	384.92
300mm	432.85	433.49
Risers		
15mm	22.41	23.22
20mm	24.73	25.38
25mm	28.82	29.49
32mm	34.11	34.78
40mm	38.02	38.52
50mm	47.15	47.93
65mm	63.46	63.96
80mm	82.29	82.73
100mm	103.29	104.05
125mm	141.82	142.30
150mm	165.87	166.61
200mm	217.72	218.07
250mm	292.04	292.60
300mm	309.13	309.61
On Floor Distribution		
15mm	34.93	36.03
20mm	38.81	39.60
25mm	43.70	44.51
32mm	51.09	51.90
40mm	58.16	58.90
50mm	71.03	72.30
65mm	-	77.15

ALL-IN-RATES

ALL-IN-RATES FOR PRICING MECHANICAL APPROXIMATE QUANTITIES *continued*

PIPEWORK *continued*

LTHW AND CHILLED WATER *excludes insulation, valves and ancillaries etc (continued)*

Black heavy weight mild steel tube to BS1387 with joints in the running length, allowance for waste, fittings and supports assuming average runs

	Cost per metre	
	LTHW £	*Chilled Water* £
Plantroom Areas etc		
15mm	37.25	38.40
20mm	41.25	42.06
25mm	46.50	47.31
32mm	54.31	55.12
40mm	61.41	62.16
50mm	74.76	75.92
65mm	83.22	83.98
80mm	110.32	110.99
100mm	139.18	140.34
125mm	177.98	176.83
150mm	207.68	208.66
200mm	299.50	299.97
250mm	339.89	340.64
300mm	448.83	449.68

ALL-IN-RATES

ALL-IN-RATES FOR PRICING ELECTRICAL APPROXIMATE QUANTITIES

HV/LV INSTALLATIONS

The cost of HV/LV equipment will vary according to the electricity supplier's requirements, the duty required and the actual location of the site. For estimating purposes the items indicated below are typical of the equipment required in a HV substation incorporated into a building.

	Cost per Unit £
RING MAIN UNIT	
Ring Main Unit , 11kv including electrical terminations...	12,000 - 18,000

	Cost per KVA £
TRANSFORMERS	
Oil filled transformers, 11kv to 415v including electrical terminations	12 to 16
Cast Resin transformers, 11kv to 415v including electrical terminations...	13 to 18

	Cost per Section £
HV SWITCHGEAR	
Cubicle section HV switchpanel, Form 4 type 6 including air circuit breakers, meters and electrical terminations...	15,000 – 0,000

	Cost per isolator £
LV SWITCHGEAR	
LV switchpanel, Form 3 including all isolators, fuses, meters and electrical terminations...	1,800 - 2,800
LV switchpanel, Form 4 type 5 including all isolators, fuses, meters and electrical terminations...	3,000 – 4,000

	£
EXTERNAL PACKAGED SUB-STATION	
Extra over cost for prefabricated packaged sub station housing excludes base and protective security fencing	20,000 – 25,000

	Cost per KVA £
STANDBY GENERATING SETS	
Diesel powered including control panel, flue, oil day tank and attenuation	
Approximate installed cost, LV	170 - 220
Approximate installed cost, HV	190 - 240

UNINTERRUPTIBLE POWER SUPPLY

Rotary UPS including control panel, automatic bypass, DC isolator and batteries for 30 minutes standby (excludes distribution)	
Approximate installed cost (range 100KVA to 1000KVA)	250 - 350
Static UPS including control panel, automatic bypass, DC isolator and batteries for 30 minutes standby (excludes distribution)	
Approximate installed cost (range 100KVA to 1000KVA)...	200 - 300

ALL-IN-RATES

ALL-IN-RATES FOR PRICING ELECTRICAL APPROXIMATE QUANTITIES *continued*

SMALL POWER

Approximate prices for wiring of power points of length not exceeding 20m, including accessories, wireways but excluding distribution boards.

Per Point
£

13 amp Accessories
Wired in PVC insulated twin and earth cable in ring main circuit
Domestic properties.. 55.00
Commercial properties.. 75.00
Industrial properties.. 75.00
Wired in PVC insulated twin and earth cable in radial circuit
Domestic properties.. 70.00
Commercial properties.. 90.00
Industrial property... 90.00
Wired in LSF insulated single cable in ring main circuit
Commercial properties.. 90.00
Industrial property 90.00
Wired in LSF insulated single cable in radial circuit
Commercial properties.. 110.00
Industrial property... 110.00
45 amp wired in PVC insulated twin and earth cable
Domestic properties.. 110.00

Low voltage power circuits
Three phase four wire radial circuit feeding an individual load, wired in LSF insulated single cable including all wireways, isolator, *not exceeding 10 metres; in commercial properties*.

Cable size mm²		£
1.5		180.00
2.5		195.00
4		210.00
6		230.00
10		265.00
16		290.00

Three phase four core radial circuit feeding an individual load item, wired in LSF/SWA/XLPE insulated cable including terminations, isolator; clipped to surface, *not exceeding 10 metres in commercial properties*.

Cable size mm²		£
1.5		133.00
2.5		148.00
4		168.00
6		180.00
10		276.00
16		357.00

ALL-IN-RATES

ALL-IN-RATES FOR PRICING ELECTRICAL APPROXIMATE QUANTITIES *continued*

LIGHTING

Approximate prices for wiring of lighting points including rose, wireways but excluding distribution boards, luminaires and switches.

	Per Point £
Final Circuits	
Wired in PVC insulated twin and earth cable	
Domestic properties	40.00
Commercial properties	50.00
Industrial properties	50.00
Wired in LSF insulated single cable	
Commercial properties	65.00
Industrial property	65.00

ELECTRICAL WORKS IN CONNECTION WITH MECHANICAL SERVICES

The cost of electrical connections to mechanical services equipment will vary depending on the type of building and complexity of the equipment. Therefore a rate of £ 5.00 per m² of gross floor area should be a useful guide to allow for power wiring, isolators and associated wireways.

FIRE ALARMS

Cost per point for two core MICC insulated wired system including all terminations, supports and wireways.

Call point	247.00
Smoke detector	216.00
Smoke/heat detector	247.00
Heat detector	238.00
Heat detector and sounder	209.00
Input/output/relay units	302.00
Alarm sounder	234.00
Alarm sounder/beacon	273.00
Speakers/voice sounders	274.00
Speakers/voice sounders (weatherproof)	288.00
Beacon/strobe	215.00
Beacon/strobe (weatherproof)	317.00
Door release units	315.00
Beam detector	844.00

ALL-IN-RATES

ALL-IN-RATES FOR PRICING ELECTRICAL APPROXIMATE QUANTITIES *continued*

FIRE ALARMS *continued*

Cost per point for wireless system

	Per Point £
Call point	188.00
Smoke detector	187.00
Smoke/heat detector	225.00
Heat detector	196.00
Heat detector and sounder	394.00
Input/output/relay units	361.00
Alarm sounder	357.00
Alarm sounder/beacon	409.00
Speakers/voice sounders	371.00
Speakers/voice sounders (weatherproof)	448.00
Beacon/strobe	382.00
Beacon/strobe (weatherproof)	379.00
Door release units	351.00
Beam detector	998.00

For costs for zone control panel, battery chargers and batteries, see 'Prices for Measured Work' section.

EXTERNAL LIGHTING

Estate road lighting
Post type road lighting lantern 70 watt CDM-T 3000k complete with 5m high column with hinged lockable door, control gear and cut-out including 2.5 mm two core butyl cable internal wiring, interconnections and earthing fed by 16 mm² four core XLPE/SWA /LSF cable and terminations. Approximate installed price *per metre road length* (based on 300 metres run) including time switch but excluding builder's work in connection

Columns erected at 30 m intervals ... ... £42.00 per m of road

Bollard lighting
Bollard lighting fitting 26 watt TC-D 3500k including control gear, all internal wiring, interconnections, earthing and 25 metres of 2.5 mm² three core XLPE/SWA/LSF cable

Approximate installed price excluding builder's work in connection ... ... £893.00 *each*

Outdoor flood lighting
Wall mounted outdoor flood light fitting complete with tungsten halogen lamp, mounting bracket, wire guard and all internal wiring; fixed to brickwork or concrete and connected.

Installed price 500 watt ... ... £84.00 - £147.00
Installed price 1000 watt... ... ... £116.00 - £168.00

Pedestal mounted outdoor floor light fitting complete with1000 watt MBF/U lamp, mounting bracket, control gear, contained in weatherproof steel box, all internal wiring, interconnections and earthing; fixed to brickwork or concrete and connected

Approximate installed price excluding builder's work in connection ... ... £1,019.00 *each*

ALL-IN-RATES

ALL-IN-RATES FOR PRICING SPECIALIST APPROXIMATE QUANTITIES

LIFT INSTALLATIONS

The cost of lift installations will vary depending upon a variety of circumstances. The following prices assume a car height of 2.2 metres, manufacturers standard car finish, brushed stainless steel 2 panel centre opening doors to BSEN81 part 1 & 2 and Lift Regulations 1997.

Passenger Lifts – Machine Room Above	8 Person £	10 Person £	13 Person £	17 Person £	21 Person £	26 Person £
Electrically operated AC drive serving 2 levels with directional collective controls and a speed of 1.0 m/s	53,281	57,632	60,576	68,358	77,511	88,932
As above serving 4 levels and a speed of 1.0m/s	62,025	66,298	69,684	78,347	88,516	101,228
As above serving 6 levels and a speed of 1.0m/s	70,581	74,941	78,602	88,148	99,300	113,242
As above serving 8 levels and a speed of 1.0m/s	79,136	83,587	87,520	97,948	110,084	125,255
As above serving 10 levels and a speed of 1.0m/s	87,690	92,229	96,436	107,749	120,867	137,268
As above serving 12 levels and a speed of 1.0m/s	96,245	100,874	105,355	117,549	131,651	149,281
As above serving 14 levels and a speed of 1.0m/s	106,505	87,417	115,977	129,053	143,383	161,293
Add to above for:						
Increase speed from 1.0 to 1.6 m/s	3,783	3,586	3,832	3,821	3,821	3,821
Increase speed from 1.6m/s to 2.0m/s	802	802	1,066	1,066	1,346	1,346
Increase speed from 2.0m/s to 2.5m/s	1,863	1,863	2,277	2,277	2,681	2,681
Enhanced finish to car – Centre mirror, flat ceiling, carpet	2,573	2,811	2,734	3,209	3,730	4,383

ALL-IN-RATES

ALL-IN-RATES FOR PRICING SPECIALIST APPROXIMATE QUANTITIES *continued*

LIFT INSTALLATIONS *continued*

Passenger Lifts Machine Room Above (cont'd)	8 Person £	10 Person £	13 Person £	16 Person £	21 Person £	26 Person £
Bottom motor room	6,676	6,676	6,676	8,021	8,021	8,280
Fire fighting control	5,330	5,330	5,330	5,330	5,330	5,330
Glass back	2,277	2,639	3,157	3,830	3,830	3,830
Glass doors	18,320	18,320	20,102	20,700	20,700	20,700
Painting to entire pit	1,974	1,974	1,974	1,974	1,974	1,974
Dual seal shaft	3,867	3,867	3,867	4,645	4,645	4,645
Dust sealing machine room	745	745	1,242	1,242	1,242	1,242
Intercom to reception desk and security room	350	350	350	350	350	350
Heating, cooling and ventilation to machine room	776	776	776	776	776	776
Shaft lighting / small power	3,705	3,705	3,705	3,705	3,705	3,705
Motor room lighting / small power	1,242	1,242	1,397	1,532	1,532	1,532
Lifting beams	1,267	1,267	1,267	1,267	1,267	1,267
10mm Equipotential bonding of all entrance metalwork	838	838	838	838	838	838
Shaft secondary steelwork	5,615	5,744	6,003	6,138	6,138	6,138
Independent insurance inspection	1,809	1,809	1,809	1,809	1,809	1,809
12 Month warranty service	1,750	1,750	1,750	1,750	1,750	1,750

ALL-IN-RATES

ALL-IN-RATES FOR PRICING SPECIALIST APPROXIMATE QUANTITIES *continued*

LIFT INSTALLATIONS *continued*

Passenger Lifts — Machine room-less	8 Person £	10 Person £	13 Person £	17 Person £	21 Person £	26 Person £
Electrically operated AC drive serving 2 levels with directional collective controls and a speed of 1.0 m/s	48,134	52,786	56,293	68,240	74,593	82,360
As above serving 4 levels and a speed of 1.0m/s	56,285	60,482	64,773	77,358	84,025	92,786
As above serving 6 levels and a speed of 1.0m/s	64,465	68,700	73,142	86,362	93,380	103,022
As above serving 8 levels and a speed of 1.0m/s	72,607	76,953	81,546	95,402	102,639	113,258
As above serving 10 levels and a speed of 1.0m/s	80,794	85,249	89,991	104,486	112,025	123,497
As above serving 12 levels and a speed of 1.0m/s	89,032	93,597	98,490	113,508	121,482	131,662
As above serving 14 levels and a speed of 1.0m/s	99,727	101,820	108,435	123,966	133,186	146,172
Add to above for:						
Increase speed from 1.0 to 1.6 m/s	2,762	2,483	2,675	4,006	4,578	5,879
Enhanced finish to car – Centre mirror, flat ceiling, carpet	2,574	2,537	2,734	3,036	4,018	4,594
Fire fighting control	6,650	6,650	6,650	-	-	-
Painting to entire pit	737	737	737	737	737	737
Dual seal shaft	683	690	716	865	917	917
Shaft lighting / small power	3,705	3,705	3,705	-	-	-

ALL-IN-RATES

ALL-IN-RATES FOR PRICING SPECIALIST APPROXIMATE QUANTITIES *continued*

LIFT INSTALLATIONS *continued*

Passenger Lifts Machine Room-less (cont'd)	8 Person £	10 Person £	13 Person £	17 Person £	21 Person £	26 Person £
Add to above for:						
Intercom to reception desk and security room	349	349	349	349	349	349
Heating, cooling and ventilation to machine room	776	776	776	776	776	776
Lifting beams	828	828	828	1,156	1,216	1,216
10mm Equipotential bonding of all entrance metalwork	397	397	397	397	397	397
Shaft secondary steelwork	5,330	5,330	5,330	-	-	-
Independent insurance inspection	1,809	1,809	1,809	1,809	1,809	1,809
12 Month warranty service	767	795	805	829	852	852

Goods Lifts Machine Room Above	2000 kg £	2250 kg £	2500 kg £	3000 kg £
Electrically operated two speed serving 2 levels to take 1000 kg load, prime coated internal finish and a speed of 1.0 m/s	88,932	98,264	99,494	107,774
As above serving 4 levels and a speed of 1.0m/s	101,228	110,955	112,184	126,619
A As above serving 6 levels and a speed of 1.0m/s	113,241	123,386	124,590	139,282
As above serving 8 levels and a speed of 1.0m/s	125,255	135,769	136,998	155,036
As above serving 10 levels and a speed of 1.0m/s	137,268	148,179	149,405	170,789
As above serving 12 levels and a speed of 1.0m/s	149,281	160,582	161,812	186,542
As above serving 14 levels and a speed of 1.0m/s	161,293	172,990	174,218	202,297

ALL-IN-RATES

ALL-IN-RATES FOR PRICING SPECIALIST APPROXIMATE QUANTITIES *continued*

LIFT INSTALLATIONS *continued*

Goods Lifts Machine Room Above (cont'd)	2000 kg £	2250 kg £	2500 kg £	3000 kg £
Add to above for:				
Increased speed of travel from 1.0 to 1.6 metres per second	1,242	-	-	-
Enhanced finish to car – Centre mirror, flat ceiling, carpet	3,353	3,353	3,353	-
Bottom motor room	8,021	-	-	-
Painting to entire pit	603	1,276	1,276	1,276
Dual seal shaft	5,423	5,423	5,423	-
Intercom to reception desk and security room	350	350	350	350
Heating, cooling and ventilation to machine room	776	776	776	776
Lifting beams	1,156	1,156	1,156	1,156
10mm Equipotential bonding of all entrance metalwork	689	838	838	838
Independent insurance inspection	2,300	2,300	2,300	2,300
12 Month warranty service	631	1,155	1,155	-

Goods Lifts Machine Room-less	2000 kg £	2250 kg £	2500 kg £
Electrically operated two speed serving 2 levels to take 1000 kg load, prime coated internal finish and a speed of 1.0 m/s	80,376	85,968	91,566
As above serving 4 levels and a speed of 1.0m/s	90,802	96,586	102,372
A As above serving 6 levels and a speed of 1.0m/s	101,038	101,038	113,177
As above serving 8 levels and a speed of 1.0m/s	111,275	117,629	123,983
As above serving 10 levels and a speed of 1.0m/s	121,513	128,152	134,790
As above serving 12 levels and a speed of 1.0m/s	131,748	138,672	145,594

ALL-IN-RATES

ALL-IN-RATES FOR PRICING SPECIALIST APPROXIMATE QUANTITIES *continued*

LIFT INSTALLATIONS *continued*

Goods Lifts Machine Room-less (cont'd)	2000 kg £	2250 kg £	2500 kg £
As above serving 14 levels and a speed of 1.0m/s	144,189	150,295	156,400

Add to above for:

Increased speed of travel from 1.0 to 1.6 metres per second	4,585	7,763	-

Add to above for:

Enhanced finish to car – Centre mirror, flat ceiling, carpet	3,782	5,175	7,764

Add to above for:

Painting to entire pit	737	-	869
Dual seal shaft	678	-	-
Intercom to reception desk and security room	350	-	-
Heating, cooling and ventilation to machine room	388	776	776
Lifting beams	1,116	1,035	1,035
10mm Equipotential bonding of all entrance metalwork	397	-	-
Independent insurance inspection	1,809	1,809	1,809
12 Month warranty service	631	631	631

ESCALATOR INSTALLATIONS

30Ø Pitch escalator with a rise of 3 to 6 metres with standard balustrades	Each £
1000mm step width	89,528

Add to above for:

Balustrade Lighting	8,280
Skirting Lighting	8,953
Emergency stop button pedestals	2,805
Truss cladding - Stainless steel..	25,565
Truss cladding - Spray painted steel	20,441

ELEMENTAL COSTS

AIRPORT TERMINAL BUILDING

New build airport terminal building, premium quality, located in the South East, handling both domestic and international flights with a gross internal floor area (GIA) of 25,000m².

Cost Summary

El. Ref.	Element	Total Cost £	Cost/m² £
5A	Sanitaryware	67,750.000	2.71
5C	Disposal Installations		
	Rainwater	122,500.00	4.90
	Soil and waste	139,000.0	5.56
	Condensate	27,750.00	1.11
5D	Water Installations		
	Hot and cold water services	343,000.00	13.72
5F	Space Heating and Air Treatment		
	LTHW Heating system	1,391,250.00	55.65
	Chilled water system	834,750.00	33.39
	Supply and extract air conditioning system...	4,173,750.00	166.95
	Allowance for services to communications rooms	83,250.00	3.33
5G	Ventilating Services		
	Mechanical ventilation to baggage handling and plantrooms	444,250.00	17.77
	Toilet extract ventilation	208,500.00	8.34
	Smoke extract installation	104,250.00	4.17
	Kitchen extract system	52,500.00	2.10
5H	Electrical Installation		
	HV/LV Switchgear	1,446,750.00	57.87
	Standby generator	556,500.00	22.26
	Mains and sub mains installation	626,000.00	25.04
	Small power installation...	347,750.00	13.91
	Lighting and luminaires...	2,226,000.00	89.04
	Emergency lighting installation...	278,250.00	11.13
	Power to mechanical services	139,000.00	5.56
5I	Gas Installation	27,750.00	1.11
5K	Protective Installations		
	Lightning protection	69,500.00	2.78
	Earthing and bonding	55,500.00	2.22
	Sprinkler installation	500,750.00	20.03
	Dry riser and hosereel installations	166,750.00	6.67
	Fire suppression installation to communications room	40,000.00	1.60
	Carried forward	14,473,000.00	578.92

ELEMENTAL COSTS

AIRPORT TERMINAL BUILDING *continued*

El. Ref.	Element	Total Cost £	Cost/m² £
	Brought forward	14,473,000.00	578.92
5L	Communications Installations		
	Fire and smoke detection and alarm system	556,500.00	22.26
	Voice/public address system	556,500.00	22.26
	Intruder detection	208,500.00	8.34
	Security, CCTV and access control	695,500.00	27.82
	Wireways for telephones, data and structured cable	139,000.00	5.56
	Structured cable installation	695,500.00	27.82
	Flight information display system	682,500.00	27.30
5M	Special Installations		
	BMS Installation	1,043,250.00	41.73
	Summary total.	19,050,250.00	762.01

ELEMENTAL COSTS

SHOPPING MALL (TENANT'S FIT OUT EXCLUDED)

Natural ventilation shopping mall with approximately 33,000m² two storey retail area and a 13,000m² above ground covered car park , , situated in a town centre in South East England

Cost Summary

El. Ref.	Element	Total Cost £	Cost/m² £
	RETAIL BUILDING 33,000m²		
5A	Sanitary appliances	25,000.00	0.76
5C	Disposal Installations		
	Rainwater	170,000.00	5.15
	Soil, waste and vent...	200,000.00	6.06
5D	Water Installations		
	Cold water installation	195,000.00	5.91
	Hot water installation	180,000.00	5.45
5E	Heat Source	Included	
5F	Space Heating and Air Treatment		
	Condenser water system	1,000,000.00	30.30
	LTHW installation	135,000.00	4.09
	Air conditioning system...	1,010,00.00	30.61
	Over-door heaters at entrances	30,000.00	0.91
5G	Ventilation Services		
	Public toilet ventilation...	20,000.00	0.61
	Plant room ventilation...	135,000.00	4.09
	Supply and extract systems to shop units...	430,000.00	13.03
	Toilet extract systems to shop units...	80,000.00	2.42
	Smoke ventilation system to Mall Area...	340,000.00	10.30
	Service corridor ventilation...	75,000.00	2.27
	Miscellaneous ventilation...	550,000.00	16.67
5H	Electrical Installation		
	LV distribution	735,000.00	22.27
	Standby power	120,000.00	3.64
	General lighting	1,940,00.00	58.79
	External lighting	165,000.00	5.00
	Emergency lighting	370,000.00	11.21
	Small power	300,000.00	9.09
	Mechanical services power supplies...	85,000.00	2.58
	General earthing	40,000.00	1.21
	UPS for security and CCTV...	22,000.00	0.67
5I	Gas Installation		
	Gas supplies to boilers	17,000.00	0.52
	Gas supplies to Anchor (major) stores	11,000.00	0.33
5K	Protective Installations		
	Lightning protection	40,000.00	1.21
	Sprinkler installation...	430,000.00	13.03
	Dry Risers...	Excluded	
	Hosereel installation...	Excluded	
	Carried forward	8,850,000.00	268.18

ELEMENTAL COSTS

SHOPPING MALL (TENANT'S FIT OUT EXCLUDED) *continued*

El. Ref.	Element	Total Cost £	Cost/m² £
	Brought forward	8,850,000.00	268.18
5L	Communications Installations		
	Fire alarm installation...	250,000.00	7.58
	Public address/ voice alarm...	165,000.00	5.00
	Security installation	360,000.00	10.91
	General containment for voice and data (excluded, by Landlord/Tennant)...	185,000.00	5.61
5M	Special Installations		
	BMS/Controls	550,000.00	16.67
	Interface to existing services...	185,000.00	5.61
	Summary total...	10,540,000.00	319.56
	CAR PARK - 13,000m²		
5C	Disposal Installations		
	Car park drainage...	75,000.00	5.77
5G	Ventilation Services		
	Car park ventilation (ducted system)...	950,000.00	73.08
5H	Electrical Installation		
	LV distribution	145,000.00	11.15
	Standby power	Included	
	General lighting	360,000.00	27.69
	External lighting	Excluded	
	Emergency lighting	75,000.00	5.77
	Small power	145,000.00	11.15
	Mechanical services power supplies...	50,000.00	3.85
	General earthing...	15,000.00	1.15
	Ramp frost protection	20,000.00	1.54
5K	Protective Installations		
	Sprinkler installation...	285,000.00	21.92
	Dry Riser and Hosereel Installation...	Excluded	
5L	Communications Installations		
	Fire alarm installation...	285,000.00	21.92
	Security installation	120,000.00	9.23
5M	Special Installations		
	BMS/Controls	75,000.00	5.77
	Interface to existing services...	75,000.00	5.77
	Entry/exit barriers...	70,000.00	5.38
	Summary total...	2,745,000.00	211.14

ELEMENTAL COSTS

OFFICE BUILDING

Speculative 14 storey office in Central London for single tenant occupancy with a gross floor area of 27,490m². A four pipe fan coil system, with roof mounted air cooled chillers, gas fired boilers.

Cost Summary

El. Ref.	Element	Total Cost £	Cost/m² £
	SHELL AND CORE 27,490m² GIA		
5A	Sanitaryware	174,286.00	6.34
5C	Disposal Installations		
	Rainwater/Soil and Waste...	256,206.00	9.32
	Condensate..	27,215.00	0.99
5D	Water Installations		
	Hot and cold water services	266,103.00	9.68
5E	Heat Source	Included in 5F	
5F	Space Heating and Air Treatment		
	LTHW Heating..	322,733.00	11.74
	Chilled water..	671,856.00	24.44
	Ductwork...	1,674,141.00	60.90
5G	Ventilating Services		
	Toilet extract ventilation..	75,047.00	2.73
	Kitchen extract..	230,641.00	8.39
	Miscellaneous ventilation systems	173,187.00	6.30
5H	Electrical Installation		
	Generator	73,673.00	2.68
	HV/LV supply/distribution..	1,161,727.00	42.26
	General lighting...	578,665.00	21.05
	General power...	79,446.00	2.89
	Electrical services for mechanical equipment..............	105,287.00	3.83
5I	Gas Installation...	46,183.00	1.68
5K	Protection		
	Dry risers	47,283.00	1.72
	Sprinklers..	349,123.00	12.70
	Earthing and bonding	31,614.00	1.15
	Lightning protection	47,833.00	1.74
5L	Communication Installation		
	Fire/Voice alarms..	281,498.00	10.24
	Voice and data (wireways)	45,633.00	1.66
	Security (wireways)...	28,865.00	1.05
	Disabled alarms..	30,239.00	1.10
	Carried forward...	6,778,484.00	246.58

ELEMENTAL COSTS

OFFICE BUILDING *continued*

El. Ref.	Element	Total Cost £	Cost/m² £
	Brought forward..	6,778,484.00	246.58
5M	Special Installation Building management systems...........................	465,406.00	16.93
	Summary total (based on Gross Internal Area - GIA)................	7,243,890.00	263.51

El. Ref.	Element	Total Cost £	Cost/m² £
	<u>CATEGORY 'A' FIT OUT - 19,186m² NIA</u>		
5C	Disposal Installations Condensate...	103,223.00	5.38
5F	Space Heating and Air Treatment LTHW Heating.. Chilled water.. Ductwork..	423,051.00 574,045.00 1,460,438.00	22.05 29.92 76.12
5H	Electrical Installation Lighting installation....................................... Electrical services in connection.................. Tenant distribution board............................	1,133,509.00 50,267.00 62,355.00	59.08 2.62 3.25
5K	Protection Sprinkler installation..................................	362,615.00	18.90
5L	Communication Installation Fire/Voice alarms.......................................	150,994.00	7.87
5M	Special Installations Building management system........................	277,238.00	14.45
	Summary total (based on Nett Internal Area – NIA)................	4,597,733.00	239.64

ELEMENTAL COSTS

BUSINESS PARK

New build office in South East within the M25 part of a speculative business park consisting of two 3 storey existing buildings and 1 new build, fitted out to Category A specification. Four pipe system, external chiller, BMS controlled with all three buildings linked with an area of 7,500m² gross internal area and nett internal area of 6,000m².

Cost Summary

El. Ref.	Element	Total Cost £	Cost/m² £
	SHELL AND CORE – 7.500m² GIA		
5A	Sanitaryware	35,550.00	4.74
5C	Disposal Installations		
	Rainwater	20,700.00	2.76
	Soil and waste	55,275.00	7.37
5D	Water Installations		
	Cold water services	37,125.00	4.95
	Hot water services	10,800.00	1.44
5E	Heat Source	Included in 5F	
5F	Space Heating and Air Treatment		
	LTHW Heating; plantroom and risers	227,625.00	30.35
	Chilled water; plantroom and risers	245,100.00	32.68
5G	Ventilating Services		
	Toilet and miscellaneous ventilation	34,500.00	4.60
5H	Electrical Installation		
	LV supply/distribution	138,000.00	18.40
	General lighting	137,250.00	18.30
	General power	28,875.00	3.85
5I	Gas Installation	9,750.00	1.30
5K	Protective Installation		
	Earthing and bonding	11,250.00	1.50
	Lightning protection	11,250.00	1.50
5L	Communication Installation		
	Fire alarms	51,600.00	6.88
	Security (wireways)	11,250.00	1.50
	Data and voice (wireways)	11,250.00	1.50
5M	Special Installation		
	Building management systems	133,875.00	17.85
	Electrical services in connection	6,000.00	0.80
	Summary total (based on Gross Internal Area – GIA)	1,217,025.00	162.27

ELEMENTAL COSTS

BUSINESS PARK *continued*

El. Ref.	Element	Total Cost £	Cost/m² £
	CATEGORY 'A' FIT OUT – 6,000m² NIA		
5C	Disposal Installation		
	Condensate...	31,980.00	5.33
5F	Space Heating and Air Treatment		
	LTHW Heating..	107,400.00	17.90
	Chilled water...	131,340.00	21.89
	Supply and extract ductwork........................	421,620.00	70.27
5H	Electrical Installation		
	Distribution boards......................................	15,840.00	2.64
	General lighting..	244,680.00	40.78
5K	Protective Installation		
	Earthing and bonding...................................	4,500.00	0.75
5L	Communication Installation		
	Fire alarms..	26,400.00	4.40
5M	Special Installations		
	Building management systems......................	61,200.00	10.20
	Electrical services in connection..................	16,500.00	2.75
	Summary total (Based on Nett Internal Area - NIA).................	1,061,460.00	176.91

ELEMENTAL COSTS

PERFORMING ARTS CENTRE (LOW SPECIFICATION)

Performing Arts centre with a Gross Internal Area (GIA) of 6,000m², on a low specification for the theatre systems and with natural ventilation.

The development comprises dance studios and a theatre auditorium. The theatre would require all the necessary stage lighting, machinery and equipment installed in a modern professional theatre (these are excluded from the model, as assumed to be FF&E, but the containment and power wiring is included). Also not included are the staff call system, audio and video recording, EPOS ticket system, production recording and relay to TV screens and enhanced lighting including dimming – for such, refer to the High Specification Model.

Cost Summary

El. Ref.	Element	Total Cost £	Cost/m² £
5A	Sanitaryware..	71,940.00	11.99
5C	Disposal Installations		
	Soil, Waste and Rainwater..	87,780.00	14.63
5D	Water Installations		
	Cold water services..	51,180.00	8.53
	Hot water services..	49,560.00	8.26
5E	Heat Source..	125,520.00	20.92
5F	Space Heating and Air Treatment		
	Heating with limited cooling...................................	337,200.00	56.20
	DX Cooling to Comms and Amps rooms......................	32,520.00	5.42
5G	Ventilating Services		
	Ventilation and extract systems..............................	676,500.00	112.75
5H	Electrical Installation		
	LV supply/distribution...	312,000.00	52.00
	General lighting..	484,000.00	80.73
	Small power...	142,860.00	23.81
5I	Gas Installation...	11,520.00	1.92
5K	Protection		
	Lighting protection...	11880.00	1.98
5L	Communication Installation		
	Fire alarms and detection.....................................	191,400.00	31.90
	Voice and Data (containment only)..........................	33,720.00	5.62
	Security, Access, Control and Disabled alarms...........	177,540.00	29.59
5M	Special Installation		
	Building management systems................................	205,200.00	34.20
	Theatre systems includes for containment and power wiring ...	273,000.00	45.50
	Summary total..	3,275,700.00	545.95

ELEMENTAL COSTS

PERFORMING ARTS CENTRE (HIGH SPECIFICATION)

Performing Arts centres with a Gross Internal Area (GIA) of 6,000m², upon which this cost analysis has been based, on a high specification for the theatre systems and with cooling to the Auditorium.

The development comprises of dance studios and a theatre auditorium. The theatre would require all the necessary stage lighting, machinery and equipment installed in a modern professional theatre (these are excluded from the model, as assumed to be FF&E, but the containment and power wiring is included). Included are the staff call system/paging, audio and video recording, EPOS ticket system, production recording and relay to TV screens and enhanced lighting including dimming (not stage)

Cost Summary

El. Ref.	Element	Total Cost £	Cost/m² £
5A	Sanitaryware..	79,140.00	13.19
5C	Disposal Installations		
	Soil, Waste and Rainwater..................................	92,160.00	15.36
5D	Water Installations		
	Cold water services..	52,200.00	8.70
	Hot water services..	49,800.00	8.30
5E	Heat Source...	131,820.00	21.97
5F	Space Heating and Air Treatment		
	Heating and ventilation.......................................	387,780.00	64.63
	Cooling to Auditorium with DX to Comms and Amps rooms ...	336,000.00	56.00
5G	Ventilating Services		
	Ventilation and extract systems to toilets, kitchen and workshop	765,000.00	127.50
5H	Electrical Installation		
	LV supply/distribution...	343,200.00	57.20
	General lighting..	554,400.00	92.40
	Small power...	157,140.00	26.19
5I	Gas Installation..	12,000.00	2.00
5K	Protection		
	Lighting protection...	11,880.00	1.98
5L	Communication Installation		
	Fire alarms and detection...................................	210,540.00	35.09
	Voice and Data complete installation (excluding active equipment)	98,940.00	16.49
	Security, Access, Control, Disabled alarms, Staff paging.......	204.180.00	34.03
5M	Special Installation		
	Building management systems............................	223,200.00	37.80
	Theatre systems includes for containment and power wiring ...	353,580.00	58.93
	Summary total..	4,066,560.00	677.76

ELEMENTAL COSTS

SPORTS HALL

Single storey sports hall, located in the South East, with a gross internal area of 1,200m² (40m x 30m).

Cost Summary

El. Ref.	Element	Total Cost £	Cost/m² £
5A	Sanitaryware...	12,036.00	10.03
5C	Disposal Installations		
	Rainwater	4,008.00	3.34
	Soil and waste	7,020.00	5.85
5D	Water Installations		
	Hot and cold water services	16,716.00	13.93
5E	Heat Source		
	Boiler, flues, pumps and controls.......................	13,368.00	11.14
5F	Space Heating and Air Treatment		
	Warm air heating to sports hall area...................	16,056.00	13.38
	Radiator heating to ancillary areas.....................	25,272.00	21.06
5G	Ventilating Services		
	Ventilation to changing, fitness and sports hall areas............	16,188.00	13.49
5H	Electrical Installations		
	Main switchgear and sub-mains...........................	13,896.00	11.58
	Small power..	12,036.00	10.03
	Lighting and luminaries to sports hall areas........	20,064.00	16.72
	Lighting and luminaries to ancillary areas...........	23,844.00	19.87
5I	Gas Installation..	Included in 5E	
5K	Protective Installations		
	Lightning protection..	4,008.00	3.34
5L	Communications Installations		
	Fire, smoke detection and alarm system, intruder detection....	12,840.00	10.70
	CCTV Installation..	15,564.00	12.97
	Public address and music systems......................	7,980.00	6.65
	Wireways for telephone and data........................	3,192.00	2.66
	Summary total...	224,088.00	186.74

ELEMENTAL COSTS

HOTELS

250 Bedroom, four star hotel, situated in Central London, with a gross internal floor area of 16,500m².

The development comprises a ten storey building with large suites on each guest floor, together with banqueting, meeting rooms and leisure facilities.

Cost Summary

El. Ref.	Element	Total Cost £	Cost/m² £
5A	Sanitaryware...	495,000.00	30.00
5C	Disposal Installations Rainwater, soil and waste...	412,500.00	25.00
5D	Water Installations Hot and cold water services...	577,500.00	35.00
5E	Heat Source Condensing boiler and pumps etc...	165,000.00	10.00
5F	Space Heating and Air Treatment Air conditioning system; chillers, pumps, air handling units, ductwork, fan coil units etc; to guest rooms, public areas, meeting and banquet rooms...	1,237,500.00	75.00
5G	Ventilating Services General toilet extract and ventilation to kitchens and bathrooms etc...	792,000.00	48.00
5H	Electrical Installation HV/LV Installation, standby power, lighting, emergency lighting and small power to guest floors and public areas including earthing and lightning protection...	2,392,500.00	145.00
5I	Gas Installation...	24,750.00	1.50
5K	Protective Installations Dry risers and sprinkler installation...	990,000.00	60.00
5L	Communications Installations Fire, smoke detection and alarm system/security CCTV	759,000.00	46.00
	Background music, AV wireways...	247,500.00	15.00
	Telecommunications, data and T.V. wiring (no hotel management and head end equipment)...	123,750.00	7.50
5M	Special Installations Building Management System...	429,000.00	26.00
	Summary total...	8,646,000.00	524.00

ELEMENTAL COSTS

STADIUM – NEW

A three storey stadium, located in Greater London with gross internal area of 85,000m² and incorporating 60,000 spectator seats

Cost Summary

El. Ref.	Element	Total Cost £	Cost/m² £
5A	Sanitaryware	800,000.00	9.41
5C	Disposal Installations		
	Rainwater	224,000.00	3.22
	Above ground drainage	909,000.00	10.69
5D	Water Installations		
	Hot and cold water	1,748,000.00	20.56
5E	Heat Source	759,000.00	8.93
5F	Space Heating and Air Treatment		
	Heating	509,000.00	5.99
	Cooling	1,179,000.00	13.87
5G	Ventilating Services		
	Ventilation	4,080,000.00	48.00
5H	Electrical Installation		
	HV/LV Supply	730,000.00	8.59
	LV Distribution	1,922,000.00	22.61
	General lighting	4,711,000.00	55.42
	Small power	1,184,000.00	13.93
	Earthing and bonding	100,000.00	1.18
	Power supply to mechanical equipment	100,000.00	1.18
	Pitch lighting	545,000.00	6.41
5I	Gas Installation	75,000.00	0.88
5K	Protective Installations		
	Lightning protection	100,000.00	1.18
	Hydrants	185,000.00	2.18
5L	Communications Installations		
	Wireways for data, TV, telecom and PA	609,000.00	7.16
	Public address	1,235,000.00	14.53
	Security	910,000.00	10.71
	Data voice installations	2,400,000.00	28.23
	Fire alarms	679,000.00	7.99
	Disabled/refuse alarm/call systems	233,000.00	2.74
5M	Special Installations		
	BMS/Controls	1,128,000.00	13.27
	Summary total	27,104,000.00	318.86

Cost per seat	£ 451.73

ELEMENTAL COSTS

PRIVATE HOSPITAL

New build project building. The works consist of a new 80 bed hospital of approximately 15,000m², eight storey with a plant room.

All heat is provided from existing steam boiler plant, medical gases are also served from existing plant. The project includes the provision of additional standby electrical generation to serve the wider site requirements.

This hospital has six operating theatres, ITU/HDU department, pathology facilities, diagnostic imaging, out patient facilities and physiotherapy.

Cost Summary

El. Ref.	Element	Total Cost £	Cost/m² £
5A	Sanitaryware..	268,500.00	17.90
5C	Disposal Installations		
	Rainwater..	31,650.00	2.11
	Soil and waste...	296,850.00	19.79
	Specialist drainage (above ground)........................	18,750.00	1.25
5D	Water Installations		
	Hot and cold water services....................................	689,850.00	45.99
5E	Heat Source...	Included in 5F	
5F	Space Heating and Air Treatment		
	LPHW Heating...	521,400.00	34.76
	Chilled Water...	486,600.00	32.44
	Steam and condensate..	297,300.00	19.92
5G	Ventilating Services		
	Ventilation, comfort cooling and air conditioning......	1,778,700.00	118.58
5H	Electrical Installation		
	HV Distribution..	25,950.00	1.73
	LV supply/distribution..	360,000.00	24.00
	Standby Power..	404,550.00	26.97
	UPS..	302,550.00	20.17
	General lighting...	528,600.00	35.24
	General power...	519,000.00	34.60
	Emergency lighting..	150,750.00	10.05
	Theatre lighting...	196,350.00	13.09
	Specialist lighting	183,450.00	12.23
	External lighting..	28,950.00	1.93
	Electrical supplies for mechanical equipment.........	161,550.00	10.77
5I	Gas Installation...	35,700.00	2.38
	Oil Installations...	77,400.00	5.16
5K	Protection		
	Dry risers...	17,400.00	1.16
	Lightning Protection..	3,600.00	0.24
	Carried forward...	7,385,400.00	492.36

ELEMENTAL COSTS

PRIVATE HOSPITAL *continued*

El. Ref.	Element	Total Cost £	Cost/m² £
	Brought forward...	7,385,400.00	492.36
5L	Communication Installation		
	Fire alarms and detection...	261,750.00	17.45
	Voice and Data..	165,900.00	11.06
	Security and CCTV...	48,900.00	3.26
	Nurse call and cardiac alarm system.........................	241,500.00	16.10
	Personnel paging...	62,100.00	4.14
	Hospital radio (entertainment)..................................	31,950.00	2.13
5M	Special Installation		
	Building management systems....................................	559,350.00	37.29
	Pneumatic tube conveying system.............................	39,600.00	2.64
	Group 1 Equipment..	533,850.00	35.59
	Summary total (based on gross floor area)..................	9,330,300.00	622.02

ELEMENTAL COSTS

SCHOOL

New build secondary school (Academy) located in Southern England, with a gross internal floor area of 10,000m².

The building comprises a three storey teaching block, including provision for music, drama, catering, sports hall, science laboratories, food technology, workshops and reception/admin (BB93 compliant). Excludes IT Cabling

Cost Summary

El. Ref.	Element	Total Cost £	Cost/m² £
5A	Sanitaryware		
	Toilet cores and changing facilities only............................	70,000.00	7.00
5C	Disposal Installations		
	Rainwater installations...	30,000.00	3.00
	Soil and waste..	100,000.00	10.00
5D	Water Installations		
	Potable hot and cold water services................................	120,000.00	12.00
	Non potable hot and cold water services to labs and art rooms	30,000.00	3.00
5E	Heat Source..	120,000.00	12.00
5F	Space Heating and Air Treatment		
	LTHW Heating system (primary)......................................	300,000.00	30.00
	LTHW Heating system (secondary)..................................	70,000.00	7.00
	DX Cooling system to ICT server rooms............................	30,000.00	3.00
	Mechanical supply and extract ventilation including DX type cooling to Music, Drama, Kitchen/Dining and Sports Hall.........	400,000.00	40.00
5G	Ventilating Services		
	Toilet extract systems..	30,000.00	3.00
	Changing area extract systems.......................................	30,000.00	3.00
	Extract ventilation from design/food technology and science labs...	40,000.00	4.00
5H	Electrical Installation		
	Mains and sub-mains distribution....................................	250,000.00	25.00
	Lighting and luminaries; including emergency fittings.............	600,000.00	60.00
	Small power installation..	325,000.00	34.00
	Earthing and bonding...	10,000.00	1.00
5I	Gas Installation...	70,000.00	7.00
5K	Protective Installations		
	Lightning protection...	30,000.00	3.00
5L	Communications Installations		
	Containment for telephone, IT data, AV and security systems .	20,000.00	2.00
	Fire, smoke detection and alarm system............................	130,000.00	13.00
	Security installations including CCTV, access control and intruder alarm..	150,000.00	15.00
	Disabled toilet, refuge and induction loop systems................	30,000.00	3.00
5M	Special Installations		
	Building Management system – To plant.............................	190,000.00	19.00
	Building Management system – To opening vents/windows...	60,000.00	6.00
	Summary total..	3,200,000.00	325.00

ELEMENTAL COSTS

AFFORDABLE RESIDENTIAL DEVELOPMENT

A twelve storey, 100 apartment affordable residential development with a gross internal area of 9,000m² and a net internal area of 7,300m², situated within the London area. The development does not include a car park and is based on 81% efficiency.

Based on an individual radiator LTHW system within each apartment, with local gas combi boilers exhausting to building façade. Kitchens and bathrooms are also ventilated to the building façade, there are pendant light fittings, an audio entry system, telephone and satellite installation. Sanitaryware is of lower quality but a disabled refuge alarm is included.

Cost Summary

El. Ref.	Element	Total Cost £	Cost/m² £
	SHELL & CORE		
5A	Sanitaryware..	5,000.00	0.56
5B	Services..	-	-
5C	Disposal Installations..	115,000.00	12.78
5D	Water Installations..	175,000.00	19.44
5E	Heat Source...	75,000.00	8.33
5F	Space Heating and Air Treatment.............................	50,000.00	5.56
5G	Ventilating Services...	80,000.00	8.89
5H	Electrical Installation...	275,000.00	30.56
5I	Gas Installation..	15,000.00	1.67
5K	Protective Installations...	45,000.00	5.00
5L	Communications Installations...................................	235,000.00	26.11
5M	Special Installations..	105,000.00	11.67
	Summary total (based on Gross Internal Area - GIA)............	1,175,000.00	130.56

ELEMENTAL COSTS

AFFORDABLE RESIDENTIAL DEVELOPMENT *continued*

Cost Summary

El. Ref.	Element	Total Cost £	Cost/m² £
	FITTING OUT		
5A	Sanitaryware...	165,000.00	22.60
5B	Services..	-	-
5C	Disposal Installations...	56,000.00	7.67
5D	Water Installations...	360,000.00	49.32
5E	Heat Source..	25,000.00	3.42
5F	Space Heating and Air Treatment..............................	185,000.00	25.34
5G	Ventilating Services...	265,000.00	36.30
5H	Electrical Installation..	220,000.00	30.14
5I	Gas Installation..	30,000.00	4.11
5K	Protective Installations...	-	-
5L	Communications Installations....................................	115,000.00	15.75
5M	Special Installations...	25,000.00	3.42
	Summary total (Net Internal Area - NIA).....................	1,446,000.00	198.08

ELEMENTAL COSTS

PRIVATE RESIDENTIAL DEVELOPMENT

A seven storey, 115 apartment private residential development with a gross internal area of 7,500m² and a net internal area of 6,900m², situated within the London area. The development does not include a car park and is based on 92% efficiency.

Included is a central boiler and hot water installation with LTHW under-floor heating to each apartment, central comfort cooling is via a Variable Refrigerant Volume system, whole house ventilation system discharging to local façade, dry riser, video entry and TV and satellite installation.

Cost Summary

El. Ref.	Element	Total Cost £	Cost/m² £
	SHELL & CORE		
5A	Sanitaryware...	7,500.00	1.00
5B	Services...	-	-
5C	Disposal Installations...	145,000.00	19.33
5D	Water Installations...	225,000.00	30.00
5E	Heat Source...	75,000.00	10.00
5F	Space Heating and Air Treatment...	400,000.00	53.33
5G	Ventilating Services...	85,000.00	11.33
5H	Electrical Installation...	260,000.00	34.67
5I	Gas Installation...	20,000.00	2.67
5K	Protective Installations...	55,000.00	7.33
5L	Communications Installations...	200,000.00	26.67
5M	Special Installations...	75,000.00	10.00
	Summary total (based on Gross Internal Area - GIA)...	1,547,500.00	206.33

ELEMENTAL COSTS

SUPERMARKET

Supermarket located in the South East with a total gross floor area of 4,000m², including a sales area of 2,350m². The building is on one level and incorporates a main sales, coffee shop, bakery, offices and amenities areas and warehouse.

Cost Summary

El. Ref.	Element	Total Cost £	Cost/m² £
5A	Sanitaryware..	6,396.00	1.59
5C	Disposal Installations		
	Soil and Waste..	10,341.00	2.58
5D	Water Installations		
	Hot and Cold water services....................................	36,910.00	9.23
5E	Heat Source...	72,698.00	18.17
5F	Space Heating and Air Treatment		
	Heating & ventilation with cooling via DX units...............	137,768.00	34.44
5G	Ventilating Services		
	Supply and extract system..	45,753.00	11.43
5H	Electrical Installation		
	Panels / Boards...	41,951.00	10.48
	LV supply/distribution	120,942.00	30.23
	General lighting	96,926.00	24.23
	Emergency lighting	24,360.00	6.09
	Small power...	72,909.00	18.22
5I	Gas Installation		
	Gas mains services to plantroom...............................	13,858.00	3.46
5K	Protection		
	Sprinklers..	106,391.00	26.59
	Lightning protection..	3,186.00	0.79
5L	Communication Installation		
	Fire alarms, detection and public address...................	25,204.00	6.30
	CCTV...	20,473.00	5.11
	Intruder alarm, detection and store security..................	45,714.00	11.42
5M	Special Installations		
	BMS Installation..	44,774.00	11.19
	Refrigeration		
	Installation...	106,201.00	26.55
	Plant..	109,087.00	27.27
	Cold Store..	40,302.00	10.07
	Cabinets...	299,047.00	74.76
	Summary total	1,480,561.00	363.20

ELEMENTAL COSTS

DISTRIBUTION CENTRE

Distribution centre located in London with a total gross floor area of 75,000m², including a refrigerated cold box of 17,500m².

The building is on one level and incorporates a office area, vehicle recovery unit, gate house and plantrooms

Cost Summary

El. Ref.	Element	Total Cost £	Cost/m² £
5C	Disposal Installations		
	Soil and Waste..	149,080.00	1.98
	Rainwater...	481,449.00	6.41
5D	Water Installations		
	Hot and Cold water services........................	128,866.00	1.71
5F	Space Heating and Air Treatment		
	Heating with ventilation to offices, displacement system to		
	main warehouse...	1,661,653.00	22.15
5G	Ventilating Services		
	Smoke extract system..................................	371,928.00	4.95
5H	Electrical Installation		
	Generator...	1,127,576.00	15.03
	Main HV installation....................................	1,302,375.00	17.36
	MV distribution...	806,882.00	10.75
	Lighting installation.....................................	740,320.00	9.87
	Small power installation...............................	907,828.00	12.10
5I	Gas Installation		
	Gas mains services to plantroom..................	41,397.00	0.55
5K	Protection		
	Sprinklers including racking protection...........	3,295,392.00	43.93
	Lightning protection.....................................	8,141.00	0.10
5L	Communication Installation		
	Fire alarms, detection and public address......	839,091.00	11.18
	CCTV..	538,693.00	7.18
5M	Special Installations		
	BMS Installation	419,347.00	5.59
	Refrigeration		
	Installation...	2,678,883.00	35.71
	Summary total..	14,760,860.00	206.45

ELEMENTAL COSTS

DATA CENTRE

New build data centre located in the London area/proximity to M25.

Net Technical Area (NTA) provided at 2,000m² with typically other areas of 250m² office space, 250m² ancillary space and 1,000m² internal plant. Total GIA 3,500m²

Power and cooling to Technical Space @ 1,000w/m². Cost/m² against Nett Technical Area - NTA.

Cost Summary

El. Ref.	Element	Total Cost £	Cost/m² £
5C	Disposal Installations		
	Soil & Waste..	30,000.00	15.00
	Rainwater..	20,000.00	10.00
	Condensate..	20,000.00	10.00
5D	Water Installations		
	Hot and Cold water services....................................	50,000.00	25.00
5F	Space Heating and Air Treatment		
	Chilled water plant with redundancy of N+1 to provide 1,000w/m² net technical space cooling............................	800,000.00	400.00
	Chilled water distribution to data centre to free standing cooling units and distribution to ancillary office and workshop/build areas...	1,230,000.00	615.00
	Freestanding cooling units with redundancy of N+20% to technical space and switchrooms. Based on single coil cooling units 30% of units with humidification....................................	650,000.00	325.00
	Floor grilles..	150,000.00	75.00
5G	Ventilating Services		
	Supply and extract ventilation systems to data centre, switchrooms and ancillary spaces including dedicated gas extract and hot aisle extract system......................................	780,000.00	390.00
5H	Electrical Installation		
	Main HV installations including transformers with redundant capacity of N+1...	290,000.00	145.00
	LV distribution including cabling and busbar installations to provide full system – System dual supplies/redundancy including supplies to mechanical services, office and ancillary areas...	1,200,000.00	600.00
	Generator Installation		
	Standby rated containerised generators with redundant capacity of N+1 and including synchronisation panel, 72 hours bulk fuel store and controls...	1,700,000.00	850.00
	Uninterruptible Power Supplies		
	Static UPS to provide 2 x (N+1) system redundancy with 10 minute battery autonomy...	1,150,000.00	575.00
	Carried Forward..	8,070,000.00	4,035.00

ELEMENTAL COSTS

DATA CENTRE *continued*

El. Ref.	Element	Total Cost £	Cost/m² £
	Brought Forward..	8,070,000.00	4,035.00
5H	Electrical Installation *continued*		
	LV Switchgear Incoming LV switchgear, UPS input and output boards for electrical and mechanical services systems........................	1,050,000.00	525.00
	Power Distribution Units (PDUs) PDUs to provide a redundant capacity of 2 Nr PDU to include static transfer switch and isolating transformer.....................	800,000.00	400.00
	Cabinet Supplies A & B supply cables from PDUs to a BS4343 socket fixed under raised floor adjacent cabinet positions........................	490,000.00	245.00
	Lighting Installation Lighting to technical, workshop and plant areas including office/ancillary spaces and external areas.............................	200,000.00	100.00
5I	Gas Installation...	20,000.00	10.00
5K	Protective Installations Gaseous suppression to technical areas and switchrooms Lightning protection..	460,000.00 30,000.00	230.00 15.00
5L	Communications Installation Fire, smoke detection and alarm system .... Very early smoke detection alarm (VESDA) to technical areas... CCTV installations... Access control installations..	 240,000.00 300,000.00 250,000.00	 120.00 150.00 125.00
5M	Special Installations Building Management System... PLC/Electrical monitoring system.................................,.	550,000.00 500,000.00	275.00 250.00
	Summary total (Based on Nett Technical Area - NTA)...............	12,960,000.00	6,480.00

ELEMENTAL COSTS

BUILDING MANAGEMENT INSTALLATION

El. Ref.	Element	Total Cost £	Cost/Point £
	Category A Fit Out		
	Option 1 – 189 nr four pipe fan coil – 756 points		
1.0	**Field Equipment** Network devices Valves/actuators Sensing devices...	48,216.00	63.78
2.0	**Cabling** Power – from local isolator to DDC controller Control – from DDC controller to field equipment...	27,243.00	36.04
3.0	**Programming** Software – central facility Software – network devices Graphics...	15,081.00	19.95
4.0	**On site testing and commissioning** Equipment Programming/graphics Power and control cabling...	15,706.00	20.78
	Total Option 1 – Four pipe fan coil (On Point Basis)...	106,246.00	140.55
	Category A Fit Out		
	Option 2 – 189 nr two pipe fan coil system with electric heating – 756 points		
1.0	**Field Equipment** Network devices Valves/actuators/thyristors Sensing devices...	61,838.00	81.80
2.0	**Cabling** Power – from local isolator to DDC controller Control – from DDC controller to field equipment...	28,526.00	37.73
3.0	**Programming** Software – central facility Software – network devices Graphics...	15,081.00	19.95
4.0	**On site testing and commissioning** Equipment Programming/graphics Power and control cabling...	15,706.00	20.78
	Total Option 2 – Two pipe fan coil with electric heating (On Point Basis)...	121,151.00	160.26

Approximate Estimating

ELEMENTAL COSTS

BUILDING MANAGEMENT INSTALLATIONS *continued*

El. Ref.	Element	Total Cost £	Cost/Point £
	Category A Fit Out		
	Option 3 – 180 Nr Chilled Beams with perimeter heating – 567 points		
1.0	**Field Equipment** Network devices Valves/actuators Sensing devices........	43,676.00	77.03
2.0	**Cabling** Power – from local isolator to DDC controller Control – from DDC controller to field equipment...	28,973.00	51.10
3.0	**Programming** Software – central facility Software – network devices Graphics...	15,081.00	26.60
4.0	**On site testing and commissioning** Equipment Programming/graphics Power and control cabling...	16,120.00	28.42
	Total Option 3 – Chilled beams with perimeter heating (On Point Basis)	103,850.00	183.15
	Combined Shell & Core and Fit Out		
1.0	**Option 1 – 4 pipe fan coil – 1196 points**		
	Shell and core and Category A Fit out...	363,246.00	303.72
2.0	**Option 2 – 2 pipe fan coil with electrical heating – 1196 points**		
	Shell and core and Category A Fit out...	378,151.00	316.18
3.0	**Option 3 – Chilled beams with perimeter heating – 1007 points**		
	Shell and core and Category A Fit out...	360,850.00	358.34

Material Costs/ Measured Work Prices

Electrical Installations *continued*

Mechanical Installations

Material Costs/Measured Work Prices

DIRECTIONS

The following explanations are given for each of the column headings and letter codes.

Unit	Prices for each unit are given as singular (i.e. 1 metre, 1 nr) unless stated otherwise
Net price	Industry tender prices, plus nominal allowance for fixings (unless measured separately), waste and applicable trade discounts.
Material cost	Net price plus percentage allowance for overheads, profit and preliminaries.
Labour norms	In man-hours for each operation.
Labour cost	Labour constant multiplied by the appropriate all-in man-hour cost based on gang rate. (See also relevant Rates of Wages Section) plus percentage allowance for overheads, profit and preliminaries
Measured work Price (total rate)	Material cost plus Labour cost.

MATERIAL COSTS

The Material Costs given are based at Third Quarter 2006 but exclude any charges in respect of VAT. At the time of going to press, the raw material cost of copper was falling but had not been reflected within factory exit prices for finished goods i.e. tube and fittings etc. Users of the book are advised to register on the SPON's website www.pricebooks.co.uk/updates to receive the free quarterly updates - alerts will then be provided by e-mail as changes arise.

MEASURED WORK PRICES

These prices are intended to apply to new work in the London area. The prices are for reasonable quantities of work and the user should make suitable adjustments if the quantities are especially small or especially large. Adjustments may also be required for locality (e.g. outside London - refer to cost indices in approximate estimating section for details of adjustment factors) and for the market conditions e.g. volume of work secured or being tendered) at the time of use.

MECHANICAL INSTALLATIONS

The labour rate has been based on average gang rates per man hour effective from 3 October 2005 (Plus a Provisional Allowance of 5% - As the wage promulgation has not been announced at the time of going to press - The labour rates will be confirmed within free updates on the aforementioned website) including allowances for all other emoluments and expenses. To this rate has been added 12.5% and 7.5% to cover preliminary items, site and head office overheads together with 5% for profit, resulting in an inclusive rate of £25.37 per man hour. The rate has been calculated on a working year of 2,016 hours; a detailed build-up of the rate is given at the end of these directions.

DUCTWORK INSTALLATIONS

The labour rate basis is as per Mechanical above and to this rate has been added 25% plus 12.5% to cover shop, site and head office overheads and preliminary items together with 10% for profit, resulting in an inclusive rate of £29.82 per man hour. The rate has been calculated on a working year of 2,016 hours; a detailed build-up of the rate is given at the end of these directions.

In calculating the 'Measured Work Prices' the following assumptions have been made:
  (a)   That the work is carried out as a sub-contract under the Standard Form of Building Contract.
  (b)   That, unless otherwise stated, the work is being carried out in open areas at a height which would not require more than simple scaffolding.
  (c)   That the building in which the work is being carried out is no more than six storey's high.

Where these assumptions are not valid, as for example where work is carried out in ducts and similar confined spaces or in multi-storey structures when additional time is needed to get to and from upper floors, then an appropriate adjustment must be made to the prices. Such adjustment will normally be to the labour element only. *Note : The rates do not include for any uplift applied if the ductwork package is procured via the Mechanical Sub Contractor*

DIRECTIONS

LABOUR RATE - MECHANICAL

The annual cost of a notional twelve man gang

		FOREMAN	SENIOR CRAFTSMAN (+2 Welding skill)	SENIOR CRAFTSMAN	CRAFTSMAN	INSTALLER	MATE (Over 18)	SUB TOTALS
		1 NR	1 NR	2 NR	4 NR	2 NR	2 NR	
Hourly Rate from 3 October 2005 Plus 5% (Provisional)		13.86	11.92	11.45	10.50	9.52	8.02	
Working hours per annum per man		1,702.40	1,702.40	1,702.40	1,702.40	1,702.40	1,702.40	
x Hourly rate x nr of men = £ per annum		23,595.26	20,288.35	38,967.94	71,500.80	32,425.61	27,313.31	214,091.27
Overtime Rate		19.53	16.80	16.14	14.81	13.43	11.31	
Overtime hours per annum per man		313.60	313.60	313.60	313.60	313.60	313.60	
x Hourly rate x nr of men = £ per annum		6,124.61	5,268.48	10,122.07	18,571.39	8,422.98	7,092.69	55,602.22
Total		29,719.87	25,556.83	49,090.00	90,072.19	40,848.60	34,406.00	269,693.49
Incentive schemes	5.00%	1,485.99	1,277.84	2,454.50	4,503.61	2,042.43	1,720.30	13,484.67
Daily Travel Time Allowance (15-20 miles each way)		8.60	8.60	8.60	8.60	8.60	8.27	
Days per annum per man		224.00	224.00	224.00	224.00	224.00	224.00	
x nr of men = £ per annum		1,926.29	1,926.29	3,852.58	7,705.15	3,852.58	3,704.96	22,968.96
Daily Travel Fare (10-20 miles each way)		8.40	8.40	8.40	8.40	8.40	8.40	
Days per annum per man		224.00	224.00	224.00	224.00	224.00	224.00	
x nr of men = £ per annum		1,881.60	1,831.60	3,763.20	7,526.40	3,763.20	3,763.20	22,579.20
Employers Contributions to EasyBuild Stakeholder Pension (Death and accident cover is provided free):								
Number of weeks		52	52	52	52	52	52	
Total weekly £ contribution each		4.00	4.00	4.00	4.00	4.00	4.00	
£ Contributions/annum		208.00	208.00	416.00	832.00	416.00	416.00	2,496.00
National Insurance Contributions:								
Wkly gross pay (subject to NI) each		35,013.75	30,642.56	59,160.28	109,807.35	50,506.80	43,596.25	
% of NI Contributions		12.8	12.8	12.8	12.8	12.8	12.8	
£ Contributions/annum		3,673.56	3,114.05	5,956.11	10,822.53	4,848.47	3,963.69	32,378.54

DIRECTIONS

LABOUR RATE - MECHANICAL

The annual cost of a notional twelve man gang

		FOREMAN	SENIOR CRAFTSMAN (+2 Welding skill)	SENIOR CRAFTSMAN	CRAFTSMAN	INSTALLER	MATE (Over 18)	SUB TOTALS
		1 NR	1 NR	2 NR	4 NR	2 NR	2 NR	
Holiday Credit and Welfare Contributions:								
Number of weeks		52	52	52	52	52	52	
Total weekly £ contribution each		72.04	62.85	60.56	56.00	51.32	44.13	
x nr of men = £ contributions/annum		3,746.11	3,268.36	6,298.66	11,647.27	5,337.70	4,589.65	34,887.76
Holiday Top-up Funding including overtime		12.81	10.92	10.51	9.69	8.79	7.41	
Cost		666.12	567.84	1,093.04	2,015.52	914.16	770.64	6,027.32
						SUB-TOTAL		404,515.95
				TRAINING (INCLUDING ANY TRADE REGISTRATIONS) – SAY			1.00%	4,045.16
				SEVERANCE PAY AND SUNDRY COSTS – SAY			1.50%	6,128.42
				EMPLOYER'S LIABILITY AND THIRD PARTY INSURANCE – SAY			2.00%	8,293.79
				ANNUAL COST OF NOTIONAL GANG				422,983.32
MEN ACTUALLY WORKING = 10.5				THEREFORE ANNUAL COST PER PRODUCTIVE MAN				40,284.13
AVERAGE NR OF HOURS WORKED PER MAN = 2016				THEREFORE ALL IN MAN HOURS				19.98
				PRELIMINARY ITEMS			12.50%	2.69
				SITE AND HEAD OFFICE OVERHEADS - SAY			7.50%	1.50
				PROFIT – SAY			5.00%	1.21
				THEREFORE INCLUSIVE MAN HOUR RATE				25.37

Notes:

(1)   The following assumptions have been made in the above calculations:-
     (a)   The working week of 38 hours i.e. the normal working week as defined by the National Agreement.
     (b)   The actual hours worked are five days of 9 hours each.
     (c)   A working year of 2016 hours, including overtime worked.
     (d)   Five days in the year are lost through sickness or similar reason.
(2)   The incentive scheme addition of 5% is intended to reflect bonus schemes typically in use.
(3)   National insurance contributions are those effective from 6 April 2006.
(4)   Weekly Holiday Credit/Welfare Stamp values are those effective from 3 October 2005 + 5% (Provisional) for October 2006.
(5)   Rates are based from 3 October 2005 + 5% (assumed) for wage increase in October 2006.
(6)   Fares (New Malden to Waterloo is £6.00 plus £4.00 Zone 1 Return) current at June 2006
(7)   Easybuild Stakeholder Pension Contributions effective from 26 June 2006.
(8)   Does not include for major project status.
(9)   Caution should be applied when utilising the labour rate and the 'all-in' rate it applies to, as the size and complexity of the project will reflect in the gang size.
(10)   Overtime rates are based on Premium Rate 1.

DIRECTIONS

LABOUR RATE - DUCTWORK

The annual cost of notional eight man gang

		FOREMAN 1 NR	SENIOR CRAFTSMAN 1 NR	CRAFTSMAN 4 NR	INSTALLER 2 NR	SUB TOTALS
Hourly Rate from 3 October 2005 Plus 5% (Provisional)		13.86	11.45	10.50	9.52	
Working hours per annum per man		1,702.40	1,702.40	1,702.40	1,702.40	
x Hourly rate x nr of men = £ per annum		23,595.26	19,483.97	71,500.80	32,425.61	147,005.64
Overtime Rate		19.53	16.14	14.81	13.43	
Overtime hours per annum per man		313.60	313.60	313.60	313.60	
x hourly rate x nr of men = £ per		6,124.61	5,061.03	18,571.39	8,422.98	38,180.02
Total		29,719.87	24,545.00	90.072.19	40,848.60	185,185.66
Incentive schemes	5.00%	1,485.99	1,227.25	4,503.61	2,042.43	9,259.28
Daily Travel Time Allowance (15-20 miles each way)		8.60	8.60	8.60	8.60	
Days per annum per man		224	224	224	224	
x nr of men = £ per annum		1,926.29	1,926.29	7,705.15	3,852.58	15,410.30
Daily Travel Fare (15-20 miles each way)		8.40	8.40	8.40	8.40	
Days per annum per man		224	224	224	224	
x nr of men = £ per annum		1,881.60	1,881.60	7,526.40	3.763.20	15,052.80
Employers Contributions to EasyBuild Stakeholder Pension (Death cover is provided free)						
Number of weeks		52	52	52	52	
Total weekly £ contribution each		4.00	4.00	4.00	4.00	
£ Contributions/annum		208.00	208.00	832.00	416.00	1,664.00
National Insurance Contributions:						
Weekly gross pay (subject to NI) each		35,013.75	29,580.14	109,807.35	50,506.80	
% of NI Contributions		12.8	12.8	12.8	12.8	
£ Contributions/annum		3,763.56	2,978.06	10,822.53	4,848.47	22,322.61
Holiday Credit and Welfare contributions:						
Number of weeks		52	52	52	52	
Total weekly £ contribution each		72.04	60.56	56.00	51.32	
x nr of men = £ Contributions/annum		3,746.11	3,149.33	11,647.27	5,337.70	23,880.40

DIRECTIONS

LABOUR RATE - DUCTWORK

The annual cost of notional eight man gang

		FOREMAN	SENIOR CRAFTSMAN	CRAFTSMAN	INSTALLER	SUB TOTALS
		1 NR	1 NR	4 NR	2 NR	
Holiday Top-up Funding including overtime		12.81	10.51	9.69	8.79	
Cost		666.12	546.52	2,015.52	914.16	4,142.32

SUB-TOTAL			276,917.38
TRAINING (INCLUDING ANY TRADE REGISTRATIONS) - SAY	1.00%		2,769.17
SEVERANCE PAY AND SUNDRY COSTS - SAY	1.50%		4,195.30
EMPLOYER'S LIABILITY AND THIRD PARTY INSURANCE - SAY	2.00%		5,677.64
ANNUAL COST OF NOTIONAL GANG			289,559.49
MEN ACTUALLY WORKING = 7.5 THEREFORE ANNUAL COST PER PRODUCTIVE MAN			38,607.93
AVERAGE NR OF HOURS WORKED PER MAN = 2016 THEREFORE ALL IN MAN HOURS			19.15
PRELIMINARY ITEMS - SAY	25.00%		5.39
SITE AND HEAD OFFICE OVERHEADS - SAY	12.50%		2.39
PROFIT - SAY	10.00%		2.69
THEREFORE INCLUSIVE MAN HOUR RATE			29.62

Notes:

(1) The following assumptions have been made in the above calculations:-
    (a) The working week of 38 hours i.e. the normal working week as defined by the National Agreement.
    (b) The actual hours worked are five days of 9 hours each.
    (c) A working year of 2016 hours, including overtime worked.
    (d) Five days in the year are lost through sickness or similar reason.
(2) The incentive scheme addition of 5% is intended to reflect bonus schemes typically in use.
(3) National insurance contributions are those effective from 6 April 2006.
(4) Weekly Holiday Credit/Welfare Stamp values are those effective from 3 October 2005 + 5% (assumed) for October 2006.
(5) Rates are based from 3 October 2006 + 5% (assumed) for wage increase in October 2006.
(6) Fares (New Malden to Waterloo is £6.00 plus £4.00 Zone 1 Return) current at June 2006
(7) Easybuild Stakeholder Pension Contributions effective from 26 June 2006.
(8) Does not include for major project status.
(9) Caution should be applied when utilising the labour rate and the 'all-in' rate it applies to, as the size and complexity of the project will reflect in the gang size.
(10) Overtime rates are based on Premium Rate 1.

The Dynamic Landscape
Naturalistic Planting in an Urban Context

Edited by
Nigel Dunnett and James Hitchmough

The last quarter of the twentieth-century witnessed a burgeoning of interest in ecological or naturally inspired use of vegetation in the designed landscape. More recently a strong aesthetic element has been added to what was formerly a movement aimed at creating nature-like landscapes.

This book advances a fusion of scientific and ecological planting design philosophy that can address the need for more sustainable designed landscapes. It is a major statement on the design, implementation and management of ecologically-inspired landscape vegetation. With contributions from people at the forefront of developments in this field, in both Europe and North America, it provides a valuable synthesis of current thinking.

Contents: 1. Introduction 2. The Historical Development of Naturalistic Planting 3. A Contemporary Overview of Naturalistic Planting 4. The Dynamic Nature of Plant Communities 5. A Naturalistic Design Process 6. Herbaceous Plantings 7. Exploring Woodland Design 8. Wetlands and Water Bodies 9. Communicating Naturalistic Plantings: Plans and Specifications 10. The Creative Management of Naturalistic Plantings 11. The Social and Cultural Context of Naturalistic Plantings. Index.

June 2004: 238x225mm: 336 pages
20 tables, 40 line drawings, 67 b+w photos and 85 colour photos
HB: 0-415-25620-8: £60.00

To Order: Tel: +44 (0) 1264 343071 Fax: +44 (0) 1264 343005, or
Post: Taylor and Francis Customer Services, Thomson Publishing Services, Cheriton House, Andover, Hants, SP10 5BE, UK Email: book.orders@tandf.co.uk

For a complete listing of all our titles visit:
www.tandf.co.uk

 Taylor & Francis
Taylor & Francis Group plc

R:DISPOSAL SYSTEMS

Item	Net Price £	Material £	Labour hours	Labour £	Unit	Total rate £
R10: RAINWATER PIPEWORK/GUTTERS						
PVC-U gutters: push fit joints; fixed with brackets to backgrounds; BS 4576 BS EN 607						
Half round gutter, with brackets measured separately						
75mm	1.30	1.67	0.69	17.51	m	**19.17**
100 mm	2.75	3.52	0.64	16.24	m	**19.76**
150mm	4.11	5.26	0.82	20.80	m	**26.07**
Brackets: including fixing to backgrounds. For minimum fixing distances, refer to the Tables and Memoranda at the rear of the book						
75mm; Fascia	0.59	0.76	0.15	3.81	nr	**4.56**
100mm; Jointing	1.51	1.93	0.16	4.06	nr	**5.99**
100mm; Support	0.65	0.83	0.16	4.06	nr	**4.89**
150mm; Fascia	1.02	1.31	0.16	4.06	nr	**5.37**
Bracket supports: including fixing to backgrounds. For minimum fixing distances, refer to the Tables and Memoranda at the rear of the book						
Side rafter	2.70	3.46	0.16	4.06	nr	**7.52**
Top rafter	2.70	3.46	0.16	4.06	nr	**7.52**
Rise and fall	2.61	3.34	0.16	4.06	nr	**7.40**
Extra over fittings half round PVC-U gutter						
Union						
75mm	0.99	1.27	0.19	4.82	nr	**6.09**
100mm	2.16	2.77	0.24	6.09	nr	**8.86**
150mm	3.00	4.03	0.28	7.10	nr	**11.14**
Rainwater pipe outlets						
Running: 75 x 53mm dia	2.27	2.91	0.12	3.04	nr	**5.95**
Running: 100 x 68mm dia	2.87	3.68	0.12	3.04	nr	**6.72**
Running: 150 x 110mm dia	5.82	7.45	0.12	3.04	nr	**10.50**
Stop end: 100 x 68mm dia	2.87	3.68	0.12	3.04	nr	**6.72**
Internal stop ends: short						
75mm	0.99	2.16	0.09	2.28	nr	**4.44**
100mm	1.15	2.32	0.09	2.28	nr	**4.60**
150mm	1.77	2.27	0.09	2.28	nr	**4.55**
External stop ends: short						
75mm	3.02	3.87	0.09	2.28	nr	**6.15**
100mm	3.02	5.02	0.09	2.28	nr	**7.30**
150mm	4.06	5.20	0.09	2.28	nr	**7.48**

R:DISPOSAL SYSTEMS

Item	Net Price £	Material £	Labour hours	Labour £	Unit	Total rate £
R10: RAINWATER PIPEWORK/GUTTERS (cont'd)						
PVC-U gutters: push fit joints (cont'd)						
Fittings; half round gutter (cont'd)						
Angles						
75mm;	2.68	3.43	0.20	5.07	nr	**8.51**
75mm; 45°	2.68	3.43	0.20	5.07	nr	**8.51**
100mm;	2.83	3.81	0.20	5.07	nr	**8.88**
100mm; 120°	3.14	4.02	0.20	5.07	nr	**9.10**
100mm; 135°	3.14	4.02	0.20	5.07	nr	**9.10**
100mm; Prefabricated to special angle	13.37	17.12	0.23	5.84	nr	**22.96**
100mm; Prefabricated to raked angle	14.47	18.53	0.23	5.84	nr	**24.37**
150mm;	5.31	6.80	0.20	5.07	nr	**11.88**
Gutter adaptors						
100mm; Stainless steel clip	1.52	1.95	0.16	4.06	nr	**6.01**
100mm; Cast iron spigot	4.17	5.34	0.23	5.84	nr	**11.18**
100mm; Cast iron socket	4.17	5.34	0.23	5.84	nr	**11.18**
100mm; Cast iron "ogee" spigot	4.22	5.40	0.23	5.84	nr	**11.24**
100mm; Cast iron "ogee" socket	4.22	5.40	0.23	5.84	nr	**11.24**
100mm; Half round to Square PVC-U	7.31	9.36	0.23	5.84	nr	**15.20**
100mm; Gutter overshoot guard	9.05	11.59	0.58	14.72	nr	**26.31**
Square gutter, with brackets measured separately						
120mm	2.23	2.86	0.82	20.80	m	**23.66**
Brackets: including fixing to backgrounds. For minimum fixing distances, refer to the Tables and Memoranda at the rear of the book						
Jointing	1.89	2.42	0.16	4.06	nr	**6.48**
Support	0.73	0.94	0.16	4.06	nr	**4.99**
Bracket support: including fixing to backgrounds. For minimum fixing distances, refer to the Tables and Memoranda at the rear of the book						
Side rafter	2.70	3.46	0.16	4.06	nr	**7.52**
Top rafter	2.70	3.46	0.16	4.06	nr	**7.52**
Rise and fall	4.22	5.40	0.16	4.06	nr	**9.46**
Extra over fittings square PVC-U gutter						
Rainwater pipe outlets						
Running: 62mm square	3.13	4.01	0.12	3.04	nr	**7.05**
Stop end: 62mm square	2.69	3.45	0.12	3.04	nr	**6.49**
Stop ends: short						
External	1.53	1.96	0.09	2.28	nr	**4.24**

R:DISPOSAL SYSTEMS

Item	Net Price £	Material £	Labour hours	Labour £	Unit	Total rate £
Angles						
90°	3.17	4.26	0.20	5.07	nr	**9.34**
120°	11.71	15.75	0.20	5.07	nr	**20.82**
135°	3.17	4.26	0.20	5.07	nr	**9.34**
Prefabricated to special angle	13.72	17.57	0.23	5.84	nr	**23.41**
Prefabricated to raked angle	14.08	18.03	0.23	5.84	nr	**23.87**
Gutter adaptors						
Cast iron	7.30	9.35	0.23	5.84	nr	**15.19**
High capacity square gutter, with brackets measured separately						
137mm	5.71	7.31	0.82	20.80	m	**28.12**
Brackets: including fixing to backgrounds. For minimum fixing distances, refer to the Tables and Memoranda at the rear of the book						
Jointing	5.18	6.97	0.16	4.06	nr	**11.03**
Support	2.16	2.90	0.16	4.06	nr	**6.96**
Overslung	2.01	2.70	0.16	4.06	nr	**6.76**
Bracket supports: including fixing to backgrounds. For minimum fixing distances, refer to the Tables and Memoranda at the rear of the book						
Side rafter	3.24	4.15	0.16	4.06	nr	**8.21**
Top rafter	3.24	4.15	0.16	4.06	nr	**8.21**
Rise and fall	4.59	5.88	0.16	4.06	nr	**9.94**
Extra over fittings high capacity square UPV-C						
Rainwater pipe outlets						
Running: 75mm square	9.09	11.64	0.12	3.04	nr	**14.69**
Running: 82mm dia	9.09	11.64	0.12	3.04	nr	**14.69**
Running: 110mm dia	7.96	10.20	0.12	3.04	nr	**13.24**
Screwed outlet adaptor						
75mm square pipe	4.78	6.12	0.23	5.84	nr	**11.96**
Stop ends: short						
External	3.02	3.87	0.09	2.28	nr	**6.15**
Angles						
90°	8.46	11.38	0.20	5.07	nr	**16.45**
135°	16.54	22.24	0.20	5.07	nr	**27.32**
Prefabricated to special angle	19.25	24.66	0.23	5.84	nr	**30.49**
Prefabricated to raked internal angle	33.51	42.92	0.23	5.84	nr	**48.75**
Prefabricated to raked external angle	33.51	42.92	0.23	5.84	nr	**48.75**
Deep elliptical gutter, with brackets measured separately						
137mm	2.71	3.47	0.82	20.80	m	**24.28**

R:DISPOSAL SYSTEMS

Item	Net Price £	Material £	Labour hours	Labour £	Unit	Total rate £
R10: RAINWATER PIPEWORK/GUTTERS (cont'd)						
PVC-U gutters: push fit joints (cont'd)						
Deep elliptical gutter (cont'd)						
Brackets: including fixing to backgrounds. For minimum fixing distances, refer to the Tables and Memoranda at the rear of the book						
Jointing	0.76	1.02	0.16	4.06	nr	**5.08**
Support	2.82	3.79	0.16	4.06	nr	**7.85**
Bracket support: including fixing to backgrounds. For minimum fixing distances, refer to the Tables and Memoranda at the rear of the book						
Side rafter	3.24	4.36	0.16	4.06	nr	**8.42**
Top rafter	3.24	4.36	0.16	4.06	nr	**8.42**
Rise and fall	4.58	6.16	0.16	4.06	nr	**10.22**
Extra over fittings deep elliptical PVC-U gutter						
Rainwater pipe outlets						
Running: 68mm dia	3.02	4.06	0.12	3.04	nr	**7.11**
Running: 82mm dia	3.02	3.87	0.12	3.04	nr	**6.91**
Stop end: 68mm dia	3.19	4.09	0.12	3.04	nr	**7.13**
Stop ends: short						
External	1.49	1.91	0.09	2.28	nr	**4.19**
Angles						
90°	3.32	4.46	0.20	5.07	nr	**9.54**
135°	3.32	4.46	0.20	5.07	nr	**9.54**
Prefabricated to special angle	9.46	12.72	0.23	5.84	nr	**18.56**
Gutter adaptors						
Stainless steel clip	2.19	2.80	0.16	4.06	nr	**6.86**
Marley deepflow	2.78	3.56	0.23	5.84	nr	**9.40**
Ogee profile PVC-U gutter, with brackets measured separately						
122mm	3.05	3.91	0.82	20.80	m	**24.71**
Brackets: including fixing to backgrounds. For minimum fixing distances, refer to the Tables and Memoranda at the rear of the book						
Jointing	2.39	3.06	0.16	4.06	nr	**7.12**
Support	0.97	1.24	0.16	4.06	nr	**5.30**
Overslung	0.97	1.24	0.16	4.06	nr	**5.30**
Extra over fittings Ogee profile PVC-U gutter						
Rainwater pipe outlets						
Running: 68mm dia	3.30	4.23	0.12	3.04	nr	**7.27**

R:DISPOSAL SYSTEMS

Item	Net Price £	Material £	Labour hours	Labour £	Unit	Total rate £
Stop ends: short						
Internal/External: left or right hand	1.68	2.15	0.09	2.28	nr	**4.44**
Angles						
90° internal or external	3.66	4.69	0.20	5.07	nr	**9.76**
135° internal or external	3.66	4.69	0.20	5.07	nr	**9.76**
PVC-U rainwater pipe: dry push fit joints; fixed with brackets to backgrounds; BS 4576/ BS EN 607						
Pipe: circular, with brackets measured separately						
53mm	3.27	4.19	0.61	15.48	m	**19.67**
68mm	3.58	4.59	0.61	15.48	m	**20.06**
Pipe clip: including fixing to backgrounds. For minimum fixing distances, refer to the Tables and Memoranda at the rear of the book						
68mm	0.97	1.24	0.16	4.06	nr	**5.30**
Pipe clip adjustable: including fixing to backgrounds. For minimum fixing distances, refer to the Tables and Memoranda at the rear of the book						
53mm	0.94	1.20	0.16	4.06	nr	**5.26**
68mm	2.10	2.69	0.16	4.06	nr	**6.75**
Pipe clip drive in: including fixing to backgrounds. For minimum fixing distances, refer to the Tables and Memoranda at the rear of the book						
68mm	2.38	3.05	0.16	4.06	nr	**7.11**
Extra over fittings circular pipework PVC-U						
Pipe coupler: PVC-U to PVC-U						
68mm	1.21	1.55	0.12	3.04	nr	**4.59**
Pipe coupler: PVC-U to Cast Iron						
68mm: to 3" cast iron	2.69	3.45	0.17	4.31	nr	**7.76**
68mm: to 3.3/4" cast iron	11.98	15.34	0.17	4.31	nr	**19.66**
Access pipe: single socket						
68mm	7.69	9.85	0.15	3.81	nr	**13.66**
Bend: short radius						
53mm: 67.5°	1.67	2.14	0.20	5.07	nr	**7.21**
68mm: 92.5°	2.10	2.69	0.20	5.07	nr	**7.76**
68mm: 112.5°	1.75	2.24	0.20	5.07	nr	**7.32**
Bend: long radius						
68mm: 112°	2.10	2.69	0.20	5.07	nr	**7.76**
Branch						
68mm:	9.72	13.07	0.23	5.84	nr	**18.91**
68mm: 112°	9.76	13.13	0.23	5.84	nr	**18.96**

R:DISPOSAL SYSTEMS

Item	Net Price £	Material £	Labour hours	Labour £	Unit	Total rate £
R10: RAINWATER PIPEWORK/GUTTERS (cont'd)						
PVC-U gutters: push fit joints (cont'd)						
Fittings; circular pipework PVC-U (cont'd)						
Double branch						
68mm: 112°	20.22	25.90	0.24	6.09	nr	**31.99**
Shoe						
53mm	1.67	2.14	0.12	3.04	nr	**5.18**
68mm	1.42	1.82	0.12	3.04	nr	**4.86**
Rainwater head: including fixing to backgrounds						
68mm	7.77	9.95	0.29	7.36	nr	**17.31**
Pipe: square, with brackets measured separately						
62mm	2.68	3.43	0.45	11.42	m	**14.85**
75mm	4.60	5.89	0.45	11.42	m	**17.31**
Pipe clip: including fixing to backgrounds. For minimum fixing distances, refer to the Tables and Memoranda at the rear of the book						
62mm	0.90	1.15	0.16	4.06	nr	**5.21**
75mm	1.81	2.32	0.16	4.06	nr	**6.38**
Pipe clip adjustable: including fixing to backgrounds. For minimum fixing distances, refer to the Tables and Memoranda at the rear of the book						
62mm	2.77	3.55	0.16	4.06	nr	**7.61**
Extra over fittings square pipework PVC-U						
Pipe coupler: PVC-U to PVC-U						
62mm	1.66	2.13	0.20	5.07	nr	**7.20**
75mm	2.16	2.77	0.20	5.07	nr	**7.84**
Square to circular adaptor: single socket						
62mm to 68mm	2.56	3.28	0.20	5.07	nr	**8.35**
Square to circular adaptor: single socket						
75mm to 62mm	3.25	4.16	0.20	5.07	nr	**9.24**
Access pipe						
62mm	10.58	13.55	0.16	4.06	nr	**17.61**
75mm	13.29	17.02	0.16	4.06	nr	**21.08**
Bends						
62mm:	2.26	2.89	0.20	5.07	nr	**7.97**
62mm: 112.5°	1.86	2.38	0.20	5.07	nr	**7.46**
75mm: 112.5°	4.33	5.55	0.20	5.07	nr	**10.62**

R:DISPOSAL SYSTEMS

Item	Net Price £	Material £	Labour hours	Labour £	Unit	Total rate £
Bends: prefabricated special angle						
62mm	10.49	13.44	0.23	5.84	nr	19.27
75mm	14.66	18.78	0.23	5.84	nr	24.61
Offset						
62mm	3.39	4.34	0.20	5.07	nr	9.42
75mm	9.59	12.28	0.20	5.07	nr	17.36
Offset: prefabricated special angle						
62mm	10.54	13.50	0.23	5.84	nr	19.34
Shoe						
62mm	1.75	2.24	0.12	3.04	nr	5.29
75mm	2.23	2.86	0.12	3.04	nr	5.90
Branch						
62mm	5.42	6.94	0.23	5.84	nr	12.78
75mm	14.30	18.32	0.23	5.84	nr	24.15
Double branch						
62mm	19.03	24.37	0.24	6.09	nr	30.46
Rainwater head						
62mm	6.68	8.56	0.29	7.36	nr	15.91
75mm	25.55	32.72	3.45	87.49	nr	120.21
PVC-U rainwater pipe: solvent welded joints; fixed with brackets to backgrounds; BS 4576/ BS EN 607						
Pipe: circular, with brackets measured separately						
82mm	7.18	9.20	0.35	8.88	m	18.08
Pipe clip: galvanised; including fixing to backgrounds. For minimum fixing distances, refer to the Tables and Memoranda at the rear of the book						
82mm	2.05	2.63	0.58	14.72	nr	17.34
Pipe clip: galvanised plastic coated; including fixing to backgrounds. For minimum fixing distances, refer to the Tables and Memoranda at the rear of the book						
82mm	2.85	3.65	0.58	14.72	nr	18.37
Pipe clip: PVC-U including fixing to backgrounds. For minimum fixing distances, refer to the Tables and Memoranda at the rear of the book						
82mm	1.54	1.97	0.58	14.72	nr	16.69
Pipe clip: PVC-U adjustable: including fixing to backgrounds. For minimum fixing distances, refer to the Tables and Memoranda at the rear of the book						
82mm	2.82	3.61	0.58	14.72	nr	18.33

R:DISPOSAL SYSTEMS

Item	Net Price £	Material £	Labour hours	Labour £	Unit	Total rate £
R10: RAINWATER PIPEWORK/GUTTERS (cont'd)						
PVC-U pipe: solvent welded joints (cont'd)						
Extra over fittings circular pipework PVC-U						
Pipe coupler: PVC-U to PVC-U						
82mm	3.01	3.86	0.21	5.33	nr	**9.18**
Access pipe						
82mm	16.30	20.88	0.23	5.84	nr	**26.71**
Bend						
82mm: 92°, 112.5° and 135°	6.72	8.61	0.29	7.36	nr	**15.96**
Shoe						
82mm	6.51	8.34	0.29	7.36	nr	**15.70**
110mm	8.19	10.49	0.32	8.12	nr	**18.61**
Branch						
82mm: 92°, 112.5° and 135°	10.47	13.41	0.35	8.88	nr	**22.29**
Rainwater head						
82mm	13.79	17.66	0.58	14.72	nr	**32.38**
110mm	12.58	16.11	0.58	14.72	nr	**30.83**
Roof outlets: 178 dia; Flat						
50mm	12.39	15.87	1.15	29.18	nr	**45.05**
82mm	12.39	15.87	1.15	29.18	nr	**45.05**
Roof outlets: 178mm dia; Domed						
50mm	12.39	15.87	1.15	29.18	nr	**45.05**
82mm	12.39	15.87	1.15	29.18	nr	**45.05**
Roof outlets: 406mm dia; Flat						
82mm	25.41	32.54	1.15	29.18	nr	**61.72**
110mm	25.41	32.54	1.15	29.18	nr	**61.72**
Roof outlets: 406mm dia; Domed						
82mm	25.41	32.54	1.15	29.18	nr	**61.72**
110mm	25.41	32.54	1.15	29.18	nr	**61.72**
Roof outlets: 406mm dia; Inverted						
82mm	53.37	68.36	1.15	29.18	nr	**97.53**
110mm	53.37	68.36	1.15	29.18	nr	**97.53**
Roof outlets: 406mm dia; Vent Pipe						
82mm	36.74	47.06	1.15	29.18	nr	**76.23**
110mm	36.74	47.06	1.15	29.18	nr	**76.23**
Balcony outlets: screed						
82mm	21.62	27.69	1.15	29.18	nr	**56.87**
Balcony outlets: asphalt						
82mm	21.63	27.70	1.15	29.18	nr	**56.88**
Adaptors						
82mm x 62mm square pipe	1.69	2.16	0.21	5.33	nr	**7.49**
82mm x 68mm circular pipe	1.56	2.00	0.21	5.33	nr	**7.33**

R:DISPOSAL SYSTEMS

Item	Net Price £	Material £	Labour hours	Labour £	Unit	Total rate £
For 110mm diameter pipework and fittings refer to R11: Above Ground Drainage						
Cast iron gutters: mastic and bolted joints; BS 460; fixed with brackets to backgrounds						
Half round gutter, with brackets measured separately						
100 mm	13.98	17.91	0.85	21.57	m	**39.47**
115 mm	15.31	19.61	0.97	24.61	m	**44.22**
125 mm	17.81	22.81	0.97	24.61	m	**47.42**
150 mm	29.70	38.04	1.12	28.42	m	**66.46**
Brackets; fixed to backgrounds. For minimum fixing distances, refer to the Tables and Memoranda at the rear of the book						
Fascia						
100 mm	1.59	2.04	0.16	4.06	nr	**6.10**
115 mm	1.59	2.04	0.16	4.06	nr	**6.10**
125 mm	1.59	2.04	0.16	4.06	nr	**6.10**
150 mm	2.00	2.56	0.16	4.06	nr	**6.62**
Rise and fall						
100 mm	2.36	3.02	0.39	9.90	nr	**12.92**
115 mm	2.36	3.02	0.39	9.90	nr	**12.92**
125 mm	2.83	3.62	0.39	9.90	nr	**13.52**
150 mm	3.32	4.25	0.39	9.90	nr	**14.15**
Top rafter						
100 mm	1.52	1.95	0.16	4.06	nr	**6.01**
115 mm	1.52	1.95	0.16	4.06	nr	**6.01**
125 mm	1.57	2.01	0.16	4.06	nr	**6.07**
150 mm	2.36	3.02	0.16	4.06	nr	**7.08**
Side rafter						
100 mm	1.52	1.95	0.16	4.06	nr	**6.01**
115 mm	1.52	1.95	0.16	4.06	nr	**6.01**
125 mm	1.57	2.01	0.16	4.06	nr	**6.07**
150 mm	2.36	3.02	0.16	4.06	nr	**7.08**
Extra over fittings half round gutter cast iron BS 460						
Union						
100 mm	3.87	4.96	0.39	9.90	nr	**14.85**
115 mm	4.72	6.05	0.48	12.18	nr	**18.22**
125 mm	5.44	6.97	0.48	12.18	nr	**19.15**
150 mm	6.12	7.84	0.55	13.95	nr	**21.79**
Stop end; internal						
100 mm	1.97	2.52	0.12	3.04	nr	**5.57**
115 mm	2.55	3.27	0.15	3.81	nr	**7.07**
125 mm	2.55	3.27	0.15	3.81	nr	**7.07**
150 mm	3.40	4.35	0.20	5.07	nr	**9.43**

R:DISPOSAL SYSTEMS

Item	Net Price £	Material £	Labour hours	Labour £	Unit	Total rate £
R10: RAINWATER PIPEWORK/GUTTERS (cont'd)						
Cast iron gutters; mastic joints (cont'd)						
Fittings; half round gutter (cont'd)						
Stop end; external						
100 mm	1.97	2.52	0.12	3.04	nr	**5.57**
115 mm	2.55	3.27	0.15	3.81	nr	**7.07**
125 mm	2.55	3.27	0.15	3.81	nr	**7.07**
150 mm	3.40	4.35	0.20	5.07	nr	**9.43**
90° angle; single socket						
100 mm	5.85	7.49	0.39	9.90	nr	**17.39**
115 mm	6.03	7.72	0.43	10.91	nr	**18.63**
125 mm	7.11	9.11	0.43	10.91	nr	**20.02**
150 mm	12.99	16.64	0.50	12.69	nr	**29.32**
90° angle; double socket						
100 mm	7.10	9.09	0.39	9.90	nr	**18.99**
115 mm	7.54	9.66	0.43	10.91	nr	**20.57**
125 mm	9.75	12.49	0.43	10.91	nr	**23.40**
135° angle; single socket						
100 mm	5.44	6.97	0.39	9.90	nr	**16.86**
115 mm	6.03	7.72	0.43	10.91	nr	**18.63**
125 mm	8.90	11.40	0.43	10.91	nr	**22.31**
150 mm	11.89	15.23	0.50	12.69	nr	**27.91**
Running outlet						
65 mm outlet						
100 mm	5.70	7.30	0.39	9.90	nr	**17.20**
115 mm	6.22	7.97	0.43	10.91	nr	**18.88**
125 mm	7.11	9.11	0.43	10.91	nr	**20.02**
75 mm outlet						
100 mm	5.70	7.30	0.39	9.90	nr	**17.20**
115 mm	6.22	7.97	0.43	10.91	nr	**18.88**
125 mm	7.11	9.11	0.43	10.91	nr	**20.02**
150 mm	12.09	15.48	0.50	12.69	nr	**28.17**
100 mm outlet						
150 mm	12.09	12.09	0.50	12.69	nr	**24.78**
Stop end outlet; socket						
65 mm outlet						
100 mm	4.54	5.81	0.39	9.90	nr	**15.71**
115 mm	4.99	6.39	0.43	10.91	nr	**17.30**
75 mm outlet						
125 mm	6.34	8.12	0.43	10.91	nr	**19.03**
150 mm	12.09	15.48	0.50	12.69	nr	**28.17**
100mm outlet						
150 mm	12.09	15.48	0.50	12.69	nr	**28.17**

R:DISPOSAL SYSTEMS

Item	Net Price £	Material £	Labour hours	Labour £	Unit	Total rate £
Stop end outlet; spigot						
65 mm outlet						
100 mm	4.54	5.81	0.39	9.90	nr	**15.71**
115 mm	4.99	6.39	0.43	10.91	nr	**17.30**
75 mm outlet						
125 mm	6.34	8.12	0.43	10.91	nr	**19.03**
150 mm	12.09	15.48	0.50	12.69	nr	**28.17**
100mm outlet						
150 mm	12.09	15.48	0.50	12.69	nr	**28.17**
Half round; 3 mm thick double beaded gutter, with brackets measured separately						
100 mm	8.52	10.91	0.85	21.57	m	**32.48**
115 mm	8.98	11.50	0.85	21.57	m	**33.07**
125 mm	10.07	12.90	0.97	24.61	m	**37.51**
Brackets; fixed to backgrounds. For minimum fixing distances, refer to the Tables and Memoranda at the rear of the book						
Fascia						
100 mm	3.59	4.60	0.16	4.06	nr	**8.66**
115 mm	3.59	4.60	0.16	4.06	nr	**8.66**
125 mm	4.39	5.62	0.16	4.06	nr	**9.68**
Extra over fittings Half Round 3mm thick Gutter BS 460						
Union						
100 mm	3.86	4.94	0.38	9.64	nr	**14.59**
115 mm	4.69	6.01	0.38	9.64	nr	**15.65**
125 mm	5.37	6.88	0.43	10.91	nr	**17.79**
Stop end; internal						
100 mm	1.76	2.25	0.12	3.04	nr	**5.30**
115 mm	2.50	3.20	0.12	3.04	nr	**6.25**
125 mm	2.55	3.27	0.15	3.81	nr	**7.07**
Stop end; external						
100 mm	1.76	2.25	0.12	3.04	nr	**5.30**
115 mm	2.50	3.20	0.12	3.04	nr	**6.25**
125 mm	2.50	3.20	0.15	3.81	nr	**7.01**
90° angle; single socket						
100 mm	5.98	7.66	0.38	9.64	nr	**17.30**
115 mm	6.17	7.90	0.38	9.64	nr	**17.54**
125 mm	7.50	9.61	0.43	10.91	nr	**20.52**
135° angle; single socket						
100 mm	5.98	7.66	0.38	9.64	nr	**17.30**
115 mm	6.17	7.90	0.38	9.64	nr	**17.54**
125 mm	7.50	9.61	0.43	10.91	nr	**20.52**

R:DISPOSAL SYSTEMS

Item	Net Price £	Material £	Labour hours	Labour £	Unit	Total rate £
R10: RAINWATER PIPEWORK/GUTTERS (cont'd)						
Cast iron gutters; mastic joints (cont'd)						
Fittings; half round 3mm thick gutter (cont'd)						
Running outlet						
65 mm outlet						
100 mm	5.98	7.66	0.38	9.64	nr	**17.30**
115 mm	6.04	7.74	0.38	9.64	nr	**17.38**
125 mm	6.58	8.43	0.43	10.91	nr	**19.34**
75 mm outlet						
115 mm	6.17	7.90	0.38	9.64	nr	**17.54**
125 mm	7.36	9.43	0.43	10.91	nr	**20.34**
Stop end outlet; socket						
65 mm outlet						
100 mm	4.62	5.92	0.38	9.64	nr	**15.56**
115 mm	5.63	7.21	0.38	9.64	nr	**16.85**
125 mm	7.36	7.36	0.43	10.91	nr	**18.27**
75 mm outlet						
125 mm	6.31	8.08	0.43	10.91	nr	**18.99**
Stop end outlet; spigot						
65 mm outlet						
100 mm	4.62	5.92	0.38	9.64	nr	**15.56**
115 mm	5.63	7.21	0.38	9.64	nr	**16.85**
125 mm	6.43	8.24	0.43	10.91	nr	**19.15**
Deep half round gutter, with brackets measured separately						
100 x 75 mm	14.52	18.60	0.85	21.57	m	**40.16**
125 x 75 mm	18.79	24.07	0.97	24.61	m	**48.68**
Brackets; fixed to backgrounds. For minimum fixing distances, refer to the Tables and Memoranda at the rear of the book						
Fascia						
100 x 75 mm	5.55	7.11	0.16	4.06	nr	**11.17**
125 x 75 mm	11.84	15.16	0.16	4.06	nr	**19.22**
Extra over fittings Deep Half Round Gutter BS 460						
Union						
100 x 75 mm	6.34	8.12	0.38	9.64	nr	**17.76**
125 x 75 mm	6.85	8.77	0.43	10.91	nr	**19.68**
Stop end; internal						
100 x 75 mm	5.55	7.11	0.12	3.04	nr	**10.15**
125 x 75 mm	6.85	8.77	0.15	3.81	nr	**12.58**
Stop end; external						
100 x 75 mm	5.55	7.11	0.12	3.04	nr	**10.15**
125 x 75 mm	6.85	8.77	0.15	3.81	nr	**12.58**

R:DISPOSAL SYSTEMS

Item	Net Price £	Material £	Labour hours	Labour £	Unit	Total rate £
90° angle; single socket						
100 x 75 mm	16.13	20.66	0.38	9.64	nr	**30.30**
125 x 75 mm	20.48	26.23	0.43	10.91	nr	**37.14**
135° angle; single socket						
100 x 75 mm	16.13	20.66	0.38	9.64	nr	**30.30**
125 x 75 mm	20.48	26.23	0.43	10.91	nr	**37.14**
Running outlet						
65 mm outlet						
100 x 75 mm	16.13	20.66	0.38	9.64	nr	**30.30**
125 x 75 mm	20.48	26.23	0.43	10.91	nr	**37.14**
75 mm outlet						
100 x 75 mm	12.10	15.50	0.38	9.64	nr	**25.14**
125 x 75 mm	14.57	18.66	0.43	10.91	nr	**29.57**
Stop end outlet; socket						
65 mm outlet						
100 x 75 mm	10.80	13.83	0.38	9.64	nr	**23.47**
75 mm outlet						
100 x 75 mm	10.80	13.83	0.38	9.64	nr	**23.47**
125 x 75 mm	13.81	17.69	0.43	10.91	nr	**28.60**
Stop end outlet; spigot						
65 mm outlet						
100 x 75 mm	10.80	13.83	0.38	9.64	nr	**23.47**
75 mm outlet						
100 x 75 mm	10.80	13.83	0.38	9.64	nr	**23.47**
125 x 75 mm	13.81	17.69	0.43	10.91	nr	**28.60**
Ogee gutter, with brackets measured separately						
100 mm	9.55	12.23	0.85	21.57	m	**33.80**
115 mm	10.67	13.67	0.97	24.61	m	**38.28**
125 mm	11.32	14.50	0.97	24.61	m	**39.11**
Brackets; fixed to backgrounds. For minimum fixing distances, refer to the Tables and Memoranda at the rear of the book						
Fascia						
100 mm	3.39	4.34	0.16	4.06	nr	**8.40**
115 mm	3.65	4.67	0.16	4.06	nr	**8.73**
125 mm	4.25	5.44	0.16	4.06	nr	**9.50**
Extra over fittings Ogee Cast Iron Gutter BS 460						
Union						
100 mm	3.71	4.75	0.38	9.64	nr	**14.39**
115 mm	3.87	4.96	0.43	10.91	nr	**15.87**
125 mm	4.87	6.24	0.43	10.91	nr	**17.15**

R:DISPOSAL SYSTEMS

Item	Net Price £	Material £	Labour hours	Labour £	Unit	Total rate £
R10: RAINWATER PIPEWORK/GUTTERS (cont'd)						
Fittings; Ogee gutters (cont'd)						
Stop end; internal						
100 mm	1.82	2.33	0.12	3.04	nr	**5.38**
115 mm	2.41	3.09	0.15	3.81	nr	**6.89**
125 mm	2.41	3.09	0.15	3.81	nr	**6.89**
Stop end; external						
100 mm	1.82	2.33	0.12	3.04	nr	**5.38**
115 mm	2.41	3.09	0.15	3.81	nr	**6.89**
125 mm	2.41	3.09	0.15	3.81	nr	**6.89**
90° angle; internal						
100 mm	6.12	7.84	0.38	9.64	nr	**17.48**
115 mm	6.63	8.49	0.43	10.91	nr	**19.40**
125 mm	7.23	9.26	0.43	10.91	nr	**20.17**
90° angle; external						
100 mm	6.23	7.98	0.38	9.64	nr	**17.62**
115 mm	6.76	8.66	0.43	10.91	nr	**19.57**
125 mm	7.38	9.45	0.43	10.91	nr	**20.36**
135° angle; internal						
100 mm	6.12	7.84	0.38	9.64	nr	**17.48**
115 mm	6.63	8.49	0.43	10.91	nr	**19.40**
125 mm	7.23	9.26	0.43	10.91	nr	**20.17**
135° angle; external						
100 mm	6.12	7.84	0.38	9.64	nr	**17.48**
115 mm	6.63	8.49	0.43	10.91	nr	**19.40**
125 mm	7.38	9.45	0.43	10.91	nr	**20.36**
Running outlet						
65 mm outlet						
100 mm	6.69	8.57	0.38	9.64	nr	**18.21**
115 mm	7.20	9.22	0.43	10.91	nr	**20.13**
125 mm	7.89	10.11	0.43	10.91	nr	**21.02**
75 mm outlet						
125 mm	7.89	10.11	0.43	10.91	nr	**21.02**
Stop end outlet; socket						
65 mm outlet						
100 mm	7.32	9.38	0.38	9.64	nr	**19.02**
115 mm	7.32	9.38	0.43	10.91	nr	**20.29**
125 mm	8.44	10.81	0.43	10.91	nr	**21.72**
75 mm outlet						
125 mm	9.40	12.04	0.43	10.91	nr	**22.95**
Stop end outlet; spigot						
65 mm outlet						
100 mm	7.32	9.38	0.38	9.64	nr	**19.02**
115 mm	7.32	9.38	0.43	10.91	nr	**20.29**
125 mm	8.44	10.81	0.43	10.91	nr	**21.72**

R:DISPOSAL SYSTEMS

Item	Net Price £	Material £	Labour hours	Labour £	Unit	Total rate £
75 mm outlet						
125 mm	3.39	4.34	0.43	10.91	nr	**15.25**
Notts Ogee Gutter, with brackets measured separately						
115 mm	15.51	19.87	0.85	21.57	m	**41.43**
Brackets; fixed to backgrounds. For minimum fixing distances, refer to the Tables and Memoranda at the rear of the book						
Fascia						
115 mm	5.43	6.95	0.16	4.06	nr	**11.01**
Extra over fittings Notts Ogee Cast Iron Gutter BS 460						
Union						
115 mm	6.35	8.13	0.38	9.64	nr	**17.77**
Stop end; internal						
115 mm	5.43	6.95	0.16	4.06	nr	**11.01**
Stop end; external						
115 mm	5.43	6.95	0.16	4.06	nr	**11.01**
90° angle; internal						
115 mm	15.17	19.43	0.43	10.91	nr	**30.34**
90° angle; external						
115 mm	15.17	19.43	0.43	10.91	nr	**30.34**
135° angle; internal						
115 mm	15.17	19.43	0.43	10.91	nr	**30.34**
135° angle; external						
115 mm	15.17	19.43	0.43	10.91	nr	**30.34**
Running outlet						
65 mm outlet						
115 mm	18.20	23.31	0.43	10.91	nr	**34.22**
75 mm outlet						
115 mm	15.19	19.46	0.43	10.91	nr	**30.37**
Stop end outlet; socket						
65 mm outlet						
115 mm	12.01	15.38	0.43	10.91	nr	**26.29**
Stop end outlet; spigot						
65 mm outlet						
115 mm	14.42	18.47	0.43	10.91	nr	**29.38**
No 46 moulded Gutter, with brackets measured separately						
100 x 75 mm	14.72	18.85	0.85	21.57	m	**40.42**
125 x 100 mm	21.18	27.13	0.97	24.61	m	**51.74**

R:DISPOSAL SYSTEMS

Item	Net Price £	Material £	Labour hours	Labour £	Unit	Total rate £
R10: RAINWATER PIPEWORK/GUTTERS (cont'd)						
No 46 moulded Gutter (cont'd)						
Brackets; fixed to backgrounds. For minimum fixing distances, refer to the Tables and Memoranda at the rear of the book						
Fascia						
100 x 75 mm	3.00	3.84	0.16	4.06	nr	**7.90**
125 x 100 mm	3.00	3.84	0.16	4.06	nr	**7.90**
Extra over fittings						
Union						
100 x 75 mm	7.85	10.05	0.38	9.64	nr	**19.70**
125 x 100 mm	8.72	11.17	0.43	10.91	nr	**22.08**
Stop end; internal						
100 x 75 mm	5.67	7.26	0.12	3.04	nr	**10.31**
125 x 100 mm	7.35	9.41	0.15	3.81	nr	**13.22**
Stop end; external						
100 x 75 mm	5.67	7.26	0.12	3.04	nr	**10.31**
125 x 100 mm	7.35	9.41	0.15	3.81	nr	**13.22**
90° angle; internal						
100 x 75 mm	14.87	19.05	0.38	9.64	nr	**28.69**
125 x 100 mm	21.37	27.37	0.43	10.91	nr	**38.28**
90° angle; external						
100 x 75 mm	14.87	19.05	0.38	9.64	nr	**28.69**
125 x 100 mm	21.37	27.37	0.43	10.91	nr	**38.28**
135° angle; internal						
100 x 75 mm	14.87	19.05	0.38	9.64	nr	**28.69**
125 x 100 mm	21.37	27.37	0.43	10.91	nr	**38.28**
135° angle; external						
100 x 75 mm	14.87	19.05	0.38	9.64	nr	**28.69**
125 x 100 mm	21.37	27.37	0.43	10.91	nr	**38.28**
Running outlet						
65 mm outlet						
100 x 75 mm	14.87	19.05	0.38	9.64	nr	**28.69**
125 x 100 mm	21.37	27.37	0.43	10.91	nr	**38.28**
75 mm outlet						
100 x 75 mm	14.87	19.05	0.38	9.64	nr	**28.69**
125 x 100 mm	21.37	27.37	0.43	10.91	nr	**38.28**
100 mm outlet						
100 x 75 mm	21.37	27.37	0.38	9.64	nr	**37.01**
125 x 100 mm	21.37	27.37	0.43	10.91	nr	**38.28**
100 x 75 mm outlet						
125 x 100 mm	21.37	27.37	0.43	10.91	nr	**38.28**

R:DISPOSAL SYSTEMS

Item	Net Price £	Material £	Labour hours	Labour £	Unit	Total rate £
Stop end outlet; socket						
65 mm outlet						
100 x 75 mm	11.79	15.10	0.38	9.64	nr	**24.74**
75 mm outlet						
125 x 100 mm	14.55	18.64	0.43	10.91	nr	**29.55**
Stop end outlet; spigot						
65 mm outlet						
100 x 75 mm	11.79	15.10	0.38	9.64	nr	**24.74**
75 mm outlet						
125 x 100 mm	14.55	18.64	0.43	10.91	nr	**29.55**
Box gutter, with brackets measured separately						
100 x 75 mm	24.70	31.64	0.85	21.57	m	**53.20**
Brackets; fixed to backgrounds. For minimum fixing distances, refer to the Tables and Memoranda at the rear of the book						
Fascia						
100 x 75 mm	3.31	4.24	0.16	4.06	nr	**8.30**
Extra over fittings Box Cast Iron Gutter BS 460						
Union						
100 x 75 mm	4.22	5.40	0.38	9.64	nr	**15.05**
Stop end; external						
100 x 75 mm	3.25	4.16	0.12	3.04	nr	**7.21**
90° angle						
100 x 75 mm	11.23	14.38	0.38	9.64	nr	**24.02**
135° angle						
100 x 75 mm	11.47	14.69	0.38	9.64	nr	**24.33**
Running outlet						
65 mm outlet						
100 x 75 mm	11.23	14.38	0.38	9.64	nr	**24.02**
75 mm outlet						
100 x 75 mm	11.23	14.38	0.38	9.64	nr	**24.02**
100 x 75 mm outlet						
100 x 75 mm	11.23	14.38	0.38	9.64	nr	**24.02**
Cast iron rainwater pipe; dry joints; BS 460; fixed to backgrounds						
Circular						
Plain socket pipe, with brackets measured separately						
65mm	14.65	18.76	0.69	17.51	m	**36.27**
75 mm	14.65	18.76	0.69	17.51	m	**36.27**
100 mm	20.00	25.62	0.69	17.51	m	**43.12**

R:DISPOSAL SYSTEMS

Item	Net Price £	Material £	Labour hours	Labour £	Unit	Total rate £
R10: RAINWATER PIPEWORK/GUTTERS (cont'd)						
Cast iron rainwater pipe circular; dry joints (cont'd)						
Bracket; fixed to backgrounds. For minimum fixing distances, refer to the Tables and Memoranda at the rear of the book						
65mm	4.67	5.98	0.29	7.36	nr	**13.34**
75 mm	4.71	6.03	0.29	7.36	nr	**13.39**
100 mm	4.71	6.03	0.29	7.36	nr	**13.39**
Eared socket pipe, with wall spacers measured separately						
65mm	15.11	19.35	0.62	15.73	m	**35.08**
75 mm	15.11	19.35	0.62	15.73	m	**35.08**
100 mm	20.28	25.97	0.62	15.73	m	**41.70**
Wall spacer plate; eared pipework						
65mm	3.25	4.16	0.16	4.06	nr	**8.22**
75 mm	3.22	4.12	0.16	4.06	nr	**8.18**
100 mm	3.37	4.32	0.16	4.06	nr	**8.38**
Extra over fittings Circular Cast Iron Pipework BS 460						
Loose sockets						
Plain socket						
65mm	10.90	13.96	0.23	5.84	nr	**19.80**
75 mm	10.90	13.96	0.23	5.84	nr	**19.80**
100 mm	14.67	18.79	0.23	5.84	nr	**24.62**
Eared socket						
65mm	14.96	19.16	0.29	7.36	nr	**26.52**
75 mm	14.96	19.16	0.29	7.36	nr	**26.52**
100 mm	20.02	25.64	0.29	7.36	nr	**33.00**
Shoe; front projection						
Plain socket						
65mm	14.96	19.16	0.23	5.84	nr	**25.00**
75 mm	14.96	19.16	0.23	5.84	nr	**25.00**
100 mm	20.02	25.64	0.23	5.84	nr	**31.48**
Eared socket						
65mm	13.33	17.07	0.29	7.36	nr	**24.43**
75 mm	13.33	17.07	0.29	7.36	nr	**24.43**
100 mm	16.64	21.31	0.29	7.36	nr	**28.67**
Access Pipe						
65mm	20.79	26.63	0.23	5.84	nr	**32.46**
75 mm	21.84	27.97	0.23	5.84	nr	**33.81**
100 mm	38.14	48.85	0.23	5.84	nr	**54.68**
100 mm; eared	43.01	55.09	0.29	7.36	nr	**62.44**
Bends; any degree						
65mm	8.16	10.45	0.23	5.84	nr	**16.29**
75 mm	9.92	12.71	0.23	5.84	nr	**18.54**
100 mm	20.00	25.62	0.23	5.84	nr	**31.45**

R:DISPOSAL SYSTEMS

Item	Net Price £	Material £	Labour hours	Labour £	Unit	Total rate £
Branch						
92.5°						
65mm	15.74	20.16	0.29	7.36	nr	**27.52**
75 mm	17.35	22.22	0.29	7.36	nr	**29.58**
100 mm	20.63	26.42	0.29	7.36	nr	**33.78**
112.5°						
65mm	12.61	16.15	0.29	7.36	nr	**23.51**
75 mm	13.90	17.80	0.29	7.36	nr	**25.16**
135°						
65mm	15.74	20.16	0.29	7.36	nr	**27.52**
75 mm	17.35	22.22	0.29	7.36	nr	**29.58**
Offsets						
75 to 150 mm projection						
65mm	12.50	16.01	0.25	6.34	nr	**22.35**
75 mm	12.50	16.01	0.25	6.34	nr	**22.35**
100 mm	23.57	30.19	0.25	6.34	nr	**36.53**
225 mm projection						
65mm	14.55	18.64	0.25	6.34	nr	**24.98**
75 mm	14.55	18.64	0.25	6.34	nr	**24.98**
100 mm	28.56	36.58	0.25	6.34	nr	**42.92**
305 mm projection						
65mm	17.03	21.81	0.25	6.34	nr	**28.15**
75 mm	17.88	22.90	0.25	6.34	nr	**29.24**
100 mm	28.56	36.58	0.25	6.34	nr	**42.92**
380 mm projection						
65mm	34.00	43.55	0.25	6.34	nr	**49.89**
75 mm	34.00	43.55	0.25	6.34	nr	**49.89**
100 mm	46.41	59.44	0.25	6.34	nr	**65.78**
455 mm projection						
65mm	39.80	50.98	0.25	6.34	nr	**57.32**
75 mm	39.80	50.98	0.25	6.34	nr	**57.32**
100 mm	56.41	72.25	0.25	6.34	nr	**78.59**
Rectangular						
Plain socket						
100 x 75 mm	59.41	76.09	1.04	26.39	m	**102.48**
Bracket; fixed to backgrounds. For minimum fixing distances, refer to the Tables and Memoranda at the rear of the book						
100 x 75mm; build in holdabat	20.68	26.49	0.35	8.88	nr	**35.37**
100 x 75mm; trefoil earband	16.18	20.72	0.29	7.36	nr	**28.08**
100 x 75mm; plain earband	15.64	20.03	0.29	7.36	nr	**27.39**
Eared Socket, with wall spacers measured separately						
100 x 75 mm	56.87	72.84	1.16	29.43	m	**102.27**
Wall spacer plate; eared pipework						
100 x 75	3.37	4.32	0.16	4.06	nr	**8.38**

R:DISPOSAL SYSTEMS

Item	Net Price £	Material £	Labour hours	Labour £	Unit	Total rate £
R10: RAINWATER PIPEWORK/GUTTERS (cont'd)						
Cast iron rainwater pipe rectangular; dry joints (cont'd)						
Extra over fittings Rectangular Cast Iron Pipework BS 460						
Loose socket						
100 x 75 mm; plain	15.12	19.37	0.23	5.84	nr	**25.20**
100 x 75 mm; eared	28.23	36.16	0.29	7.36	nr	**43.51**
Shoe; front						
100 x 75 mm; plain	15.18	19.44	0.23	5.84	nr	**25.28**
100 x 75 mm; eared	25.15	32.21	0.29	7.36	nr	**39.57**
Shoe; side						
100 x 75 mm; plain	40.09	51.35	0.23	5.84	nr	**57.18**
100 x 75 mm; eared	50.99	65.31	0.29	7.36	nr	**72.67**
Bends; side; any degree						
100 x 75 mm; plain	36.31	46.51	0.25	6.34	nr	**52.85**
100 x 75 mm; 135°; plain	38.44	49.23	0.25	6.34	nr	**55.58**
Bends; side; any degree						
100 x 75 mm; eared	45.39	58.14	0.25	6.34	nr	**64.48**
Bends; front; any degree						
100 x 75 mm; plain	36.31	46.51	0.25	6.34	nr	**52.85**
100 x 75 mm; eared	45.39	58.14	0.25	6.34	nr	**64.48**
Offset; side; Plain socket						
75mm projection	49.57	63.49	0.25	6.34	nr	**69.83**
115mm projection	51.55	66.02	0.25	6.34	nr	**72.37**
225mm projection	64.21	82.24	0.25	6.34	nr	**88.58**
305mm projection	74.01	94.79	0.25	6.34	nr	**101.13**
Offset; front Plain socket						
75mm projection	49.57	63.49	0.25	6.34	nr	**69.83**
150mm projection	52.44	67.16	0.25	6.34	nr	**73.51**
225mm projection	64.21	82.24	0.25	6.34	nr	**88.58**
305mm projection	74.01	94.79	0.25	6.34	nr	**101.13**
Eared socket						
75mm projection	47.33	60.62	0.25	6.34	nr	**66.96**
150mm projection	52.44	67.16	0.25	6.34	nr	**73.51**
225mm projection	66.97	85.77	0.25	6.34	nr	**92.12**
305mm projection	71.58	91.68	0.25	6.34	nr	**98.02**
Offset; plinth						
115mm projection; plain	38.91	49.84	0.25	6.34	nr	**56.18**
115mm projection; eared	49.16	62.96	0.25	6.34	nr	**69.31**

R:DISPOSAL SYSTEMS

Item	Net Price £	Material £	Labour hours	Labour £	Unit	Total rate £
Rainwater heads						
Flat hopper						
210 x 160 x 185 mm; 65 mm outlet	10.42	13.35	0.40	10.15	nr	**23.49**
210 x 160 x 185 mm; 75 mm outlet	11.84	15.16	0.40	10.15	nr	**25.31**
250 x 215 x 215 mm; 100 mm outlet	26.22	33.58	0.40	10.15	nr	**43.73**
Flat rectangular						
225 x 125 x 125 mm; 65 mm outlet	18.20	23.31	0.40	10.15	nr	**33.46**
225 x 125 x 125 mm; 75 mm outlet	18.20	23.31	0.40	10.15	nr	**33.46**
280 x 150 x 130 mm; 100 mm outlet	25.13	32.19	0.40	10.15	nr	**42.34**
Rectangular						
250 x 180 x 175mm; 75 mm outlet	26.13	33.47	0.40	10.15	nr	**43.62**
250 x 180 x 175mm; 100 mm outlet	27.54	35.27	0.40	10.15	nr	**45.42**
300 x 250 x 200mm; 65 mm outlet	48.96	62.71	0.40	10.15	nr	**72.86**
300 x 250 x 200mm; 75 mm outlet	48.96	62.71	0.40	10.15	nr	**72.86**
300 x 250 x 200mm; 100 mm outlet	48.96	62.71	0.40	10.15	nr	**72.86**
300 x 250 x 200mm; 100 x 75 mm outlet	48.96	62.71	0.40	10.15	nr	**72.86**
Castellated rectangular						
250 x 180 x 175mm; 65 mm outlet	48.96	62.71	0.40	10.15	nr	**72.86**

R:DISPOSAL SYSTEMS

Item	Net Price £	Material £	Labour hours	Labour £	Unit	Total rate £
R11: ABOVE GROUND DRAINAGE						
Pricing note: degree angles are only indicated where material rates differ PVC-U overflow pipe; solvent welded joints; fixed with clips to backgrounds						
Pipe, with brackets measured separately						
19mm	1.92	2.46	0.21	5.33	m	**7.79**
Fixings						
Pipe clip: including fixing to backgrounds. For minimum fixing distances, refer to the Tables and Memoranda at the rear of the book						
19mm	0.32	0.41	0.18	4.57	nr	**4.98**
Extra over fittings overflow pipework PVC-U						
Straight coupler						
19mm	0.75	0.96	0.17	4.31	nr	**5.27**
Bend						
19mm	0.90	1.15	0.17	4.31	nr	**5.47**
19mm: 135°	0.91	1.17	0.17	4.31	nr	**5.48**
Tee						
19mm	0.98	1.26	0.18	4.57	nr	**5.82**
Reverse nut connector						
19mm	0.38	0.49	0.15	3.81	nr	**4.29**
BSP adaptor: solvent welded socket to threaded socket						
19mm x 3/4"	1.25	1.60	0.14	3.55	nr	**5.15**
Straight tank connector						
19mm	1.19	1.52	0.21	5.33	nr	**6.85**
32mm	2.20	2.82	0.28	7.10	nr	**9.92**
40mm	2.39	3.06	0.30	7.61	nr	**10.67**
Bent tank connector						
19mm	1.41	1.81	0.21	5.33	nr	**7.13**
Tundish						
19mm	20.22	25.90	0.38	9.64	nr	**35.54**
MuPVC waste pipe; solvent welded joints; fixed with clips to backgrounds; BS 5255						
Pipe, with brackets measured separately						
32mm	1.69	2.16	0.23	5.84	m	**8.00**
40mm	2.09	2.68	0.23	5.84	m	**8.51**
50mm	3.23	4.14	0.26	6.60	m	**10.73**

R:DISPOSAL SYSTEMS

Item	Net Price £	Material £	Labour hours	Labour £	Unit	Total rate £
Fixings						
Pipe clip: including fixing to backgrounds. For minimum fixing distances, refer to the Tables and Memoranda at the rear of the book						
32mm	0.26	0.33	0.13	3.30	nr	**3.63**
40mm	0.34	0.44	0.13	3.30	nr	**3.73**
50mm	0.63	0.81	0.13	3.30	nr	**4.11**
Pipe clip: expansion: including fixing to backgrounds. For minimum fixing distances, refer to the Tables and Memoranda at the rear of the book						
32mm	0.29	0.37	0.13	3.30	nr	**3.67**
40mm	0.37	0.47	0.13	3.30	nr	**3.77**
50mm	1.00	1.28	0.13	3.30	nr	**4.58**
Pipe clip: metal; including fixing to backgrounds. For minimum fixing distances, refer to the Tables and Memoranda at the rear of the book						
32mm	1.17	1.50	0.13	3.30	nr	**4.80**
40mm	1.39	1.78	0.13	3.30	nr	**5.08**
50mm	1.76	2.25	0.13	3.30	nr	**5.55**
Extra over fittings waste pipework MuPVC						
Screwed access plug						
32mm	1.23	1.58	0.18	4.57	nr	**6.14**
40mm	1.30	1.67	0.18	4.57	nr	**6.23**
50mm	2.13	2.73	0.25	6.34	nr	**9.07**
Straight coupling						
32mm	0.81	1.04	0.27	6.85	nr	**7.89**
40mm	0.96	1.23	0.27	6.85	nr	**8.08**
50mm	1.48	1.90	0.27	6.85	nr	**8.75**
Expansion coupling						
32mm	1.42	1.82	0.27	6.85	nr	**8.67**
40mm	1.72	2.20	0.27	6.85	nr	**9.05**
50mm	2.32	2.97	0.27	6.85	nr	**9.82**
MuPVC to copper coupling						
32mm	1.31	1.68	0.27	6.85	nr	**8.53**
40mm	1.47	1.88	0.27	6.85	nr	**8.73**
50mm	1.97	2.52	0.27	6.85	nr	**9.37**
Spigot and socket coupling						
32mm	1.42	1.82	0.27	6.85	nr	**8.67**
40mm	1.72	2.20	0.27	6.85	nr	**9.05**
50mm	2.32	2.97	0.27	6.85	nr	**9.82**
Union						
32mm	3.39	4.34	0.28	7.10	nr	**11.45**
40mm	4.45	5.70	0.28	7.10	nr	**12.80**
50mm	6.94	8.89	0.28	7.10	nr	**15.99**

R:DISPOSAL SYSTEMS

Item	Net Price £	Material £	Labour hours	Labour £	Unit	Total rate £
R11: ABOVE GROUND DRAINAGE (cont'd)						
MuPVC waste pipe; solvent welded joints (cont'd)						
Fittings; waste pipework MuPVC (cont'd)						
Reducer: socket						
32 x 19mm	1.02	1.31	0.27	6.85	nr	8.16
40 x 32mm	0.91	1.17	0.27	6.85	nr	8.02
50 x 32mm	1.56	2.00	0.27	6.85	nr	8.85
50 x 40mm	1.58	2.02	0.27	6.85	nr	8.87
Reducer: level invert						
40 x 32mm	1.39	1.78	0.27	6.85	nr	8.63
50 x 32mm	1.94	2.48	0.27	6.85	nr	9.34
50 x 40mm	1.84	2.36	0.27	6.85	nr	9.21
Swept bend						
32mm	1.27	1.63	0.27	6.85	nr	8.48
32mm: 165°	1.53	1.96	0.27	6.85	nr	8.81
40mm	1.41	1.81	0.27	6.85	nr	8.66
40mm: 165°	1.99	2.55	0.27	6.85	nr	9.40
50mm	2.34	3.00	0.30	7.61	nr	10.61
50mm: 165°	2.61	3.34	0.30	7.61	nr	10.95
Knuckle bend						
32mm	1.17	1.50	0.27	6.85	nr	8.35
40mm	1.28	1.64	0.27	6.85	nr	8.49
Spigot and socket bend						
32mm	1.44	1.84	0.27	6.85	nr	8.69
32mm: 150°	1.40	1.79	0.27	6.85	nr	8.64
40mm	1.60	2.05	0.27	6.85	nr	8.90
50mm	3.79	4.85	0.30	7.61	nr	12.47
Swept tee						
32mm:	1.80	2.31	0.31	7.87	nr	10.17
32mm: 135°.	2.13	2.73	0.31	7.87	nr	10.59
40mm:	2.26	2.89	0.31	7.87	nr	10.76
40mm: 135°	2.68	3.43	0.31	7.87	nr	11.30
50mm:	4.12	5.28	0.31	7.87	nr	13.14
Swept cross						
40mm:	6.39	8.18	0.31	7.87	nr	16.05
50mm:	7.34	9.40	0.43	10.91	nr	20.31
50mm: 135°	15.76	20.19	0.31	7.87	nr	28.05
Male iron adaptor						
32mm	1.30	1.67	0.28	7.10	nr	8.77
40mm	1.53	1.96	0.28	7.10	nr	9.06
Female iron adaptor						
32mm	1.30	1.67	0.28	7.10	nr	8.77
40mm	1.53	1.96	0.28	7.10	nr	9.06
50mm	2.20	2.82	0.31	7.87	nr	10.68
Reverse nut adaptor						
32mm	1.85	2.37	0.20	5.07	nr	7.44
40mm	1.85	2.37	0.20	5.07	nr	7.44

R:DISPOSAL SYSTEMS

Item	Net Price £	Material £	Labour hours	Labour £	Unit	Total rate £
Automatic air admittance valve						
32mm	10.34	13.24	0.27	6.85	nr	**20.09**
40mm	10.34	13.24	0.28	7.10	nr	**20.35**
50mm	10.34	13.24	0.31	7.87	nr	**21.11**
MuPVC to metal adaptor: including heat shrunk joint to metal						
50mm	4.45	5.70	0.38	9.64	nr	**15.34**
Caulking bush: including joint to metal						
32mm	1.94	2.48	0.31	7.87	nr	**10.35**
40mm	1.94	2.48	0.31	7.87	nr	**10.35**
50mm	1.94	2.48	0.32	8.12	nr	**10.60**
Weathering apron						
50mm	1.62	2.07	0.65	16.49	nr	**18.57**
Vent Cowl						
50mm	1.66	2.13	0.19	4.82	nr	**6.95**
ABS waste pipe; solvent welded joints; fixed with clips to backgrounds; BS 5255						
Pipe, with brackets measured separately						
32mm	1.16	1.49	0.23	5.84	m	**7.32**
40mm	1.39	1.78	0.23	5.84	m	**7.62**
50mm	1.83	2.34	0.26	6.60	m	**8.94**
Fixings						
Pipe clip: including fixing to backgrounds. For minimum fixing distances, refer to the Tables and Memoranda at the rear of the book						
32mm	0.25	0.32	0.17	4.31	nr	**4.63**
40mm	0.32	0.41	0.17	4.31	nr	**4.72**
50mm	0.61	0.78	0.17	4.31	nr	**5.09**
Pipe clip: expansion: including fixing to backgrounds. For minimum fixing distances, refer to the Tables and Memoranda at the rear of the book						
32mm	0.28	0.36	0.17	4.31	nr	**4.67**
40mm	0.34	0.44	0.17	4.31	nr	**4.75**
50mm	0.87	1.11	0.17	4.31	nr	**5.43**
Pipe clip: metal; including fixing to backgrounds. For minimum fixing distances, refer to the Tables and Memoranda at the rear of the book						
32mm	1.13	1.45	0.17	4.31	nr	**5.76**
40mm	1.34	1.72	0.17	4.31	nr	**6.03**
50mm	1.71	2.19	0.17	4.31	nr	**6.50**
Extra over fittings waste pipework ABS						
Screwed access plug						
32mm	0.57	0.73	0.18	4.57	nr	**5.30**
40mm	0.57	0.73	0.18	4.57	nr	**5.30**
50mm	1.32	1.69	0.25	6.34	nr	**8.03**

R:DISPOSAL SYSTEMS

Item	Net Price £	Material £	Labour hours	Labour £	Unit	Total rate £
R11: ABOVE GROUND DRAINAGE (cont'd)						
ABS waste pipe; solvent welded joints (cont'd)						
Fittings; waste pipework ABS (cont'd)						
Straight coupling						
32mm	0.57	0.73	0.27	6.85	nr	**7.58**
40mm	0.57	0.73	0.27	6.85	nr	**7.58**
50mm	1.32	1.69	0.27	6.85	nr	**8.54**
Expansion coupling						
32mm	1.32	1.69	0.27	6.85	nr	**8.54**
40mm	1.32	1.69	0.27	6.85	nr	**8.54**
50mm	2.64	3.38	0.27	6.85	nr	**10.23**
ABS to Copper coupling						
32mm	1.32	1.69	0.27	6.85	nr	**8.54**
40mm	1.32	1.69	0.27	6.85	nr	**8.54**
50mm	2.64	3.38	0.27	6.85	nr	**10.23**
Reducer: socket						
40 x 32mm	0.57	0.73	0.27	6.85	nr	**7.58**
50 x 32mm	0.57	0.73	0.27	6.85	nr	**7.58**
50 x 40mm	1.32	1.69	0.27	6.85	nr	**8.54**
Swept bend						
32mm	0.57	0.73	0.27	6.85	nr	**7.58**
40mm	0.57	0.73	0.27	6.85	nr	**7.58**
50mm	1.32	1.69	0.30	7.61	nr	**9.30**
Knuckle bend						
32mm	0.57	0.73	0.27	6.85	nr	**7.58**
40mm	0.57	0.73	0.27	6.85	nr	**7.58**
Swept tee						
32mm	0.57	0.73	0.31	7.87	nr	**8.60**
40mm	0.57	0.73	0.31	7.87	nr	**8.60**
50mm	1.32	1.69	0.31	7.87	nr	**9.56**
Swept cross						
40mm	3.77	4.83	0.23	5.84	nr	**10.66**
50mm	4.84	6.20	0.43	10.91	nr	**17.11**
Male iron adaptor						
32mm	1.14	1.46	0.28	7.10	nr	**8.56**
40mm	1.39	1.78	0.28	7.10	nr	**8.88**
Female iron adaptor						
32mm	1.10	1.41	0.28	7.10	nr	**8.51**
40mm	1.39	1.78	0.28	7.10	nr	**8.88**
50mm	1.83	2.34	0.31	7.87	nr	**10.21**
Tank connectors						
32mm	2.14	2.74	0.29	7.36	nr	**10.10**
40mm	2.33	2.98	0.29	7.36	nr	**10.34**
Caulking bush: including joint to pipework						
50mm	2.14	2.74	0.50	12.69	nr	**15.43**

R:DISPOSAL SYSTEMS

Item	Net Price £	Material £	Labour hours	Labour £	Unit	Total rate £
Polypropylene waste pipe; push fit joints; fixed with clips to backgrounds; BS 5254						
Pipe, with brackets measured separately						
32mm	0.93	1.19	0.21	5.33	m	**6.52**
40mm	1.07	1.37	0.21	5.33	m	**6.70**
50mm	1.69	2.16	0.38	9.64	m	**11.81**
Pipe clip: saddle; including fixing to backgrounds. For minimum fixing distances, refer to the Tables and Memoranda at the rear of the book						
32mm	0.20	0.26	0.17	4.31	nr	**4.57**
40mm	0.20	0.26	0.17	4.31	nr	**4.57**
Pipe clip: including fixing to backgrounds. For minimum fixing distances, refer to the Tables and Memoranda at the rear of the book						
50mm	0.49	0.63	0.17	4.31	nr	**4.94**
Extra over fittings waste pipework polypropylene						
Screwed access plug						
32mm	0.60	0.77	0.16	4.06	nr	**4.83**
40mm	0.60	0.77	0.16	4.06	nr	**4.83**
50mm	1.04	1.33	0.20	5.07	nr	**6.41**
Straight coupling						
32mm	0.60	0.77	0.19	4.82	nr	**5.59**
40mm	0.60	0.77	0.19	4.82	nr	**5.59**
50mm	1.04	1.33	0.20	5.07	nr	**6.41**
Universal waste pipe coupler						
32mm dia.	1.41	1.81	0.20	5.07	nr	**6.88**
40mm dia.	1.48	1.90	0.20	5.07	nr	**6.97**
Reducer						
40 x 32mm	0.60	0.77	0.19	4.82	nr	**5.59**
50 x 32mm	0.60	0.77	0.19	4.82	nr	**5.59**
50 x 40mm	1.04	1.33	0.20	5.07	nr	**6.41**
Swept bend						
32mm	0.60	0.77	0.19	4.82	nr	**5.59**
40mm	0.60	0.77	0.19	4.82	nr	**5.59**
50mm	1.04	1.33	0.20	5.07	nr	**6.41**
Knuckle bend						
32mm	0.60	0.77	0.19	4.82	nr	**5.59**
40mm	0.60	0.77	0.19	4.82	nr	**5.59**
50mm	1.04	1.33	0.20	5.07	nr	**6.41**
Spigot and socket bend						
32mm	0.60	0.77	0.19	4.82	nr	**5.59**
40mm	0.60	0.77	0.19	4.82	nr	**5.59**

R:DISPOSAL SYSTEMS

Item	Net Price £	Material £	Labour hours	Labour £	Unit	Total rate £
R11: ABOVE GROUND DRAINAGE (cont'd)						
Polypropylene waste pipe; push fit joints; (cont'd)						
Fittings; polypropylene waste pipework (cont'd)						
Swept tee						
32mm	0.60	0.77	0.22	5.58	nr	**6.35**
40mm	0.60	0.77	0.22	5.58	nr	**6.35**
50mm	1.08	1.38	0.23	5.84	nr	**7.22**
Male iron adaptor						
32mm	1.14	1.46	0.13	3.30	nr	**4.76**
40mm	1.31	1.68	0.19	4.82	nr	**6.50**
50mm	1.65	2.11	0.15	3.81	nr	**5.92**
Tank connector						
32mm	0.60	0.77	0.24	6.09	nr	**6.86**
40mm	0.60	0.77	0.24	6.09	nr	**6.86**
50mm	0.98	1.26	0.35	8.88	nr	**10.14**
Polypropylene traps; including fixing to appliance and connection to pipework; BS 3943						
Tubular P trap; 75mm seal						
32mm dia.	3.09	3.96	0.20	5.07	nr	**9.03**
40mm dia.	3.57	4.57	0.20	5.07	nr	**9.65**
Tubular S trap; 75mm seal						
32mm dia.	3.91	5.01	0.20	5.07	nr	**10.08**
40mm dia.	4.59	5.88	0.20	5.07	nr	**10.95**
Running tubular P trap; 75mm seal						
32mm dia.	4.74	6.07	0.20	5.07	nr	**11.15**
40mm dia.	5.18	6.63	0.20	5.07	nr	**11.71**
Running tubular S trap; 75mm seal						
32mm dia.	5.70	7.30	0.20	5.07	nr	**12.37**
40mm dia.	6.15	7.88	0.20	5.07	nr	**12.95**
Spigot and socket bend; converter from P to S Trap						
32mm	1.21	1.55	0.20	5.07	nr	**6.62**
40mm	1.30	1.67	0.21	5.33	nr	**6.99**
Bottle P trap; 75mm seal						
32mm dia.	3.44	4.41	0.20	5.07	nr	**9.48**
40mm dia.	4.11	5.26	0.20	5.07	nr	**10.34**
Bottle S trap; 75mm seal						
32mm dia.	4.15	5.32	0.20	5.07	nr	**10.39**
40mm dia.	5.04	6.46	0.25	6.34	nr	**12.80**
Bottle P trap; resealing; 75mm seal						
32mm dia.	4.28	5.48	0.20	5.07	nr	**10.56**
40mm dia.	5.01	6.42	0.25	6.34	nr	**12.76**

R:DISPOSAL SYSTEMS

Item	Net Price £	Material £	Labour hours	Labour £	Unit	Total rate £
Bottle S trap; resealing; 75mm seal						
32mm dia.	4.90	6.28	0.20	5.07	nr	**11.35**
40mm dia.	5.68	7.27	0.25	6.34	nr	**13.62**
Bath trap, low level; 38mm seal						
40mm dia.	4.29	5.49	0.25	6.34	nr	**11.84**
Bath trap, low level; 38mm seal complete with overflow hose						
40mm dia.	6.64	8.50	0.25	6.34	nr	**14.85**
Bath trap; 75mm seal complete with overflow hose						
40mm dia.	6.61	8.47	0.25	6.34	nr	**14.81**
Bath trap; 75mm seal complete with overflow hose and overflow outlet						
40mm dia.	11.28	14.45	0.20	5.07	nr	**19.52**
Bath trap; 75mm seal complete with overflow hose, overflow outlet and ABS chrome waste						
40mm dia.	15.60	19.98	0.20	5.07	nr	**25.05**
Washing machine trap; 75mm seal including stand pipe						
40mm dia.	8.98	11.50	0.25	6.34	nr	**17.84**
Washing machine standpipe						
40mm dia.	4.69	6.01	0.25	6.34	nr	**12.35**
Plastic unslotted chrome plated basin/sink waste including plug						
32mm	5.64	7.22	0.34	8.63	nr	**15.85**
40mm	7.54	9.66	0.34	8.63	nr	**18.28**
Plastic slotted chrome plated basin/sink waste including plug						
32mm	4.47	5.73	0.34	8.63	nr	**14.35**
40mm	7.49	9.59	0.34	8.63	nr	**18.22**
Bath overflow outlet; plastic; white						
42mm	3.92	5.02	0.37	9.39	nr	**14.41**
Bath overflow outlet; plastic; chrome plated						
42mm	5.20	6.66	0.37	9.39	nr	**16.05**
Combined cistern and bath overflow outlet; plastic; white						
42mm	8.07	10.34	0.39	9.90	nr	**20.23**
Combined cistern and bath overflow outlet; plastic; chrome plated						
42mm	8.06	10.32	0.39	9.90	nr	**20.22**
Cistern overflow outlet; plastic; white						
42mm	6.50	8.33	0.15	3.81	nr	**12.13**
Cistern overflow outlet; plastic; chrome plated						
42mm	5.86	7.51	0.15	3.81	nr	**11.31**

R:DISPOSAL SYSTEMS

Item	Net Price £	Material £	Labour hours	Labour £	Unit	Total rate £
R11: ABOVE GROUND DRAINAGE (cont'd)						
PVC-U soil and waste pipe; solvent welded joints; fixed with clips to backgrounds; BS 4514/ BS EN 607						
Pipe, with brackets measured separately						
82mm	6.25	8.00	0.35	8.88	m	**16.89**
110mm	6.30	8.07	0.41	10.40	m	**18.47**
160mm	17.06	21.85	0.51	12.94	m	**34.79**
Fixings						
Galvanised steel pipe clip: including fixing to backgrounds. For minimum fixing distances, refer to the Tables and Memoranda at the rear of the book						
82mm	2.10	2.69	0.18	4.57	nr	**7.26**
110mm	2.17	2.78	0.18	4.57	nr	**7.35**
160mm	5.25	6.72	0.18	4.57	nr	**11.29**
Plastic coated steel pipe clip: including fixing to backgrounds. For minimum fixing distances, refer to the Tables and Memoranda at the rear of the book						
82mm	2.89	3.70	0.18	4.57	nr	**8.27**
110mm	2.89	3.70	0.18	4.57	nr	**8.27**
160mm	5.05	6.47	0.18	4.57	nr	**11.03**
Plastic pipe clip: including fixing to backgrounds. For minimum fixing distances, refer to the Tables and Memoranda at the rear of the book						
82mm	2.89	3.70	0.18	4.57	nr	**8.27**
110mm	2.89	3.70	0.18	4.57	nr	**8.27**
Plastic coated steel pipe clip: adjustable; including fixing to backgrounds. For minimum fixing distances, refer to the Tables and Memoranda at the rear of the book						
82mm	2.87	3.68	0.20	5.07	nr	**8.75**
110mm	2.96	3.79	0.20	5.07	nr	**8.87**
Galvanised steel pipe clip: drive in; including fixing to backgrounds. For minimum fixing distances, refer to the Tables and Memoranda at the rear of the book						
110mm	4.52	5.79	0.22	5.58	nr	**11.37**
Extra over fittings solvent welded pipework PVC-U						
Straight coupling						
82mm	2.81	3.60	0.21	5.33	nr	**8.93**
110mm	2.88	3.69	0.22	5.58	nr	**9.27**
160mm	8.48	10.86	0.24	6.09	nr	**16.95**

R:DISPOSAL SYSTEMS

Item	Net Price £	Material £	Labour hours	Labour £	Unit	Total rate £
Expansion coupling						
82mm	2.81	3.60	0.21	5.33	nr	8.93
110mm	3.39	4.34	0.22	5.58	nr	9.92
160mm	5.48	7.02	0.24	6.09	nr	13.11
Slip coupling; double ring socket						
82mm	8.35	10.69	0.21	5.33	nr	16.02
110mm	10.44	13.37	0.22	5.58	nr	18.95
160mm	25.65	32.85	0.24	6.09	nr	38.94
Puddle flanges						
110mm	80.12	102.62	0.45	11.42	nr	114.03
160mm	129.47	165.82	0.55	13.95	nr	179.78
Socket reducer						
82 to 50mm	4.07	5.21	0.18	4.57	nr	9.78
110 to 50mm	5.10	6.53	0.18	4.57	nr	11.10
110 to 82mm	5.25	6.72	0.22	5.58	nr	12.31
160 to 110mm	10.63	13.61	0.26	6.60	nr	20.21
Socket plugs						
82mm	3.98	5.10	0.15	3.81	nr	8.90
110mm	4.28	5.48	0.20	5.07	nr	10.56
160mm	8.85	11.34	0.27	6.85	nr	18.19
Access door; including cutting into pipe						
82mm	8.98	11.50	0.28	7.10	nr	18.61
110mm	8.98	11.50	0.34	8.63	nr	20.13
160mm	16.04	20.54	0.46	11.67	nr	32.22
Screwed access cap						
82mm	6.35	8.13	0.15	3.81	nr	11.94
110mm	7.48	9.58	0.20	5.07	nr	14.65
160mm	14.08	18.03	0.27	6.85	nr	24.88
Access pipe: spigot and socket						
110mm	11.06	14.17	0.22	5.58	nr	19.75
Access pipe: double socket						
110mm	11.06	14.17	0.22	5.58	nr	19.75
Swept bend						
82mm	6.74	8.63	0.29	7.36	nr	15.99
110mm	7.89	10.11	0.32	8.12	nr	18.22
160mm	19.65	25.17	0.49	12.43	nr	37.60
Bend; special angle						
82mm	13.04	16.70	0.29	7.36	nr	24.06
110mm	15.56	19.93	0.32	8.12	nr	28.05
160mm	26.25	33.62	0.49	12.43	nr	46.05
Spigot and socket bend						
82mm	6.53	8.36	0.26	6.60	nr	14.96
110mm	7.64	9.79	0.32	8.12	nr	17.90
110mm: 135°	8.59	11.00	0.32	8.12	nr	19.12
160mm: 135°	19.01	24.35	0.44	11.16	nr	35.51
Variable bend: single socket						
110mm	14.76	18.90	0.33	8.37	nr	27.28

R:DISPOSAL SYSTEMS

Item	Net Price £	Material £	Labour hours	Labour £	Unit	Total rate £
R11: ABOVE GROUND DRAINAGE (cont'd)						
PVC-U soil and waste pipe; solvent welded Joints (cont'd)						
Fittings; solvent welded pipework PVC-U (cont'd)						
Variable bend: double socket						
110mm	14.78	18.93	0.33	8.37	nr	**27.30**
Access bend						
110mm	21.89	28.04	0.33	8.37	nr	**36.41**
Single branch: two bosses						
82mm	9.43	12.08	0.35	8.88	nr	**20.96**
82mm: 104°	9.43	12.08	0.35	8.88	nr	**20.96**
110mm	9.92	12.71	0.42	10.66	nr	**23.36**
110mm: 135°	10.88	13.94	0.42	10.66	nr	**24.59**
160mm	23.71	30.37	0.50	12.69	nr	**43.05**
160mm: 135°	42.29	54.16	0.50	12.69	nr	**66.85**
Single branch; four bosses						
110mm	13.06	16.73	0.42	10.66	nr	**27.38**
Single access branch						
82mm	42.75	54.75	0.35	8.88	nr	**63.63**
110mm	25.07	32.11	0.42	10.66	nr	**42.77**
Unequal single branch						
160 x 160 x 110mm	25.02	32.05	0.50	12.69	nr	**44.73**
160 x 160 x 110mm: 135°	26.89	34.44	0.50	12.69	nr	**47.13**
Double branch						
110mm	24.80	31.76	0.42	10.66	nr	**42.42**
110mm: 135°	26.95	34.52	0.42	10.66	nr	**45.17**
Corner branch						
110mm	45.55	58.34	0.42	10.66	nr	**69.00**
Unequal double branch						
160 x 160 x 110mm	46.55	59.62	0.50	12.69	nr	**72.31**
Single boss pipe; single socket						
110 x 110 x 32mm	6.39	8.18	0.24	6.09	nr	**14.27**
110 x 110 x 40mm	6.39	8.18	0.24	6.09	nr	**14.27**
110 x 110 x 50mm	6.16	7.89	0.24	6.09	nr	**13.98**
Single boss pipe; triple socket						
110 x 110 x 40mm	4.40	5.64	0.24	6.09	nr	**11.72**
Waste boss; including cutting into pipe						
82 to 32mm	3.68	4.71	0.29	7.36	nr	**12.07**
82 to 40mm	3.68	4.71	0.29	7.36	nr	**12.07**
110 to 32mm	3.68	4.71	0.29	7.36	nr	**12.07**
110 to 40mm	3.68	4.71	0.29	7.36	nr	**12.07**
110 to 50mm	3.82	4.89	0.29	7.36	nr	**12.25**
160 to 32mm	5.21	6.67	0.30	7.61	nr	**14.28**
160 to 40mm	5.21	6.67	0.35	8.88	nr	**15.55**
160 to 50mm	5.21	6.67	0.40	10.15	nr	**16.82**

R:DISPOSAL SYSTEMS

Item	Net Price £	Material £	Labour hours	Labour £	Unit	Total rate £
Self locking waste boss; including cutting into pipe						
110 to 32mm	4.92	6.30	0.30	7.61	nr	**13.91**
110 to 40mm	5.16	6.61	0.30	7.61	nr	**14.22**
110 to 50mm	5.83	7.47	0.30	7.61	nr	**15.08**
Adaptor saddle; including cutting to pipe						
82 to 32mm	2.25	2.88	0.29	7.36	nr	**10.24**
110 to 40mm	2.77	3.55	0.29	7.36	nr	**10.91**
160 to 50mm	5.02	6.43	0.29	7.36	nr	**13.79**
Branch boss adaptor						
32mm	1.46	1.87	0.26	6.60	nr	**8.47**
40 mm	1.46	1.87	0.26	6.60	nr	**8.47**
50 mm	2.07	2.65	0.26	6.60	nr	**9.25**
Branch boss adaptor bend						
32mm	2.20	2.82	0.26	6.60	nr	**9.41**
40 mm	2.40	3.07	0.26	6.60	nr	**9.67**
50 mm	2.35	3.01	0.26	6.60	nr	**9.61**
Automatic air admittance valve						
82 to 110mm	23.81	30.50	0.19	4.82	nr	**35.32**
PVC-U to metal adaptor: including heat shrunk joint to metal						
110mm	6.16	7.89	0.57	14.46	nr	**22.35**
Caulking bush: including joint to pipework						
82mm	6.31	8.08	0.46	11.67	nr	**19.75**
110mm	6.31	8.08	0.46	11.67	nr	**19.75**
Vent cowl						
82mm	1.90	2.43	0.13	3.30	nr	**5.73**
110mm	1.91	2.45	0.13	3.30	nr	**5.74**
160mm	5.02	6.43	0.13	3.30	nr	**9.73**
Weathering apron; to lead slates						
82mm	1.90	2.43	1.15	29.18	nr	**31.61**
110mm	2.17	2.78	1.15	29.18	nr	**31.96**
160mm	6.55	8.39	1.15	29.18	nr	**37.57**
Weathering apron; to asphalt						
82mm	8.02	10.27	1.10	27.91	nr	**38.18**
110mm	8.02	10.27	1.10	27.91	nr	**38.18**
Weathering slate; flat; 406 x 406mm						
82mm	22.60	28.95	1.04	26.39	nr	**55.33**
110mm	22.60	28.95	1.04	26.39	nr	**55.33**
Weathering slate; flat; 457 x 457mm						
82mm	23.17	29.68	1.04	26.39	nr	**56.06**
110mm	23.17	29.68	1.04	26.39	nr	**56.06**
Weathering slate; angled; 610 x 610mm						
82mm	31.30	40.09	1.04	26.39	nr	**66.48**
110mm	31.30	40.09	1.04	26.39	nr	**66.48**

R:DISPOSAL SYSTEMS

Item	Net Price £	Material £	Labour hours	Labour £	Unit	Total rate £
R11: ABOVE GROUND DRAINAGE (cont'd)						
PVC-U soil and waste pipe; ring seal joints; fixed with clips to backgrounds; BS 4514/ BS EN 607						
Pipe, with brackets measured separately						
82mm dia.	6.81	8.72	0.35	8.88	m	**17.60**
110mm dia.	6.50	8.33	0.41	10.40	m	**18.73**
160mm dia.	16.02	20.52	0.51	12.94	m	**33.46**
Fixings						
Galvanised steel pipe clip: including fixing to backgrounds. For minimum fixing distances, refer to the Tables and Memoranda at the rear of the book						
82mm	2.10	2.69	0.18	4.57	nr	**7.26**
110mm	2.17	2.78	0.18	4.57	nr	**7.35**
160mm	5.25	6.72	0.18	4.57	nr	**11.29**
Plastic coated steel pipe clip: including fixing to backgrounds. For minimum fixing distances, refer to the Tables and Memoranda at the rear of the book						
82mm	2.89	3.70	0.18	4.57	nr	**8.27**
110mm	2.89	3.70	0.18	4.57	nr	**8.27**
160mm	5.05	6.47	0.18	4.57	nr	**11.03**
Plastic pipe clip: including fixing to backgrounds. For minimum fixing distances, refer to the Tables and Memoranda at the rear of the book						
82mm	2.89	3.70	0.18	4.57	nr	**8.27**
110mm	2.89	3.70	0.18	4.57	nr	**8.27**
Plastic coated steel pipe clip: adjustable; including fixing to backgrounds. For minimum fixing distances, refer to the Tables and Memoranda at the rear of the book						
82mm	2.87	3.68	0.20	5.07	nr	**8.75**
110mm	2.96	3.79	0.20	5.07	nr	**8.87**
Galvanised steel pipe clip: drive in; including fixing to backgrounds. For minimum fixing distances, refer to the Tables and Memoranda at the rear of the book						
110mm	4.52	5.79	0.22	5.58	nr	**11.37**
Extra over fittings ring seal pipework PVC-U						
Straight coupling						
82mm	3.73	4.78	0.21	5.33	nr	**10.11**
110mm	3.98	5.10	0.22	5.58	nr	**10.68**
160mm	8.75	11.21	0.24	6.09	nr	**17.30**

R:DISPOSAL SYSTEMS

Item	Net Price £	Material £	Labour hours	Labour £	Unit	Total rate £
Straight coupling; double socket						
82mm	3.78	4.84	0.21	5.33	nr	**10.17**
110mm	6.78	8.68	0.22	5.58	nr	**14.27**
160mm	14.01	17.94	0.24	6.09	nr	**24.03**
Reducer; socket						
82 to 50mm	5.57	7.13	0.15	3.81	nr	**10.94**
110 to 50mm	3.73	4.78	0.15	3.81	nr	**8.58**
110 to 82mm	7.59	9.72	0.19	4.82	nr	**14.54**
160 to 110mm	10.87	13.92	0.31	7.87	nr	**21.79**
Access Cap						
82mm	6.80	8.71	0.15	3.81	nr	**12.52**
110mm	7.10	9.09	0.17	4.31	nr	**13.41**
Access Cap; pressure plug						
160mm	19.71	25.24	0.33	8.37	nr	**33.62**
Access pipe						
82mm	13.16	16.86	0.22	5.58	nr	**22.44**
110mm	13.43	17.20	0.22	5.58	nr	**22.78**
160mm	31.95	40.92	0.24	6.09	nr	**47.01**
Bend						
82mm	7.71	9.87	0.29	7.36	nr	**17.23**
82mm; adjustable radius	12.50	16.01	0.29	7.36	nr	**23.37**
110mm	9.04	11.58	0.32	8.12	nr	**19.70**
110mm; adjustable radius	11.91	15.25	0.32	8.12	nr	**23.37**
160mm	22.48	28.79	0.49	12.43	nr	**41.22**
160mm; adjustable radius	23.82	30.51	0.49	12.43	nr	**42.94**
Bend; spigot and socket						
110mm	9.04	11.58	0.32	8.12	nr	**19.70**
Bend; offset						
82mm	7.54	9.66	0.21	5.33	nr	**14.99**
110mm	8.99	11.51	0.32	8.12	nr	**19.63**
160mm	24.32	31.15	-	-	nr	**31.15**
Bend; access						
110mm	18.82	24.10	0.33	8.37	nr	**32.48**
Single branch						
82mm	10.98	14.06	0.35	8.88	nr	**22.94**
110mm	13.82	17.70	0.42	10.66	nr	**28.36**
110mm; 45°	14.65	18.76	0.31	7.87	nr	**26.63**
160mm	40.58	51.97	0.50	12.69	nr	**64.66**
Single branch; access						
82mm	17.01	21.79	0.35	8.88	nr	**30.67**
110mm	24.48	31.35	0.42	10.66	nr	**42.01**
Unequal single branch						
160 x 160 x 110mm	26.89	34.44	0.50	12.69	nr	**47.13**
160 x 160 x 110mm; 45°	23.14	29.64	0.50	12.69	nr	**42.32**
Double branch; 4 bosses						
110mm	21.56	27.61	0.49	12.43	nr	**40.05**

R:DISPOSAL SYSTEMS

Item	Net Price £	Material £	Labour hours	Labour £	Unit	Total rate £
R11: ABOVE GROUND DRAINAGE (cont'd)						
PVC-U soil and waste pipe; ring seal joints (cont'd)						
Fittings; ring seal pipework PVC-U (cont'd)						
Corner branch; 2 bosses						
110mm	51.09	65.44	0.49	12.43	nr	**77.87**
Multibranch; 4 bosses						
110mm	40.94	52.44	0.52	13.19	nr	**65.63**
Boss Branch						
110 x 32mm	2.89	3.70	0.34	8.63	nr	**12.33**
110 x 40mm	2.89	3.70	0.34	8.63	nr	**12.33**
Strap on boss						
110 x 32mm	3.76	4.82	0.30	7.61	nr	**12.43**
110 x 40mm	3.76	4.82	0.30	7.61	nr	**12.43**
110 x 50mm	3.76	4.82	0.30	7.61	nr	**12.43**
Patch boss						
82 x 32mm	3.07	3.93	0.31	7.87	nr	**11.80**
82 x 40mm	3.07	3.93	0.31	7.87	nr	**11.80**
82 x 50mm	3.07	3.93	0.31	7.87	nr	**11.80**
Boss Pipe; collar 4 boss						
110mm	15.61	19.99	0.35	8.88	nr	**28.87**
Boss adaptor; rubber; push fit						
32mm	1.43	1.83	0.26	6.60	nr	**8.43**
40 mm	1.43	1.83	0.26	6.60	nr	**8.43**
50 mm	2.05	2.63	0.26	6.60	nr	**9.22**
WC connector; cap and seal; solvent socket						
110mm	3.55	4.55	0.23	5.84	nr	**10.38**
110mm; 90°	9.30	11.91	0.27	6.85	nr	**18.76**
Vent terminal						
82mm	2.29	2.93	0.13	3.30	nr	**6.23**
110mm	2.45	3.14	0.13	3.30	nr	**6.44**
160mm	7.84	10.04	0.13	3.30	nr	**13.34**
Weathering slate; inclined; 610 x 610mm						
82mm	26.13	33.47	1.04	26.39	nr	**59.85**
110mm	26.13	33.47	1.04	26.39	nr	**59.85**
Weathering slate; inclined; 450 x 450mm						
82mm	16.33	20.92	1.04	26.39	nr	**47.30**
110mm	16.33	20.92	1.04	26.39	nr	**47.30**
Weathering slate; flat; 400 x 400mm						
82mm	16.24	20.80	1.04	26.39	nr	**47.19**
110mm	16.24	20.80	1.04	26.39	nr	**47.19**
Air admittance valve						
82mm	35.87	45.94	0.19	4.82	nr	**50.76**
110mm	36.89	47.25	0.19	4.82	nr	**52.07**

R:DISPOSAL SYSTEMS

Item	Net Price £	Material £	Labour hours	Labour £	Unit	Total rate £
Cast iron pipe; nitrile rubber gasket joint with continuity clip BS 416/6087; fixed vertically to backgrounds						
Pipe, with brackets and jointing couplings measured separately						
50 mm	13.84	17.73	0.25	6.34	m	**24.07**
75 mm	15.29	19.58	0.45	11.42	m	**31.00**
100 mm	18.22	23.34	0.60	15.23	m	**38.56**
150 mm	37.58	48.13	0.70	17.77	m	**65.90**
Fixings						
Brackets; fixed to backgrounds. For minimum fixing distances, refer to the Tables and Memoranda at the rear of the book						
50 mm	3.42	4.38	0.15	3.81	nr	**8.19**
75 mm	3.42	4.38	0.18	4.57	nr	**8.95**
100 mm	3.80	4.87	0.18	4.57	nr	**9.43**
150 mm	7.11	9.11	0.20	5.07	nr	**14.18**
Extra over fittings nitrile gasket cast iron pipework BS 416/6087, with jointing couplings measured separately						
Standard coupling						
50 mm	5.63	7.21	0.50	12.69	nr	**19.90**
75 mm	6.22	7.97	0.60	15.23	nr	**23.19**
100 mm	8.12	10.40	0.67	17.00	nr	**27.40**
150 mm	16.23	20.79	0.83	21.07	nr	**41.86**
Conversion coupling						
65 x 75 mm	6.58	8.43	0.60	15.23	nr	**23.66**
70 x 75 mm	6.58	8.43	0.60	15.23	nr	**23.66**
90 x 100 mm	6.58	8.43	0.67	17.00	nr	**25.43**
Access pipe; round door						
50 mm	23.41	29.98	0.41	10.40	nr	**40.39**
75 mm	24.00	30.74	0.46	11.67	nr	**42.41**
100 mm	25.11	32.16	0.67	17.03	nr	**49.19**
150 mm	41.76	53.49	0.83	21.07	nr	**74.56**
Access pipe; square door						
100 mm	49.46	63.35	0.67	17.03	nr	**80.38**
150 mm	75.72	96.98	0.83	21.07	nr	**118.06**
Taper reducer						
75 mm	13.03	16.69	0.60	15.23	nr	**31.92**
100 mm	17.32	22.18	0.67	17.00	nr	**39.19**
150 mm	33.75	43.23	0.83	21.07	nr	**64.30**
Blank cap						
50 mm	3.43	4.39	0.24	6.09	nr	**10.48**
75 mm	4.18	5.35	0.26	6.60	nr	**11.95**
100 mm	4.18	5.35	0.32	8.12	nr	**13.47**
150 mm	-	-	0.40	10.15	nr	**10.15**

R:DISPOSAL SYSTEMS

Item	Net Price £	Material £	Labour hours	Labour £	Unit	Total rate £
R11: ABOVE GROUND DRAINAGE (cont'd)						
Cast iron pipe; nitrile rubber gasket joint (cont'd)						
Fittings; cast iron nitrile						
Blank cap; 50 mm screwed tapping						
75 mm	8.47	10.85	0.26	6.60	nr	**17.45**
100 mm	9.98	12.78	0.32	8.12	nr	**20.90**
150 mm	10.73	13.74	0.40	10.15	nr	**23.89**
Universal connector						
50 x 56/48/40 mm	3.59	4.60	0.33	8.37	nr	**12.97**
Change piece; BS416						
100 mm	10.73	13.74	0.47	11.92	nr	**25.67**
WC connector						
100 mm	16.35	20.94	0.49	12.43	nr	**33.37**
Boss pipe; 2 " BSPT socket						
50 mm	20.55	26.32	0.58	14.72	nr	**41.04**
75 mm	20.55	26.32	0.65	16.49	nr	**42.81**
100 mm	24.55	31.44	0.79	20.04	nr	**51.49**
150 mm	40.05	51.30	0.86	21.82	nr	**73.12**
Boss pipe; 2 " BSPT socket; 135°						
100 mm	29.57	37.87	0.79	20.04	nr	**57.92**
Boss pipe; 2 x 2 " BSPT socket; opposed						
75 mm	27.21	34.85	0.65	16.49	nr	**51.34**
100 mm	31.74	40.65	0.79	20.04	nr	**60.70**
Boss pipe; 2 x 2 " BSPT socket; in line						
100 mm	32.06	41.06	0.79	20.04	nr	**61.11**
Boss pipe; 2 x 2 " BSPT socket; 90°						
100 mm	31.74	40.65	0.79	20.04	nr	**60.70**
Bend; short radius						
50 mm	9.92	12.71	0.50	12.69	nr	**25.39**
75 mm	9.92	12.71	0.60	15.23	nr	**27.93**
100 mm	13.75	17.61	0.67	17.00	nr	**34.62**
100 mm; 11°	11.85	15.18	0.67	17.00	nr	**32.18**
100 mm; 67°	13.75	17.61	0.67	17.00	nr	**34.62**
150 mm	22.98	29.43	0.83	21.07	nr	**50.51**
Access bend; short radius						
50 mm	24.44	31.30	0.50	12.69	nr	**43.99**
75 mm	24.44	31.30	0.60	15.23	nr	**46.53**
100 mm	29.05	37.21	0.67	17.00	nr	**54.21**
100 mm; 45°	29.05	37.21	0.67	17.00	nr	**54.21**
150 mm	41.27	52.86	0.83	21.07	nr	**73.93**
150 mm; 45°	41.27	52.86	0.83	21.07	nr	**73.93**
Long radius bend						
75 mm	18.75	24.01	0.60	15.23	nr	**39.24**
100 mm	22.25	28.50	0.67	17.00	nr	**45.50**
100 mm; 5°	13.75	17.61	0.67	17.00	nr	**34.62**
150 mm	44.85	57.44	0.83	21.07	nr	**78.52**
150 mm; 22.5°	39.31	50.35	0.83	21.07	nr	**71.42**

R:DISPOSAL SYSTEMS

Item	Net Price £	Material £	Labour hours	Labour £	Unit	Total rate £
Access bend; long radius						
75 mm	33.25	42.59	0.60	15.23	nr	**57.81**
100 mm	37.59	48.14	0.67	17.00	nr	**65.15**
150 mm	66.21	84.80	0.83	21.07	nr	**105.88**
Long tail bend						
100 x 250 mm long	17.75	22.73	0.70	17.77	nr	**40.50**
100 x 815 mm long	43.20	55.33	0.70	17.77	nr	**73.10**
Offset						
75 mm projection						
75 mm	10.34	13.24	0.53	13.45	nr	**26.69**
100 mm	14.45	18.51	0.66	16.75	nr	**35.25**
115 mm projection						
75 mm	12.23	15.66	0.53	13.45	nr	**29.11**
100 mm	17.23	22.07	0.66	16.75	nr	**38.81**
150 mm projection						
75 mm	12.23	15.66	0.53	13.45	nr	**29.11**
100 mm	17.23	22.07	0.66	16.75	nr	**38.81**
225 mm projection						
100 mm	19.74	25.28	0.66	16.75	nr	**42.03**
300 mm projection						
100 mm	22.25	28.50	0.66	16.75	nr	**45.24**
Branch; equal and unequal						
50 mm	14.94	19.14	0.78	19.79	nr	**38.93**
75 mm	14.94	19.14	0.85	21.57	nr	**40.71**
100 mm	21.24	27.20	1.00	25.37	nr	**52.58**
150 mm	52.64	67.42	1.20	30.46	nr	**97.88**
150 x 100 mm	40.28	51.59	1.21	30.57	nr	**82.16**
150 x 100 mm; 45°	57.52	73.67	1.21	30.57	nr	**104.24**
Branch; 2" BSPT screwed socket						
100 mm	28.95	37.08	1.00	25.37	nr	**62.45**
Branch; long tail						
100 x 915 mm long	56.16	71.93	1.00	25.37	nr	**97.30**
Access branch; equal and unequal						
50 mm	29.48	37.76	0.78	19.79	nr	**57.55**
75 mm	29.48	37.76	0.85	21.57	nr	**59.33**
100 mm	36.57	46.84	1.02	25.89	nr	**72.73**
150 mm	72.29	92.59	1.20	30.46	nr	**123.05**
150 x 100 mm	57.01	73.02	1.20	30.46	nr	**103.48**
150 x 100 mm; 45°	70.94	90.86	1.20	30.46	nr	**121.32**
Parallel branch						
100 mm	22.25	28.50	1.00	25.37	nr	**53.87**
Double branch						
75 mm	25.11	32.16	0.95	24.10	nr	**56.26**
100 mm	26.27	33.65	1.30	32.99	nr	**66.64**
150 x 100 mm	73.99	94.77	1.56	39.58	nr	**134.35**
Double access branch						
100 mm	41.59	53.27	1.43	36.30	nr	**89.57**

R:DISPOSAL SYSTEMS

Item	Net Price £	Material £	Labour hours	Labour £	Unit	Total rate £
R11: ABOVE GROUND DRAINAGE (cont'd)						
Cast iron pipe; nitrile rubber gasket joint (cont'd)						
Fittings; cast iron nitrile rubber (cont'd)						
Corner branch						
100 mm	35.30	45.21	1.30	32.99	nr	**78.21**
Puddle flange; grey epoxy coated						
100 mm	25.11	32.16	1.00	25.37	nr	**57.53**
Roof vent connector; asphalt						
75 mm	30.39	38.92	0.90	22.83	nr	**61.76**
100 mm	23.72	30.38	0.97	24.61	nr	**54.99**
P trap						
100 mm	22.01	28.19	1.00	25.37	nr	**53.56**
P trap with access						
50 mm	33.70	43.16	0.77	19.54	nr	**62.70**
75 mm	33.70	43.16	0.90	22.83	nr	**66.00**
100 mm	37.35	47.84	1.16	29.43	nr	**77.27**
150 mm	62.49	80.04	1.77	44.91	nr	**124.94**
Bellmouth gully inlet						
100 mm	31.29	40.08	1.08	27.40	nr	**67.48**
Balcony gully inlet						
100 mm	67.40	86.33	1.08	27.40	nr	**113.73**
Roof outlet						
Flat grate						
75 mm	66.50	85.17	0.83	21.06	nr	**106.23**
100 mm	77.02	98.65	1.08	27.40	nr	**126.05**
Dome grate						
75 mm	67.12	85.97	0.83	21.06	nr	**107.03**
100 mm	78.27	100.25	1.08	27.40	nr	**127.65**
Top Hat						
100 mm	146.48	187.61	1.08	27.40	nr	**215.01**
Cast iron pipe; EPDM rubber gasket joint with continuity clip; BS EN877; fixed to backgrounds						
Pipe, with brackets and jointing couplings measured separately						
50 mm	10.10	13.32	0.25	6.34	m	**19.67**
70 mm	11.42	15.07	0.45	11.42	m	**26.48**
100 mm	13.44	17.73	0.60	15.22	m	**32.95**
125 mm	21.88	28.86	0.65	16.49	m	**45.36**
150 mm	26.50	34.96	0.70	17.76	m	**52.72**
200 mm	51.32	67.70	1.14	28.92	m	**96.63**
250 mm	60.89	80.33	1.25	31.71	m	**112.04**
300 mm	74.86	98.76	1.53	38.82	m	**137.58**

R:DISPOSAL SYSTEMS

Item	Net Price £	Material £	Labour hours	Labour £	Unit	Total rate £
Fixings						
Brackets; fixed to backgrounds. For minimum fixing distances, refer to the Tables and Memoranda at the rear of the book						
Ductile iron						
50mm	3.24	4.15	0.10	2.54	nr	**6.69**
70 mm	3.24	4.15	0.10	2.54	nr	**6.69**
100 mm	3.56	4.56	0.15	3.81	nr	**8.37**
150 mm	6.74	8.63	0.20	5.07	nr	**13.71**
200 mm	25.69	32.90	0.25	6.34	nr	**39.25**
Mild steel; vertical						
125 mm	6.31	8.08	0.15	3.81	nr	**11.89**
Mild steel; stand off						
250 mm	13.74	17.60	0.25	6.34	nr	**23.94**
300 mm	15.12	19.37	0.25	6.34	nr	**25.71**
Stack support; rubber seal						
70 mm	7.50	9.61	0.74	18.78	nr	**28.38**
100 mm	9.94	12.73	0.88	22.33	nr	**35.06**
125 mm	11.02	14.11	1.00	25.37	nr	**39.49**
150 mm	14.90	19.08	1.19	30.19	nr	**49.28**
200 mm	30.58	39.17	1.31	33.24	nr	**72.40**
Wall spacer plate; cast iron (eared sockets)						
50 mm	2.83	3.62	0.10	2.54	nr	**6.16**
70 mm	2.83	3.62	0.10	2.54	nr	**6.16**
100 mm	2.83	3.62	0.10	2.54	nr	**6.16**
Extra over fittings EPDM rubber jointed cast iron pipework BS EN 877, with jointing couplings measured separately						
Coupling						
50 mm	3.46	4.43	0.50	12.69	nr	**17.12**
70 mm	3.80	4.87	0.60	15.23	nr	**20.10**
100 mm	4.97	6.37	0.67	17.00	nr	**23.37**
125 mm	6.15	7.88	0.75	19.04	nr	**26.92**
150mm	9.92	12.71	0.83	21.07	nr	**33.78**
200 mm	22.21	28.45	1.21	30.70	nr	**59.15**
250 mm	30.93	39.61	1.33	33.74	nr	**73.36**
300 mm	35.80	45.85	1.63	41.36	nr	**87.21**
Plain socket						
50 mm	8.30	10.63	0.25	6.34	nr	**16.97**
70 mm	8.30	10.63	0.25	6.34	nr	**16.97**
100 mm	9.46	12.12	0.25	6.34	nr	**18.46**
Eared socket						
50 mm	8.57	10.98	0.25	6.34	nr	**17.32**
70 mm	8.57	10.98	0.25	6.34	nr	**17.32**
100 mm	10.30	13.19	0.25	6.34	nr	**19.54**

R:DISPOSAL SYSTEMS

Item	Net Price £	Material £	Labour hours	Labour £	Unit	Total rate £
R11: ABOVE GROUND DRAINAGE (cont'd)						
Cast iron pipe; EPDM rubber gasket joint (cont'd)						
Fittings; cast iron EPDM (cont'd)						
Slip socket						
50 mm	10.67	13.67	0.25	6.34	nr	20.01
70 mm	10.67	13.67	0.25	6.34	nr	20.01
100 mm	12.53	16.05	0.25	6.34	nr	22.39
Stack support pipe						
70 mm	12.84	16.45	0.74	18.78	nr	35.22
100 mm	14.27	18.28	0.88	22.33	nr	40.60
125 mm	15.84	20.29	1.00	25.37	nr	45.66
150 mm	22.60	28.95	1.19	30.19	nr	59.14
200 mm	38.25	48.99	1.31	33.24	nr	82.23
Access pipe; round door						
50 mm	15.95	20.43	0.54	13.70	nr	34.13
70 mm	16.88	21.62	0.64	16.24	nr	37.86
100 mm	18.56	23.77	0.60	15.22	nr	38.99
150 mm	33.57	43.00	0.83	21.06	nr	64.05
Access pipe; square door						
100 mm	35.89	45.97	0.60	15.22	nr	61.19
125 mm	37.34	47.82	0.67	17.00	nr	64.82
150 mm	56.16	71.93	0.71	18.01	nr	89.94
200 mm	111.57	142.90	1.21	30.70	nr	173.60
250 mm	175.53	224.82	1.31	33.24	nr	258.05
300 mm	218.97	280.45	1.43	36.28	nr	316.74
Taper reducer						
70 mm	9.20	11.78	0.51	12.94	nr	24.72
100 mm	10.83	13.87	0.58	14.72	nr	28.59
125 mm	10.89	13.95	0.64	16.24	nr	30.19
150 mm	20.77	26.60	0.67	17.00	nr	43.60
200 mm	33.75	43.23	1.15	29.18	nr	72.40
250 mm	69.71	89.28	1.25	31.71	nr	121.00
300 mm	95.86	122.78	1.37	34.76	nr	157.54
Blank cap						
50 mm	2.41	3.09	0.24	6.09	nr	9.18
70 mm	2.54	3.25	0.26	6.60	nr	9.85
100 mm	2.96	3.79	0.32	8.12	nr	11.91
125 mm	4.18	5.35	0.35	8.88	unit	14.23
150 mm	4.27	5.47	0.40	10.15	nr	15.62
200 mm	18.90	24.21	0.60	15.22	nr	39.43
250 mm	38.60	49.44	0.65	16.49	nr	65.93
300 mm	50.55	64.74	0.72	18.27	nr	83.01
Blank cap; 50 mm screwed tapping						
70 mm	5.68	7.27	0.26	6.60	nr	13.87
100 mm	6.12	7.84	0.32	8.12	nr	15.96
150 mm	7.35	9.41	0.40	10.15	nr	19.56
Universal connector; EPDM rubber						
50 x 56/48/40 mm	3.59	4.60	0.30	7.61	nr	12.21

R:DISPOSAL SYSTEMS

Item	Net Price £	Material £	Labour hours	Labour £	Unit	Total rate £
Blank end; push fit						
100 x 38/32 mm	5.19	6.65	0.39	9.90	nr	**16.54**
Boss pipe; 2 " BSPT socket						
50 mm	13.48	17.27	0.54	13.70	nr	**30.97**
75 mm	13.48	17.27	0.64	16.24	nr	**33.50**
100 mm	16.48	21.11	0.78	19.79	nr	**40.90**
150 mm	26.88	34.43	1.09	27.66	nr	**62.08**
Boss pipe; 2 x 2 " BSPT socket; opposed						
100 mm	21.30	27.28	0.78	19.79	nr	**47.07**
Boss pipe; 2 x 2 " BSPT socket; 90°						
100 mm	21.30	27.28	0.78	19.79	nr	**47.07**
Manifold connector						
100 mm	32.87	42.10	0.78	19.79	nr	**61.89**
Bend; short radius						
50mm	5.98	7.66	0.50	12.69	nr	**20.35**
70mm	6.74	8.63	0.60	15.22	nr	**23.86**
100mm	7.98	10.22	0.67	17.00	nr	**27.22**
125 mm	14.14	18.11	0.78	19.79	nr	**37.90**
150mm	14.34	18.37	0.83	21.07	nr	**39.44**
200 mm; 45°	42.65	54.63	1.21	30.70	nr	**85.33**
250 mm; 45°	83.24	106.61	1.31	33.24	nr	**139.85**
300 mm; 45°	117.07	149.94	1.43	36.28	nr	**186.22**
Access bend; short radius						
70 mm	13.06	16.73	0.64	16.24	nr	**32.97**
100mm	19.08	24.44	0.78	19.79	nr	**44.23**
150mm	29.66	37.99	0.83	21.07	nr	**59.06**
Bend; long radius bend						
100mm	18.90	24.21	0.67	17.00	nr	**41.21**
100 mm; 22°	14.93	19.12	0.78	19.79	nr	**38.91**
150mm	56.50	72.36	0.83	21.07	nr	**93.44**
Access bend; long radius						
100mm	24.68	31.61	0.67	17.00	nr	**48.61**
Bend; long tail						
100 mm	13.99	17.92	0.78	19.79	nr	**37.71**
Bend; long tail double						
70 mm	21.66	27.74	0.64	16.24	nr	**43.98**
100 mm	23.90	30.61	0.78	19.79	nr	**50.40**
Bend; air pipe						
100 mm	25.60	32.79	0.78	19.79	nr	**52.58**
Offset						
75 mm projection						
100 mm	12.27	15.72	0.78	19.79	nr	**35.51**
130 mm projection						
50 mm	10.14	12.99	0.54	13.70	nr	**26.69**
70 mm	15.39	19.71	0.64	16.24	nr	**35.95**
100 mm	20.21	25.88	0.78	19.79	nr	**45.67**
125 mm	25.64	32.84	0.78	19.79	nr	**52.63**

R:DISPOSAL SYSTEMS

Item	Net Price £	Material £	Labour hours	Labour £	Unit	Total rate £
R11: ABOVE GROUND DRAINAGE (cont'd)						
Cast iron pipe; EPDM rubber gasket joint (cont'd)						
Fittings; cast iron EPDM (cont'd)						
Branch; equal and unequal						
50 mm	9.61	12.31	0.78	19.79	nr	**32.10**
70 mm	10.14	12.99	0.85	21.57	nr	**34.56**
100 mm	13.90	17.80	1.00	25.37	nr	**43.17**
125 mm	27.88	35.71	1.16	29.43	nr	**65.14**
150 mm	34.63	44.35	1.37	34.76	nr	**79.11**
200 mm	80.73	103.40	1.51	38.31	nr	**141.71**
250 mm	186.62	239.02	1.63	41.36	nr	**280.38**
300mm	284.91	364.91	1.77	44.91	nr	**409.82**
Branch; radius; equal and unequal						
70 mm	12.36	15.83	0.79	20.04	nr	**35.87**
100 mm	14.27	18.28	0.96	24.36	nr	**42.63**
125 mm	34.96	44.78	1.16	29.43	nr	**74.21**
150 mm	39.25	50.27	1.37	34.76	nr	**85.03**
200 mm	108.70	139.22	1.51	38.31	nr	**177.53**
Branch; long tail						
100 mm	45.37	58.11	0.96	24.36	nr	**82.47**
Access branch; radius; equal and unequal						
70 mm	18.11	23.20	0.79	20.04	nr	**43.24**
100 mm	24.56	31.46	0.96	24.36	nr	**55.81**
150 mm	52.49	67.23	1.20	30.46	nr	**97.69**
Double branch; equal and unequal						
100mm	18.58	23.80	1.30	32.99	nr	**56.79**
100 mm; 69°	20.45	26.19	1.30	32.99	nr	**59.19**
150 mm	59.90	76.72	1.37	34.76	nr	**111.48**
200 mm	109.71	140.52	1.51	38.31	nr	**178.83**
Double branch; radius; equal and unequal						
100 mm	17.67	22.63	1.30	32.99	nr	**55.63**
150 mm	63.54	81.38	1.37	34.76	nr	**116.14**
200 mm	109.71	140.52	1.51	38.31	nr	**178.83**
Corner branch						
100 mm	33.84	43.34	1.30	32.99	nr	**76.34**
Corner branch; long tail						
100 mm	53.31	68.28	1.30	32.99	nr	**101.27**
Roof vent connector; asphalt						
100 mm	22.39	28.68	0.78	19.79	nr	**48.47**
Roof vent connector; felt						
100 mm	50.56	64.76	0.78	19.79	nr	**84.55**
Movement connector						
100 mm	25.29	32.39	0.78	19.79	nr	**52.18**
150 mm	43.89	56.21	0.78	19.79	nr	**76.00**

R:DISPOSAL SYSTEMS

Item	Net Price £	Material £	Labour hours	Labour £	Unit	Total rate £
Expansion plugs						
70 mm	11.62	14.88	0.32	8.12	nr	**23.00**
100 mm	12.55	16.07	0.39	9.90	nr	**25.97**
150 mm	18.77	24.04	0.55	13.95	nr	**37.99**
P trap						
100mm dia.	14.81	18.97	1.00	25.37	nr	**44.34**
P trap with access						
50 mm	22.61	28.96	0.54	13.70	nr	**42.66**
70 mm	22.61	28.96	0.64	16.24	nr	**45.20**
100mm	24.49	31.37	1.00	25.37	nr	**56.74**
150 mm	43.79	56.09	1.09	27.66	nr	**83.74**
Branch trap						
100 mm	54.12	69.32	1.17	29.69	nr	**99.00**
Balcony gully inlet						
100 mm	47.30	60.58	1.00	25.37	nr	**85.95**
Roof outlet						
Flat grate						
70 mm	65.50	83.89	1.00	25.37	nr	**109.26**
100 mm	76.99	98.61	1.00	25.37	nr	**123.98**
Dome grate						
70 mm	67.12	85.97	1.00	25.37	nr	**111.34**
100 mm	-	-	1.00	25.37	nr	**25.37**
Top Hat						
100 mm	152.08	194.78	1.00	25.37	nr	**220.15**
Floor drains; for cast iron pipework BS 416 and BS EN877						
Adjustable clamp plate body						
100 mm; 165mm nickel bronze grate and frame	52.39	67.10	0.50	12.69	nr	**79.79**
100 mm; 165mm nickel bronze rodding eye	61.58	78.87	0.50	12.69	nr	**91.56**
100 mm; 150 x 150mm nickel bronze grate and frame	56.24	72.03	0.50	12.69	nr	**84.72**
100 mm; 150 x 150 nickel bronze rodding eye	64.44	82.53	0.50	12.69	nr	**95.22**
Deck plate body						
100 mm; 165mm nickel bronze grate and frame	53.70	68.78	0.50	12.69	nr	**81.46**
100 mm; 165mm nickel bronze rodding eye	64.01	81.98	0.50	12.69	nr	**94.67**
100 mm; 150 x 150mm nickel bronze grate and frame	59.55	76.27	0.50	12.69	nr	**88.96**
100 mm; 150 x 150mm nickel bronze rodding eye	65.16	83.46	0.50	12.69	nr	**96.14**
Extra for						
100 mm; Screwed extension piece	19.86	25.44	0.30	7.61	nr	**33.05**
100 mm; Grating extension piece; screwed or spigot	16.54	21.18	0.30	7.61	nr	**28.80**
100 mm; Brewery trap	630.30	807.28	2.00	50.74	nr	**858.03**

S:PIPED SUPPLY SYSTEMS

Item	Net Price £	Material £	Labour hours	Labour £	Unit	Total rate £
S10 : COLD WATER						
Y10 - PIPELINES						
COPPER PIPEWORK						
Copper pipe; capillary or compression joints in the running length; EN1057 R250 (TX) formerly BS 2871 Table X						
Fixed vertically or at low level, with brackets measured separately						
12mm dia.	1.22	1.56	0.39	9.90	m	**11.46**
15mm dia.	1.35	1.73	0.40	10.15	m	**11.88**
22mm dia.	2.65	3.39	0.47	11.92	m	**15.32**
28mm dia.	3.48	4.46	0.51	12.94	m	**17.40**
35mm dia.	7.09	9.08	0.58	14.72	m	**23.80**
42mm dia.	8.82	11.30	0.66	16.75	m	**28.04**
54mm dia.	11.74	15.04	0.72	18.27	m	**33.30**
67mm dia.	15.32	19.62	0.75	19.03	m	**38.65**
76mm dia.	21.53	27.58	0.76	19.28	m	**46.86**
108mm dia.	30.94	39.63	0.78	19.79	m	**59.42**
133mm dia.	40.61	52.01	1.05	26.64	m	**78.65**
159mm dia.	61.89	79.27	1.15	29.18	m	**108.45**
Fixed horizontally at high level or suspended, with brackets measured separately						
12mm dia.	1.22	1.56	0.45	11.42	m	**12.98**
15mm dia.	1.35	1.73	0.46	11.67	m	**13.40**
22mm dia.	2.65	3.39	0.54	13.70	m	**17.09**
28mm dia.	-	4.46	0.59	14.97	m	**19.43**
35mm dia.	7.09	9.08	0.67	17.00	m	**26.08**
42mm dia.	8.82	11.30	0.76	19.28	m	**30.58**
54mm dia.	11.74	15.04	0.83	21.06	m	**36.10**
67mm dia.	15.32	19.62	0.86	21.82	m	**41.44**
76mm dia.	21.53	27.58	0.87	22.07	m	**49.65**
108mm dia.	30.94	39.63	0.90	22.83	m	**62.46**
133mm dia.	40.61	52.01	1.21	30.70	m	**82.71**
159mm dia.	61.89	79.27	1.32	33.49	m	**112.76**
Copper pipe; capillary or compression joints in the running length; EN1057 R250 (TY) formerly BS 2871 Table Y						
Fixed vertically or at low level with brackets measured separately						
12mm dia.	1.64	2.10	0.41	10.40	m	**12.50**
15mm dia.	2.40	3.07	0.43	10.91	m	**13.98**
22mm dia.	4.25	5.44	0.50	12.69	m	**18.13**
28mm dia.	5.55	7.11	0.54	13.70	m	**20.81**
35mm dia.	9.47	12.13	0.62	15.73	m	**27.86**
42mm dia.	11.67	14.95	0.71	18.01	m	**32.96**
54mm dia.	20.04	25.67	0.78	19.79	m	**45.46**
67mm dia.	26.00	33.30	0.82	20.80	m	**54.11**
76mm dia.	30.18	38.65	0.60	15.22	m	**53.88**
108mm dia.	53.00	67.88	0.88	22.33	m	**90.21**

S:PIPED SUPPLY SYSTEMS

Item	Net Price £	Material £	Labour hours	Labour £	Unit	Total rate £
Copper pipe; capillary or compression joints in the running length; EN1057 R250 (TX) formerly BS 2871 Table X						
Plastic coated gas and cold water service pipe for corrosive environments, fixed vertically or at low level with brackets measured separately						
15mm dia. (white)	1.71	2.19	0.59	14.97	m	**17.16**
22mm dia. (white)	3.27	4.19	0.68	17.26	m	**21.45**
28mm dia. (white)	4.14	5.30	0.74	18.78	m	**24.08**
FIXINGS						
For copper pipework						
Saddle band						
6mm dia.	0.09	0.12	0.11	2.79	nr	**2.91**
8mm dia.	0.09	0.12	0.12	3.04	nr	**3.16**
10mm dia.	0.09	0.12	0.12	3.04	nr	**3.16**
12mm dia.	0.09	0.12	0.12	3.04	nr	**3.16**
15mm dia.	0.10	0.13	0.13	3.30	nr	**3.43**
22mm dia.	0.10	0.13	0.13	3.30	nr	**3.43**
28mm dia.	0.11	0.14	0.16	4.06	nr	**4.20**
35mm dia.	0.17	0.22	0.18	4.57	nr	**4.78**
42mm dia.	0.37	0.47	0.21	5.33	nr	**5.80**
54mm dia.	0.50	0.64	0.21	5.33	nr	**5.97**
Single spacing clip						
15mm dia.	0.11	0.14	0.14	3.55	nr	**3.69**
22mm dia.	0.11	0.14	0.15	3.81	nr	**3.95**
28mm dia.	0.27	0.35	0.17	4.31	nr	**4.66**
Two piece spacing clip						
8mm dia. Bottom	0.09	0.12	0.11	2.79	nr	**2.91**
8mm dia. Top	0.09	0.12	0.11	2.79	nr	**2.91**
12mm dia. Bottom	0.09	0.12	0.13	3.30	nr	**3.41**
12mm dia. Top	0.09	0.12	0.13	3.30	nr	**3.41**
15mm dia. Bottom	0.10	0.13	0.13	3.30	nr	**3.43**
15mm dia. Top	0.10	0.13	0.13	3.30	nr	**3.43**
22mm dia. Bottom	0.10	0.13	0.14	3.55	nr	**3.68**
22mm dia. Top	0.10	0.13	0.14	3.55	nr	**3.68**
28mm dia. Bottom	0.10	0.13	0.16	4.06	nr	**4.19**
28mm dia. Top	0.17	0.22	0.16	4.06	nr	**4.28**
35mm dia. Bottom	0.11	0.14	0.21	5.33	nr	**5.47**
35mm dia. Top	0.27	0.35	0.21	5.33	nr	**5.67**
Single pipe bracket						
15mm dia.	1.23	1.58	0.14	3.55	nr	**5.13**
22mm dia.	1.41	1.81	0.14	3.55	nr	**5.36**
28mm dia.	1.68	2.15	0.17	4.31	nr	**6.46**

S:PIPED SUPPLY SYSTEMS

Item	Net Price £	Material £	Labour hours	Labour £	Unit	Total rate £
S10 : COLD WATER (cont'd)						
COPPER PIPEWORK (cont'd)						
Fixings (cont'd)						
Copper pipe; capillary or compression joints (cont'd)						
Single pipe ring						
15mm dia.	1.86	2.38	0.26	6.60	nr	8.98
22mm dia.	1.99	2.55	0.26	6.60	nr	9.15
28mm dia.	2.37	3.04	0.31	7.87	nr	10.90
35mm dia.	2.50	3.20	0.32	8.12	nr	11.32
42mm dia.	2.80	3.59	0.32	8.12	nr	11.71
54mm dia.	3.33	4.27	0.34	8.63	nr	12.89
67mm dia.	7.82	10.02	0.35	8.88	nr	18.90
76mm dia.	9.87	12.64	0.42	10.66	nr	23.30
108mm dia.	15.27	19.56	0.42	10.66	nr	30.21
Double pipe ring						
15mm dia.	2.50	3.20	0.26	6.60	nr	9.80
22mm dia.	2.66	3.41	0.26	6.60	nr	10.00
28mm dia.	3.56	4.56	0.31	7.87	nr	12.42
35mm dia.	3.69	4.73	0.32	8.12	nr	12.85
42mm dia.	4.02	5.15	0.32	8.12	nr	13.27
54mm dia.	4.84	6.20	0.34	8.63	nr	14.83
67mm dia.	10.34	13.24	0.35	8.88	nr	22.12
76mm dia.	12.90	16.52	0.42	10.66	nr	27.18
108mm dia.	21.71	27.81	0.42	10.66	nr	38.46
Wall bracket						
15mm dia.	3.14	4.02	0.05	1.27	nr	5.29
22mm dia.	3.69	4.73	0.05	1.27	nr	5.99
28mm dia.	4.50	5.76	0.05	1.27	nr	7.03
35mm dia.	5.78	7.40	0.05	1.27	nr	8.67
42mm dia.	7.66	9.81	0.05	1.27	nr	11.08
54mm dia.	9.67	12.39	0.05	1.27	nr	13.65
Hospital bracket						
15mm dia.	2.66	3.41	0.26	6.60	nr	10.00
22mm dia.	2.80	3.59	0.26	6.60	nr	10.18
28mm dia.	3.42	4.38	0.31	7.87	nr	12.25
35mm dia.	3.69	4.73	0.32	8.12	nr	12.85
42mm dia.	5.18	6.63	0.32	8.12	nr	14.75
54mm dia.	7.04	9.02	0.34	8.63	nr	17.64
Screw on backplate, female						
All sizes 15mm to 54mm x 10mm	1.18	1.51	0.10	2.54	nr	4.05
Screw on backplate, male						
All sizes 15mm to 54mm x 10mm	1.04	1.33	0.10	2.54	nr	3.87
Pipe joist clips, single						
15mm dia.	0.54	0.69	0.08	2.03	nr	2.72
22mm dia.	0.54	0.69	0.08	2.03	nr	2.72
Pipe joist clips, double						
15mm dia.	0.76	0.97	0.08	2.03	nr	3.00
22mm dia.	0.76	0.97	0.08	2.03	nr	3.00

S:PIPED SUPPLY SYSTEMS

Item	Net Price £	Material £	Labour hours	Labour £	Unit	Total rate £
Extra Over channel sections for fabricated hangers and brackets						
Galvanised steel; including inserts, bolts, nuts, washers; fixed to backgrounds						
41 x 21mm	2.40	3.07	0.15	3.81	m	**6.88**
41 x 41mm	3.75	4.80	0.15	3.81	m	**8.61**
Threaded rods; metric thread; including nuts, washers etc						
10mm dia x 600mm long for ring clips up to 54mm	1.25	1.60	0.10	2.54	nr	**4.14**
12mm dia x 600mm long for ring clips from 54mm	1.30	1.66	0.10	2.54	nr	**4.20**
Extra over copper pipes; capillary fittings; BS 864						
Stop end						
15mm dia.	0.87	1.11	0.13	3.30	nr	**4.41**
22mm dia.	1.62	2.07	0.14	3.55	nr	**5.63**
28mm dia.	2.90	3.71	0.17	4.31	nr	**8.03**
35mm dia.	6.40	8.20	0.19	4.82	nr	**13.02**
42mm dia.	11.02	14.11	0.22	5.58	nr	**19.70**
54mm dia.	15.38	19.70	0.23	5.84	nr	**25.53**
Straight coupling; copper to copper						
6mm dia.	0.73	0.94	0.23	5.84	nr	**6.77**
8mm dia.	0.75	0.96	0.23	5.84	nr	**6.80**
10mm dia.	0.38	0.49	0.23	5.84	nr	**6.32**
15mm dia.	0.12	0.15	0.23	5.84	nr	**5.99**
22mm dia.	0.32	0.41	0.26	6.60	nr	**7.01**
28mm dia.	0.80	1.02	0.30	7.61	nr	**8.64**
35mm dia.	2.61	3.34	0.34	8.63	nr	**11.97**
42mm dia.	4.37	5.60	0.38	9.64	nr	**15.24**
54mm dia.	8.05	10.31	0.42	10.66	nr	**20.97**
67mm dia.	23.93	30.65	0.53	13.45	nr	**44.10**
Adaptor coupling; imperial to metric						
1/2" x 15mm dia.	1.89	2.42	0.27	6.85	nr	**9.27**
3/4" x 22mm dia.	1.65	2.11	0.31	7.87	nr	**9.98**
1" x 28mm dia.	3.26	4.18	0.36	9.13	nr	**13.31**
1 1/4" x 35mm dia.	5.45	6.98	0.41	10.40	nr	**17.38**
1 1/2" x 42mm dia.	6.93	8.88	0.46	11.67	nr	**20.55**
Reducing coupling						
15 x 10mm dia.	1.61	2.06	0.23	5.84	nr	**7.90**
22 x 10mm dia.	2.34	3.00	0.26	6.60	nr	**9.59**
22 x 15mm dia.	1.57	2.01	0.27	6.85	nr	**8.86**
28 x 15mm dia.	3.59	4.60	0.28	7.10	nr	**11.70**
28 x 22mm dia.	2.20	2.82	0.30	7.61	nr	**10.43**
35 x 28mm dia.	5.19	6.65	0.34	8.63	nr	**15.27**
42 x 35mm dia.	7.61	9.75	0.38	9.64	nr	**19.39**
54 x 35mm dia.	13.35	17.10	0.42	10.66	nr	**27.75**
54 x 42mm dia.	14.57	18.66	0.42	10.66	nr	**29.32**

S:PIPED SUPPLY SYSTEMS

Item	Net Price £	Material £	Labour hours	Labour £	Unit	Total rate £
S10 : COLD WATER (cont'd)						
COPPER PIPEWORK (cont'd)						
Extra over copper pipes; capillary fittings (cont'd)						
Straight female connector						
15mm x 1/2" dia.	1.98	2.54	0.27	6.85	nr	**9.39**
22mm x 3/4" dia.	2.87	3.68	0.31	7.87	nr	**11.54**
28mm x 1" dia.	5.41	6.93	0.36	9.13	nr	**16.06**
35mm x 1 1/4" dia.	9.36	11.99	0.41	10.40	nr	**22.39**
42mm x 1 1/2" dia.	12.15	15.56	0.46	11.67	nr	**27.23**
54mm x 2" dia.	19.27	24.68	0.52	13.19	nr	**37.87**
Straight male connector						
15mm x 1/2" dia.	1.69	2.16	0.27	6.85	nr	**9.01**
22mm x 3/4" dia.	3.01	3.86	0.31	7.87	nr	**11.72**
28mm x 1" dia.	4.85	6.21	0.36	9.13	nr	**15.35**
35mm x 1 1/4" dia.	8.54	10.94	0.41	10.40	nr	**21.34**
42mm x 1 1/2" dia.	10.99	14.08	0.46	11.67	nr	**25.75**
54mm x 2" dia.	16.68	21.36	0.52	13.19	nr	**34.56**
67mm x 2 1/2" dia.	26.64	34.12	0.63	15.98	nr	**50.10**
Female reducing connector						
15mm x 3/4" dia.	4.92	6.30	0.27	6.85	nr	**13.15**
Male reducing connector						
15mm x 3/4" dia.	4.39	5.62	0.27	6.85	nr	**12.47**
22mm x 1" dia.	6.69	8.57	0.31	7.87	nr	**16.43**
Flanged connector						
28mm dia.	31.41	40.23	0.36	9.13	nr	**49.36**
35mm dia.	39.76	50.92	0.41	10.40	nr	**61.33**
42mm dia.	47.53	60.88	0.46	11.67	nr	**72.55**
54mm dia.	71.85	92.02	0.52	13.19	nr	**105.22**
67mm dia.	88.72	113.63	0.61	15.48	nr	**129.11**
Tank connector						
15mm x 1/2" dia.	4.21	5.39	0.25	6.34	nr	**11.74**
22mm x 3/4" dia.	6.42	8.22	0.28	7.10	nr	**15.33**
28mm x 1" dia.	8.43	10.80	0.32	8.12	nr	**18.92**
35mm x 1 1/4" dia.	10.81	13.85	0.37	9.39	nr	**23.23**
42mm x 1 1/2" dia.	14.17	18.15	0.43	10.91	nr	**29.06**
54mm x 2" dia.	21.66	27.74	0.46	11.67	nr	**39.41**
Tank connector with long thread						
15mm x 1/2" dia.	5.45	6.98	0.30	7.61	nr	**14.59**
22mm x 3/4" dia.	7.77	9.95	0.33	8.37	nr	**18.32**
28mm x 1" dia.	9.59	12.28	0.39	9.90	nr	**22.18**

S:PIPED SUPPLY SYSTEMS

Item	Net Price £	Material £	Labour hours	Labour £	Unit	Total rate £
Reducer						
15 x 10mm dia.	0.54	0.69	0.23	5.84	nr	**6.53**
22 x 15mm dia.	1.26	1.61	0.26	6.60	nr	**8.21**
28 x 15mm dia.	1.77	2.27	0.28	7.10	nr	**9.37**
28 x 22mm dia.	1.36	1.74	0.30	7.61	nr	**9.35**
35 x 22mm dia.	5.04	6.46	0.34	8.63	nr	**15.08**
42 x 22mm dia.	9.09	11.64	0.36	9.13	nr	**20.78**
42 x 35mm dia.	7.04	9.02	0.38	9.64	nr	**18.66**
54 x 35mm dia.	14.76	18.90	0.40	10.15	nr	**29.05**
54 x 42mm dia.	12.73	16.30	0.42	10.66	nr	**26.96**
67 x 54mm dia.	17.31	22.17	0.53	13.45	nr	**35.62**
Adaptor; copper to female iron						
15mm x 1/2" dia.	3.41	4.37	0.27	6.85	nr	**11.22**
22mm x 3/4" dia.	5.19	6.65	0.31	7.87	nr	**14.51**
28mm x 1" dia.	7.32	9.38	0.36	9.13	nr	**18.51**
35mm x 1 1/4" dia.	13.24	16.96	0.41	10.40	nr	**27.36**
42mm x 1 1/2" dia.	16.68	21.36	0.46	11.67	nr	**33.03**
54mm x 2" dia.	20.08	25.72	0.52	13.19	nr	**38.91**
Adaptor; copper to male iron						
15mm x 1/2" dia.	3.48	4.46	0.27	6.85	nr	**11.31**
22mm x 3/4" dia.	4.45	5.70	0.31	7.87	nr	**13.56**
28mm x 1" dia.	7.43	9.52	0.36	9.13	nr	**18.65**
35mm x 1 1/4" dia.	10.82	13.86	0.41	10.40	nr	**24.26**
42mm x 1 1/2" dia.	14.95	19.15	0.46	11.67	nr	**30.82**
54mm x 2" dia.	20.08	25.72	0.52	13.19	nr	**38.91**
Union coupling						
15mm dia.	4.70	6.02	0.41	10.40	nr	**16.42**
22mm dia.	7.54	9.66	0.45	11.42	nr	**21.07**
28mm dia.	10.99	14.08	0.51	12.94	nr	**27.02**
35mm dia.	14.42	18.47	0.64	16.24	nr	**34.71**
42mm dia.	21.07	26.99	0.68	17.25	nr	**44.24**
54mm dia.	40.10	51.36	0.78	19.79	nr	**71.15**
67mm dia.	67.89	86.95	0.96	24.36	nr	**111.31**
Elbow						
15mm dia.	0.22	0.28	0.23	5.84	nr	**6.12**
22mm dia.	0.56	0.72	0.26	6.60	nr	**7.31**
28mm dia.	1.29	1.65	0.31	7.87	nr	**9.52**
35mm dia.	5.60	7.17	0.35	8.88	nr	**16.05**
42mm dia.	9.26	11.86	0.41	10.40	nr	**22.26**
54mm dia.	19.12	24.49	0.44	11.16	nr	**35.65**
67mm dia.	49.62	63.55	0.54	13.70	nr	**77.25**
Backplate elbow						
15mm dia.	3.54	4.53	0.51	12.94	nr	**17.47**
22mm dia.	7.59	9.72	0.54	13.70	nr	**23.42**
Overflow bend						
22mm dia.	10.70	13.70	0.26	6.60	nr	**20.30**
Return bend						
15mm dia.	5.29	6.78	0.23	5.84	nr	**12.61**
22mm dia.	10.41	13.33	0.26	6.60	nr	**19.93**
28mm dia.	13.29	17.02	0.31	7.87	nr	**24.89**

S:PIPED SUPPLY SYSTEMS

Item	Net Price £	Material £	Labour hours	Labour £	Unit	Total rate £
S10 : COLD WATER (cont'd)						
COPPER PIPEWORK (cont'd)						
Extra over copper pipes; capillary fittings (cont'd)						
Obtuse elbow						
15mm dia.	0.67	0.86	0.23	5.84	nr	**6.69**
22mm dia.	1.41	1.81	0.26	6.60	nr	**8.40**
28mm dia.	2.72	3.48	0.31	7.87	nr	**11.35**
35mm dia.	8.45	10.82	0.36	9.13	nr	**19.96**
42mm dia.	15.05	19.28	0.41	10.40	nr	**29.68**
54mm dia.	27.22	34.86	0.44	11.16	nr	**46.03**
67mm dia.	49.39	63.26	0.54	13.70	nr	**76.96**
Straight tap connector						
15mm x 1/2" dia.	1.05	1.34	0.13	3.30	nr	**4.64**
22mm x 3/4" dia.	1.35	1.73	0.14	3.55	nr	**5.28**
Bent tap connector						
15mm x 1/2" dia.	1.34	1.72	0.13	3.30	nr	**5.01**
22mm x 3/4" dia.	4.12	5.28	0.14	3.55	nr	**8.83**
Bent male union connector						
15mm x 1/2" dia.	6.88	8.81	0.41	10.40	nr	**19.21**
22mm x 3/4" dia.	8.96	11.48	0.45	11.42	nr	**22.89**
28mm x 1" dia.	12.81	16.41	0.51	12.94	nr	**29.35**
35mm x 1 1/4" dia.	20.89	26.76	0.64	16.24	nr	**42.99**
42mm x 1 1/2" dia.	33.97	43.51	0.68	17.25	nr	**60.76**
54mm x 2" dia.	53.65	68.71	0.78	19.79	nr	**88.50**
Bent female union connector						
15mm dia.	6.88	8.81	0.41	10.40	nr	**19.21**
22mm x 3/4" dia.	8.96	11.48	0.45	11.42	nr	**22.89**
28mm x 1" dia.	12.81	16.41	0.51	12.94	nr	**29.35**
35mm x 1 1/4" dia.	20.89	26.76	0.64	16.24	nr	**42.99**
42mm x 1 1/2" dia.	33.97	43.51	0.68	17.25	nr	**60.76**
54mm x 2" dia.	53.65	68.71	0.78	19.79	nr	**88.50**
Straight union adaptor						
15mm x 3/4" dia.	2.94	3.77	0.41	10.40	nr	**14.17**
22mm x 1" dia.	4.18	5.35	0.45	11.42	nr	**16.77**
28mm x 1 1/4" dia.	6.76	8.66	0.51	12.94	nr	**21.60**
35mm x 1 1/2" dia.	10.41	13.33	0.64	16.24	nr	**29.57**
42mm x 2" dia.	13.14	16.83	0.68	17.25	nr	**34.08**
54mm x 2 1/2" dia.	20.28	25.97	0.78	19.79	nr	**45.76**
Straight male union connector						
15mm x 1/2" dia.	5.86	7.51	0.41	10.40	nr	**17.91**
22mm x 3/4" dia.	7.61	9.75	0.45	11.42	nr	**21.16**
28mm x 1" dia.	11.34	14.52	0.51	12.94	nr	**27.46**
35mm x 1 1/4" dia.	16.34	20.93	0.64	16.24	nr	**37.17**
42mm x 1 1/2" dia.	25.67	32.88	0.68	17.25	nr	**50.13**
54mm x 2" dia.	36.88	47.24	0.78	19.79	nr	**67.03**

S:PIPED SUPPLY SYSTEMS

Item	Net Price £	Material £	Labour hours	Labour £	Unit	Total rate £
Straight female union connector						
15mm x 1/2" dia.	5.86	7.51	0.41	10.40	nr	**17.91**
22mm x 3/4" dia.	7.61	9.75	0.45	11.42	nr	**21.16**
28mm x 1" dia.	11.34	14.52	0.51	12.94	nr	**27.46**
35mm x 1 1/4" dia.	16.34	20.93	0.64	16.24	nr	**37.17**
42mm x 1 1/2" dia.	25.67	32.88	0.68	17.25	nr	**50.13**
54mm x 2" dia.	36.88	47.24	0.78	19.79	nr	**67.03**
Male nipple						
3/4 x 1/2" dia.	2.90	3.71	0.28	7.10	nr	**10.82**
1 x 3/4" dia.	3.36	4.30	0.32	8.12	nr	**12.42**
1 1/4 x 1" dia.	4.58	5.87	0.37	9.39	nr	**15.25**
1 1/2 x 1 1/4" dia.	6.78	8.68	0.42	10.66	nr	**19.34**
2 x 1 1/2" dia.	13.87	17.76	0.46	11.67	nr	**29.44**
2 1/2 x 2" dia.	18.53	23.73	0.56	14.21	nr	**37.94**
Female nipple						
3/4 x 1/2" dia.	2.90	3.71	0.28	7.10	nr	**10.82**
1 x 3/4" dia.	3.36	4.30	0.32	8.12	nr	**12.42**
1 1/4 x 1" dia.	4.58	5.87	0.37	9.39	nr	**15.26**
1 1/2 x 1 1/4" dia.	6.78	8.68	0.42	10.66	nr	**19.34**
2 x 1 1/2" dia.	13.87	17.76	0.46	11.67	nr	**29.44**
2 1/2 x 2" dia.	18.53	23.73	0.56	14.21	nr	**37.94**
Equal tee						
10mm dia.	1.47	1.88	0.25	6.34	nr	**8.23**
15mm dia.	0.41	0.53	0.36	9.13	nr	**9.66**
22mm dia.	1.29	1.65	0.39	9.90	nr	**11.55**
28mm dia.	3.57	4.57	0.43	10.91	nr	**15.48**
35mm dia.	9.12	11.68	0.57	14.46	nr	**26.14**
42mm dia.	14.63	18.74	0.60	15.22	nr	**33.96**
54mm dia.	29.51	37.80	0.65	16.49	nr	**54.29**
67mm dia.	64.25	82.29	0.78	19.79	nr	**102.08**
Female tee, reducing branch Fl						
15 x 15mm x 1/4" dia.	4.41	5.65	0.36	9.13	nr	**14.78**
22 x 22mm x 1/2" dia.	3.07	3.93	0.39	9.90	nr	**13.83**
28 x 28mm x 3/4" dia.	10.59	13.56	0.43	10.91	nr	**24.47**
35 x 35mm x 3/4" dia.	15.27	19.56	0.47	11.92	nr	**31.48**
42 x 42mm x 1/2" dia.	18.35	23.50	0.60	15.22	nr	**38.73**
Backplate tee						
15 x 15mm x 1/2" dia.	8.36	10.71	0.62	15.73	nr	**26.44**
Heater tee						
1/2 x 1/2" x 15mm dia.	7.52	9.63	0.36	9.13	nr	**18.77**
Union heater tee						
1/2 x 1/2" x 15mm dia.	-	-	0.36	9.13	nr	**9.13**
Sweep tee - equal						
15mm dia.	5.98	7.66	0.36	9.13	nr	**16.79**
22mm dia.	7.69	9.85	0.39	9.90	nr	**19.74**
28mm dia.	12.94	16.57	0.43	10.91	nr	**27.48**
35mm dia.	18.35	23.50	0.57	14.46	nr	**37.96**
42mm dia.	27.20	34.84	0.60	15.22	nr	**50.06**
54mm dia.	30.12	38.58	0.65	16.49	nr	**55.07**
67mm dia.	52.43	67.15	0.78	19.79	nr	**86.94**

S:PIPED SUPPLY SYSTEMS

Item	Net Price £	Material £	Labour hours	Labour £	Unit	Total rate £
S10 : COLD WATER (cont'd)						
COPPER PIPEWORK (cont'd)						
Extra over copper pipes; capillary fittings (cont'd)						
Sweep tee - reducing						
22 x 22 x 15mm dia.	6.44	8.25	0.39	9.90	nr	**18.14**
28 x 28 x 22mm dia.	10.94	14.01	0.43	10.91	nr	**24.92**
35 x 35 x 22mm dia.	18.35	23.50	0.57	14.46	nr	**37.96**
Sweep tee - double						
15mm dia.	6.76	8.66	0.36	9.13	nr	**17.79**
22mm dia.	9.20	11.78	0.39	9.90	nr	**21.68**
28mm dia.	13.97	17.89	0.43	10.91	nr	**28.80**
Cross						
15mm dia.	8.94	11.45	0.48	12.18	nr	**23.63**
22mm dia.	9.99	12.80	0.53	13.45	nr	**26.24**
28mm dia.	14.33	18.35	0.61	15.48	nr	**33.83**
Extra over copper pipes; high duty capillary fittings; BS 864						
Stop end						
15mm dia.	4.41	5.65	0.16	4.06	nr	**9.71**
Straight coupling; copper to copper						
15mm dia.	2.02	2.59	0.27	6.85	nr	**9.44**
22mm dia.	3.24	4.15	0.32	8.12	nr	**12.27**
28mm dia.	4.58	5.87	0.37	9.39	nr	**15.25**
35mm dia.	8.08	10.35	0.43	10.91	nr	**21.26**
42mm dia.	8.84	11.32	0.50	12.69	nr	**24.01**
54mm dia.	13.00	16.65	0.54	13.70	nr	**30.35**
Reducing coupling						
15 x 12mm dia.	3.81	4.88	0.27	6.85	nr	**11.73**
22 x 15mm dia.	4.41	5.65	0.32	8.12	nr	**13.77**
28 x 22mm dia.	6.08	7.79	0.37	9.39	nr	**17.17**
Straight female connector						
15mm x 1/2" dia.	4.97	6.37	0.32	8.12	nr	**14.48**
22mm x 3/4" dia.	5.60	7.17	0.36	9.13	nr	**16.31**
28mm x 1" dia.	8.26	10.58	0.42	10.66	nr	**21.24**
Straight male connector						
15mm x 1/2" dia.	4.84	6.20	0.32	8.12	nr	**14.32**
22mm x 3/4" dia.	5.60	7.17	0.36	9.13	nr	**16.31**
28mm x 1" dia.	8.26	10.58	0.42	10.66	nr	**21.24**
42mm x 1 1/2" dia.	16.13	20.66	0.53	13.45	nr	**34.11**
54mm x 2" dia.	26.20	33.56	0.62	15.73	nr	**49.29**
Reducer						
15 x 12mm dia.	2.50	3.20	0.27	6.85	nr	**10.05**
22 x 15mm dia.	2.44	3.13	0.32	8.12	nr	**11.24**
28 x 22mm dia.	4.41	5.65	0.37	9.39	nr	**15.04**
35 x 28mm dia.	5.60	7.17	0.43	10.91	nr	**18.08**
42 x 35mm dia.	7.22	9.25	0.50	12.69	nr	**21.93**
54 x 42mm dia.	11.65	14.92	0.39	9.90	nr	**24.82**

S:PIPED SUPPLY SYSTEMS

Item	Net Price £	Material £	Labour hours	Labour £	Unit	Total rate £
Straight union adaptor						
15mm x 3/4" dia.	4.03	5.16	0.27	6.85	nr	**12.01**
22mm x 1" dia.	5.46	6.99	0.32	8.12	nr	**15.11**
28mm x 1 1/4" dia.	7.22	9.25	0.37	9.39	nr	**18.63**
35mm x 1 1/2" dia.	13.09	16.77	0.43	10.91	nr	**27.68**
42mm x 2" dia.	16.57	21.22	0.50	12.69	nr	**33.91**
Bent union adaptor						
15mm x 3/4" dia.	10.49	13.44	0.27	6.85	nr	**20.29**
22mm x 1" dia.	14.16	18.14	0.32	8.12	nr	**26.25**
28mm x 1 1/4" dia.	19.07	24.42	0.37	9.39	nr	**33.81**
Adaptor; male copper to FI						
15mm x 1/2" dia.	7.48	9.58	0.27	6.85	nr	**16.43**
22mm x 3/4" dia.	7.63	9.77	0.32	8.12	nr	**17.89**
Union coupling						
15mm dia.	9.10	11.66	0.54	13.70	nr	**25.36**
22mm dia.	11.65	14.92	0.60	15.22	nr	**30.14**
28mm dia.	16.17	20.71	0.68	17.25	nr	**37.96**
35mm dia.	28.21	36.13	0.83	21.06	nr	**57.19**
42mm dia.	33.23	42.56	0.89	22.58	nr	**65.14**
Elbow						
15mm dia.	5.66	7.25	0.27	6.85	nr	**14.10**
22mm dia.	6.25	8.00	0.32	8.12	nr	**16.12**
28mm dia.	9.30	11.91	0.37	9.39	nr	**21.30**
35mm dia.	14.53	18.61	0.43	10.91	nr	**29.52**
42mm dia.	18.10	23.18	0.50	12.69	nr	**35.87**
54mm dia.	31.49	40.33	0.52	13.19	nr	**53.53**
Return bend						
28mm dia.	19.07	24.42	0.37	9.39	nr	**33.81**
35mm dia.	22.15	28.37	0.43	10.91	nr	**39.28**
Bent male union connector						
15mm x 1/2" dia.	13.57	17.38	0.54	13.70	nr	**31.08**
22mm x 3/4" dia.	18.29	23.43	0.60	15.22	nr	**38.65**
28mm x 1" dia.	33.23	42.56	0.68	17.25	nr	**59.81**
Composite flange						
35mm dia.	-	-	0.38	9.64	nr	**9.64**
42mm dia.	-	-	0.41	10.40	nr	**10.40**
54mm dia.	-	-	0.43	10.91	nr	**10.91**
Equal tee						
15mm dia.	6.72	8.61	0.44	11.16	nr	**19.77**
22mm dia.	8.45	10.82	0.47	11.92	nr	**22.75**
28mm dia.	11.15	14.28	0.53	13.45	nr	**27.73**
35mm dia.	19.07	24.42	0.70	17.76	nr	**42.19**
42mm dia.	24.27	31.08	0.84	21.31	nr	**52.40**
54mm dia.	38.23	48.96	0.79	20.04	nr	**69.01**

S:PIPED SUPPLY SYSTEMS

Item	Net Price £	Material £	Labour hours	Labour £	Unit	Total rate £
S10 : COLD WATER (cont'd)						
COPPER PIPEWORK (cont'd)						
Extra over copper pipes; high duty capillary fittings (cont'd)						
Reducing tee						
15 x 12mm dia.	-	-	0.44	11.16	nr	**11.16**
22 x 15mm dia.	-	-	0.47	11.92	nr	**11.92**
28 x 22mm dia.	-	-	0.53	13.45	nr	**13.45**
35 x 28mm dia.	-	-	0.73	18.52	nr	**18.52**
42 x 28mm dia.	-	-	0.84	21.31	nr	**21.31**
54 x 28mm dia.	-	-	1.01	25.63	nr	**25.63**
Extra over copper pipes; compression fittings; BS 864						
Stop end						
15mm dia.	1.06	1.36	0.10	2.54	nr	**3.89**
22mm dia.	1.27	1.63	0.12	3.04	nr	**4.67**
28mm dia.	4.01	5.14	0.15	3.81	nr	**8.94**
Straight connector; copper to copper						
15mm dia.	0.69	0.88	0.18	4.57	nr	**5.45**
22mm dia.	1.18	1.51	0.21	5.33	nr	**6.84**
28mm dia.	3.61	4.62	0.24	6.09	nr	**10.71**
Straight connector; copper to imperial copper						
22mm dia.	3.39	4.34	0.21	5.33	nr	**9.67**
Male coupling; copper to MI (BSP)						
15mm x 1/2" dia.	0.63	0.81	0.19	4.82	nr	**5.63**
22mm x 3/4" dia.	0.99	1.27	0.23	5.84	nr	**7.10**
28mm x 1" dia.	2.14	2.74	0.26	6.60	nr	**9.34**
Male coupling with long thread and backnut						
15mm dia.	3.76	4.82	0.19	4.82	nr	**9.64**
22mm dia.	4.77	6.11	0.23	5.84	nr	**11.94**
Female coupling; copper to FI (BSP)						
15mm x 1/2" dia.	0.77	0.99	0.19	4.82	nr	**5.81**
22mm x 3/4" dia.	1.11	1.42	0.23	5.84	nr	**7.26**
28mm x 1" dia.	4.74	6.07	0.27	6.85	nr	**12.92**
Elbow						
15mm dia.	0.83	1.06	0.18	4.57	nr	**5.63**
22mm dia.	1.41	1.81	0.21	5.33	nr	**7.13**
28mm dia.	4.42	5.66	0.24	6.09	nr	**11.75**
Male elbow; copper to FI (BSP)						
15mm x 1/2" dia.	1.47	1.88	0.19	4.82	nr	**6.70**
22mm x 3/4" dia.	1.90	2.43	0.23	5.84	nr	**8.27**
28mm x 1" dia.	4.62	5.92	0.27	6.85	nr	**12.77**
Female elbow; copper to FI (BSP)						
15mm x 1/2" dia.	2.26	2.89	0.19	4.82	nr	**7.72**
22mm x 3/4" dia.	3.26	4.18	0.23	5.84	nr	**10.01**
28mm x 1" dia.	5.75	7.36	0.27	6.85	nr	**14.21**

S:PIPED SUPPLY SYSTEMS

Item	Net Price £	Material £	Labour hours	Labour £	Unit	Total rate £
Backplate elbow						
15mm x 1/2" dia.	3.26	4.18	0.50	12.69	nr	**16.86**
Tank coupling; long thread						
22mm dia.	3.76	4.82	0.46	11.67	nr	**16.49**
Tee equal						
15mm dia.	1.18	1.51	0.28	7.10	nr	**8.62**
22mm dia.	1.97	2.52	0.30	7.61	nr	**10.13**
28mm dia.	8.31	10.64	0.34	8.63	nr	**19.27**
Tee reducing						
22mm dia.	4.70	6.02	0.30	7.61	nr	**13.63**
Backplate tee						
15mm dia.	6.20	7.94	0.62	15.73	nr	**23.67**
Extra over fittings; silver brazed welded joints						
Reducer						
76 x 67mm dia	26.66	34.15	1.40	35.52	nr	**69.67**
108 x 76mm dia	37.89	48.53	1.80	45.67	nr	**94.20**
133 x 108mm dia	58.26	74.62	2.20	55.82	nr	**130.44**
159 x 133mm dia	68.63	87.90	2.60	65.97	nr	**153.87**
90° elbow						
76mm dia	43.52	55.74	1.60	40.60	nr	**96.34**
108mm dia	73.80	94.52	2.00	50.74	nr	**145.27**
133mm dia	194.40	248.99	2.40	60.89	nr	**309.88**
159mm dia	210.92	270.14	2.80	71.04	nr	**341.19**
45° elbow						
76mm dia	46.87	60.03	1.60	40.60	nr	**100.63**
108mm dia	80.35	102.91	2.00	50.74	nr	**153.66**
133mm dia	160.92	206.10	2.40	60.89	nr	**267.00**
159mm dia	187.49	240.14	2.80	71.04	nr	**311.18**
Equal tee						
76mm dia	73.66	94.34	2.40	60.89	nr	**155.24**
108mm dia	100.44	128.64	3.00	76.12	nr	**204.76**
133mm dia	183.10	234.51	3.60	91.34	nr	**325.85**
159mm dia	227.66	291.58	4.20	106.56	nr	**398.15**
Extra over copper pipes; dezincification resistant compression fittings; BS 864						
Stop end						
15mm dia.	1.46	1.87	0.10	2.54	nr	**4.41**
22mm dia.	2.11	2.70	0.13	3.30	nr	**6.00**
28mm dia.	4.40	5.64	0.15	3.81	nr	**9.44**
35mm dia.	6.76	8.66	0.18	4.57	nr	**13.23**
42mm dia.	11.26	14.42	0.20	5.07	nr	**19.50**

S:PIPED SUPPLY SYSTEMS

Item	Net Price £	Material £	Labour hours	Labour £	Unit	Total rate £
S10 : COLD WATER (cont'd)						
COPPER PIPEWORK (cont'd)						
Extra over copper pipes; dezincification resistant compression fittings (cont'd)						
Straight coupling; copper to copper						
15mm dia.	1.17	1.50	0.18	4.57	nr	**6.07**
22mm dia.	1.91	2.45	0.21	5.33	nr	**7.77**
28mm dia.	4.20	5.38	0.24	6.09	nr	**11.47**
35mm dia.	8.73	11.18	0.29	7.36	nr	**18.54**
42mm dia.	11.48	14.70	0.33	8.37	nr	**23.08**
54mm dia.	17.17	21.99	0.38	9.64	nr	**31.63**
Straight swivel connector; copper to imperial copper						
22mm x 3/4" dia.	4.08	5.23	0.20	5.07	nr	**10.30**
Male coupling; copper to MI (BSP)						
15mm x 1/2" dia.	1.04	1.33	0.19	4.82	nr	**6.15**
22mm x 3/4" dia.	1.58	2.02	0.23	5.84	nr	**7.86**
28mm x 1" dia.	2.98	3.82	0.26	6.60	nr	**10.41**
35mm x 1 1/4" dia.	6.63	8.49	0.32	8.12	nr	**16.61**
42mm x 1 1/2" dia.	9.95	12.74	0.37	9.39	nr	**22.13**
54mm x 2" dia.	14.70	18.83	0.57	14.46	nr	**33.29**
Male coupling with long thread and backnuts						
22mm dia.	5.75	7.36	0.23	5.84	nr	**13.20**
28mm dia.	6.37	8.16	0.24	6.09	nr	**14.25**
Female coupling; copper to FI (BSP)						
15mm x 1/2" dia.	1.25	1.60	0.19	4.82	nr	**6.42**
22mm x 3/4" dia.	1.84	2.36	0.23	5.84	nr	**8.19**
28mm x 1" dia.	3.86	4.94	0.27	6.85	nr	**11.79**
35mm x 1 1/4" dia.	7.97	10.21	0.32	8.12	nr	**18.33**
42mm x 1 1/2" dia.	10.71	13.72	0.37	9.39	nr	**23.10**
54mm x 2" dia.	15.72	20.13	0.42	10.66	nr	**30.79**
Elbow						
15mm dia.	1.41	1.81	0.18	4.57	nr	**6.37**
22mm dia.	2.24	2.87	0.21	5.33	nr	**8.20**
28mm dia.	5.43	6.95	0.24	6.09	nr	**13.04**
35mm dia.	9.20	11.78	0.29	7.36	nr	**19.14**
42mm dia.	15.96	20.44	0.33	8.37	nr	**28.81**
54mm dia.	27.46	35.17	0.38	9.64	nr	**44.81**
Male elbow; copper to MI (BSP)						
15mm x 1/2" dia.	2.37	3.04	0.19	4.82	nr	**7.86**
22mm x 3/4" dia.	2.66	3.41	0.23	5.84	nr	**9.24**
28mm x 1" dia.	4.98	6.38	0.27	6.85	nr	**13.23**
Female elbow; copper to FI (BSP)						
15mm x 1/2" dia.	2.54	3.25	0.19	4.82	nr	**8.07**
22mm x 3/4" dia.	3.66	4.69	0.23	5.84	nr	**10.52**
28mm x 1" dia.	6.06	7.76	0.27	6.85	nr	**14.61**
Backplate elbow						
15mm x 1/2" dia.	3.70	4.74	0.50	12.69	nr	**17.42**

S:PIPED SUPPLY SYSTEMS

Item	Net Price £	Material £	Labour hours	Labour £	Unit	Total rate £
Straight tap connector						
15mm dia.	2.16	2.77	0.13	3.30	nr	**6.06**
22mm dia.	4.56	5.84	0.15	3.81	nr	**9.65**
Tank coupling						
15mm dia.	2.97	3.80	0.19	4.82	nr	**8.62**
22mm dia.	3.30	4.23	0.23	5.84	nr	**10.06**
28mm dia.	6.76	8.66	0.27	6.85	nr	**15.51**
35mm dia.	11.70	14.99	0.32	8.12	nr	**23.10**
42mm dia.	19.03	24.37	0.37	9.39	nr	**33.76**
54mm dia.	24.51	31.39	0.31	7.87	nr	**39.26**
Reducing set; internal						
22mm dia.	-	-	0.23	5.84	nr	**5.84**
Tee equal						
15mm dia.	1.97	2.52	0.28	7.10	nr	**9.63**
22mm dia.	3.27	4.19	0.30	7.61	nr	**11.80**
28mm dia.	8.65	11.08	0.34	8.63	nr	**19.71**
35mm dia.	15.33	19.63	0.43	10.91	nr	**30.54**
42mm dia.	24.10	30.87	0.46	11.67	nr	**42.54**
54mm dia.	38.71	49.58	0.54	13.70	nr	**63.28**
Tee reducing						
22mm dia.	5.07	6.49	0.30	7.61	nr	**14.11**
28mm dia.	8.35	10.69	0.34	8.63	nr	**19.32**
35mm dia.	14.98	19.19	0.43	10.91	nr	**30.10**
42mm dia.	23.15	29.65	0.46	11.67	nr	**41.32**
54mm dia.	38.71	49.58	0.54	13.70	nr	**63.28**
Extra over copper pipes; bronze one piece brazing flanges; metric, including jointing ring and bolts						
Bronze flange; PN6						
15mm dia.	14.08	18.03	0.27	6.85	nr	**24.88**
22mm dia.	16.63	21.30	0.32	8.12	nr	**29.42**
28mm dia.	19.07	24.42	0.36	9.13	nr	**33.56**
35mm dia.	27.23	34.88	0.47	11.92	nr	**46.80**
42mm dia.	32.94	42.19	0.54	13.70	nr	**55.89**
54mm dia.	46.41	59.44	0.63	15.98	nr	**75.43**
67mm dia.	53.49	68.51	0.77	19.54	nr	**88.05**
76mm dia.	62.01	79.42	0.93	23.60	nr	**103.02**
108mm dia.	83.06	106.38	1.14	28.92	nr	**135.31**
133mm dia.	100.25	128.40	1.41	35.77	nr	**164.17**
159mm dia.	142.83	182.94	1.74	44.15	nr	**227.08**
Bronze flange; PN10						
15mm dia.	18.17	23.27	0.27	6.85	nr	**30.12**
22mm dia.	21.10	27.02	0.32	8.12	nr	**35.14**
28mm dia.	21.33	27.32	0.38	9.64	nr	**36.96**
35mm dia.	29.55	37.85	0.47	11.92	nr	**49.77**
42mm dia.	34.94	44.75	0.54	13.70	nr	**58.45**
54mm dia.	49.22	63.04	0.63	15.98	nr	**79.02**
67mm dia.	53.49	68.51	0.77	19.54	nr	**88.05**
76mm dia.	68.56	87.81	0.93	23.60	nr	**111.41**
108mm dia.	100.77	129.07	1.14	28.92	nr	**157.99**
133mm dia.	115.53	147.97	1.41	35.77	nr	**183.74**
159mm dia.	176.92	226.60	1.74	44.15	nr	**270.74**

S:PIPED SUPPLY SYSTEMS

Item	Net Price £	Material £	Labour hours	Labour £	Unit	Total rate £
S10 : COLD WATER (cont'd)						
COPPER PIPEWORK (cont'd)						
Extra over copper pipes; bronze one piece brazing flanges (cont'd)						
Bronze flange; PN16						
15mm dia.	18.17	23.27	0.27	6.85	nr	**30.12**
22mm dia.	21.10	27.02	0.32	8.12	nr	**35.14**
28mm dia.	21.97	28.14	0.38	9.64	nr	**37.78**
35mm dia.	29.55	37.85	0.47	11.92	nr	**49.77**
42mm dia.	34.94	44.75	0.54	13.70	nr	**58.45**
54mm dia.	51.89	66.46	0.63	15.98	nr	**82.44**
67mm dia.	63.00	80.69	0.77	19.54	nr	**100.23**
76mm dia.	80.91	103.63	0.93	23.60	nr	**127.22**
108mm dia.	103.92	133.10	1.14	28.92	nr	**162.02**
133mm dia.	169.37	216.93	1.41	35.77	nr	**252.70**
159mm dia.	212.53	272.21	1.74	44.15	nr	**316.35**
Extra Over copper pipes; bronze blank flanges; metric, including jointing ring and bolts						
Bronze blank flange; PN6						
15mm dia.	11.84	15.16	0.27	6.85	nr	**22.02**
22mm dia.	15.11	19.35	0.27	6.85	nr	**26.20**
28mm dia.	15.53	19.89	0.27	6.85	nr	**26.74**
35mm dia.	25.41	32.54	0.32	8.12	nr	**40.66**
42mm dia.	34.28	43.91	0.32	8.12	nr	**52.02**
54mm dia.	37.21	47.66	0.34	8.63	nr	**56.28**
67mm dia.	46.51	59.57	0.36	9.13	nr	**68.70**
76mm dia.	59.91	76.73	0.37	9.39	nr	**86.12**
108mm dia.	95.19	121.92	0.41	10.40	nr	**132.32**
133mm dia.	112.30	143.83	0.58	14.72	nr	**158.55**
159mm dia.	141.49	181.22	0.61	15.48	nr	**196.70**
Bronze blank flange; PN10						
15mm dia.	14.32	18.34	0.27	6.85	nr	**25.19**
22mm dia.	18.53	23.73	0.27	6.85	nr	**30.58**
28mm dia.	20.52	26.28	0.27	6.85	nr	**33.13**
35mm dia.	25.41	32.54	0.32	8.12	nr	**40.66**
42mm dia.	47.63	61.00	0.32	8.12	nr	**69.12**
54mm dia.	54.30	69.55	0.34	8.63	nr	**78.17**
67mm dia.	58.25	74.61	0.46	11.67	nr	**86.28**
76mm dia.	74.84	95.85	0.47	11.92	nr	**107.78**
108mm dia.	114.30	146.39	0.51	12.94	nr	**159.33**
133mm dia.	120.68	154.57	0.58	14.72	nr	**169.28**
159mm dia.	220.70	282.67	0.71	18.01	nr	**300.68**

S:PIPED SUPPLY SYSTEMS

Item	Net Price £	Material £	Labour hours	Labour £	Unit	Total rate £
Bronze blank flange; PN16						
15mm dia.	14.32	18.34	0.27	6.85	nr	**25.19**
22mm dia.	18.81	24.09	0.27	6.85	nr	**30.94**
28mm dia.	20.52	26.28	0.27	6.85	nr	**33.13**
35mm dia.	25.41	32.54	0.32	8.12	nr	**40.66**
42mm dia.	47.63	61.00	0.32	8.12	nr	**69.12**
54mm dia.	54.30	69.55	0.34	8.63	nr	**78.17**
67mm dia.	84.45	108.16	0.46	11.67	nr	**119.83**
76mm dia.	100.01	128.09	0.47	11.92	nr	**140.02**
108mm dia.	127.14	162.84	0.51	12.94	nr	**175.78**
133mm dia.	208.65	267.24	0.58	14.72	nr	**281.95**
159mm dia.	267.07	342.06	0.71	18.01	nr	**360.07**
Extra Over copper pipes; bronze screwed flanges; metric, including jointing ring and bolts						
Bronze screwed flange; 6 BSP						
15mm dia.	12.59	16.13	0.35	8.88	nr	**25.01**
22mm dia.	14.50	18.57	0.47	11.92	nr	**30.50**
28mm dia.	15.19	19.46	0.52	13.19	nr	**32.65**
35mm dia.	20.91	26.78	0.62	15.73	nr	**42.51**
42mm dia.	24.77	31.73	0.70	17.76	nr	**49.49**
54mm dia.	33.27	42.61	0.84	21.31	nr	**63.92**
67mm dia.	41.51	53.17	1.03	26.13	nr	**79.30**
76mm dia.	51.09	65.44	1.22	30.95	nr	**96.39**
108mm dia.	79.79	102.19	1.41	35.77	nr	**137.97**
133mm dia.	94.12	120.55	1.75	44.40	nr	**164.95**
159mm dia.	121.28	155.33	2.21	56.07	nr	**211.41**
Bronze screwed flange; 10 BSP						
15mm dia.	15.09	19.33	0.35	8.88	nr	**28.21**
22mm dia.	17.51	22.43	0.47	11.92	nr	**34.35**
28mm dia.	19.36	24.80	0.52	13.19	nr	**37.99**
35mm dia.	27.96	35.81	0.62	15.73	nr	**51.54**
42mm dia.	33.77	43.25	0.70	17.76	nr	**61.01**
54mm dia.	47.53	60.88	0.84	21.31	nr	**82.19**
67mm dia.	55.55	71.15	1.03	26.13	nr	**97.28**
76mm dia.	63.86	81.79	1.22	30.95	nr	**112.74**
108mm dia.	84.57	108.32	1.41	35.77	nr	**144.09**
133mm dia.	101.91	130.53	1.75	44.40	nr	**174.93**
159mm dia.	179.86	230.36	2.21	56.07	nr	**286.43**
Bronze screwed flange; 16 BSP						
15mm dia.	15.09	19.33	0.35	8.88	nr	**28.21**
22mm dia.	17.51	22.43	0.47	11.92	nr	**34.35**
28mm dia.	19.36	24.80	0.52	13.19	nr	**37.99**
35mm dia.	27.96	35.81	0.62	15.73	nr	**51.54**
42mm dia.	33.77	43.25	0.70	17.76	nr	**61.01**
54mm dia.	47.53	60.88	0.84	21.31	nr	**82.19**
67mm dia.	65.02	83.28	1.03	26.13	nr	**109.41**
76mm dia.	82.95	106.24	1.22	30.95	nr	**137.20**
108mm dia.	105.92	135.66	1.41	35.77	nr	**171.44**
133mm dia.	172.24	220.60	1.75	44.40	nr	**265.00**
159mm dia.	215.15	275.56	2.21	56.07	nr	**331.63**

S:PIPED SUPPLY SYSTEMS

Item	Net Price £	Material £	Labour hours	Labour £	Unit	Total rate £
S10 : COLD WATER (cont'd)						
COPPER PIPEWORK (cont'd)						
Extra over copper pipes; labour						
Made bend						
15mm dia.	-	-	0.26	6.60	nr	**6.60**
22mm dia.	-	-	0.28	7.10	nr	**7.10**
28mm dia.	-	-	0.31	7.87	nr	**7.87**
35mm dia.	-	-	0.42	10.66	nr	**10.66**
42mm dia.	-	-	0.51	12.94	nr	**12.94**
54mm dia.	-	-	0.58	14.72	nr	**14.72**
67mm dia.	-	-	0.69	17.51	nr	**17.51**
76mm dia.	-	-	0.80	20.30	nr	**20.30**
Bronze butt weld						
15mm dia.	-	-	0.25	6.34	nr	**6.34**
22mm dia.	-	-	0.31	7.87	nr	**7.87**
28mm dia.	-	-	0.37	9.39	nr	**9.39**
35mm dia.	-	-	0.49	12.43	nr	**12.43**
42mm dia.	-	-	0.58	14.72	nr	**14.72**
54mm dia.	-	-	0.72	18.27	nr	**18.27**
67mm dia.	-	-	0.88	22.33	nr	**22.33**
76mm dia.	-	-	1.08	27.40	nr	**27.40**
108mm dia.	-	-	1.37	34.76	nr	**34.76**
133mm dia.	-	-	1.73	43.89	nr	**43.89**
159mm dia.	-	-	2.03	51.51	nr	**51.51**
PRESSFIT (copper fittings)						
Mechanical pressfit joints; butyl rubber O ring						
Coupler						
15mm dia	0.71	0.91	0.36	9.13	nr	**10.04**
22mm dia	1.11	1.42	0.36	9.13	nr	**10.56**
28mm dia	2.21	2.83	0.44	11.16	nr	**13.99**
35mm dia	2.79	3.57	0.44	11.16	nr	**14.74**
42mm dia	5.02	6.43	0.52	13.19	nr	**19.62**
54mm dia	6.41	8.21	0.60	15.22	nr	**23.43**
Stop end						
22mm dia	1.73	2.22	0.18	4.57	nr	**6.78**
28mm dia	2.73	3.50	0.22	5.58	nr	**9.08**
35mm dia	4.69	6.01	0.22	5.58	nr	**11.59**
42mm dia	7.06	9.04	0.26	6.60	nr	**15.64**
54mm dia	8.51	10.90	0.30	7.61	nr	**18.51**
Reducer						
22 x 15mm dia	0.82	1.05	0.36	9.13	nr	**10.18**
28 x 15mm dia	2.11	2.70	0.40	10.15	nr	**12.85**
28 x 22mm dia	2.19	2.80	0.40	10.15	nr	**12.95**
35 x 22mm dia	2.62	3.36	0.40	10.15	nr	**13.50**
35 x 28mm dia	2.90	3.71	0.44	11.16	nr	**14.88**
42 x 22mm dia	4.53	5.80	0.44	11.16	nr	**16.97**
42 x 28mm dia	4.31	5.52	0.48	12.18	nr	**17.70**
42 x 35mm dia	4.31	5.52	0.48	12.18	nr	**17.70**
54 x 35mm dia	5.84	7.48	0.52	13.19	nr	**20.67**
54 x 42mm dia	5.84	7.48	0.56	14.21	nr	**21.69**

S:PIPED SUPPLY SYSTEMS

Item	Net Price £	Material £	Labour hours	Labour £	Unit	Total rate £
90° elbow						
15mm dia	0.76	0.97	0.36	9.13	nr	**10.11**
22mm dia	1.28	1.64	0.36	9.13	nr	**10.77**
28mm dia	2.76	3.54	0.44	11.16	nr	**14.70**
35mm dia	5.52	7.07	0.44	11.16	nr	**18.23**
42mm dia	10.38	13.29	0.52	13.19	nr	**26.49**
54mm dia	14.39	18.43	0.60	15.22	nr	**33.65**
45° elbow						
15mm dia	0.91	1.17	0.36	9.13	nr	**10.30**
22mm dia	1.26	1.61	0.36	9.13	nr	**10.75**
28mm dia	3.79	4.85	0.44	11.16	nr	**16.02**
35mm dia	5.41	6.93	0.44	11.16	nr	**18.09**
42mm dia	9.00	11.53	0.52	13.19	nr	**24.72**
54mm dia	12.80	16.39	0.60	15.22	nr	**31.62**
Equal tee						
15mm dia	1.22	1.56	0.54	13.70	nr	**15.26**
22mm dia	2.21	2.83	0.54	13.70	nr	**16.53**
28mm dia	3.86	4.94	0.66	16.75	nr	**21.69**
35mm dia	6.68	8.56	0.66	16.75	nr	**25.30**
42mm dia	13.19	16.89	0.78	19.79	nr	**36.68**
54mm dia	16.45	21.07	0.90	22.83	nr	**43.90**
Reducing tee						
22 x 15mm dia	1.80	2.31	0.54	13.70	nr	**16.01**
28 x 15mm dia	7.39	9.47	0.62	15.73	nr	**25.20**
28 x 22mm dia	4.56	5.84	0.62	15.73	nr	**21.57**
35 x 22mm dia	5.94	7.61	0.62	15.73	nr	**23.34**
35 x 28mm dia	6.62	8.48	0.62	15.73	nr	**24.21**
42 x 28mm dia	11.97	15.33	0.70	17.76	nr	**33.09**
42 x 35mm dia	11.97	15.33	0.70	17.76	nr	**33.09**
54 x 35mm dia	20.22	25.90	0.82	20.80	nr	**46.70**
54 x 42mm dia	20.22	25.90	0.82	20.80	nr	**46.70**
Male iron connector; BSP thread						
15mm dia x 1/2" dia.	2.58	3.30	0.18	4.57	nr	**7.87**
22mm x 3/4" dia.	3.70	4.74	0.18	4.57	nr	**9.31**
28mm x 1" dia.	4.95	6.34	0.22	5.58	nr	**11.92**
35mm dia x 1 1/4" dia.	8.96	11.48	0.22	5.58	nr	**17.06**
42mm dia x 1 1/2" dia.	12.01	15.38	0.26	6.60	nr	**21.98**
54mm dia x 2" dia.	23.17	29.68	0.30	7.61	nr	**37.29**
90° elbow; male iron BSP thread						
15mm dia x 1/2" dia.	4.04	5.17	0.36	9.13	nr	**14.31**
22mm x 3/4" dia.	6.32	8.09	0.36	9.13	nr	**17.23**
28mm x 1" dia.	9.69	12.41	0.44	11.16	nr	**23.57**
35mm dia x 1 1/4" dia.	12.60	16.14	0.44	11.16	nr	**27.30**
42mm dia x 1 1/2" dia.	16.43	21.04	0.52	13.19	nr	**34.24**
54mm dia x 2" dia.	24.01	30.75	0.60	15.22	nr	**45.97**
Female iron connector; BSP thread						
15mm dia x 1/2" dia.	2.86	3.66	0.18	4.57	nr	**8.23**
22mm x 3/4" dia.	3.76	4.82	0.18	4.57	nr	**9.38**
28mm x 1" dia.	5.08	6.51	0.22	5.58	nr	**12.09**
35mm dia x 1 1/4" dia.	9.92	12.71	0.22	5.58	nr	**18.29**
42mm dia x 1 1/2" dia.	14.17	18.15	0.26	6.60	nr	**24.75**
54mm dia x 2" dia.	24.30	31.12	0.30	7.61	nr	**38.73**

S:PIPED SUPPLY SYSTEMS

Item	Net Price £	Material £	Labour hours	Labour £	Unit	Total rate £
S10 : COLD WATER (cont'd)						
PRESSFIT (copper fittings) (cont'd)						
Mechanical pressfit joints; butyl rubber O ring (cont'd)						
90° elbow; female iron BSP thread						
15mm dia x 1/2" dia.	3.42	4.38	0.36	9.13	nr	**13.51**
22mm x 3/4" dia.	5.02	6.43	0.36	9.13	nr	**15.56**
28mm dia x 1" dia.	8.30	10.63	0.44	11.16	nr	**21.79**
35mm dia x 1 1/4" dia.	10.71	13.72	0.44	11.16	nr	**24.88**
42mm dia x 1 1/2" dia.	14.58	18.67	0.52	13.19	nr	**31.87**
54mm dia x 2" dia.	21.45	27.47	0.60	15.22	nr	**42.70**
STAINLESS STEEL PIPEWORK						
Stainless steel pipes; capillary or compression joints; BS 4127, vertical or at low level, with brackets measured separately						
Grade 304; satin finish						
15mm dia.	2.06	2.64	0.41	10.40	m	**13.04**
22mm dia.	2.89	3.70	0.51	12.94	m	**16.65**
28mm dia.	3.94	5.05	0.58	14.72	m	**19.76**
35mm dia.	5.96	7.63	0.65	16.50	m	**24.13**
42mm dia.	7.57	9.70	0.71	18.02	m	**27.71**
54mm dia.	10.53	13.49	0.80	20.30	m	**33.78**
Grade 316 satin finish						
15mm dia.	2.67	3.42	0.61	15.48	m	**18.90**
22mm dia.	3.85	4.93	0.76	19.30	m	**24.23**
28mm dia.	5.88	7.53	0.87	22.08	m	**29.61**
35mm dia.	6.94	8.89	0.98	24.87	m	**33.76**
42mm dia.	8.90	11.40	1.06	26.90	m	**38.30**
54mm dia.	12.10	15.50	1.16	29.43	m	**44.93**
FIXINGS						
For Stainless Steel pipework						
Single pipe ring						
15mm dia.	6.86	8.79	0.26	6.60	nr	**15.38**
22mm dia.	7.99	10.23	0.26	6.60	nr	**16.83**
28mm dia.	8.37	10.72	0.31	7.87	nr	**18.59**
35mm dia.	9.50	12.17	0.32	8.12	nr	**20.29**
42mm dia.	10.80	13.83	0.32	8.12	nr	**21.95**
54mm dia.	12.33	15.79	0.34	8.63	nr	**24.42**
Screw on backplate, female						
All sizes 15mm to 54mm dia.	6.18	7.92	0.10	2.54	nr	**10.45**
Screw on backplate, male						
All sizes 15mm to 54mm dia.	7.04	9.02	0.10	2.54	nr	**11.55**
Stainless steel threaded rods; metric thread; including nuts, washers etc						
10mm dia x 600mm long	7.36	9.42	0.10	2.54	nr	**11.96**

S:PIPED SUPPLY SYSTEMS

Item	Net Price £	Material £	Labour hours	Labour £	Unit	Total rate £
Extra over stainless steel pipes; capillary fittings						
Straight coupling						
15mm dia.	0.83	1.06	0.25	6.34	nr	**7.41**
22mm dia.	1.34	1.72	0.28	7.10	nr	**8.82**
28mm dia.	1.77	2.27	0.33	8.37	nr	**10.64**
35mm dia.	4.09	5.24	0.37	9.39	nr	**14.63**
42mm dia.	4.71	6.03	0.42	10.66	nr	**16.69**
54mm dia.	7.09	9.08	0.45	11.42	nr	**20.50**
45° bend						
15mm dia.	4.46	5.71	0.25	6.34	nr	**12.06**
22mm dia.	5.86	7.51	0.30	7.53	nr	**15.03**
28mm dia.	7.19	9.21	0.33	8.37	nr	**17.58**
35mm dia.	-	-	0.37	9.39	nr	**9.39**
42mm dia.	-	-	0.42	10.66	nr	**10.66**
54mm dia.	-	-	0.45	11.42	nr	**11.42**
90° bend						
15mm dia.	2.30	2.95	0.28	7.10	nr	**10.05**
22mm dia.	3.11	3.98	0.28	7.10	nr	**11.09**
28mm dia.	4.39	5.62	0.33	8.37	nr	**14.00**
35mm dia.	10.70	13.70	0.37	9.39	nr	**23.09**
42mm dia.	14.73	18.87	0.42	10.66	nr	**29.52**
54mm dia.	19.97	25.58	0.45	11.42	nr	**36.99**
Reducer						
22 x 15mm dia.	5.31	6.80	0.28	7.10	nr	**13.91**
28 x 22mm dia.	5.93	7.60	0.33	8.37	nr	**15.97**
35 x 28mm dia.	7.25	9.29	0.37	9.39	nr	**18.68**
42 x 35mm dia.	7.82	10.02	0.42	10.66	nr	**20.67**
54 x 42mm dia.	23.18	29.69	0.48	12.20	nr	**41.89**
Tap connector						
15mm dia.	11.22	14.37	0.13	3.30	nr	**17.67**
22mm dia.	14.83	18.99	0.14	3.55	nr	**22.55**
28mm dia.	20.57	26.35	0.17	4.31	nr	**30.66**
Tank connector						
15mm dia.	14.49	18.56	0.13	3.30	nr	**21.86**
22mm dia.	21.56	27.61	0.13	3.30	nr	**30.91**
28mm dia	26.54	33.99	0.15	3.81	nr	**37.80**
35mm dia.	38.39	49.17	0.18	4.57	nr	**53.74**
42mm dia.	50.69	64.92	0.21	5.33	nr	**70.25**
54mm dia.	76.73	98.28	0.24	6.09	nr	**104.36**
Tee equal						
15mm dia.	4.13	5.29	0.37	9.39	nr	**14.68**
22mm dia.	5.14	6.58	0.40	10.15	nr	**16.73**
28mm dia.	6.21	7.95	0.45	11.42	nr	**19.37**
35mm dia.	14.93	19.12	0.59	14.98	nr	**34.10**
42mm dia.	18.41	23.58	0.62	15.74	nr	**39.32**
54mm dia.	37.19	47.63	0.67	17.00	nr	**64.64**

S:PIPED SUPPLY SYSTEMS

Item	Net Price £	Material £	Labour hours	Labour £	Unit	Total rate £
S10 : COLD WATER (cont'd)						
FIXINGS (cont'd)						
Extra over stainless steel pipes; capillary Fittings (cont'd)						
Unequal tee						
22 x 15mm dia.	8.38	10.73	0.37	9.39	nr	**20.12**
28 x 15mm dia.	9.44	12.09	0.45	11.42	nr	**23.51**
28 x 22mm dia.	9.44	12.09	0.45	11.42	nr	**23.51**
35 x 22mm dia.	16.48	21.11	0.59	14.98	nr	**36.08**
35 x 28mm dia.	16.48	21.11	0.59	14.98	nr	**36.08**
42 x 28mm dia.	20.26	25.95	0.62	15.74	nr	**41.69**
42 x 35mm dia.	20.26	25.95	0.62	15.74	nr	**41.69**
54 x 35mm dia.	41.95	53.73	0.67	17.00	nr	**70.73**
54 x 42mm dia.	41.95	53.73	0.67	17.00	nr	**70.73**
Union, conical seat						
15mm dia.	18.57	23.78	0.25	6.34	nr	**30.13**
22mm dia.	29.25	37.46	0.28	7.10	nr	**44.57**
28mm dia.	37.82	48.44	0.33	8.37	nr	**56.81**
35mm dia.	49.64	63.58	0.37	9.39	nr	**72.97**
42mm dia.	62.61	80.19	0.42	10.66	nr	**90.85**
54mm dia.	82.83	106.09	0.45	11.42	nr	**117.51**
Union, flat seat						
15mm dia.	19.40	24.85	0.25	6.34	nr	**31.19**
22mm dia.	30.20	38.68	0.28	7.10	nr	**45.78**
28mm dia.	39.05	50.01	0.33	8.37	nr	**58.39**
35mm dia.	51.02	65.35	0.37	9.39	nr	**74.74**
42mm dia.	64.26	82.30	0.42	10.66	nr	**92.96**
54mm dia.	86.18	110.38	0.45	11.42	nr	**121.80**
Extra over stainless steel pipes; compression fittings						
Straight coupling						
15mm dia.	11.73	15.02	0.18	4.57	nr	**19.59**
22mm dia.	22.35	28.63	0.22	5.58	nr	**34.21**
28mm dia.	30.09	38.54	0.25	6.34	nr	**44.88**
35mm dia.	46.45	59.49	0.30	7.61	nr	**67.10**
42mm dia.	54.22	69.44	0.40	10.15	nr	**79.59**
90° bend						
15mm dia.	14.79	18.94	0.18	4.57	nr	**23.51**
22mm dia.	29.38	37.63	0.22	5.58	nr	**43.21**
28mm dia.	40.07	51.32	0.25	6.34	nr	**57.66**
35mm dia.	81.13	103.91	0.33	8.37	nr	**112.28**
42mm dia.	118.56	151.85	0.35	8.88	nr	**160.73**
Reducer						
22 x 15mm dia.	21.29	27.27	0.28	7.10	nr	**34.37**
28 x 22mm dia.	29.16	37.35	0.28	7.10	nr	**44.45**
35 x 28mm dia.	42.60	54.56	0.30	7.61	nr	**62.17**
42 x 35mm dia.	56.68	72.60	0.37	9.39	nr	**81.99**

S:PIPED SUPPLY SYSTEMS

Item	Net Price £	Material £	Labour hours	Labour £	Unit	Total rate £
Stud coupling						
15mm dia.	12.20	15.63	0.42	10.66	nr	**26.28**
22mm dia.	20.63	26.42	0.25	6.34	nr	**32.77**
28mm dia.	28.63	36.67	0.25	6.34	nr	**43.01**
35mm dia.	45.87	58.75	0.37	9.39	nr	**68.14**
42mm dia.	54.22	69.44	0.42	10.66	nr	**80.10**
Equal tee						
15mm dia.	20.83	26.68	0.37	9.39	nr	**36.07**
22mm dia.	43.02	55.10	0.40	10.15	nr	**65.25**
28mm dia.	58.84	75.36	0.45	11.42	nr	**86.78**
35mm dia.	117.01	149.87	0.59	14.98	nr	**164.84**
42mm dia.	162.25	207.81	0.62	15.74	nr	**223.55**
Running tee						
15mm dia.	25.64	32.84	0.37	9.39	nr	**42.23**
22mm dia.	46.19	59.16	0.40	10.15	nr	**69.31**
28mm dia.	78.20	100.16	0.59	14.98	nr	**115.13**
PRESS FIT (stainless steel)						
Press fit jointing system; butyl rubber O ring mechanical joint						
Pipework						
15mm dia	2.56	3.28	0.46	11.67	m	**14.95**
22mm dia	4.00	5.12	0.48	12.18	m	**17.30**
28mm dia	4.90	6.28	0.52	13.19	m	**19.47**
35mm dia	7.10	9.09	0.56	14.21	m	**23.30**
42mm dia	8.81	11.28	0.58	14.72	m	**26.00**
54mm dia	11.13	14.26	0.66	16.75	m	**31.00**
FIXINGS						
For stainless steel pipes						
Refer to fixings for stainless steel pipes; capillary or compression joints; BS 4127						
Extra over stainless steel pipes; Press fit jointing system						
Coupling						
15mm dia	1.83	2.34	0.36	9.13	nr	**11.48**
22mm dia	2.61	3.34	0.36	9.13	nr	**12.48**
28mm dia	3.24	4.15	0.44	11.16	nr	**15.31**
35mm dia	4.42	5.66	0.44	11.16	nr	**16.82**
42mm dia	5.31	6.80	0.52	13.19	nr	**19.99**
54mm dia	6.13	7.85	0.60	15.22	nr	**23.07**
Stop end						
22mm dia	2.03	2.60	0.18	4.57	nr	**7.17**
28mm dia	2.87	3.68	0.22	5.58	nr	**9.26**
35mm dia	3.32	4.25	0.22	5.58	nr	**9.83**
42mm dia	4.66	5.97	0.26	6.60	nr	**12.57**
54mm dia	5.40	6.92	0.30	7.61	nr	**14.53**

S:PIPED SUPPLY SYSTEMS

Item	Net Price £	Material £	Labour hours	Labour £	Unit	Total rate £
S10 : COLD WATER (cont'd)						
FIXINGS (cont'd)						
Extra over stainless steel pipes; Press fit jointing system (cont'd)						
Reducer						
22 x 15mm dia	2.19	2.80	0.36	9.13	nr	**11.94**
28 x 15mm dia	2.48	3.18	0.40	10.15	nr	**13.33**
28 x 22mm dia	2.56	3.28	0.40	10.15	nr	**13.43**
35 x 22mm dia	3.13	4.01	0.40	10.15	nr	**14.16**
35 x 28mm dia	3.88	4.97	0.44	11.16	nr	**16.13**
42 x 35mm dia	4.09	5.24	0.48	12.18	nr	**17.42**
54 x 42mm dia	4.68	5.99	0.56	14.21	nr	**20.20**
90° bend						
15mm dia	2.63	3.37	0.36	9.13	nr	**12.50**
22mm dia	3.67	4.70	0.36	9.13	nr	**13.83**
28mm dia	4.64	5.94	0.44	11.16	nr	**17.11**
35mm dia	7.29	9.34	0.44	11.16	nr	**20.50**
42mm dia	12.18	15.60	0.52	13.19	nr	**28.79**
54mm dia	16.81	21.53	0.60	15.22	nr	**36.75**
45° bend						
15mm dia	3.57	4.57	0.36	9.13	nr	**13.71**
22mm dia	4.43	5.67	0.36	9.13	nr	**14.81**
28mm dia	5.16	6.61	0.44	11.16	nr	**17.77**
35mm dia	6.06	7.76	0.44	11.16	nr	**18.93**
42mm dia	9.74	12.47	0.52	13.19	nr	**25.67**
54mm dia	12.66	16.21	0.60	15.22	nr	**31.44**
Equal tee						
15mm dia	4.31	5.52	0.54	13.70	nr	**19.22**
22mm dia	5.28	6.76	0.54	13.70	nr	**20.46**
28mm dia	6.18	7.92	0.66	16.75	nr	**24.66**
35mm dia	7.83	10.03	0.66	16.75	nr	**26.77**
42mm dia	11.11	14.23	0.78	19.79	nr	**34.02**
54mm dia	13.29	17.02	0.90	22.83	nr	**39.86**
Reducing tee						
22 x 15mm dia	4.52	5.79	0.54	13.70	nr	**19.49**
28 x 15mm dia	5.48	7.02	0.62	15.73	nr	**22.75**
28 x 22mm dia	5.93	7.60	0.62	15.73	nr	**23.33**
35 x 22mm dia	7.03	9.00	0.62	15.73	nr	**24.73**
35 x 28mm dia	7.34	9.40	0.62	15.73	nr	**25.13**
42 x 28mm dia	10.43	13.36	0.70	17.76	nr	**31.12**
42 x 35mm dia	10.75	13.77	0.70	17.76	nr	**31.53**
54 x 35mm dia	12.14	15.55	0.82	20.80	nr	**36.35**
54 x 42mm dia	12.48	15.98	0.82	20.80	nr	**36.79**
FIXINGS						
For stainless steel pipes						
Refer to fixings for stainless steel pipes; capillary or compression joints; BS 4127						

S:PIPED SUPPLY SYSTEMS

Item	Net Price £	Material £	Labour hours	Labour £	Unit	Total rate £
MEDIUM DENSITY POLYETHYLENE - BLUE						
Note : MDPE is sized on Outside Diameter i.e. OD not ID						
Pipes for water distribution; laid underground; electrofusion joints in the running length; BS 6572						
Coiled service pipe						
20mm dia.	0.50	0.66	0.37	9.39	m	10.05
25mm dia.	0.58	0.77	0.41	10.40	m	11.17
32mm dia.	0.97	1.28	0.47	11.93	m	13.21
50mm dia.	2.32	3.06	0.53	13.45	m	16.51
63mm dia.	3.68	4.85	0.60	15.23	m	20.08
Mains service pipe						
90mm dia.	7.41	9.78	0.90	22.84	m	32.61
110mm dia	11.04	14.14	1.10	27.91	m	42.05
125mm dia.	14.07	18.56	1.20	30.46	m	49.02
160mm dia	24.30	31.12	1.48	37.55	m	68.67
180mm dia.	30.54	40.29	1.50	38.10	m	78.38
225mm dia	34.53	44.23	1.77	44.91	m	89.13
250mm dia.	42.49	56.05	1.75	44.43	m	100.49
315mm dia	67.45	86.39	1.90	48.21	m	134.60
Extra over fittings; MDPE blue; electrofusion joints						
Coupler						
20mm dia	3.79	4.85	0.36	9.13	nr	13.99
25mm dia	3.79	4.85	0.40	10.15	nr	15.00
32mm dia	3.79	4.85	0.44	11.16	nr	16.02
40mm dia	3.79	4.85	0.48	12.18	nr	17.03
50mm dia	6.89	8.82	0.52	13.19	nr	22.02
63mm dia.	7.12	9.12	0.58	14.72	nr	23.83
90mm dia.	10.51	13.46	0.67	17.00	nr	30.46
110mm dia	16.96	21.72	0.74	18.78	nr	40.50
125mm dia.	19.07	24.42	0.83	21.06	nr	45.48
160mm dia	30.38	38.91	1.00	25.37	nr	64.28
180mm dia.	35.42	45.37	1.25	31.71	nr	77.08
225mm dia	56.87	72.84	1.35	34.25	nr	107.09
250mm dia.	83.02	106.33	1.50	38.06	nr	144.39
315mm dia	137.77	176.45	1.80	45.67	nr	222.12

S:PIPED SUPPLY SYSTEMS

Item	Net Price £	Material £	Labour hours	Labour £	Unit	Total rate £
S10 : COLD WATER (cont'd)						
MEDIUM DENSITY POLYETHYLENE - BLUE (cont'd)						
Extra over fittings; MDPE blue; butt fused joints						
Cap						
25mm dia	7.52	9.63	0.20	5.07	nr	**14.71**
32mm dia	7.52	9.63	0.22	5.58	nr	**15.21**
40mm dia	12.68	16.24	0.24	6.09	nr	**22.33**
50mm dia	13.64	17.47	0.26	6.60	nr	**24.07**
63mm dia.	18.01	23.07	0.32	8.12	nr	**31.19**
90mm dia.	32.86	42.09	0.37	9.39	nr	**51.47**
110mm dia	51.49	65.95	0.40	10.15	nr	**76.10**
125mm dia.	62.88	80.54	0.46	11.67	nr	**92.21**
160mm dia	83.67	107.16	0.50	12.69	nr	**119.85**
180mm dia.	106.25	136.08	0.60	15.22	nr	**151.31**
225mm dia	132.76	170.04	0.68	17.25	nr	**187.29**
250mm dia	132.76	170.04	0.75	19.03	nr	**189.07**
315mm dia	295.35	378.28	0.90	22.83	nr	**401.12**
Reducer						
63 x 32mm dia	9.84	12.60	0.54	13.70	nr	**26.30**
63 x 50mm dia	11.56	14.81	0.60	15.22	nr	**30.03**
90 x 63mm dia.	11.56	14.81	0.67	17.00	nr	**31.81**
110 x 90mm dia	14.67	18.79	0.74	18.78	nr	**37.56**
125 x 90mm dia.	19.05	24.40	0.83	21.06	nr	**45.46**
125 x 110mm dia	29.41	37.67	1.00	25.37	nr	**63.04**
160 x 110mm dia	49.71	63.67	1.10	27.91	nr	**91.58**
180 x 125mm dia.	53.96	69.11	1.25	31.71	nr	**100.83**
225 x 160mm dia	76.29	97.71	1.40	35.52	nr	**133.23**
250 x 180mm dia	85.01	108.88	1.80	45.67	nr	**154.55**
315 x 250mm dia	139.26	178.36	2.40	60.89	nr	**239.26**
Bend; 45°						
50mm dia	18.37	23.53	0.50	12.69	nr	**36.21**
63mm dia.	18.37	23.53	0.58	14.72	nr	**38.24**
90mm dia.	28.34	36.30	0.67	17.00	nr	**53.30**
110mm dia	41.19	52.76	0.74	18.78	nr	**71.53**
125mm dia.	46.36	59.38	0.83	21.06	nr	**80.44**
160mm dia	86.10	110.28	1.00	25.37	nr	**135.65**
180mm dia.	105.19	134.73	1.25	31.71	nr	**166.44**
225mm dia	130.91	167.67	1.40	35.52	nr	**203.19**
250mm dia	161.43	206.76	1.80	45.67	nr	**252.43**
315mm dia	290.73	372.36	2.40	60.89	nr	**433.26**
Bend; 90°						
50mm dia	18.37	23.53	0.50	12.69	nr	**36.21**
63mm dia.	18.37	23.53	0.58	14.72	nr	**38.24**
90mm dia.	28.34	36.30	0.67	17.00	nr	**53.30**
110mm dia	35.29	45.20	0.74	18.78	nr	**63.97**
125mm dia.	46.36	59.38	0.83	21.06	nr	**80.44**
160mm dia	86.10	110.28	1.00	25.37	nr	**135.65**
180mm dia.	105.19	134.73	1.25	31.71	nr	**166.44**
225mm dia	166.66	213.46	1.40	35.52	nr	**248.98**
250mm dia	194.81	249.51	1.80	45.67	nr	**295.18**
315mm dia	395.32	506.32	2.40	60.89	nr	**567.21**

S:PIPED SUPPLY SYSTEMS

Item	Net Price £	Material £	Labour hours	Labour £	Unit	Total rate £
Equal tee						
50mm dia	16.69	21.38	0.70	17.76	nr	39.14
63mm dia.	18.01	23.07	0.75	19.03	nr	42.10
90mm dia.	32.86	42.09	0.87	22.07	nr	64.16
110mm dia	49.01	62.77	1.00	25.37	nr	88.14
125mm dia.	62.88	80.54	1.08	27.40	nr	107.94
160mm dia.	103.60	132.69	1.35	34.25	nr	166.94
180mm dia.	106.25	136.08	1.63	41.36	nr	177.44
225mm dia	443.33	567.81	1.90	48.21	nr	616.02
250mm dia	538.91	690.23	2.70	68.50	nr	758.73
315mm dia	811.17	1038.94	3.60	91.34	nr	1130.28
Extra over plastic fittings, compression joints						
Straight connector						
20mm dia.	2.02	2.59	0.38	9.64	nr	12.23
25mm dia.	2.14	2.74	0.45	11.42	nr	14.16
32mm dia.	5.06	6.48	0.50	12.69	nr	19.17
50mm dia.	11.14	14.27	0.68	17.26	nr	31.53
63mm dia.	11.65	14.92	0.85	21.57	nr	36.49
Reducing connector						
25mm dia.	4.18	5.35	0.38	9.64	nr	15.00
32mm dia.	6.74	8.63	0.45	11.42	nr	20.05
50mm dia.	11.65	14.92	0.50	12.69	nr	27.61
63mm dia.	26.06	33.38	0.62	15.74	nr	49.12
Straight connector; polyethylene to MI						
20mm dia.	1.83	2.34	0.31	7.87	nr	10.21
25mm dia.	2.33	2.98	0.35	8.88	nr	11.86
32mm dia.	3.40	4.35	0.40	10.15	nr	14.50
50mm dia.	8.67	11.10	0.55	13.96	nr	25.06
63mm dia.	12.21	15.64	0.65	16.50	nr	32.14
Straight connector; polyethylene to FI						
20mm dia.	2.48	3.18	0.31	7.87	nr	11.04
25mm dia.	2.68	3.43	0.35	8.88	nr	12.31
32mm dia.	3.21	4.11	0.40	10.15	nr	14.26
50mm dia.	10.17	13.03	0.55	13.96	nr	26.98
63mm dia.	14.26	18.26	0.75	19.03	nr	37.30
Elbow						
20mm dia.	2.71	3.47	0.38	9.64	nr	13.11
25mm dia.	4.00	5.12	0.45	11.42	nr	16.54
32mm dia.	5.82	7.45	0.50	12.69	nr	20.14
50mm dia.	12.47	15.97	0.68	17.26	nr	33.23
63mm dia.	18.40	23.57	0.80	20.30	nr	43.86
Elbow; polyethylene to MI						
25mm dia.	3.44	4.41	0.35	8.88	nr	13.29
Elbow; polyethylene to FI						
20mm dia.	2.46	3.15	0.31	7.87	nr	11.02
25mm dia.	3.35	4.29	0.35	8.88	nr	13.17
32mm dia.	5.01	6.42	0.42	10.66	nr	17.07
50mm dia.	11.91	15.25	0.50	12.69	nr	27.94
63mm dia.	15.62	20.01	0.55	13.96	nr	33.96

S:PIPED SUPPLY SYSTEMS

Item	Net Price £	Material £	Labour hours	Labour £	Unit	Total rate £
S10 : COLD WATER (cont'd)						
MEDIUM DENSITY POLYETHYLENE - BLUE (cont'd)						
Extra over plastic fittings, compression joints (cont'd)						
Tank coupling						
25mm dia.	5.17	6.62	0.42	10.66	nr	**17.28**
Equal tee						
20mm dia.	5.24	6.71	0.53	13.45	nr	**20.16**
25mm dia.	7.86	10.07	0.55	13.96	nr	**24.02**
32mm dia.	9.72	12.45	0.64	16.24	nr	**28.69**
50mm dia.	21.61	27.68	0.75	19.03	nr	**46.71**
63mm dia.	33.15	42.46	0.87	22.08	nr	**64.54**
Equal tee; FI branch						
20mm dia.	3.53	4.52	0.45	11.42	nr	**15.94**
25mm dia.	5.62	7.20	0.50	12.69	nr	**19.88**
32mm dia.	6.84	8.76	0.60	15.23	nr	**23.99**
50mm dia.	15.76	20.19	0.68	17.26	nr	**37.45**
63mm dia.	22.13	28.34	0.81	20.56	nr	**48.91**
Equal tee; MI branch						
25mm dia.	5.52	7.07	0.50	12.69	nr	**19.76**
ABS PIPEWORK						
Pipes; solvent welded joints in the running length, brackets measured separately						
Class C (9 bar pressure)						
1" dia.	2.13	2.73	0.30	7.61	m	**10.34**
1 1/4" dia.	3.51	4.50	0.33	8.37	m	**12.87**
1 1/2" dia.	4.44	5.69	0.36	9.13	m	**14.82**
2" dia.	6.05	7.75	0.39	9.90	m	**17.64**
3" dia.	13.68	17.52	0.46	11.67	m	**29.19**
4" dia.	22.15	28.37	0.53	13.45	m	**41.82**
6" dia.	45.08	57.74	0.76	19.28	m	**77.02**
8" dia.	79.14	101.36	0.97	24.61	m	**125.97**
Class E (15 bar pressure)						
1/2" dia.	1.65	2.11	0.24	6.09	m	**8.20**
3/4" dia.	2.49	3.19	0.27	6.85	m	**10.04**
1" dia.	3.27	4.19	0.30	7.61	m	**11.80**
1 1/4" dia.	4.91	6.29	0.33	8.37	m	**14.66**
1 1/2" dia.	6.45	8.26	0.36	9.13	m	**17.40**
2" dia.	8.12	10.40	0.39	9.90	m	**20.30**
3" dia.	17.20	22.03	0.49	12.43	m	**34.46**
4" dia.	27.32	34.99	0.57	14.46	m	**49.45**

S:PIPED SUPPLY SYSTEMS

Item	Net Price £	Material £	Labour hours	Labour £	Unit	Total rate £
FIXINGS						
Refer to steel pipes; galvanised iron. For minimum fixing dimensions, refer to the Tables and Memoranda at the rear of the book						
Extra over fittings; solvent welded joints						
Cap						
1/2" dia.	0.66	0.85	0.16	4.06	nr	**4.90**
3/4" dia.	0.76	0.97	0.19	4.82	nr	**5.79**
1" dia.	0.87	1.11	0.22	5.58	nr	**6.70**
1 1/4" dia.	1.46	1.87	0.25	6.34	nr	**8.21**
1 1/2" dia.	2.24	2.87	0.28	7.10	nr	**9.97**
2" dia.	2.85	3.65	0.31	7.87	nr	**11.52**
3" dia.	8.56	10.96	0.36	9.13	nr	**20.10**
4" dia.	13.08	16.75	0.44	11.16	nr	**27.92**
Elbow 90°						
1/2" dia.	0.92	1.18	0.29	7.36	nr	**8.54**
3/4" dia.	1.10	1.41	0.34	8.63	nr	**10.04**
1" dia.	1.54	1.97	0.40	10.15	nr	**12.12**
1 1/4" dia.	2.60	3.33	0.45	11.42	nr	**14.75**
1 1/2" dia.	3.38	4.33	0.51	12.94	nr	**17.27**
2" dia.	5.14	6.58	0.56	14.21	nr	**20.79**
3" dia.	14.75	18.89	0.65	16.49	nr	**35.38**
4" dia.	22.04	28.23	0.80	20.30	nr	**48.53**
6" dia.	88.70	113.61	1.21	30.70	nr	**144.31**
8" dia.	135.38	173.39	1.45	36.79	nr	**210.18**
Elbow 45°						
1/2" dia.	1.77	2.27	0.29	7.36	nr	**9.62**
3/4" dia.	1.80	2.31	0.34	8.63	nr	**10.93**
1" dia.	2.24	2.87	0.40	10.15	nr	**13.02**
1 1/4" dia.	3.29	4.21	0.45	11.42	nr	**15.63**
1 1/2" dia.	4.08	5.23	0.51	12.94	nr	**18.17**
2" dia.	5.67	7.26	0.56	14.21	nr	**21.47**
3" dia.	13.34	17.09	0.65	16.49	nr	**33.58**
4" dia.	27.66	35.43	0.80	20.30	nr	**55.72**
6" dia.	57.34	73.44	1.21	30.70	nr	**104.14**
8" dia.	123.39	158.04	1.45	36.79	nr	**194.83**
Reducing bush						
3/4" x 1/2" dia.	0.68	0.87	0.42	10.66	nr	**11.53**
1" x 1/2" dia.	0.88	1.13	0.45	11.42	nr	**12.54**
1" x 3/4" dia.	0.88	1.13	0.45	11.42	nr	**12.54**
1 1/4" x 1" dia.	1.18	1.51	0.48	12.18	nr	**13.69**
1 1/2" x 3/4" dia.	1.54	1.97	0.51	12.94	nr	**14.91**
1 1/2" x 1" dia.	1.54	1.97	0.51	12.94	nr	**14.91**
1 1/2" x 1 1/4" dia.	1.54	1.97	0.51	12.94	nr	**14.91**
2" x 1" dia.	2.01	2.57	0.56	14.21	nr	**16.78**
2" x 1 1/4" dia.	2.01	2.57	0.56	14.21	nr	**16.78**
2" x 1 1/2" dia.	2.01	2.57	0.56	14.21	nr	**16.78**
3" x 1 1/2" dia.	5.67	7.26	0.65	16.49	nr	**23.75**
3" x 2" dia.	5.67	7.26	0.65	16.49	nr	**23.75**
4" x 3" dia.	7.81	10.00	0.80	20.30	nr	**30.30**
6" x 4" dia.	24.06	30.82	1.21	30.70	nr	**61.52**

S:PIPED SUPPLY SYSTEMS

Item	Net Price £	Material £	Labour hours	Labour £	Unit	Total rate £
S10 : COLD WATER (cont'd)						
ABS PIPEWORK (cont'd)						
Extra over fittings; solvent welded joints (cont'd)						
Union						
1/2" dia.	3.65	4.67	0.34	8.63	nr	**13.30**
3/4" dia.	3.94	5.05	0.39	9.90	nr	**14.94**
1" dia.	5.31	6.80	0.43	10.91	nr	**17.71**
1 1/4" dia.	6.50	8.33	0.50	12.69	nr	**21.01**
1 1/2" dia.	8.96	11.48	0.57	14.46	nr	**25.94**
2" dia.	11.68	14.96	0.62	15.73	nr	**30.69**
Sockets						
1/2" dia.	0.68	0.87	0.34	8.63	nr	**9.50**
3/4" dia.	0.76	0.97	0.39	9.90	nr	**10.87**
1" dia.	0.88	1.13	0.43	10.91	nr	**12.04**
1 1/4" dia.	1.54	1.97	0.50	12.69	nr	**14.66**
1 1/2" dia.	1.85	2.37	0.57	14.46	nr	**16.83**
2" dia.	2.60	3.33	0.62	15.73	nr	**19.06**
3" dia.	10.45	13.38	0.70	17.76	nr	**31.14**
4" dia.	14.83	18.99	0.70	17.76	nr	**36.75**
6" dia.	37.06	47.47	1.26	31.97	nr	**79.43**
8" dia.	74.01	94.79	1.55	39.33	nr	**134.12**
Barrel nipple						
1/2" dia.	1.28	1.64	0.34	8.63	nr	**10.27**
3/4" dia.	1.66	2.13	0.39	9.90	nr	**12.02**
1" dia.	2.15	2.75	0.43	10.91	nr	**13.66**
1 1/4" dia.	2.98	3.82	0.50	12.69	nr	**16.50**
1 1/2" dia.	3.51	4.50	0.57	14.46	nr	**18.96**
2" dia.	4.26	5.46	0.62	15.73	nr	**21.19**
3" dia.	11.33	14.51	0.70	17.76	nr	**32.27**
Tee, 90°						
1/2" dia.	1.04	1.33	0.41	10.40	nr	**11.73**
3/4" dia.	1.46	1.87	0.47	11.92	nr	**13.79**
1" dia.	2.01	2.57	0.55	13.95	nr	**16.53**
1 1/4" dia.	2.90	3.71	0.64	16.24	nr	**19.95**
1 1/2" dia.	4.26	5.46	0.71	18.01	nr	**23.47**
2" dia.	6.50	8.33	0.78	19.79	nr	**28.12**
3" dia.	18.97	24.30	0.91	23.09	nr	**47.39**
4" dia.	27.84	35.66	1.12	28.42	nr	**64.07**
6" dia.	97.31	124.63	1.69	42.88	nr	**167.51**
8" dia.	151.72	194.32	2.03	51.51	nr	**245.83**
Full face flange						
1/2" dia.	8.59	11.00	0.10	2.54	nr	**13.54**
3/4" dia.	8.83	11.31	0.13	3.30	nr	**14.61**
1" dia.	9.66	12.37	0.15	3.81	nr	**16.18**
1 1/4" dia.	10.80	13.83	0.18	4.57	nr	**18.40**
1 1/2" dia.	13.21	16.92	0.21	5.33	nr	**22.25**
2" dia.	17.58	22.52	0.29	7.36	nr	**29.87**
3" dia.	29.15	37.34	0.37	9.39	nr	**46.72**
4" dia.	38.53	49.35	0.41	10.40	nr	**59.75**

S:PIPED SUPPLY SYSTEMS

Item	Net Price £	Material £	Labour hours	Labour £	Unit	Total rate £
PVC-U PIPEWORK						
Pipes; solvent welded joints in the running length, brackets Measured separately						
Class C (9 bar pressure)						
2" dia.	6.84	8.76	0.41	10.40	m	**19.16**
3" dia.	13.92	17.83	0.47	11.92	m	**29.75**
4" dia.	24.17	30.96	0.50	12.69	m	**43.64**
6" dia.	53.12	68.04	1.76	44.65	m	**112.69**
Class D (12 bar pressure)						
1 1/4" dia.	3.99	5.11	0.41	10.40	m	**15.51**
1 1/2" dia.	5.44	6.97	0.42	10.66	m	**17.62**
2" dia.	8.40	10.76	0.45	11.42	m	**22.18**
3" dia.	18.69	23.94	0.48	12.18	m	**36.12**
4" dia.	30.94	39.63	0.53	13.45	m	**53.07**
6" dia.	58.93	75.48	0.58	14.72	m	**90.19**
Class E (15 bar pressure)						
1/2" dia.	1.94	2.48	0.38	9.64	m	**12.13**
3/4" dia.	2.74	3.51	0.40	10.15	m	**13.66**
1" dia.	3.19	4.09	0.41	10.40	m	**14.49**
1 1/4" dia.	4.72	6.05	0.41	10.40	m	**16.45**
1 1/2" dia.	6.11	7.83	0.42	10.66	m	**18.48**
2" dia.	9.50	12.17	0.45	11.42	m	**23.58**
3" dia.	21.28	27.26	0.47	11.92	m	**39.18**
4" dia.	34.63	44.35	0.50	12.69	m	**57.04**
6" dia.	75.82	97.11	0.53	13.45	m	**110.56**
Class 7						
1/2" dia.	3.35	4.29	0.32	8.12	m	**12.41**
3/4" dia.	4.65	5.96	0.33	8.37	m	**14.33**
1" dia.	7.06	9.04	0.40	10.15	m	**19.19**
1 1/4" dia.	9.76	12.50	0.40	10.15	m	**22.65**
1 1/2" dia.	12.06	15.45	0.41	10.40	m	**25.85**
2" dia.	19.98	25.59	0.43	10.91	m	**36.50**
FIXINGS						
Refer to steel pipes; galvanised iron. For minimum fixing dimensions, refer to the Tables and Memoranda at the rear of the book						
Extra over fittings; solvent welded joints						
End cap						
1/2" dia.	0.62	0.79	0.17	4.31	nr	**5.11**
3/4" dia.	0.73	0.94	0.19	4.82	nr	**5.76**
1" dia.	0.82	1.05	0.22	5.58	nr	**6.63**
1 1/4" dia.	1.28	1.64	0.25	6.34	nr	**7.98**
1 1/2" dia.	2.15	2.75	0.28	7.10	nr	**9.86**
2" dia.	2.63	3.37	0.31	7.87	nr	**11.23**
3" dia.	8.07	10.34	0.36	9.13	nr	**19.47**
4" dia.	12.47	15.97	0.44	11.16	nr	**27.14**
6" dia.	30.12	38.58	0.67	17.00	nr	**55.58**

S:PIPED SUPPLY SYSTEMS

Item	Net Price £	Material £	Labour hours	Labour £	Unit	Total rate £
S10 : COLD WATER (cont'd)						
PVC-U PIPEWORK (cont'd)						
Extra over fittings; solvent welded joints (cont'd)						
Socket						
1/2" dia.	0.66	0.85	0.31	7.87	nr	**8.71**
3/4" dia.	0.73	0.94	0.35	8.88	nr	**9.82**
1" dia.	0.85	1.09	0.42	10.66	nr	**11.74**
1 1/4" dia.	1.54	1.97	0.45	11.42	nr	**13.39**
1 1/2" dia.	1.80	2.31	0.51	12.94	nr	**15.25**
2" dia.	2.55	3.27	0.56	14.21	nr	**17.47**
3" dia.	9.76	12.50	0.65	16.49	nr	**28.99**
4" dia.	14.14	18.11	0.80	20.30	nr	**38.41**
6" dia.	35.48	45.44	1.21	30.70	nr	**76.14**
Reducing socket						
3/4 x 1/2" dia.	0.78	1.00	0.31	7.87	nr	**8.86**
1 x 3/4" dia.	0.96	1.23	0.35	8.88	nr	**10.11**
1 1/4 x 1" dia.	1.85	2.37	0.42	10.66	nr	**13.03**
1 1/2 x 1 1/4" dia.	2.07	2.65	0.45	11.42	nr	**14.07**
2 x 1 1/2" dia.	3.12	4.00	0.51	12.94	nr	**16.94**
3 x 2" dia.	9.49	12.15	0.56	14.21	nr	**26.36**
4 x 3" dia.	14.05	18.00	0.65	16.49	nr	**34.49**
6 x 4" dia.	51.20	65.58	0.80	20.30	nr	**85.87**
8 x 6" dia.	79.31	101.58	1.21	30.70	nr	**132.28**
Elbow, 90°						
1/2" dia.	0.88	1.13	0.31	7.87	nr	**8.99**
3/4" dia.	1.04	1.33	0.35	8.88	nr	**10.21**
1" dia.	1.46	1.87	0.42	10.66	nr	**12.53**
1 1/4" dia.	2.55	3.27	0.45	11.42	nr	**14.68**
1 1/2" dia.	3.29	4.21	0.45	11.42	nr	**15.63**
2" dia.	4.87	6.24	0.56	14.21	nr	**20.45**
3" dia.	14.05	18.00	0.65	16.49	nr	**34.49**
4" dia.	21.16	27.10	0.80	20.30	nr	**47.40**
6" dia.	83.78	107.30	1.21	30.70	nr	**138.00**
Elbow 45°						
1/2" dia.	1.66	2.13	0.31	7.87	nr	**9.99**
3/4" dia.	1.77	2.27	0.35	8.88	nr	**11.15**
1" dia.	2.15	2.75	0.45	11.42	nr	**14.17**
1 1/4" dia.	3.07	3.93	0.45	11.42	nr	**15.35**
1 1/2" dia.	3.87	4.96	0.51	12.94	nr	**17.90**
2" dia.	5.45	6.98	0.56	14.21	nr	**21.19**
3" dia.	12.82	16.42	0.65	16.49	nr	**32.91**
4" dia.	26.34	33.74	0.80	20.30	nr	**54.03**
6" dia.	54.36	69.62	1.21	30.70	nr	**100.32**
Bend 90° (long radius)						
3" dia.	41.05	52.58	0.65	16.49	nr	**69.07**
4" dia.	79.12	101.34	0.80	20.30	nr	**121.63**
6" dia.	173.88	222.70	1.21	30.70	nr	**253.40**

S:PIPED SUPPLY SYSTEMS

Item	Net Price £	Material £	Labour hours	Labour £	Unit	Total rate £
Bend 45° (long radius)						
1 1/2" dia.	9.30	11.91	0.51	12.94	nr	**24.85**
2" dia.	15.20	19.47	0.56	14.21	nr	**33.68**
3" dia.	32.49	41.61	0.65	16.49	nr	**58.10**
4" dia.	63.22	80.97	0.80	20.30	nr	**101.27**
Socket union						
1/2" dia.	3.38	4.33	0.34	8.63	nr	**12.96**
3/4" dia.	3.87	4.96	0.39	9.90	nr	**14.85**
1" dia.	5.01	6.42	0.45	11.42	nr	**17.83**
1 1/4" dia.	6.24	7.99	0.50	12.69	nr	**20.68**
1 1/2" dia.	8.56	10.96	0.57	14.46	nr	**25.43**
2" dia.	11.06	14.17	0.62	15.73	nr	**29.90**
3" dia.	41.19	52.76	0.70	17.76	nr	**70.52**
4" dia.	55.76	71.42	0.89	22.58	nr	**94.00**
Saddle plain						
2" x 1 1/4" dia.	8.70	11.14	0.42	10.66	nr	**21.80**
3" x 1 1/2" dia.	12.21	15.64	0.48	12.18	nr	**27.82**
4" x 2" dia.	13.76	17.62	0.68	17.25	nr	**34.88**
6" x 2" dia.	16.15	20.68	0.91	23.09	nr	**43.77**
Straight tank connector						
1/2" dia.	2.24	2.87	0.13	3.30	nr	**6.17**
3/4" dia.	2.54	3.25	0.14	3.55	nr	**6.81**
1" dia.	5.39	6.90	0.14	3.55	nr	**10.46**
1 1/4" dia.	13.70	17.55	0.16	4.06	nr	**21.61**
1 1/2" dia.	15.02	19.24	0.18	4.57	nr	**23.80**
2" dia.	18.00	23.05	0.24	6.09	nr	**29.14**
3" dia.	18.45	23.63	0.29	7.36	nr	**30.99**
Equal tee						
1/2" dia.	1.02	1.31	0.44	11.16	nr	**12.47**
3/4" dia.	1.28	1.64	0.48	12.18	nr	**13.82**
1" dia.	1.93	2.47	0.54	13.70	nr	**16.17**
1 1/4" dia.	2.73	3.50	0.70	17.76	nr	**21.26**
1 1/2" dia.	3.94	5.05	0.74	18.78	nr	**23.82**
2" dia.	6.24	7.99	0.80	20.30	nr	**28.29**
3" dia.	18.09	23.17	1.04	26.39	nr	**49.56**
4" dia.	26.53	33.98	1.28	32.48	nr	**66.46**
6" dia.	92.39	118.33	1.93	48.97	nr	**167.30**
PVC - C						
Pipes; solvent welded in the running length, brackets measured separately						
Pipe; 3m long; PN25						
16 x 2.0mm	1.78	2.28	0.20	5.07	m	**7.35**
20 x 2.3mm	2.71	3.47	0.20	5.07	m	**8.55**
25 x 2.8mm	3.50	4.48	0.20	5.07	m	**9.56**
32 x 3.6mm	5.20	6.66	0.20	5.07	m	**11.73**
Pipe; 5m long; PN25						
40 x 4.5mm	6.34	8.12	0.20	5.07	m	**13.19**
50 x 5.6mm	9.54	12.22	0.20	5.07	m	**17.29**
63 x 7.0mm	14.74	18.88	0.20	5.07	m	**23.95**

S:PIPED SUPPLY SYSTEMS

Item	Net Price £	Material £	Labour hours	Labour £	Unit	Total rate £
S10 : COLD WATER (cont'd)						
PVC-C Pipe (cont'd)						
FIXINGS						
Refer to steel pipes; galvanised iron. For minimum fixing dimensions, refer to the Tables and Memoranda at the rear of the book						
Extra over fittings; solvent welded joints						
Straight coupling; PN25						
16mm	0.27	0.35	0.20	5.07	nr	**5.42**
20mm	0.39	0.50	0.20	5.07	nr	**5.57**
25mm	0.47	0.60	0.20	5.07	nr	**5.67**
32mm	1.48	1.90	0.20	5.07	nr	**6.97**
40mm	1.90	2.43	0.20	5.07	nr	**7.51**
50mm	2.55	3.27	0.20	5.07	nr	**8.34**
63mm	4.50	5.76	0.20	5.07	nr	**10.84**
Elbow; 90°; PN25						
16mm	0.45	0.58	0.20	5.07	nr	**5.65**
20mm	0.68	0.87	0.20	5.07	nr	**5.95**
25mm	0.85	1.09	0.20	5.07	nr	**6.16**
32mm	1.77	2.27	0.20	5.07	nr	**7.34**
40mm	2.73	3.50	0.20	5.07	nr	**8.57**
50mm	3.79	4.85	0.20	5.07	nr	**9.93**
63mm	6.47	8.29	0.20	5.07	nr	**13.36**
Elbow; 45°; PN25						
20mm	0.68	0.87	0.20	5.07	nr	**5.95**
25mm	0.85	1.09	0.20	5.07	nr	**6.16**
32mm	1.77	2.27	0.20	5.07	nr	**7.34**
40mm	2.73	3.50	0.20	5.07	nr	**8.57**
50mm	3.79	4.85	0.20	5.07	nr	**9.93**
63mm	6.47	8.29	0.20	5.07	nr	**13.36**
Reducer fitting; single stage reduction						
20/16mm	0.48	0.61	0.20	5.07	nr	**5.69**
25/20mm	0.58	0.74	0.20	5.07	nr	**5.82**
32/25mm	1.16	1.49	0.20	5.07	nr	**6.56**
40/32mm	1.54	1.97	0.20	5.07	nr	**7.05**
50/40mm	1.77	2.27	0.20	5.07	nr	**7.34**
63/50mm	2.69	3.45	0.20	5.07	nr	**8.52**
Equal tee; 90°; PN25						
16mm	0.75	0.96	0.20	5.07	nr	**6.04**
20mm	1.02	1.31	0.20	5.07	nr	**6.38**
25mm	1.30	1.67	0.20	5.07	nr	**6.74**
32mm	2.11	2.70	0.20	5.07	nr	**7.78**
40mm	3.65	4.67	0.20	5.07	nr	**9.75**
50mm	5.45	6.98	0.20	5.07	nr	**12.05**
63mm	9.20	11.78	0.20	5.07	nr	**16.86**
Cap; PN25						
20mm	0.51	0.65	0.20	5.07	nr	**5.73**
25mm	0.68	0.87	0.20	5.07	nr	**5.95**
32mm	0.99	1.27	0.20	5.07	nr	**6.34**

S:PIPED SUPPLY SYSTEMS

Item	Net Price £	Material £	Labour hours	Labour £	Unit	Total rate £
40mm	1.37	1.75	0.20	5.07	nr	6.83
50mm	1.91	2.45	0.20	5.07	nr	7.52
63mm	3.03	3.88	0.20	5.07	nr	8.96
SCREWED STEEL PIPEWORK						
Galvanised steel pipes; screwed and socketed joints; BS 1387: 1985						
Galvanised; medium, fixed vertically, with brackets measured separately, screwed joints are within the running length, but any flanges are additional						
10mm dia.	2.73	3.50	0.51	12.94	m	16.44
15mm dia.	2.50	3.20	0.52	13.19	m	16.40
20mm dia.	2.82	3.61	0.55	13.95	m	17.57
25mm dia.	3.95	5.06	0.60	15.22	m	20.28
32mm dia.	4.93	6.31	0.67	17.00	m	23.31
40mm dia.	5.74	7.35	0.75	19.03	m	26.38
50mm dia.	8.10	10.37	0.85	21.57	m	31.94
65mm dia.	11.29	14.46	0.93	23.60	m	38.06
80mm dia.	14.63	18.74	1.07	27.15	m	45.89
100mm dia.	21.28	27.26	1.46	37.04	m	64.30
125mm dia.	35.80	45.85	1.72	43.64	m	89.49
150mm dia.	40.63	52.04	1.96	49.73	m	101.77
Galvanised; heavy, fixed vertically, with brackets measured separately, screwed joints are within the running length, but any flanges are additional						
15mm dia.	2.94	3.77	0.52	13.19	m	16.96
20mm dia.	3.33	4.27	0.55	13.95	m	18.22
25mm dia.	4.75	6.08	0.60	15.22	m	21.31
32mm dia.	5.94	7.61	0.67	17.00	m	24.61
40mm dia.	6.94	8.89	0.75	19.03	m	27.92
50mm dia.	9.67	12.39	0.85	21.57	m	33.95
65mm dia.	13.44	17.21	0.93	23.60	m	40.81
80mm dia.	17.09	21.89	1.07	27.15	m	49.04
100mm dia.	24.46	31.33	1.46	37.04	m	68.37
125mm dia.	37.73	48.32	1.72	43.64	m	91.96
150mm dia.	43.14	55.25	1.96	49.73	m	104.98
Galvanised; medium, fixed horizontally or suspended at high level, with brackets measured separately, screwed joints are within the running length, but any flanges are additional						
10mm dia.	2.73	3.50	0.51	12.94	m	16.44
15mm dia.	2.50	3.20	0.52	13.19	m	16.40
20mm dia.	2.82	3.61	0.55	13.95	m	17.57
25mm dia.	3.95	5.06	0.60	15.22	m	20.28
32mm dia.	4.93	6.31	0.67	17.00	m	23.31
40mm dia.	5.74	7.35	0.75	19.03	m	26.38
50mm dia.	8.10	10.37	0.85	21.57	m	31.94
65mm dia.	11.29	14.46	0.93	23.60	m	38.06
80mm dia.	14.63	18.74	1.07	27.15	m	45.89
100mm dia.	21.28	27.26	1.46	37.04	m	64.30
125mm dia.	35.80	45.85	1.72	43.64	m	89.49
150mm dia.	40.63	52.04	1.96	49.73	m	101.77

S:PIPED SUPPLY SYSTEMS

Item	Net Price £	Material £	Labour hours	Labour £	Unit	Total rate £
S10 : COLD WATER (cont'd)						
SCREWED STEEL PIPEWORK (cont'd)						
Galvanised steel pipes; screwed and socketed joints (cont'd)						
Galvanised; heavy, fixed horizontally or suspended at high level, with brackets measured separately, screwed joints are within the running length, but any flanges are additional						
15mm dia.	2.94	3.77	0.52	13.19	m	**16.96**
20mm dia.	3.33	4.27	0.55	13.95	m	**18.22**
25mm dia.	4.75	6.08	0.60	15.22	m	**21.31**
32mm dia.	5.94	7.61	0.67	17.00	m	**24.61**
40mm dia.	6.94	8.89	0.75	19.03	m	**27.92**
50mm dia.	9.67	12.39	0.85	21.57	m	**33.95**
65mm dia.	13.44	17.21	0.93	23.60	m	**40.81**
80mm dia.	17.09	21.89	1.07	27.15	m	**49.04**
100mm dia.	24.46	31.33	1.46	37.04	m	**68.37**
125mm dia.	37.73	48.32	1.72	43.64	m	**91.96**
150mm dia.	43.14	55.25	1.96	49.73	m	**104.98**
FIXINGS						
For steel pipes; galvanised iron. For minimum fixing dimensions, refer to the Tables and Memoranda at the rear of the book						
Single pipe bracket, screw on, galvanised iron; screwed to wood						
15mm dia	0.93	1.19	0.14	3.55	nr	**4.74**
20mm dia	1.04	1.33	0.14	3.55	nr	**4.88**
25mm dia	1.21	1.55	0.17	4.31	nr	**5.86**
32mm dia	1.65	2.11	0.19	4.82	nr	**6.93**
40mm dia	2.20	2.82	0.22	5.58	nr	**8.40**
50mm dia	2.91	3.73	0.22	5.58	nr	**9.31**
65mm dia	3.84	4.92	0.28	7.10	nr	**12.02**
80mm dia	5.28	6.76	0.32	8.12	nr	**14.88**
100mm dia	7.69	9.85	0.35	8.88	nr	**18.73**
Single pipe bracket, screw on, galvanised iron; plugged and screwed						
15mm dia	0.93	1.19	0.25	6.34	nr	**7.53**
20mm dia	1.04	1.33	0.25	6.34	nr	**7.68**
25mm dia	1.21	1.55	0.30	7.61	nr	**9.16**
32mm dia	1.65	2.11	0.32	8.12	nr	**10.23**
40mm dia	2.20	2.82	0.32	8.12	nr	**10.94**
50mm dia	2.91	3.73	0.32	8.12	nr	**11.85**
65mm dia	3.84	4.92	0.35	8.88	nr	**13.80**
80mm dia	5.28	6.76	0.42	10.66	nr	**17.42**
100mm dia	7.69	9.85	0.42	10.66	nr	**20.51**

S:PIPED SUPPLY SYSTEMS

Item	Net Price £	Material £	Labour hours	Labour £	Unit	Total rate £
Single pipe bracket for building in, galvanised iron						
15mm dia	1.68	2.15	0.10	2.54	nr	**4.69**
20mm dia	1.68	2.15	0.11	2.79	nr	**4.94**
25mm dia	1.68	2.15	0.12	3.04	nr	**5.20**
32mm dia	1.90	2.43	0.14	3.55	nr	**5.99**
40mm dia	1.92	2.46	0.15	3.81	nr	**6.26**
50mm dia	1.99	2.55	0.16	4.06	nr	**6.61**
Pipe ring, single socket, galvanised iron						
15mm dia.	0.98	1.26	0.10	2.54	nr	**3.79**
20mm dia.	1.10	1.41	0.11	2.79	nr	**4.20**
25mm dia	1.21	1.55	0.12	3.04	nr	**4.59**
32mm dia	1.27	1.63	0.15	3.81	nr	**5.43**
40mm dia	1.65	2.11	0.15	3.81	nr	**5.92**
50mm dia	2.09	2.68	0.16	4.06	nr	**6.74**
65mm dia	3.03	3.88	0.30	7.61	nr	**11.49**
80mm dia	3.63	4.65	0.35	8.88	nr	**13.53**
100mm dia	5.49	7.03	0.40	10.15	nr	**17.18**
125mm dia	11.11	14.23	0.60	15.23	nr	**29.46**
150mm dia	12.42	15.91	0.77	19.54	nr	**35.44**
Pipe ring, double socket, galvanised iron						
15mm dia	1.15	1.47	0.10	2.54	nr	**4.01**
20mm dia	1.32	1.69	0.11	2.79	nr	**4.48**
25mm dia	1.49	1.91	0.12	3.04	nr	**4.95**
32mm dia	1.76	2.25	0.14	3.55	nr	**5.81**
40mm dia	2.03	2.60	0.15	3.81	nr	**6.41**
50mm dia	2.31	2.96	0.16	4.06	nr	**7.02**
Screw on backplate (Male), galvanised iron; plugged and screwed						
All sizes 15mm to 50mm x M12	1.40	1.79	0.10	2.54	nr	**4.33**
Screw on backplate (Female), galvanised iron; plugged and screwed						
All sizes 15mm to 50mm x M12	1.40	1.79	0.10	2.54	nr	**4.33**
Extra Over channel sections for fabricated hangers and brackets						
Galvanised steel; including inserts, bolts, nuts, washers; fixed to backgrounds						
41 x 21mm	2.40	3.07	0.15	3.81	m	**6.88**
41 x 41mm	3.75	4.80	0.15	3.81	m	**8.61**
Threaded rods; metric thread; including nuts, washers etc						
10mm dia x 600mm long for ring clips up to 50mm	1.25	1.60	0.10	2.54	nr	**4.14**
12mm dia x 600mm long for ring clips from 50mm	1.30	1.66	0.10	2.54	nr	**4.20**

S:PIPED SUPPLY SYSTEMS

Item	Net Price £	Material £	Labour hours	Labour £	Unit	Total rate £
S10 : COLD WATER (cont'd)						
SCREWED STEEL PIPEWORK (cont'd)						
Extra over steel flanges, screwed and drilled; metric; BS 4504						
Screwed flanges; PN6						
15mm dia.	10.82	13.86	0.35	8.88	nr	**22.74**
20mm dia.	11.24	14.40	0.47	11.92	nr	**26.32**
25mm dia.	11.35	14.54	0.53	13.45	nr	**27.98**
32mm dia.	11.35	14.54	0.62	15.73	nr	**30.27**
40mm dia.	11.35	14.54	0.70	17.76	nr	**32.30**
50mm dia.	12.24	15.68	0.84	21.31	nr	**36.99**
65mm dia.	16.72	21.41	1.03	26.13	nr	**47.55**
80mm dia.	20.17	25.83	1.23	31.21	nr	**57.04**
100mm dia.	24.17	30.96	1.41	35.77	nr	**66.73**
125mm dia.	51.19	65.56	1.77	44.91	nr	**110.47**
150mm dia.	51.19	65.56	2.21	56.07	nr	**121.64**
Screwed flanges; PN16						
15mm dia.	14.71	18.84	0.35	8.88	nr	**27.72**
20mm dia.	14.77	18.92	0.47	11.92	nr	**30.84**
25mm dia.	14.88	19.06	0.53	13.45	nr	**32.51**
32mm dia.	15.64	20.03	0.62	15.73	nr	**35.76**
40mm dia.	15.64	20.03	0.70	17.76	nr	**37.79**
50mm dia.	16.52	21.16	0.84	21.31	nr	**42.47**
65mm dia.	20.66	26.46	1.03	26.13	nr	**52.59**
80mm dia.	24.30	31.12	1.23	31.21	nr	**62.33**
100mm dia.	29.11	37.28	1.41	35.77	nr	**73.06**
125mm dia.	53.08	67.98	1.77	44.91	nr	**112.89**
150mm dia.	55.96	71.67	2.21	56.07	nr	**127.74**
Extra over steel flanges, screwed and drilled; imperial; BS 10						
Screwed flanges; table E						
1/2" dia.	15.61	19.99	0.35	8.88	nr	**28.87**
3/4" dia.	15.67	20.07	0.47	11.92	nr	**31.99**
1" dia.	15.78	20.21	0.53	13.45	nr	**33.66**
1 1/4" dia.	16.54	21.18	0.62	15.73	nr	**36.91**
1 1/2" dia.	16.54	21.18	0.70	17.76	nr	**38.94**
2" dia.	16.76	21.47	0.84	21.31	nr	**42.78**
2 1/2" dia.	19.79	25.35	1.03	26.13	nr	**51.48**
3" dia.	23.69	30.34	1.23	31.21	nr	**61.55**
4" dia.	31.64	40.52	1.41	35.77	nr	**76.30**
5" dia.	62.12	79.56	1.77	44.91	nr	**124.47**
Extra over steel flange connections						
Bolted connection between pair of flanges; including gasket, bolts, nuts and washers						
50mm dia.	31.02	39.73	0.53	13.45	nr	**53.18**
65mm dia.	39.07	50.04	0.53	13.45	nr	**63.49**
80mm dia.	44.80	57.38	0.53	13.45	nr	**70.83**
100mm dia.	53.46	68.47	0.53	13.45	nr	**81.92**
125mm dia.	101.02	129.39	0.61	15.48	nr	**144.86**
150mm dia.	103.90	133.07	0.90	22.83	nr	**155.91**

S:PIPED SUPPLY SYSTEMS

Item	Net Price £	Material £	Labour hours	Labour £	Unit	Total rate £
Extra over heavy steel tubular fittings; BS 1387						
Long screw connection with socket and backnut						
15mm dia.	4.64	5.94	0.63	15.98	nr	**21.93**
20mm dia.	5.43	6.95	0.84	21.31	nr	**28.27**
25mm dia.	8.01	10.26	0.95	24.10	nr	**34.36**
32mm dia.	11.06	14.17	1.11	28.16	nr	**42.33**
40mm dia.	9.32	11.94	1.28	32.48	nr	**44.41**
50mm dia.	13.70	17.55	1.53	38.82	nr	**56.37**
65mm dia.	31.13	39.87	1.87	47.45	nr	**87.32**
80mm dia.	39.59	50.71	2.21	56.07	nr	**106.78**
100mm dia.	104.46	133.79	3.05	77.38	nr	**211.18**
Running nipple						
15mm dia.	0.92	1.18	0.50	12.69	nr	**13.86**
20mm dia.	1.15	1.47	0.68	17.25	nr	**18.73**
25mm dia.	1.24	1.59	0.77	19.54	nr	**21.12**
32mm dia.	1.99	2.55	0.90	22.83	nr	**25.38**
40mm dia.	2.69	3.45	1.03	26.13	nr	**29.58**
50mm dia.	4.09	5.24	1.23	31.21	nr	**36.45**
65mm dia.	8.80	11.27	1.50	38.06	nr	**49.33**
80mm dia.	13.72	17.57	1.78	45.16	nr	**62.73**
100mm dia.	21.49	27.52	2.38	60.39	nr	**87.91**
Barrel nipple						
15mm dia.	0.78	1.00	0.50	12.69	nr	**13.69**
20mm dia.	0.87	1.11	0.68	17.25	nr	**18.37**
25mm dia.	1.24	1.59	0.77	19.54	nr	**21.12**
32mm dia.	1.57	2.01	0.90	22.83	nr	**24.85**
40mm dia.	1.94	2.48	1.03	26.13	nr	**28.62**
50mm dia.	2.76	3.54	1.23	31.21	nr	**34.74**
65mm dia.	5.01	6.42	1.50	38.06	nr	**44.47**
80mm dia.	6.98	8.94	1.78	45.16	nr	**54.10**
100mm dia.	12.63	16.18	2.38	60.39	nr	**76.56**
125mm dia.	23.50	30.10	2.87	72.82	nr	**102.92**
150mm dia.	36.94	47.31	3.39	86.01	nr	**133.32**
Close taper nipple						
15mm dia.	1.09	1.40	0.50	12.69	nr	**14.08**
20mm dia.	1.42	1.82	0.68	17.25	nr	**19.07**
25mm dia.	1.86	2.38	0.77	19.54	nr	**21.92**
32mm dia.	2.77	3.55	0.90	22.83	nr	**26.38**
40mm dia.	3.44	4.41	1.03	26.13	nr	**30.54**
50mm dia.	5.29	6.78	1.23	31.21	nr	**37.98**
65mm dia.	8.35	10.69	1.50	38.06	nr	**48.75**
80mm dia.	13.68	17.52	1.78	45.16	nr	**62.68**
100mm dia.	26.02	33.33	2.38	60.39	nr	**93.71**
90° bend with socket						
15mm dia.	2.98	3.82	0.64	16.24	nr	**20.05**
20mm dia.	4.00	5.12	0.85	21.57	nr	**26.69**
25mm dia.	6.13	7.85	0.97	24.61	nr	**32.46**
32mm dia.	8.80	11.27	1.12	28.42	nr	**39.69**
40mm dia.	10.75	13.77	1.29	32.73	nr	**46.50**
50mm dia.	16.72	21.41	1.55	39.33	nr	**60.74**
65mm dia.	33.74	43.21	1.89	47.95	nr	**91.17**
80mm dia.	50.16	64.24	2.24	56.83	nr	**121.08**
100mm dia.	88.88	113.84	3.09	78.40	nr	**192.24**
125mm dia.	218.82	280.26	3.92	99.46	nr	**379.72**
150mm dia.	329.05	421.44	4.74	120.26	nr	**541.71**

S:PIPED SUPPLY SYSTEMS

Item	Net Price £	Material £	Labour hours	Labour £	Unit	Total rate £
S10 : COLD WATER (cont'd)						
SCREWED STEEL PIPEWORK (cont'd)						
Extra over heavy steel fittings; BS 1740						
Plug						
15mm dia.	0.74	0.95	0.28	7.10	nr	**8.05**
20mm dia.	1.15	1.47	0.38	9.64	nr	**11.11**
25mm dia.	2.02	2.59	0.44	11.16	nr	**13.75**
32mm dia.	3.13	4.01	0.51	12.94	nr	**16.95**
40mm dia.	3.46	4.43	0.59	14.97	nr	**19.40**
50mm dia.	4.95	6.34	0.70	17.76	nr	**24.10**
65mm dia.	11.83	15.15	0.85	21.57	nr	**36.72**
80mm dia.	22.14	28.36	1.00	25.37	nr	**53.73**
100mm dia.	42.52	54.46	1.44	36.54	nr	**90.99**
Socket						
15mm dia.	0.71	0.91	0.64	16.24	nr	**17.15**
20mm dia.	0.76	0.97	0.85	21.57	nr	**22.54**
25mm dia.	1.12	1.43	0.97	24.61	nr	**26.05**
32mm dia.	1.65	2.11	1.12	28.42	nr	**30.53**
40mm dia.	2.00	2.56	1.29	32.73	nr	**35.29**
50mm dia.	3.06	3.92	1.55	39.33	nr	**43.25**
65mm dia.	6.06	7.76	1.89	47.95	nr	**55.71**
80mm dia.	7.82	10.02	2.24	56.83	nr	**66.85**
100mm dia.	14.71	18.84	3.09	78.40	nr	**97.24**
150mm dia.	35.06	44.90	4.74	120.26	nr	**165.17**
Elbow, female/female						
15mm dia.	3.92	5.02	0.64	16.24	nr	**21.26**
20mm dia.	5.11	6.54	0.85	21.57	nr	**28.11**
25mm dia.	6.94	8.89	0.97	24.61	nr	**33.50**
32mm dia.	12.92	16.55	1.12	28.42	nr	**44.96**
40mm dia.	15.40	19.72	1.29	32.73	nr	**52.45**
50mm dia.	25.25	32.34	1.55	39.33	nr	**71.67**
65mm dia.	61.69	79.01	1.89	47.95	nr	**126.96**
80mm dia.	73.57	94.23	2.24	56.83	nr	**151.06**
100mm dia.	127.39	163.16	3.09	78.40	nr	**241.56**
Equal tee						
15mm dia.	4.87	6.24	0.91	23.09	nr	**29.33**
20mm dia.	5.66	7.25	1.22	30.95	nr	**38.20**
25mm dia.	8.34	10.68	1.40	35.52	nr	**46.20**
32mm dia.	17.23	22.07	1.62	41.10	nr	**63.17**
40mm dia.	18.76	24.03	1.86	47.19	nr	**71.22**
50mm dia.	30.51	39.08	2.21	56.07	nr	**95.15**
65mm dia.	73.77	94.48	2.72	69.01	nr	**163.50**
80mm dia.	79.17	101.40	3.21	81.44	nr	**182.84**
100mm dia.	127.41	163.19	4.44	112.65	nr	**275.84**

S:PIPED SUPPLY SYSTEMS

Item	Net Price £	Material £	Labour hours	Labour £	Unit	Total rate £
Extra over malleable iron fittings; BS 143						
Cap						
15mm dia.	0.63	0.81	0.32	8.12	nr	8.93
20mm dia.	0.72	0.92	0.43	10.91	nr	11.83
25mm dia.	0.97	1.24	0.49	12.43	nr	13.67
32mm dia.	1.49	1.91	0.58	14.72	nr	16.62
40mm dia.	1.90	2.43	0.66	16.75	nr	19.18
50mm dia.	3.54	4.53	0.78	19.79	nr	24.32
65mm dia.	6.08	7.79	0.96	24.36	nr	32.14
80mm dia.	6.88	8.81	1.13	28.67	nr	37.48
100mm dia.	15.07	19.30	1.70	43.13	nr	62.43
Plain plug, hollow						
15mm dia.	0.49	0.63	0.28	7.10	nr	7.73
20mm dia.	0.62	0.79	0.38	9.64	nr	10.44
25mm dia.	0.76	0.97	0.44	11.16	nr	12.14
32mm dia.	1.18	1.51	0.51	12.94	nr	14.45
40mm dia.	1.83	2.34	0.59	14.97	nr	17.31
50mm dia.	2.56	3.28	0.70	17.76	nr	21.04
65mm dia.	4.47	5.73	0.85	21.57	nr	27.29
80mm dia.	6.68	8.56	1.00	25.37	nr	33.93
100mm dia.	12.31	15.77	1.44	36.54	nr	52.30
Plain plug, solid						
15mm dia.	1.37	1.75	0.29	7.36	nr	9.11
20mm dia.	1.47	1.88	0.38	9.64	nr	11.52
25mm dia.	1.94	2.48	0.44	11.16	nr	13.65
32mm dia.	2.67	3.42	0.51	12.94	nr	16.36
40mm dia.	3.60	4.61	0.59	14.97	nr	19.58
50mm dia.	4.73	6.06	0.70	17.76	nr	23.82
Elbow, male/female						
15mm dia.	0.72	0.92	0.64	16.24	nr	17.16
20mm dia.	0.96	1.23	0.85	21.57	nr	22.80
25mm dia.	1.60	2.05	0.97	24.61	nr	26.66
32mm dia.	3.25	4.16	1.12	28.42	nr	32.58
40mm dia.	4.48	5.74	1.29	32.73	nr	38.47
50mm dia.	5.77	7.39	1.55	39.33	nr	46.72
65mm dia.	12.86	16.47	1.89	47.95	nr	64.42
80mm dia.	17.59	22.53	2.24	56.83	nr	79.36
100mm dia.	30.75	39.38	3.09	78.40	nr	117.78
Elbow						
15mm dia.	0.64	0.82	0.64	16.24	nr	17.06
20mm dia.	0.87	1.11	0.85	21.57	nr	22.68
25mm dia.	1.35	1.73	0.97	24.61	nr	26.34
32mm dia.	2.54	3.25	1.12	28.42	nr	31.67
40mm dia.	3.80	4.87	1.29	32.73	nr	37.60
50mm dia.	4.45	5.70	1.55	39.33	nr	45.03
65mm dia.	9.93	12.72	1.89	47.95	nr	60.67
80mm dia.	14.59	18.69	2.24	56.83	nr	75.52
100mm dia.	25.06	32.10	3.09	78.40	nr	110.50
125mm dia.	60.30	77.23	4.44	112.65	nr	189.88
150mm dia.	112.25	143.77	5.79	146.90	nr	290.67

S:PIPED SUPPLY SYSTEMS

Item	Net Price £	Material £	Labour hours	Labour £	Unit	Total rate £
S10 : COLD WATER (cont'd)						
SCREWED STEEL PIPEWORK (cont'd)						
Extra over malleable iron fittings; BS 143 (cont'd)						
45° elbow						
15mm dia.	1.65	2.11	0.64	16.24	nr	**18.35**
20mm dia.	2.03	2.60	0.85	21.57	nr	**24.17**
25mm dia.	2.80	3.59	0.97	24.61	nr	**28.20**
32mm dia.	5.86	7.51	1.12	28.42	nr	**35.92**
40mm dia.	6.88	8.81	1.29	32.73	nr	**41.54**
50mm dia.	9.44	12.09	1.55	39.33	nr	**51.42**
65mm dia.	13.27	17.00	1.89	47.95	nr	**64.95**
80mm dia.	19.95	25.55	2.24	56.83	nr	**82.38**
100mm dia.	38.54	49.36	3.09	78.40	nr	**127.76**
150mm dia.	103.15	132.11	5.79	146.90	nr	**279.02**
Bend, male/female						
15mm dia.	1.26	1.61	0.64	16.24	nr	**17.85**
20mm dia.	2.08	2.66	0.85	21.57	nr	**24.23**
25mm dia.	2.91	3.73	0.97	24.61	nr	**28.34**
32mm dia.	4.42	5.66	1.12	28.42	nr	**34.08**
40mm dia.	6.48	8.30	1.29	32.73	nr	**41.03**
50mm dia.	12.17	15.59	1.55	39.33	nr	**54.91**
65mm dia.	18.64	23.87	1.89	47.95	nr	**71.83**
80mm dia.	25.28	32.38	2.24	56.83	nr	**89.21**
100mm dia.	62.61	80.19	3.09	78.40	nr	**158.59**
Bend, male						
15mm dia.	2.88	3.69	0.64	16.24	nr	**19.93**
20mm dia.	3.24	4.15	0.85	21.57	nr	**25.72**
25mm dia.	4.76	6.10	0.97	24.61	nr	**30.71**
32mm dia.	9.45	12.10	1.12	28.42	nr	**40.52**
40mm dia.	13.26	16.98	1.29	32.73	nr	**49.71**
50mm dia.	17.73	22.71	1.55	39.33	nr	**62.03**
Bend, female						
15mm dia.	1.29	1.65	0.64	16.24	nr	**17.89**
20mm dia.	1.85	2.37	0.85	21.57	nr	**23.94**
25mm dia.	2.59	3.32	0.97	24.61	nr	**27.93**
32mm dia.	4.53	5.80	1.12	28.42	nr	**34.22**
40mm dia.	5.40	6.92	1.29	32.73	nr	**39.65**
50mm dia.	8.51	10.90	1.55	39.33	nr	**50.23**
65mm dia.	18.64	23.87	1.89	47.95	nr	**71.83**
80mm dia.	27.64	35.40	2.24	56.83	nr	**92.23**
100mm dia.	57.98	74.26	3.09	78.40	nr	**152.66**
125mm dia.	154.67	198.10	4.44	112.65	nr	**310.75**
150mm dia.	226.41	289.98	5.79	146.90	nr	**436.89**
Return bend						
15mm dia.	5.82	7.45	0.64	16.24	nr	**23.69**
20mm dia.	9.41	12.05	0.85	21.57	nr	**33.62**
25mm dia.	11.74	15.04	0.97	24.61	nr	**39.65**
32mm dia.	16.65	21.33	1.12	28.42	nr	**49.74**
40mm dia.	19.83	25.40	1.29	32.73	nr	**58.13**
50mm dia.	30.27	38.77	1.55	39.33	nr	**78.10**

S:PIPED SUPPLY SYSTEMS

Item	Net Price £	Material £	Labour hours	Labour £	Unit	Total rate £
Equal socket, parallel thread						
15mm dia.	0.68	0.87	0.64	16.24	nr	**17.11**
20mm dia.	0.81	1.04	0.85	21.57	nr	**22.60**
25mm dia.	1.08	1.38	0.97	24.61	nr	**25.99**
32mm dia.	2.02	2.59	1.12	28.42	nr	**31.00**
40mm dia.	2.88	3.69	1.29	32.73	nr	**36.42**
50mm dia.	4.13	5.29	1.55	39.33	nr	**44.62**
65mm dia.	7.24	9.27	1.89	47.95	nr	**57.23**
80mm dia.	9.95	12.74	2.24	56.83	nr	**69.58**
100mm dia.	16.88	21.62	3.09	78.40	nr	**100.02**
Concentric reducing socket						
20 x 15mm dia.	0.99	1.27	0.76	19.28	nr	**20.55**
25 x 15mm dia.	1.29	1.65	0.86	21.82	nr	**23.47**
25 x 20mm dia.	1.22	1.56	0.86	21.82	nr	**23.38**
32 x 25mm dia.	2.11	2.70	1.01	25.63	nr	**28.33**
40 x 25mm dia.	2.78	3.56	1.16	29.43	nr	**32.99**
40 x 32mm dia.	3.08	3.94	1.16	29.43	nr	**33.38**
50 x 25mm dia.	5.34	6.84	1.38	35.01	nr	**41.85**
50 x 40mm dia.	4.32	5.53	1.38	35.01	nr	**40.55**
65 x 50mm dia.	7.54	9.66	1.69	42.88	nr	**52.54**
80 x 50mm dia.	9.39	12.03	2.00	50.74	nr	**62.77**
100 x 50mm dia.	18.74	24.00	2.75	69.77	nr	**93.77**
100 x 80mm dia.	17.39	22.27	2.75	69.77	nr	**92.05**
150 x 100mm dia.	45.88	58.76	4.10	104.02	nr	**162.79**
Eccentric reducing socket						
20 x 15mm dia.	1.97	2.52	0.76	19.28	nr	**21.81**
25 x 15mm dia.	5.62	7.20	0.86	21.82	nr	**29.02**
25 x 20mm dia.	6.38	8.17	0.86	21.82	nr	**29.99**
32 x 25mm dia.	7.40	9.48	1.01	25.63	nr	**35.10**
40 x 25mm dia.	8.48	10.86	1.16	29.43	nr	**40.29**
40 x 32mm dia.	4.57	5.85	1.16	29.43	nr	**35.28**
50 x 25mm dia.	5.50	7.04	1.38	35.01	nr	**42.06**
50 x 40mm dia.	5.50	7.04	1.38	35.01	nr	**42.06**
65 x 50mm dia.	9.39	12.03	1.69	42.88	nr	**54.91**
80 x 50mm dia.	15.27	19.56	2.00	50.74	nr	**70.30**
Hexagon bush						
20 x 15mm dia.	0.56	0.72	0.37	9.39	nr	**10.10**
25 x 15mm dia.	0.76	0.97	0.43	10.91	nr	**11.88**
25 x 20mm dia.	0.72	0.92	0.43	10.91	nr	**11.83**
32 x 25mm dia.	0.87	1.11	0.51	12.94	nr	**14.05**
40 x 25mm dia.	1.30	1.67	0.58	14.72	nr	**16.38**
40 x 32mm dia.	1.37	1.75	0.58	14.72	nr	**16.47**
50 x 25mm dia.	2.75	3.52	0.71	18.01	nr	**21.54**
50 x 40mm dia.	2.56	3.28	0.71	18.01	nr	**21.29**
65 x 50mm dia.	4.71	6.03	0.84	21.31	nr	**27.34**
80 x 50mm dia.	7.12	9.12	1.00	25.37	nr	**34.49**
100 x 50mm dia.	16.48	21.11	1.52	38.57	nr	**59.67**
100 x 80mm dia.	13.71	17.56	1.52	38.57	nr	**56.12**
150 x 100mm dia.	43.41	55.60	2.48	62.92	nr	**118.52**

S:PIPED SUPPLY SYSTEMS

Item	Net Price £	Material £	Labour hours	Labour £	Unit	Total rate £
S10 : COLD WATER (cont'd)						
SCREWED STEEL PIPEWORK (cont'd)						
Extra over malleable iron fittings; BS 143 (cont'd)						
Hexagon nipple						
15mm dia.	0.60	0.77	0.28	7.10	nr	**7.87**
20mm dia.	0.68	0.87	0.38	9.64	nr	**10.51**
25mm dia.	0.96	1.23	0.44	11.16	nr	**12.39**
32mm dia.	1.83	2.34	0.51	12.94	nr	**15.28**
40mm dia.	2.11	2.70	0.59	14.97	nr	**17.67**
50mm dia.	3.84	4.92	0.70	17.76	nr	**22.68**
65mm dia.	6.43	8.24	0.85	21.57	nr	**29.80**
80mm dia.	8.90	11.40	1.00	25.37	nr	**36.77**
100mm dia.	15.78	20.21	1.44	36.54	nr	**56.75**
150mm dia.	44.62	57.15	2.32	58.86	nr	**116.01**
Union, male/female						
15mm dia.	3.01	3.86	0.64	16.24	nr	**20.09**
20mm dia.	3.68	4.71	0.85	21.57	nr	**26.28**
25mm dia.	4.28	5.48	0.97	24.61	nr	**30.09**
32mm dia.	6.78	8.68	1.12	28.42	nr	**37.10**
40mm dia.	8.68	11.12	1.29	32.73	nr	**43.85**
50mm dia.	13.67	17.51	1.55	39.33	nr	**56.83**
65mm dia.	26.93	34.49	1.89	47.95	nr	**82.44**
Union, female						
15mm dia.	6.85	8.77	0.64	16.24	nr	**25.01**
20mm dia.	8.14	10.43	0.85	21.57	nr	**31.99**
25mm dia.	10.01	12.82	0.97	24.61	nr	**37.43**
32mm dia.	15.05	19.28	1.12	28.42	nr	**47.69**
40mm dia.	18.20	23.31	1.29	32.73	nr	**56.04**
50mm dia.	21.51	27.55	1.55	39.33	nr	**66.88**
65mm dia.	48.24	61.79	1.89	47.95	nr	**109.74**
80mm dia.	78.78	100.90	2.24	56.83	nr	**157.73**
100mm dia.	113.46	145.32	3.09	78.40	nr	**223.72**
Union elbow, male/female						
15mm dia.	4.00	5.12	0.64	16.24	nr	**21.36**
20mm dia.	5.01	6.42	0.85	21.57	nr	**27.98**
25mm dia.	7.03	9.00	0.97	24.61	nr	**33.61**
Twin elbow						
15mm dia.	3.85	4.93	0.91	23.09	nr	**28.02**
20mm dia.	4.25	5.44	1.22	30.95	nr	**36.40**
25mm dia.	6.88	8.81	1.39	35.27	nr	**44.08**
32mm dia.	12.38	15.86	1.62	41.10	nr	**56.96**
40mm dia.	15.67	20.07	1.86	47.19	nr	**67.26**
50mm dia.	20.14	25.80	2.21	56.07	nr	**81.87**
65mm dia.	32.56	41.70	2.72	69.01	nr	**110.71**
80mm dia.	55.47	71.05	3.21	81.44	nr	**152.49**

S:PIPED SUPPLY SYSTEMS

Item	Net Price £	Material £	Labour hours	Labour £	Unit	Total rate £
Equal tee						
15mm dia.	0.87	1.11	0.91	23.09	nr	24.20
20mm dia.	1.28	1.64	1.22	30.95	nr	32.59
25mm dia.	1.84	2.36	1.39	35.27	nr	37.62
32mm dia.	3.48	4.46	1.62	41.10	nr	45.56
40mm dia.	4.76	6.10	1.86	47.19	nr	53.29
50mm dia.	6.86	8.79	2.21	56.07	nr	64.86
65mm dia.	16.08	20.60	2.72	69.01	nr	89.61
80mm dia.	18.74	24.00	3.21	81.44	nr	105.45
100mm dia.	33.97	43.51	4.44	112.65	nr	156.16
125mm dia.	83.31	106.70	5.38	136.50	nr	243.20
150mm dia.	132.75	170.02	6.31	160.10	nr	330.12
Tee reducing on branch						
20 x 15mm dia.	1.31	1.68	1.22	30.95	nr	32.63
25 x 15mm dia.	1.78	2.28	1.39	35.27	nr	37.55
25 x 20mm dia.	2.03	2.60	1.39	35.27	nr	37.87
32 x 25mm dia.	3.54	4.53	1.62	41.10	nr	45.64
40 x 25mm dia.	4.48	5.74	1.86	47.19	nr	52.93
40 x 32mm dia.	6.58	8.43	1.86	47.19	nr	55.62
50 x 25mm dia.	5.95	7.62	2.21	56.07	nr	63.69
50 x 40mm dia.	9.25	11.85	2.21	56.07	nr	67.92
65 x 50mm dia.	14.27	18.28	2.72	69.01	nr	87.29
80 x 50mm dia.	19.29	24.71	3.21	81.44	nr	106.15
100 x 50mm dia.	28.14	36.04	4.44	112.65	nr	148.69
100 x 80mm dia.	43.14	55.25	4.44	112.65	nr	167.90
150 x 100mm dia.	97.78	125.24	6.31	160.10	nr	285.33
Equal pitcher tee						
15mm dia.	3.04	3.89	0.91	23.09	nr	26.98
20mm dia.	3.74	4.79	1.22	30.95	nr	35.74
25mm dia.	5.62	7.20	1.39	35.27	nr	42.46
32mm dia.	7.81	10.00	1.62	41.10	nr	51.11
40mm dia.	12.08	15.47	1.86	47.19	nr	62.66
50mm dia.	16.96	21.72	2.21	56.07	nr	77.79
65mm dia.	24.12	30.89	2.72	69.01	nr	99.90
80mm dia.	33.16	42.47	3.21	81.44	nr	123.91
100mm dia.	74.61	95.56	4.44	112.65	nr	208.21
Cross						
15mm dia.	2.52	3.23	1.00	25.37	nr	28.60
20mm dia.	3.95	5.06	1.33	33.74	nr	38.80
25mm dia.	5.01	6.42	1.51	38.31	nr	44.73
32mm dia.	6.68	8.56	1.77	44.91	nr	53.46
40mm dia.	8.99	11.51	2.02	51.25	nr	62.77
50mm dia.	13.98	17.91	2.42	61.40	nr	79.31
65mm dia.	19.95	25.55	2.97	75.35	nr	100.91
80mm dia.	26.53	33.98	3.50	88.80	nr	122.78
100mm dia.	48.24	61.79	4.84	122.80	nr	184.59

S:PIPED SUPPLY SYSTEMS

Item	Net Price £	Material £	Labour hours	Labour £	Unit	Total rate £
S10 : COLD WATER (cont'd)						
Y11 - PIPELINE ANCILLARIES						
VALVES						
Regulators						
Gunmetal; self-acting two port thermostat; single seat; screwed; normally closed; with adjustable or fixed bleed device						
25mm dia.	337.04	431.68	1.46	37.04	nr	**468.72**
32mm dia.	346.77	444.15	1.45	36.79	nr	**480.94**
40mm dia.	370.65	474.72	1.55	39.34	nr	**514.05**
50mm dia.	446.19	571.48	1.68	42.62	nr	**614.10**
Self acting temperature regulator for storage calorifier; integral sensing element and pocket; screwed ends						
15mm dia.	331.61	424.72	1.32	33.49	nr	**458.21**
25mm dia.	363.99	466.19	1.52	38.57	nr	**504.75**
32mm dia.	470.05	602.04	1.79	45.42	nr	**647.46**
40mm dia.	575.01	736.46	1.99	50.49	nr	**786.95**
50mm dia.	672.15	860.88	2.26	57.34	nr	**918.22**
Self acting temperature regulator for storage calorifier; integral sensing element and pocket; flanged ends; bolted connection						
15mm dia.	486.80	623.49	0.61	15.48	nr	**638.97**
25mm dia.	557.14	713.58	0.72	18.27	nr	**731.85**
32mm dia.	702.29	899.49	0.94	23.85	nr	**923.34**
40mm dia.	831.81	1065.37	1.03	26.13	nr	**1091.50**
50mm dia.	965.79	1236.97	1.18	29.94	nr	**1266.91**
Chrome plated thermostatic mixing valves including non-return valves and inlet swivel connections with strainers; copper compression fittings						
15mm dia.	122.82	157.30	0.69	17.51	nr	**174.81**
Chrome plated thermostatic mixing valves including non-return valves and inlet swivel connections with angle pattern combined isolating valves and strainers; copper compression fittings						
15mm dia.	128.87	165.06	0.69	17.51	nr	**182.57**
Gunmetal thermostatic mixing valves including non-return valves and inlet swivel connections with strainers; copper compression fittings						
15mm dia.	111.50	142.81	0.69	17.51	nr	**160.32**
Gunmetal thermostatic mixing valves including non-return valves and inlet swivel connections with angle pattern combined isolating valves and strainers; copper compression fittings						
15mm dia.	117.23	150.15	0.69	17.51	nr	**167.66**

S:PIPED SUPPLY SYSTEMS

Item	Net Price £	Material £	Labour hours	Labour £	Unit	Total rate £
Ball float valves						
Bronze, equilibrium; copper float; working pressure cold services up to 16 bar; flanged ends; BS 4504 Table 16/21; bolted connections						
25mm dia.	289.20	370.41	1.04	26.39	nr	**396.79**
32mm dia.	401.21	513.87	1.22	30.95	nr	**544.83**
40mm dia.	544.13	696.91	1.38	35.01	nr	**731.93**
50mm dia.	876.66	1122.82	1.66	42.12	nr	**1164.93**
65mm dia.	916.98	1174.46	1.93	48.97	nr	**1223.43**
80mm dia.	1133.62	1451.92	2.16	54.80	nr	**1506.73**
Heavy, equilibrium; with long tail and backnut; copper float; screwed for iron						
25mm dia.	158.86	203.46	1.58	40.09	nr	**243.55**
32mm dia.	246.83	316.14	1.78	45.16	nr	**361.30**
40mm dia.	268.06	343.33	1.90	48.21	nr	**391.53**
50mm dia.	438.72	561.91	2.65	67.24	nr	**629.15**
Brass, ball valve; BS 1212; copper float; screwed						
15mm dia.	8.99	11.51	0.25	6.34	nr	**17.85**
22mm dia	15.15	19.40	0.29	7.36	nr	**26.76**
28mm dia	57.04	73.06	0.35	8.88	nr	**81.94**
Gate valves						
DZR copper alloy wedge non-rising stem; capillary joint to copper						
15mm dia.	9.95	12.75	0.84	21.31	nr	**34.06**
22mm dia.	12.22	15.65	1.01	25.63	nr	**41.28**
28mm dia.	16.52	21.16	1.19	30.19	nr	**51.35**
35mm dia.	29.65	37.97	1.38	35.01	nr	**72.98**
42mm dia.	50.36	64.50	1.62	41.10	nr	**105.60**
54mm dia.	70.56	90.37	1.94	49.22	nr	**139.59**
Cocks; capillary joints to copper						
Stopcock; brass head with gun metal body						
15mm dia.	2.94	3.76	0.45	11.42	nr	**15.18**
22mm dia.	5.67	7.27	0.46	11.67	nr	**18.94**
28mm dia.	16.13	20.66	0.54	13.70	nr	**34.36**
Lockshield stop cocks; brass head with gun metal body						
15mm dia.	7.68	9.84	0.45	11.42	nr	**21.25**
22mm dia.	11.04	14.15	0.46	11.67	nr	**25.82**
28mm dia.	19.55	25.04	0.54	13.70	nr	**38.74**
DZR stopcock; brass head with gun metal body						
15mm dia.	7.63	9.77	0.45	11.42	nr	**21.19**
22mm dia.	13.22	16.93	0.46	11.67	nr	**28.60**
28mm dia.	22.03	28.21	0.54	13.70	nr	**41.91**
Gunmetal stopcock						
35mm dia.	34.55	44.25	0.69	17.51	nr	**61.76**
42mm dia.	45.87	58.75	0.71	18.01	nr	**76.76**
54mm dia.	68.52	87.76	0.81	20.55	nr	**108.31**

S:PIPED SUPPLY SYSTEMS

Item	Net Price £	Material £	Labour hours	Labour £	Unit	Total rate £
S10 : COLD WATER (cont'd)						
VALVES (cont'd)						
Cocks; capillary joints to copper (cont'd)						
Double union stopcock						
15mm dia.	14.71	18.84	0.60	15.22	nr	**34.06**
22mm dia.	18.09	23.17	0.60	15.22	nr	**38.39**
28mm dia.	33.45	42.85	0.69	17.51	nr	**60.35**
Double union DZR stopcock						
15mm dia.	15.99	20.48	0.60	15.22	nr	**35.70**
22mm dia.	19.66	25.18	0.61	15.48	nr	**40.66**
28mm dia.	36.36	46.57	0.69	17.51	nr	**64.08**
Double union gun metal stopcock						
35mm dia.	60.58	77.58	0.63	15.98	nr	**93.57**
42mm dia.	83.13	106.47	0.67	17.00	nr	**123.46**
54mm dia.	130.78	167.50	0.85	21.57	nr	**189.06**
Double union stopcock with easy clean cover						
15mm dia.	17.09	21.89	0.60	15.22	nr	**37.11**
22mm dia.	21.33	27.31	0.61	15.48	nr	**42.79**
28mm dia.	39.65	50.78	0.69	17.51	nr	**68.29**
Combined stopcock and drain						
15mm dia.	16.85	21.58	0.67	17.00	nr	**38.58**
22mm dia.	20.69	26.50	0.68	17.25	nr	**43.75**
Combined DZR stopcock and drain						
15mm dia.	22.26	28.51	0.67	17.00	nr	**45.51**
Gate valve						
DZR copper alloy wedge non-rising stem; compression joint to copper						
15mm dia.	9.95	12.75	0.84	21.31	nr	**34.06**
22mm dia.	12.22	15.65	1.01	25.63	nr	**41.28**
28mm dia.	16.52	21.16	1.19	30.19	nr	**51.35**
35mm dia.	29.65	37.97	1.38	35.01	nr	**72.98**
42mm dia.	50.36	64.50	1.62	41.10	nr	**105.60**
54mm dia.	70.56	90.37	1.94	49.22	nr	**139.59**
Cocks; compression joints to copper						
Stopcock; brass head gun metal body						
15mm dia.	3.69	4.72	0.42	10.66	nr	**15.38**
22mm dia.	6.48	8.29	0.42	10.66	nr	**18.95**
28mm dia.	16.89	21.63	0.45	11.42	nr	**33.05**
Lockshield stopcock; brass head gun metal body						
15mm dia.	8.01	10.26	0.42	10.66	nr	**20.92**
22mm dia.	11.30	14.47	0.42	10.66	nr	**25.13**
28mm dia.	23.98	30.71	0.45	11.42	nr	**42.12**

S:PIPED SUPPLY SYSTEMS

Item	Net Price £	Material £	Labour hours	Labour £	Unit	Total rate £
DZR Stopcock						
15mm dia.	8.83	11.31	0.38	9.64	nr	**20.95**
22mm dia.	14.49	18.56	0.39	9.90	nr	**28.46**
28mm dia.	23.99	30.73	0.40	10.15	nr	**40.88**
35mm dia.	46.41	59.45	0.52	13.19	nr	**72.64**
42mm dia.	65.83	84.32	0.54	13.70	nr	**98.02**
54mm dia.	89.80	115.01	0.63	15.98	nr	**130.99**
DZR Lockshield stopcock						
15mm dia.	10.88	13.93	0.38	9.64	nr	**23.57**
22mm dia.	17.05	21.84	0.39	9.90	nr	**31.73**
Combined stop/draincock						
15mm dia.	13.63	17.45	0.22	5.58	nr	**23.03**
22mm dia.	17.59	22.53	0.45	11.42	nr	**33.94**
DZR Combined stop/draincock						
15mm dia.	17.08	21.87	0.41	10.40	nr	**32.27**
22mm dia.	23.40	29.97	0.42	10.66	nr	**40.63**
Stopcock to polyethylene						
15mm dia.	8.68	11.11	0.38	9.64	nr	**20.75**
20mm dia.	13.48	17.26	0.39	9.90	nr	**27.15**
25mm dia.	16.51	21.15	0.40	10.15	nr	**31.30**
Draw off coupling						
15mm dia.	5.35	6.85	0.38	9.64	nr	**16.49**
DZR Draw off coupling						
15mm dia.	7.38	9.45	0.38	9.64	nr	**19.09**
22mm dia.	8.54	10.93	0.39	9.90	nr	**20.83**
Draw off elbow						
15mm dia.	5.47	7.01	0.38	9.64	nr	**16.65**
22mm dia.	6.28	8.04	0.39	9.90	nr	**17.93**
Lockshield drain cock						
15mm dia.	4.65	5.96	0.41	10.40	nr	**16.36**
Check valves						
DZR copper alloy and bronze, WRC approved cartridge double check valve; BS 6282; working pressure cold services up to 10 bar at 65°C; screwed ends						
32mm dia.	57.48	73.62	1.38	35.01	nr	**108.63**
40mm dia.	70.86	90.76	1.62	41.10	nr	**131.86**
50mm dia.	100.72	129.00	1.94	49.22	nr	**178.22**

S:PIPED SUPPLY SYSTEMS

Item	Net Price £	Material £	Labour hours	Labour £	Unit	Total rate £
S10 : COLD WATER (cont'd)						
Y20 - PUMPS						
Packaged cold water pressure booster set; fully automatic; 3 phase supply; includes fixing in position; electrical work elsewhere.						
Pressure booster set						
0.75 l/s at 30m head	3253.15	4166.60	9.38	237.99	nr	**4404.59**
1.5 l/s at 30m head	3863.12	4947.85	9.38	237.99	nr	**5185.84**
3 l/s at 30m head	4719.97	6045.29	10.38	263.36	nr	**6308.65**
6 l/s at 30m head	10659.89	13653.08	10.38	263.36	nr	**13916.44**
12 l/s at 30m head	13317.59	17057.03	12.38	314.10	nr	**17371.14**
0.75 l/s at 50m head	3703.37	4743.24	9.38	237.99	nr	**4981.23**
1.5 l/s at 50m head	4719.97	6045.29	9.38	237.99	nr	**6283.28**
3 l/s at 50m head	5213.76	6677.73	10.38	263.36	nr	**6941.09**
6 l/s at 50m head	11905.96	15249.03	10.38	263.36	nr	**15512.39**
12 l/s at 50m head	14624.66	18731.12	12.38	314.10	nr	**19045.23**
0.75 l/s at 70m head	4051.92	5189.66	9.38	237.99	nr	**5427.65**
1.5 l/s at 70m head	5210.86	6674.01	9.38	237.99	nr	**6912.00**
3 l/s at 70m head	5547.79	7105.56	10.38	263.36	nr	**7368.92**
6 l/s at 70m head	12998.09	16647.82	10.38	263.36	nr	**16911.18**
12 l/s at 70m head	15801.02	20237.79	12.38	314.10	nr	**20551.90**
Automatic sump pump for clear and drainage water; single stage centrifugal pump, pressure tight electric motor; single phase supply; includes fixing in position; electrical work elsewhere						
Single pump						
1 l/s at 2.68m total head	162.42	208.03	3.50	88.80	nr	**296.83**
1 l/s at 4.68m total head	178.78	228.98	3.50	88.80	nr	**317.78**
1 l/s at 6.68m total head	236.04	302.31	3.50	88.80	nr	**391.12**
2 l/s at 4.38m total head	236.04	302.31	4.00	101.49	nr	**403.80**
2 l/s at 6.38m total head	236.04	302.31	4.00	101.49	nr	**403.80**
2 l/s at 8.38m total head	303.81	389.12	4.00	101.49	nr	**490.60**
3 l/s at 3.7m total head	236.04	302.31	4.50	114.17	nr	**416.49**
3 l/s at 5.7m total head	303.81	389.12	4.50	114.17	nr	**503.29**
4 l/s at 2.9m total head	236.04	302.31	5.00	126.86	nr	**429.17**
4 l/s at 4.9m total head	303.81	389.12	5.00	126.86	nr	**515.98**
4 l/s at 6.9m total head	842.49	1079.05	5.00	126.86	nr	**1205.91**
Extra for high level alarm box with single float switch, local alarm and volt free contacts for remote alarm.	247.23	316.65	-	-	nr	**316.65**

S:PIPED SUPPLY SYSTEMS

Item	Net Price £	Material £	Labour hours	Labour £	Unit	Total rate £
Duty/standby pump unit						
1 l/s at 2.68m total head	308.48	395.10	5.00	126.86	nr	**521.96**
1 l/s at 4.68m total head	338.87	434.01	5.00	126.86	nr	**560.87**
1 l/s at 6.68 total head	458.05	586.67	5.00	126.86	nr	**713.53**
2 l/s at 4.38m total head	458.05	586.67	5.50	139.55	nr	**726.21**
2 l/s at 6.38m total head	458.05	586.67	5.50	139.55	nr	**726.21**
2 l/s at 8.38m total head	586.59	751.29	5.50	139.55	nr	**890.84**
3 l/s at 3.7m total head	458.05	586.67	6.00	152.23	nr	**738.90**
3 l/s at 5.7m total head	586.59	751.29	6.00	152.23	nr	**903.53**
4 l/s at 2.9m total head	458.05	586.67	6.50	164.92	nr	**751.59**
4 l/s at 4.9m total head	586.59	751.29	6.50	164.92	nr	**916.21**
4 /s at 6.9m total head	1558.78	1996.47	7.00	177.60	nr	**2174.07**
Extra for 4nr float switches for pump on, off and high level alarm	258.30	330.83	-	-	nr	**330.83**
Extra for dual pump control panel, internal wall mounted IP54, including volt free contacts	1230.00	1575.37	4.00	101.49	nr	**1676.86**

S:PIPED SUPPLY SYSTEMS

Item	Net Price £	Material £	Labour hours	Labour £	Unit	Total rate £
S10 : COLD WATER (cont'd)						
Y21 - TANKS						
Cisterns; fibreglass; complete with ball valve, fixing plate and fitted covers						
Rectangular						
70 litres capacity	182.40	261.01	1.33	33.74	nr	**294.75**
110 litres capacity	189.60	271.31	1.40	35.52	nr	**306.83**
170 litres capacity	200.40	286.76	1.61	40.85	nr	**327.61**
280 litres capacity	326.40	467.06	1.61	40.85	nr	**507.91**
420 litres capacity	342.00	489.39	1.99	50.49	nr	**539.88**
710 litres capacity	564.00	807.06	3.31	83.98	nr	**891.04**
840 litres capacity	685.20	980.49	3.60	91.34	nr	**1071.83**
1590 litres capacity	958.80	1372.00	13.32	337.95	nr	**1709.95**
2275 litres capacity	1198.80	1715.43	20.18	512.00	nr	**2227.43**
3365 litres capacity	1492.80	2136.13	24.50	621.61	nr	**2757.74**
4545 litres capacity	1786.80	2556.83	29.91	758.87	nr	**3315.70**
Cisterns; polypropylene; complete with ball valve, fixing plate and cover; includes placing in position						
Rectangular						
18 litres capacity	8.79	11.16	1.00	25.37	nr	**36.53**
68 litres capacity	24.11	30.61	1.00	25.37	nr	**55.98**
91 litres capacity	24.58	31.22	1.00	25.37	nr	**56.59**
114 litres capacity	32.66	41.48	1.00	25.37	nr	**66.85**
182 litres capacity	58.40	74.16	1.00	25.37	nr	**99.53**
227 litres capacity	58.87	74.75	1.00	25.37	nr	**100.12**
Circular						
114 litres capacity	27.45	34.85	1.00	25.37	nr	**60.23**
227 litres capacity	41.65	52.88	1.00	25.37	nr	**78.26**
318 litres capacity	111.78	141.94	1.00	25.37	nr	**167.32**
455 litres capacity	127.04	161.31	1.00	25.37	nr	**186.68**
Steel sectional water storage tank; hot pressed steel tank to BS 1564 TYPE 1; 5mm plate; pre-insulated and complete with all connections and fittings to comply with BS EN 13280; 2001 and WRAS water supply (water fittings) regulations 1999; externally flanged base and sides; cost of erection (on prepared base) is included within the net price, labour cost allows for offloading and positioning materials						
Note - Rates are based on the most economical tank size for each volume, and the cost will vary with differing tank dimensions, for the same volume						
Volume, size						
4,900 litres, 3.66m x 1.22m x 1,22m (h)	4778.29	5970.47	6.00	152.23	nr	**6122.70**
20,300 litres, 3.66m x 2.4m x 2.4m (h)	10178.52	12718.06	12.00	304.46	nr	**13022.52**
52,000 litres, 6.1m x 3.6m x 2.4m (h)	18148.79	22676.91	19.00	482.07	nr	**23158.98**
94,000 litres, 7.3m x 3.6m x 3.6m (h)	27303.65	34115.91	28.00	710.41	nr	**34826.32**
140,000 litres, 9.7m x 6.1m x 2.44m (h)	35658.70	44555.55	28.00	710.41	nr	**45265.96**

S:PIPED SUPPLY SYSTEMS

Item	Net Price £	Material £	Labour hours	Labour £	Unit	Total rate £
GRP sectional water storage tank; pre-insulated and complete with all connections and fittings to comply with BS EN 13280; 2001 and WRAS water supply (water fittings) regulations 1999; externally flanged base and sides; cost of erection (on prepared base) is included within the net price, labour cost allows for offloading and positioning materials						
Note - Rates are based on the most economical tank size for each volume, and the cost will vary with differing tank dimensions, for the same volume						
Volume, size						
4,500 litres, 3m x 1m x 1.5m (h)	3359.51	4197.71	5.00	126.86	nr	**4324.57**
10,000 litres, 2.5m x 2m x 2m (h)	4465.60	5579.76	7.00	177.60	nr	**5757.36**
20,000 litres, 4m x 2.5m x 2m (h)	6423.30	8025.92	10.00	253.72	nr	**8279.64**
30,000 litres 5m x 3m x 2m (h)	8015.38	10015.21	12.00	304.46	nr	**10319.68**
40,000 litres, 5m x 4m x 2m (h)	9391.96	11735.26	12.00	304.46	nr	**12039.72**
50,000 litres, 5m x 4m x 2.5m (h)	11578.91	14467.85	14.00	355.21	nr	**14823.06**
60,000 litres, 6m x 4m x 2.5m (h)	13340.62	16669.11	16.00	405.95	nr	**17075.06**
70,000 litres, 7m x 4m x 2.5m (h)	14782.54	18470.79	16.00	405.95	nr	**18876.74**
80,000 litres, 8m x 4m x 2.5m (h)	16802.15	20994.28	16.00	405.95	nr	**21400.23**
90,000 litres, 6m x 5m x 3m (h)	17123.08	21395.29	16.00	405.95	nr	**21801.24**
105,000 litres, 7m x 5m x 3m (h)	18927.20	23649.54	24.00	608.93	nr	**24258.46**
120,000 litres, 8m x 5m x 3m (h)	19971.39	24954.25	24.00	608.93	nr	**25563.18**
135,000 litres, 9m x 6m x 2.5m (h)	23019.13	28762.41	24.00	608.93	nr	**29371.33**
144,000 litres, 8m x 6m x 3m (h)	22228.26	27774.21	24.00	608.93	nr	**28383.13**

S:PIPED SUPPLY SYSTEMS

Item	Net Price £	Material £	Labour hours	Labour £	Unit	Total rate £
S10 : COLD WATER (cont'd)						
Y25 - CLEANING AND CHEMICAL TREATMENT						
Electromagnetic water conditioner, complete with control box; maximum inlet pressure 16 bar; electrical work elsewhere						
Connection size, nominal flow rate at 50mbar						
20mm dia, 0.3l/s	1386.75	1776.14	1.25	31.71	nr	1807.85
25mm dia, 0.6l/s	1961.88	2512.75	1.45	36.79	nr	2549.54
32mm dia, 1.2l/s	2848.75	3648.65	1.55	39.33	nr	3687.98
40mm dia, 1.7l/s	3466.88	4440.34	1.65	41.86	nr	4482.20
50mm dia, 3.4l/s	4649.38	5954.87	1.75	44.40	nr	5999.27
65mm dia, 5.2l/s	5149.25	6595.11	1.90	48.21	nr	6643.31
100mm dia, 30.5l/s	16866.75	21602.76	3.00	76.12	nr	21678.88
Ultraviolet water sterilising unit, complete with control unit; UV lamp housed in quartz tube; unit complete with UV intensity sensor, flushing and discharge valve and facilities for remote alarm; electrical work elsewhere						
Maximum flow rate (at 250J/m^2 exposure), connection size						
0.82l/s, 40mm dia	3149.75	4034.17	1.98	50.24	nr	4084.40
1.28l/s, 40mm dia	4063.50	5204.49	1.98	50.24	nr	5254.73
2.00l/s, 40mm dia	4536.50	5810.30	1.98	50.24	nr	5860.54
4.14l/s, 50mm dia	7310.00	9362.57	2.10	53.28	nr	9415.86
1.28l/s, 40mm dia	6181.25	7916.88	1.98	50.24	nr	7967.12
2.00l/s, 40mm dia	10121.12	12963.04	1.98	50.24	nr	13013.27
4.14l/s, 50mm dia	13104.25	16783.79	2.10	53.28	nr	16837.07
7.4l/s, 80mm dia	16313.13	20893.69	3.60	91.34	nr	20985.03
16.8l/s 100mm dia	18602.88	23826.38	3.60	91.34	nr	23917.72
32.3l/s 100mm dia	23596.25	30221.84	3.60	91.34	nr	30313.18
Base exchange water softener complete with resin tank, brine tank and consumption data monitoring facilities Capacities of softeners are based on 300ppm hardness and quoted in m^3 of softened water produced. Design flow rates are recommended for continuous use						
Simplex configuration						
Design flow rate, min-max softened water produced						
1l/s, 5.8m^3-11.2m^3	1591.00	2037.74	8.00	202.98	nr	2240.71
1.3l/s, 11.7m^3-21.4m^3	2214.50	2836.31	8.00	202.98	nr	3039.28
1.3l/s, 15.5m^3-28.5m^3	2537.00	3249.36	10.00	253.72	nr	3503.08
1.6l/s, 23.3m^3-42.7m^3	3251.88	4164.97	10.00	253.72	nr	4418.69
1.6l/s, 38.8m^3-71.2m^3	3676.50	4708.82	12.00	304.46	nr	5013.29
1.9l/s, 11.7m^3-21.4m^3	3676.50	4708.82	12.00	304.46	nr	5013.29
3.2l/s, 19.4m^3-35.6m^3	3988.25	5108.11	12.00	304.46	nr	5412.57
4.4l/s, 31m^3-57m^3	4719.25	6044.37	15.00	380.58	nr	6424.95
5.1l/s, 46.6m^3-85.4m^3	7084.25	9073.44	15.00	380.58	nr	9454.02
5.1l/s, 77.7m^3-142.4m^3	8686.00	11124.94	18.00	456.69	nr	11581.64

S:PIPED SUPPLY SYSTEMS

Item	Net Price £	Material £	Labour hours	Labour £	Unit	Total rate £
Duplex configuration						
Design flow rate, min-max softened water produced						
2l/s, 5.8m³-22.4m³	2612.25	3345.74	12.00	304.46	nr	**3650.21**
2.6l/s, 11.7m³-42.8m³	3622.75	4639.98	12.00	304.46	nr	**4944.44**
2.6l/s, 15.5m³-57m³	3805.50	4874.05	15.00	380.58	nr	**5254.62**
3.2l/s, 23.3m³-85.4m³	4977.25	6374.81	15.00	380.58	nr	**6755.39**
3.2l/s, 38.8m³-142.4m³	5987.75	7669.05	18.00	456.69	nr	**8125.74**
3.8l/s, 11.7m³-42.8m³	6471.50	8288.63	18.00	456.69	nr	**8745.33**
6.4l/s, 19.4m³-71.2m³	6976.75	8935.75	18.00	456.69	nr	**9392.45**
8.8l/s, 31.1m³-114m³	8030.25	10285.06	23.00	583.55	nr	**10868.62**
10.2l/s, 46.6m³-170.8m³	12996.75	16646.11	23.00	583.55	nr	**17229.66**
10.2l/s, 77.7m³-284.8m³	16168.00	20707.81	27.00	685.04	nr	**21392.85**
Triplex configuration						
Design flow rate, min-max softened water produced						
3l/s, 5.8m³-33.6m³	3606.63	4619.33	15.00	380.58	nr	**4999.91**
3.9l/s, 11.7m³-64.2m³	5246.00	6719.02	15.00	380.58	nr	**7099.60**
3.9l/s, 15.5m³-85.5m³	5557.75	7118.31	18.00	456.69	nr	**7575.00**
4.8l/s, 23.3m³-128.1m³	7331.50	9390.11	18.00	456.69	nr	**9846.81**
4.8l/s, 38.8m³-213.6m³	8471.00	10849.57	22.00	558.18	nr	**11407.75**
5.7l/s, 11.7m³-64.2m³	9524.50	12198.88	22.00	558.18	nr	**12757.07**
9.6l/s, 19.4m³-106.8m³	10352.25	13259.06	22.00	558.18	nr	**13817.24**
13.2l/s, 31.1m³-171.0m³	11921.75	15269.26	27.00	685.04	nr	**15954.30**
15.3l/s, 46.6m³-256.2m³	19371.50	24810.82	27.00	685.04	nr	**25495.86**
15.3l/s, 77.7m³-427.2m³	24284.25	31103.02	32.00	811.90	nr	**31914.93**

S:PIPED SUPPLY SYSTEMS

Item	Net Price £	Material £	Labour hours	Labour £	Unit	Total rate £
S10 : COLD WATER (cont'd)						
Y50 -THERMAL INSULATION						
Flexible closed cell walled insulation; Class 1/ Class O; adhesive joints; including around fittings						
6mm wall thickness						
15mm diameter	0.68	0.89	0.15	3.08	m	3.97
22mm diameter	0.80	1.05	0.15	3.08	m	4.13
28mm diameter	1.02	1.33	0.15	3.08	m	4.41
9mm wall thickness						
15mm diameter	0.72	0.94	0.15	3.08	m	4.02
22mm diameter	0.87	1.14	0.15	3.08	m	4.22
28mm diameter	0.96	1.25	0.15	3.08	m	4.33
35mm diameter	1.11	1.46	0.15	3.08	m	4.53
42mm diameter	1.29	1.69	0.15	3.08	m	4.77
54mm diameter	1.86	2.43	0.15	3.08	m	5.51
13mm wall thickness						
15mm diameter	0.93	1.22	0.15	3.08	m	4.30
22mm diameter	1.14	1.49	0.15	3.08	m	4.56
28mm diameter	1.39	1.81	0.15	3.08	m	4.89
35mm diameter	1.50	1.96	0.15	3.08	m	5.04
42mm diameter	1.79	2.34	0.15	3.08	m	5.41
54mm diameter	2.22	2.90	0.15	3.08	m	5.98
67mm diameter	3.50	4.58	0.15	3.08	m	7.65
76mm diameter	4.06	5.31	0.15	3.08	m	8.38
108mm diameter	5.94	7.77	0.15	3.08	m	10.85
19mm wall thickness						
15mm diameter	1.56	2.04	0.15	3.08	m	5.12
22mm diameter	1.90	2.49	0.15	3.08	m	5.56
28mm diameter	2.61	3.41	0.15	3.08	m	6.49
35mm diameter	3.04	3.98	0.15	3.08	m	7.05
42mm diameter	3.60	4.71	0.15	3.08	m	7.79
54mm diameter	4.58	5.99	0.15	3.08	m	9.07
67mm diameter	5.48	7.16	0.15	3.08	m	10.24
76mm diameter	6.30	8.24	0.22	4.51	m	12.75
108mm diameter	9.29	12.15	0.22	4.51	m	16.66
25mm wall thickness						
15mm diameter	3.10	4.05	0.15	3.08	m	7.13
22mm diameter	3.42	4.47	0.15	3.08	m	7.55
28mm diameter	3.88	5.07	0.15	3.08	m	8.15
35mm diameter	4.31	5.64	0.15	3.08	m	8.71
42mm diameter	4.59	6.01	0.15	3.08	m	9.08
54mm diameter	5.42	7.09	0.15	3.08	m	10.17
67mm diameter	6.61	8.65	0.15	3.08	m	11.73
76mm diameter	7.68	10.04	0.22	4.51	m	14.56

S:PIPED SUPPLY SYSTEMS

Item	Net Price £	Material £	Labour hours	Labour £	Unit	Total rate £
32mm wall thickness						
15mm diameter	3.93	5.13	0.15	3.08	m	**8.21**
22mm diameter	4.31	5.64	0.15	3.08	m	**8.71**
28mm diameter	5.01	6.55	0.15	3.08	m	**9.63**
35mm diameter	5.08	6.64	0.15	3.08	m	**9.72**
42mm diameter	5.91	7.73	0.15	3.08	m	**10.80**
54mm diameter	7.44	9.73	0.15	3.08	m	**12.81**
76mm diameter	11.13	14.56	0.22	4.51	m	**19.07**

Note : For mineral fibre sectional insulation; bright class O foil faced; bright class O foil taped joints; 19mm aluminium bands rates, refer to section T31 - Low Temperature Hot Water Heating, Y50 Thermal Insulation

Note : For mineral fibre sectional insulation; bright class O foil faced; bright class O foil taped joints; 22 swg plain/embossed aluminium cladding; pop riveted rates, refer to section T31 - Low Temperature Hot Water Heating, Y50 Thermal Insulation

Note : For mineral fibre sectional insulation; bright class O foil faced; bright class O foil taped joints; 0.8mm polyisobutylene sheeting; welded joints rates, refer to section T31 - Low Temperature Hot Water Heating, Y50 Thermal Insulation

S:PIPED SUPPLY SYSTEMS

Item	Net Price £	Material £	Labour hours	Labour £	Unit	Total rate £
S11 : HOT WATER						
Y10 - PIPELINES						
Note : For pipework rates refer to section S10 - Cold Water						
Y11 - PIPELINE ANCILLARIES						
Note : For rates for ancillaries refer to section S10 - Cold Water						
Y23 - STORAGE CYLINDERS/CALORIFIERS/ WATER HEATERS						
CYLINDERS						
Insulated copper storage cylinders; BS 699; includes placing in position						
Grade 3 (maximum 10m working head)						
BS size 6; 115 litres capacity; 400mm dia.; 1050mm high	103.96	133.15	1.50	38.10	nr	**171.25**
BS size 7; 120 litres capacity; 450mm dia.; 900mm high	122.92	157.44	2.00	50.74	nr	**208.18**
BS size 8; 144 litres capacity; 450mm dia.; 1050mm high	130.66	167.35	2.80	71.07	nr	**238.42**
Grade 4 (maximum 6m working head)						
BS size 2; 96 litres capacity; 400mm dia.; 900mm high	80.49	103.09	1.50	38.10	nr	**141.18**
BS size 7; 120 litres capacity; 450mm dia.; 900mm high	84.59	108.34	1.50	38.10	nr	**146.44**
BS size 8; 144 litres capacity; 450mm dia.; 1050mm high	100.98	129.34	1.50	38.10	nr	**167.43**
BS size 9; 166 litres capacity; 450mm dia.; 1200mm high	147.65	189.11	1.50	38.10	nr	**227.20**
Storage cylinders; brazed copper construction; to BS 699; screwed bosses; includes placing in position						
Tested to 2.2 bar, 15m maximum head						
144 litres	457.62	586.11	3.00	76.19	nr	**662.31**
160 litres	517.32	662.57	3.00	76.19	nr	**738.77**
200 litres	533.23	682.96	3.76	95.38	nr	**778.34**
255 litres	606.84	777.24	3.76	95.38	nr	**872.62**
290 litres	819.73	1049.90	3.76	95.38	nr	**1145.29**
370 litres	939.12	1202.82	4.50	114.29	nr	**1317.11**
450 litres	1279.27	1638.48	5.00	126.86	nr	**1765.34**
Tested to 2.55 bar, 17m maximum head						
550 litres	1385.20	1774.15	5.00	126.86	nr	**1901.01**
700 litres	1619.43	2074.15	6.02	152.84	nr	**2226.99**
800 litres	1873.07	2399.01	6.54	165.83	nr	**2564.84**
900 litres	2026.85	2595.97	8.00	202.98	nr	**2798.95**
1000 litres	2138.90	2739.48	8.00	202.98	nr	**2942.45**
1250 litres	2342.58	3000.36	13.16	333.84	nr	**3334.20**
1500 litres	3564.80	4565.77	15.15	384.42	nr	**4950.19**
2000 litres	4278.07	5479.31	17.24	437.45	nr	**5916.76**
3000 litres	6009.23	7696.57	24.39	618.83	nr	**8315.39**

S:PIPED SUPPLY SYSTEMS

Item	Net Price £	Material £	Labour hours	Labour £	Unit	Total rate £
Indirect cylinders; copper; bolted top; up to 5 tappings for connections; BS 1586; includes placing in position						
Grade 3, tested to 1.45 bar, 10m maximum head						
74 litres capacity	208.45	266.98	1.50	38.10	nr	**305.08**
96 litres capacity	212.23	271.83	1.50	38.10	nr	**309.92**
114 litres capacity	217.91	279.10	1.50	38.10	nr	**317.20**
117 litres capacity	226.17	289.68	2.00	50.74	nr	**340.42**
140 litres capacity	233.06	298.50	2.50	63.43	nr	**361.93**
162 litres capacity	325.92	417.44	3.00	76.19	nr	**493.63**
190 litres capacity	356.25	456.28	3.51	89.02	nr	**545.30**
245 litres capacity	416.89	533.94	3.80	96.47	nr	**630.42**
280 litres capacity	739.01	946.52	4.00	101.49	nr	**1048.01**
360 litres capacity	799.65	1024.19	4.50	114.29	nr	**1138.47**
440 litres capacity	928.51	1189.23	4.50	114.29	nr	**1303.51**
Grade 2, tested to 2.2 bar, 15m maximum head						
117 litres capacity	301.28	385.87	2.00	50.74	nr	**436.62**
140 litres capacity	327.83	419.88	2.50	63.43	nr	**483.31**
162 litres capacity	375.19	480.54	2.80	71.07	nr	**551.61**
190 litres capacity	435.85	558.23	3.00	76.19	nr	**634.42**
245 litres capacity	526.80	674.72	4.00	101.49	nr	**776.21**
280 litres capacity	841.34	1077.58	4.00	101.49	nr	**1179.07**
360 litres capacity	928.51	1189.23	4.50	114.29	nr	**1303.51**
440 litres capacity	1099.04	1407.64	4.50	114.29	nr	**1521.93**
Grade 1, tested 3.65 bar, 25m maximum head						
190 litres capacity	648.07	830.04	3.00	76.19	nr	**906.23**
245 litres capacity	737.10	944.08	3.00	76.19	nr	**1020.27**
280 litres capacity	1042.21	1334.85	4.00	101.49	nr	**1436.34**
360 litres capacity	1318.88	1689.21	4.50	114.29	nr	**1803.49**
440 litres capacity	1601.19	2050.79	4.50	114.29	nr	**2165.08**
Indirect cylinders, including manhole; BS 853						
Grade 3, tested to 1.5 bar, 10m maximum head						
550 litres capacity	1347.82	1726.28	5.21	132.14	nr	**1858.42**
700 litres capacity	1492.24	1911.25	6.02	152.84	nr	**2064.09**
800 litres capacity	1732.92	2219.51	6.54	165.83	nr	**2385.34**
1000 litres capacity	2166.16	2774.40	7.04	178.68	nr	**2953.08**
1500 litres capacity	2503.11	3205.96	10.00	253.72	nr	**3459.68**
2000 litres capacity	3465.86	4439.04	16.13	409.22	nr	**4848.26**
Grade 2, tested to 2.55 bar, 15m maximum head						
550 litres capacity	1494.31	1913.90	5.21	132.14	nr	**2046.04**
700 litres capacity	1870.23	2395.38	6.02	152.84	nr	**2548.22**
800 litres capacity	1973.61	2527.78	6.54	165.83	nr	**2693.60**
1000 litres capacity	2443.52	3129.64	7.04	178.68	nr	**3308.31**
1500 litres capacity	3007.41	3851.86	10.00	253.72	nr	**4105.58**
2000 litres capacity	3759.27	4814.83	16.13	409.22	nr	**5224.06**
Grade 1, tested to 4 bar, 25m maximum head						
550 litres capacity	1738.65	2226.85	5.21	132.14	nr	**2358.99**
700 litres capacity	1973.61	2527.78	6.02	152.84	nr	**2680.62**
800 litres capacity	2114.59	2708.34	6.54	165.83	nr	**2874.17**
1000 litres capacity	2819.46	3611.13	7.04	178.68	nr	**3789.81**
1500 litres capacity	3383.32	4333.33	10.00	253.72	nr	**4587.05**
2000 litres capacity	4135.18	5296.30	16.13	409.22	nr	**5705.52**

S:PIPED SUPPLY SYSTEMS

Item	Net Price £	Material £	Labour hours	Labour £	Unit	Total rate £
S11 : HOT WATER (cont'd)						
CYLINDERS (cont'd)						
Indirect cylinders; mild steel, welded throughout, galvanised; with bolted connections; includes placing in position						
3.2mm plate						
136 litres capacity	744.97	954.15	2.50	63.43	nr	1017.58
159 litres capacity	802.29	1027.56	2.80	71.07	nr	1098.63
182 litres capacity	985.69	1262.46	3.00	76.19	nr	1338.65
227 litres capacity	1214.89	1556.02	3.00	76.19	nr	1632.22
273 litres capacity	1417.17	1815.09	4.00	101.49	nr	1916.58
364 litres capacity	1650.42	2113.84	4.50	114.29	nr	2228.13
455 litres capacity	1704.68	2183.34	5.00	126.86	nr	2310.20
683 litres capacity	2758.69	3533.30	6.02	152.84	nr	3686.14
910 litres capacity	3204.00	4103.66	7.04	178.68	nr	4282.33
CALORIFIERS						
Storage calorifiers; copper; heater battery capable of raising temperature of contents from 10°C to 65°C in one hour; static head not exceeding 1.35 bar; BS 853						
Horizontal; primary LPHW at 82°C/71°C						
400 litres capacity	1929.46	2471.23	7.04	178.68	nr	2649.90
1000 litres capacity	3087.13	3953.96	8.00	202.98	nr	4156.94
2000 litres capacity	6174.29	7907.97	14.08	357.35	nr	8265.32
3000 litres capacity	7621.37	9761.38	25.00	634.30	nr	10395.67
4000 litres capacity	9261.41	11861.92	40.00	1014.88	nr	12876.79
4500 litres capacity	10437.20	13367.86	50.00	1268.59	nr	14636.45
Vertical; primary LPHW at 82°C/71°C						
400 litres capacity	2174.51	2785.09	7.04	178.68	nr	2963.77
1000 litres capacity	3494.76	4476.06	8.00	202.98	nr	4679.04
2000 litres capacity	6656.65	8525.76	14.08	357.35	nr	8883.11
3000 litres capacity	8320.82	10657.22	25.00	634.30	nr	11291.52
4000 litres capacity	10206.84	13072.82	40.00	1014.88	nr	14087.70
4500 litres capacity	11538.18	14777.98	50.00	1268.59	nr	16046.58
Storage calorifiers; galvanised mild steel; heater battery capable of raising temperature of contents from 10°C to 65°C in one hour; static head not exceeding 1.35 bar; BS 853						
Horizontal; primary LPHW at 82°C/71°C						
400 litres capacity	1929.46	2471.23	7.04	178.68	nr	2649.90
1000 litres capacity	3087.13	3953.96	8.00	202.98	nr	4156.94
2000 litres capacity	6174.29	7907.97	14.08	357.35	nr	8265.32
3000 litres capacity	7621.37	9761.38	25.00	634.30	nr	10395.67
4000 litres capacity	9261.41	11861.92	40.00	1014.88	nr	12876.79
4500 litres capacity	10437.20	13367.86	50.00	1268.59	nr	14636.45
Vertical; primary LPHW at 82°C/71°C						
400 litres capacity	2174.51	2785.09	7.04	178.68	nr	2963.77
1000 litres capacity	3494.76	4476.06	8.00	202.98	nr	4679.04
2000 litres capacity	6656.65	8525.76	14.08	357.35	nr	8883.11
3000 litres capacity	8320.82	10657.22	25.00	634.30	nr	11291.52
4000 litres capacity	10206.84	13072.82	40.00	1014.88	nr	14087.70
4500 litres capacity	11538.18	14777.98	50.00	1268.59	nr	16046.58

S:PIPED SUPPLY SYSTEMS

Item	Net Price £	Material £	Labour hours	Labour £	Unit	Total rate £
LOCAL ELECTRICAL HOT WATER HEATERS						
Unvented multi-point water heater; providing hot water for one or more outlets: Used with conventional taps or mixers: Factory fitted temperature and pressure relief valve: Externally adjustable thermostat: Elemental 'on' indicator: Fitted with 1 metre of 3 core cable: Electrical supply and connection excluded						
5 litre capacity, 2.2kW rating	153.90	197.11	1.50	38.06	nr	**235.17**
10 litre capacity, 2.2kW rating	171.64	219.83	1.50	38.06	nr	**257.89**
15 litre capacity, 2.2kW rating	190.73	244.29	1.50	38.06	nr	**282.34**
30 litre capacity, 3kW rating	339.93	435.38	2.00	50.74	nr	**486.12**
50 litre capacity, 3kW rating	357.88	458.37	2.00	50.74	nr	**509.11**
80 litre capacity, 3kW rating	704.20	901.93	2.00	50.74	nr	**952.68**
100 litre capacity, 3kW rating	756.00	968.28	2.00	50.74	nr	**1019.02**
Accessories						
Pressure reducing valve and expansion kit	128.55	164.65	2.00	50.74	nr	**215.39**
Thermostatic blending valve	68.41	87.62	1.00	25.37	nr	**112.99**
Y50 - INSULATION						
Refer to sections S10 - Cold Water and T31 - Low Temperature Hot Water Heating for details						

S:PIPED SUPPLY SYSTEMS

Item	Net Price £	Material £	Labour hours	Labour £	Unit	Total rate £
S32 : NATURAL GAS						
Y10 - PIPELINES						
MEDIUM DENSITY POLYETHELENE - YELLOW						
Pipe; laid underground; electrofusion joints in the running length; BS 6572; BGT PL2 standards						
Coiled service pipe						
20mm dia.	0.75	0.99	0.37	9.39	m	**10.38**
25mm dia.	0.98	1.29	0.41	10.40	m	**11.70**
32mm dia.	1.61	2.12	0.47	11.93	m	**14.05**
63mm dia.	6.14	8.10	0.60	15.22	m	**23.32**
90mm dia.	8.07	10.65	0.90	22.84	m	**33.48**
Mains service pipe						
63mm dia.	6.14	7.86	0.60	15.22	m	**23.09**
90mm dia.	8.07	10.34	0.90	22.83	m	**33.17**
125mm dia.	15.58	19.95	1.20	30.45	m	**50.40**
180mm dia.	32.17	41.20	1.50	38.06	m	**79.26**
250mm dia.	59.21	75.84	1.75	44.40	m	**120.24**
Extra over fittings, electrofusion joints						
Straight connector						
32mm dia.	4.82	6.17	0.47	11.92	nr	**18.10**
63mm dia.	9.05	11.59	0.58	14.72	nr	**26.31**
90mm dia.	13.36	17.11	0.67	17.00	nr	**34.11**
125mm dia.	25.00	32.02	0.83	21.06	nr	**53.08**
180mm dia.	45.02	57.66	1.25	31.71	nr	**89.38**
Reducing connector						
90 x 63mm dia.	18.65	23.89	0.67	17.00	nr	**40.89**
125 x 90mm dia.	37.38	47.88	0.83	21.06	nr	**68.93**
180 x 125mm dia.	68.60	87.86	1.25	31.71	nr	**119.58**
Bend; 45°						
90mm dia.	36.03	46.15	0.67	17.00	nr	**63.15**
125mm dia.	58.94	75.49	0.83	21.06	nr	**96.55**
180mm dia.	133.71	171.25	1.25	31.71	nr	**202.97**
Bend; 90°						
63mm dia.	23.35	29.91	0.58	14.72	nr	**44.62**
90mm dia.	36.03	46.15	0.67	17.00	nr	**63.15**
125mm dia.	58.94	75.49	0.83	21.06	nr	**96.55**
180mm dia.	133.71	171.25	1.25	31.71	nr	**202.97**
Extra over malleable iron fittings, compression joints						
Straight connector						
20mm dia.	6.87	8.80	0.38	9.64	nr	**18.44**
25mm dia.	7.50	9.61	0.45	11.42	nr	**21.02**
32mm dia.	8.41	10.77	0.50	12.69	nr	**23.46**
63mm dia.	16.88	21.62	0.85	21.57	nr	**43.19**

S:PIPED SUPPLY SYSTEMS

Item	Net Price £	Material £	Labour hours	Labour £	Unit	Total rate £
Straight connector; polyethylene to MI						
20mm dia.	5.84	7.48	0.31	7.87	nr	**15.35**
25mm dia.	6.37	8.16	0.35	8.88	nr	**17.04**
32mm dia.	7.11	9.11	0.40	10.15	nr	**19.26**
63mm dia.	11.92	15.27	0.65	16.50	nr	**31.76**
Straight connector; polyethylene to FI						
20mm dia.	5.84	7.48	0.31	7.87	nr	**15.35**
25mm dia.	6.37	8.16	0.35	8.88	nr	**17.04**
32mm dia.	7.11	9.11	0.40	10.15	nr	**19.26**
63mm dia.	11.92	15.27	0.75	19.03	nr	**34.30**
Elbow						
20mm dia.	8.94	11.45	0.38	9.64	nr	**21.09**
25mm dia.	9.75	12.49	0.45	11.42	nr	**23.91**
32mm dia.	10.93	14.00	0.50	12.69	nr	**26.69**
63mm dia.	21.95	28.11	0.80	20.30	nr	**48.41**
Equal tee						
20mm dia.	11.92	15.27	0.53	13.45	nr	**28.72**
25mm dia.	13.94	17.85	0.55	13.96	nr	**31.81**
32mm dia.	17.55	22.48	0.64	16.24	nr	**38.72**

SCREWED STEEL

For rates for steel pipework refer to section T31 - Low Temperature Hot Water Heating

PIPE IN PIPE

Note - for pipe in pipe, a sleeve size two pipe sizes bigger than actual pipe size has been allowed. All rates refer to actual pipe size.

Black steel pipes –
Screwed and socketed joints; BS 1387: 1985 up to 50mm pipe size. Butt welded joints; BS 1387: 1985 65mm pipe size and above.

Item	Net Price £	Material £	Labour hours	Labour £	Unit	Total rate £
Pipe						
25mm	9.19	11.77	1.73	43.89	m	**55.66**
32mm	12.30	15.75	1.95	49.48	m	**65.23**
40mm	13.92	17.83	2.16	54.80	m	**72.63**
50mm	18.39	23.55	2.44	61.91	m	**85.46**
65mm	23.51	30.11	2.95	74.85	m	**104.96**
80mm	30.96	39.65	3.42	86.77	m	**126.43**
100mm	38.64	49.49	4.00	101.49	m	**150.98**

FIXINGS

Refer to steel pipes Section T31; For pipe in pipe, a bracket Size two pipe sizes bigger than actual pipe size has to be allowed. For minimum fixing dimensions, refer to the Tables and Memoranda at the rear of the book.

S:PIPED SUPPLY SYSTEMS

Item	Net Price £	Material £	Labour hours	Labour £	Unit	Total rate £
S32 : NATURAL GAS (cont'd)						
PIPE IN PIPE (cont'd)						
Extra over black steel pipes - Screwed pipework; black malleable iron fittings; BS 143. Welded pipework; butt welded steel fittings; BS 1965						
Bend, 90°						
25mm	3.66	4.69	2.91	73.83	m	**78.52**
32mm	4.79	6.14	3.45	87.53	m	**93.67**
40mm	8.03	10.28	5.34	135.49	m	**145.77**
50mm	9.25	11.85	6.53	165.68	m	**177.53**
65mm	14.67	18.79	8.84	224.29	m	**243.08**
80mm	24.02	30.76	10.73	272.24	m	**303.01**
100mm	33.55	42.97	12.76	323.75	m	**366.72**
Bend, 45°						
25mm	6.52	8.35	2.91	73.83	m	**82.18**
32mm	9.98	12.78	3.45	87.53	m	**100.32**
40mm	9.88	12.65	5.34	135.49	m	**148.14**
50mm	11.42	14.63	6.53	165.68	m	**180.31**
65mm	13.27	17.00	8.84	224.29	m	**241.28**
80mm	20.41	26.14	10.73	272.24	m	**298.38**
100mm	28.51	36.52	12.76	323.75	m	**360.26**
Equal tee						
25mm	4.64	5.94	4.18	106.05	m	**112.00**
32mm	7.04	9.02	4.94	125.34	m	**134.35**
40mm	29.05	37.21	7.28	184.71	m	**221.91**
50mm	30.73	39.36	8.48	215.15	m	**254.51**
65mm	34.04	43.60	11.47	291.02	m	**334.61**
80mm	99.63	127.61	14.23	361.04	m	**488.65**
100mm	112.76	144.42	17.92	454.66	m	**599.09**
Copper pipe; capillary or compression joints in the running length; EN1057 R250 (TX) formerly BS 2871 Table X						
Plastic coated gas service pipe for corrosive environments, fixed vertically or at low level with brackets measured separately						
15mm dia. (yellow)	1.69	2.16	0.85	21.57	m	**23.74**
22mm dia. (yellow)	3.23	4.14	0.96	24.37	m	**28.51**
28mm dia. (yellow)	4.11	5.26	1.06	26.89	m	**32.16**
Copper pipe; capillary or compression joints in the running length; EN1057 R250 (TY) formerly BS 2871 Table Y						
Plastic coated gas and cold water service pipe for corrosive environments, fixed vertically or at low level with brackets measured separately (Refer to Copper Pipe Table X Section)						
15mm dia. (yellow)	2.53	3.24	0.61	15.48	m	**18.72**
22mm dia. (yellow)	4.46	5.71	0.69	17.51	m	**23.22**

S:PIPED SUPPLY SYSTEMS

Item	Net Price £	Material £	Labour hours	Labour £	Unit	Total rate £
FIXINGS						
Refer to Section S10 Cold Water						
Extra over copper pipes; capillary fittings; BS 864						
Refer to Section S10 Cold Water						
GAS BOOSTERS						
Complete skid mounted gas booster set, including AV mounts, flexible connections, low pressure switch, control panel and NRV (for run/standby unit); 3 phase supply; in accordance with IGE/UP/2; includes delivery, offloading and positioning						
Single unit						
Flow, pressure range						
0-200 m³/hour, 0.1-2.6 kPa	1175.27	1505.27	10.00	253.72	nr	**1758.99**
0-200 m³/hour, 0.1-4.0 kPa	1341.01	1717.55	10.00	253.72	nr	**1971.27**
0-200 m³/hour, 0.1-7 kPa	1560.56	1998.75	10.00	253.72	nr	**2252.47**
0-200 m³/hour, 0.1-9.5 kPa	1625.14	2081.46	10.00	253.72	nr	**2335.18**
0-200 m³/hour 0.1-11.0 kPa	1815.64	2325.45	10.00	253.72	nr	**2579.17**
0-400 m³/hour, 0.1-4.0 kPa	1480.92	1896.75	10.00	253.72	nr	**2150.47**
0-1000 m³/hour, 0.1-7.4 kPa	2051.33	2627.33	10.00	253.72	nr	**2881.05**
50-1000 m³/hour, 0.1-16.0 kPa	3832.53	4908.67	20.00	507.44	nr	**5416.11**
50-1000 m³/hour, 0.1-24.5 kPa	4332.98	5549.64	20.00	507.44	nr	**6057.08**
50-1000 m³/hour, 0.1-31.0 kPa	4928.15	6311.93	20.00	507.44	nr	**6819.37**
50-1000 m³/hour, 0.1-41.0 kPa	5439.37	6966.69	20.00	507.44	nr	**7474.13**
50-1000 m³/hour, 0.1-51.0 kPa	5643.85	7228.59	20.00	507.44	nr	**7736.03**
100-1800 m³/hour, 3.5-23.5 kPa	5791.31	7417.45	20.00	507.44	nr	**7924.89**
100-1800 m³/hour, 4.5-27.0 kPa	6246.56	8000.53	20.00	507.44	nr	**8507.96**
100-1800 m³/hour, 6.0-32.5 kPa	6918.14	8860.68	20.00	507.44	nr	**9368.12**
100-1800 m³/hour, 7.2-39.0 kPa	7958.87	10193.65	20.00	507.44	nr	**10701.08**
100-1800 m³/hour, 9.0-42.0 kPa	8365.70	10714.70	20.00	507.44	nr	**11222.14**
Run/Standby unit						
Flow, pressure range						
0-200 m³/hour, 0.1-2.6 kPa	6712.58	8597.40	16.00	405.95	nr	**9003.35**
0-200 m³/hour, 0.1-4.0 kPa	6877.24	8808.30	16.00	405.95	nr	**9214.25**
0-200 m³/hour, 0.1-7 kPa	7036.52	9012.31	16.00	405.95	nr	**9418.26**
0-200 m³/hour, 0.1-9.5 kPa	7210.88	9235.62	16.00	405.95	nr	**9641.57**
0-200 m³/hour 0.1-11.0 kPa	7330.34	9388.63	16.00	405.95	nr	**9794.58**
0-400 m³/hour, 0.1-4.0 kPa	8083.72	10353.55	16.00	405.95	nr	**10759.50**
0-1000 m³/hour, 0.1-7.4 kPa	10304.02	13197.28	25.00	634.30	nr	**13831.58**
50-1000 m³/hour, 0.1-16.0 kPa	14122.55	18088.02	25.00	634.30	nr	**18722.32**
50-1000 m³/hour, 0.1-24.5 kPa	15966.17	20449.32	25.00	634.30	nr	**21083.61**
50-1000 m³/hour, 0.1-31.0 kPa	17805.48	22805.08	25.00	634.30	nr	**23439.38**
50-1000 m³/hour, 0.1-41.0 kPa	19650.17	25167.74	25.00	634.30	nr	**25802.04**
50-1000 m³/hour, 0.1-51.0 kPa	19418.63	24871.19	25.00	634.30	nr	**25505.48**
100-1800 m³/hour, 3.5-23.5 kPa	17338.39	22206.83	25.00	634.30	nr	**22841.13**
100-1800 m³/hour, 4.5-27.0 kPa	18700.92	23951.95	25.00	634.30	nr	**24586.25**
100-1800 m³/hour, 6.0-32.5 kPa	19617.89	25126.39	25.00	634.30	nr	**25760.69**
100-1800 m³/hour, 7.2-39.0 kPa	23826.02	30516.13	25.00	634.30	nr	**31150.43**
100-1800 m³/hour, 9.0-42.0 kPa	25044.34	32076.54	25.00	634.30	nr	**32710.83**

S:PIPED SUPPLY SYSTEMS

Item	Net Price £	Material £	Labour hours	Labour £	Unit	Total rate £
S41 : FUEL OIL STORAGE/DISTRIBUTION						
Y10 - PIPELINES						
For pipework rates refer to Section T31 - Low Temperature Hot Water Heating						
Y21 - TANKS						
Fuel storage tanks; mild steel; with all necessary screwed bosses; oil resistant joint rings; includes placing in position						
Rectangular						
1360 litres (300 gallon) capacity; 2mm plate	299.17	383.18	12.03	305.22	nr	**688.40**
2730 litres (600 gallon) capacity; 2.5mm plate	399.77	512.02	18.60	471.92	nr	**983.94**
4550 litres (1000 gallon) capacity; 3mm plate	838.14	1073.48	25.00	634.30	nr	**1707.78**
Fuel storage tanks; 5mm plate mild steel to BS 799 type J; complete with raised neck manhole with bolted cover, screwed connections, vent and fill connections, drain valve, gauge and overfill alarm; includes placing in position; excludes pumps and control panel						
Nominal capacity, size						
5,600 litres, 2.5m x 1.5m x 1.5m high	1956.25	2505.55	20.00	507.44	nr	**3012.98**
Extra for bund unit (internal use)	1125.00	1440.89	30.00	761.16	nr	**2202.05**
Extra for external use with bund (watertight)	712.50	912.56	2.00	50.74	nr	**963.31**
10,200 litres, 3.05m x 1.83m x 1.83m high	2431.25	3113.92	30.00	761.16	nr	**3875.08**
Extra for bund unit (internal use)	1637.50	2097.29	40.00	1014.88	nr	**3112.17**
Extra for external use with bund (watertight)	875.00	1120.69	2.00	50.74	nr	**1171.44**
15,000 litres, 3.75m x 2m x 2m high	3075.00	3938.43	40.00	1014.88	nr	**4953.31**
Extra for bund unit (internal use)	2200.00	2817.74	55.00	1395.45	nr	**4213.19**
Extra for external use with bund (watertight)	1025.00	1312.81	2.00	50.74	nr	**1363.55**
20,000 litres, 4m x 2.5m x 2m high	4056.25	5195.20	50.00	1268.59	nr	**6463.80**
Extra for bund unit (internal use)	2668.75	3418.11	65.00	1649.17	nr	**5067.28**
Extra for external use with bund (watertight)	1212.50	1552.96	2.00	50.74	nr	**1603.70**
Extra for BMS output (all tank sizes)	425.00	544.34	-	-	nr	**544.34**
Fuel storage tanks; plastic; with all necessary screwed bosses; oil resistant joint rings; includes placing in position						
Cylindrical; horizontal						
1250 litres (285 gallon) capacity	242.05	310.01	3.73	94.64	nr	**404.65**
1350 litres (300 gallon) capacity	222.38	284.82	4.30	109.10	nr	**393.92**
2500 litres (550 gallon) capacity	380.82	487.75	4.88	123.81	nr	**611.56**
Cylindrical; vertical						
1365 litres (300 gallon) capacity	153.54	196.66	3.73	94.64	nr	**291.30**
2600 litres (570 gallon) capacity	233.67	299.29	4.88	123.81	nr	**423.10**
3635 litres (800 gallon) capacity	364.30	466.59	4.88	123.81	nr	**590.41**
5455 litres (1200 gallon) capacity	530.70	679.72	5.95	150.96	nr	**830.68**
Bunded tanks						
1135 litres (250 gallon) capacity	492.19	630.40	4.30	109.10	nr	**739.50**
1590 litres (350 gallon) capacity	585.54	749.96	4.88	123.81	nr	**873.77**
2500 litres (550 gallon) capacity	695.99	891.42	5.95	150.96	nr	**1042.38**
5000 litres (1100 gallon) capacity	1404.44	1798.80	6.53	165.68	nr	**1964.48**

S:PIPED SUPPLY SYSTEMS

Item	Net Price £	Material £	Labour hours	Labour £	Unit	Total rate £
S60 : FIRE HOSE REELS						
Y10 - PIPELINES						
For pipework rates refer to section S10 - Cold water						
Y11 - PIPELINE ANCILLARIES						
For rates for ancillaries refer to section S10 - Cold water						
Hose reels; automatic; connection to 25mm screwed joint; reel with 30.5 metres, 19mm rubber hose; suitable for working pressure up to 7 bar						
Reels						
Non-swing pattern	171.50	213.96	3.75	91.18	nr	**305.15**
Recessed non-swing pattern	219.94	274.39	3.75	91.18	nr	**365.58**
Swinging pattern	229.02	285.73	3.75	91.18	nr	**376.91**
Recessed swinging pattern	236.27	294.77	3.75	91.18	nr	**385.95**
Hose reels; manual; connection to 25mm screwed joint; reel with 30.5 metres, 19mm rubber hose; suitable for working pressure up to 7 bar						
Reels						
Non-swing pattern	172.73	215.50	3.25	79.03	nr	**294.52**
Recessed non-swing pattern	202.75	252.94	3.25	79.03	nr	**331.97**
Swinging pattern	218.79	272.96	3.25	79.03	nr	**351.98**
Recessed swinging pattern	226.66	282.79	3.25	79.03	nr	**361.81**

S:PIPED SUPPLY SYSTEMS

Item	Net Price £	Material £	Labour hours	Labour £	Unit	Total rate £
S61 : DRY RISERS						
Y10 - PIPELINES						
For pipework rates refer to section S10 - Cold water						
Y11 - PIPELINE ANCILLARIES VALVES (BS 5041, parts 2 and 3)						
Bronze/gunmetal inlet breeching for pumping in with 65mm dia. instantaneous male coupling; with cap, chain and 25mm drain valve						
Double inlet with back pressure valve, flanged to steel	155.37	193.83	1.75	42.55	nr	**236.38**
Quadruple inlet with back pressure valve, flanged to steel	347.88	434.01	1.75	42.55	nr	**476.56**
Bronze/gunmetal gate type outlet valve with 65mm dia. instantaneous female coupling; cap and chain; wheel head secured by padlock and leather strap						
Flanged to BS 4504 PN6 (bolted connection to counter flanges measured separately)	124.60	155.45	1.75	42.55	nr	**198.01**
Bronze/gunmetal landing type outlet valve, with 65mm dia. instantaneous female coupling; cap and chain; wheelhead secured by padlock and leather strap; bolted connections to counter flanges measured separately						
Horizontal, flanged to BS 4504 PN6	135.70	169.30	1.50	36.47	nr	**205.77**
Oblique, flanged to BS 4504 PN6	135.70	169.30	1.50	36.47	nr	**205.77**
Air valve, screwed joints to steel						
25mm dia.	26.97	33.64	0.55	13.37	nr	**47.02**
INLET BOXES (BS 5041, part 5)						
Steel dry riser inlet box with hinged wire glazed door suitably lettered (fixing by others)						
610 x 460 x 325mm; double inlet	165.36	206.30	3.00	72.95	nr	**279.25**
610 x 610 x 356mm; quadruple inlet	308.61	385.02	3.00	72.95	nr	**457.97**
OUTLET BOXES (BS 5041, part 5)						
Steel dry riser outlet box with hinged wire glazed door suitably lettered (fixing by others)						
610 x 460 x 325; single outlet	161.48	201.46	3.00	72.95	nr	**274.41**

S:PIPED SUPPLY SYSTEMS

Item	Net Price £	Material £	Labour hours	Labour £	Unit	Total rate £
S63 : SPRINKLERS						
Y10 - PIPELINES						
Prefabricated black steel pipework; screwed joints, including all couplings, unions and the like to BS 1387:1985; includes fixing to backgrounds, with brackets measured separately						
Heavy weight						
25mm dia	3.74	4.67	0.47	11.43	m	**16.09**
32mm dia	4.68	5.84	0.53	12.89	m	**18.73**
40mm dia	5.45	6.80	0.58	14.10	m	**20.90**
50mm dia	7.62	9.51	0.63	15.32	m	**24.83**
FIXINGS						
For steel pipes; black malleable iron. For minimum fixing distances, refer to the Tables and Memoranda at the rear of the book						
Pipe ring, single socket, black malleable iron						
25mm dia	0.90	1.15	0.12	3.04	nr	**4.20**
32mm dia	0.95	1.22	0.14	3.55	nr	**4.77**
40mm dia	1.23	1.58	0.15	3.81	nr	**5.38**
50mm dia	1.55	1.99	0.16	4.06	nr	**6.04**
Extra Over channel sections for fabricated hangers and brackets						
Galvanised steel; including inserts, bolts, nuts, washers; fixed to backgrounds						
41 x 21mm	2.40	3.07	0.15	3.81	m	**6.88**
41 x 41mm	3.75	4.80	0.15	3.81	m	**8.61**
Threaded rods; metric thread; including nuts, washers etc						
12mm dia x 600mm long for ring clips	1.30	1.66	0.10	2.54	nr	**4.20**
Extra over for black malleable iron fittings; BS 143						
Plain plug, solid						
25mm dia	0.95	1.19	0.40	9.73	nr	**10.91**
32mm dia	1.02	1.27	0.44	10.70	nr	**11.97**
40mm dia	1.36	1.70	0.48	11.67	nr	**13.37**
50mm dia	1.82	2.27	0.56	13.62	nr	**15.89**
Concentric reducing socket						
32mm dia	1.50	1.87	0.48	11.67	nr	**13.54**
40mm dia	1.95	2.43	0.55	13.37	nr	**15.81**
50mm dia	3.65	4.55	0.60	14.59	nr	**19.14**
Elbow; 90° female/female						
25mm dia	0.95	1.19	0.44	10.70	nr	**11.88**
32mm dia	1.61	2.01	0.53	12.89	nr	**14.90**
40mm dia	2.71	3.38	0.60	14.59	nr	**17.97**
50mm dia	3.17	3.95	0.65	15.81	nr	**19.76**

S:PIPED SUPPLY SYSTEMS

Item	Net Price £	Material £	Labour hours	Labour £	Unit	Total rate £
S63 : SPRINKLERS (cont'd)						
Fittings; Prefabricated black steel screwed joints (cont'd)						
Tee						
25mm dia equal	1.28	1.60	0.51	12.40	nr	**14.00**
32mm dia reducing to 25mm dia	3.01	3.76	0.54	13.13	nr	**16.89**
40mm dia	5.21	6.50	0.65	15.81	nr	**22.31**
50mm dia	5.92	7.39	0.78	18.97	nr	**26.35**
Cross tee						
25mm dia equal	4.59	5.73	1.16	28.21	nr	**33.93**
32mm dia	6.65	8.30	1.40	34.04	nr	**42.34**
40mm dia	8.57	10.69	1.60	38.91	nr	**49.60**
50mm dia	13.91	17.35	1.68	40.85	nr	**58.20**
Prefabricated black steel pipework; welded joints, including all couplings, unions and the like to BS 1387:1985; fixing to backgrounds						
Heavy weight						
65mm dia	8.47	10.57	0.65	15.81	m	**26.37**
80mm dia	10.78	13.45	0.70	17.02	m	**30.47**
100mm dia	15.04	18.76	0.85	20.67	m	**39.43**
150mm dia	23.60	29.44	1.15	27.96	m	**57.41**
FIXINGS						
For steel pipes; black malleable iron. For minimum fixing distances, refer to the Tables and Memoranda at the rear of the book						
Pipe ring, single socket, black malleable iron						
65mm dia	2.24	2.87	0.30	7.61	nr	**10.48**
80mm dia	2.69	3.45	0.35	8.88	nr	**12.33**
100mm dia	4.07	5.21	0.40	10.15	nr	**15.36**
150mm dia	9.23	11.82	0.77	19.54	nr	**31.36**
Extra over fittings						
Reducer (one size down)						
65mm dia	7.08	8.83	2.70	65.65	nr	**74.49**
80mm dia	7.19	8.97	2.86	69.54	nr	**78.51**
100mm dia	8.16	10.18	3.22	78.30	nr	**88.48**
150mm dia	19.38	24.18	4.20	102.13	nr	**126.30**
Elbow; 90°						
65mm dia	5.32	6.64	3.06	74.41	nr	**81.04**
80mm dia	6.08	7.59	3.40	82.67	nr	**90.26**
100mm dia	9.35	11.67	3.70	89.97	nr	**101.63**
150mm dia	24.20	30.19	5.20	126.44	nr	**156.63**
Branch bend						
65mm dia	5.32	6.64	3.60	87.54	nr	**94.17**
80mm dia	6.08	7.59	3.80	92.40	nr	**99.98**
100mm dia	9.35	11.67	5.10	124.01	nr	**135.67**
150mm dia	24.20	30.19	7.50	182.37	nr	**212.56**

S:PIPED SUPPLY SYSTEMS

Item	Net Price £	Material £	Labour hours	Labour £	Unit	Total rate £
Prefabricated black steel pipe; victaulic joints; including all couplings and the like to BS 1387: 1985; fixing to backgrounds						
Heavy weight						
65mm dia	8.85	11.04	0.70	17.02	m	**28.06**
80mm dia	11.50	14.35	0.78	18.97	m	**33.31**
100mm dia	16.76	20.91	0.93	22.61	m	**43.52**
150mm dia	33.38	41.64	1.25	30.39	m	**72.04**
FIXINGS						
For fixings refer to For steel pipes; black malleable iron.						
Extra over fittings						
Coupling						
65mm dia	7.27	9.07	0.26	6.32	nr	**15.39**
80mm dia	7.80	9.73	0.26	6.32	nr	**16.05**
100mm dia	10.22	12.75	0.32	7.78	nr	**20.53**
150mm dia	17.99	22.44	0.35	8.51	nr	**30.95**
Reducer						
65mm dia	7.86	9.81	0.48	11.67	nr	**21.48**
80mm dia	9.50	11.85	0.43	10.46	nr	**22.31**
100mm dia	20.57	25.66	0.46	11.19	nr	**36.85**
150mm dia	22.19	27.68	0.45	10.94	nr	**38.63**
Elbow; any degree						
65mm dia	9.44	11.78	0.56	13.62	nr	**25.39**
80mm dia	9.64	12.03	0.63	15.32	nr	**27.35**
100mm dia	12.91	16.11	0.71	17.26	nr	**33.37**
150mm dia	27.63	34.47	0.80	19.45	nr	**53.92**
Equal tee						
65mm dia	17.01	21.22	0.74	17.99	nr	**39.22**
80mm dia	18.02	22.48	0.83	20.18	nr	**42.66**
100mm dia	20.15	25.14	0.94	22.86	nr	**48.00**
150mm dia	49.54	61.81	1.05	25.53	nr	**87.34**

S:PIPED SUPPLY SYSTEMS

Item	Net Price £	Material £	Labour hours	Labour £	Unit	Total rate £
S63 : SPRINKLERS (cont'd)						
Y11 - PIPELINE ANCILLARIES						
SPRINKLER HEADS						
Sprinkler heads; brass body; frangible glass bulb; manufactured to standard operating temperature of 57-141°C; quick response; RTI<50						
Conventional pattern; 15mm dia.	3.18	3.96	0.15	3.65	nr	**7.61**
Sidewall pattern; 15mm dia.	4.67	5.82	0.15	3.65	nr	**9.47**
Conventional pattern; 15mm dia.; satin chrome plated	3.73	4.65	0.15	3.65	nr	**8.30**
Sidewall pattern; 15mm dia.; satin chrome plated	4.85	6.05	0.15	3.65	nr	**9.70**
Fully concealed; fusible link; 15mm dia	11.65	14.53	0.15	3.65	nr	**18.18**
VALVES						
Wet system alarm valves; including internal non-return valve; working pressure up to 12.5 bar; BS4504 PN16 flanged ends; bolted connections						
100mm dia.	1008.53	1258.23	25.00	607.89	nr	**1866.12**
150mm dia.	1231.94	1536.96	25.00	607.89	nr	**2144.85**
Wet system by-pass alarm valves; including internal non-return valve; working pressure up to 12.5 bar; BS4504 PN16 flanged ends; bolted connections						
100mm dia.	1777.75	2217.91	25.00	607.89	nr	**2825.80**
150mm dia.	2247.68	2804.20	25.00	607.89	nr	**3412.09**
Alternate system wet/dry alarm station; including butterfly valve, wet alarm valve, dry pipe differential pressure valve and pressure gauges; working pressure up to 10.5 bar; BS4505 PN16 flanged ends; bolted connections						
100mm dia.	2203.25	2748.76	40.00	972.62	nr	**3721.39**
150mm dia.	2567.46	3203.15	40.00	972.62	nr	**4175.77**
Alternate system wet/dry alarm station; including electrically operated butterfly valve, water supply accelerator set, wet alarm valve, dry pipe differential pressure valve and pressure gauges; working pressure up to 10.5 bar; BS4505 PN16 flanged ends; bolted connections						
100mm dia.	2511.44	3133.25	45.00	1094.20	nr	**4227.45**
150mm dia.	2879.29	3592.18	45.00	1094.20	nr	**4686.38**
ALARM/GONGS						
Water operated motor alarm and gong; stainless steel and aluminium body and gong; screwed connections						
Connection to sprinkler system and drain pipework	307.42	383.53	6.00	145.89	nr	**529.42**

S:PIPED SUPPLY SYSTEMS

Item	Net Price £	Material £	Labour hours	Labour £	Unit	Total rate £
Y21 - WATER TANKS						
Note - Rates are based on the most economical tank size for each volume, and the cost will vary with differing tank dimensions, for the same volume						
Steel sectional sprinkler tank; ordinary hazard, life safety classification; two compartment tank, complete with all fittings and accessories to comply with LPCB type A requirements; cost of erection (on prepared supports) is included within net price, labour cost allows for offloading and positioning of materials						
Volume, size						
70m³, 6.1m x 4.88m x 2.44m (h)	21178.70	26462.79	24.00	608.93	nr	**27071.71**
105m³, 7.3m x 6.1m x 2.44m (h)	24852.00	31052.57	28.00	710.41	nr	**31762.99**
168m³, 9.76m x 4.88m x 3.66m (h)	30705.30	38366.27	28.00	710.41	nr	**39076.69**
211m³, 9.76m x 6.1m x 3.66m (h)	33899.00	42356.80	32.00	811.90	nr	**43168.70**
Steel sectional sprinkler tank; ordinary hazard, property protection classification; single compartment tank, complete with all fittings and accessories to comply with LPCB type A requirements; cost of erection (on prepared supports) is included within net price, labour cost allows for offloading and positioning of materials						
Volume, size						
70m³, 6.1m x 4.88m x 2.44m (h)	14933.00	18658.78	24.00	608.93	nr	**19267.71**
105m³, 7.3m x 6.1m x 2.44m (h)	18448.25	23051.09	28.00	710.41	nr	**23761.50**
168m³, 9.76m x 4.88m x 3.66m (h)	23838.30	29785.96	28.00	710.41	nr	**30496.37**
211m³, 9.76m x 6.1m x 3.66m (h)	26748.60	33422.38	32.00	811.90	nr	**34234.28**
GRP sectional sprinkler tank; ordinary hazard, life safety classification; two compartment tank, complete with all fittings and accessories to comply with LPCB type A requirements; cost of erection (on prepared supports) is included within net price, labour cost allows for offloading and positioning of materials						
Volume, size						
55m³, 6m x 4m x 3m (h)	23633.60	29530.19	14.00	355.21	nr	**29885.39**
70m³, 6m x 5m x 3m (h)	26424.08	33016.89	16.00	405.95	nr	**33422.84**
80m³, 8m x 4m x 3m (h)	27384.80	34217.31	16.00	405.95	nr	**34623.26**
105m³, 10m x 4m x 3m (h)	30988.04	38719.55	24.00	608.93	nr	**39328.48**
125m³, 10m x 5m x 3m (h)	34947.64	43667.07	24.00	608.93	nr	**44276.00**
140m³, 8m x 7m x 3m (h)	37495.33	46850.41	24.00	608.93	nr	**47459.34**
135m³, 9m x 6m x 3m (h)	36394.98	45475.52	24.00	608.93	nr	**46084.45**
160m³, 13m x 5m x 3m (h)	41166.29	51437.28	24.00	608.93	nr	**52046.21**
185m³, 12m x 6m x 3m (h)	43546.22	54411.00	24.00	608.93	nr	**55019.93**

S:PIPED SUPPLY SYSTEMS

Item	Net Price £	Material £	Labour hours	Labour £	Unit	Total rate £
S63 : SPRINKLERS (cont'd)						
Y21 - WATER TANKS (cont'd)						
GRP sectional sprinkler tank; ordinary hazard, property protection classification; single compartment tank, complete with all fittings and accessories to comply with LPCB type A requirements; cost of erection (on prepared supports) is within net price, labour cost allows for offloading and positioning of materials						
Volume, size						
55m³, 6m x 4m x 3m (h)	18181.86	22718.23	14.00	355.21	nr	**23073.44**
70m³, 6m x 5m x 3m (h)	20540.95	25665.91	16.00	405.95	nr	**26071.86**
80m³, 8m x 4m x 3m (h)	21924.72	27394.94	16.00	405.95	nr	**27800.89**
105m³, 10m x 4m x 3m (h)	25685.30	32093.78	24.00	608.93	nr	**32702.71**
125m³, 10m x 5m x 3m (h)	29044.71	36291.36	24.00	608.93	nr	**36900.29**
140m³, 8m x 7m x 3m (h)	30744.21	38414.89	24.00	608.93	nr	**39023.82**
135m³, 9m x 6m x 3m (h)	30067.95	37569.91	24.00	608.93	nr	**38178.83**
160m³, 13m x 5m x 3m (h)	35416.54	44252.96	24.00	608.93	nr	**44861.89**
185m³, 12m x 6m x 3m (h)	37205.65	46488.46	24.00	608.93	nr	**47097.39**

S:PIPED SUPPLY SYSTEMS

Item	Net Price £	Material £	Labour hours	Labour £	Unit	Total rate £
S65 : FIRE HYDRANTS						
EXTINGUISHERS						
Fire extinguishers; hand held; BS 5423; placed in position						
Water type; cartridge operated; for Class A fires						
Water type, 9 litres capacity; 55gm CO_2 cartridge; Class A fires (fire rating 13A)	72.44	90.37	1.00	24.32	nr	**114.69**
Foam type, 9 litres capacity; 75gm CO_2 Class A & B fires (fire rating 13A:183B)	85.91	107.18	1.00	24.32	nr	**131.50**
Dry powder type; cartridge operated; for Class A, B & C fires and electrical equipment fires						
Dry powder type, 1kg capacity; 12gm CO_2 cartridge; Class A, B & C fires (fire rating 5A:34B)	30.64	38.22	1.00	24.32	nr	**62.54**
Dry powder type, 2kg capacity; 28gm CO_2 cartridge; Class A, B & C fires (fire rating 13A:55B)	41.25	51.46	1.00	24.32	nr	**75.78**
Dry powder type, 4kg capacity; 90gm CO_2 cartridge; Class A, B & C fires (fire rating 21A:183B)	73.48	91.67	1.00	24.32	nr	**115.99**
Dry powder type, 9kg capacity; 190gm CO_2 cartridge; Class A, B & C fires (fire rating 43A:233B)	98.45	122.83	1.00	24.32	nr	**147.14**
Dry powder type; stored pressure type; for Class A, B & C fires and electrical equipment fires						
Dry powder type, 1kg capacity; Class A, B & C fires (fire rating 5A:34B)	28.22	35.20	1.00	24.32	nr	**59.52**
Dry powder type, 2kg capacity; Class A, B & C fires (fire rating 13A:55B)	33.50	41.79	1.00	24.32	nr	**66.10**
Dry powder type, 4kg capacity; Class A, B & C fires (fire rating 21A:183B)	62.57	78.06	1.00	24.32	nr	**102.38**
Dry powder type, 9kg capacity; Class A, B & C fires (fire rating 43A:233B)	81.07	101.14	1.00	24.32	nr	**125.46**
Carbon dioxide type; for Class B fires and electrical equipment fires						
CO2 type with hose and horn, 2kg capacity, Class B fires (fire rating 34B)	77.50	96.68	1.00	24.32	nr	**121.00**
CO2 type with hose and horn, 5kg capacity, Class B fires (fire rating 55B)	121.44	151.51	1.00	24.32	nr	**175.82**
Glass fibre blanket, in GRP container						
1100 x 1100mm	24.59	30.67	0.50	12.16	nr	**42.83**
1200 x 1200mm	26.95	33.62	0.50	12.16	nr	**45.78**
1800 x 1200mm	35.95	44.85	0.50	12.16	nr	**57.01**

S:PIPED SUPPLY SYSTEMS

Item	Net Price £	Material £	Labour hours	Labour £	Unit	Total rate £
S65 : FIRE HYDRANTS (cont'd)						
HYDRANTS						
Fire hydrants; bolted connections						
Underground hydrants, complete with frost plug to BS 750						
Sluice valve pattern type 1	202.27	252.35	4.50	109.42	nr	**361.77**
Screw down pattern type 2	146.92	183.29	4.50	109.42	nr	**292.71**
Stand pipe for underground hydrant; screwed base; light alloy						
Single outlet	116.33	145.13	1.00	24.32	nr	**169.44**
Double outlet	170.21	212.36	1.00	24.32	nr	**236.67**
64mm diameter bronze/gunmetal outlet valves						
Oblique flanged landing valve	127.33	158.85	1.00	24.32	nr	**183.17**
Oblique screwed landing valve	127.33	158.85	1.00	24.32	nr	**183.17**
Cast iron surface box; fixing by others						
400 x 200 x 100mm	111.49	139.09	1.00	24.32	nr	**163.40**
500 x 200 x 150mm	150.22	187.41	1.00	24.32	nr	**211.72**
Frost Plug	27.06	33.76	0.25	6.08	nr	**39.84**

T:MECHANICAL/COOLING/HEATING SYSTEMS

Item	Net Price £	Material £	Labour hours	Labour £	Unit	Total rate £
T10 : GAS/OIL FIRED BOILERS						
DOMESTIC						
Domestic water boilers; stove enamelled casing; electric controls; placing in position; assembling and connecting; electrical work elsewhere						
Gas fired; floor standing; connected to conventional flue						
9 to 12 kW	511.91	655.65	8.59	217.94	nr	**873.60**
12 to 15 kW	538.54	689.75	8.59	217.94	nr	**907.70**
15 to 18 kW	573.60	734.66	8.88	225.30	nr	**959.97**
18 to 21 kW	666.39	853.50	9.92	251.69	nr	**1105.19**
21 to 23 kW	738.77	946.21	10.66	270.46	nr	**1216.67**
23 to 29 kW	958.17	1227.21	11.81	299.64	nr	**1526.85**
29 to 37 kW	1133.28	1451.50	11.81	299.64	nr	**1751.14**
37 to 41 kW	1178.56	1509.49	12.68	321.72	nr	**1831.20**
Gas fired; wall hung; connected to conventional flue						
9 to 12 kW	449.92	576.26	8.59	217.94	nr	**794.20**
12 to 15 kW	580.88	743.98	8.59	217.94	nr	**961.92**
13 to 18 kW	737.19	944.19	8.59	217.94	nr	**1162.13**
Gas fired; floor standing; connected to balanced flue						
9 to 12 kW	639.47	819.03	9.16	232.41	nr	**1051.44**
12 to 15 kW	665.34	852.16	10.95	277.82	nr	**1129.99**
15 to 18 kW	716.18	917.28	11.98	303.96	nr	**1221.24**
18 to 21 kW	843.95	1080.92	12.78	324.25	nr	**1405.17**
21 to 23 kW	973.39	1246.71	12.78	324.25	nr	**1570.96**
23 to 29 kW	1240.69	1589.07	15.45	392.00	nr	**1981.06**
29 to 37 kW	2385.11	3054.83	17.65	447.81	nr	**3502.64**
Gas fired; wall hung; connected to balanced flue						
6 to 9 kW	482.82	618.39	9.16	232.41	nr	**850.79**
9 to 12 kW	552.51	707.65	9.16	232.41	nr	**940.06**
12 to 15kW	625.64	801.31	9.45	239.76	nr	**1041.08**
15 to 18kW	750.17	960.81	9.74	247.12	nr	**1207.93**
18 to 22kW	796.28	1019.87	9.74	247.12	nr	**1266.99**
Gas fired; wall hung; connected to fan flue (including flue kit)						
6 to 9kW	552.21	707.27	9.16	232.41	nr	**939.68**
9 to 12kW	615.39	788.19	9.16	232.41	nr	**1020.59**
12 to 15kW	668.03	855.60	10.95	277.82	nr	**1133.43**
15 to 18kW	719.57	921.62	11.98	303.96	nr	**1225.58**
18 to 23kW	937.57	1200.83	12.78	324.25	nr	**1525.08**
23 to 29kW	1214.13	1555.04	15.45	392.00	nr	**1947.04**
29 to 35kW	1490.61	1909.15	17.65	447.81	nr	**2356.97**
Oil fired; floor standing; connected to conventional flue						
12 to 15	1001.94	1283.28	10.38	263.36	nr	**1546.64**
15 to 19	1045.18	1338.66	12.20	309.54	nr	**1648.20**
21 to 25	1192.85	1527.78	14.30	362.82	nr	**1890.60**
26 to 32	1307.80	1675.02	15.80	400.88	nr	**2075.89**
35 to 50	1476.55	1891.15	20.46	519.11	nr	**2410.26**

T:MECHANICAL/COOLING/HEATING SYSTEMS

Item	Net Price £	Material £	Labour hours	Labour £	Unit	Total rate £
T10 : GAS/OIL FIRED BOILERS (cont'd)						
DOMESTIC (cont'd)						
Domestic water boilers; stove enamelled casing (cont'd)						
Fire place mounted natural gas fire and back boiler; cast iron water boiler; electric control box; fire output 3kW with wood surround						
10.50kW	279.71	358.24	8.88	225.30	nr	**583.55**
FORCED DRAFT						
Commercial cast iron sectional floor standing boilers; pressure jet burner; including controls, enamelled jacket, insulation, assembly and commissioning; electrical work elsewhere						
Gas fired (on/off type), connected to conventional flue						
16-26 kW; 3 sections; 125mm dia. flue	1933.10	2475.90	8.00	202.98	nr	**2678.88**
26-33 kW; 4 sections; 125mm dia. flue	2061.03	2639.75	8.00	202.98	nr	**2842.72**
33-40 kW; 5 sections; 125mm dia. flue	2331.10	2985.64	8.00	202.98	nr	**3188.62**
35-50 kW; 3 sections; 153mm dia. flue	2558.52	3276.93	8.00	202.98	nr	**3479.90**
50-65 kW; 4 sections; 153mm dia. flue	2757.52	3531.80	8.00	202.98	nr	**3734.77**
65-80 kW; 5 sections; 153mm dia. flue	2999.15	3841.29	8.00	202.98	nr	**4044.26**
80-100 kW; 6 sections; 180mm dia. flue	3485.27	4463.90	8.00	202.98	nr	**4666.88**
100-120 kW; 7 sections; 180mm dia. flue	4406.34	5643.60	8.00	202.98	nr	**5846.57**
105-140 kW; 5 sections; 180mm dia. flue	5117.04	6553.85	8.00	202.98	nr	**6756.83**
140-180 kW; 6 sections; 180mm dia. flue	5827.74	7464.11	8.00	202.98	nr	**7667.09**
180-230 kW; 7 sections; 200mm dia. flue	6538.44	8374.37	8.00	202.98	nr	**8577.34**
230-280 kW; 8 sections; 200mm dia. flue	6964.86	8920.52	8.00	202.98	nr	**9123.50**
280-330 kW; 9 sections; 200mm dia. flue	7817.70	10012.83	8.00	202.98	nr	**10215.81**
Gas fired (high/low type), connected to conventional flue						
105-140 kW; 5 sections; 180mm dia. flue	6112.02	7828.21	8.00	202.98	nr	**8031.19**
140-180 kW; 6 sections; 180mm dia. flue	6566.87	8410.78	8.00	202.98	nr	**8613.75**
180-230 kW; 7 sections; 200mm dia. flue	7448.14	9539.50	8.00	202.98	nr	**9742.47**
230-280 kW; 8 sections; 200mm dia. flue	7959.84	10194.88	8.00	202.98	nr	**10397.86**
280-330 kW; 9 sections; 200mm dia. flue	8528.40	10923.09	8.00	202.98	nr	**11126.06**
300-390 kW; 8 sections; 250mm dia. flue	10234.08	13107.71	12.00	304.46	nr	**13412.17**
390-450 kW; 9 sections; 250mm dia. flue	11271.70	14436.68	12.00	304.46	nr	**14741.15**
450-540 kW; 10 sections; 250mm dia. flue	12366.18	15838.48	12.00	304.46	nr	**16142.94**
540-600 kW; 11 sections; 300mm dia. flue	12650.46	16202.58	12.00	304.46	nr	**16507.05**
600-670 kW; 12 sections; 300mm dia. flue	15529.31	19889.79	12.00	304.46	nr	**20194.25**
670-720 kW; 13 sections; 300mm dia. flue	15813.59	20253.89	12.00	304.46	nr	**20558.35**
720-780 kW; 14 sections; 300mm dia. flue	16026.80	20526.97	12.00	304.46	nr	**20831.43**
754-812 kW; 14 sections; 400mm dia. flue	16168.94	20709.02	12.00	304.46	nr	**21013.48**
812-870 kW; 15 sections; 400mm dia. flue	20433.14	26170.56	12.00	304.46	nr	**26475.02**
870-928 kW; 16 sections; 400mm dia. flue	22138.82	28355.18	12.00	304.46	nr	**28659.64**
928-986 kW; 17 sections; 400mm dia. flue	22707.38	29083.39	12.00	304.46	nr	**29387.85**
986-1044 kW; 18 sections; 400mm dia. flue	23702.36	30357.75	12.00	304.46	nr	**30662.21**
1044-1102 kW; 19 sections; 400mm dia. flue	24270.92	31085.95	12.00	304.46	nr	**31390.41**
1102-1160 kW; 20 sections; 400mm dia. flue	24896.34	31886.98	12.00	304.46	nr	**32191.44**
1160-1218 kW; 21 sections; exceeding 400mm dia. flue	26260.88	33634.67	12.00	304.46	nr	**33939.14**

T:MECHANICAL/COOLING/HEATING SYSTEMS

Item	Net Price £	Material £	Labour hours	Labour £	Unit	Total rate £
1218-1276 kW; 22 sections; exceeding 400mm dia. flue	27607.09	35358.88	12.00	304.46	nr	**35663.35**
1276-1334 kW; 23 sections; exceeding 400mm dia. flue	28587.86	36615.04	12.00	304.46	nr	**36919.50**
1334-1392 kW; 24 sections; exceeding 400mm dia. flue	29867.12	38253.50	12.00	304.46	nr	**38557.97**
1392-1450 kW; 25 sections; exceeding 400mm dia. flue	30719.96	39345.81	12.00	304.46	nr	**39650.28**
Oil fired (on/off type), connected to conventional flue						
16-26 kW; 3 sections; 125mm dia. flue	1421.40	1820.51	8.00	202.98	nr	**2023.49**
26-33 kW; 4 sections; 125mm dia. flue	1563.54	2002.57	8.00	202.98	nr	**2205.54**
33-40 kW; 5 sections; 125mm dia. flue	1705.68	2184.62	8.00	202.98	nr	**2387.59**
35-50 kW; 3 sections; 153mm dia. flue	2004.17	2566.93	8.00	202.98	nr	**2769.90**
50-65 kW; 4 sections; 153mm dia. flue	2103.67	2694.36	8.00	202.98	nr	**2897.34**
65-80 kW; 5 sections; 153mm dia. flue	2188.96	2803.59	8.00	202.98	nr	**3006.57**
80-100 kW; 6 sections; 180mm dia. flue	3269.22	4187.18	8.00	202.98	nr	**4390.16**
100-120 kW; 7 sections; 180mm dia. flue	3695.64	4733.34	8.00	202.98	nr	**4936.31**
105-140 kW; 5 sections; 180mm dia. flue	4264.20	5461.54	8.00	202.98	nr	**5664.52**
140-180 kW; 6 sections; 180mm dia. flue	4832.76	6189.75	8.00	202.98	nr	**6392.73**
180-230 kW; 7 sections; 200mm dia. flue	5969.88	7646.16	8.00	202.98	nr	**7849.14**
230-280 kW; 8 sections; 200mm dia. flue	6538.44	8374.37	8.00	202.98	nr	**8577.34**
280-330 kW; 9 sections; 200mm dia. flue	7249.14	9284.63	8.00	202.98	nr	**9487.60**
Oil fired (high/low type), connected to conventional flue						
105-140 kW; 5 sections; 180mm dia. flue	3425.57	4387.44	8.00	202.98	nr	**4590.42**
140-180 kW; 6 sections; 180mm dia. flue	3695.64	4733.34	8.00	202.98	nr	**4936.31**
180-230 kW; 7 sections; 200mm dia. flue	4122.06	5279.49	8.00	202.98	nr	**5482.47**
230-280 kW; 8 sections; 200mm dia. flue	4861.19	6226.16	8.00	202.98	nr	**6429.14**
280-330 kW; 9 sections; 200mm dia. flue	7675.56	9830.78	8.00	202.98	nr	**10033.76**
300-390 kW; 8 sections; 250mm dia. flue	8528.40	10923.09	12.00	304.46	nr	**11227.55**
390-450 kW; 9 sections; 250mm dia. flue	9168.03	11742.32	12.00	304.46	nr	**12046.78**
450-540 kW; 10 sections; 250mm dia. flue	10376.22	13289.76	12.00	304.46	nr	**13594.22**
540-600 kW; 11 sections; 300mm dia. flue	10902.14	13963.35	12.00	304.46	nr	**14267.81**
600-670 kW; 12 sections; 300mm dia. flue	11655.48	14928.22	12.00	304.46	nr	**15232.69**
670-720 kW; 13 sections; 300mm dia. flue	13219.02	16930.79	12.00	304.46	nr	**17235.25**
720-780 kW; 14 sections; 300mm dia. flue	13432.23	17203.87	12.00	304.46	nr	**17508.33**
754-812 kW; 14 sections; 400mm dia. flue	13617.01	17440.53	12.00	304.46	nr	**17745.00**
812-870 kW; 15 sections; 400mm dia. flue	18193.92	23302.59	12.00	304.46	nr	**23607.05**
870-928 kW; 16 sections; 400mm dia. flue	18890.41	24194.64	12.00	304.46	nr	**24499.11**
928-986 kW; 17 sections; 400mm dia. flue	19544.25	25032.08	12.00	304.46	nr	**25336.54**
986-1044 kW; 18 sections; 400mm dia. flue	20326.02	26033.36	12.00	304.46	nr	**26337.83**
1044-1102 kW; 19 sections; 400mm dia. flue	21321.00	27307.72	12.00	304.46	nr	**27612.19**
1102-1160 kW; 20 sections; 400 mm dia. flue	22514.98	28836.96	12.00	304.46	nr	**29141.42**
1160-1218 kW; 21 sections; exceeding 400mm dia. flue	23765.81	30439.01	12.00	304.46	nr	**30743.47**
1218-1276 kW; 22 sections; exceeding 400mm dia. flue	24945.57	31950.04	12.00	304.46	nr	**32254.50**
1276-1334 kW; 23 sections; exceeding 400mm dia. flue	25826.84	33078.76	12.00	304.46	nr	**33383.22**
1334-1392 kW; 24 sections; exceeding 400mm dia. flue	27340.63	35017.60	12.00	304.46	nr	**35322.07**
1392-1450 kW; 25 sections; exceeding 400mm dia. flue	28342.72	36301.07	12.00	304.46	nr	**36605.53**

T:MECHANICAL/COOLING/HEATING SYSTEMS

Item	Net Price £	Material £	Labour hours	Labour £	Unit	Total rate £
T10 : GAS/OIL FIRED BOILERS (cont'd)						
FORCED DRAFT (cont'd)						
Commercial steel shell floor standing boilers; pressure jet burner; including controls, enamelled jacket, insulation, placing in position and commissioning; electrical work elsewhere						
Gas fired (on/off type), connected to conventional flue						
130-190kW	5096.34	6527.34	8.00	202.98	nr	**6730.32**
200-250kW	5772.54	7393.41	8.00	202.98	nr	**7596.39**
280-360kW	6913.80	8855.13	8.00	202.98	nr	**9058.10**
375-500kW	8426.28	10792.30	8.00	202.98	nr	**10995.27**
Gas fired (high/low type), connected to conventional flue						
130-190kW	6207.24	7950.17	8.00	202.98	nr	**8153.15**
200-250kW	6629.52	8491.02	8.00	202.98	nr	**8694.00**
280-360kW	7923.96	10148.93	8.00	202.98	nr	**10351.90**
375-500kW	8840.28	11322.54	8.00	202.98	nr	**11525.52**
580-730kW	11158.68	14291.93	10.00	253.72	nr	**14545.64**
655-820kW	11386.38	14583.56	10.00	253.72	nr	**14837.28**
830-1040kW	11426.40	14634.82	12.00	304.46	nr	**14939.28**
1070-1400kW	15112.38	19355.79	12.00	304.46	nr	**19660.25**
1420-1850kW	18819.06	24103.26	12.00	304.46	nr	**24407.73**
1850-2350kW	21374.82	27376.66	14.00	355.21	nr	**27731.86**
2300-3000kW	25220.88	32302.65	14.00	355.21	nr	**32657.86**
2800-3500kW	32708.76	41893.05	14.00	355.21	nr	**42248.26**
Oil fired (on/off type), connected to conventional flue						
130-190kW	4665.78	5975.88	8.00	202.98	nr	**6178.86**
200-250kW	5094.96	6525.57	8.00	202.98	nr	**6728.55**
Oil fired (high/low type), connected to conventional flue						
130-190kW	5078.40	6504.36	8.00	202.98	nr	**6707.34**
200-250kW	5507.58	7054.05	8.00	202.98	nr	**7257.03**
280-360kW	6349.38	8132.22	8.00	202.98	nr	**8335.20**
375-500kW	7591.38	9722.96	8.00	202.98	nr	**9925.94**
580-730kW	8444.22	10815.27	10.00	253.72	nr	**11068.99**
655-820kW	8670.54	11105.14	10.00	253.72	nr	**11358.86**
830-1040kW	10031.22	12847.89	12.00	304.46	nr	**13152.35**
1070-1400kW	12257.16	15698.85	12.00	304.46	nr	**16003.31**
1420-1850kW	14841.90	19009.36	12.00	304.46	nr	**19313.82**
1850-2350kW	19006.74	24343.64	14.00	355.21	nr	**24698.85**
2300-3000kW	22201.44	28435.38	14.00	355.21	nr	**28790.59**
2800-3500kW	29689.32	38025.78	14.00	355.21	nr	**38380.99**

T:MECHANICAL/COOLING/HEATING SYSTEMS

Item	Net Price £	Material £	Labour hours	Labour £	Unit	Total rate £
ATMOSPHERIC						
Commercial cast iron sectional floor standing boilers; atmospheric; including controls, enamelled jacket, insulation, assembly and commissioning; electrical work elsewhere						
Gas (on/off type), connected to conventional flue						
30-40kW	1821.04	2332.37	8.00	202.98	nr	**2535.35**
40-50kW	1996.48	2557.07	8.00	202.98	nr	**2760.05**
50-60kW	2121.60	2717.32	8.00	202.98	nr	**2920.30**
60-70kW	2316.08	2966.41	8.00	202.98	nr	**3169.39**
70-80kW	2454.80	3144.08	8.00	202.98	nr	**3347.06**
80-90kW	2868.24	3673.61	8.00	202.98	nr	**3876.59**
90-100kW	2983.84	3821.67	8.00	202.98	nr	**4024.65**
100-110kW	3162.00	4049.86	8.00	202.98	nr	**4252.83**
110-120kW	3344.24	4283.27	8.00	202.98	nr	**4486.24**
Gas (high/low type), connected to conventional flue						
30-40kW	2067.20	2647.65	8.00	202.98	nr	**2850.62**
40-50kW	2192.32	2807.90	8.00	202.98	nr	**3010.88**
50-60kW	2356.88	3018.67	8.00	202.98	nr	**3221.64**
60-70kW	2570.40	3292.14	8.00	202.98	nr	**3495.12**
70-80kW	2721.36	3485.49	8.00	202.98	nr	**3688.47**
80-90kW	3024.64	3873.93	8.00	202.98	nr	**4076.90**
90-100kW	3189.20	4084.70	8.00	202.98	nr	**4287.67**
100-110kW	3376.88	4325.07	8.00	202.98	nr	**4528.05**
110-120kW	3587.68	4595.06	8.00	202.98	nr	**4798.04**
120-140kW	5381.52	6892.60	8.00	202.98	nr	**7095.57**
140-160kW	5517.52	7066.78	8.00	202.98	nr	**7269.76**
160-180kW	5611.36	7186.97	8.00	202.98	nr	**7389.95**
180-200kW	6136.32	7859.34	8.00	202.98	nr	**8062.31**
200-220kW	6401.52	8199.00	10.00	253.72	nr	**8452.72**
220-260kW	6956.40	8909.69	10.00	253.72	nr	**9163.41**
260-300kW	7586.08	9716.18	10.00	253.72	nr	**9969.89**
300-340kW	8090.64	10362.41	12.00	304.46	nr	**10666.87**
CONDENSING						
Low Nox wall mounted condensing boiler with high efficiency modulating pre-mix burner: Aluminium heat exchanger: Placing in position						
Maximum output						
35kW	1520.00	1946.80	11.00	279.09	nr	**2225.89**
45kW	1571.00	2012.12	11.00	279.09	nr	**2291.21**
60kW	1775.00	2273.40	11.00	279.09	nr	**2552.49**
80kW	2136.00	2735.77	11.00	279.09	nr	**3014.86**

T:MECHANICAL/COOLING/HEATING SYSTEMS

Item	Net Price £	Material £	Labour hours	Labour £	Unit	Total rate £
T10 : GAS/OIL FIRED BOILERS (cont'd)						
CONDENSING (cont'd)						
Low Nox floor standing condensing boiler with high efficiency modulating pre-mix burner: Stainless steel heat exchanger: Including controls: Placing in position.						
Maximum output						
50kW	2498.00	3199.41	8.00	202.98	nr	**3402.39**
60kW	2670.00	3419.71	8.00	202.98	nr	**3622.68**
80kW	2915.00	3733.50	8.00	202.98	nr	**3936.48**
100kW	3352.00	4293.21	8.00	202.98	nr	**4496.18**
125kW	4166.00	5335.77	10.00	253.72	nr	**5589.49**
150kW	4610.00	5904.44	10.00	253.72	nr	**6158.16**
200kW	6373.00	8162.47	10.00	253.72	nr	**8416.19**
250kW	6977.00	8936.07	10.00	253.72	nr	**9189.79**
300kW	8833.00	11313.22	10.00	253.72	nr	**11566.94**
350kW	9624.00	12326.32	10.00	253.72	nr	**12580.04**
400kW	10353.00	13260.02	10.00	253.72	nr	**13513.74**
450kW	10526.00	13481.60	10.00	253.72	nr	**13735.31**
500kW	10699.00	13703.17	10.00	253.72	nr	**13956.89**
650kW	11250.00	14408.89	10.00	253.72	nr	**14662.61**
700kW	17972.00	23018.36	10.00	253.72	nr	**23272.08**
800kW	19219.00	24615.50	12.00	304.46	nr	**24919.97**
900kW	19881.00	25463.39	12.00	304.46	nr	**25767.85**
1000kW	20545.00	26313.83	12.00	304.46	nr	**26618.29**
1300kW	21795.00	27914.82	12.00	304.46	nr	**28219.28**
FLUE SYSTEMS						
Flues; suitable for domestic, medium sized industrial and commercial oil and gas appliances; stainless steel, twin wall, insulated; for use internally or externally						
Straight length; 120mm long; including one locking band						
127mm dia.	36.29	46.48	0.49	12.43	nr	**58.91**
152mm dia.	40.67	52.09	0.51	12.94	nr	**65.03**
175mm dia.	47.21	60.47	0.54	13.70	nr	**74.17**
203mm dia.	53.72	68.81	0.58	14.72	nr	**83.52**
254mm dia.	63.53	81.37	0.70	17.76	nr	**99.13**
304mm dia.	79.58	101.92	0.74	18.78	nr	**120.70**
355mm dia.	113.96	145.96	0.80	20.30	nr	**166.26**
Straight length; 300mm long; including one locking band						
127mm dia.	55.99	71.71	0.52	13.19	nr	**84.91**
152mm dia.	63.30	81.08	0.52	13.19	nr	**94.27**
178mm dia.	72.73	93.15	0.55	13.95	nr	**107.11**
203mm dia.	82.31	105.43	0.64	16.24	nr	**121.66**
254mm dia.	91.28	116.91	0.79	20.04	nr	**136.95**
304mm dia.	109.51	140.26	0.86	21.82	nr	**162.08**
355mm dia.	120.16	153.90	0.94	23.85	nr	**177.75**

T:MECHANICAL/COOLING/HEATING SYSTEMS

Item	Net Price £	Material £	Labour hours	Labour £	Unit	Total rate £
400mm dia.	128.59	164.69	1.03	26.13	nr	**190.82**
450mm dia.	147.08	188.38	1.03	26.13	nr	**214.51**
500mm dia.	157.75	202.05	1.10	27.91	nr	**229.96**
550mm dia.	174.09	222.98	1.10	27.91	nr	**250.89**
600mm dia.	192.10	246.04	1.10	27.91	nr	**273.95**
Straight length; 500mm long; including one locking band						
127mm dia.	65.81	84.29	0.55	13.95	nr	**98.24**
152mm dia.	73.36	93.96	0.55	13.95	nr	**107.91**
178mm dia.	82.63	105.83	0.63	15.98	nr	**121.81**
203mm dia.	96.58	123.70	0.63	15.98	nr	**139.69**
254mm dia.	112.17	143.67	0.86	21.82	nr	**165.49**
304mm dia.	134.39	172.13	0.95	24.10	nr	**196.23**
355mm dia.	150.92	193.30	1.03	26.13	nr	**219.43**
400mm dia.	164.80	211.08	1.12	28.42	nr	**239.49**
450mm dia.	190.32	243.76	1.12	28.42	nr	**272.18**
500mm dia.	205.07	262.65	1.19	30.19	nr	**292.84**
550mm dia.	226.05	289.52	1.19	30.19	nr	**319.71**
600mm dia.	234.16	299.91	1.19	30.19	nr	**330.10**
Straight length; 1000mm long; including one locking band						
127mm dia.	117.64	150.67	0.62	15.73	nr	**166.40**
152mm dia.	131.11	167.92	0.68	17.25	nr	**185.17**
178mm dia.	147.52	188.94	0.74	18.78	nr	**207.72**
203mm dia.	173.75	222.54	0.80	20.30	nr	**242.84**
254mm dia.	196.92	252.21	0.87	22.07	nr	**274.28**
304mm dia.	227.30	291.12	1.06	26.89	nr	**318.02**
355mm dia.	260.83	334.07	1.16	29.43	nr	**363.50**
400mm dia.	279.33	357.76	1.26	31.97	nr	**389.73**
450mm dia.	294.74	377.50	1.26	31.97	nr	**409.47**
500mm dia.	319.85	409.65	1.33	33.74	nr	**443.40**
550mm dia.	351.77	450.55	1.33	33.74	nr	**484.29**
600mm dia.	369.55	473.32	1.33	33.74	nr	**507.07**
Adjustable length; boiler removal; internal use only; including one locking band						
127mm dia.	53.92	69.06	0.52	13.19	nr	**82.26**
152mm dia.	60.94	78.06	0.55	13.95	nr	**92.01**
178mm dia.	69.05	88.43	0.59	14.97	nr	**103.40**
203mm dia.	79.19	101.43	0.64	16.24	nr	**117.66**
254mm dia.	117.42	150.40	0.79	20.04	nr	**170.44**
304mm dia.	140.64	180.13	0.86	21.82	nr	**201.95**
355mm dia.	157.72	202.00	0.99	25.12	nr	**227.12**
400mm dia.	275.98	353.47	0.91	23.09	nr	**376.56**
450mm dia.	295.42	378.37	0.91	23.09	nr	**401.46**
500mm dia.	322.20	412.67	0.99	25.12	nr	**437.79**
550mm dia.	351.50	450.19	0.99	25.12	nr	**475.31**
600mm dia.	367.96	471.28	0.99	25.12	nr	**496.40**

T:MECHANICAL/COOLING/HEATING SYSTEMS

Item	Net Price £	Material £	Labour hours	Labour £	Unit	Total rate £
T10 : GAS/OIL FIRED BOILERS (cont'd)						
FLUE SYSTEMS (cont'd)						
Flues; suitable for domestic, medium sized (cont'd)						
Inspection length; 500mm long; including one locking band						
127mm dia.	138.60	177.52	0.55	13.95	nr	**191.48**
152mm dia.	143.52	183.82	0.55	13.95	nr	**197.77**
178mm dia.	151.16	193.60	0.63	15.98	nr	**209.58**
203mm dia.	160.07	205.02	0.63	15.98	nr	**221.00**
254mm dia.	206.96	265.07	0.86	21.82	nr	**286.89**
304mm dia.	224.30	287.29	0.95	24.10	nr	**311.39**
355mm dia.	253.26	324.37	1.03	26.13	nr	**350.50**
400mm dia.	406.33	520.43	1.12	28.42	nr	**548.84**
450mm dia.	417.38	534.58	1.12	28.42	nr	**562.99**
500mm dia.	457.99	586.58	1.19	30.19	nr	**616.78**
550mm dia.	478.92	613.39	1.19	30.19	nr	**643.58**
600mm dia.	487.05	623.81	1.19	30.19	nr	**654.01**
Adapters						
127mm dia.	11.59	14.84	0.49	12.43	nr	**27.27**
152mm dia.	12.68	16.24	0.51	12.94	nr	**29.18**
178mm dia.	13.56	17.36	0.54	13.70	nr	**31.06**
203mm dia.	15.44	19.77	0.58	14.72	nr	**34.49**
254mm dia.	16.99	21.76	0.70	17.76	nr	**39.52**
304mm dia.	21.19	27.14	0.74	18.78	nr	**45.92**
355mm dia.	25.74	32.97	0.80	20.30	nr	**53.27**
400mm dia.	28.87	36.97	0.89	22.58	nr	**59.55**
450mm dia.	30.84	39.49	0.89	22.58	nr	**62.08**
500mm dia.	32.73	41.92	0.96	24.36	nr	**66.28**
550mm dia.	38.06	48.75	0.96	24.36	nr	**73.10**
600mm dia.	45.73	58.57	0.96	24.36	nr	**82.93**
Fittings for flue system						
90° insulated tee; including two locking bands						
127mm dia.	134.75	172.59	1.89	47.95	nr	**220.55**
152mm dia.	155.53	199.21	2.04	51.76	nr	**250.96**
178mm dia.	169.98	217.71	2.39	60.64	nr	**278.35**
203mm dia.	199.03	254.92	2.56	64.95	nr	**319.87**
254mm dia.	201.34	257.88	2.95	74.85	nr	**332.72**
304mm dia.	250.74	321.14	3.41	86.52	nr	**407.66**
355mm dia.	319.66	409.41	3.77	95.65	nr	**505.07**
400mm dia.	421.22	539.49	4.25	107.83	nr	**647.32**
450mm dia.	439.03	562.30	4.76	120.77	nr	**683.07**
500mm dia.	496.80	636.29	5.12	129.90	nr	**766.20**
550mm dia.	534.02	683.96	5.61	142.34	nr	**826.30**
600mm dia.	555.65	711.67	5.98	151.72	nr	**863.39**
135° insulated tee; including two locking bands						
127mm dia.	174.77	223.84	1.89	47.95	nr	**271.80**
152mm dia.	188.97	242.02	2.04	51.76	nr	**293.78**
178mm dia.	206.61	264.62	2.39	60.64	nr	**325.26**
203mm dia.	262.21	335.84	2.56	64.95	nr	**400.79**
254mm dia.	297.94	381.60	2.95	74.85	nr	**456.44**
304mm dia.	224.48	287.51	3.41	86.52	nr	**374.03**
355mm dia.	441.71	565.74	3.77	95.65	nr	**661.39**

T:MECHANICAL/COOLING/HEATING SYSTEMS

Item	Net Price £	Material £	Labour hours	Labour £	Unit	Total rate £
400mm dia.	581.45	744.72	4.25	107.83	nr	**852.55**
450mm dia.	629.45	806.19	4.76	120.77	nr	**926.96**
500mm dia.	733.31	939.22	5.12	129.90	nr	**1069.13**
550mm dia.	752.37	963.63	5.61	142.34	nr	**1105.97**
600mm dia.	794.05	1017.01	5.98	151.72	nr	**1168.73**
Wall sleeve; for 135° tee through wall						
127mm dia.	17.89	22.92	1.89	47.95	nr	**70.87**
152mm dia.	24.00	30.74	2.04	51.76	nr	**82.50**
178mm dia.	25.14	32.20	2.39	60.64	nr	**92.84**
203mm dia.	28.27	36.20	2.56	64.95	nr	**101.15**
254mm dia.	31.54	40.39	2.95	74.85	nr	**115.24**
304mm dia.	36.75	47.08	3.41	86.52	nr	**133.59**
355mm dia.	40.83	52.30	3.77	95.65	nr	**147.95**
15° insulated elbow; including two locking bands						
127mm dia.	93.52	119.78	1.57	39.83	nr	**159.62**
152mm dia.	103.93	133.11	1.79	45.42	nr	**178.53**
178mm dia.	111.15	142.37	2.05	52.01	nr	**194.38**
203mm dia.	117.84	150.93	2.33	59.12	nr	**210.04**
254mm dia.	121.36	155.44	2.45	62.16	nr	**217.60**
304mm dia.	153.44	196.52	3.43	87.03	nr	**283.55**
355mm dia.	205.27	262.90	4.71	119.50	nr	**382.41**
30° insulated elbow; including two locking bands						
127mm dia.	93.52	119.78	1.44	36.54	nr	**156.32**
152mm dia.	103.93	133.11	1.62	41.10	nr	**174.22**
178mm dia.	111.15	142.37	1.89	47.95	nr	**190.32**
203mm dia.	117.84	150.93	2.17	55.06	nr	**205.98**
254mm dia.	121.36	155.44	2.16	54.80	nr	**210.24**
304mm dia.	153.44	196.52	2.74	69.52	nr	**266.04**
355mm dia.	204.88	262.41	3.17	80.43	nr	**342.84**
400mm dia.	205.27	262.90	3.53	89.56	nr	**352.47**
450mm dia.	237.90	304.70	3.88	98.44	nr	**403.14**
500mm dia.	249.41	319.44	4.24	107.58	nr	**427.02**
550mm dia.	267.67	342.82	4.61	116.96	nr	**459.79**
600mm dia.	291.87	373.82	4.96	125.84	nr	**499.67**
45° insulated elbow; including two locking bands						
127mm dia.	93.52	119.78	1.44	36.54	nr	**156.32**
152mm dia.	103.93	133.11	1.51	38.31	nr	**171.43**
178mm dia.	111.15	142.37	1.58	40.09	nr	**182.45**
203mm dia.	117.84	150.93	1.66	42.12	nr	**193.04**
254mm dia.	121.36	155.44	1.72	43.64	nr	**199.08**
304mm dia.	153.44	196.52	1.80	45.67	nr	**242.19**
355mm dia.	205.27	262.90	1.94	49.22	nr	**312.13**
400mm dia.	261.76	335.26	2.01	51.00	nr	**386.26**
450mm dia.	274.19	351.18	2.09	53.03	nr	**404.20**
500mm dia.	294.11	376.70	2.16	54.80	nr	**431.50**
550mm dia.	320.77	410.84	2.23	56.58	nr	**467.42**
600mm dia.	328.90	421.25	2.30	58.36	nr	**479.61**

T:MECHANICAL/COOLING/HEATING SYSTEMS

Item	Net Price £	Material £	Labour hours	Labour £	Unit	Total rate £
T10 : GAS/OIL FIRED BOILERS (cont'd)						
FLUE SYSTEMS (cont'd)						
Flue supports						
Wall support, galvanised; including plate and brackets						
127mm dia.	56.77	72.71	2.24	56.83	nr	**129.55**
152mm dia.	62.42	79.95	2.44	61.91	nr	**141.86**
178mm dia.	68.91	88.26	2.52	63.94	nr	**152.20**
203mm dia.	71.75	91.89	2.77	70.28	nr	**162.17**
254mm dia.	85.57	109.59	2.98	75.61	nr	**185.20**
304mm dia.	96.75	123.92	3.46	87.79	nr	**211.71**
355mm dia.	129.51	165.87	4.08	103.52	nr	**269.39**
400mm dia.; with 300mm support length and collar	332.66	426.06	4.80	121.79	nr	**547.85**
450mm dia.; with 300mm support length and collar	354.67	454.26	5.62	142.59	nr	**596.85**
500mm dia.; with 300mm support length and collar	387.47	496.27	6.24	158.32	nr	**654.59**
550mm dia.; with 300mm support length and collar	423.92	542.95	6.97	176.84	nr	**719.80**
600mm dia.; with 300mm support length and collar	447.52	573.17	7.49	190.04	nr	**763.21**
Ceiling/floor support						
127mm dia.	17.57	22.50	1.86	47.19	nr	**69.69**
152mm dia.	19.54	25.03	2.14	54.30	nr	**79.32**
178mm dia.	23.39	29.96	1.93	48.97	nr	**78.92**
203mm dia.	35.22	45.11	2.74	69.52	nr	**114.63**
254mm dia.	40.18	51.46	3.21	81.44	nr	**132.90**
304mm dia.	45.92	58.82	3.68	93.37	nr	**152.19**
355mm dia.	54.73	70.09	4.28	108.59	nr	**178.69**
400mm dia.	277.70	355.67	4.86	123.31	nr	**478.98**
450mm dia.	292.36	374.45	5.46	138.53	nr	**512.98**
500mm dia.	308.62	395.28	6.04	153.25	nr	**548.53**
550mm dia.	337.86	432.72	6.65	168.72	nr	**601.45**
600mm dia.	343.63	440.11	7.24	183.69	nr	**623.81**
Ceiling/floor firestop spacer						
127mm dia.	3.42	4.38	0.66	16.75	nr	**21.13**
152mm dia.	3.81	4.88	0.69	17.51	nr	**22.39**
178mm dia.	4.30	5.50	0.70	17.76	nr	**23.26**
203mm dia.	5.10	6.53	0.87	22.07	nr	**28.61**
254mm dia.	5.26	6.74	0.91	23.09	nr	**29.82**
304mm dia.	6.33	8.11	0.95	24.10	nr	**32.21**
355mm dia.	11.53	14.77	0.99	25.12	nr	**39.89**
Wall band; internal or external use						
127mm dia.	20.70	26.51	1.03	26.13	nr	**52.64**
152mm dia.	21.61	27.68	1.07	27.15	nr	**54.83**
178mm dia.	22.46	28.77	1.11	28.16	nr	**56.93**
203mm dia.	23.72	30.38	1.18	29.94	nr	**60.32**
254mm dia.	24.68	31.61	1.30	32.98	nr	**64.59**
304mm dia.	26.66	34.15	1.45	36.79	nr	**70.94**
355mm dia.	28.38	36.35	1.65	41.86	nr	**78.21**
400mm dia.	35.19	45.07	1.85	46.94	nr	**92.00**
450mm dia.	37.50	48.03	2.39	60.64	nr	**108.67**

T:MECHANICAL/COOLING/HEATING SYSTEMS

Item	Net Price £	Material £	Labour hours	Labour £	Unit	Total rate £
500mm dia.	45.07	57.73	2.25	57.09	nr	**114.81**
550mm dia.	47.21	60.47	2.45	62.16	nr	**122.63**
600mm dia.	49.91	63.92	2.66	67.49	nr	**131.41**
Flashings and terminals						
Insulated top stub; including one locking band						
127mm dia.	50.99	65.31	1.49	37.80	nr	**103.11**
152mm dia.	57.65	73.84	1.90	48.21	nr	**122.04**
178mm dia.	62.07	79.50	1.92	48.71	nr	**128.21**
203mm dia.	66.05	84.60	2.20	55.82	nr	**140.41**
254mm dia.	70.13	89.82	2.49	63.18	nr	**153.00**
304mm dia.	95.87	122.78	2.79	70.79	nr	**193.57**
355mm dia.	126.56	162.10	3.19	80.94	nr	**243.03**
400mm dia.	122.03	156.29	3.59	91.09	nr	**247.38**
450mm dia.	132.57	169.80	3.97	100.73	nr	**270.53**
500mm dia.	153.36	196.42	4.38	111.13	nr	**307.55**
550mm dia.	160.47	205.52	4.78	121.28	nr	**326.80**
600mm dia.	166.08	212.71	5.17	131.17	nr	**343.89**
Rain cap; including one locking band						
127mm dia.	27.25	34.90	1.49	37.80	nr	**72.70**
152mm dia.	28.48	36.47	1.54	39.07	nr	**75.55**
178mm dia.	31.36	40.17	1.72	43.64	nr	**83.81**
203mm dia.	37.51	48.04	2.00	50.74	nr	**98.79**
254mm dia.	49.32	63.17	2.49	63.18	nr	**126.35**
304mm dia.	66.49	85.16	2.80	71.04	nr	**156.20**
355mm dia.	89.08	114.10	3.19	80.94	nr	**195.03**
400mm dia.	88.87	113.82	3.45	87.53	nr	**201.35**
450mm dia.	96.65	123.78	3.97	100.73	nr	**224.51**
500mm dia.	104.51	133.86	4.38	111.13	nr	**244.98**
550mm dia.	112.14	143.63	4.78	121.28	nr	**264.91**
600mm dia.	119.89	153.55	5.17	131.17	nr	**284.72**
Round top; including one locking band						
127mm dia	52.11	66.74	1.49	37.80	nr	**104.54**
152mm dia	56.82	72.78	1.65	41.86	nr	**114.64**
178mm dia	64.78	82.97	1.92	48.71	nr	**131.69**
203mm dia	76.24	97.65	2.20	55.82	nr	**153.47**
254mm dia	90.13	115.44	2.49	63.18	nr	**178.62**
304mm dia	118.38	151.62	2.80	71.04	nr	**222.66**
355mm dia	157.93	202.28	3.19	80.94	nr	**283.22**
Coping cap; including one locking band						
127mm dia.	29.18	37.38	1.49	37.80	nr	**75.18**
152mm dia.	30.57	39.16	1.65	41.86	nr	**81.02**
178mm dia.	33.64	43.09	1.92	48.71	nr	**91.80**
203mm dia.	40.36	51.69	2.20	55.82	nr	**107.51**
254mm dia.	49.32	63.17	2.49	63.18	nr	**126.35**
304mm dia.	66.49	85.16	2.79	70.79	nr	**155.94**
355mm dia.	89.08	114.10	3.19	80.94	nr	**195.03**

T:MECHANICAL/COOLING/HEATING SYSTEMS

Item	Net Price £	Material £	Labour hours	Labour £	Unit	Total rate £
T10 : GAS/OIL FIRED BOILERS (cont'd)						
FLUE SYSTEMS (cont'd)						
Flashings and terminals (cont'd)						
Storm collar						
127mm dia.	5.54	7.10	0.52	13.19	nr	**20.29**
152mm dia.	5.92	7.58	0.55	13.95	nr	**21.53**
178mm dia.	6.56	8.41	0.57	14.46	nr	**22.87**
203mm dia.	6.87	8.79	0.66	16.75	nr	**25.54**
254mm dia.	8.61	11.02	0.66	16.75	nr	**27.77**
304mm dia.	8.97	11.49	0.72	18.27	nr	**29.76**
355mm dia.	9.58	12.27	0.77	19.54	nr	**31.81**
400mm dia.	25.42	32.56	0.82	20.80	nr	**53.36**
450mm dia.	27.98	35.83	0.87	22.07	nr	**57.91**
500mm dia.	30.51	39.08	0.92	23.34	nr	**62.42**
550mm dia.	33.06	42.34	0.98	24.86	nr	**67.20**
600mm dia.	35.60	45.60	1.03	26.13	nr	**71.73**
Flat flashing; including storm collar and sealant						
127mm dia.	31.07	39.79	1.49	37.80	nr	**77.60**
152mm dia.	32.11	41.12	1.65	41.86	nr	**82.98**
178mm dia.	33.69	43.15	1.92	48.71	nr	**91.86**
203mm dia.	36.89	47.25	2.20	55.82	nr	**103.07**
254mm dia.	50.42	64.57	2.49	63.18	nr	**127.75**
304mm dia.	60.24	77.16	2.80	71.04	nr	**148.20**
355mm dia.	94.93	121.58	3.20	81.19	nr	**202.77**
400mm dia.	132.43	169.61	3.59	91.09	nr	**260.70**
450mm dia.	152.13	194.84	3.97	100.73	nr	**295.57**
500mm dia.	164.81	211.09	4.38	111.13	nr	**322.22**
550mm dia.	174.99	224.13	4.78	121.28	nr	**345.40**
600mm dia.	181.31	232.22	5.17	131.17	nr	**363.39**
5°-30° rigid adjustable flashing; including storm collar and sealant						
127mm dia.	52.92	67.79	1.49	37.80	nr	**105.59**
152mm dia.	55.76	71.42	1.65	41.86	nr	**113.28**
178mm dia.	59.30	75.96	1.92	48.71	nr	**124.67**
203mm dia.	62.35	79.85	2.20	55.82	nr	**135.67**
254mm dia.	65.64	84.07	2.49	63.18	nr	**147.24**
304mm dia.	81.23	104.04	2.80	71.04	nr	**175.08**
355mm dia.	92.38	118.32	3.19	80.94	nr	**199.26**
400mm dia.	297.84	381.47	3.59	91.09	nr	**472.55**
450mm dia.	347.58	445.18	3.97	100.73	nr	**545.91**
500mm dia.	371.07	475.26	4.38	111.13	nr	**586.39**
550mm dia.	388.30	497.33	4.77	121.02	nr	**618.35**
600mm dia.	422.47	541.10	5.17	131.17	nr	**672.27**
Domestic and small commercial; twin walled gas vent system suitable for gas fired appliances; domestic gas boilers; small commercial boilers with internal or external flues						
152mm long						
100mm dia.	5.66	7.25	0.52	13.19	nr	**20.44**
125mm dia.	6.96	8.92	0.52	13.19	nr	**22.11**
150mm dia.	7.54	9.66	0.52	13.19	nr	**22.85**

T:MECHANICAL/COOLING/HEATING SYSTEMS

Item	Net Price £	Material £	Labour hours	Labour £	Unit	Total rate £
305mm long						
100mm dia.	8.59	11.00	0.52	13.19	nr	24.20
125mm dia.	10.08	12.91	0.52	13.19	nr	26.11
150mm dia.	11.95	15.30	0.52	13.19	nr	28.49
457mm long						
100mm dia.	9.51	12.17	0.55	13.95	nr	26.13
125mm dia.	10.68	13.68	0.55	13.95	nr	27.64
150mm dia.	13.22	16.94	0.55	13.95	nr	30.89
914mm long						
100mm dia.	16.98	21.75	0.62	15.73	nr	37.48
125mm dia.	19.79	25.34	0.62	15.73	nr	41.07
150mm dia.	22.69	29.06	0.62	15.73	nr	44.79
1524mm long						
100mm dia.	24.53	31.42	0.82	20.80	nr	52.23
125mm dia.	30.18	38.65	0.84	21.31	nr	59.97
150mm dia.	32.41	41.51	0.84	21.31	nr	62.82
Adjustable length 305mm long						
100mm dia.	10.87	13.92	0.56	14.21	nr	28.13
125mm dia.	12.21	15.64	0.56	14.21	nr	29.85
150mm dia.	15.39	19.71	0.56	14.21	nr	33.92
Adjustable length 457mm long						
100mm dia.	14.65	18.77	0.56	14.21	nr	32.97
125mm dia.	17.77	22.76	0.56	14.21	nr	36.97
150mm dia.	19.77	25.33	0.56	14.21	nr	39.53
Adjustable elbow 0°-90°						
100mm dia.	12.40	15.88	0.48	12.18	nr	28.06
125mm dia.	14.65	18.77	0.48	12.18	nr	30.94
150mm dia.	18.36	23.51	0.48	12.18	nr	35.69
Draughthood connector						
100mm dia.	3.83	4.91	0.48	12.18	nr	17.09
125mm dia.	4.31	5.52	0.48	12.18	nr	17.70
150mm dia.	4.68	6.00	0.48	12.18	nr	18.18
Adaptor						
100mm dia.	9.28	11.89	0.48	12.18	nr	24.06
125mm dia.	9.48	12.14	0.48	12.18	nr	24.32
150mm dia.	9.68	12.40	0.48	12.18	nr	24.58
Support plate						
100mm dia.	6.70	8.58	0.48	12.18	nr	20.76
125mm dia.	7.13	9.13	0.48	12.18	nr	21.30
150mm dia.	7.63	9.77	0.48	12.18	nr	21.95
Wall band						
100mm dia.	6.05	7.75	0.48	12.18	nr	19.93
125mm dia.	6.45	8.26	0.48	12.18	nr	20.44
150mm dia.	8.18	10.47	0.48	12.18	nr	22.65
Firestop						
100mm dia.	2.63	3.37	0.48	12.18	nr	15.55
125mm dia.	2.63	3.37	0.48	12.18	nr	15.55
150mm dia.	3.01	3.85	0.48	12.18	nr	16.03

T:MECHANICAL/COOLING/HEATING SYSTEMS

Item	Net Price £	Material £	Labour hours	Labour £	Unit	Total rate £
T10 : GAS/OIL FIRED BOILERS (cont'd)						
FLUE SYSTEMS (cont'd)						
Domestic and small commercial; twin walled gas vent (cont'd)						
Flat flashing						
125mm dia.	17.85	22.86	0.55	13.95	nr	**36.81**
150mm dia.	24.83	31.81	0.55	13.95	nr	**45.76**
Adjustable flashing 5°-30°						
100mm dia.	46.52	59.59	0.55	13.95	nr	**73.54**
125mm dia.	72.63	93.03	0.55	13.95	nr	**106.98**
Storm collar						
100mm dia.	3.71	4.75	0.55	13.95	nr	**18.70**
125mm dia.	3.79	4.86	0.55	13.95	nr	**18.81**
150mm dia.	3.89	4.99	0.55	13.95	nr	**18.94**
Gas vent terminal						
100mm dia.	14.15	18.12	0.55	13.95	nr	**32.08**
125mm dia.	15.55	19.92	0.55	13.95	nr	**33.88**
150mm dia.	19.95	25.55	0.55	13.95	nr	**39.51**
Twin wall galvanised steel flue box, 125mm dia.; fitted for gas fire, where no chimney exists						
Free standing	89.95	115.21	2.15	54.55	nr	**169.76**
Recess	89.95	115.21	2.15	54.55	nr	**169.76**
Back boiler	66.27	84.88	2.40	60.89	nr	**145.77**

T:MECHANICAL/COOLING/HEATING SYSTEMS

Item	Net Price £	Material £	Labour hours	Labour £	Unit	Total rate £
T13 : PACKAGED STEAM GENERATORS						
Packaged steam boilers; boiler mountings centrifugal water feed pump; insulation; and sheet steel wrap around casing; plastic coated						
Gas fired						
293 kW rating	17554.39	22483.49	86.45	2193.40	nr	**24676.89**
1465 kW rating	37443.07	47956.72	148.22	3760.62	nr	**51717.34**
2930 kW rating	53769.61	68867.57	207.50	5264.67	nr	**74132.24**
Oil fired						
293 kW rating	15921.74	20392.41	86.45	2193.40	nr	**22585.81**
1465 kW rating	34636.53	44362.12	148.22	3760.62	nr	**48122.75**
2930 kW rating	52420.31	67139.40	207.50	5264.67	nr	**72404.07**

T:MECHANICAL/COOLING/HEATING SYSTEMS

Item	Net Price £	Material £	Labour hours	Labour £	Unit	Total rate £
T31 : LOW TEMPERATURE HOT WATER HEATING						
Y10 - PIPELINES						
SCREWED STEEL						
Black steel pipes; screwed and socketed joints; BS 1387: 1985. Fixed vertically, brackets measured separately. Screwed joints are within the running length, but any flanges are additional						
Medium weight						
10mm dia	1.69	2.16	0.37	9.39	m	**11.55**
15mm dia	1.83	2.34	0.37	9.39	m	**11.73**
20mm dia	2.16	2.77	0.37	9.39	m	**12.15**
25mm dia	3.09	3.96	0.41	10.40	m	**14.36**
32mm dia	3.86	4.94	0.48	12.18	m	**17.12**
40mm dia	4.49	5.75	0.52	13.19	m	**18.94**
50mm dia	6.35	8.13	0.62	15.73	m	**23.86**
65mm dia	8.85	11.34	0.65	16.49	m	**27.83**
80mm dia	11.50	14.73	1.10	27.91	m	**42.64**
100mm dia	16.76	21.47	1.31	33.24	m	**54.70**
125mm dia	27.93	35.77	1.66	42.12	m	**77.89**
150mm dia	33.58	43.01	1.88	47.70	m	**90.71**
Heavy weight						
15mm dia	2.16	2.77	0.37	9.39	m	**12.15**
20mm dia	2.57	3.29	0.37	9.39	m	**12.68**
25mm dia	3.74	4.79	0.41	10.40	m	**15.19**
32mm dia	4.68	5.99	0.48	12.18	m	**18.17**
40mm dia	5.45	6.98	0.52	13.19	m	**20.17**
50mm dia	7.62	9.76	0.62	15.73	m	**25.49**
65mm dia	10.59	13.56	0.64	16.24	m	**29.80**
80mm dia	13.50	17.29	1.10	27.91	m	**45.20**
100mm dia	19.33	24.76	1.31	33.24	m	**57.99**
125mm dia	29.51	37.80	1.66	42.12	m	**79.91**
150mm dia	35.61	45.61	1.88	47.70	m	**93.31**
200mm dia	45.17	57.85	2.99	75.86	m	**133.72**
250mm dia	58.38	74.77	3.49	88.55	m	**163.32**
300mm dia	96.43	123.51	3.91	99.20	m	**222.71**

T:MECHANICAL/COOLING/HEATING SYSTEMS

Item	Net Price £	Material £	Labour hours	Labour £	Unit	Total rate £
Black steel pipes; screwed and socketed joints; BS 1387: 1985. Fixed at high level or suspended, brackets measured separately. Screwed joints are within the running length, but any flanges are additional						
Medium weight						
10mm dia	1.69	2.16	0.58	14.72	m	**16.88**
15mm dia	1.83	2.34	0.58	14.72	m	**17.06**
20mm dia	2.16	2.77	0.58	14.72	m	**17.48**
25mm dia	3.09	3.96	0.60	15.22	m	**19.18**
32mm dia	3.86	4.94	0.68	17.25	m	**22.20**
40mm dia	4.49	5.75	0.73	18.52	m	**24.27**
50mm dia	6.35	8.13	0.85	21.57	m	**29.70**
65mm dia	8.85	11.34	0.88	22.33	m	**33.66**
80mm dia	11.50	14.73	1.45	36.79	m	**51.52**
100mm dia	16.76	21.47	1.74	44.15	m	**65.61**
125mm dia	27.93	35.77	2.21	56.07	m	**91.84**
150mm dia	33.58	43.01	2.50	63.43	m	**106.44**
Heavy weight						
15mm dia	2.16	2.77	0.58	14.72	m	**17.48**
20mm dia	2.57	3.29	0.58	14.72	m	**18.01**
25mm dia	3.74	4.79	0.60	15.22	m	**20.01**
32mm dia	4.68	5.99	0.68	17.25	m	**23.25**
40mm dia	5.45	6.98	0.73	18.52	m	**25.50**
50mm dia	7.62	9.76	0.85	21.57	m	**31.33**
65mm dia	10.59	13.56	0.88	22.33	m	**35.89**
80mm dia	13.50	17.29	1.45	36.79	m	**54.08**
100mm dia	19.33	24.76	1.74	44.15	m	**68.90**
125mm dia	29.51	37.80	2.21	56.07	m	**93.87**
150mm dia	35.61	45.61	2.50	63.43	m	**109.04**
200mm dia	45.17	57.85	2.99	75.86	m	**133.72**
250mm dia	58.38	74.77	3.49	88.55	m	**163.32**
300mm dia	96.43	123.51	3.91	99.20	m	**222.71**
FIXINGS						
For steel pipes; black malleable iron. For minimum fixing distances, refer to the Tables and Memoranda to the rear of the book						
Single pipe bracket, screw on, black malleable iron; screwed to wood						
15mm dia	0.69	0.88	0.14	3.55	nr	**4.44**
20mm dia	0.77	0.99	0.14	3.55	nr	**4.54**
25mm dia	0.90	1.15	0.17	4.31	nr	**5.47**
32mm dia	1.23	1.58	0.19	4.82	nr	**6.40**
40mm dia	1.63	2.09	0.22	5.58	nr	**7.67**
50mm dia	2.17	2.78	0.22	5.58	nr	**8.36**
65mm dia	2.86	3.66	0.28	7.10	nr	**10.77**
80mm dia	3.92	5.02	0.32	8.12	nr	**13.14**
100mm dia	5.71	7.31	0.35	8.88	nr	**16.19**

T:MECHANICAL/COOLING/HEATING SYSTEMS

Item	Net Price £	Material £	Labour hours	Labour £	Unit	Total rate £
T31 : LOW TEMPERATURE HOT WATER HEATING (cont'd)						
Y10 - PIPELINES (cont'd)						
Fixings; steel pipes; black malleable iron (cont'd)						
Single pipe bracket, screw on, black malleable iron; plugged and screwed						
15mm dia	0.69	0.88	0.25	6.34	nr	**7.23**
20mm dia	0.77	0.99	0.25	6.34	nr	**7.33**
25mm dia	0.90	1.15	0.30	7.61	nr	**8.76**
32mm dia	1.23	1.58	0.32	8.12	nr	**9.69**
40mm dia	1.63	2.09	0.32	8.12	nr	**10.21**
50mm dia	2.17	2.78	0.32	8.12	nr	**10.90**
65mm dia	2.86	3.66	0.35	8.88	nr	**12.54**
80mm dia	3.92	5.02	0.42	10.66	nr	**15.68**
100mm dia	5.71	7.31	0.42	10.66	nr	**17.97**
Single pipe bracket for building in, black malleable iron						
15mm dia	1.12	1.43	0.10	2.54	nr	**3.97**
20mm dia	1.12	1.43	0.11	2.79	nr	**4.23**
25mm dia	1.12	1.43	0.12	3.04	nr	**4.48**
32mm dia	1.27	1.63	0.14	3.55	nr	**5.18**
40mm dia	1.28	1.64	0.15	3.81	nr	**5.45**
50mm dia	1.33	1.70	0.16	4.06	nr	**5.76**
Pipe ring, single socket, black malleable iron						
15mm dia	0.73	0.94	0.10	2.54	nr	**3.47**
20mm dia	0.82	1.05	0.11	2.79	nr	**3.84**
25mm dia	0.90	1.15	0.12	3.04	nr	**4.20**
32mm dia	0.95	1.22	0.14	3.55	nr	**4.77**
40mm dia	1.23	1.58	0.15	3.81	nr	**5.38**
50mm dia	1.55	1.99	0.16	4.06	nr	**6.04**
65mm dia	2.24	2.87	0.30	7.61	nr	**10.48**
80mm dia	2.69	3.45	0.35	8.88	nr	**12.33**
100mm dia	4.07	5.21	0.40	10.15	nr	**15.36**
125mm dia	8.24	10.55	0.60	15.23	nr	**25.78**
150mm dia	9.23	11.82	0.77	19.54	nr	**31.36**
200mm dia	13.49	17.28	0.90	22.83	nr	**40.11**
250mm dia	16.86	21.59	1.10	27.91	nr	**49.50**
300mm dia	20.23	25.91	1.25	31.71	nr	**57.63**
350mm dia	23.61	30.24	1.50	38.06	nr	**68.30**
400mm dia	26.98	34.56	1.75	44.40	nr	**78.96**
Pipe ring, double socket, black malleable iron						
15mm dia	0.85	1.09	0.10	2.54	nr	**3.63**
20mm dia	0.98	1.26	0.11	2.79	nr	**4.05**
25mm dia	1.10	1.41	0.12	3.04	nr	**4.45**
32mm dia	1.30	1.67	0.14	3.55	nr	**5.22**
40mm dia	1.51	1.93	0.15	3.81	nr	**5.74**
50mm dia	1.72	2.20	0.16	4.06	nr	**6.26**
Screw on backplate (Male), black malleable iron; plugged and screwed						
M12	0.57	0.73	0.10	2.54	nr	**3.27**

T:MECHANICAL/COOLING/HEATING SYSTEMS

Item	Net Price £	Material £	Labour hours	Labour £	Unit	Total rate £
Screw on backplate (Female), black malleable iron; plugged and screwed						
M12	0.57	0.73	0.10	2.54	nr	3.27
Extra Over channel sections for fabricated hangers and brackets						
Galvanised steel; including inserts, bolts, nuts, washers; fixed to backgrounds						
41 x 21mm	2.40	3.07	0.15	3.81	m	6.88
41 x 41mm	3.75	4.80	0.15	3.81	m	8.61
Threaded rods; metric thread; including nuts, washers etc						
12mm dia x 600mm long for ring clips	1.30	1.66	0.10	2.54	nr	4.20
Pipe roller and chair						
Roller and chair; black malleable						
Up to 50mm dia	2.93	3.75	0.20	5.07	nr	8.83
65mm dia	2.93	3.75	0.20	5.07	nr	8.83
80mm dia	4.19	5.37	0.20	5.07	nr	10.44
100mm dia	4.50	5.76	0.20	5.07	nr	10.84
125mm dia	4.92	6.30	0.20	5.07	nr	11.38
150mm dia	5.46	6.99	0.30	7.61	nr	14.60
175mm dia	13.69	17.53	0.30	7.61	nr	25.15
200mm dia	13.69	17.53	0.30	7.61	nr	25.15
250mm dia	19.56	25.05	0.30	7.61	nr	32.66
300mm dia	20.54	26.31	0.30	7.61	nr	33.92
Roller bracket; black malleable						
25mm dia	1.78	2.28	0.20	5.07	nr	7.35
32mm dia	1.87	2.40	0.20	5.07	nr	7.47
40mm dia	2.00	2.56	0.20	5.07	nr	7.64
50mm dia	2.12	2.72	0.20	5.07	nr	7.79
65mm dia	2.78	3.56	0.20	5.07	nr	8.64
80mm dia	4.01	5.14	0.20	5.07	nr	10.21
100mm dia	4.45	5.70	0.20	5.07	nr	10.77
125mm dia	7.35	9.41	0.20	5.07	nr	14.49
150mm dia	7.35	9.41	0.30	7.61	nr	17.03
175mm dia	16.38	20.98	0.30	7.61	nr	28.59
200mm dia	16.38	20.98	0.30	7.61	nr	28.59
250mm dia	21.77	27.88	0.30	7.61	nr	35.49
300mm dia	26.98	34.56	0.30	7.61	nr	42.17
350mm dia	43.37	55.55	0.30	7.61	nr	63.16
400mm dia	49.57	63.49	0.30	7.61	nr	71.10
Extra over black steel screwed pipes; black steel flanges, screwed and drilled; metric; BS 4504						
Screwed flanges; PN6						
15mm dia	5.73	7.34	0.35	8.88	nr	16.22
20mm dia	6.09	7.80	0.47	11.92	nr	19.72
25mm dia	6.26	8.02	0.53	13.45	nr	21.46
32mm dia	6.26	8.02	0.62	15.73	nr	23.75
40mm dia	6.26	8.02	0.70	17.76	nr	25.78
50mm dia	6.81	8.72	0.84	21.31	nr	30.03

T:MECHANICAL/COOLING/HEATING SYSTEMS

Item	Net Price £	Material £	Labour hours	Labour £	Unit	Total rate £
T31 : LOW TEMPERATURE HOT WATER HEATING (cont'd)						
Y10 - PIPELINES (cont'd)						
SCREWED STEEL (cont'd)						
Extra over black steel screwed pipes (cont'd)						
Screwed flanges; PN6						
65mm dia	9.17	11.74	1.03	26.13	nr	**37.88**
80mm dia	11.30	14.47	1.23	31.21	nr	**45.68**
100mm dia	13.62	17.44	1.41	35.77	nr	**53.22**
125mm dia	28.04	35.91	1.77	44.91	nr	**80.82**
150mm dia	28.04	35.91	2.21	56.07	nr	**91.99**
Screwed flanges; PN16						
15mm dia	7.79	9.98	0.35	8.88	nr	**18.86**
20mm dia	7.85	10.05	0.47	11.92	nr	**21.98**
25mm dia	7.96	10.20	0.53	13.45	nr	**23.64**
32mm dia	8.72	11.17	0.62	15.73	nr	**26.90**
40mm dia	8.72	11.17	0.70	17.76	nr	**28.93**
50mm dia	9.27	11.87	0.84	21.31	nr	**33.19**
65mm dia	11.45	14.67	1.03	26.13	nr	**40.80**
80mm dia	14.05	18.00	1.23	31.21	nr	**49.20**
100mm dia	16.94	21.70	1.41	35.77	nr	**57.47**
125mm dia	29.11	37.28	1.77	44.91	nr	**82.19**
150mm dia	31.99	40.97	2.21	56.07	nr	**97.04**
Extra over black steel screwed pipes; black steel flanges, screwed and drilled; imperial; BS 10						
Screwed flanges; Table E						
1/2" dia	8.24	10.55	0.35	8.88	nr	**19.43**
3/4" dia	8.30	10.63	0.47	11.92	nr	**22.56**
1" dia	8.41	10.77	0.53	13.45	nr	**24.22**
1 1/4" dia	9.17	11.74	0.62	15.73	nr	**27.48**
1 1/2" dia	9.17	11.74	0.70	17.76	nr	**29.51**
2" dia	9.39	12.03	0.84	21.31	nr	**33.34**
2 1/2" dia	11.02	14.11	1.03	26.13	nr	**40.25**
3" dia	13.74	17.60	1.23	31.21	nr	**48.81**
4" dia	18.20	23.31	1.41	35.77	nr	**59.08**
5" dia	33.63	43.07	1.77	44.91	nr	**87.98**
6" dia	36.51	46.76	2.21	56.07	nr	**102.83**
Extra over black steel screwed pipes; black steel flange connections						
Bolted connection between pair of flanges; including gasket, bolts, nuts and washers						
50mm dia	9.90	12.68	0.53	13.45	nr	**26.13**
65mm dia	11.70	14.99	0.53	13.45	nr	**28.43**
80mm dia	15.24	19.52	0.53	13.45	nr	**32.97**
100mm dia	18.22	23.34	0.61	15.48	nr	**38.81**
125mm dia	28.27	36.21	0.61	15.48	nr	**51.68**
150mm dia	34.04	43.60	0.90	22.83	nr	**66.43**

T:MECHANICAL/COOLING/HEATING SYSTEMS

Item	Net Price £	Material £	Labour hours	Labour £	Unit	Total rate £
Extra over black steel screwed pipes; black heavy steel tubular fittings; BS 1387						
Long screw connection with socket and backnut						
15mm dia	2.80	3.59	0.63	15.98	nr	**19.57**
20mm dia	3.52	4.51	0.84	21.31	nr	**25.82**
25mm dia	4.62	5.92	0.95	24.10	nr	**30.02**
32mm dia	6.05	7.75	1.11	28.16	nr	**35.91**
40mm dia	7.39	9.47	1.28	32.48	nr	**41.94**
50mm dia	10.85	13.90	1.53	38.82	nr	**52.72**
65mm dia	24.74	31.69	1.87	47.45	nr	**79.13**
80mm dia	34.02	43.57	2.21	56.07	nr	**99.64**
100mm dia	54.97	70.41	3.05	77.38	nr	**147.79**
Running nipple						
15mm dia	0.71	0.91	0.50	12.69	nr	**13.60**
20mm dia	0.90	1.15	0.68	17.25	nr	**18.41**
25mm dia	1.10	1.41	0.77	19.54	nr	**20.95**
32mm dia	1.55	1.99	0.90	22.83	nr	**24.82**
40mm dia	2.09	2.68	1.03	26.13	nr	**28.81**
50mm dia	3.18	4.07	1.23	31.21	nr	**35.28**
65mm dia	6.84	8.76	1.50	38.06	nr	**46.82**
80mm dia	10.67	13.67	1.78	45.16	nr	**58.83**
100mm dia	16.72	21.41	2.38	60.39	nr	**81.80**
Barrel nipple						
15mm dia	0.41	0.53	0.50	12.69	nr	**13.21**
20mm dia	0.65	0.83	0.68	17.25	nr	**18.09**
25mm dia	0.85	1.09	0.77	19.54	nr	**20.63**
32mm dia	1.27	1.63	0.90	22.83	nr	**24.46**
40mm dia	1.57	2.01	1.03	26.13	nr	**28.14**
50mm dia	2.23	2.86	1.23	31.21	nr	**34.06**
65mm dia	4.78	6.12	1.50	38.06	nr	**44.18**
80mm dia	6.66	8.53	1.78	45.16	nr	**53.69**
100mm dia	12.07	15.46	2.38	60.39	nr	**75.84**
125mm dia	22.41	28.70	2.87	72.82	nr	**101.52**
150mm dia	35.32	45.24	3.39	86.01	nr	**131.25**
Close taper nipple						
15mm dia	0.85	1.09	0.50	12.69	nr	**13.77**
20mm dia	1.10	1.41	0.68	17.25	nr	**18.66**
25mm dia	1.44	1.84	0.77	19.54	nr	**21.38**
32mm dia	2.16	2.77	0.90	22.83	nr	**25.60**
40mm dia	2.68	3.43	1.03	26.13	nr	**29.57**
50mm dia	4.12	5.28	1.23	31.21	nr	**36.48**
65mm dia	7.99	10.23	1.50	38.06	nr	**48.29**
80mm dia	9.85	12.62	1.78	45.16	nr	**57.78**
100mm dia	20.23	25.91	2.38	60.39	nr	**86.30**
Extra over black steel screwed pipes; black malleable iron fittings; BS 143						
Cap						
15mm dia	0.44	0.56	0.32	8.12	nr	**8.68**
20mm dia	0.50	0.64	0.43	10.91	nr	**11.55**
25mm dia	0.63	0.81	0.49	12.43	nr	**13.24**
32mm dia	1.02	1.31	0.58	14.72	nr	**16.02**
40mm dia	1.20	1.54	0.66	16.75	nr	**18.28**

T:MECHANICAL/COOLING/HEATING SYSTEMS

Item	Net Price £	Material £	Labour hours	Labour £	Unit	Total rate £
T31 : LOW TEMPERATURE HOT WATER HEATING (cont'd)						
Y10 - PIPELINES (cont'd)						
SCREWED STEEL (cont'd)						
Fittings; black steel screwed pipes (cont'd)						
Cap						
50mm dia	2.52	3.23	0.78	19.79	nr	**23.02**
65mm dia	4.20	5.38	0.96	24.36	nr	**29.74**
80mm dia	4.75	6.08	1.13	28.67	nr	**34.75**
100mm dia	10.40	13.32	1.70	43.13	nr	**56.45**
Plain plug, hollow						
15mm dia	0.31	0.40	0.28	7.10	nr	**7.50**
20mm dia	0.39	0.50	0.38	9.64	nr	**10.14**
25mm dia	0.53	0.68	0.44	11.16	nr	**11.84**
32mm dia	0.69	0.88	0.51	12.94	nr	**13.82**
40mm dia	1.29	1.65	0.59	14.97	nr	**16.62**
50mm dia	1.81	2.32	0.70	17.76	nr	**20.08**
65mm dia	2.96	3.79	0.85	21.57	nr	**25.36**
80mm dia	4.61	5.90	1.00	25.37	nr	**31.28**
100mm dia	8.49	10.87	1.44	36.54	nr	**47.41**
Plain plug, solid						
15mm dia	0.95	1.22	0.28	7.10	nr	**8.32**
20mm dia	0.92	1.18	0.38	9.64	nr	**10.82**
25mm dia	1.36	1.74	0.44	11.16	nr	**12.91**
32mm dia	1.82	2.33	0.51	12.94	nr	**15.27**
40mm dia	2.56	3.28	0.59	14.97	nr	**18.25**
50mm dia	3.37	4.32	0.70	17.76	nr	**22.08**
90° Elbow, male/female						
15mm dia	0.50	0.64	0.64	16.24	nr	**16.88**
20mm dia	0.67	0.86	0.85	21.57	nr	**22.42**
25mm dia	1.11	1.42	0.97	24.61	nr	**26.03**
32mm dia	1.91	2.45	1.12	28.42	nr	**30.86**
40mm dia	3.19	4.09	1.29	32.73	nr	**36.82**
50mm dia	4.11	5.26	1.55	39.33	nr	**44.59**
65mm dia	8.87	11.36	1.89	47.95	nr	**59.31**
80mm dia	12.13	15.54	2.24	56.83	nr	**72.37**
100mm dia	21.20	27.15	3.09	78.40	nr	**105.55**
90° Elbow						
15mm dia	0.45	0.58	0.64	16.24	nr	**16.81**
20mm dia	0.61	0.78	0.85	21.57	nr	**22.35**
25mm dia	0.95	1.22	0.97	24.61	nr	**25.83**
32mm dia	1.61	2.06	1.12	28.42	nr	**30.48**
40mm dia	2.71	3.47	1.29	32.73	nr	**36.20**
50mm dia	3.17	4.06	1.55	39.33	nr	**43.39**
65mm dia	6.85	8.77	1.89	47.95	nr	**56.73**
80mm dia	10.06	12.88	2.24	56.83	nr	**69.72**
100mm dia	19.41	24.86	3.09	78.40	nr	**103.26**
125mm dia	41.59	53.27	4.44	112.65	nr	**165.92**
150mm dia	77.42	99.16	5.79	146.90	nr	**246.06**

T:MECHANICAL/COOLING/HEATING SYSTEMS

Item	Net Price £	Material £	Labour hours	Labour £	Unit	Total rate £
45° Elbow						
15mm dia	1.07	1.37	0.64	16.24	nr	**17.61**
20mm dia	1.31	1.68	0.85	21.57	nr	**23.24**
25mm dia	1.96	2.51	0.97	24.61	nr	**27.12**
32mm dia	3.72	4.76	1.12	28.42	nr	**33.18**
40mm dia	4.56	5.84	1.29	32.73	nr	**38.57**
50mm dia	6.26	8.02	1.55	39.33	nr	**47.34**
65mm dia	9.15	11.72	1.89	47.95	nr	**59.67**
80mm dia	13.76	17.62	2.24	56.83	nr	**74.46**
100mm dia	26.58	34.04	3.09	78.40	nr	**112.44**
150mm dia	74.24	95.09	5.79	146.90	nr	**241.99**
90° Bend, male/female						
15mm dia	0.88	1.13	0.64	16.24	nr	**17.37**
20mm dia	1.28	1.64	0.85	21.57	nr	**23.21**
25mm dia	1.89	2.42	0.97	24.61	nr	**27.03**
32mm dia	2.80	3.59	1.12	28.42	nr	**32.00**
40mm dia	4.42	5.66	1.29	32.73	nr	**38.39**
50mm dia	7.72	9.89	1.55	39.33	nr	**49.21**
65mm dia	12.86	16.47	1.89	47.95	nr	**64.42**
80mm dia	17.43	22.32	2.24	56.83	nr	**79.16**
100mm dia	43.18	55.30	3.09	78.40	nr	**133.70**
90° Bend, male						
15mm dia	2.02	2.59	0.64	16.24	nr	**18.83**
20mm dia	2.27	2.91	0.85	21.57	nr	**24.47**
25mm dia	3.33	4.27	0.97	24.61	nr	**28.88**
32mm dia	6.73	8.62	1.12	28.42	nr	**37.04**
40mm dia	9.44	12.09	1.29	32.73	nr	**44.82**
50mm dia	12.63	16.18	1.55	39.33	nr	**55.50**
90° Bend, female						
15mm dia	0.81	1.04	0.64	16.24	nr	**17.28**
20mm dia	1.15	1.47	0.85	21.57	nr	**23.04**
25mm dia	1.62	2.07	0.97	24.61	nr	**26.69**
32mm dia	2.86	3.66	1.12	28.42	nr	**32.08**
40mm dia	3.81	4.88	1.29	32.73	nr	**37.61**
50mm dia	5.36	6.87	1.55	39.33	nr	**46.19**
65mm dia	11.46	14.68	1.89	47.95	nr	**62.63**
80mm dia	18.25	23.37	2.24	56.83	nr	**80.21**
100mm dia	39.99	51.22	3.09	78.40	nr	**129.62**
125mm dia	106.68	136.63	4.44	112.65	nr	**249.29**
150mm dia	162.94	208.69	5.79	146.90	nr	**355.60**
Return bend						
15mm dia	4.07	5.21	0.64	16.24	nr	**21.45**
20mm dia	6.59	8.44	0.85	21.57	nr	**30.01**
25mm dia	8.22	10.53	0.97	24.61	nr	**35.14**
32mm dia	11.86	15.19	1.12	28.42	nr	**43.61**
40mm dia	14.12	18.08	1.29	32.73	nr	**50.81**
50mm dia	21.55	27.60	1.55	39.33	nr	**66.93**
Equal socket, parallel thread						
15mm dia	0.47	0.60	0.64	16.24	nr	**16.84**
20mm dia	0.57	0.73	0.85	21.57	nr	**22.30**
25mm dia	0.76	0.97	0.97	24.61	nr	**25.58**
32mm dia	1.34	1.72	1.12	28.42	nr	**30.13**
40mm dia	1.96	2.51	1.29	32.73	nr	**35.24**
50mm dia	2.65	3.39	1.55	39.33	nr	**42.72**

T:MECHANICAL/COOLING/HEATING SYSTEMS

Item	Net Price £	Material £	Labour hours	Labour £	Unit	Total rate £
T31 : LOW TEMPERATURE HOT WATER HEATING (cont'd)						
Y10 - PIPELINES (cont'd)						
SCREWED STEEL (cont'd)						
Fittings; black steel screwed pipes (cont'd)						
Equal socket, parallel thread (cont'd)						
65mm dia	4.99	6.39	1.89	47.95	nr	**54.34**
80mm dia	6.86	8.79	2.24	56.83	nr	**65.62**
100mm dia	11.64	14.91	3.09	78.40	nr	**93.31**
Concentric reducing socket						
20 x 15mm dia	0.69	0.88	0.76	19.28	nr	**20.17**
25 x 15mm dia	0.81	1.04	0.85	21.57	nr	**22.60**
25 x 20mm dia	0.85	1.09	0.86	21.82	nr	**22.91**
32 x 25mm dia	1.50	1.92	1.01	25.63	nr	**27.55**
40 x 25mm dia	1.89	2.42	1.16	29.43	nr	**31.85**
40 x 32mm dia	1.95	2.50	1.16	29.43	nr	**31.93**
50 x 25mm dia	3.65	4.67	1.38	35.01	nr	**39.69**
50 x 40mm dia	2.74	3.51	1.38	35.01	nr	**38.52**
65 x 50mm dia	4.98	6.38	1.69	42.88	nr	**49.26**
80 x 50mm dia	6.48	8.30	2.00	50.74	nr	**59.04**
100 x 50mm dia	12.92	16.55	2.75	69.77	nr	**86.32**
100 x 80mm dia	11.99	15.36	2.75	69.77	nr	**85.13**
150 x 100mm dia	31.64	40.52	4.10	104.02	nr	**144.55**
Eccentric reducing socket						
20 x 15mm dia	1.22	1.56	0.73	18.52	nr	**20.08**
25 x 15mm dia	3.50	4.48	0.85	21.57	nr	**26.05**
25 x 20mm dia	3.98	5.10	0.85	21.57	nr	**26.66**
32 x 25mm dia	4.69	6.01	1.01	25.63	nr	**31.63**
40 x 25mm dia	5.78	7.40	1.16	29.43	nr	**36.83**
40 x 32mm dia	2.90	3.71	1.16	29.43	nr	**33.15**
50 x 25mm dia	3.75	4.80	1.38	35.01	nr	**39.82**
50 x 40mm dia	3.49	4.47	1.38	35.01	nr	**39.48**
65 x 50mm dia	6.48	8.30	1.69	42.88	nr	**51.18**
80 x 50mm dia	10.53	13.49	2.00	50.74	nr	**64.23**
Hexagon bush						
20 x 15mm dia	0.39	0.50	0.37	9.39	nr	**9.89**
25 x 15mm dia	0.47	0.60	0.43	10.91	nr	**11.51**
25 x 20mm dia	0.50	0.64	0.43	10.91	nr	**11.55**
32 x 25mm dia	0.60	0.77	0.51	12.94	nr	**13.71**
40 x 25mm dia	0.95	1.22	0.58	14.72	nr	**15.93**
40 x 32mm dia	0.87	1.11	0.58	14.72	nr	**15.83**
50 x 25mm dia	1.95	2.50	0.71	18.01	nr	**20.51**
50 x 40mm dia	1.81	2.32	0.71	18.01	nr	**20.33**
65 x 50mm dia	3.03	3.88	0.85	21.57	nr	**25.45**
80 x 50mm dia	4.91	6.29	1.00	25.37	nr	**31.66**
100 x 50mm dia	11.36	14.55	1.52	38.57	nr	**53.12**
100 x 80mm dia	9.46	12.12	1.52	38.57	nr	**50.68**
150 x 100mm dia	29.94	38.35	2.57	65.21	nr	**103.55**

T:MECHANICAL/COOLING/HEATING SYSTEMS

Item	Net Price £	Material £	Labour hours	Labour £	Unit	Total rate £
Hexagon nipple						
15mm dia	0.42	0.54	0.28	7.10	nr	**7.64**
20mm dia	0.47	0.60	0.38	9.64	nr	**10.24**
25mm dia	0.67	0.86	0.44	11.16	nr	**12.02**
32mm dia	1.29	1.65	0.51	12.94	nr	**14.59**
40mm dia	1.50	1.92	0.59	14.97	nr	**16.89**
50mm dia	2.71	3.47	0.70	17.76	nr	**21.23**
65mm dia	4.43	5.67	0.85	21.57	nr	**27.24**
80mm dia	6.41	8.21	1.00	25.37	nr	**33.58**
100mm dia	10.88	13.94	1.44	36.54	nr	**50.47**
150mm dia	32.11	41.13	2.32	58.86	nr	**99.99**
Union, male/female						
15mm dia	2.04	2.61	0.64	16.24	nr	**18.85**
20mm dia	2.50	3.20	0.85	21.57	nr	**24.77**
25mm dia	3.13	4.01	0.97	24.61	nr	**28.62**
32mm dia	4.83	6.19	1.12	28.42	nr	**34.60**
40mm dia	6.18	7.92	1.29	32.73	nr	**40.65**
50mm dia	9.73	12.46	1.55	39.33	nr	**51.79**
65mm dia	18.58	23.80	1.89	47.95	nr	**71.75**
80mm dia	25.54	32.71	2.24	56.83	nr	**89.54**
Union, female						
15mm dia	4.75	6.08	0.64	16.24	nr	**22.32**
20mm dia	5.25	6.72	0.85	21.57	nr	**28.29**
25mm dia	6.94	8.89	0.97	24.61	nr	**33.50**
32mm dia	9.88	12.65	1.12	28.42	nr	**41.07**
40mm dia	11.95	15.31	1.29	32.73	nr	**48.04**
50mm dia	15.17	19.43	1.55	39.33	nr	**58.76**
65mm dia	33.27	42.61	1.89	47.95	nr	**90.56**
80mm dia	54.34	69.60	2.24	56.83	nr	**126.43**
100mm dia	78.25	100.22	3.09	78.40	nr	**178.62**
Union elbow, male/female						
15mm dia	8.15	10.44	0.55	13.95	nr	**24.39**
20mm dia	10.98	14.06	0.85	21.57	nr	**35.63**
25mm dia	13.24	16.96	0.97	24.61	nr	**41.57**
Twin elbow						
15mm dia	2.57	3.29	0.91	23.09	nr	**26.38**
20mm dia	2.85	3.65	1.22	30.95	nr	**34.60**
25mm dia	4.61	5.90	1.39	35.27	nr	**41.17**
32mm dia	8.82	11.30	1.62	41.10	nr	**52.40**
40mm dia	11.16	14.29	1.86	47.19	nr	**61.49**
50mm dia	14.35	18.38	2.21	56.07	nr	**74.45**
65mm dia	22.46	28.77	2.72	69.01	nr	**97.78**
80mm dia	38.26	49.00	3.21	81.44	nr	**130.45**
Equal tee						
15mm dia	0.61	0.78	0.91	23.09	nr	**23.87**
20mm dia	0.89	1.14	1.22	30.95	nr	**32.09**
25mm dia	1.28	1.64	1.39	35.27	nr	**36.91**
32mm dia	2.19	2.80	1.62	41.10	nr	**43.91**
40mm dia	3.36	4.30	1.86	47.19	nr	**51.50**
50mm dia	4.84	6.20	2.21	56.07	nr	**62.27**
65mm dia	10.62	13.60	2.72	69.01	nr	**82.61**
80mm dia	12.92	16.55	3.21	81.44	nr	**97.99**
100mm dia	23.43	30.01	4.44	112.65	nr	**142.66**
125mm dia	59.96	76.80	5.38	136.50	nr	**213.30**
150mm dia	95.54	122.37	6.31	160.10	nr	**282.46**

T:MECHANICAL/COOLING/HEATING SYSTEMS

Item	Net Price £	Material £	Labour hours	Labour £	Unit	Total rate £
T31 : LOW TEMPERATURE HOT WATER HEATING (cont'd)						
Y10 - PIPELINES (cont'd)						
SCREWED STEEL (cont'd)						
Fittings; black steel screwed pipes (cont'd)						
Tee reducing on branch						
20 x 15mm dia	1.06	1.36	1.22	30.95	nr	32.31
25 x 15mm dia	1.46	1.87	1.39	35.27	nr	37.14
25 x 20mm dia	1.74	2.23	1.39	35.27	nr	37.50
32 x 25mm dia	3.01	3.86	1.62	41.10	nr	44.96
40 x 25mm dia	3.98	5.10	1.86	47.19	nr	52.29
40 x 32mm dia	5.21	6.67	1.86	47.19	nr	53.86
50 x 25mm dia	5.92	7.58	2.21	56.07	nr	63.65
50 x 40mm dia	8.81	11.28	2.21	56.07	nr	67.36
65 x 50mm dia	13.75	17.61	2.72	69.01	nr	86.62
80 x 50mm dia	29.63	37.95	3.21	81.44	nr	119.39
100 x 50mm dia	27.12	34.74	4.44	112.65	nr	147.39
100 x 80mm dia	41.84	53.59	4.44	112.65	nr	166.24
150 x 100mm dia	98.34	125.95	6.31	160.10	nr	286.05
Equal pitcher tee						
15mm dia	2.46	3.15	0.91	23.09	nr	26.24
20mm dia	3.06	3.92	1.22	30.95	nr	34.87
25mm dia	4.59	5.88	1.39	35.27	nr	41.15
32mm dia	6.93	8.88	1.62	41.10	nr	49.98
40mm dia	10.71	13.72	1.86	47.19	nr	60.91
50mm dia	15.03	19.25	2.21	56.07	nr	75.32
65mm dia	23.25	29.78	2.72	69.01	nr	98.79
80mm dia	31.96	40.93	3.21	81.44	nr	122.38
100mm dia	71.91	92.10	4.44	112.65	nr	204.75
Cross						
15mm dia	2.41	3.09	1.00	25.37	nr	28.46
20mm dia	3.62	4.64	1.33	33.74	nr	38.38
25mm dia	4.59	5.88	1.51	38.31	nr	44.19
32mm dia	6.65	8.52	1.77	44.91	nr	53.43
40mm dia	8.57	10.98	2.02	51.25	nr	62.23
50mm dia	13.91	17.82	2.42	61.40	nr	79.22
65mm dia	19.23	24.63	2.97	75.35	nr	99.98
80mm dia	25.57	32.75	3.50	88.80	nr	121.55
100mm dia	46.49	59.54	4.84	122.80	nr	182.34
BLACK WELDED STEEL						
Black steel pipes; butt welded joints; BS 1387: 1985; including protective painting. Fixed Vertical with brackets measured separately (Refer to Screwed Steel Section). Welded butt joints are within the running length, but any flanges are additional						
Medium weight						
10mm dia	1.41	1.81	0.37	9.39	m	11.19
15mm dia	1.50	1.92	0.37	9.39	m	11.31
20mm dia	1.72	2.20	0.37	9.39	m	11.59

T:MECHANICAL/COOLING/HEATING SYSTEMS

Item	Net Price £	Material £	Labour hours	Labour £	Unit	Total rate £
25mm dia	2.46	3.15	0.41	10.40	m	**13.55**
32mm dia	3.05	3.91	0.48	12.18	m	**16.08**
40mm dia	3.54	4.53	0.52	13.19	m	**17.73**
50mm dia	4.99	6.39	0.62	15.73	m	**22.12**
65mm dia	6.77	8.67	0.64	16.24	m	**24.91**
80mm dia	8.79	11.26	1.10	27.91	m	**39.17**
100mm dia	12.45	15.95	1.31	33.24	m	**49.18**
125mm dia	18.93	24.25	1.66	42.12	m	**66.36**
150mm dia	21.98	28.15	1.88	47.70	m	**75.85**
Heavy weight						
15mm dia	1.86	2.38	0.37	9.39	m	**11.77**
20mm dia	2.12	2.72	0.37	9.39	m	**12.10**
25mm dia	3.10	3.97	0.41	10.40	m	**14.37**
32mm dia	3.84	4.92	0.48	12.18	m	**17.10**
40mm dia	4.48	5.74	0.52	13.19	m	**18.93**
50mm dia	6.23	7.98	0.62	15.73	m	**23.71**
65mm dia	8.47	10.85	0.64	16.24	m	**27.09**
80mm dia	10.78	13.81	1.10	27.91	m	**41.72**
100mm dia	15.04	19.26	1.31	33.24	m	**52.50**
125mm dia	20.19	25.86	1.66	42.12	m	**67.98**
150mm dia	23.60	30.23	1.88	47.70	m	**77.93**
Black steel pipes; butt welded joints; BS 1387: 1985; including protective painting. Fixed at High Level or Suspended with brackets measured separately (Refer to Screwed Steel Section). Welded butt joints are within the running length, but any flanges are additional						
Medium weight						
10mm dia	1.41	1.81	0.58	14.72	m	**16.52**
15mm dia	1.50	1.92	0.58	14.72	m	**16.64**
20mm dia	1.72	2.20	0.58	14.72	m	**16.92**
25mm dia	2.46	3.15	0.60	15.22	m	**18.37**
32mm dia	3.05	3.91	0.68	17.25	m	**21.16**
40mm dia	3.54	4.53	0.73	18.52	m	**23.06**
50mm dia	4.99	6.39	0.85	21.57	m	**27.96**
65mm dia	6.77	8.67	0.88	22.33	m	**31.00**
80mm dia	8.79	11.26	1.45	36.79	m	**48.05**
100mm dia	12.45	15.95	1.74	44.15	m	**60.09**
125mm dia	18.93	24.25	2.21	56.07	m	**80.32**
150mm dia	21.98	28.15	2.50	63.43	m	**91.58**
Heavy weight						
15mm dia	1.86	2.38	0.58	14.72	m	**17.10**
20mm dia	2.12	2.72	0.58	14.72	m	**17.43**
25mm dia	3.10	3.97	0.60	15.22	m	**19.19**
32mm dia	3.84	4.92	0.68	17.25	m	**22.17**
40mm dia	4.48	5.74	0.73	18.52	m	**24.26**
50mm dia	6.23	7.98	0.85	21.57	m	**29.55**
65mm dia	8.47	10.85	0.88	22.33	m	**33.18**
80mm dia	10.78	13.81	1.45	36.79	m	**50.60**
100mm dia	15.04	19.26	1.74	44.15	m	**63.41**
125mm dia	20.19	25.86	2.21	56.07	m	**81.93**
150mm dia	23.60	30.23	2.50	63.43	m	**93.66**

T:MECHANICAL/COOLING/HEATING SYSTEMS

Item	Net Price £	Material £	Labour hours	Labour £	Unit	Total rate £
T31 : LOW TEMPERATURE HOT WATER HEATING (cont'd)						
Y10 - PIPELINES (cont'd)						
BLACK WELDED STEEL (cont'd)						
FIXINGS						
Refer to steel pipes; black malleable iron. For minimum fixing distances, refer to the Tables and Memoranda to the rear of the book						
Extra over black steel butt welded pipes; black steel flanges, welded and drilled; metric; BS 4504						
Welded flanges; PN6						
15mm dia	2.13	2.73	0.59	14.97	nr	**17.70**
20mm dia	2.13	2.73	0.69	17.51	nr	**20.23**
25mm dia	2.13	2.73	0.84	21.31	nr	**24.04**
32mm dia	2.29	2.93	1.00	25.37	nr	**28.30**
40mm dia	2.29	2.93	1.11	28.16	nr	**31.10**
50mm dia	2.62	3.36	1.37	34.76	nr	**38.12**
65mm dia	3.24	4.15	1.54	39.07	nr	**43.22**
80mm dia	4.69	6.01	1.67	42.37	nr	**48.38**
100mm dia	5.44	6.97	2.22	56.33	nr	**63.29**
125mm dia	9.34	11.96	2.61	66.22	nr	**78.18**
150mm dia	10.50	13.45	2.99	75.86	nr	**89.31**
Welded flanges; PN16						
15mm dia	2.54	3.25	0.59	14.97	nr	**18.22**
20mm dia	2.54	3.25	0.69	17.51	nr	**20.76**
25mm dia	2.54	3.25	0.84	21.31	nr	**24.57**
32mm dia	3.84	4.92	1.00	25.37	nr	**30.29**
40mm dia	3.84	4.92	1.11	28.16	nr	**33.08**
50mm dia	3.94	5.05	1.37	34.76	nr	**39.81**
65mm dia	4.72	6.05	1.54	39.07	nr	**45.12**
80mm dia	5.72	7.33	1.67	42.37	nr	**49.70**
100mm dia	6.73	8.62	2.22	56.33	nr	**64.95**
125mm dia	11.56	14.81	2.61	66.22	nr	**81.03**
150mm dia	13.01	16.66	2.99	75.86	nr	**92.53**
Blank flanges, slip on for welding; PN6						
15mm dia	1.01	1.29	0.48	12.18	nr	**13.47**
20mm dia	1.01	1.29	0.55	13.95	nr	**15.25**
25mm dia	1.01	1.29	0.64	16.24	nr	**17.53**
32mm dia	1.68	2.15	0.76	19.28	nr	**21.43**
40mm dia	1.70	2.18	0.84	21.31	nr	**23.49**
50mm dia	1.87	2.40	1.01	25.63	nr	**28.02**
65mm dia	3.72	4.76	1.30	32.98	nr	**37.75**
80mm dia	3.77	4.83	1.41	35.77	nr	**40.60**
100mm dia	3.98	5.10	1.78	45.16	nr	**50.26**
125mm dia	6.50	8.33	2.06	52.27	nr	**60.59**
150mm dia	7.13	9.13	2.35	59.62	nr	**68.76**

T:MECHANICAL/COOLING/HEATING SYSTEMS

Item	Net Price £	Material £	Labour hours	Labour £	Unit	Total rate £
Blank flanges, slip on for welding; PN16						
15mm dia	0.93	1.19	0.48	12.18	nr	**13.37**
20mm dia	1.06	1.36	0.55	13.95	nr	**15.31**
25mm dia	1.15	1.47	0.64	16.24	nr	**17.71**
32mm dia	2.21	2.83	0.76	19.28	nr	**22.11**
40mm dia	2.41	3.09	0.84	21.31	nr	**24.40**
50mm dia	2.79	3.57	1.01	25.63	nr	**29.20**
65mm dia	3.65	4.67	1.30	32.98	nr	**37.66**
80mm dia	4.35	5.57	1.41	35.77	nr	**41.35**
100mm dia	5.64	7.22	1.78	45.16	nr	**52.39**
125mm dia	8.39	10.75	2.06	52.27	nr	**63.01**
150mm dia	9.85	12.62	2.35	59.62	nr	**72.24**
Extra over black steel butt welded pipes; **black steel flanges, welding and drilled;** **imperial; BS 10**						
Welded flanges; Table E						
15mm dia	3.69	4.73	0.59	14.97	nr	**19.70**
20mm dia	3.69	4.73	0.69	17.51	nr	**22.23**
25mm dia	3.69	4.73	0.84	21.31	nr	**26.04**
32mm dia	3.69	4.73	1.00	25.37	nr	**30.10**
40mm dia	3.69	4.73	1.11	28.16	nr	**32.89**
50mm dia	3.69	4.73	1.37	34.76	nr	**39.49**
65mm dia	4.57	5.85	1.54	39.07	nr	**44.93**
80mm dia	4.66	5.97	1.67	42.37	nr	**48.34**
100mm dia	6.94	8.89	2.22	56.33	nr	**65.21**
125mm dia	15.20	19.47	2.61	66.22	nr	**85.69**
150mm dia	16.93	21.68	2.99	75.86	nr	**97.55**
Blank flanges, slip on for welding; Table E						
15mm dia	3.78	4.84	0.48	12.18	nr	**17.02**
20mm dia	3.78	4.84	0.55	13.95	nr	**18.80**
25mm dia	3.78	4.84	0.64	16.24	nr	**21.08**
32mm dia	4.77	6.11	0.76	19.28	nr	**25.39**
40mm dia	5.19	6.65	0.84	21.31	nr	**27.96**
50mm dia	5.73	7.34	1.01	25.63	nr	**32.96**
65mm dia	6.60	8.45	1.30	32.98	nr	**41.44**
80mm dia	7.80	9.99	1.41	35.77	nr	**45.76**
100mm dia	11.34	14.52	1.78	45.16	nr	**59.69**
125mm dia	18.09	23.17	2.06	52.27	nr	**75.44**
150mm dia	25.92	33.20	2.35	59.62	nr	**92.82**
Extra over black steel butt welded pipes; **black steel flange connections**						
Bolted connection between pair of flanges; **including gasket, bolts, nuts and washers**						
50mm dia	9.90	12.68	0.50	12.69	nr	**25.37**
65mm dia	11.70	14.99	0.50	12.69	nr	**27.67**
80mm dia	15.24	19.52	0.50	12.69	nr	**32.21**
100mm dia	18.22	23.34	0.50	12.69	nr	**36.02**
125mm dia	28.27	36.21	0.50	12.69	nr	**48.89**
150mm dia	34.04	43.60	0.88	22.33	nr	**65.93**

T:MECHANICAL/COOLING/HEATING SYSTEMS

Item	Net Price £	Material £	Labour hours	Labour £	Unit	Total rate £
T31 : LOW TEMPERATURE HOT WATER HEATING (cont'd)						
Y10 - PIPELINES (cont'd)						
BLACK WELDED STEEL (cont'd)						
Extra over fittings; BS 1965; butt welded						
Cap						
25mm dia	8.03	10.28	0.47	11.92	nr	22.21
32mm dia	8.03	10.28	0.59	14.97	nr	25.25
40mm dia	8.03	10.28	0.70	17.76	nr	28.05
50mm dia	9.54	12.22	0.99	25.12	nr	37.34
65mm dia	11.14	14.27	1.35	34.25	nr	48.52
80mm dia	11.33	14.51	1.66	42.12	nr	56.63
100mm dia	14.72	18.85	2.23	56.58	nr	75.43
125mm dia	20.86	26.72	3.03	76.88	nr	103.59
150mm dia	23.87	30.57	3.79	96.16	nr	126.73
Concentric reducer						
20 x 15mm dia	4.47	5.73	0.69	17.51	nr	23.23
25 x 15mm dia	4.28	5.48	0.87	22.07	nr	27.56
25 x 20mm dia	5.56	7.12	0.87	22.07	nr	29.19
32 x 25mm dia	6.08	7.79	1.08	27.40	nr	35.19
40 x 25mm dia	7.92	10.14	1.38	35.01	nr	45.16
40 x 32mm dia	5.41	6.93	1.38	35.01	nr	41.94
50 x 25mm dia	7.15	9.16	1.82	46.18	nr	55.33
50 x 40mm dia	5.18	6.63	1.82	46.18	nr	52.81
65 x 50mm dia	7.08	9.07	2.52	63.94	nr	73.01
80 x 50mm dia	7.19	9.21	3.24	82.21	nr	91.41
100 x 50mm dia	11.95	15.31	4.08	103.52	nr	118.82
100 x 80mm dia	8.16	10.45	4.08	103.52	nr	113.97
125 x 80mm dia	17.99	23.04	4.71	119.50	nr	142.54
150 x 100mm dia	19.38	24.82	5.33	135.23	nr	160.05
Eccentric reducer						
20 x 15mm dia	6.63	8.49	0.69	17.51	nr	26.00
25 x 15mm dia	8.74	11.19	0.87	22.07	nr	33.27
25 x 20mm dia	7.30	9.35	0.87	22.07	nr	31.42
32 x 25mm dia	8.27	10.59	1.08	27.40	nr	37.99
40 x 25mm dia	10.18	13.04	1.38	35.01	nr	48.05
40 x 32mm dia	9.73	12.46	1.38	35.01	nr	47.48
50 x 25mm dia	12.52	16.04	1.82	46.18	nr	62.21
50 x 40mm dia	8.20	10.50	1.82	46.18	nr	56.68
65 x 50mm dia	9.60	12.30	2.52	63.94	nr	76.23
80 x 50mm dia	11.71	15.00	3.24	82.21	nr	97.20
100 x 50mm dia	19.92	25.51	4.08	103.52	nr	129.03
100 x 80mm dia	14.60	18.70	4.08	103.52	nr	122.22
125 x 80mm dia	39.32	50.36	4.71	119.50	nr	169.86
150 x 100mm dia	29.67	38.00	5.33	135.23	nr	173.23
45° elbow, long radius						
15mm dia	1.73	2.22	0.56	14.21	nr	16.42
20mm dia	1.73	2.22	0.75	19.03	nr	21.24
25mm dia	2.21	2.83	0.93	23.60	nr	26.43
32mm dia	3.32	4.25	1.17	29.69	nr	33.94
40mm dia	3.06	3.92	1.46	37.04	nr	40.96
50mm dia	4.10	5.25	1.97	49.98	nr	55.23
65mm dia	5.32	6.81	2.70	68.50	nr	75.32
80mm dia	5.16	6.61	3.32	84.23	nr	90.84

T:MECHANICAL/COOLING/HEATING SYSTEMS

Item	Net Price £	Material £	Labour hours	Labour £	Unit	Total rate £
100mm dia	7.95	10.18	4.09	103.77	nr	113.95
125mm dia	15.25	19.53	4.94	125.34	nr	144.87
150mm dia	20.56	26.33	5.78	146.65	nr	172.98
90° elbow, long radius						
15mm dia	1.73	2.22	0.56	14.21	nr	16.42
20mm dia	1.73	2.22	0.75	19.03	nr	21.24
25mm dia	2.21	2.83	0.93	23.60	nr	26.43
32mm dia	3.32	4.25	1.17	29.69	nr	33.94
40mm dia	3.06	3.92	1.46	37.04	nr	40.96
50mm dia	4.10	5.25	1.97	49.98	nr	55.23
65mm dia	5.32	6.81	2.70	68.50	nr	75.32
80mm dia	6.08	7.79	3.32	84.23	nr	92.02
100mm dia	9.35	11.98	4.09	103.77	nr	115.75
125mm dia	17.94	22.98	4.94	125.34	nr	148.31
150mm dia	24.20	31.00	5.78	146.65	nr	177.64
Branch Bend (based on branch and pipe sizes being the same)						
15mm dia	6.90	8.84	0.85	21.57	nr	30.40
20mm dia	6.90	8.84	0.85	21.57	nr	30.40
25mm dia	6.90	8.84	1.02	25.88	nr	34.72
32mm dia	6.65	8.52	1.11	28.16	nr	36.68
40mm dia	6.59	8.44	1.36	34.51	nr	42.95
50mm dia	6.27	8.03	1.70	43.13	nr	51.16
65mm dia	9.20	11.78	1.78	45.16	nr	56.95
80mm dia	14.80	18.96	1.82	46.18	nr	65.13
100mm dia	20.00	25.62	1.87	47.45	nr	73.06
125mm dia	37.31	47.79	2.21	56.07	nr	103.86
150mm dia	57.31	73.40	2.65	67.24	nr	140.64
Equal tee						
15mm dia	16.83	21.56	0.82	20.80	nr	42.36
20mm dia	16.83	21.56	1.10	27.91	nr	49.46
25mm dia	16.83	21.56	1.35	34.25	nr	55.81
32mm dia	16.83	21.56	1.63	41.36	nr	62.91
40mm dia	16.83	21.56	2.14	54.30	nr	75.85
50mm dia	17.72	22.70	3.02	76.62	nr	99.32
65mm dia	25.69	32.90	3.61	91.59	nr	124.50
80mm dia	25.89	33.16	4.18	106.05	nr	139.21
100mm dia	32.21	41.25	5.24	132.95	nr	174.20
125mm dia	73.74	94.45	6.70	169.99	nr	264.44
150mm dia	80.54	103.15	8.45	214.39	nr	317.55
Extra over black steel butt welded pipes; labour						
Made bend						
15mm dia	-	-	0.42	10.66	nr	10.66
20mm dia	-	-	0.42	10.66	nr	10.66
25mm dia	-	-	0.50	12.69	nr	12.69
32mm dia	-	-	0.62	15.73	nr	15.73
40mm dia	-	-	0.74	18.78	nr	18.78
50mm dia	-	-	0.89	22.58	nr	22.58
65mm dia	-	-	1.05	26.64	nr	26.64
80mm dia	-	-	1.13	28.67	nr	28.67
100mm dia	-	-	2.90	73.58	nr	73.58
125mm dia	-	-	3.56	90.32	nr	90.32
150mm dia	-	-	4.18	106.05	nr	106.05

T:MECHANICAL/COOLING/HEATING SYSTEMS

Item	Net Price £	Material £	Labour hours	Labour £	Unit	Total rate £
T31 : LOW TEMPERATURE HOT WATER HEATING (cont'd)						
Y10 - PIPELINES (cont'd)						
BLACK WELDED STEEL (cont'd)						
Extra over black steel butt welded pipes; Labour (cont'd)						
Splay cut end						
15mm dia	-	-	0.14	3.55	nr	3.55
20mm dia	-	-	0.16	4.06	nr	4.06
25mm dia	-	-	0.18	4.57	nr	4.57
32mm dia	-	-	0.25	6.34	nr	6.34
40mm dia	-	-	0.27	6.85	nr	6.85
50mm dia	-	-	0.31	7.87	nr	7.87
65mm dia	-	-	0.35	8.88	nr	8.88
80mm dia	-	-	0.40	10.15	nr	10.15
100mm dia	-	-	0.48	12.18	nr	12.18
125mm dia	-	-	0.56	14.21	nr	14.21
150mm dia	-	-	0.64	16.24	nr	16.24
Screwed joint to fitting						
15mm dia	-	-	0.30	7.61	nr	7.61
20mm dia	-	-	0.40	10.15	nr	10.15
25mm dia	-	-	0.46	11.67	nr	11.67
32mm dia	-	-	0.53	13.45	nr	13.45
40mm dia	-	-	0.61	15.48	nr	15.48
50mm dia	-	-	0.73	18.52	nr	18.52
65mm dia	-	-	0.89	22.58	nr	22.58
80mm dia	-	-	1.05	26.64	nr	26.64
100mm dia	-	-	1.46	37.04	nr	37.04
125mm dia	-	-	2.10	53.28	nr	53.28
150mm dia	-	-	2.73	69.27	nr	69.27
Straight butt weld						
15mm dia	-	-	0.31	7.87	nr	7.87
20mm dia	-	-	0.42	10.66	nr	10.66
25mm dia	-	-	0.52	13.19	nr	13.19
32mm dia	-	-	0.69	17.51	nr	17.51
40mm dia	-	-	0.83	21.06	nr	21.06
50mm dia	-	-	1.22	30.95	nr	30.95
65mm dia	-	-	1.57	39.83	nr	39.83
80mm dia	-	-	1.95	49.48	nr	49.48
100mm dia	-	-	2.38	60.39	nr	60.39
125mm dia	-	-	2.83	71.80	nr	71.80
150mm dia	-	-	3.27	82.97	nr	82.97

T:MECHANICAL/COOLING/HEATING SYSTEMS

Item	Net Price £	Material £	Labour hours	Labour £	Unit	Total rate £
CARBON WELDED STEEL						
Hot finished seamless carbon steel pipe; BS 806 and BS 3601; wall thickness to BS 3600; butt welded joints; including protective painting, fixed vertically or at low level, brackets measured separately (Refer to Screwed Pipework Section). Welded butt joints are within the running length, but any flanges are additional						
Pipework						
200mm dia	39.56	50.67	2.04	51.76	m	**102.43**
250mm dia	49.33	63.18	2.59	65.71	m	**128.89**
300mm dia	54.53	69.84	2.99	75.86	m	**145.70**
350mm dia	80.50	103.10	3.52	89.31	m	**192.41**
400mm dia	178.85	229.07	4.08	103.52	m	**332.59**
Hot finished seamless carbon steel pipe; BS 806 and BS 3601; wall thickness to BS 3600; butt welded joints; including protective painting, fixed at high level or suspended, brackets measured separately (Refer to Screwed Pipework Section). Welded butt joints are within the running length, but any flanges are additional						
Pipework						
200mm dia	39.56	50.67	3.70	93.88	m	**144.54**
250mm dia	49.33	63.18	4.73	120.01	m	**183.19**
300mm dia	54.53	69.84	5.65	143.35	m	**213.19**
350mm dia	80.50	103.10	6.68	169.48	m	**272.59**
400mm dia	178.85	229.07	7.70	195.36	m	**424.43**
FIXINGS						
Refer to steel pipes; black malleable iron. For minimum fixing distances, refer to the Tables and Memoranda to the rear of the book						
Extra over fittings; BS 1965 part 1; butt welded						
Cap						
200mm dia	34.60	44.32	3.70	93.88	nr	**138.19**
250mm dia	66.82	85.58	4.73	120.01	nr	**205.59**
300mm dia	31.26	40.04	5.65	143.35	nr	**183.39**
350mm dia	40.27	51.58	6.68	169.48	nr	**221.06**
400mm dia	45.71	58.54	7.70	195.36	nr	**253.91**

T:MECHANICAL/COOLING/HEATING SYSTEMS

Item	Net Price £	Material £	Labour hours	Labour £	Unit	Total rate £
T31 : LOW TEMPERATURE HOT WATER HEATING (cont'd)						
Y10 - PIPELINES (cont'd)						
CARBON WELDED STEEL (cont'd)						
Extra over fittings; BS 1965 part 1; butt welded (cont'd)						
Concentric reducer						
200mm x 150mm dia	36.01	46.12	7.27	184.45	nr	**230.57**
250mm x 150mm dia	54.88	70.29	9.05	229.62	nr	**299.91**
250mm x 200mm dia	33.76	43.24	9.10	230.88	nr	**274.12**
300mm x 150mm dia	115.80	148.32	10.75	272.75	nr	**421.06**
300mm x 200mm dia	65.99	84.52	10.75	272.75	nr	**357.27**
300mm x 250mm dia	58.66	75.13	11.15	282.90	nr	**358.03**
350mm x 200mm dia	-	-	12.50	317.15	nr	**317.15**
350mm x 250mm dia	-	-	12.70	322.22	nr	**322.22**
350mm x 300mm dia	-	-	13.00	329.83	nr	**329.83**
400mm x 250mm dia	-	-	14.46	366.88	nr	**366.88**
400mm x 300mm dia	-	-	14.51	368.15	nr	**368.15**
400mm x 350mm dia	-	-	15.16	384.64	nr	**384.64**
Eccentric reducer						
200mm x 150mm dia	65.89	84.39	7.27	184.45	nr	**268.85**
250mm x 150mm dia	91.54	117.24	9.05	229.62	nr	**346.86**
250mm x 200mm dia	59.90	76.72	9.10	230.88	nr	**307.60**
300mm x 150mm dia	132.88	170.19	10.75	272.75	nr	**442.94**
300mm x 200mm dia	126.67	162.24	10.75	272.75	nr	**434.99**
300mm x 250mm dia	101.96	130.59	11.15	282.90	nr	**413.49**
350mm x 200mm dia	-	-	12.50	317.15	nr	**317.15**
350mm x 250mm dia	-	-	12.70	322.22	nr	**322.22**
350mm x 300mm dia	66.60	85.30	13.00	329.83	nr	**415.14**
400mm x 250mm dia	-	-	14.46	366.88	nr	**366.88**
400mm x 300mm dia	96.81	123.99	14.51	368.15	nr	**492.14**
400mm x 350mm dia	85.90	110.02	15.16	384.64	nr	**494.66**
45° elbow						
200mm dia	35.53	45.51	7.75	196.63	nr	**242.14**
250mm dia	68.33	87.52	10.05	254.99	nr	**342.50**
300mm dia	100.75	129.04	12.20	309.54	nr	**438.58**
350mm dia	70.64	90.48	14.65	371.70	nr	**462.17**
400mm dia	91.63	117.36	17.12	434.37	nr	**551.73**
90° elbow						
200mm dia	41.80	53.54	7.75	196.63	nr	**250.17**
250mm dia	81.20	104.00	10.05	254.99	nr	**358.99**
300mm dia	118.53	151.81	12.20	309.54	nr	**461.35**
350mm dia	88.30	113.09	14.65	371.70	nr	**484.79**
400mm dia	114.53	146.69	17.12	434.37	nr	**581.06**
Equal tee						
200mm dia	113.08	144.83	11.25	285.43	nr	**430.27**
250mm dia	193.90	248.35	14.53	368.65	nr	**617.00**
300mm dia	122.54	156.95	17.55	445.28	nr	**602.22**
350mm dia	-	-	20.98	532.30	nr	**532.30**
400mm dia	-	-	24.38	618.57	nr	**618.57**

·T:MECHANICAL/COOLING/HEATING SYSTEMS

Item	Net Price £	Material £	Labour hours	Labour £	Unit	Total rate £
Extra over black steel butt welded pipes; labour						
Straight butt weld						
200mm dia	-	-	4.08	103.52	nr	**103.52**
250mm dia	-	-	5.20	131.93	nr	**131.93**
300mm dia	-	-	6.22	157.81	nr	**157.81**
350mm dia	-	-	7.33	185.98	nr	**185.98**
400mm dia	-	-	8.41	213.38	nr	**213.38**
Branch weld						
100mm dia	-	-	3.46	87.79	nr	**87.79**
125mm dia	-	-	4.23	107.32	nr	**107.32**
150mm dia	-	-	5.00	126.86	nr	**126.86**
Extra over black steel butt welded pipes; black steel flanges, welding and drilled; metric; BS 4504						
Welded flanges; PN16						
200mm dia	15.38	19.70	4.10	104.02	nr	**123.72**
250mm dia	27.46	35.17	5.33	135.23	nr	**170.40**
300mm dia	32.18	41.22	6.40	162.38	nr	**203.60**
350mm dia	69.33	88.80	7.43	188.51	nr	**277.31**
400mm dia	90.57	116.00	8.45	214.39	nr	**330.39**
Welded flanges; PN25						
200mm dia	58.62	75.08	4.10	104.02	nr	**179.10**
250mm dia	70.27	90.00	5.33	135.23	nr	**225.23**
300mm dia	95.03	121.71	6.40	162.38	nr	**284.09**
Blank flanges, slip on for welding; PN16						
200mm dia	16.17	20.71	2.70	68.50	nr	**89.21**
250mm dia	23.12	29.61	3.48	88.29	nr	**117.91**
300mm dia	32.27	41.33	4.20	106.56	nr	**147.89**
350mm dia	61.71	79.04	4.78	121.28	nr	**200.32**
400mm dia	84.86	108.69	5.35	135.74	nr	**244.43**
Blank flanges, slip on for welding; PN25						
200mm dia	123.20	157.79	2.70	68.50	nr	**226.30**
250mm dia	183.18	234.62	3.48	88.29	nr	**322.91**
300mm dia	252.64	323.58	4.20	106.56	nr	**430.14**
Extra over black steel butt welded pipes; black steel flange connections						
Bolted connection between pair of flanges; including gasket, bolts, nuts and washers						
200mm dia	49.00	62.76	3.83	97.17	nr	**159.93**
250mm dia	81.67	104.60	4.93	125.08	nr	**229.69**
300mm dia	94.49	121.02	5.90	149.69	nr	**270.72**

T:MECHANICAL/COOLING/HEATING SYSTEMS

Item	Net Price £	Material £	Labour hours	Labour £	Unit	Total rate £
T31 : LOW TEMPERATURE HOT WATER HEATING (cont'd)						
Y10 - PIPELINES (cont'd)						
PRESS FIT						
Press fit jointing system; operating temperature -20°C to +120°C; operating pressure 16 bar; butyl rubber 'O' ring mechanical joint. With brackets measured separately (Refer to Screwed Steel Section)						
Carbon steel						
Pipework						
15mm dia	0.96	1.23	0.46	11.67	m	**12.90**
20mm dia	1.49	1.91	0.48	12.18	m	**14.09**
25mm dia	2.07	2.65	0.52	13.19	m	**15.84**
32mm dia	2.75	3.52	0.56	14.21	m	**17.73**
40mm dia	3.75	4.80	0.58	14.72	m	**19.52**
50mm dia	4.82	6.17	0.66	16.75	m	**22.92**
Extra over for Carbon Steel pressfit fittings						
Coupling						
15mm dia	0.65	0.83	0.36	9.13	nr	**9.97**
20mm dia	0.78	1.00	0.36	9.13	nr	**10.13**
25mm dia	0.99	1.27	0.44	11.16	nr	**12.43**
32mm dia	1.66	2.13	0.44	11.16	nr	**13.29**
40mm dia	2.22	2.84	0.52	13.19	nr	**16.04**
50mm dia	2.62	3.36	0.60	15.22	nr	**18.58**
Reducer						
20 x 15mm dia	0.60	0.77	0.36	9.13	nr	**9.90**
25 x 15mm dia	0.78	1.00	0.40	10.15	nr	**11.15**
25 x 20mm dia	0.82	1.05	0.40	10.15	nr	**11.20**
32 x 20mm dia	0.92	1.18	0.40	10.15	nr	**11.33**
32 x 25mm dia	0.98	1.26	0.44	11.16	nr	**12.42**
40 x 32mm dia	2.12	2.72	0.48	12.18	nr	**14.89**
50 x 20mm dia	6.14	7.86	0.48	12.18	nr	**20.04**
50 x 25mm dia	6.18	7.92	0.52	13.19	nr	**21.11**
50 x 40mm dia	6.50	8.33	0.56	14.21	nr	**22.53**
90° Elbow						
15mm dia	0.94	1.20	0.36	9.13	nr	**10.34**
20mm dia	1.22	1.56	0.36	9.13	nr	**10.70**
25mm dia	1.68	2.15	0.44	11.16	nr	**13.32**
32mm dia	4.19	5.37	0.44	11.16	nr	**16.53**
40mm dia	6.71	8.59	0.52	13.19	nr	**21.79**
50mm dia	8.02	10.27	0.60	15.22	nr	**25.50**
45° Elbow						
15mm dia	1.11	1.42	0.36	9.13	nr	**10.56**
20mm dia	1.24	1.59	0.36	9.13	nr	**10.72**
25mm dia	1.69	2.16	0.44	11.16	nr	**13.33**
32mm dia	3.32	4.25	0.44	11.16	nr	**15.42**
40mm dia	4.17	5.34	0.52	13.19	nr	**18.53**
50mm dia	4.71	6.03	0.60	15.22	nr	**21.26**

T:MECHANICAL/COOLING/HEATING SYSTEMS

Item	Net Price £	Material £	Labour hours	Labour £	Unit	Total rate £
Equal tee						
15mm dia	1.79	2.29	0.54	13.70	nr	**15.99**
20mm dia	2.06	2.64	0.54	13.70	nr	**16.34**
25mm dia	2.77	3.55	0.66	16.75	nr	**20.29**
32mm dia	4.31	5.52	0.66	16.75	nr	**22.27**
40mm dia	6.37	8.16	0.78	19.79	nr	**27.95**
50mm dia	7.63	9.77	0.90	22.83	nr	**32.61**
Reducing tee						
20 x 15mm dia	2.03	2.60	0.54	13.70	nr	**16.30**
25 x 15mm dia	2.74	3.51	0.62	15.73	nr	**19.24**
25 x 20mm dia	2.97	3.80	0.62	15.73	nr	**19.53**
32 x 15mm dia	4.02	5.15	0.62	15.73	nr	**20.88**
32 x 20mm dia	4.34	5.56	0.62	15.73	nr	**21.29**
32 x 25mm dia	4.40	5.64	0.62	15.73	nr	**21.37**
40 x 20mm dia	5.81	7.44	0.70	17.76	nr	**25.20**
40 x 25mm dia	6.02	7.71	0.70	17.76	nr	**25.47**
40 x 32mm dia	5.89	7.54	0.70	17.76	nr	**25.30**
50 x 20mm dia	6.93	8.88	0.82	20.80	nr	**29.68**
50 x 25mm dia	7.06	9.04	0.82	20.80	nr	**29.85**
50 x 32mm dia	7.27	9.31	0.82	20.80	nr	**30.12**
50 x 40mm dia	7.62	9.76	0.82	20.80	nr	**30.56**

MECHANICAL GROOVED

Mechanical grooved jointing system; working temperature not exceeding 82°s C BS 5750; pipework complete with grooved joints; painted finish. With brackets measured separately (Refer to Screwed Steel Section)

Item	Net Price £	Material £	Labour hours	Labour £	Unit	Total rate £
Grooved Joints						
65 mm	9.68	12.40	0.58	14.72	m	**27.11**
80 mm	12.08	15.47	0.68	17.25	m	**32.72**
100 mm	16.75	21.45	0.79	20.04	m	**41.50**
125 mm	22.57	28.91	1.02	25.88	m	**54.79**
150mm	26.60	34.07	1.15	29.18	m	**63.25**

Extra over mechanical grooved system fittings

Item	Net Price £	Material £	Labour hours	Labour £	Unit	Total rate £
Couplings						
65mm	7.27	9.31	0.41	10.40	nr	**19.71**
80mm	7.80	9.99	0.41	10.40	nr	**20.39**
100mm	10.22	13.09	0.66	16.75	nr	**29.84**
125mm	14.31	18.33	0.68	17.25	nr	**35.58**
150mm	17.99	23.04	0.80	20.30	nr	**43.34**
Concentric reducers (one size down)						
80mm	9.50	12.17	0.59	14.97	nr	**27.14**
100mm	20.51	26.27	0.71	18.01	nr	**44.28**
125mm	90.10	115.40	0.85	21.57	nr	**136.97**
150mm	22.19	28.42	0.98	24.86	nr	**53.29**
Short radius elbow; 90°						
65mm	9.44	12.09	0.53	13.45	nr	**25.54**
80mm	9.64	12.35	0.61	15.48	nr	**27.82**
100mm	12.91	16.54	0.80	20.30	nr	**36.83**
125mm	21.29	27.27	0.90	22.83	nr	**50.10**
150mm	27.63	35.39	0.94	23.85	nr	**59.24**

T:MECHANICAL/COOLING/HEATING SYSTEMS

Item	Net Price £	Material £	Labour hours	Labour £	Unit	Total rate £
T31 : LOW TEMPERATURE HOT WATER HEATING (cont'd)						
Y10 - PIPELINES (cont'd)						
MECHANICAL GROOVED (cont'd)						
Extra over mechanical grooved system Fittings (cont'd)						
Short radius elbow; 45°						
65mm	8.10	10.37	0.53	13.45	nr	**23.82**
80mm	9.09	11.64	0.61	15.48	nr	**27.12**
100mm	11.30	14.47	0.80	20.30	nr	**34.77**
125mm	19.10	24.46	0.90	22.83	nr	**47.30**
150mm	21.03	26.93	0.94	23.85	nr	**50.78**
Equal tee						
65mm	17.01	21.79	0.83	21.06	nr	**42.84**
80mm	18.02	23.08	0.93	23.60	nr	**46.68**
100mm	20.15	25.81	1.18	29.94	nr	**55.75**
125mm	53.43	68.43	1.37	34.76	nr	**103.19**
150mm	49.54	63.45	1.43	36.28	nr	**99.73**
PLASTIC PIPEWORK						
Polypropylene PP-R 80 pipe, mechanically stabilised by fibre compound mixture in middle layer; suitable for continuous working temperatures of 0-90°C; thermal fused joints in the running length						
Pipe; 4m long; PN 20						
20mm dia	1.39	1.78	0.35	8.88	m	**10.66**
25mm dia	2.10	2.69	0.39	9.90	m	**12.58**
32mm dia	2.39	3.06	0.43	10.91	m	**13.97**
40mm dia	3.19	4.09	0.47	11.92	m	**16.01**
50mm dia	4.65	5.96	0.51	12.94	m	**18.90**
63mm dia	7.66	9.81	0.52	13.19	m	**23.00**
75mm dia	9.91	12.69	0.60	15.22	m	**27.92**
90mm dia	15.29	19.58	0.69	17.51	m	**37.09**
110mm dia	22.98	29.43	0.69	17.51	m	**46.94**
125mm dia	24.62	31.53	0.85	21.57	m	**53.10**
FIXINGS						
Refer to steel pipes; black malleable iron. For minimum fixing distances, refer to the Tables and Memoranda to the rear of the book						
Extra over fittings; thermal fused joints						
Overbridge bow						
20mm dia	0.92	1.18	0.51	12.94	nr	**14.12**
25mm dia	1.70	2.18	0.56	14.21	nr	**16.39**
32mm dia	3.39	4.34	0.65	16.49	nr	**20.83**

T:MECHANICAL/COOLING/HEATING SYSTEMS

Item	Net Price £	Material £	Labour hours	Labour £	Unit	Total rate £
Elbow; 90°						
20mm dia	0.32	0.41	0.44	11.16	nr	**11.57**
25mm dia	0.42	0.54	0.52	13.19	nr	**13.73**
32mm dia	0.60	0.77	0.59	14.97	nr	**15.74**
40mm dia	0.93	1.19	0.66	16.75	nr	**17.94**
50mm dia	2.01	2.57	0.73	18.52	nr	**21.10**
63mm dia	3.08	3.94	0.85	21.57	nr	**25.51**
75mm dia	6.80	8.71	0.85	21.57	nr	**30.28**
90mm dia	12.56	16.09	1.04	26.39	nr	**42.47**
110mm dia	17.88	22.90	1.04	26.39	nr	**49.29**
125mm dia	27.53	35.26	1.30	32.98	nr	**68.24**
Long bend; 90°						
20mm dia	1.66	2.13	0.48	12.18	nr	**14.30**
25mm dia	1.73	2.22	0.57	14.46	nr	**16.68**
32mm dia	2.03	2.60	0.65	16.49	nr	**19.09**
40mm dia	3.71	4.75	0.73	18.52	nr	**23.27**
Elbow female/male; 90°						
20mm dia	0.32	0.41	0.44	11.16	nr	**11.57**
25mm dia	0.42	0.54	0.52	13.19	nr	**13.73**
32mm dia	0.60	0.77	0.59	14.97	nr	**15.74**
Elbow 45°						
20mm dia	0.32	0.41	0.44	11.16	nr	**11.57**
25mm dia	0.42	0.54	0.52	13.19	nr	**13.73**
32mm dia	0.60	0.77	0.59	14.97	nr	**15.74**
40mm dia	0.93	1.19	0.66	16.75	nr	**17.94**
50mm dia	2.01	2.57	0.73	18.52	nr	**21.10**
63mm dia	3.08	3.94	0.85	21.57	nr	**25.51**
75mm dia	6.80	8.71	0.85	21.57	nr	**30.28**
90mm dia	12.56	16.09	1.04	26.39	nr	**42.47**
110mm dia	17.88	22.90	1.04	26.39	nr	**49.29**
125mm dia	27.53	35.26	1.30	32.98	nr	**68.24**
Elbow female/male; 45°						
20mm dia	0.32	0.41	0.44	11.16	nr	**11.57**
25mm dia	0.42	0.54	0.52	13.19	nr	**13.73**
32mm dia	0.60	0.77	0.59	14.97	nr	**15.74**
T-Piece; 90°						
20mm dia	0.43	0.55	0.61	15.48	nr	**16.03**
25mm dia	0.59	0.76	0.72	18.27	nr	**19.02**
32mm dia	0.77	0.99	0.83	21.06	nr	**22.04**
40mm dia	1.17	1.50	0.92	23.34	nr	**24.84**
50mm dia	3.34	4.28	1.01	25.63	nr	**29.90**
63mm dia	4.80	6.15	1.11	28.16	nr	**34.31**
75mm dia	7.99	10.23	1.18	29.94	nr	**40.17**
90mm dia	14.71	18.84	1.46	37.04	nr	**55.88**
110mm dia	22.96	29.41	1.46	37.04	nr	**66.45**
125mm dia	30.49	39.05	1.82	46.18	nr	**85.23**

T:MECHANICAL/COOLING/HEATING SYSTEMS

Item	Net Price £	Material £	Labour hours	Labour £	Unit	Total rate £
T31 : LOW TEMPERATURE HOT WATER HEATING (cont'd)						
Y10 - PIPELINES (cont'd)						
PLASTIC PIPEWORK (cont'd)						
Extra over fittings; thermal fused joints (cont'd)						
T-Piece reducing; 90°						
25 x 20 x 25mm	0.60	0.77	0.72	18.27	nr	**19.04**
32 x 20 x 32mm	0.77	0.99	0.83	21.06	nr	**22.04**
32 x 25 x 32mm	0.77	0.99	0.83	21.06	nr	**22.04**
40 x 20 x 40mm	1.17	1.50	0.92	23.34	nr	**24.84**
40 x 25 x 40mm	1.17	1.50	0.92	23.34	nr	**24.84**
40 x 32 x 40mm	1.17	1.50	0.92	23.34	nr	**24.84**
50 x 25 x 50mm	3.34	4.28	1.01	25.63	nr	**29.90**
50 x 32 x 50mm	3.34	4.28	1.01	25.63	nr	**29.90**
50 x 40 x 50mm	3.34	4.28	1.01	25.63	nr	**29.90**
63 x 20 x 63mm	4.51	5.78	1.11	28.16	nr	**33.94**
63 x 25 x 63mm	4.51	5.78	1.11	28.16	nr	**33.94**
63 x 32 x 63mm	4.51	5.78	1.11	28.16	nr	**33.94**
63 x 40 x 63mm	4.51	5.78	1.01	25.63	nr	**31.40**
63 x 50 x 63mm	4.51	5.78	1.01	25.63	nr	**31.40**
75 x 20 x 75mm	7.33	9.39	1.18	29.94	nr	**39.33**
75 x 25 x 75mm	7.33	9.39	1.18	29.94	nr	**39.33**
75 x 32 x 75mm	7.33	9.39	1.18	29.94	nr	**39.33**
75 x 40 x 75mm	7.33	9.39	1.18	29.94	nr	**39.33**
75 x 50 x 75mm	7.33	9.39	1.18	29.94	nr	**39.33**
75 x 63 x 75mm	7.33	9.39	1.18	29.94	nr	**39.33**
32 x 32 x 25mm	-	-	0.83	21.06	nr	**21.06**
25 x 20 x 20mm	0.60	0.77	0.72	18.27	nr	**19.04**
20 x 25 x 25mm	-	-	0.72	18.27	nr	**18.27**
32 x 20 x 25mm	-	-	0.72	18.27	nr	**18.27**
32 x 25 x 25mm	-	-	0.72	18.27	nr	**18.27**
90 x 63 x 90mm	14.71	18.84	1.46	37.04	nr	**55.88**
110 x 75 x 110mm	22.96	29.41	1.46	37.04	nr	**66.45**
110 x 90 x 110mm	22.96	29.41	1.46	37.04	nr	**66.45**
125 x 75 x 125mm	-	-	1.82	46.18	nr	**46.18**
125 x 90 x 125mm	27.30	34.97	1.82	46.18	nr	**81.14**
125 x 110 x 125mm	27.86	35.68	1.82	46.18	nr	**81.86**
Reducer						
25 x 20mm	0.44	0.56	0.59	14.97	nr	**15.53**
32 x 20mm	0.44	0.56	0.62	15.73	nr	**16.29**
32 x 25mm	0.70	0.90	0.62	15.73	nr	**16.63**
40 x 20mm	0.70	0.90	0.66	16.75	nr	**17.64**
40 x 25mm	0.70	0.90	0.66	16.75	nr	**17.64**
40 x 32mm	0.70	0.90	0.66	16.75	nr	**17.64**
50 x 20mm	1.13	1.45	0.73	18.52	nr	**19.97**
50 x 25mm	1.13	1.45	0.73	18.52	nr	**19.97**
50 x 32mm	1.13	1.45	0.73	18.52	nr	**19.97**
50 x 40mm	1.13	1.45	0.73	18.52	nr	**19.97**
63 x 40mm	2.28	2.92	0.78	19.79	nr	**22.71**
63 x 25mm	2.28	2.92	0.78	19.79	nr	**22.71**
63 x 32mm	2.28	2.92	0.78	19.79	nr	**22.71**
63 x 50mm	2.28	2.92	0.78	19.79	nr	**22.71**
75 x 50mm	2.54	3.25	0.85	21.57	nr	**24.82**
75 x 63mm	2.54	3.25	0.85	21.57	nr	**24.82**

T:MECHANICAL/COOLING/HEATING SYSTEMS

Item	Net Price £	Material £	Labour hours	Labour £	Unit	Total rate £
90 x 63mm	5.68	7.27	1.04	26.39	nr	**33.66**
90 x 75mm	5.68	7.27	1.04	26.39	nr	**33.66**
110 x 90mm	9.17	11.74	1.17	29.69	nr	**41.43**
125 x 110mm	14.34	18.37	1.43	36.28	nr	**54.65**
Socket						
20mm dia	0.31	0.40	0.51	12.94	nr	**13.34**
25mm dia	0.35	0.45	0.56	14.21	nr	**14.66**
32mm dia	0.44	0.56	0.65	16.49	nr	**17.06**
40mm dia	0.55	0.70	0.74	18.78	nr	**19.48**
50mm dia	1.13	1.45	0.81	20.55	nr	**22.00**
63mm dia	2.28	2.92	0.86	21.82	nr	**24.74**
75mm dia	2.54	3.25	0.91	23.09	nr	**26.34**
90mm dia	6.56	8.40	0.91	23.09	nr	**31.49**
110mm dia	11.14	14.27	0.91	23.09	nr	**37.36**
125mm dia	15.53	19.89	1.30	32.98	nr	**52.87**
End Cap						
20mm dia	0.48	0.61	0.25	6.34	nr	**6.96**
25mm dia	0.59	0.76	0.29	7.36	nr	**8.11**
32mm dia	0.72	0.92	0.33	8.37	nr	**9.29**
40mm dia	1.15	1.47	0.36	9.13	nr	**10.61**
50mm dia	1.61	2.06	0.40	10.15	nr	**12.21**
63mm dia	2.69	3.45	0.44	11.16	nr	**14.61**
75mm dia	3.89	4.98	0.47	11.92	nr	**16.91**
90mm dia	8.81	11.28	0.57	14.46	nr	**25.75**
110mm dia	10.59	13.56	0.57	14.46	nr	**28.03**
125mm dia	16.14	20.67	0.85	21.57	nr	**42.24**
Stub flange with gasket						
32mm dia	10.70	13.70	0.23	5.84	nr	**19.54**
40mm dia	13.47	17.25	0.27	6.85	nr	**24.10**
50mm dia	16.29	20.86	0.38	9.64	nr	**30.51**
63mm dia	19.55	25.04	0.43	10.91	nr	**35.95**
75mm dia	22.95	29.39	0.48	12.18	nr	**41.57**
90mm dia	31.05	39.77	0.53	13.45	nr	**53.22**
110mm dia	43.49	55.70	0.53	13.45	nr	**69.15**
125mm dia	62.34	79.84	0.75	19.03	nr	**98.87**
Weld in saddle with female thread						
40 - 1/2"	0.78	1.00	0.36	9.13	nr	**10.13**
50 - 1/2"	0.78	1.00	0.36	9.13	nr	**10.13**
63 - 1/2"	0.78	1.00	0.40	10.15	nr	**11.15**
75 - 1/2"	0.78	1.00	0.40	10.15	nr	**11.15**
90 - 1/2"	0.78	1.00	0.46	11.67	nr	**12.67**
110 - 1/2"	0.78	1.00	0.46	11.67	nr	**12.67**
Weld in saddle with male thread						
50 - 1/2"	0.78	1.00	0.36	9.13	nr	**10.13**
63 - 1/2"	0.78	1.00	0.40	10.15	nr	**11.15**
75 - 1/2"	0.78	1.00	0.40	10.15	nr	**11.15**
90 - 1/2"	0.78	1.00	0.46	11.67	nr	**12.67**
110 - 1/2"	0.78	1.00	0.46	11.67	nr	**12.67**
Transition piece, round with female thread						
20 x 1/2"	1.80	2.31	0.29	7.36	nr	**9.66**
20 x 3/4"	2.37	3.04	0.29	7.36	nr	**10.39**
25 x 1/2"	1.80	2.31	0.33	8.37	nr	**10.68**
25 x 3/4"	2.37	3.04	0.33	8.37	nr	**11.41**

T:MECHANICAL/COOLING/HEATING SYSTEMS

Item	Net Price £	Material £	Labour hours	Labour £	Unit	Total rate £
T31 : LOW TEMPERATURE HOT WATER HEATING (cont'd)						
Y10 - PIPELINES (cont'd)						
PLASTIC PIPEWORK (cont'd)						
Extra over fittings; thermal fused joints (cont'd)						
Transition piece, hexagon with female thread						
32 x 1"	6.70	8.58	0.36	9.13	nr	**17.72**
40 x 1 1/4"	10.60	13.58	0.36	9.13	nr	**22.71**
50 x 1 1/2"	12.30	15.75	0.36	9.13	nr	**24.89**
63 x 2"	19.07	24.42	0.40	10.15	nr	**34.57**
75 x 2"	19.89	25.47	0.40	10.15	nr	**35.62**
125 x 5"	112.82	144.50	0.51	12.94	nr	**157.44**
Stop valve for surface assembly						
20mm dia	7.39	9.47	0.25	6.34	nr	**15.81**
25mm dia	7.39	9.47	0.29	7.36	nr	**16.82**
32mm dia	13.93	17.84	0.33	8.37	nr	**26.21**
Ball valve						
20mm dia	26.05	33.36	0.25	6.34	nr	**39.71**
25mm dia	27.91	35.75	0.29	7.36	nr	**43.10**
32mm dia	33.53	42.94	0.33	8.37	nr	**51.32**
40mm dia	42.81	54.83	0.36	9.13	nr	**63.96**
50mm dia	58.76	75.26	0.40	10.15	nr	**85.41**
63mm dia	66.25	84.85	0.44	11.16	nr	**96.02**
Floor or ceiling cover plates						
Plastic						
15mm dia	0.30	0.38	0.16	4.06	nr	**4.44**
20mm dia	0.30	0.38	0.22	5.58	nr	**5.97**
25mm dia	0.32	0.41	0.22	5.58	nr	**5.99**
32mm dia	0.36	0.46	0.24	6.09	nr	**6.55**
40mm dia	0.80	1.02	0.26	6.60	nr	**7.62**
50mm dia	0.89	1.14	0.26	6.60	nr	**7.74**
Chromium plated						
15mm dia	1.86	2.38	0.16	4.06	nr	**6.44**
20mm dia	1.98	2.54	0.17	4.31	nr	**6.85**
25mm dia	2.06	2.64	0.21	5.33	nr	**7.97**
32mm dia	2.10	2.69	0.22	5.58	nr	**8.27**
40mm dia	2.37	3.04	0.26	6.60	nr	**9.63**
50mm dia	2.84	3.64	0.26	6.60	nr	**10.23**

T:MECHANICAL/COOLING/HEATING SYSTEMS

Item	Net Price £	Material £	Labour hours	Labour £	Unit	Total rate £
Y11 - PIPELINE ANCILLARIES						
EXPANSION JOINTS						
Axial movement bellows expansion joints; stainless steel						
Screwed ends for steel pipework; up to 6 bar G at 100°C						
15mm dia	67.31	86.20	0.68	17.25	nr	**103.46**
20mm dia	86.53	110.83	0.81	20.55	nr	**131.38**
25mm dia	90.66	116.11	0.93	23.60	nr	**139.71**
32mm dia	109.57	140.33	1.06	26.89	nr	**167.23**
40mm dia	128.54	164.63	1.16	29.43	nr	**194.07**
50mm dia	141.48	181.20	1.19	30.19	nr	**211.39**
Screwed ends for steel pipework; aluminium and steel outer sleeves; up to 16 bar G at 120°C						
20mm dia	138.74	177.69	1.32	33.49	nr	**211.18**
25mm dia	144.86	185.53	1.52	38.57	nr	**224.10**
32mm dia	164.83	211.11	1.80	45.67	nr	**256.78**
40mm dia	184.05	235.74	2.03	51.51	nr	**287.24**
50mm dia	203.28	260.36	2.26	57.34	nr	**317.70**
Flanged ends for steel pipework; aluminium and steel outer sleeves; up to 16 bar G at 120°C						
20mm dia	263.73	337.78	0.53	13.45	nr	**351.23**
25mm dia	265.10	339.54	0.64	16.24	nr	**355.78**
32mm dia	274.71	351.85	0.74	18.78	nr	**370.62**
40mm dia	281.59	360.66	0.82	20.80	nr	**381.46**
50mm dia	291.20	372.96	0.89	22.58	nr	**395.54**
Flanged ends for steel pipework; up to 16 bar G at 120°C						
65mm dia	206.03	263.88	1.10	27.91	nr	**291.79**
80mm dia	251.36	321.94	1.31	33.24	nr	**355.18**
100mm dia	287.28	367.95	1.78	45.16	nr	**413.11**
150mm dia	432.67	554.16	3.08	78.15	nr	**632.31**
Screwed ends for non-ferrous pipework; up to 6 bar G at 100°C						
20mm dia	94.83	121.45	0.72	18.27	nr	**139.72**
25mm dia	100.16	128.28	0.84	21.31	nr	**149.60**
32mm dia	120.18	153.93	1.02	25.88	nr	**179.81**
40mm dia	138.88	177.88	1.11	28.16	nr	**206.04**
50mm dia	157.58	201.83	1.18	29.94	nr	**231.77**
Flanged ends for steel, copper or non-ferrous pipework; up to 16 bar G at 120°C						
65mm dia	271.97	348.34	0.87	22.07	nr	**370.41**
80mm dia	317.31	406.41	0.95	24.10	nr	**430.51**
100mm dia	368.12	471.48	1.15	29.18	nr	**500.66**
150mm dia	545.33	698.45	1.36	34.51	nr	**732.96**

T:MECHANICAL/COOLING/HEATING SYSTEMS

Item	Net Price £	Material £	Labour hours	Labour £	Unit	Total rate £
T31 : LOW TEMPERATURE HOT WATER HEATING (cont'd)						
Y11 - PIPELINE ANCILLARIES (cont'd)						
EXPANSION JOINTS (cont'd)						
Angular movement bellows expansion joints; stainless steel						
Flanged ends for steel pipework; up to 16 bar G at 120°C						
50mm dia	386.07	494.48	0.71	18.01	nr	**512.49**
65mm dia	409.57	524.58	0.83	21.06	nr	**545.64**
80mm dia	484.98	621.16	0.91	23.09	nr	**644.25**
100mm dia	579.90	742.74	0.97	24.61	nr	**767.35**
125mm dia	678.72	869.30	1.16	29.43	nr	**898.73**
150mm dia	709.93	909.27	1.18	29.94	nr	**939.20**
Universal lateral movement bellows expansion joints; stainless steel						
Flanged ends for steel pipework; up to 16 bar G at 120°C						
50mm dia	543.50	696.11	0.89	22.58	nr	**718.69**
65mm dia	568.21	727.75	1.10	27.91	nr	**755.66**
80mm dia	616.31	789.36	1.31	33.24	nr	**822.60**
100mm dia	735.93	942.58	1.78	45.16	nr	**987.74**
125mm dia	1196.21	1532.10	3.06	77.64	nr	**1609.74**
150mm dia	1352.24	1731.94	3.08	78.15	nr	**1810.08**
Universal movement expansion joints; reinforced neoprene flexible connector						
Spherical expansion joints; flanged to BS 10, Table E; up to 10 bar at 100°C						
40mm dia	122.22	156.54	0.82	20.80	nr	**177.34**
50mm dia	124.81	159.86	0.89	22.58	nr	**182.44**
65mm dia	135.23	173.20	1.10	27.91	nr	**201.11**
80mm dia	154.73	198.17	1.31	33.24	nr	**231.41**
100mm dia	183.33	234.81	1.78	45.16	nr	**279.97**
150mm dia	262.65	336.40	3.08	78.15	nr	**414.54**
Hose connector; BSP threaded union ends; up to 8 bar at 100°C						
20mm dia	34.06	43.62	1.32	33.49	nr	**77.11**
25mm dia	46.31	59.32	1.52	38.57	nr	**97.88**
32mm dia	51.77	66.30	1.80	45.67	nr	**111.97**
40mm dia	68.11	87.24	2.03	51.51	nr	**138.74**
50mm dia	88.54	113.40	2.26	57.34	nr	**170.75**

T:MECHANICAL/COOLING/HEATING SYSTEMS

Item	Net Price £	Material £	Labour hours	Labour £	Unit	Total rate £
VALVES						
Isolating valves						
Bronze gate valve; non-rising stem; BS 5154, series B, PN 32; working pressure up to 14 bar for saturated steam, 32 bar from -10°C to 100°C; screwed ends to steel						
15mm dia	19.28	24.70	1.11	28.16	nr	**52.86**
20mm dia	18.52	23.72	1.28	32.48	nr	**56.19**
25mm dia	24.26	31.08	1.49	37.80	nr	**68.88**
32mm dia	34.58	44.29	1.88	47.70	nr	**91.99**
40mm dia	65.15	83.45	2.31	58.61	nr	**142.06**
50mm dia	94.15	120.59	2.80	71.04	nr	**191.63**
Bronze gate valve; non-rising stem; BS 5154, series B, PN 20; working pressure up to 9 bar for saturated steam, 20 bar from -10°C to 100°C; screwed ends to steel						
15mm dia	13.05	16.72	0.84	21.31	nr	**38.03**
20mm dia	18.52	23.72	1.01	25.63	nr	**49.34**
25mm dia	24.26	31.08	1.19	30.19	nr	**61.27**
32mm dia	34.58	44.29	1.38	35.01	nr	**79.30**
40mm dia	47.86	61.30	1.62	41.10	nr	**102.40**
50mm dia	69.48	88.99	1.94	49.22	nr	**138.21**
Bronze gate valve; non-rising stem; BS 5154, series B, PN 16; working pressure up to 7 bar for saturated steam, 16 bar from -10°C to 100°C; BS4504 flanged ends; bolted connections						
15mm dia	58.68	75.16	1.24	31.46	nr	**106.62**
20mm dia	75.83	97.13	1.31	33.24	nr	**130.37**
25mm dia	99.48	127.41	1.43	36.28	nr	**163.69**
32mm dia	129.60	165.99	1.53	38.82	nr	**204.81**
40mm dia	155.25	198.85	1.63	41.36	nr	**240.20**
50mm dia	216.51	277.30	1.71	43.39	nr	**320.68**
65mm dia	329.00	421.38	1.88	47.70	nr	**469.08**
80mm dia	465.30	595.95	2.03	51.51	nr	**647.46**
100mm dia	825.21	1056.93	2.81	71.30	nr	**1128.22**
Cast iron gate valve; bronze trim; non rising stem; BS 5150, PN6; working pressure 6 bar from -10°C to 120°C; BS4504 flanged ends; bolted connections						
50mm dia	156.58	200.55	1.85	46.94	nr	**247.48**
65mm dia	156.58	200.55	2.00	50.74	nr	**251.29**
80mm dia	180.96	231.77	2.27	57.59	nr	**289.37**
100mm dia	239.09	306.22	2.76	70.03	nr	**376.25**
125mm dia	335.03	429.10	6.05	153.50	nr	**582.60**
150mm dia	403.04	516.21	8.03	203.74	nr	**719.94**
200mm dia	734.40	940.61	9.17	232.66	nr	**1173.27**
250mm dia	1130.02	1447.32	10.72	271.99	nr	**1719.31**
300mm dia	1340.29	1716.63	11.75	298.12	nr	**2014.75**

T:MECHANICAL/COOLING/HEATING SYSTEMS

Item	Net Price £	Material £	Labour hours	Labour £	Unit	Total rate £
T31 : LOW TEMPERATURE HOT WATER HEATING (cont'd)						
Y11 - PIPELINE ANCILLARIES (cont'd)						
VALVES (cont'd)						
Isolating valves (cont'd)						
Cast iron gate valve; bronze trim; non rising stem; BS 5150, PN10; working pressure up to 8.4 bar for saturated steam, 10 bar from -10°C to 120°C; BS4504 flanged ends; bolted connections						
50mm dia	167.75	214.86	1.85	46.94	nr	**261.80**
65mm dia	194.49	249.10	2.00	50.74	nr	**299.84**
80mm dia	258.08	330.54	2.27	57.59	nr	**388.14**
100mm dia	364.09	466.32	2.76	70.03	nr	**536.34**
125mm dia	419.38	537.14	6.05	153.50	nr	**690.64**
150mm dia	760.44	973.96	8.03	203.74	nr	**1177.70**
200mm dia	1133.73	1452.08	9.17	232.66	nr	**1684.74**
250mm dia	1442.51	1847.56	10.72	271.99	nr	**2119.54**
300mm dia	1916.53	2454.67	11.75	298.12	nr	**2752.79**
350mm dia	3220.78	4125.14	12.67	321.46	nr	**4446.61**
Cast iron gate valve; bronze trim; non rising stem; BS 5163 series A, PN16; working pressure for cold water services up to 16 bar; BS4504 flanged ends; bolted connections						
50mm dia	326.22	417.81	1.85	46.94	nr	**464.75**
65mm dia	338.83	433.98	2.00	50.74	nr	**484.72**
80mm dia	348.56	446.44	2.27	57.59	nr	**504.03**
100mm dia	452.62	579.71	2.76	70.03	nr	**649.74**
125mm dia	590.20	755.93	6.05	153.50	nr	**909.43**
150mm dia	711.78	911.64	8.03	203.74	nr	**1115.38**
Ball valves						
Malleable iron body; lever operated stainless steel ball and stem; working pressure up to 12 bar; flanged ends to BS 4504 16/11; bolted connections						
40mm dia	140.36	179.77	1.54	39.07	nr	**218.85**
50mm dia	176.59	226.18	1.64	41.61	nr	**267.79**
80mm dia	295.33	378.25	1.92	48.71	nr	**426.97**
100mm dia	545.88	699.16	2.80	71.04	nr	**770.20**
150mm dia	742.62	951.14	12.05	305.73	nr	**1256.87**
Malleable iron body; lever operated stainless steel ball and stem; working pressure up to 16 bar; screwed ends to steel						
20mm dia	30.88	39.55	1.34	34.01	nr	**73.56**
25mm dia	31.76	40.67	1.40	35.54	nr	**76.21**
32mm dia	43.74	56.02	1.46	37.09	nr	**93.11**
40mm dia	43.74	56.02	1.54	39.09	nr	**95.11**
50mm dia	52.31	67.00	1.64	41.66	nr	**108.66**

T:MECHANICAL/COOLING/HEATING SYSTEMS

Item	Net Price £	Material £	Labour hours	Labour £	Unit	Total rate £
Carbon steel body; lever operated stainless steel ball and stem; Class 150; working pressure up to 19 bar; screwed ends to steel						
15mm dia	22.68	29.05	0.84	21.32	nr	**50.37**
20mm dia	23.81	30.50	1.14	28.93	nr	**59.43**
25mm dia	27.26	34.92	1.30	32.99	nr	**67.91**
Globe valves						
Bronze; rising stem; renewable disc; BS 5154 series B, PN32; working pressure up to 14 bar for saturated steam, 32 bar from -10°C to 100°C; screwed ends to steel						
15mm dia	15.46	19.80	0.77	19.54	nr	**39.34**
20mm dia	21.06	26.97	1.03	26.13	nr	**53.10**
25mm dia	32.20	41.24	1.19	30.19	nr	**71.43**
32mm dia	45.44	58.20	1.38	35.01	nr	**93.21**
40mm dia	56.41	72.25	1.62	41.10	nr	**113.35**
50mm dia	89.37	114.46	1.61	40.85	nr	**155.31**
Bronze; needle valve; rising stem; BS 5154, series B, PN32; working pressure up to 14 bar for saturated steam, 32 bar from -10°C to 100°C; screwed ends to steel						
15mm dia	21.87	28.01	1.07	27.15	nr	**55.15**
20mm dia	37.02	47.42	1.18	29.94	nr	**77.36**
25mm dia	52.22	66.88	1.27	32.22	nr	**99.10**
32mm dia	108.33	138.75	1.35	34.25	nr	**173.01**
40mm dia	170.68	218.61	1.47	37.30	nr	**255.90**
50mm dia	230.70	295.48	1.61	40.85	nr	**336.33**
Bronze; rising stem; renewable disc; BS 5154, series B, PN16; working pressure upto 7 bar for saturated steam, 16 bar from -10°C to 100°C; BS4504 flanged ends; bolted connections						
15mm dia	56.32	72.13	1.16	29.43	nr	**101.56**
20mm dia	65.12	83.40	1.26	31.97	nr	**115.37**
25mm dia	114.22	146.29	1.38	35.01	nr	**181.31**
32mm dia	144.18	184.67	1.47	37.30	nr	**221.97**
40mm dia	186.46	238.82	1.56	39.58	nr	**278.40**
50mm dia	235.24	301.29	1.71	43.39	nr	**344.67**
Bronze; rising stem; renewable disc; BS 2060, class 250; working pressure up to 24 bar for saturated steam, 38 bar from -10°C to 100°C; flanged ends (BS 10 table H); bolted connections						
15mm dia	143.68	184.02	1.16	29.43	nr	**213.45**
20mm dia	167.44	214.45	1.26	31.97	nr	**246.42**
25mm dia	229.45	293.88	1.38	35.01	nr	**328.90**
32mm dia	304.44	389.93	1.47	37.30	nr	**427.22**
40mm dia	358.95	459.74	1.56	39.58	nr	**499.32**
50mm dia	558.16	714.89	1.71	43.39	nr	**758.27**
65mm dia	804.75	1030.71	1.88	47.70	nr	**1078.41**
80mm dia	1947.96	2494.93	2.03	51.51	nr	**2546.44**

T:MECHANICAL/COOLING/HEATING SYSTEMS

Item	Net Price £	Material £	Labour hours	Labour £	Unit	Total rate £
T31 : LOW TEMPERATURE HOT WATER HEATING (cont'd)						
Y11 - PIPELINE ANCILLARIES (cont'd)						
VALVES (cont'd)						
Check valves						
Bronze; swing pattern; BS 5154 series B, PN 25; working pressure up to 10.5 bar for saturated steam, 25 bar from -10°C to 100°C; screwed ends to steel						
15mm dia	13.71	17.56	0.77	19.54	nr	**37.10**
20mm dia	16.32	20.90	1.03	26.13	nr	**47.04**
25mm dia	22.60	28.94	1.19	30.19	nr	**59.13**
32mm dia	38.31	49.06	1.38	35.01	nr	**84.07**
40mm dia	47.66	61.04	1.62	41.10	nr	**102.14**
50mm dia	73.09	93.61	1.94	49.22	nr	**142.83**
65mm dia	123.52	158.20	2.45	62.16	nr	**220.36**
80mm dia	174.65	223.69	2.83	71.80	nr	**295.49**
Bronze; vertical lift pattern; BS 5154 series B, PN32; working pressure up to 14 bar for saturated steam, 32 bar from -10°C to 100°C; screwed ends to steel						
15mm dia	25.43	32.57	0.96	24.36	nr	**56.92**
20mm dia	29.20	37.40	1.07	27.15	nr	**64.55**
25mm dia	32.29	41.36	1.17	29.69	nr	**71.04**
32mm dia	40.30	51.62	1.33	33.74	nr	**85.36**
40mm dia	53.76	68.85	1.41	35.77	nr	**104.62**
50mm dia	82.68	105.90	1.55	39.33	nr	**145.23**
65mm dia	299.44	383.52	1.80	45.67	nr	**429.19**
80mm dia	447.82	573.56	1.99	50.49	nr	**624.05**
Bronze; oblique swing pattern; BS 5154 series A, PN32; working pressure up to 14 bar for saturated steam, 32 bar from -10°C to 120°C; screwed connections to steel						
15mm dia	39.53	50.63	0.96	24.36	nr	**74.98**
20mm dia	43.40	55.59	1.07	27.15	nr	**82.74**
25mm dia	66.14	84.71	1.17	29.69	nr	**114.40**
32mm dia	95.60	122.44	1.33	33.74	nr	**156.19**
40mm dia	105.95	135.70	1.41	35.77	nr	**171.47**
50mm dia	151.93	194.59	1.55	39.33	nr	**233.91**
Cast iron; swing pattern; BS 5153 PN6; working pressure up to 6 bar from -10°C to 120°C; BS 4504 flanged ends; bolted connections						
50mm dia	177.67	227.56	1.86	47.19	nr	**274.75**
65mm dia	177.67	227.56	2.00	50.74	nr	**278.30**
80mm dia	220.01	281.78	2.56	64.95	nr	**346.73**
100mm dia	256.20	328.13	2.76	70.03	nr	**398.16**
125mm dia	382.23	489.55	6.05	153.50	nr	**643.05**
150mm dia	429.78	550.46	8.11	205.77	nr	**756.22**
200mm dia	950.43	1217.31	9.26	234.94	nr	**1452.25**
250mm dia	1446.32	1852.43	10.72	271.99	nr	**2124.42**
300mm dia	1921.54	2461.08	11.75	298.12	nr	**2759.20**

T:MECHANICAL/COOLING/HEATING SYSTEMS

Item	Net Price £	Material £	Labour hours	Labour £	Unit	Total rate £
Cast iron; horizontal lift pattern; BS 5153 PN16; working pressure up to 13 bar for saturated steam, 16 bar from -10°C to 120°C; BS 4504 flanged ends; bolted connections						
50mm dia	204.74	262.23	1.86	47.19	nr	309.43
65mm dia	321.15	411.33	2.00	50.74	nr	462.07
80mm dia	390.89	500.64	2.56	64.95	nr	565.60
100mm dia	506.12	648.23	2.96	75.10	nr	723.33
125mm dia	765.16	980.01	7.76	196.89	nr	1176.90
150mm dia	841.56	1077.86	10.50	266.40	nr	1344.26
Cast iron; semi lugged butterfly valve; BS5155 PN16; working pressure 16 bar from -10°C to 120°C; BS 4504 flanged ends; bolted connections						
50mm dia	56.83	72.79	2.20	55.82	nr	128.61
65mm dia	56.83	72.79	2.31	58.61	nr	131.40
80mm dia	68.20	87.35	2.88	73.07	nr	160.42
100mm dia	150.13	192.28	3.11	78.91	nr	271.19
125mm dia	150.13	192.28	5.02	127.37	nr	319.65
150mm dia	174.21	223.12	6.98	177.10	nr	400.22
200mm dia	304.24	389.67	8.25	209.32	nr	598.99
250mm dia	607.40	777.96	10.47	265.64	nr	1043.60
300mm dia	770.47	986.81	11.48	291.27	nr	1278.08
Commissioning valves						
Bronze commissioning set; metering station; double regulating valve; BS5154 PN20 Series B; working pressure 20 bar from -10°C to 100°C; screwed ends to steel						
15mm dia	35.45	45.40	1.08	27.40	nr	72.81
20mm dia	60.35	77.29	1.46	37.04	nr	114.34
25mm dia	69.61	89.15	1.68	42.62	nr	131.78
32mm dia	96.49	123.58	1.95	49.48	nr	173.06
40mm dia	137.06	175.55	2.27	57.59	nr	233.14
50mm dia	226.97	290.71	2.73	69.27	nr	359.97
Cast iron commissioning set; metering station; double regulating valve; BS5152 PN16; working pressure 16 bar from -10°C to 90°C; flanged ends (BS 4504, Part 1, Table 16); bolted connections						
65mm dia	348.04	445.76	1.80	45.67	nr	491.43
80mm dia	418.29	535.74	2.56	64.95	nr	600.69
100mm dia	566.17	725.15	2.30	58.36	nr	783.51
125mm dia	921.55	1180.32	2.44	61.91	nr	1242.22
150mm dia	1156.43	1481.15	2.90	73.58	nr	1554.73
200mm dia	3375.70	4323.57	8.26	209.57	nr	4533.14
250mm dia	4866.36	6232.78	10.49	266.15	nr	6498.93
300mm dia	7667.94	9821.03	11.49	291.52	nr	10112.55

T:MECHANICAL/COOLING/HEATING SYSTEMS

Item	Net Price £	Material £	Labour hours	Labour £	Unit	Total rate £
T31 : LOW TEMPERATURE HOT WATER HEATING (cont'd)						
Y11 - PIPELINE ANCILLARIES (cont'd)						
VALVES (cont'd)						
Commissioning valves (cont'd)						
Cast iron variable orifice double regulating valve; orifice valve; BS5152 PN16; working pressure 16 bar from -10° to 90°C; flanged ends (BS 4504, Part 1, Table 16); bolted connections						
65mm dia	307.98	394.46	2.00	50.74	nr	445.20
80mm dia	399.98	512.29	2.56	64.95	nr	577.24
100mm dia	532.79	682.39	2.96	75.10	nr	757.50
125mm dia	779.97	998.97	7.76	196.89	nr	1195.86
150mm dia	968.24	1240.11	10.50	266.40	nr	1506.51
200mm dia	2506.67	3210.52	8.26	209.57	nr	3420.10
250mm dia	3792.56	4857.47	10.49	266.15	nr	5123.62
300mm dia	5958.65	7631.78	11.49	291.52	nr	7923.30
Cast iron globe valve with double regulating feature; BS5152 PN16; working pressure 16 bar from -10°C to 120°C; flanged ends (BS 4504, Part 1, Table 16); bolted connections						
65mm dia	246.75	316.03	2.00	50.74	nr	366.78
80mm dia	302.74	387.74	2.56	64.95	nr	452.70
100mm dia	425.23	544.63	2.96	75.10	nr	619.73
125mm dia	652.62	835.87	7.76	196.89	nr	1032.75
150mm dia	842.67	1079.29	10.50	266.40	nr	1345.69
200mm dia	2537.73	3250.30	8.26	209.57	nr	3459.87
250mm dia	3935.18	5040.14	10.49	266.15	nr	5306.29
300mm dia	6226.67	7975.05	11.49	291.52	nr	8266.57
Bronze autoflow commissioning valve; PN25 ; working pressure 25 bar up to 100°C; screwed ends to steel						
15mm dia	73.00	93.50	0.82	20.80	nr	114.31
20mm dia	97.14	124.41	1.08	27.40	nr	151.81
25mm dia	110.41	141.42	1.27	32.22	nr	173.64
32mm dia	143.93	184.34	1.50	38.06	nr	222.40
40mm dia	171.28	219.37	1.76	44.65	nr	264.03
50mm dia	227.72	291.66	2.13	54.04	nr	345.70
Ductile iron autoflow commissioning valves; PN16; working pressure 16 bar from -10°C to 120°C; for ANSI 150 flanged ends						
65mm dia	547.13	700.75	2.31	58.61	nr	759.36
80mm dia	604.90	774.75	2.88	73.07	nr	847.82
100mm dia	943.33	1208.20	3.11	78.91	nr	1287.11
150mm dia	1591.89	2038.88	6.98	177.10	nr	2215.97
200mm dia	2297.49	2942.60	8.26	209.57	nr	3152.17
250mm dia	3190.84	4086.80	10.49	266.15	nr	4352.95
300mm dia	4075.22	5219.50	11.49	291.52	nr	5511.02

T:MECHANICAL/COOLING/HEATING SYSTEMS

Item	Net Price £	Material £	Labour hours	Labour £	Unit	Total rate £
Strainers						
Bronze strainer; Y type; PN32 ; working pressure 32 bar from -10°C to 100°C; screwed ends to steel						
15mm dia	14.99	19.20	0.82	20.80	nr	**40.01**
20mm dia	19.06	24.41	1.08	27.40	nr	**51.81**
25mm dia	26.86	34.40	1.27	32.22	nr	**66.63**
32mm dia	45.19	57.88	1.50	38.06	nr	**95.94**
40mm dia	61.39	78.63	1.76	44.65	nr	**123.29**
50mm dia	241.13	308.84	2.13	54.04	nr	**362.88**
Cast iron strainer; Y type; PN16; working pressure 16 bar from -10°C to 120°C; BS 4504 flanged ends						
65mm dia	96.83	124.02	2.31	58.61	nr	**182.63**
80mm dia	112.15	143.65	2.88	73.07	nr	**216.72**
100mm dia	166.06	212.69	3.11	78.91	nr	**291.60**
125mm dia	342.85	439.12	5.02	127.37	nr	**566.49**
150mm dia	444.20	568.92	6.98	177.10	nr	**746.02**
200mm dia	725.96	929.80	8.26	209.57	nr	**1139.38**
250mm dia	1097.31	1405.42	10.49	266.15	nr	**1671.57**
300mm dia	1841.15	2358.13	11.49	291.52	nr	**2649.66**
Regulators						
Gunmetal; self-acting two port thermostatic regulator; single seat; screwed ends; complete with sensing element, 2m long capillary tube						
15mm dia	439.63	563.07	1.37	34.76	nr	**597.83**
20mm dia	450.96	577.58	1.24	31.46	nr	**609.04**
25mm dia	465.68	596.44	1.34	34.00	nr	**630.44**
Gunmetal; self-acting two port thermostatic regulator; double seat; flanged ends (BS 4504 PN25); with sensing element, 2m long capillary tube; steel body						
65mm dia	1630.34	2088.12	1.23	31.23	nr	**2119.36**
80mm dia	1924.28	2464.60	1.62	41.13	nr	**2505.74**
Control valves; electrically operated (electrical work elsewhere)						
Cast iron; butterfly type; two position electrically controlled 240V motor and linkage mechanism; for low pressure hot water; maximum pressure 6 bar at 120°C; flanged ends						
25mm dia	345.15	442.06	1.47	37.30	nr	**479.36**
32mm dia	356.64	456.78	1.52	38.57	nr	**495.35**
40mm dia	498.34	638.27	1.61	40.85	nr	**679.12**
50mm dia	512.50	656.40	1.71	43.39	nr	**699.79**
65mm dia	525.34	672.84	2.51	63.68	nr	**736.53**
80mm dia	545.88	699.16	2.69	68.25	nr	**767.41**
100mm dia	572.85	733.70	2.81	71.30	nr	**805.00**
125mm dia	634.50	812.66	2.94	74.59	nr	**887.25**
150mm dia	687.59	880.66	3.33	84.49	nr	**965.14**
200mm dia	852.84	1092.31	3.67	93.11	nr	**1185.42**

T:MECHANICAL/COOLING/HEATING SYSTEMS

Item	Net Price £	Material £	Labour hours	Labour £	Unit	Total rate £
T31 : LOW TEMPERATURE HOT WATER HEATING (cont'd)						
Y11 - PIPELINE ANCILLARIES (cont'd)						
VALVES (cont'd)						
Control valves; electrically operated (cont'd)						
Cast iron; three way 240V motorized; for low pressure hot water; maximum pressure 6 bar 120°C; flanged ends, drilled (BS 10, Table F)						
25mm dia	385.76	494.08	1.99	50.49	nr	**544.57**
40mm dia	404.60	518.21	2.13	54.04	nr	**572.25**
50mm dia	399.46	511.62	3.21	81.44	nr	**593.07**
65mm dia	446.33	571.65	3.23	81.95	nr	**653.61**
80mm dia	500.11	640.54	3.50	88.80	nr	**729.34**
Two port normally closed motorised valve; electric actuator; spring return; domestic usage						
22mm dia	54.69	70.04	1.18	29.94	nr	**99.98**
28mm dia	74.13	94.95	1.35	34.25	nr	**129.20**
Two port on/off motorised valve; electric actuator; spring return; domestic usage						
22mm dia	54.69	70.04	1.18	29.94	nr	**99.98**
Three port motorised valve; electric actuator; spring return; domestic usage						
22mm dia	79.80	102.21	1.18	29.94	nr	**132.15**
Safety and relief valves						
Bronze relief valve; spring type; side outlet; working pressure up to 20.7 bar at 120°C; screwed ends to steel						
15mm dia	76.59	98.10	0.26	6.60	nr	**104.70**
20mm dia	103.37	132.40	0.36	9.13	nr	**141.53**
Bronze relief valve; spring type; side outlet; working pressure up to 17.2 bar at 120°C; screwed ends to steel						
25mm dia	123.63	158.35	0.38	9.64	nr	**167.99**
32mm dia	188.22	241.07	0.48	12.18	nr	**253.25**
Bronze relief valve; spring type; side outlet; working pressure up to 13.8 bar at 120°C; screwed ends to steel						
40mm dia	164.38	210.54	0.64	16.24	nr	**226.78**
50mm dia	250.02	320.22	0.76	19.28	nr	**339.50**
65mm dia	396.97	508.44	0.94	23.85	nr	**532.29**
80mm dia	520.82	667.06	1.10	27.91	nr	**694.97**

T:MECHANICAL/COOLING/HEATING SYSTEMS

Item	Net Price £	Material £	Labour hours	Labour £	Unit	Total rate £
Cocks; screwed joints to steel						
Bronze gland cock; complete with malleable iron lever; working pressure up to 10 bar at 100°C; screwed ends to steel						
15mm dia	19.85	25.43	0.77	19.54	nr	**44.96**
20mm dia	28.93	37.05	1.03	26.13	nr	**63.18**
25mm dia	41.55	53.21	1.19	30.19	nr	**83.40**
32mm dia	150.13	192.29	1.38	35.01	nr	**227.30**
40mm dia	213.19	273.05	1.62	41.10	nr	**314.16**
50mm dia	303.37	388.55	1.94	49.22	nr	**437.77**
Bronze three-way plug cock; complete with malleable iron lever; working pressure up to 10 bar at 100°C; screwed ends to steel						
15mm dia	44.15	56.55	0.77	19.54	nr	**76.09**
20mm dia	51.10	65.45	1.03	26.13	nr	**91.58**
25mm dia	71.44	91.50	1.19	30.19	nr	**121.69**
32mm dia	101.32	129.77	1.38	35.01	nr	**164.79**
40mm dia	122.16	156.46	1.62	41.10	nr	**197.56**
Air vents; including regulating, adjusting and testing						
Automatic air vent; maximum pressure up to 7 bar at 93°C; screwed ends to steel						
15mm dia	84.08	107.69	0.80	20.30	nr	**127.99**
Automatic air vent; maximum pressure up to 7 bar at 93°C; lockhead isolating valve; screwed ends to steel						
15mm dia	92.08	117.94	0.83	21.06	nr	**138.99**
Automatic air vent; maximum pressure up to 17 bar at 200°C; flanged ends (BS10, Table H); bolted connections to counter flange (measured separately)						
15mm dia	373.74	478.68	0.83	21.06	nr	**499.74**
Radiator valves						
Bronze; wheelhead or lockshield; chromium plated finish; screwed joints to steel						
Straight						
15mm dia	28.43	36.42	0.59	14.97	nr	**51.39**
20mm dia	36.96	47.34	0.73	18.52	nr	**65.86**
25mm dia	46.20	59.18	0.85	21.57	nr	**80.74**
Angled						
15mm dia	20.21	25.88	0.59	14.97	nr	**40.85**
20mm dia	26.63	34.11	0.73	18.52	nr	**52.63**
25mm dia	34.34	43.98	0.85	21.57	nr	**65.54**

T:MECHANICAL/COOLING/HEATING SYSTEMS

Item	Net Price £	Material £	Labour hours	Labour £	Unit	Total rate £
T31 : LOW TEMPERATURE HOT WATER HEATING (cont'd)						
Y11 - PIPELINE ANCILLARIES (cont'd)						
VALVES (cont'd)						
Radiator valves (cont'd)						
Bronze; wheelhead or lockshield; chromium plated finish; compression joints to copper						
Straight						
15mm dia	30.55	39.13	0.59	14.97	nr	**54.10**
20mm dia	38.12	48.82	0.73	18.52	nr	**67.35**
25mm dia	47.91	61.37	0.85	21.57	nr	**82.93**
Angled						
15mm dia	20.21	25.88	0.59	14.97	nr	**40.85**
20mm dia	28.05	35.92	0.73	18.52	nr	**54.44**
25mm dia	35.75	45.79	0.85	21.57	nr	**67.36**
Twin entry						
8mm dia	31.42	40.64	0.23	5.84	nr	**46.48**
10mm dia	35.66	46.13	0.23	5.84	nr	**51.97**
Bronze; thermostatic head; chromium plated finish; compression joints to copper						
Straight						
15mm dia	25.81	33.05	0.59	14.97	nr	**48.02**
20mm dia	30.11	38.56	0.73	18.52	nr	**57.08**
Angled						
15mm dia	27.19	34.83	0.59	14.97	nr	**49.79**
20mm dia	35.25	45.15	0.73	18.52	nr	**63.67**
TEMPERATURE & PRESSURE GAUGES						
Dial thermometer; coated steel case and dial; glass window; brass pocket; BS 5235; pocket length 100mm; screwed end						
Back/bottom entry						
100mm dia face	53.91	69.05	0.81	20.56	nr	**89.61**
150mm dia face	61.49	78.76	0.81	20.56	nr	**99.32**
Dial pressure/altitude gauge; bronze bourdon tube type; coated steel case and dial; glass window BS 1780; screwed end						
100mm dia face	30.89	39.57	0.81	20.56	nr	**60.13**
150mm dia face	33.54	42.96	0.81	20.56	nr	**63.52**

T:MECHANICAL/COOLING/HEATING SYSTEMS

Item	Net Price £	Material £	Labour hours	Labour £	Unit	Total rate £
EQUIPMENT						
PRESSURISATION UNITS						
LTHW pressurisation unit complete with expansion vessel(s), interconnecting pipework and all necessary isolating and drain valves; includes placing in position; electrical work elsewhere. Selection based on a final working pressure of 4 bar, a 3m static head and system operating temperatures of 82/71°s C						
System volume						
2,400 litres	1503.41	1925.55	15.00	380.58	nr	**2306.13**
6,000 - 20,000 litres	1648.39	2111.24	22.00	558.18	nr	**2669.43**
25,000 litres	2046.57	2621.22	22.00	558.18	nr	**3179.41**
DIRT SEPARATORS						
Dirt separator; maximum operating temperature and pressure of 110°C and 10 bar; fitted with drain valve						
Bore size, flow rate (at 1.0m/s velocity); threaded connections						
32mm dia, 3.7m³/h	95.04	121.73	2.29	58.10	nr	**179.83**
40mm dia, 5.0m³/h	114.05	146.07	2.45	62.16	nr	**208.23**
Bore size, flow rate (at 1.5m/s velocity); flanged connections to PN16						
50mm dia, 13.0m³/h	789.47	1011.15	3.00	76.12	nr	**1087.26**
65mm dia, 21.0m³/h	818.61	1048.47	3.00	76.12	nr	**1124.58**
80mm dia, 29.0m³/h	1146.82	1468.83	3.84	97.43	nr	**1566.26**
100mm dia, 49.0m³/h	1188.64	1522.40	4.44	112.65	nr	**1635.05**
125mm dia, 74.0m³/h	2278.43	2918.19	11.64	295.33	nr	**3213.52**
150mm dia, 109.0m³/h	2377.27	3044.78	15.75	399.61	nr	**3444.38**
200mm dia, 181.0m³/h	3326.40	4260.42	15.75	399.61	nr	**4660.03**
250mm dia, 288.0m³/h	5033.31	6446.62	15.75	399.61	nr	**6846.23**
300mm dia, 407.0m³/h	8021.38	10273.70	17.24	437.41	nr	**10711.11**
MICROBUBBLE DEAERATORS						
Microbubble deaerator; maximum operating temperature and pressure of 110°C and 10 bar; fitted with drain valve						
Bore size, flow rate (at 1.0m/s velocity); threaded connections						
32mm dia, 3.7m³/h	95.04	121.73	2.29	58.10	nr	**179.83**
40mm dia, 5.0m³/h	114.05	146.07	2.45	62.16	nr	**208.23**

T:MECHANICAL/COOLING/HEATING SYSTEMS

Item	Net Price £	Material £	Labour hours	Labour £	Unit	Total rate £
T31 : LOW TEMPERATURE HOT WATER HEATING (cont'd)						
Y11 - PIPELINE ANCILLARIES (cont'd)						
MICROBUBBLE DEAERATORS (cont'd)						
Microbubble deaerator; maximum operating Temperature (cont'd)						
Bore size, flow rate (at 1.5m/s velocity); flanged connections to PN16						
50mm dia, 13.0m³/h	909.34	1164.68	3.00	76.12	nr	**1240.79**
65mm dia, 21.0m³/h	938.33	1201.81	3.00	76.12	nr	**1277.92**
80mm dia, 29.0m³/h	1270.45	1627.17	3.84	97.43	nr	**1724.60**
100mm dia, 49.0m³/h	1320.52	1691.31	4.44	112.65	nr	**1803.96**
125mm dia, 74.0m³/h	2617.32	3352.24	11.64	295.33	nr	**3647.57**
150mm dia, 109.0m³/h	2731.99	3499.10	15.75	399.61	nr	**3898.71**
200mm dia, 181.0m³/h	3338.21	4275.55	15.75	399.61	nr	**4675.16**
250mm dia, 288.0m³/h	4906.50	6284.20	15.75	399.61	nr	**6683.81**
300mm dia, 407.0m³/h	8719.14	11167.39	17.24	437.41	nr	**11604.81**
Pressure step deaerator for high pressure systems						
Heating (where static head of water above boiler exceeds 15m); maximum working pressure						
6 bar (single phase supply)	4658.73	5966.86	8.00	202.98	nr	**6169.84**
10 bar (3 phase supply)	10023.85	12838.45	10.00	253.72	nr	**13092.17**
15 bar (3 phase supply)	10728.92	13741.50	12.00	304.46	nr	**14045.96**
Cooling (where static head of water above chiller exceeds 5m); maximum working pressure						
6 bar (single phase supply)	5222.79	6689.30	8.00	202.98	nr	**6892.28**
10 bar (3 phase supply)	10587.91	13560.90	10.00	253.72	nr	**13814.61**
15 bar (3 phase supply)	11294.30	14465.63	12.00	304.46	nr	**14770.09**
COMBINED MICROBUBBLE DEAERATORS AND DIRT SEPARATORS						
Combined deaerator and dirt separators; maximum operating temperature and pressure of 110°C and 10 bar; fitted with drain valve						
Bore size, flow rate (at 1.5m/s velocity); threaded connections						
25mm dia, 2.0m³/h	117.83	150.92	2.75	69.77	nr	**220.69**
Bore size, flow rate (at 1.5m/s velocity); flanged connections to PN16						
50mm dia, 13.0m³/h	983.84	1260.09	3.60	91.34	nr	**1351.43**
65mm dia, 21.0m³/h	1035.32	1326.03	3.60	91.34	nr	**1417.37**
80mm dia, 29.0m³/h	1261.83	1616.14	4.61	116.96	nr	**1733.11**
100mm dia, 49.0m³/h	1387.67	1777.32	5.33	135.23	nr	**1912.55**
125mm dia, 74.0m³/h	2397.82	3071.11	13.97	354.45	nr	**3425.55**
150mm dia, 109.0m³/h	2587.73	3314.34	18.90	479.53	nr	**3793.87**
200mm dia, 181.0m³/h	3755.75	4810.33	18.90	479.53	nr	**5289.86**
250mm dia, 288.0m³/h	5633.06	7214.76	18.90	479.53	nr	**7694.29**
300mm dia, 407.0m³/h	9939.07	12729.86	20.69	524.94	nr	**13254.81**

T:MECHANICAL/COOLING/HEATING SYSTEMS

Item	Net Price £	Material £	Labour hours	Labour £	Unit	Total rate £
Y20 - PUMPS						
Centrifugal heating and chilled water pump; belt drive; 3 phase, 1450 rpm motor; max. pressure 1000kN/m²; max. temperature 125°C; bed plate; coupling guard; bolted connections; supply only mating flanges; includes fixing on prepared concrete base; electrical work elsewhere						
40mm pump size; 4.0 l/s at 70 kPa max head; 0.25kW max motor rating	1043.21	1336.14	7.59	192.57	nr	**1528.71**
40mm pump size; 4.0 l/s at 130 kPa max head; 1.5kW max motor rating	1305.56	1672.15	8.09	205.26	nr	**1877.41**
50mm pump size; 8.5 l/s at 90 kPa max head; 2.2kW max motor rating	1346.40	1724.46	8.67	219.97	nr	**1944.43**
50mm pump size; 8.5 l/s at 190 kPa max head; 3kW max motor rating	2340.11	2997.19	11.20	284.17	nr	**3281.36**
50mm pump size; 8.5 l/s at 215 kPa max head; 4 kW max motor rating	1529.55	1959.03	11.70	296.85	nr	**2255.88**
65mm pump size; 14.0 l/s at 90 kPa max head; 3kW max motor rating	1341.45	1718.12	11.70	296.85	nr	**2014.97**
65mm pump size; 14.0 l/s at 160 kPa max head; 4 kW max motor rating	1487.48	1905.14	11.70	296.85	nr	**2201.99**
80mm pump size; 14.5 l/s at 210 kPa max head; 5.5 kW max motor rating	2281.95	2922.70	11.70	296.85	nr	**3219.55**
80mm pump size; 22.0 l/s at 130 kPa max head; 5.5 kW max motor rating	2281.95	2922.70	13.64	346.07	nr	**3268.77**
80mm pump size; 22.0 l/s at 200 kPa max head; 7.5 kW max motor rating	2389.61	3060.59	13.64	346.07	nr	**3406.66**
100mm pump size; 22.0 l/s at 250 kPa max head; 11kW max motor rating	3353.63	4295.29	13.64	346.07	nr	**4641.36**
100mm pump size; 30.0 l/s at 100 kPa max head; 4.0 kW max motor rating	1877.29	2404.41	19.15	485.87	nr	**2890.28**
100mm pump size; 36.0 l/s at 250 kPa max head; 15.0kW max motor rating	3551.63	4548.89	19.15	485.87	nr	**5034.76**
100mm pump size; 36.0 l/s at 550 kPa max head; 30.0 kW max motor rating	4820.06	6173.49	19.15	485.87	nr	**6659.36**
Centrifugal heating and chilled water pump; TWIN HEAD BELT DRIVE; 3 phase, 1450 rpm motor; max. pressure 1000kN/m²; max. temperature 125°C; bed plate; coupling guard; bolted connections; supply only mating flanges; includes fixing on prepared concrete base; electrical work elsewhere						
40mm pump size; 4.0 l/s at 70 kPa max head; 0.75kW max motor rating	2234.93	2862.47	7.59	192.57	nr	**3055.04**
40mm pump size; 4.0 l/s at 130 kPa max head; 1.5kW max motor rating	2690.32	3445.74	8.09	205.26	nr	**3651.00**
50mm pump size; 8.5 l/s at 90 kPa max head; 2.2kW max motor rating	2748.49	3520.24	8.67	219.97	nr	**3740.21**
50mm pump size; 8.5 l/s at 190 kPa max head; 4kW max motor rating	3168.00	4057.54	11.20	284.17	nr	**4341.71**
65mm pump size; 8.5 l/s at 215 kPa max head; 4 kW max motor rating	3304.13	4231.89	11.70	296.85	nr	**4528.74**

T:MECHANICAL/COOLING/HEATING SYSTEMS

Item	Net Price £	Material £	Labour hours	Labour £	Unit	Total rate £
T31 : LOW TEMPERATURE HOT WATER HEATING (cont'd)						
Y20 – PUMPS (cont'd)						
Centrifugal heating and chilled water pump; TWIN HEAD BELT DRIVE (cont'd)						
65mm pump size; 14.0 l/s at 90 kPa max head; 3kW max motor rating	2958.86	3789.68	11.70	296.85	nr	**4086.53**
65mm pump size; 14.0 l/s at 160 kPa max head; 4 kW max motor rating	3304.13	4231.89	11.70	296.85	nr	**4528.74**
80mm pump size; 14.5 l/s at 210 kPa max head; 7.5 kW max motor rating	4963.61	6357.35	13.64	346.07	nr	**6703.42**
Centrifugal heating and chilled water pump; CLOSE COUPLED; 3 phase, 1450 rpm motor; max. pressure 1000kN/m²; max. temperature 110°C; bed plate; coupling guard; bolted connections; supply only mating flanges; includes fixing on prepared concrete base; electrical work elsewhere						
40mm pump size; 4.0 l/s at 23 kPa max head; 0.55kW max motor rating	627.41	803.58	7.31	185.47	nr	**989.05**
50mm pump size; 4.0 l/s at 75 kPa max head; 0.75 kW max motor rating	695.48	890.76	7.31	185.47	nr	**1076.23**
50mm pump size; 7.0 l/s at 65 kPa max head; 0.75 kW max motor rating	695.48	890.76	8.01	203.23	nr	**1093.99**
65mm pump size; 10.0 l/s at 33 kPa max head; 0.75kW max motor rating	788.29	1009.63	8.01	203.23	nr	**1212.86**
50mm pump size; 4.0 l/s at 120 kPa max head; 1.5 kW max motor rating	949.16	1215.68	8.01	203.23	nr	**1418.91**
80mm pump size; 16.0 l/s at 80 kPa max head; 2.2 kW max motor rating	1228.84	1573.88	12.35	313.34	nr	**1887.23**
80mm pump size; 16.0 l/s at 120 kPa max head; 4.0 kW max motor rating	1197.90	1534.26	12.35	313.34	nr	**1847.60**
100mm pump size; 28.0 l/s at 40 kPa max head; 2.2 kW max motor rating	1213.99	1554.86	17.86	453.14	nr	**2008.01**
100mm pump size; 28.0 l/s at 90 kPa max head; 4.0 kW max motor rating	1340.21	1716.53	17.86	453.14	nr	**2169.67**
125mm pump size; 40.0 l/s at 50 kPa max head; 3.0 kW max motor rating	1278.34	1637.28	25.85	655.86	nr	**2293.15**
125mm pump size; 40.0 l/s at 120 kPa max head; 7.5 kW max motor rating	1577.81	2020.85	25.85	655.86	nr	**2676.71**
150mm pump size; 70.0 l/s at 75 kPa max head; 11 kW max motor rating	2319.07	2970.25	30.43	772.07	nr	**3742.31**
150mm pump size; 70.0 l/s at 120 kPa max head; 15.0 kW max motor rating	2478.71	3174.71	30.43	772.07	nr	**3946.78**
150mm pump size; 70.0 l/s at 150 kPa max head; 15.0 kW max motor rating	2478.71	3174.71	30.43	772.07	nr	**3946.78**

T:MECHANICAL/COOLING/HEATING SYSTEMS

Item	Net Price £	Material £	Labour hours	Labour £	Unit	Total rate £
Centrifugal heating & chilled water pump; close coupled; 3 phase, VARIABLE SPEED motor; max. system pressure 1000 kN/m²; max. temperature 110°C; bed plate; coupling guard; bolted connections; supply only mating flanges; includes fixing on prepared concrete base; electrical work elsewhere.						
40mm pump size; 4.0 l/s at 23 kPa max head; 0.55kW max motor rating	1132.31	1450.25	7.31	185.47	nr	**1635.72**
40mm pump size; 4.0 l/s at 75 kPa max head; 0.75kW max motor rating	1249.88	1600.83	7.31	185.47	nr	**1786.30**
50mm pump size; 7.0 l/s at 65 kPa max head; 1.5kW max motor rating	1533.26	1963.79	8.01	203.23	nr	**2167.02**
50mm pump size; 10.0 l/s at 33 kPa max head; 1.5kW max motor rating	1533.26	1963.79	8.01	203.23	nr	**2167.02**
50mm pump size; 4.0 l/s at 120 kPa max head; 1.5kW max motor rating	1533.26	1963.79	8.01	203.23	nr	**2167.02**
80mm pump size; 16.0 l/s at 80 kPa max head; 2.2kW max motor rating	1998.56	2559.74	12.35	313.34	nr	**2873.08**
80mm pump size; 16.0 l/s at 120 kPa max head; 3.0kW max motor rating	2155.73	2761.03	12.35	313.34	nr	**3074.37**
100mm pump size; 28.0 l/s at 40 kPa max head; 2.2kW max motor rating	2048.06	2623.14	17.86	453.14	nr	**3076.28**
100mm pump size; 28.0 l/s at 90 kPa max head; 4.0.kW max motor rating	2442.83	3128.75	17.86	453.14	nr	**3581.89**
125mm pump size; 40.0 l/s at 50 kPa max head; 3.0kW max motor rating	2345.06	3003.53	25.85	655.86	nr	**3659.40**
125mm pump size; 40.0 l/s at 120 kPa max head; 7.5kW max motor rating	3300.41	4227.14	25.85	655.86	nr	**4883.00**
150mm pump size; 70.0 l/s at 75 kPa max head; 7.5kW max motor rating	3815.21	4886.49	30.43	772.07	nr	**5658.55**
Glandless domestic heating pump; for low pressure domestic hot water heating systems; 240 volt; 50Hz electric motor; max working pressure 1000N/m² and max temperature of 130°C; includes fixing in position; electrical work elsewhere						
1" BSP unions - 2 speed	112.86	144.55	1.58	40.09	nr	**184.64**
1.25" BSP unions - 3 speed	166.32	213.02	1.58	40.09	nr	**253.11**
Glandless pumps; for hot water secondary supply; silent running; 3 phase; max pressure 1000kN/m²; max temperature 130°C ; bolted connections; supply only mating flanges; including fixing in position; electrical elsewhere						
1" BSP unions - 3 speed	219.04	280.54	1.58	40.09	nr	**320.63**

T:MECHANICAL/COOLING/HEATING SYSTEMS

Item	Net Price £	Material £	Labour hours	Labour £	Unit	Total rate £
T31 : LOW TEMPERATURE HOT WATER HEATING (cont'd)						
Y20 – PUMPS (cont'd)						
Pipeline mounted circulator; for heating and chilled water; silent running; 3 phase; 1450 rpm motor; max pressure 1000 kN/m²; max temperature 120°C; bolted connections; supply only mating flanges; includes fixing in position; electrical elsewhere						
32mm pump size; 2.0 l/s at 17 kPa max head; 0.2kW max motor rating	438.08	561.08	6.44	163.40	nr	**724.48**
50mm pump size; 3.0 l/s at 20 kPa max head; 0.2kW max motor rating	435.60	557.91	6.86	174.05	nr	**731.96**
65mm pump size; 5.0 l/s at 30 kPa max head; 0.37 kW max motor rating	693.00	887.59	7.48	189.78	nr	**1077.37**
65mm pump size; 8.0 l/s at 37 kPa max head; 0.75 kW max motor rating	775.91	993.78	7.48	189.78	nr	**1183.56**
80mm pump size; 12.0 l/s at 42 kPa max head; 1.1 kW max motor rating	902.14	1155.45	8.01	203.23	nr	**1358.68**
100mm pump size; 25.0 l/s at 37 kPa max head; 2.2 kW max motor rating	1285.76	1646.79	9.11	231.14	nr	**1877.93**
Dual pipeline mounted circulator; for heating & chilled water; silent running; 3 phase; 1450 rpm motor; max pressure 1000 kN/m²; max temperature 120°C; bolted connections; supply only mating flanges; includes fixing in position; electrical work elsewhere						
40mm pump size; 2.0 l/s at 17 kPa max head; 0.8kW max motor rating	808.09	1034.99	7.88	199.93	nr	**1234.92**
50mm pump size; 3.0 l/s at 20 kPa max head; 0.2 kW max motor rating	810.56	1038.16	8.01	203.23	nr	**1241.39**
65mm pump size; 5.0 l/s at 30 kPa max head; 0.37 kW max motor rating	1315.46	1684.83	9.20	233.42	nr	**1918.25**
65mm pump size; 8.0 l/s at 37 kPa max head; 0.75 kW max motor rating	1520.89	1947.94	9.20	233.42	nr	**2181.36**
100mm pump size; 12.0 l/s at 42 kPa max head; 1.1 kW max motor rating	1720.13	2203.12	9.45	239.76	nr	**2442.88**
Glandless accelerator pumps; for low and medium pressure heating services; silent running; 3 phase; 1450 rpm motor; max pressure 1000 kN/m²; max temperature 130°C; bolted connections; supply only mating flanges; includes fixing in position; electrical work elsewhere						
40mm pump size; 4.0 l/s at 15 kPa max head; 0.35kW max motor rating	393.75	504.31	6.94	176.08	nr	**680.39**
50mm pump size; 6.0 l/s at 20 kPa max head; 0.45kW max motor rating	419.63	537.45	7.35	186.48	nr	**723.93**
80mm pump size; 13.0 l/s at 28 kPa max head; 0.58kW max motor rating	826.88	1059.05	7.76	196.89	nr	**1255.94**

T:MECHANICAL/COOLING/HEATING SYSTEMS

Item	Net Price £	Material £	Labour hours	Labour £	Unit	Total rate £
Y22 - HEAT EXCHANGERS						
Plate heat exchanger; for use in LTHW systems; painted carbon steel frame; stainless steel plates, nitrile rubber gaskets; design pressure of 10 bar and operating temperature of 110/135°C						
Primary side; 80°C in, 69°C out; secondary side; 82°C in, 71°C out						
107 KW, 2.38 l/s	1910.00	2446.31	10.00	253.72	nr	**2700.03**
245 kW, 5.46 l/s	2935.00	3759.12	10.00	253.72	nr	**4012.84**
287 kW, 6.38 l/s	3258.00	4172.81	10.00	253.72	nr	**4426.53**
328 kW, 7.31 l/s	3555.00	4553.21	10.00	253.72	nr	**4806.93**
364 kW, 8.11 l/s	4219.00	5403.65	10.00	253.72	nr	**5657.37**
403 kW, 8.96 l/s	4429.00	5672.62	10.00	253.72	nr	**5926.34**
453 kW, 10.09 l/s	4608.00	5901.88	10.00	253.72	nr	**6155.60**
490 kW, 10.89 l/s	4777.00	6118.33	10.00	253.72	nr	**6372.05**
1000 kW, 21.7 l/s	3554.00	4551.93	12.00	304.46	nr	**4856.39**
1500 kW, 32.6 l/s	4528.00	5799.42	12.00	304.46	nr	**6103.88**
2000 kW, 43.4 l/s	5432.00	6957.25	15.00	380.58	nr	**7337.83**
2500 kW, 54.3 l/s	6677.00	8551.83	15.00	380.58	nr	**8932.41**
Note - For temperature conditions different to those above, the cost of the units can vary significantly, and so manufacturers advice should be sought.						

T:MECHANICAL/COOLING/HEATING SYSTEMS

Item	Net Price £	Material £	Labour hours	Labour £	Unit	Total rate £
T31 : LOW TEMPERATURE HOT WATER HEATING (cont'd)						
Y23 - CALORIFIERS						
Non-storage calorifiers; mild steel; heater battery duty 82°C/71°C to BS 853, maximum test on shell 11.55 bar, tubes 26.25 bar						
Horizontal or vertical; primary water at 116°C on, 90°C off						
40 kW capacity	471.76	604.23	3.00	76.12	nr	**680.34**
88 kW capacity	549.31	703.55	5.00	126.86	nr	**830.41**
176 kW capacity	851.76	1090.92	7.04	178.68	nr	**1269.60**
293 kW capacity	1265.36	1620.66	9.01	228.58	nr	**1849.23**
586 kW capacity	1859.91	2382.15	22.22	563.82	nr	**2945.97**
879 kW capacity	2424.73	3105.57	28.57	724.91	nr	**3830.48**
1465 kW capacity	3833.56	4909.98	50.00	1268.59	nr	**6178.57**
2000 kW capacity	5299.25	6787.23	60.00	1522.31	nr	**8309.54**
HEAT EMITTERS						
Perimeter convector heating; metal casing with standard finish; aluminium extruded grille; including backplates						
Top/sloping/flat front outlet						
60 x 200mm	31.85	40.79	2.00	50.74	m	**91.54**
60 x 300mm	34.64	44.37	2.00	50.74	m	**95.11**
60 x 450mm	42.94	55.00	2.00	50.74	m	**105.74**
60 x 525mm	45.71	58.54	2.00	50.74	m	**109.29**
60 x 600mm	49.86	63.86	2.00	50.74	m	**114.60**
90 x 260mm	34.64	44.37	2.00	50.74	m	**95.11**
90 x 300mm	36.01	46.12	2.00	50.74	m	**96.87**
90 x 450mm	44.34	56.79	2.00	50.74	m	**107.53**
90 x 525mm	48.49	62.10	2.00	50.74	m	**112.85**
90 x 600mm	51.26	65.65	2.00	50.74	m	**116.39**
Extra over for dampers						
Damper	15.23	19.51	0.25	6.34	nr	**25.85**
Extra over for fittings						
60mm End caps	12.19	15.61	0.25	6.34	nr	**21.96**
90mm End caps	19.81	25.38	0.25	6.34	nr	**31.72**
60mm Corners	25.91	33.19	0.25	6.34	nr	**39.53**
90mm Corners	38.10	48.80	0.25	6.34	nr	**55.14**
Radiant Strip Heaters						
Suitable for connection to hot water system; aluminium sheet panels with steel pipe clamped to upper surface; including insulation, sliding brackets, cover plates, end closures; weld or screwed BSP ends						
One pipe						
1500mm long	64.41	82.50	3.11	78.91	nr	**161.40**
3000mm long	101.88	130.49	3.11	78.91	nr	**209.40**
4500mm long	138.19	176.99	3.11	78.91	nr	**255.90**
6000mm long	190.50	243.99	3.11	78.91	nr	**322.90**

T:MECHANICAL/COOLING/HEATING SYSTEMS

Item	Net Price £	Material £	Labour hours	Labour £	Unit	Total rate £
Two pipe						
1500mm long	120.25	154.01	4.15	105.29	nr	**259.31**
3000mm long	189.94	243.28	4.15	105.29	nr	**348.57**
4500mm long	259.38	332.21	4.15	105.29	nr	**437.50**
6000mm long	349.45	447.57	4.15	105.29	nr	**552.86**
Pressed steel panel type radiators; fixed with and including brackets; taking down once for decoration; refixing						
300mm high; single panel						
500mm length	16.43	21.05	2.03	51.51	nr	**72.55**
1000mm length	32.84	42.07	2.03	51.51	nr	**93.57**
1500mm length	43.03	55.11	2.03	51.51	nr	**106.61**
2000mm length	48.27	61.83	2.47	62.67	nr	**124.50**
2500mm length	53.50	68.52	2.97	75.35	nr	**143.87**
3000mm length	64.03	82.01	3.22	81.70	nr	**163.70**
300mm high; double panel; convector						
500mm length	31.59	40.46	2.13	54.04	nr	**94.51**
1000mm length	63.22	80.98	2.13	54.04	nr	**135.02**
1500mm length	94.82	121.44	2.13	54.04	nr	**175.48**
2000mm length	126.44	161.94	2.57	65.21	nr	**227.14**
2500mm length	158.03	202.40	3.07	77.89	nr	**280.29**
3000mm length	189.65	242.90	3.31	83.98	nr	**326.88**
450mm high; single panel						
500mm length	15.33	19.64	2.08	52.77	nr	**72.41**
1000mm length	30.68	39.29	2.08	52.77	nr	**92.07**
1600mm length	49.08	62.86	2.53	64.19	nr	**127.05**
2000mm length	61.36	78.58	2.97	75.35	nr	**153.94**
2400mm length	73.62	94.29	3.47	88.04	nr	**182.33**
3000mm length	92.03	117.87	3.82	96.92	nr	**214.80**
450mm high; double panel; convector						
500mm length	28.09	35.98	2.18	55.31	nr	**91.29**
1000mm length	56.18	71.96	2.18	55.31	nr	**127.27**
1600mm length	102.89	131.78	2.63	66.73	nr	**198.50**
2000mm length	172.97	221.53	3.06	77.64	nr	**299.17**
2400mm length	207.58	265.86	3.37	85.50	nr	**351.37**
3000mm length	259.47	332.32	3.92	99.46	nr	**431.78**
600mm high; single panel						
500mm length	20.54	26.31	2.18	55.31	nr	**81.62**
1000mm length	41.09	52.62	2.43	61.65	nr	**114.28**
1600mm length	65.74	84.19	3.13	79.41	nr	**163.61**
2000mm length	82.17	105.24	3.77	95.65	nr	**200.89**
2400mm length	98.62	126.31	4.07	103.26	nr	**229.57**
3000mm length	123.27	157.88	5.11	129.65	nr	**287.53**
600mm high; double panel; convector						
500mm length	35.37	45.30	2.28	57.85	nr	**103.15**
1000mm length	70.74	90.60	2.28	57.85	nr	**148.45**
1600mm length	129.57	165.95	3.23	81.95	nr	**247.90**
2000mm length	217.82	278.99	3.87	98.19	nr	**377.18**
2400mm length	261.38	334.78	4.17	105.80	nr	**440.58**
3000mm length	326.72	418.47	5.24	132.95	nr	**551.41**

T:MECHANICAL/COOLING/HEATING SYSTEMS

Item	Net Price £	Material £	Labour hours	Labour £	Unit	Total rate £
T31 : LOW TEMPERATURE HOT WATER HEATING (cont'd)						
Y23 – CALORIFIERS (cont'd)						
HEAT EMITTERS (cont'd)						
Pressed steel panel type radiators; fixed with and including brackets (cont'd)						
700mm high; single panel						
500mm length	24.04	30.80	2.23	56.58	nr	**87.38**
1000mm length	48.06	61.56	2.83	71.80	nr	**133.36**
1600mm length	76.90	98.49	3.73	94.64	nr	**193.13**
2000mm length	96.13	123.12	4.46	113.16	nr	**236.28**
2400mm length	115.37	147.77	4.48	113.67	nr	**261.43**
3000mm length	144.19	184.68	5.24	132.95	nr	**317.63**
700mm high; double panel; convector						
500mm length	45.97	58.88	2.33	59.12	nr	**118.00**
1000mm length	123.68	158.40	3.08	78.15	nr	**236.55**
1600mm length	197.88	253.44	3.83	97.17	nr	**350.61**
2000mm length	247.35	316.81	4.17	105.80	nr	**422.61**
2400mm length	296.83	380.17	4.37	110.88	nr	**491.05**
3000mm length	371.03	475.21	4.82	122.29	nr	**597.50**
Flat panel type steel radiators; fixed with and including brackets; taking down once for decoration; refixing						
300mm high; single panel (44mm deep)						
500mm length	46.36	59.38	2.03	51.51	nr	**110.88**
1000mm length	75.31	96.45	2.03	51.51	nr	**147.96**
1500mm length	104.26	133.53	2.03	51.51	nr	**185.04**
2000mm length	133.20	170.61	2.47	62.67	nr	**233.27**
2400mm length	156.36	200.27	2.97	75.35	nr	**275.62**
3000mm length	191.10	244.76	3.22	81.70	nr	**326.46**
300mm high; double panel (100mm deep)						
500mm length	127.64	163.48	2.03	51.51	nr	**214.98**
1000mm length	184.19	235.91	2.03	51.51	nr	**287.41**
1500mm length	240.74	308.33	2.03	51.51	nr	**359.84**
2000mm length	297.29	380.76	2.47	62.67	nr	**443.43**
2400mm length	342.52	438.70	2.97	75.35	nr	**514.06**
3000mm length	410.38	525.61	3.22	81.70	nr	**607.31**
500mm high; single panel (44mm deep)						
500mm length	57.27	73.35	2.13	54.04	nr	**127.39**
1000mm length	96.31	123.36	2.13	54.04	nr	**177.40**
1500mm length	135.36	173.37	2.13	54.04	nr	**227.41**
2000mm length	174.40	223.37	2.57	65.21	nr	**288.58**
2400mm length	205.64	263.38	3.07	77.89	nr	**341.27**
3000mm length	252.49	323.39	3.31	83.98	nr	**407.37**
500mm high; double panel (100mm deep)						
500mm length	148.60	190.32	2.08	52.77	nr	**243.10**
1000mm length	223.32	286.03	2.08	52.77	nr	**338.80**
1500mm length	298.05	381.74	2.53	64.19	nr	**445.93**
2000mm length	372.77	477.44	2.97	75.35	nr	**552.80**

T:MECHANICAL/COOLING/HEATING SYSTEMS

Item	Net Price £	Material £	Labour hours	Labour £	Unit	Total rate £
2400mm length	432.55	554.01	3.47	88.04	nr	**642.05**
3000mm length	522.22	668.86	3.82	96.92	nr	**765.78**
600mm high; single panel (44mm deep)						
500mm length	62.38	79.90	2.18	55.31	nr	**135.21**
1000mm length	106.14	135.94	2.18	55.31	nr	**191.26**
1500mm length	106.14	135.94	2.63	66.73	nr	**202.67**
2000mm length	193.66	248.03	3.06	77.64	nr	**325.67**
2400mm length	228.66	292.87	3.37	85.50	nr	**378.37**
3000mm length	-	360.12	3.92	99.46	nr	**459.58**
600mm high; double panel (100mm deep)						
500mm length	159.91	204.81	2.18	55.31	nr	**260.12**
1000mm length	244.06	312.59	2.43	61.65	nr	**374.24**
1500mm length	-	420.36	3.13	79.41	nr	**499.78**
2000mm length	412.36	528.14	3.77	95.65	nr	**623.80**
2400mm length	479.68	614.37	4.07	103.26	nr	**717.63**
3000mm length	580.66	743.70	5.11	129.65	nr	**873.35**
700mm high; single panel (44mm deep)						
500mm length	67.50	86.45	2.28	57.85	nr	**144.30**
1000mm length	115.97	148.53	2.28	57.85	nr	**206.38**
1500mm length	164.44	210.61	3.23	81.95	nr	**292.56**
2000mm length	212.91	272.69	3.87	98.19	nr	**370.88**
2400mm length	251.69	322.36	4.17	105.80	nr	**428.16**
3000mm length	309.85	396.85	5.24	132.95	nr	**529.80**
700mm high; double panel (100mm deep)						
500mm length	170.54	218.43	2.23	56.58	nr	**275.01**
1000mm length	263.45	337.42	2.83	71.80	nr	**409.22**
1500mm length	356.35	456.41	3.73	94.64	nr	**551.04**
2000mm length	449.25	575.39	4.46	113.16	nr	**688.55**
2400mm length	523.57	670.58	4.48	113.67	nr	**784.25**
3000mm length	635.05	813.37	5.24	132.95	nr	**946.32**
Fan convector; sheet metal casing with lockable access panel; centrifugal fan; air filter; LPHW heating coil; extruded aluminium grilles; 3 speed; includes fixing in position; electrical work elsewhere						
Free standing flat top, 695mm high, medium speed rating						
Entering air temperature, 18°C						
695mm long, 1 row 1.94 kW, 75 l/sec	633.08	810.85	2.73	69.27	nr	**880.11**
695mm long, 2 row 2.64 kW, 75 l/sec	633.08	810.85	2.73	69.27	nr	**880.11**
895mm long, 1 row 4.02 kW, 150 l/sec	713.38	913.68	2.73	69.27	nr	**982.95**
895mm long, 2 row 5.62 kW, 150 l/sec	713.38	913.68	2.73	69.27	nr	**982.95**
1195mm long, 1 row 6.58 kW, 250 l/sec	812.19	1040.25	3.00	76.12	nr	**1116.36**
1195mm long, 2 row 9.27 kW, 250 l/sec	812.19	1040.25	3.00	76.12	nr	**1116.36**
1495mm long, 1 row 9.04 kW, 340 l/sec	906.39	1160.90	3.26	82.71	nr	**1243.61**
1495mm long, 2 row 12.73 kW, 340 l/sec	906.39	1160.90	3.26	82.71	nr	**1243.61**

T:MECHANICAL/COOLING/HEATING SYSTEMS

Item	Net Price £	Material £	Labour hours	Labour £	Unit	Total rate £
T31 : LOW TEMPERATURE HOT WATER HEATING (cont'd)						
Y23 – CALORIFIERS (cont'd)						
HEAT EMITTERS (cont'd)						
Fan convector; sheet metal casing with lockable access panel (cont'd)						
Free standing flat top, 695mm high, medium speed rating, c/w floor plinth						
695mm long, 1 row 1.94 kW, 75 l/sec	664.73	851.38	2.73	69.27	nr	**920.65**
695mm long, 2 row 2.64 kW, 75 l/sec	664.73	851.38	2.73	69.27	nr	**920.65**
895mm long, 1 row 4.02 kW, 150 l/sec	749.04	959.36	2.73	69.27	nr	**1028.63**
895mm long, 2 row 5.62 kW, 150 l/sec	749.04	959.36	2.73	69.27	nr	**1028.63**
1195mm long, 1 row 6.58 kW, 250 l/sec	852.81	1092.27	3.00	76.12	nr	**1168.38**
1195mm long, 2 row 9.27 kW, 250 l/sec	852.81	1092.27	3.00	76.12	nr	**1168.38**
1495mm long, 1 row 9.04 kW, 340 l/sec	951.70	1218.93	3.26	82.71	nr	**1301.64**
1495mm long, 2 row 12.73 kW, 340 l/sec	951.70	1218.93	3.26	82.71	nr	**1301.64**
Free standing sloping top, 695mm high, medium speed rating, c/w floor plinth						
695mm long, 1 row 1.94 kW, 75 l/sec	687.89	881.05	2.73	69.27	nr	**950.31**
695mm long, 2 row 2.64 kW, 75 l/sec	687.89	881.05	2.73	69.27	nr	**950.31**
895mm long, 1 row 4.02 kW, 150 l/sec	772.20	989.03	2.73	69.27	nr	**1058.29**
895mm long, 2 row 5.62 kW, 150 l/sec	772.20	989.03	2.73	69.27	nr	**1058.29**
1195mm long, 1 row 6.58 kW, 250 l/sec	875.97	1121.93	3.00	76.12	nr	**1198.05**
1195mm long, 2 row 9.27 kW, 250 l/sec	875.97	1121.93	3.00	76.12	nr	**1198.05**
1495mm long, 1 row 9.04 kW, 340 l/sec	974.86	1248.59	3.26	82.71	nr	**1331.30**
1495mm long, 2 row 12.73 kW, 340 l/sec	974.86	1248.59	3.26	82.71	nr	**1331.30**
Wall mounted high level sloping discharge						
695mm long, 1 row 1.94 kW, 75 l/sec	694.84	889.95	2.73	69.27	nr	**959.21**
695mm long, 2 row 2.64 kW, 75 l/sec	694.84	889.95	2.73	69.27	nr	**959.21**
895mm long, 1 row 4.02 kW, 150 l/sec	714.93	915.68	2.73	69.27	nr	**984.94**
895mm long, 2 row 5.62 kW, 150 l/sec	714.93	915.68	2.73	69.27	nr	**984.94**
1195mm long, 1 row 6.58 kW, 250 l/sec	870.08	1114.38	3.00	76.12	nr	**1190.50**
1195mm long, 2 row 9.27 kW, 250 l/sec	870.08	1114.38	3.00	76.12	nr	**1190.50**
1495mm long, 1 row 9.04 kW, 340 l/sec	943.43	1208.34	3.26	82.71	nr	**1291.05**
1495mm long, 2 row 12.73 kW, 340 l/sec	943.43	1208.34	3.26	82.71	nr	**1291.05**
Ceiling mounted sloping inlet/outlet 665mm wide						
895mm long, 1 row 4.02 kW, 150 l/sec	785.94	1006.63	4.15	105.29	nr	**1111.92**
895mm long, 2 row 5.62 kW, 150 l/sec	785.94	1006.63	4.15	105.29	nr	**1111.92**
1195mm long, 1 row 6.58 kW, 250 l/sec	883.23	1131.23	4.15	105.29	nr	**1236.53**
1195mm long, 2 row 9.27 kW, 250 l/sec	883.23	1131.23	4.15	105.29	nr	**1236.53**
1495mm long, 1 row 9.04 kW, 340 l/sec	969.69	1241.97	4.15	105.29	nr	**1347.27**
1495mm long, 2 row 12.73 kW, 340 l/sec	969.69	1241.97	4.15	105.29	nr	**1347.27**
Free standing unit, extended height 1700/1900/2100mm						
895mm long, 1 row 4.02 kW, 150 l/sec	909.48	1164.85	3.11	78.91	nr	**1243.75**
895mm long, 2 row 5.62 kW, 150 l/sec	909.48	1164.85	3.11	78.91	nr	**1243.75**
1195mm long, 1 row 6.58 kW, 250 l/sec	1062.34	1360.63	3.11	78.91	nr	**1439.54**
1195mm long, 2 row 9.27 kW, 250 l/sec	1062.34	1360.63	3.11	78.91	nr	**1439.54**
1495mm long, 1 row 9.04 kW, 340 l/sec	1168.89	1497.10	3.11	78.91	nr	**1576.01**
1495mm long, 2 row 12.73 kW, 340 l/sec	1168.89	1497.10	3.11	78.91	nr	**1576.01**

T:MECHANICAL/COOLING/HEATING SYSTEMS

Item	Net Price £	Material £	Labour hours	Labour £	Unit	Total rate £
LTHW trench heating; water temperatures 90°C/70°C; room air temperature 20°C; convector with copper tubes and aluminium fins within steel duct; Includes fixing within floor screed; electrical work elsewhere						
Natural convection type						
Normal capacity, 182mm width, complete with linear, natural anodised aluminium grille (grille also costed separately below)						
92mm deep						
1250mm long, 234 W output	202.13	258.88	2.00	50.74	nr	309.62
2250mm long, 471 W output	323.40	414.21	4.00	101.49	nr	515.70
3250mm long, 709 W output	442.37	566.58	5.00	126.86	nr	693.44
4250mm long, 946 W output	564.80	723.38	7.00	177.60	nr	900.99
5000mm long, 1124 W output	771.54	988.18	8.00	202.98	nr	1191.16
120mm deep						
1250mm long, 294 W output	235.62	301.78	2.00	50.74	nr	352.52
2250mm long, 471 W output	370.76	474.86	4.00	101.49	nr	576.35
3250mm long, 891 W output	504.74	646.46	5.00	126.86	nr	773.32
4250mm long, 1190 W output	638.72	818.06	7.00	177.60	nr	995.66
5000mm long, 1414 W output	860.48	1102.09	8.00	202.98	nr	1305.06
150mm deep						
1250mm long, 329 W output	241.40	309.18	2.00	50.74	nr	359.92
2250mm long, 664 W output	381.15	488.17	4.00	101.49	nr	589.66
3250mm long, 998 W output	518.60	664.21	5.00	126.86	nr	791.07
4250mm long, 1333 W output	659.51	844.69	7.00	177.60	nr	1022.29
5000mm long, 1584 W output	884.73	1133.15	8.00	202.98	nr	1336.13
200mm deep						
1250mm long, 396 W output	251.79	322.49	2.00	50.74	nr	373.23
2250mm long, 799 W output	460.85	590.25	4.00	101.49	nr	691.73
3250mm long, 1201 W output	548.63	702.67	5.00	126.86	nr	829.53
4250mm long, 1603 W output	702.24	899.42	7.00	177.60	nr	1077.03
5000mm long, 1905 W output	933.24	1195.28	8.00	202.98	nr	1398.26
Fan assisted type (outputs assume fan at 50%)						
Normal capacity, 182mm width, complete with Natural anodised aluminium grille						
112mm deep						
1250mm long, 437 W output	533.61	683.44	2.00	50.74	nr	734.19
2250mm long, 1019 W output	645.65	826.94	4.00	101.49	nr	928.42
3250mm long, 1488 W output	758.84	971.91	5.00	126.86	nr	1098.77
4250mm long, 1845 W output	870.87	1115.40	7.00	177.60	nr	1293.00
5000mm long, 2038 W output	982.91	1258.89	8.00	202.98	nr	1461.87
Linear grille anodised aluminium, 170mm width (if supplied as a separate item)	125.90	161.25	-	-	m	161.25
Roll up grille, natural anodised aluminium	125.90	161.25	-	-	m	161.25
Thermostatic valve with remote regulator (c/w valve body)	51.98	66.57	4.00	101.49	nr	168.06
Fan speed controller	25.41	32.54	2.00	50.74	nr	83.29

T:MECHANICAL/COOLING/HEATING SYSTEMS

Item	Net Price £	Material £	Labour hours	Labour £	Unit	Total rate £
T31 : LOW TEMPERATURE HOT WATER HEATING (cont'd)						
Y23 – CALORIFIERS (cont'd)						
HEAT EMITTERS (cont'd)						
LTHW trench heating; Fan assisted type (outputs assume fan at 50%) (cont'd)						
Note - as an alternative to thermostatic control, the system can be controlled via two port valves. Refer to valve section in T31 for valve rates						
LTHW underfloor heating; water flow and return temperatures of 60°C and 70°C; pipework at 300mm centres; pipe fixings; flow and return manifolds and zone actuators; wiring block; insulation; includes fixing in position; excludes secondary pump, mixing valve, zone thermostats and floor finishes; electrical work elsewhere						
Note - All rates are expressed on a m² basis, for the following example areas						
Screeded floor with 15-25mm stone/marble finish (producing 80-100W/m²)						
250m² area (single zone)	16.26	20.82	0.14	3.55	m²	**24.37**
1000m² area (single zone)	15.84	20.29	0.12	3.04	m²	**23.34**
5000m² area (multi-zone)	15.80	20.23	0.10	2.54	m²	**22.77**
Screeded floor with 10mm carpet tile (producing 80-100W/m²)						
250m² area (single zone)	16.26	20.82	0.14	3.55	m²	**24.37**
1000m² area (single zone)	15.84	20.29	0.12	3.04	m²	**23.34**
5000m² area (multi-zone)	15.80	20.23	0.10	2.54	m²	**22.77**
Floating timber floor with 20mm timber finish (producing 70-80Wm²)						
250m² are (single zone)	23.22	29.74	0.14	3.55	m²	**33.30**
1000m² area(single zone)	21.96	28.13	0.12	3.04	m²	**31.18**
5000m² area (multi-zone)	20.98	26.87	0.10	2.54	m²	**29.41**
Floating timber floor with 10mm carpet tile (producing 70-80W/m²)						
250m² are (single zone)	26.66	34.14	0.16	4.06	m²	**38.20**
1000m² area(single zone)	24.73	31.68	0.12	3.04	m²	**34.72**
5000m² area (multi-zone)	23.65	30.29	0.10	2.54	m²	**32.82**

T:MECHANICAL/COOLING/HEATING SYSTEMS

Item	Net Price £	Material £	Labour hours	Labour £	Unit	Total rate £
PIPE FREEZING						
Freeze isolation of carbon steel or copper pipelines containing static water, either side of work location, freeze duration not exceeding 4 hours assuming that flow and return circuits are treated concurrently and activities undertaken during normal working hours						
Up to 4 freezes						
50mm dia	345.69	442.76	-	-	nr	**442.76**
65mm dia	345.69	442.76	-	-	nr	**442.76**
80mm dia	395.20	506.17	-	-	nr	**506.17**
100mm dia	444.72	569.59	-	-	nr	**569.59**
150mm dia	739.98	947.75	-	-	nr	**947.75**
200mm dia	1134.26	1452.75	-	-	nr	**1452.75**
ENERGY METERS						
Ultrasonic						
Energy meter for measuring energy use in LTHW systems; includes ultrasonic flow meter (with sensor and signal converter), energy calculator, pair of temperature sensors with brass pockets, and 3m of interconnecting cable; includes fixing in position; electrical work elsewhere						
Pipe size (flanged connections to PN16); maximum flow rate						
50mm, 36m³/hr	1020.76	1307.38	1.80	45.67	nr	**1353.05**
65mm, 60m³/hr	1125.19	1441.14	2.32	58.86	nr	**1500.00**
80mm, 100m³/hr	1261.25	1615.39	2.56	64.95	nr	**1680.34**
125mm, 250m³/hr	1461.28	1871.59	3.60	91.34	nr	**1962.93**
150mm, 360m³/hr	1586.30	2031.72	4.80	121.79	nr	**2153.50**
200mm, 600m³/hr	1772.36	2270.02	6.24	158.32	nr	**2428.34**
250mm, 1000m³/hr	2045.20	2619.48	9.60	243.57	nr	**2863.05**
300mm, 1500m³/hr	2404.09	3079.13	10.80	274.02	nr	**3353.15**
350mm, 2000m³/hr	2896.82	3710.22	13.20	334.91	nr	**4045.13**
400mm, 2500m³/hr	3316.01	4247.11	15.60	395.80	nr	**4642.91**
500mm, 3000m³/hr	3763.14	4819.80	24.00	608.93	nr	**5428.72**
600mm, 3500m³/hr	4227.93	5415.09	28.00	710.41	nr	**6125.50**

T:MECHANICAL/COOLING/HEATING SYSTEMS

Item	Net Price £	Material £	Labour hours	Labour £	Unit	Total rate £
T31 : LOW TEMPERATURE HOT WATER HEATING (cont'd)						
Y53 - CONTROL COMPONENTS - MECHANICAL						
Room thermostats; light and medium duty; installed and connected						
Range 3°C to 27°C; 240 Volt						
1 amp; on/off type	22.39	28.68	0.30	7.61	nr	**36.29**
Range 0°C to +15°C; 240 Volt						
6 amp; frost thermostat	15.10	19.34	0.30	7.61	nr	**26.95**
Range 3°C to 27°C; 250 Volt						
2 amp; changeover type; dead zone	37.45	47.97	0.30	7.61	nr	**55.58**
2 amp; changeover type	17.37	22.24	0.30	7.61	nr	**29.85**
2 amp; changeover type; concealed setting	21.91	28.06	0.30	7.61	nr	**35.67**
6 amp; on/off type	13.53	17.33	0.30	7.61	nr	**24.94**
6 amp; temperature set-back	27.99	35.84	0.30	7.61	nr	**43.46**
16 amp; on/off type	20.83	26.68	0.30	7.61	nr	**34.29**
16 amp; on/off type; concealed setting	22.71	29.09	0.30	7.61	nr	**36.70**
20 amp; on/off type; concealed setting	24.80	31.77	0.30	7.61	nr	**39.38**
20 amp; indicated "off" position	24.39	31.24	0.30	7.61	nr	**38.85**
20 amp; manual; double pole on/off and neon indicator	44.40	56.87	0.30	7.61	nr	**64.48**
20 amp; indicated "off" position	29.05	37.21	0.30	7.61	nr	**44.82**
Range 10°C to 40°C; 240 Volt						
20 amp; changeover contacts	26.47	33.90	0.30	7.61	nr	**41.51**
2 amp; 'heating-cooling' switch	57.26	73.34	0.30	7.61	nr	**80.95**
Surface thermostats						
Cylinder thermostat						
6 amp; changeover type; with cable	14.48	19.01	0.25	6.34	nr	**25.35**
Electrical thermostats; installed and connected						
Range 5°C to 30°C; 230 Volt standard port single time						
10 amp with sensor	21.71	27.80	0.30	7.61	nr	**35.42**
Range 5°C to 30°C; 230 Volt standard port double time						
10 amp with sensor	24.60	31.51	0.30	7.61	nr	**39.12**
10 amp with sensor and on/off switch	37.53	48.06	0.30	7.61	nr	**55.67**
Radiator thermostats						
Angled valve body; thermostatic head; built in sensor						
15mm; liquid filled	13.54	17.34	0.84	21.32	nr	**38.66**
15mm; wax filled	13.54	17.34	0.84	21.32	nr	**38.66**

T:MECHANICAL/COOLING/HEATING SYSTEMS

Item	Net Price £	Material £	Labour hours	Labour £	Unit	Total rate £
Immersion thermostats; stem type; domestic water boilers; fitted; electrical work elsewhere						
Temperature range 0°C to 40°C						
Non standard; 280mm stem	8.24	10.55	0.25	6.34	nr	**16.89**
Temperature range 18°C to 88°C						
13 amp; 178mm stem	5.47	7.00	0.25	6.34	nr	**13.35**
20 amp; 178mm stem	8.50	10.89	0.25	6.34	nr	**17.23**
Non standard; pocket clip; 280mm stem	7.89	10.10	0.25	6.34	nr	**16.44**
Temperature range 40°C to 80°C						
13 amp; 178mm stem	3.00	3.84	0.25	6.34	nr	**10.18**
20 amp; 178mm stem	5.68	7.28	0.25	6.34	nr	**13.62**
Non standard; pocket clip; 280mm stem	8.84	11.32	0.25	6.34	nr	**17.66**
13 amp; 457mm stem	3.55	4.54	0.25	6.34	nr	**10.88**
20 amp; 457mm stem	6.22	7.96	0.25	6.34	nr	**14.31**
Temperature range 50°C to 100°C						
Non standard; 1780mm stem	7.44	9.53	0.25	6.34	nr	**15.88**
Non standard; 280mm stem	7.75	9.93	0.25	6.34	nr	**16.27**
Pockets for thermostats						
For 178mm stem	9.73	12.46	0.25	6.34	nr	**18.80**
For 280mm stem	9.64	12.35	0.25	6.34	nr	**18.69**
Immersion thermostats; stem type; industrial installations; fitted; electrical work elsewhere						
Temperature range 5°C to 105°C						
For 305mm stem	122.24	156.56	0.50	12.69	nr	**169.25**

T:MECHANICAL/COOLING/HEATING SYSTEMS

Item	Net Price £	Material £	Labour hours	Labour £	Unit	Total rate £
T31 : LOW TEMPERATURE HOT WATER HEATING (cont'd)						
Y50 -THERMAL INSULATION						
For flexible closed cell insulation see Section S10 - Cold Water						
CONCEALED PIPEWORK						
Mineral fibre sectional insulation; bright class O foil faced; bright class O foil taped joints; 19mm aluminium bands						
20mm thick						
15mm diameter	2.64	3.45	0.15	3.08	m	**6.53**
20mm diameter	2.81	3.68	0.15	3.08	m	**6.76**
25mm diameter	3.02	3.96	0.15	3.08	m	**7.03**
32mm diameter	3.37	4.41	0.15	3.08	m	**7.49**
40mm diameter	3.61	4.72	0.15	3.08	m	**7.79**
50mm diameter	4.12	5.38	0.15	3.08	m	**8.46**
Extra over for fittings concealed insulation						
Flange/union						
15mm diameter	1.31	1.72	0.13	2.67	nr	**4.39**
20mm diameter	1.41	1.85	0.13	2.67	nr	**4.52**
25mm diameter	1.51	1.98	0.13	2.67	nr	**4.64**
32mm diameter	1.69	2.20	0.13	2.67	nr	**4.87**
40mm diameter	1.80	2.35	0.13	2.67	nr	**5.02**
50mm diameter	2.06	2.69	0.13	2.67	nr	**5.36**
Valves						
15mm diameter	2.64	3.45	0.15	3.08	nr	**6.53**
20mm diameter	3.02	3.96	0.15	3.08	nr	**7.03**
25mm diameter	3.02	3.96	0.15	3.08	nr	**7.03**
32mm diameter	3.37	4.41	0.15	3.08	nr	**7.49**
40mm diameter	3.61	4.72	0.15	3.08	nr	**7.79**
50mm diameter	4.12	5.38	0.15	3.08	nr	**8.46**
Expansion bellows						
15mm diameter	5.27	6.89	0.22	4.51	nr	**11.40**
20mm diameter	5.63	7.36	0.22	4.51	nr	**11.87**
25mm diameter	6.05	7.91	0.22	4.51	nr	**12.42**
32mm diameter	6.73	8.80	0.22	4.51	nr	**13.32**
40mm diameter	7.21	9.43	0.22	4.51	nr	**13.95**
50mm diameter	8.24	10.78	0.22	4.51	nr	**15.29**
25mm thick						
15mm diameter	2.90	3.79	0.15	3.08	m	**6.87**
20mm diameter	3.12	4.08	0.15	3.08	m	**7.16**
25mm diameter	3.50	4.57	0.15	3.08	m	**7.65**
32mm diameter	3.81	4.98	0.15	3.08	m	**8.05**
40mm diameter	4.08	5.33	0.15	3.08	m	**8.41**
50mm diameter	4.67	6.11	0.15	3.08	m	**9.19**

T:MECHANICAL/COOLING/HEATING SYSTEMS

Item	Net Price £	Material £	Labour hours	Labour £	Unit	Total rate £
65mm diameter	5.33	6.97	0.15	3.08	m	**10.05**
80mm diameter	5.85	7.65	0.22	4.51	m	**12.16**
100mm diameter	7.71	10.08	0.22	4.51	m	**14.60**
125mm diameter	8.91	11.65	0.22	4.51	m	**16.17**
150mm diameter	10.62	13.89	0.22	4.51	m	**18.41**
200mm diameter	15.01	19.63	0.25	5.13	m	**24.76**
250mm diameter	17.97	23.50	0.25	5.13	m	**28.63**
300mm diameter	19.21	25.13	0.25	5.13	m	**30.25**
Extra over for fittings concealed insulation						
Flange/union						
15mm diameter	1.45	1.90	0.13	2.67	nr	**4.56**
20mm diameter	1.56	2.04	0.13	2.67	nr	**4.71**
25mm diameter	1.75	2.29	0.13	2.67	nr	**4.95**
32mm diameter	1.91	2.50	0.13	2.67	nr	**5.16**
40mm diameter	2.03	2.66	0.13	2.67	nr	**5.33**
50mm diameter	2.33	3.05	0.13	2.67	nr	**5.71**
65mm diameter	2.67	3.49	0.13	2.67	nr	**6.15**
80mm diameter	2.93	3.83	0.18	3.69	nr	**7.52**
100mm diameter	3.86	5.04	0.18	3.69	nr	**8.73**
125mm diameter	4.46	5.84	0.18	3.69	nr	**9.53**
150mm diameter	5.31	6.94	0.18	3.69	nr	**10.63**
200mm diameter	7.50	9.81	0.22	4.51	nr	**14.32**
250mm diameter	8.99	11.75	0.22	4.51	nr	**16.27**
300mm diameter	9.61	12.56	0.22	4.51	nr	**17.08**
Valves						
15mm diameter	2.90	3.79	0.15	3.08	nr	**6.87**
20mm diameter	3.12	4.08	0.15	3.08	nr	**7.16**
25mm diameter	3.50	4.57	0.15	3.08	nr	**7.65**
32mm diameter	3.81	4.98	0.15	3.08	nr	**8.05**
40mm diameter	4.08	5.33	0.15	3.08	nr	**8.41**
50mm diameter	4.67	6.11	0.15	3.08	nr	**9.19**
65mm diameter	5.33	6.97	0.15	3.08	nr	**10.05**
80mm diameter	5.85	7.65	0.20	4.10	nr	**11.75**
100mm diameter	7.71	10.08	0.20	4.10	nr	**14.19**
125mm diameter	8.91	11.65	0.20	4.10	nr	**15.76**
150mm diameter	10.62	13.89	0.20	4.10	nr	**18.00**
200mm diameter	15.01	19.63	0.25	5.13	nr	**24.76**
250mm diameter	17.97	23.50	0.25	5.13	nr	**28.63**
300mm diameter	19.21	25.13	0.25	5.13	nr	**30.25**
Expansion bellows						
15mm diameter	5.80	7.59	0.22	4.51	nr	**12.10**
20mm diameter	6.26	8.19	0.22	4.51	nr	**12.70**
25mm diameter	6.98	9.13	0.22	4.51	Unit	**13.64**
32mm diameter	7.61	9.95	0.22	4.51	nr	**14.47**
40mm diameter	8.16	10.67	0.22	4.51	nr	**15.18**
50mm diameter	9.35	12.22	0.22	4.51	nr	**16.74**
65mm diameter	10.67	13.96	0.22	4.51	nr	**18.47**
80mm diameter	11.69	15.29	0.29	5.95	nr	**21.24**
100mm diameter	15.41	20.15	0.29	5.95	nr	**26.10**
125mm diameter	17.83	23.31	0.29	5.95	nr	**29.26**
150mm diameter	21.25	27.78	0.29	5.95	nr	**33.73**
200mm diameter	30.01	39.24	0.36	7.39	nr	**46.63**
250mm diameter	35.94	46.99	0.36	7.39	nr	**54.38**
300mm diameter	38.43	50.25	0.36	7.39	nr	**57.64**

T:MECHANICAL/COOLING/HEATING SYSTEMS

Item	Net Price £	Material £	Labour hours	Labour £	Unit	Total rate £
T31 : LOW TEMPERATURE HOT WATER HEATING (cont'd)						
Y50 - THERMAL INSULATION (cont'd)						
CONCEALED PIPEWORK (cont'd)						
Mineral fibre sectional insulation; concealed pipework (cont'd)						
30mm thick						
15mm diameter	3.77	4.93	0.15	3.08	m	**8.01**
20mm diameter	4.04	5.28	0.15	3.08	m	**8.36**
25mm diameter	4.28	5.59	0.15	3.08	m	**8.67**
32mm diameter	4.66	6.09	0.15	3.08	m	**9.17**
40mm diameter	4.93	6.45	0.15	3.08	m	**9.53**
50mm diameter	5.64	7.38	0.15	3.08	m	**10.45**
65mm diameter	6.38	8.35	0.15	3.08	m	**11.43**
80mm diameter	6.97	9.11	0.22	4.51	m	**13.62**
100mm diameter	9.01	11.78	0.22	4.51	m	**16.30**
125mm diameter	10.38	13.57	0.22	4.51	m	**18.08**
150mm diameter	12.19	15.93	0.22	4.51	m	**20.45**
200mm diameter	17.04	22.29	0.25	5.13	m	**27.42**
250mm diameter	20.27	26.50	0.25	5.13	m	**31.63**
300mm diameter	21.48	28.09	0.25	5.13	m	**33.22**
350mm diameter	23.60	30.86	0.25	5.13	m	**35.99**
Extra over for fittings concealed insulation						
Flange/union						
15mm diameter	1.88	2.46	0.13	2.67	nr	**5.13**
20mm diameter	2.02	2.64	0.13	2.67	nr	**5.31**
25mm diameter	2.13	2.79	0.13	2.67	nr	**5.46**
32mm diameter	2.33	3.05	0.13	2.67	nr	**5.71**
40mm diameter	2.47	3.23	0.13	2.67	nr	**5.89**
50mm diameter	2.83	3.70	0.13	2.67	nr	**6.36**
65mm diameter	3.20	4.18	0.13	2.67	nr	**6.85**
80mm diameter	3.48	4.56	0.18	3.69	nr	**8.25**
100mm diameter	4.51	5.90	0.18	3.69	nr	**9.59**
125mm diameter	5.18	6.78	0.18	3.69	nr	**10.47**
150mm diameter	6.10	7.98	0.18	3.69	nr	**11.67**
200mm diameter	8.52	11.14	0.22	4.51	nr	**15.65**
250mm diameter	10.14	13.26	0.22	4.51	nr	**17.77**
300mm diameter	10.75	14.05	0.22	4.51	nr	**18.57**
350mm diameter	11.80	15.43	0.22	4.51	nr	**19.95**
Valves						
15mm diameter	4.04	5.28	0.15	3.08	nr	**8.36**
20mm diameter	4.04	5.28	0.15	3.08	nr	**8.36**
25mm diameter	4.28	5.59	0.15	3.08	nr	**8.67**
32mm diameter	4.66	6.09	0.15	3.08	nr	**9.17**
40mm diameter	4.93	6.45	0.15	3.08	nr	**9.53**
50mm diameter	5.64	7.38	0.15	3.08	nr	**10.45**
65mm diameter	6.38	8.35	0.15	3.08	nr	**11.43**
80mm diameter	6.97	9.11	0.20	4.10	nr	**13.21**
100mm diameter	9.01	11.78	0.20	4.10	nr	**15.89**
125mm diameter	10.38	13.57	0.20	4.10	nr	**17.67**
150mm diameter	12.19	15.93	0.20	4.10	nr	**20.04**
200mm diameter	17.04	22.29	0.25	5.13	nr	**27.42**
250mm diameter	20.27	26.50	0.25	5.13	nr	**31.63**
300mm diameter	21.48	28.09	0.25	5.13	nr	**33.22**
350mm diameter	23.60	30.86	0.25	5.13	nr	**35.99**

T:MECHANICAL/COOLING/HEATING SYSTEMS

Item	Net Price £	Material £	Labour hours	Labour £	Unit	Total rate £
Expansion bellows						
15mm diameter	7.55	9.87	0.22	4.51	nr	**14.39**
20mm diameter	8.07	10.55	0.22	4.51	nr	**15.07**
25mm diameter	8.54	11.17	0.22	4.51	nr	**15.68**
32mm diameter	9.33	12.21	0.22	4.51	nr	**16.72**
40mm diameter	9.87	12.90	0.22	4.51	nr	**17.42**
50mm diameter	11.29	14.77	0.22	4.51	nr	**19.28**
65mm diameter	12.77	16.70	0.22	4.51	nr	**21.21**
80mm diameter	13.93	18.22	0.29	5.95	nr	**24.17**
100mm diameter	18.04	23.59	0.29	5.95	nr	**29.54**
125mm diameter	20.74	27.12	0.29	5.95	nr	**33.07**
150mm diameter	24.37	31.87	0.29	5.95	nr	**37.82**
200mm diameter	34.08	44.56	0.36	7.39	nr	**51.95**
250mm diameter	40.53	53.01	0.36	7.39	nr	**60.39**
300mm diameter	42.98	56.20	0.36	7.39	nr	**63.59**
350mm diameter	47.19	61.71	0.36	7.39	nr	**69.10**
40mm thick						
15mm diameter	4.83	6.32	0.15	3.08	m	**9.40**
20mm diameter	4.98	6.52	0.15	3.08	m	**9.59**
25mm diameter	5.36	7.00	0.15	3.08	m	**10.08**
32mm diameter	5.71	7.47	0.15	3.08	m	**10.55**
40mm diameter	6.00	7.85	0.15	3.08	m	**10.92**
50mm diameter	6.79	8.88	0.15	3.08	m	**11.96**
65mm diameter	7.62	9.97	0.15	3.08	m	**13.05**
80mm diameter	8.29	10.84	0.22	4.51	m	**15.36**
100mm diameter	10.76	14.07	0.22	4.51	m	**18.58**
125mm diameter	12.16	15.90	0.22	4.51	m	**20.42**
150mm diameter	14.19	18.56	0.22	4.51	m	**23.07**
200mm diameter	19.61	25.64	0.25	5.13	m	**30.77**
250mm diameter	22.98	30.05	0.25	5.13	m	**35.18**
300mm diameter	24.57	32.13	0.25	5.13	m	**37.26**
350mm diameter	27.17	35.53	0.25	5.13	m	**40.66**
400mm diameter	30.28	39.60	0.25	5.13	m	**44.73**
Extra over for fittings concealed insulation						
Flange/union						
15mm diameter	2.42	3.16	0.13	2.67	nr	**5.83**
20mm diameter	2.49	3.26	0.13	2.67	nr	**5.93**
25mm diameter	2.68	3.50	0.13	2.67	nr	**6.17**
32mm diameter	2.85	3.73	0.13	2.67	nr	**6.40**
40mm diameter	3.00	3.92	0.13	2.67	nr	**6.59**
50mm diameter	3.40	4.44	0.13	2.67	nr	**7.11**
65mm diameter	3.81	4.98	0.13	2.67	nr	**7.64**
80mm diameter	4.15	5.43	0.18	3.69	nr	**9.12**
100mm diameter	5.38	7.04	0.18	3.69	nr	**10.73**
125mm diameter	6.07	7.94	0.18	3.69	nr	**11.64**
150mm diameter	7.10	9.29	0.18	3.69	nr	**12.98**
200mm diameter	9.81	12.82	0.22	4.51	nr	**17.34**
250mm diameter	11.49	15.03	0.22	4.51	nr	**19.54**
300mm diameter	12.28	16.06	0.22	4.51	nr	**20.58**
350mm diameter	13.59	17.77	0.22	4.51	nr	**22.28**
400mm diameter	15.15	19.81	0.22	4.51	nr	**24.32**

T:MECHANICAL/COOLING/HEATING SYSTEMS

Item	Net Price £	Material £	Labour hours	Labour £	Unit	Total rate £
T31 : LOW TEMPERATURE HOT WATER HEATING (cont'd)						
Y50 - THERMAL INSULATION (cont'd)						
CONCEALED PIPEWORK (cont'd)						
Mineral fibre sectional insulation; concealed pipework (cont'd)						
40mm thick Extra over for fittings concealed insulation (cont'd)						
Valves						
15mm diameter	4.83	6.32	0.15	3.08	nr	9.40
20mm diameter	4.98	6.52	0.15	3.08	nr	9.59
25mm diameter	5.36	7.00	0.15	3.08	nr	10.08
32mm diameter	5.71	7.47	0.15	3.08	nr	10.55
40mm diameter	6.00	7.85	0.15	3.08	nr	10.92
50mm diameter	6.79	8.88	0.15	3.08	nr	11.96
65mm diameter	7.62	9.97	0.15	3.08	nr	13.05
80mm diameter	8.29	10.84	0.20	4.10	nr	14.95
100mm diameter	10.76	14.07	0.20	4.10	nr	18.17
125mm diameter	12.16	15.90	0.20	4.10	nr	20.01
150mm diameter	14.19	18.56	0.20	4.10	nr	22.66
200mm diameter	19.61	25.64	0.25	5.13	nr	30.77
250mm diameter	22.98	30.05	0.25	5.13	nr	35.18
300mm diameter	24.57	32.13	0.25	5.13	nr	37.26
350mm diameter	27.17	35.53	0.25	5.13	nr	40.66
400mm diameter	30.28	39.60	0.25	5.13	nr	44.73
Expansion bellows						
15mm diameter	9.68	12.66	0.22	4.51	nr	17.17
20mm diameter	9.95	13.02	0.22	4.51	nr	17.53
25mm diameter	10.72	14.02	0.22	4.51	nr	18.54
32mm diameter	11.42	14.93	0.22	4.51	nr	19.44
40mm diameter	12.01	15.71	0.22	4.51	nr	20.22
50mm diameter	13.59	17.77	0.22	4.51	nr	22.28
65mm diameter	15.23	19.92	0.22	4.51	nr	24.44
80mm diameter	16.59	21.69	0.29	5.95	nr	27.64
100mm diameter	21.51	28.12	0.29	5.95	nr	34.07
125mm diameter	24.31	31.79	0.29	5.95	nr	37.74
150mm diameter	28.39	37.12	0.29	5.95	nr	43.07
200mm diameter	39.22	51.29	0.36	7.39	nr	58.67
250mm diameter	45.96	60.11	0.36	7.39	nr	67.49
300mm diameter	49.14	64.26	0.36	7.39	nr	71.64
350mm diameter	27.17	35.53	0.36	7.39	nr	42.92
400mm diameter	60.57	79.20	0.36	7.39	nr	86.59
50mm thick						
15mm diameter	6.72	8.79	0.15	3.08	m	11.86
20mm diameter	7.07	9.24	0.15	3.08	m	12.32
25mm diameter	7.51	9.82	0.15	3.08	m	12.90
32mm diameter	7.85	10.26	0.15	3.08	m	13.34
40mm diameter	8.26	10.80	0.15	3.08	m	13.87
50mm diameter	9.26	12.11	0.15	3.08	m	15.19
65mm diameter	10.12	13.23	0.15	3.08	m	16.30
80mm diameter	10.83	14.17	0.22	4.51	m	18.68
100mm diameter	13.85	18.11	0.22	4.51	m	22.62
125mm diameter	15.54	20.33	0.22	4.51	m	24.84
150mm diameter	17.94	23.46	0.22	4.51	m	27.97

T:MECHANICAL/COOLING/HEATING SYSTEMS

Item	Net Price £	Material £	Labour hours	Labour £	Unit	Total rate £
200mm diameter	24.49	32.03	0.25	5.13	m	**37.16**
250mm diameter	28.29	36.99	0.25	5.13	m	**42.12**
300mm diameter	29.97	39.20	0.25	5.13	m	**44.32**
350mm diameter	33.08	43.26	0.25	5.13	m	**48.39**
400mm diameter	36.69	47.98	0.25	5.13	m	**53.11**
Extra over for fittings concealed insulation						
Flange/union						
15mm diameter	3.36	4.39	0.13	2.67	nr	**7.06**
20mm diameter	3.53	4.62	0.13	2.67	nr	**7.29**
25mm diameter	3.76	4.91	0.13	2.67	nr	**7.58**
32mm diameter	3.93	5.14	0.13	2.67	nr	**7.81**
40mm diameter	4.13	5.40	0.13	2.67	nr	**8.07**
50mm diameter	4.62	6.05	0.13	2.67	nr	**8.71**
65mm diameter	5.06	6.61	0.13	2.67	nr	**9.28**
80mm diameter	5.42	7.08	0.18	3.69	nr	**10.78**
100mm diameter	6.93	9.06	0.18	3.69	nr	**12.75**
125mm diameter	7.77	10.16	0.18	3.69	nr	**13.86**
150mm diameter	8.97	11.74	0.18	3.69	nr	**15.43**
200mm diameter	12.25	16.02	0.22	4.51	nr	**20.53**
250mm diameter	14.14	18.50	0.22	4.51	nr	**23.01**
300mm diameter	14.99	19.60	0.22	4.51	nr	**24.11**
350mm diameter	16.55	21.64	0.22	4.51	nr	**26.15**
400mm diameter	18.35	23.99	0.22	4.51	nr	**28.50**
Valves						
15mm diameter	6.72	8.79	0.15	3.08	nr	**11.86**
20mm diameter	7.07	9.24	0.15	3.08	nr	**12.32**
25mm diameter	7.51	9.82	0.15	3.08	nr	**12.90**
32mm diameter	7.85	10.26	0.15	3.08	nr	**13.34**
40mm diameter	8.26	10.80	0.15	3.08	nr	**13.87**
50mm diameter	9.26	12.11	0.15	3.08	nr	**15.19**
65mm diameter	10.12	13.23	0.15	3.08	nr	**16.30**
80mm diameter	10.83	14.17	0.20	4.10	nr	**18.27**
100mm diameter	13.85	18.11	0.20	4.10	nr	**22.21**
125mm diameter	15.54	20.33	0.20	4.10	nr	**24.43**
150mm diameter	17.94	23.46	0.20	4.10	nr	**27.56**
200mm diameter	24.49	32.03	0.25	5.13	nr	**37.16**
250mm diameter	28.29	36.99	0.25	5.13	nr	**42.12**
300mm diameter	29.97	39.20	0.25	5.13	nr	**44.32**
350mm diameter	33.08	43.26	0.25	5.13	nr	**48.39**
400mm diameter	36.69	47.98	0.25	5.13	nr	**53.11**
Expansion bellows						
15mm diameter	13.42	17.56	0.22	4.51	nr	**22.07**
20mm diameter	14.13	18.48	0.22	4.51	nr	**22.99**
25mm diameter	15.02	19.65	0.22	4.51	nr	**24.16**
32mm diameter	15.71	20.54	0.22	4.51	nr	**25.05**
40mm diameter	16.51	21.59	0.22	4.51	nr	**26.11**
50mm diameter	18.51	24.20	0.22	4.51	nr	**28.72**
65mm diameter	20.23	26.45	0.22	4.51	nr	**30.97**
80mm diameter	21.67	28.33	0.29	5.95	nr	**34.28**
100mm diameter	27.69	36.21	0.29	5.95	nr	**42.16**
125mm diameter	31.08	40.64	0.29	5.95	nr	**46.59**
150mm diameter	35.89	46.93	0.29	5.95	nr	**52.88**
200mm diameter	49.00	64.08	0.36	7.39	nr	**71.46**
250mm diameter	56.56	73.97	0.36	7.39	nr	**81.35**
300mm diameter	59.96	78.41	0.36	7.39	nr	**85.79**
350mm diameter	66.17	86.53	0.36	7.39	nr	**93.91**
400mm diameter	73.37	95.95	0.36	7.39	nr	**103.33**

T:MECHANICAL/COOLING/HEATING SYSTEMS

Item	Net Price £	Material £	Labour hours	Labour £	Unit	Total rate £
T31 : LOW TEMPERATURE HOT WATER HEATING (cont'd)						
Y50 - THERMAL INSULATION (cont'd)						
PLANTROOM PIPEWORK						
Mineral fibre sectional insulation; bright class O foil faced; bright class O foil taped joints; 22 swg plain/embossed aluminium cladding; pop riveted						
20mm thick						
15mm diameter	4.30	5.62	0.44	9.03	m	**14.65**
20mm diameter	4.55	5.95	0.44	9.03	m	**14.97**
25mm diameter	4.87	6.37	0.44	9.03	m	**15.40**
32mm diameter	5.28	6.90	0.44	9.03	m	**15.93**
40mm diameter	5.56	7.27	0.44	9.03	m	**16.29**
50mm diameter	6.19	8.10	0.44	9.03	m	**17.12**
Extra over for fittings plantroom insulation						
Flange/union						
15mm diameter	4.81	6.29	0.58	11.90	nr	**18.19**
20mm diameter	5.10	6.67	0.58	11.90	nr	**18.57**
25mm diameter	5.47	7.16	0.58	11.90	nr	**19.06**
32mm diameter	5.98	7.81	0.58	11.90	nr	**19.71**
40mm diameter	6.31	8.25	0.58	11.90	nr	**20.15**
50mm diameter	7.09	9.27	0.58	11.90	nr	**21.17**
Bends						
15mm diameter	2.36	3.09	0.44	9.03	nr	**12.12**
20mm diameter	2.51	3.28	0.44	9.03	nr	**12.31**
25mm diameter	2.69	3.52	0.44	9.03	nr	**12.54**
32mm diameter	2.90	3.80	0.44	9.03	nr	**12.83**
40mm diameter	3.05	3.99	0.44	9.03	nr	**13.01**
50mm diameter	3.41	4.46	0.44	9.03	nr	**13.48**
Tees						
15mm diameter	1.42	1.85	0.44	9.03	nr	**10.88**
20mm diameter	1.50	1.96	0.44	9.03	nr	**10.99**
25mm diameter	1.61	2.10	0.44	9.03	nr	**11.13**
32mm diameter	1.74	2.28	0.44	9.03	nr	**11.30**
40mm diameter	1.84	2.40	0.44	9.03	nr	**11.43**
50mm diameter	2.04	2.67	0.44	9.03	nr	**11.70**
Valves						
15mm diameter	2.64	3.45	0.78	16.00	nr	**19.46**
20mm diameter	3.02	3.96	0.78	16.00	nr	**19.96**
25mm diameter	3.02	3.96	0.78	16.00	nr	**19.96**
32mm diameter	3.37	4.41	0.78	16.00	nr	**20.41**
40mm diameter	3.61	4.72	0.78	16.00	nr	**20.72**
50mm diameter	4.12	5.38	0.78	16.00	nr	**21.39**
Pumps						
15mm diameter	14.14	18.49	2.34	48.01	nr	**66.50**
20mm diameter	15.00	19.62	2.34	48.01	nr	**67.63**
25mm diameter	16.10	21.06	2.34	48.01	nr	**69.07**
32mm diameter	17.57	22.97	2.34	48.01	nr	**70.98**

T:MECHANICAL/COOLING/HEATING SYSTEMS

Item	Net Price £	Material £	Labour hours	Labour £	Unit	Total rate £
40mm diameter	18.56	24.28	2.34	48.01	nr	**72.29**
50mm diameter	20.87	27.29	2.34	48.01	nr	**75.30**
Expansion bellows						
15mm diameter	11.30	14.78	1.05	21.54	nr	**36.33**
20mm diameter	12.00	15.69	1.05	21.54	nr	**37.24**
25mm diameter	12.89	16.85	1.05	21.54	nr	**38.40**
32mm diameter	14.05	18.38	1.05	21.54	nr	**39.92**
40mm diameter	14.84	19.41	1.05	21.54	nr	**40.95**
50mm diameter	16.69	21.83	1.05	21.54	nr	**43.37**
25mm thick						
15mm diameter	4.72	6.17	0.44	9.03	m	**15.19**
20mm diameter	5.10	6.67	0.44	9.03	m	**15.70**
25mm diameter	5.60	7.33	0.44	9.03	m	**16.36**
32mm diameter	5.96	7.80	0.44	9.03	m	**16.83**
40mm diameter	6.41	8.38	0.44	9.03	m	**17.41**
50mm diameter	7.18	9.38	0.44	9.03	m	**18.41**
65mm diameter	8.11	10.61	0.44	9.03	m	**19.64**
80mm diameter	8.80	11.50	0.52	10.67	m	**22.17**
100mm diameter	10.92	14.28	0.52	10.67	m	**24.95**
125mm diameter	12.68	16.59	0.52	10.67	m	**27.26**
150mm diameter	14.76	19.30	0.52	10.67	m	**29.97**
200mm diameter	20.16	26.36	0.60	12.31	m	**38.67**
250mm diameter	23.95	31.32	0.60	12.31	m	**43.63**
300mm diameter	26.82	35.07	0.60	12.31	m	**47.38**
Extra over for fittings plantroom insulation						
Flange/union						
15mm diameter	5.28	6.90	0.58	11.90	nr	**18.80**
20mm diameter	5.71	7.47	0.58	11.90	nr	**19.37**
25mm diameter	6.30	8.24	0.58	11.90	nr	**20.14**
32mm diameter	6.76	8.83	0.58	11.90	nr	**20.73**
40mm diameter	7.26	9.49	0.58	11.90	nr	**21.39**
50mm diameter	8.18	10.70	0.58	11.90	nr	**22.60**
65mm diameter	9.28	12.13	0.58	11.90	nr	**24.03**
80mm diameter	10.09	13.20	0.67	13.75	nr	**26.94**
100mm diameter	12.77	16.70	0.67	13.75	nr	**30.44**
125mm diameter	14.83	19.40	0.67	13.75	nr	**33.14**
150mm diameter	17.38	22.72	0.67	13.75	nr	**36.47**
200mm diameter	24.01	31.40	0.87	17.85	nr	**49.25**
250mm diameter	28.61	37.41	0.87	17.85	nr	**55.26**
300mm diameter	31.61	41.33	0.87	17.85	nr	**59.18**
Bends						
15mm diameter	2.59	3.39	0.44	9.03	nr	**12.42**
20mm diameter	2.81	3.67	0.44	9.03	nr	**12.70**
25mm diameter	3.08	4.03	0.44	9.03	nr	**13.06**
32mm diameter	3.29	4.30	0.44	9.03	nr	**13.33**
40mm diameter	3.53	4.61	0.44	9.03	nr	**13.64**
50mm diameter	3.95	5.16	0.44	9.03	nr	**14.19**
65mm diameter	4.46	5.84	0.44	9.03	nr	**14.86**
80mm diameter	4.84	6.32	0.52	10.67	nr	**16.99**
100mm diameter	6.00	7.85	0.52	10.67	nr	**18.52**
125mm diameter	6.98	9.13	0.52	10.67	nr	**19.80**
150mm diameter	8.11	10.61	0.52	10.67	nr	**21.28**
200mm diameter	11.09	14.50	0.60	12.31	nr	**26.81**
250mm diameter	13.18	17.23	0.60	12.31	nr	**29.54**
300mm diameter	14.75	19.29	0.60	12.31	nr	**31.60**

T:MECHANICAL/COOLING/HEATING SYSTEMS

Item	Net Price £	Material £	Labour hours	Labour £	Unit	Total rate £
T31 : LOW TEMPERATURE HOT WATER HEATING (cont'd)						
Y50 - THERMAL INSULATION (cont'd)						
Mineral fibre insulation; Plantroom pipework (cont'd)						
25mm thick; Extra over for fittings plantroom Insulation (cont'd)						
Tees						
15mm diameter	1.56	2.04	0.44	9.03	nr	11.07
20mm diameter	1.68	2.20	0.44	9.03	nr	11.22
25mm diameter	1.85	2.42	0.44	9.03	nr	11.44
32mm diameter	1.97	2.57	0.44	9.03	nr	11.60
40mm diameter	2.11	2.76	0.44	9.03	nr	11.79
50mm diameter	2.36	3.09	0.44	9.03	nr	12.12
65mm diameter	2.68	3.50	0.44	9.03	nr	12.53
80mm diameter	2.90	3.80	0.52	10.67	nr	14.47
100mm diameter	3.60	4.71	0.52	10.67	nr	15.38
125mm diameter	4.19	5.48	0.52	10.67	nr	16.15
150mm diameter	4.87	6.37	0.52	10.67	nr	17.04
200mm diameter	6.65	8.69	0.60	12.31	nr	21.00
250mm diameter	7.91	10.34	0.60	12.31	nr	22.65
300mm diameter	8.86	11.58	0.60	12.31	nr	23.89
Valves						
15mm diameter	8.39	10.97	0.78	16.00	nr	26.97
20mm diameter	9.07	11.86	0.78	16.00	nr	27.87
25mm diameter	10.01	13.09	0.78	16.00	nr	29.09
32mm diameter	10.74	14.04	0.78	16.00	nr	30.05
40mm diameter	11.52	15.06	0.78	16.00	nr	31.07
50mm diameter	12.98	16.98	0.78	16.00	nr	32.98
65mm diameter	14.74	19.27	0.78	16.00	nr	35.27
80mm diameter	16.03	20.96	0.92	18.88	nr	39.84
100mm diameter	20.28	26.52	0.92	18.88	nr	45.40
125mm diameter	23.56	30.80	0.92	18.88	nr	49.68
150mm diameter	27.60	36.09	0.92	18.88	nr	54.97
200mm diameter	38.14	49.87	1.12	22.98	nr	72.85
250mm diameter	45.44	59.43	1.12	22.98	nr	82.41
300mm diameter	50.20	65.64	1.12	22.98	nr	88.62
Pumps						
15mm diameter	15.53	20.31	2.34	48.01	nr	68.32
20mm diameter	16.79	21.95	2.34	48.01	nr	69.96
25mm diameter	18.54	24.24	2.34	48.01	nr	72.25
32mm diameter	19.88	26.00	2.34	48.01	nr	74.01
40mm diameter	21.34	27.90	2.34	48.01	nr	75.91
50mm diameter	24.06	31.46	2.34	48.01	nr	79.47
65mm diameter	27.29	35.68	2.34	48.01	nr	83.69
80mm diameter	29.69	38.82	2.76	56.63	nr	95.45
100mm diameter	37.55	49.10	2.76	56.63	nr	105.73
125mm diameter	43.61	57.02	2.76	56.63	nr	113.65
150mm diameter	51.11	66.83	2.76	56.63	nr	123.46
200mm diameter	70.63	92.36	3.36	68.94	nr	161.30
250mm diameter	84.14	110.03	3.36	68.94	nr	178.97
300mm diameter	92.95	121.55	3.36	68.94	nr	190.49

T:MECHANICAL/COOLING/HEATING SYSTEMS

Item	Net Price £	Material £	Labour hours	Labour £	Unit	Total rate £
Expansion bellows						
15mm diameter	12.42	16.24	1.05	21.54	nr	37.78
20mm diameter	13.43	17.56	1.05	21.54	nr	39.10
25mm diameter	14.83	19.40	1.05	21.54	nr	40.94
32mm diameter	15.91	20.81	1.05	21.54	nr	42.35
40mm diameter	17.08	22.33	1.05	21.54	nr	43.87
50mm diameter	19.25	25.17	1.05	21.54	nr	46.71
65mm diameter	21.83	28.54	1.05	21.54	nr	50.09
80mm diameter	23.75	31.05	1.26	25.85	nr	56.91
100mm diameter	30.04	39.28	1.26	25.85	nr	65.13
125mm diameter	34.88	45.62	1.26	25.85	nr	71.47
150mm diameter	40.88	53.46	1.26	25.85	nr	79.31
200mm diameter	56.51	73.89	1.53	31.39	nr	105.29
250mm diameter	67.32	88.03	1.53	31.39	nr	119.42
300mm diameter	74.36	97.24	1.53	31.39	nr	128.64
30mm thick						
15mm diameter	5.82	7.61	0.44	9.03	m	16.64
20mm diameter	6.16	8.05	0.44	9.03	m	17.08
25mm diameter	6.52	8.52	0.44	9.03	m	17.55
32mm diameter	7.14	9.34	0.44	9.03	m	18.36
40mm diameter	7.44	9.73	0.44	9.03	m	18.76
50mm diameter	8.26	10.80	0.44	9.03	m	19.82
65mm diameter	9.31	12.18	0.44	9.03	m	21.20
80mm diameter	10.18	13.31	0.52	10.67	m	23.98
100mm diameter	12.50	16.35	0.52	10.67	m	27.02
125mm diameter	14.29	18.69	0.52	10.67	m	29.36
150mm diameter	16.66	21.78	0.52	10.67	m	32.45
200mm diameter	22.38	29.27	0.60	12.31	m	41.58
250mm diameter	26.48	34.63	0.60	12.31	m	46.94
300mm diameter	29.15	38.12	0.60	12.31	m	50.43
350mm diameter	32.46	42.45	0.60	12.31	m	54.76
Extra over for fittings plantroom insulation						
Flange/union						
15mm diameter	6.62	8.66	0.58	11.90	nr	20.56
20mm diameter	7.03	9.20	0.58	11.90	nr	21.10
25mm diameter	7.45	9.74	0.58	11.90	nr	21.64
32mm diameter	8.16	10.67	0.58	11.90	nr	22.57
40mm diameter	8.53	11.16	0.58	11.90	nr	23.06
50mm diameter	9.56	12.51	0.58	11.90	nr	24.41
65mm diameter	10.80	14.12	0.58	11.90	nr	26.02
80mm diameter	11.80	15.43	0.67	13.75	nr	29.17
100mm diameter	14.74	19.27	0.67	13.75	nr	33.02
125mm diameter	16.87	22.06	0.67	13.75	nr	35.81
150mm diameter	19.73	25.80	0.67	13.75	nr	39.54
200mm diameter	26.87	35.13	0.87	17.85	nr	52.98
250mm diameter	31.86	41.66	0.87	17.85	nr	59.51
300mm diameter	34.66	45.32	0.87	17.85	nr	63.17
350mm diameter	38.45	50.28	0.87	17.85	nr	68.13

T:MECHANICAL/COOLING/HEATING SYSTEMS

Item	Net Price £	Material £	Labour hours	Labour £	Unit	Total rate £
T31 : LOW TEMPERATURE HOT WATER HEATING (cont'd)						
Y50 - THERMAL INSULATION (cont'd)						
Mineral fibre insulation; Plantroom pipework (cont'd)						
30mm thick; Extra over for fittings plantroom insulation (cont'd)						
Bends						
15mm diameter	3.20	4.19	0.44	9.03	nr	**13.22**
20mm diameter	3.38	4.43	0.44	9.03	nr	**13.45**
25mm diameter	3.59	4.69	0.44	9.03	nr	**13.72**
32mm diameter	3.92	5.13	0.44	9.03	nr	**14.16**
40mm diameter	4.09	5.35	0.44	9.03	nr	**14.38**
50mm diameter	4.55	5.95	0.44	9.03	nr	**14.97**
65mm diameter	5.12	6.70	0.44	9.03	nr	**15.73**
80mm diameter	5.59	7.31	0.52	10.67	nr	**17.98**
100mm diameter	6.88	8.99	0.52	10.67	nr	**19.66**
125mm diameter	7.86	10.28	0.52	10.67	nr	**20.95**
150mm diameter	9.17	11.99	0.52	10.67	nr	**22.66**
200mm diameter	12.31	16.10	0.60	12.31	nr	**28.41**
250mm diameter	14.57	19.05	0.60	12.31	nr	**31.36**
300mm diameter	16.03	20.96	0.60	12.31	nr	**33.28**
350mm diameter	17.86	23.35	0.60	12.31	nr	**35.66**
Tees						
15mm diameter	1.92	2.51	0.44	9.03	nr	**11.54**
20mm diameter	2.03	2.65	0.44	9.03	nr	**11.68**
25mm diameter	2.15	2.81	0.44	9.03	nr	**11.84**
32mm diameter	2.35	3.08	0.44	9.03	nr	**12.10**
40mm diameter	2.46	3.22	0.44	9.03	nr	**12.24**
50mm diameter	2.72	3.56	0.44	9.03	nr	**12.59**
65mm diameter	3.07	4.02	0.44	9.03	nr	**13.04**
80mm diameter	3.36	4.39	0.52	10.67	nr	**15.06**
100mm diameter	4.13	5.40	0.52	10.67	nr	**16.07**
125mm diameter	4.72	6.17	0.52	10.67	nr	**16.84**
150mm diameter	5.50	7.19	0.52	10.67	nr	**17.86**
200mm diameter	7.39	9.67	0.60	12.31	nr	**21.98**
250mm diameter	8.74	11.42	0.60	12.31	nr	**23.73**
300mm diameter	9.62	12.59	0.60	12.31	nr	**24.90**
350mm diameter	10.72	14.01	0.60	12.31	nr	**26.32**
Valves						
15mm diameter	10.52	13.76	0.78	16.00	nr	**29.77**
20mm diameter	11.16	14.59	0.78	16.00	nr	**30.60**
25mm diameter	11.83	15.47	0.78	16.00	nr	**31.48**
32mm diameter	12.96	16.95	0.78	16.00	nr	**32.95**
40mm diameter	13.56	17.73	0.78	16.00	nr	**33.74**
50mm diameter	15.18	19.85	0.78	16.00	nr	**35.85**
65mm diameter	17.16	22.44	0.78	16.00	nr	**38.44**
80mm diameter	18.73	24.50	0.92	18.88	nr	**43.37**
100mm diameter	23.40	30.60	0.92	18.88	nr	**49.48**
125mm diameter	26.80	35.04	0.92	18.88	nr	**53.92**
150mm diameter	31.33	40.97	0.92	18.88	nr	**59.85**
200mm diameter	42.67	55.80	1.12	22.98	nr	**78.78**
250mm diameter	50.59	66.16	1.12	22.98	nr	**89.14**
300mm diameter	55.04	71.98	1.12	22.98	nr	**94.96**
350mm diameter	61.07	79.86	1.12	22.98	nr	**102.84**

T:MECHANICAL/COOLING/HEATING SYSTEMS

Item	Net Price £	Material £	Labour hours	Labour £	Unit	Total rate £
Pumps						
15mm diameter	19.49	25.48	2.34	48.01	nr	**73.49**
20mm diameter	20.68	27.04	2.34	48.01	nr	**75.05**
25mm diameter	21.90	28.64	2.34	48.01	nr	**76.65**
32mm diameter	24.00	31.38	2.34	48.01	nr	**79.39**
40mm diameter	25.10	32.83	2.34	48.01	nr	**80.84**
50mm diameter	28.12	36.77	2.34	48.01	nr	**84.78**
65mm diameter	31.76	41.54	2.34	48.01	nr	**89.55**
80mm diameter	34.69	45.37	2.76	56.63	nr	**101.99**
100mm diameter	43.33	56.66	2.76	56.63	nr	**113.29**
125mm diameter	49.63	64.90	2.76	56.63	nr	**121.53**
150mm diameter	58.03	75.89	2.76	56.63	nr	**132.51**
200mm diameter	79.02	103.33	3.36	68.94	nr	**172.27**
250mm diameter	93.70	122.52	3.36	68.94	nr	**191.46**
300mm diameter	101.94	133.30	3.36	68.94	nr	**202.24**
350mm diameter	113.08	147.87	3.36	68.94	nr	**216.80**
Expansion bellows						
15mm diameter	15.59	20.38	1.05	21.54	nr	**41.93**
20mm diameter	16.54	21.62	1.05	21.54	nr	**43.17**
25mm diameter	17.52	22.91	1.05	21.54	nr	**44.45**
32mm diameter	19.20	25.11	1.05	21.54	nr	**46.65**
40mm diameter	20.09	26.27	1.05	21.54	nr	**47.81**
50mm diameter	22.49	29.41	1.05	21.54	nr	**50.95**
65mm diameter	25.42	33.24	1.05	21.54	nr	**54.78**
80mm diameter	27.76	36.30	1.26	25.85	nr	**62.15**
100mm diameter	34.67	45.33	1.26	25.85	nr	**71.19**
125mm diameter	39.71	51.93	1.26	25.85	nr	**77.78**
150mm diameter	46.43	60.71	1.26	25.85	nr	**86.56**
200mm diameter	63.22	82.67	1.53	31.39	nr	**114.06**
250mm diameter	74.95	98.01	1.53	31.39	nr	**129.40**
300mm diameter	81.55	106.64	1.53	31.39	nr	**138.03**
350mm diameter	90.47	118.30	1.53	31.39	nr	**149.69**
40mm thick						
15mm diameter	7.12	9.31	0.44	9.03	m	**18.33**
20mm diameter	7.48	9.78	0.44	9.03	m	**18.80**
25mm diameter	7.96	10.40	0.44	9.03	m	**19.43**
32mm diameter	8.34	10.91	0.44	9.03	m	**19.93**
40mm diameter	8.87	11.60	0.44	9.03	m	**20.62**
50mm diameter	9.82	12.84	0.44	9.03	m	**21.86**
65mm diameter	10.94	14.31	0.44	9.03	m	**23.34**
80mm diameter	11.77	15.39	0.52	10.67	m	**26.06**
100mm diameter	17.24	22.55	0.52	10.67	m	**33.22**
125mm diameter	16.15	21.12	0.52	10.67	m	**31.79**
150mm diameter	18.84	24.64	0.52	10.67	m	**35.31**
200mm diameter	24.90	32.56	0.60	12.31	m	**44.87**
250mm diameter	29.18	38.16	0.60	12.31	m	**50.47**
300mm diameter	32.27	42.20	0.60	12.31	m	**54.51**
350mm diameter	36.06	47.15	0.60	12.31	m	**59.47**
400mm diameter	40.54	53.01	0.60	12.31	m	**65.32**

T:MECHANICAL/COOLING/HEATING SYSTEMS

Item	Net Price £	Material £	Labour hours	Labour £	Unit	Total rate £
T31 : LOW TEMPERATURE HOT WATER HEATING (cont'd)						
Y50 - THERMAL INSULATION (cont'd)						
Mineral fibre insulation; Plantroom pipework (cont'd)						
40mm thick; Extra over for fittings plantroom Insulation (cont'd)						
Flange/union						
15mm diameter	8.22	10.75	0.58	11.90	nr	**22.65**
20mm diameter	8.58	11.22	0.58	11.90	nr	**23.12**
25mm diameter	9.18	12.00	0.58	11.90	nr	**23.90**
32mm diameter	9.66	12.63	0.58	11.90	nr	**24.53**
40mm diameter	10.25	13.40	0.58	11.90	nr	**25.30**
50mm diameter	11.42	14.94	0.58	11.90	nr	**26.84**
65mm diameter	12.76	16.68	0.58	11.90	nr	**28.58**
80mm diameter	13.76	18.00	0.67	13.75	nr	**31.75**
100mm diameter	17.24	22.55	0.67	13.75	nr	**36.30**
125mm diameter	19.30	25.23	0.67	13.75	nr	**38.98**
150mm diameter	22.52	29.45	0.67	13.75	nr	**43.20**
200mm diameter	30.22	39.51	0.87	17.85	nr	**57.36**
250mm diameter	35.42	46.32	0.87	17.85	nr	**64.17**
300mm diameter	38.76	50.69	0.87	17.85	nr	**68.54**
350mm diameter	43.20	56.49	0.87	17.85	nr	**74.34**
400mm diameter	48.44	63.35	0.87	17.85	nr	**81.20**
Bends						
15mm diameter	3.91	5.12	0.44	9.03	nr	**14.14**
20mm diameter	4.10	5.37	0.44	9.03	nr	**14.39**
25mm diameter	4.38	5.73	0.44	9.03	nr	**14.76**
32mm diameter	4.58	5.99	0.44	9.03	nr	**15.02**
40mm diameter	4.88	6.39	0.44	9.03	nr	**15.41**
50mm diameter	5.40	7.06	0.44	9.03	nr	**16.09**
65mm diameter	6.02	7.88	0.44	9.03	nr	**16.91**
80mm diameter	6.47	8.46	0.52	10.67	nr	**19.13**
100mm diameter	7.98	10.44	0.52	10.67	nr	**21.10**
125mm diameter	8.88	11.61	0.52	10.67	nr	**22.28**
150mm diameter	10.36	13.54	0.52	10.67	nr	**24.21**
200mm diameter	13.69	17.90	0.60	12.31	nr	**30.21**
250mm diameter	15.94	20.84	0.60	12.31	nr	**33.15**
300mm diameter	17.75	23.21	0.60	12.31	nr	**35.52**
350mm diameter	19.84	25.94	0.60	12.31	nr	**38.25**
400mm diameter	22.30	29.16	0.60	12.31	nr	**41.47**
Tees						
15mm diameter	2.35	3.08	0.44	9.03	nr	**12.10**
20mm diameter	2.46	3.22	0.44	9.03	nr	**12.24**
25mm diameter	2.63	3.44	0.44	9.03	nr	**12.46**
32mm diameter	2.75	3.59	0.44	9.03	nr	**12.62**
40mm diameter	2.93	3.83	0.44	9.03	nr	**12.86**
50mm diameter	3.24	4.24	0.44	9.03	nr	**13.26**
65mm diameter	3.61	4.72	0.44	9.03	nr	**13.75**
80mm diameter	3.89	5.08	0.52	10.67	nr	**15.75**
100mm diameter	4.79	6.26	0.52	10.67	nr	**16.93**
125mm diameter	5.33	6.97	0.52	10.67	nr	**17.64**
150mm diameter	6.22	8.13	0.52	10.67	nr	**18.80**

T:MECHANICAL/COOLING/HEATING SYSTEMS

Item	Net Price £	Material £	Labour hours	Labour £	Unit	Total rate £
200mm diameter	8.22	10.75	0.60	12.31	nr	**23.06**
250mm diameter	9.64	12.60	0.60	12.31	nr	**24.91**
300mm diameter	10.64	13.92	0.60	12.31	nr	**26.23**
350mm diameter	11.90	15.57	0.60	12.31	nr	**27.88**
400mm diameter	13.38	17.50	0.60	12.31	nr	**29.81**
Valves						
15mm diameter	13.06	17.07	0.78	16.00	nr	**33.08**
20mm diameter	13.63	17.83	0.78	16.00	nr	**33.83**
25mm diameter	14.57	19.05	0.78	16.00	nr	**35.05**
32mm diameter	15.34	20.05	0.78	16.00	nr	**36.06**
40mm diameter	16.27	21.28	0.78	16.00	nr	**37.28**
50mm diameter	18.13	23.71	0.78	16.00	nr	**39.71**
65mm diameter	20.26	26.49	0.78	16.00	nr	**42.49**
80mm diameter	21.86	28.59	0.92	18.88	nr	**47.47**
100mm diameter	27.40	35.82	0.92	18.88	nr	**54.70**
125mm diameter	30.64	40.06	0.92	18.88	nr	**58.94**
150mm diameter	35.77	46.78	0.92	18.88	nr	**65.65**
200mm diameter	47.99	62.75	1.12	22.98	nr	**85.73**
250mm diameter	56.26	73.56	1.12	22.98	nr	**96.54**
300mm diameter	61.56	80.50	1.12	22.98	nr	**103.48**
350mm diameter	68.62	89.73	1.12	22.98	nr	**112.71**
400mm diameter	76.94	100.62	1.12	22.98	nr	**123.60**
Pumps						
15mm diameter	24.17	31.60	2.34	48.01	nr	**79.61**
20mm diameter	25.24	33.00	2.34	48.01	nr	**81.01**
25mm diameter	26.99	35.29	2.34	48.01	nr	**83.30**
32mm diameter	28.40	37.14	2.34	48.01	nr	**85.15**
40mm diameter	30.13	39.40	2.34	48.01	nr	**87.41**
50mm diameter	33.59	43.92	2.34	48.01	nr	**91.93**
65mm diameter	37.51	49.05	2.34	48.01	nr	**97.06**
80mm diameter	40.49	52.94	2.76	56.63	nr	**109.57**
100mm diameter	50.72	66.33	2.76	56.63	nr	**122.96**
125mm diameter	56.74	74.19	2.76	56.63	nr	**130.82**
150mm diameter	66.25	86.64	2.76	56.63	nr	**143.26**
200mm diameter	88.87	116.22	3.36	68.94	nr	**185.15**
250mm diameter	104.18	136.24	3.36	68.94	nr	**205.18**
300mm diameter	113.99	149.06	3.36	68.94	nr	**218.00**
350mm diameter	127.06	166.15	3.36	68.94	nr	**235.09**
400mm diameter	142.49	186.33	3.36	68.94	nr	**255.27**
Expansion bellows						
15mm diameter	19.33	25.28	1.05	21.54	nr	**46.82**
20mm diameter	20.20	26.41	1.05	21.54	nr	**47.95**
25mm diameter	21.59	28.23	1.05	21.54	nr	**49.77**
32mm diameter	22.72	29.71	1.05	21.54	nr	**51.25**
40mm diameter	24.10	31.51	1.05	21.54	nr	**53.05**
50mm diameter	26.87	35.13	1.05	21.54	nr	**56.68**
65mm diameter	30.01	39.25	1.05	21.54	nr	**60.79**
80mm diameter	32.39	42.35	1.26	25.85	nr	**68.20**
100mm diameter	40.58	53.07	1.26	25.85	nr	**78.92**
125mm diameter	45.38	59.35	1.26	25.85	nr	**85.20**
150mm diameter	52.99	69.30	1.26	25.85	nr	**95.15**
200mm diameter	71.09	92.96	1.53	31.39	nr	**124.35**
250mm diameter	83.35	109.00	1.53	31.39	nr	**140.39**
300mm diameter	91.19	119.24	1.53	31.39	nr	**150.64**
350mm diameter	101.64	132.91	1.53	31.39	nr	**164.30**
400mm diameter	113.99	149.06	1.53	31.39	nr	**180.45**

T:MECHANICAL/COOLING/HEATING SYSTEMS

Item	Net Price £	Material £	Labour hours	Labour £	Unit	Total rate £
T31 : LOW TEMPERATURE HOT WATER HEATING (cont'd)						
Y50 - THERMAL INSULATION (cont'd)						
Mineral fibre insulation; Plantroom pipework (cont'd)						
50mm thick						
15mm diameter	9.29	12.15	0.44	9.03	m	**21.17**
20mm diameter	9.79	12.80	0.44	9.03	m	**21.83**
25mm diameter	10.37	13.56	0.44	9.03	m	**22.59**
32mm diameter	10.90	14.25	0.44	9.03	m	**23.28**
40mm diameter	11.40	14.91	0.44	9.03	m	**23.94**
50mm diameter	12.58	16.45	0.44	9.03	m	**25.47**
65mm diameter	13.52	17.68	0.44	9.03	m	**26.71**
80mm diameter	14.52	18.99	0.52	10.67	m	**29.66**
100mm diameter	17.94	23.46	0.52	10.67	m	**34.13**
125mm diameter	19.86	25.97	0.52	10.67	m	**36.64**
150mm diameter	22.81	29.83	0.52	10.67	m	**40.50**
200mm diameter	29.90	39.10	0.60	12.31	m	**51.42**
250mm diameter	34.56	45.19	0.60	12.31	m	**57.50**
300mm diameter	37.62	49.19	0.60	12.31	m	**61.50**
350mm diameter	41.93	54.83	0.60	12.31	m	**67.14**
400mm diameter	46.90	61.32	0.60	12.31	m	**73.63**
Extra over for fittings plantroom insulation						
Flange/union						
15mm diameter	10.94	14.31	0.58	11.90	nr	**26.21**
20mm diameter	11.54	15.10	0.58	11.90	nr	**27.00**
25mm diameter	12.23	15.99	0.58	11.90	nr	**27.89**
32mm diameter	12.84	16.79	0.58	11.90	nr	**28.69**
40mm diameter	13.45	17.59	0.58	11.90	nr	**29.49**
50mm diameter	14.93	19.52	0.58	11.90	nr	**31.42**
65mm diameter	16.13	21.09	0.58	11.90	nr	**32.99**
80mm diameter	17.29	22.61	0.67	13.75	nr	**36.36**
100mm diameter	21.62	28.28	0.67	13.75	nr	**42.02**
125mm diameter	24.05	31.45	0.67	13.75	nr	**45.19**
150mm diameter	27.67	36.19	0.67	13.75	nr	**49.93**
200mm diameter	36.78	48.10	0.87	17.85	nr	**65.95**
250mm diameter	42.50	55.58	0.87	17.85	nr	**73.43**
300mm diameter	45.86	59.98	0.87	17.85	nr	**77.83**
350mm diameter	50.99	66.68	0.87	17.85	nr	**84.53**
400mm diameter	56.89	74.40	0.87	17.85	nr	**92.25**
Bend						
15mm diameter	5.11	6.68	0.44	9.03	nr	**15.71**
20mm diameter	5.39	7.05	0.44	9.03	nr	**16.07**
25mm diameter	5.70	7.45	0.44	9.03	nr	**16.48**
32mm diameter	6.00	7.85	0.44	9.03	nr	**16.87**
40mm diameter	6.26	8.19	0.44	9.03	nr	**17.22**
50mm diameter	6.92	9.05	0.44	9.03	nr	**18.08**
65mm diameter	7.44	9.73	0.44	9.03	nr	**18.76**
80mm diameter	7.98	10.44	0.52	10.67	nr	**21.10**
100mm diameter	9.86	12.90	0.52	10.67	nr	**23.57**
125mm diameter	10.92	14.28	0.52	10.67	nr	**24.95**
150mm diameter	12.55	16.41	0.52	10.67	nr	**27.08**
200mm diameter	16.45	21.51	0.60	12.31	nr	**33.82**

T:MECHANICAL/COOLING/HEATING SYSTEMS

Item	Net Price £	Material £	Labour hours	Labour £	Unit	Total rate £
250mm diameter	19.01	24.86	0.60	12.31	nr	**37.17**
300mm diameter	20.69	27.05	0.60	12.31	nr	**39.36**
350mm diameter	23.06	30.16	0.60	12.31	nr	**42.47**
400mm diameter	25.79	33.72	0.60	12.31	nr	**46.03**
Tee						
15mm diameter	3.07	4.02	0.44	9.03	nr	**13.04**
20mm diameter	3.23	4.22	0.44	9.03	nr	**13.25**
25mm diameter	3.42	4.47	0.44	9.03	nr	**13.50**
32mm diameter	3.60	4.71	0.44	9.03	nr	**13.74**
40mm diameter	3.76	4.91	0.44	9.03	nr	**13.94**
50mm diameter	4.15	5.43	0.44	9.03	nr	**14.46**
65mm diameter	4.46	5.84	0.44	9.03	nr	**14.86**
80mm diameter	4.79	6.26	0.52	10.67	nr	**16.93**
100mm diameter	5.92	7.74	0.52	10.67	nr	**18.41**
125mm diameter	6.55	8.57	0.52	10.67	nr	**19.24**
150mm diameter	7.52	9.84	0.52	10.67	nr	**20.51**
200mm diameter	9.86	12.90	0.60	12.31	nr	**25.21**
250mm diameter	11.41	14.92	0.60	12.31	nr	**27.23**
300mm diameter	12.42	16.24	0.60	12.31	nr	**28.55**
350mm diameter	13.84	18.09	0.60	12.31	nr	**30.40**
400mm diameter	15.48	20.24	0.60	12.31	nr	**32.55**
Valves						
15mm diameter	17.39	22.74	0.78	16.00	nr	**38.74**
20mm diameter	18.32	23.96	0.78	16.00	nr	**39.97**
25mm diameter	19.42	25.39	0.78	16.00	nr	**41.39**
32mm diameter	20.39	26.66	0.78	16.00	nr	**42.66**
40mm diameter	21.36	27.93	0.78	16.00	nr	**43.94**
50mm diameter	23.70	30.99	0.78	16.00	nr	**47.00**
65mm diameter	25.61	33.49	0.78	16.00	nr	**49.49**
80mm diameter	27.47	35.92	0.92	18.88	nr	**54.80**
100mm diameter	34.34	44.91	0.92	18.88	nr	**63.79**
125mm diameter	38.18	49.93	0.92	18.88	nr	**68.81**
150mm diameter	43.94	57.46	0.92	18.88	nr	**76.34**
200mm diameter	58.43	76.40	1.12	22.98	nr	**99.38**
250mm diameter	67.51	88.28	1.12	22.98	nr	**111.26**
300mm diameter	72.85	95.27	1.12	22.98	nr	**118.25**
350mm diameter	80.98	105.89	1.12	22.98	nr	**128.87**
400mm diameter	90.35	118.15	1.12	22.98	nr	**141.12**
Pumps						
15mm diameter	32.20	42.10	2.34	48.01	nr	**90.11**
20mm diameter	33.94	44.38	2.34	48.01	nr	**92.39**
25mm diameter	35.95	47.01	2.34	48.01	nr	**95.02**
32mm diameter	37.76	49.38	2.34	48.01	nr	**97.39**
40mm diameter	39.56	51.74	2.34	48.01	nr	**99.75**
50mm diameter	43.88	57.39	2.34	48.01	nr	**105.40**
65mm diameter	35.42	46.32	2.34	48.01	nr	**94.33**
80mm diameter	50.88	66.53	2.76	56.63	nr	**123.16**
100mm diameter	63.59	83.15	2.76	56.63	nr	**139.78**
125mm diameter	70.72	92.47	2.76	56.63	nr	**149.10**
150mm diameter	81.38	106.42	2.76	56.63	nr	**163.05**
200mm diameter	108.19	141.48	3.36	68.94	nr	**210.42**
250mm diameter	125.02	163.48	3.36	68.94	nr	**232.42**
300mm diameter	134.90	176.41	3.36	68.94	nr	**245.35**
350mm diameter	149.96	196.10	3.36	68.94	nr	**265.04**
400mm diameter	167.32	218.79	3.36	68.94	nr	**287.73**

T:MECHANICAL/COOLING/HEATING SYSTEMS

Item	Net Price £	Material £	Labour hours	Labour £	Unit	Total rate £
T31 : LOW TEMPERATURE HOT WATER HEATING (cont'd)						
Y50 - THERMAL INSULATION (cont'd)						
Mineral fibre insulation; Plantroom pipework (cont'd)						
50mm thick; Extra over for fittings plantroom insulation (cont'd)						
Expansion bellows						
15mm diameter	25.75	33.68	1.05	21.54	nr	**55.22**
20mm diameter	27.14	35.50	1.05	21.54	nr	**57.04**
25mm diameter	28.76	37.61	1.05	21.54	nr	**59.16**
32mm diameter	30.20	39.50	1.05	21.54	nr	**61.04**
40mm diameter	31.64	41.38	1.05	21.54	nr	**62.92**
50mm diameter	35.11	45.91	1.05	21.54	nr	**67.46**
65mm diameter	37.93	49.60	1.05	21.54	nr	**71.15**
80mm diameter	40.70	53.23	1.26	25.85	nr	**79.08**
100mm diameter	50.87	66.52	1.26	25.85	nr	**92.37**
125mm diameter	56.57	73.97	1.26	25.85	nr	**99.82**
150mm diameter	65.11	85.15	1.26	25.85	nr	**111.00**
200mm diameter	86.56	113.19	1.53	31.39	nr	**144.58**
250mm diameter	100.01	130.78	1.53	31.39	nr	**162.17**
300mm diameter	107.93	141.13	1.53	31.39	nr	**172.53**
350mm diameter	119.96	156.87	1.53	31.39	nr	**188.26**
400mm diameter	133.85	175.03	1.53	31.39	nr	**206.42**
Mineral fibre sectional insulation; bright class O foil faced; bright class O foil taped joints; 0.8mm polyisobutylene sheeting; welded joints						
External pipework						
20mm thick						
15mm diameter	3.96	5.18	0.30	6.16	m	**11.33**
20mm diameter	4.22	5.52	0.30	6.16	m	**11.68**
25mm diameter	4.55	5.95	0.30	6.16	m	**12.10**
32mm diameter	5.00	6.54	0.30	6.16	m	**12.70**
40mm diameter	5.33	6.97	0.30	6.16	m	**13.12**
50mm diameter	6.02	7.88	0.30	6.16	m	**14.03**
Extra over for fittings external insulation						
Flange/union						
15mm diameter	6.07	7.94	0.75	15.39	nr	**23.33**
20mm diameter	6.44	8.43	0.75	15.39	nr	**23.81**
25mm diameter	6.92	9.05	0.75	15.39	nr	**24.44**
32mm diameter	7.54	9.85	0.75	15.39	nr	**25.24**
40mm diameter	7.96	10.40	0.75	15.39	nr	**25.79**
50mm diameter	8.92	11.66	0.75	15.39	nr	**27.05**
Bends						
15mm diameter	1.00	1.30	0.30	6.16	nr	**7.46**
20mm diameter	1.06	1.38	0.30	6.16	nr	**7.54**
25mm diameter	1.14	1.49	0.30	6.16	nr	**7.65**
32mm diameter	1.25	1.63	0.30	6.16	nr	**7.79**
40mm diameter	1.33	1.74	0.30	6.16	nr	**7.90**
50mm diameter	1.50	1.96	0.30	6.16	nr	**8.12**

T:MECHANICAL/COOLING/HEATING SYSTEMS

Item	Net Price £	Material £	Labour hours	Labour £	Unit	Total rate £
Tees						
15mm diameter	1.00	1.30	0.30	6.16	nr	**7.46**
20mm diameter	1.06	1.38	0.30	6.16	nr	**7.54**
25mm diameter	1.14	1.49	0.30	6.16	nr	**7.65**
32mm diameter	1.25	1.63	0.30	6.16	nr	**7.79**
40mm diameter	1.33	1.74	0.30	6.16	nr	**7.90**
50mm diameter	1.50	1.96	0.30	6.16	nr	**8.12**
Valves						
15mm diameter	9.64	12.60	1.03	21.13	nr	**33.73**
20mm diameter	10.24	13.39	1.03	21.13	nr	**34.52**
25mm diameter	10.99	14.37	1.03	21.13	nr	**35.51**
32mm diameter	11.96	15.65	1.03	21.13	nr	**36.78**
40mm diameter	12.64	16.52	1.03	21.13	nr	**37.66**
50mm diameter	14.16	18.52	1.03	21.13	nr	**39.65**
Expansion bellows						
15mm diameter	14.28	18.67	1.42	29.13	nr	**47.81**
20mm diameter	15.17	19.83	1.42	29.13	nr	**48.97**
25mm diameter	16.30	21.31	1.42	29.13	nr	**50.44**
32mm diameter	17.74	23.19	1.42	29.13	nr	**52.33**
40mm diameter	18.72	24.48	1.42	29.13	nr	**53.61**
50mm diameter	20.98	27.43	1.42	29.13	nr	**56.56**
25mm thick						
15mm diameter	4.37	5.71	0.30	6.16	m	**11.87**
20mm diameter	4.69	6.14	0.30	6.16	m	**12.29**
25mm diameter	5.16	6.75	0.30	6.16	m	**12.90**
32mm diameter	5.59	7.31	0.30	6.16	m	**13.47**
40mm diameter	5.94	7.77	0.30	6.16	m	**13.92**
50mm diameter	6.71	8.77	0.30	6.16	m	**14.93**
65mm diameter	7.61	9.95	0.30	6.16	m	**16.10**
80mm diameter	8.32	10.87	0.40	8.21	m	**19.08**
100mm diameter	10.51	13.75	0.40	8.21	m	**21.95**
125mm diameter	12.10	15.82	0.40	8.21	m	**24.02**
150mm diameter	14.20	18.56	0.40	8.21	m	**26.77**
200mm diameter	19.26	25.19	0.50	10.26	m	**35.44**
250mm diameter	22.99	30.07	0.50	10.26	m	**40.32**
300mm diameter	25.01	32.70	0.50	10.26	m	**42.96**
Extra over for fittings external insulation						
Flange/union						
15mm diameter	6.68	8.74	0.75	15.39	nr	**24.13**
20mm diameter	7.20	9.42	0.75	15.39	nr	**24.80**
25mm diameter	7.90	10.33	0.75	15.39	nr	**25.71**
32mm diameter	8.46	11.06	0.75	15.39	nr	**26.45**
40mm diameter	9.05	11.83	0.75	15.39	nr	**27.22**
50mm diameter	10.14	13.26	0.75	15.39	nr	**28.65**
65mm diameter	11.47	15.00	0.75	15.39	nr	**30.39**
80mm diameter	12.48	16.32	0.89	18.26	nr	**34.58**
100mm diameter	15.50	20.27	0.89	18.26	nr	**38.53**
125mm diameter	17.95	23.48	0.89	18.26	nr	**41.74**
150mm diameter	20.89	27.32	0.89	18.26	nr	**45.58**
200mm diameter	28.27	36.97	1.15	23.59	nr	**60.57**
250mm diameter	33.65	44.00	1.15	23.59	nr	**67.60**
300mm diameter	37.38	48.88	1.15	23.59	nr	**72.48**

T:MECHANICAL/COOLING/HEATING SYSTEMS

Item	Net Price £	Material £	Labour hours	Labour £	Unit	Total rate £
T31 : LOW TEMPERATURE HOT WATER HEATING (cont'd)						
Y50 - THERMAL INSULATION (cont'd)						
Mineral fibre PIB clad; External pipework (cont'd)						
25mm thick; Extra over for fittings external Insulation (cont'd)						
Bends						
15mm diameter	1.09	1.43	0.30	6.16	nr	7.58
20mm diameter	1.18	1.54	0.30	6.16	nr	7.69
25mm diameter	1.28	1.68	0.30	6.16	nr	7.83
32mm diameter	1.39	1.82	0.30	6.16	nr	7.98
40mm diameter	1.49	1.95	0.30	6.16	nr	8.10
50mm diameter	1.68	2.20	0.30	6.16	nr	8.35
65mm diameter	1.91	2.50	0.30	6.16	nr	8.65
80mm diameter	2.08	2.71	0.40	8.21	nr	10.92
100mm diameter	2.63	3.44	0.40	8.21	nr	11.64
125mm diameter	3.02	3.95	0.40	8.21	nr	12.16
150mm diameter	3.55	4.64	0.40	8.21	nr	12.85
200mm diameter	4.81	6.29	0.50	10.26	nr	16.55
250mm diameter	5.75	7.52	0.50	10.26	nr	17.78
300mm diameter	6.25	8.18	0.50	10.26	nr	18.43
Tees						
15mm diameter	1.09	1.43	0.30	6.16	nr	7.58
20mm diameter	1.18	1.54	0.30	6.16	nr	7.69
25mm diameter	1.28	1.68	0.30	6.16	nr	7.83
32mm diameter	1.39	1.82	0.30	6.16	nr	7.98
40mm diameter	1.49	1.95	0.30	6.16	nr	8.10
50mm diameter	1.68	2.20	0.30	6.16	nr	8.35
65mm diameter	1.91	2.50	0.30	6.16	nr	8.65
80mm diameter	2.08	2.71	0.40	8.21	nr	10.92
100mm diameter	2.63	3.44	0.40	8.21	nr	11.64
125mm diameter	3.02	3.95	0.40	8.21	nr	12.16
150mm diameter	3.55	4.64	0.40	8.21	nr	12.85
200mm diameter	4.81	6.29	0.50	10.26	nr	16.55
250mm diameter	5.75	7.52	0.50	10.26	nr	17.78
300mm diameter	6.25	8.18	0.50	10.26	nr	18.43
Valves						
15mm diameter	10.61	13.87	1.03	21.13	nr	35.00
20mm diameter	11.44	14.95	1.03	21.13	nr	36.09
25mm diameter	12.54	16.40	1.03	21.13	nr	37.53
32mm diameter	13.44	17.58	1.03	21.13	nr	38.71
40mm diameter	14.36	18.78	1.03	21.13	nr	39.92
50mm diameter	16.12	21.07	1.03	21.13	nr	42.21
65mm diameter	18.23	23.84	1.03	21.13	nr	44.97
80mm diameter	19.81	25.91	1.25	25.65	nr	51.55
100mm diameter	24.64	32.22	1.25	25.65	nr	57.86
125mm diameter	28.50	37.27	1.25	25.65	nr	62.92
150mm diameter	33.19	43.40	1.25	25.65	nr	69.05
200mm diameter	44.90	58.72	1.55	31.80	nr	90.52
250mm diameter	53.44	69.88	1.55	31.80	nr	101.68
300mm diameter	59.36	77.63	1.55	31.80	nr	109.43

T:MECHANICAL/COOLING/HEATING SYSTEMS

Item	Net Price £	Material £	Labour hours	Labour £	Unit	Total rate £
Expansion bellows						
15mm diameter	15.72	20.56	1.42	29.13	nr	49.69
20mm diameter	16.94	22.16	1.42	29.13	nr	51.29
25mm diameter	18.58	24.29	1.42	29.13	nr	53.43
32mm diameter	19.92	26.05	1.42	29.13	nr	55.18
40mm diameter	21.29	27.84	1.42	29.13	nr	56.97
50mm diameter	23.87	31.21	1.42	29.13	nr	60.35
65mm diameter	27.00	35.31	1.42	29.13	nr	64.44
80mm diameter	29.35	38.38	1.75	35.91	nr	74.29
100mm diameter	36.49	47.72	1.75	35.91	nr	83.62
125mm diameter	42.23	55.22	1.75	35.91	nr	91.13
150mm diameter	49.16	64.29	1.75	35.91	nr	100.20
200mm diameter	66.52	86.98	2.17	44.52	nr	131.50
250mm diameter	79.16	103.52	2.17	44.52	nr	148.04
300mm diameter	87.94	114.99	3.17	65.04	nr	180.03
30mm thick						
15mm diameter	5.38	7.03	0.30	6.16	m	13.19
20mm diameter	5.72	7.49	0.30	6.16	m	13.64
25mm diameter	6.07	7.94	0.30	6.16	m	14.10
32mm diameter	6.58	8.60	0.30	6.16	m	14.75
40mm diameter	6.94	9.07	0.30	6.16	m	15.23
50mm diameter	7.81	10.22	0.30	6.16	m	16.37
65mm diameter	8.78	11.49	0.30	6.16	m	17.64
80mm diameter	9.55	12.49	0.40	8.21	m	20.70
100mm diameter	11.94	15.61	0.40	8.21	m	23.82
125mm diameter	13.67	17.87	0.40	8.21	m	26.08
150mm diameter	15.88	20.76	0.40	8.21	m	28.97
200mm diameter	21.38	27.96	0.50	10.26	m	38.22
250mm diameter	25.37	33.17	0.50	10.26	m	43.43
300mm diameter	27.36	35.78	0.50	10.26	m	46.04
350mm diameter	29.92	39.12	0.50	10.26	m	49.38
Extra over for fittings external insulation						
Flange/union						
15mm diameter	8.17	10.69	0.75	15.39	nr	26.07
20mm diameter	8.66	11.33	0.75	15.39	nr	26.72
25mm diameter	9.18	12.00	0.75	15.39	nr	27.39
32mm diameter	10.01	13.09	0.75	15.39	nr	28.48
40mm diameter	10.46	13.68	0.75	15.39	nr	29.07
50mm diameter	11.68	15.27	0.75	15.39	nr	30.66
65mm diameter	13.14	17.18	0.75	15.39	nr	32.57
80mm diameter	14.32	18.72	0.89	18.26	nr	36.98
100mm diameter	17.63	23.05	0.89	18.26	nr	41.31
125mm diameter	20.14	26.33	0.89	18.26	nr	44.59
150mm diameter	23.40	30.60	0.89	18.26	nr	48.86
200mm diameter	31.27	40.89	1.15	23.59	nr	64.49
250mm diameter	37.03	48.43	1.15	23.59	nr	72.02
300mm diameter	40.57	53.05	1.15	23.59	nr	76.65
350mm diameter	44.82	58.61	1.15	23.59	nr	82.20

T:MECHANICAL/COOLING/HEATING SYSTEMS

Item	Net Price £	Material £	Labour hours	Labour £	Unit	Total rate £
T31 : LOW TEMPERATURE HOT WATER HEATING (cont'd)						
Y50 - THERMAL INSULATION (cont'd)						
Mineral fibre PIB clad; External pipework (cont'd)						
30mm thick; Extra over for fittings external Insulation (cont'd)						
Bends						
15mm diameter	1.34	1.76	0.30	6.16	nr	7.91
20mm diameter	1.43	1.87	0.30	6.16	nr	8.02
25mm diameter	1.52	1.99	0.30	6.16	nr	8.15
32mm diameter	1.64	2.15	0.30	6.16	nr	8.31
40mm diameter	1.73	2.26	0.30	6.16	nr	8.41
50mm diameter	1.96	2.56	0.30	6.16	nr	8.71
65mm diameter	2.20	2.87	0.30	6.16	nr	9.03
80mm diameter	2.39	3.12	0.40	8.21	nr	11.33
100mm diameter	2.99	3.91	0.40	8.21	nr	12.11
125mm diameter	3.42	4.47	0.40	8.21	nr	12.68
150mm diameter	3.97	5.19	0.40	8.21	nr	13.40
200mm diameter	5.35	7.00	0.50	10.26	nr	17.26
250mm diameter	6.35	8.30	0.50	10.26	nr	18.56
300mm diameter	6.84	8.94	0.50	10.26	nr	19.20
350mm diameter	7.48	9.78	0.50	10.26	nr	20.03
Tees						
15mm diameter	1.34	1.76	0.30	6.16	nr	7.91
20mm diameter	1.43	1.87	0.30	6.16	nr	8.02
25mm diameter	1.52	1.99	0.30	6.16	nr	8.15
32mm diameter	1.64	2.15	0.30	6.16	nr	8.31
40mm diameter	1.73	2.26	0.30	6.16	nr	8.41
50mm diameter	1.96	2.56	0.30	6.16	nr	8.71
65mm diameter	2.20	2.87	0.30	6.16	nr	9.03
80mm diameter	2.39	3.12	0.40	8.21	nr	11.33
100mm diameter	2.99	3.91	0.40	8.21	nr	12.11
125mm diameter	3.42	4.47	0.40	8.21	nr	12.68
150mm diameter	3.97	5.19	0.40	8.21	nr	13.40
200mm diameter	5.35	7.00	0.50	10.26	nr	17.26
250mm diameter	6.35	8.30	0.50	10.26	nr	18.56
300mm diameter	6.84	8.94	0.50	10.26	nr	19.20
350mm diameter	7.48	9.78	0.50	10.26	nr	20.03
Valves						
15mm diameter	12.98	16.98	1.03	21.13	nr	38.11
20mm diameter	13.75	17.98	1.03	21.13	nr	39.12
25mm diameter	14.58	19.07	1.03	21.13	nr	40.20
32mm diameter	15.89	20.78	1.03	21.13	nr	41.91
40mm diameter	16.63	21.75	1.03	21.13	nr	42.88
50mm diameter	18.54	24.24	1.03	21.13	nr	45.38
65mm diameter	20.87	27.29	1.03	21.13	nr	48.42
80mm diameter	22.74	29.74	1.25	25.65	nr	55.38
100mm diameter	27.98	36.59	1.25	25.65	nr	62.24
125mm diameter	31.98	41.82	1.25	25.65	nr	67.47
150mm diameter	37.15	48.58	1.25	25.65	nr	74.23
200mm diameter	49.67	64.95	1.55	31.80	nr	96.75
250mm diameter	58.81	76.91	1.55	31.80	nr	108.71
300mm diameter	64.44	84.27	1.55	31.80	nr	116.07
350mm diameter	71.18	93.09	1.55	31.80	nr	124.89

T:MECHANICAL/COOLING/HEATING SYSTEMS

Item	Net Price £	Material £	Labour hours	Labour £	Unit	Total rate £
Expansion bellows						
15mm diameter	19.22	25.14	1.42	29.13	nr	**54.27**
20mm diameter	20.38	26.65	1.42	29.13	nr	**55.78**
25mm diameter	21.60	28.25	1.42	29.13	nr	**57.38**
32mm diameter	23.54	30.79	1.42	29.13	nr	**59.92**
40mm diameter	24.64	32.22	1.42	29.13	nr	**61.35**
50mm diameter	27.46	35.90	1.42	29.13	nr	**65.04**
65mm diameter	30.92	40.44	1.42	29.13	nr	**69.57**
80mm diameter	33.70	44.06	1.75	35.91	nr	**79.97**
100mm diameter	41.46	54.22	1.75	35.91	nr	**90.12**
125mm diameter	47.38	61.95	1.75	35.91	nr	**97.86**
150mm diameter	55.04	71.98	1.75	35.91	nr	**107.88**
200mm diameter	73.58	96.22	2.17	44.52	nr	**140.75**
250mm diameter	87.13	113.94	2.17	44.52	nr	**158.46**
300mm diameter	95.46	124.83	2.17	44.52	nr	**169.35**
350mm diameter	105.46	137.90	2.17	44.52	nr	**182.42**
40mm thick						
15mm diameter	6.73	8.80	0.30	6.16	m	**14.96**
20mm diameter	6.96	9.10	0.30	6.16	m	**15.26**
25mm diameter	7.44	9.73	0.30	6.16	m	**15.88**
32mm diameter	7.91	10.34	0.30	6.16	m	**16.50**
40mm diameter	8.29	10.84	0.30	6.16	m	**17.00**
50mm diameter	9.24	12.08	0.30	6.16	m	**18.24**
65mm diameter	10.30	13.46	0.30	6.16	m	**19.62**
80mm diameter	11.16	14.59	0.40	8.21	m	**22.80**
100mm diameter	13.94	18.23	0.40	8.21	m	**26.44**
125mm diameter	15.71	20.54	0.40	8.21	m	**28.75**
150mm diameter	18.13	23.71	0.40	8.21	m	**31.92**
200mm diameter	24.19	31.64	0.50	10.26	m	**41.89**
250mm diameter	28.32	37.03	0.50	10.26	m	**47.29**
300mm diameter	30.67	40.11	0.50	10.26	m	**50.37**
350mm diameter	33.70	44.06	0.50	10.26	m	**54.32**
400mm diameter	37.51	49.05	0.50	10.26	m	**59.31**
Extra over for fittings external insulation						
Flange/union						
15mm diameter	10.06	13.15	0.75	15.39	nr	**28.54**
20mm diameter	10.50	13.73	0.75	15.39	nr	**29.12**
25mm diameter	11.20	14.64	0.75	15.39	nr	**30.03**
32mm diameter	11.80	15.43	0.75	15.39	nr	**30.81**
40mm diameter	12.47	16.30	0.75	15.39	nr	**31.69**
50mm diameter	13.81	18.06	0.75	15.39	nr	**33.45**
65mm diameter	15.38	20.12	0.75	15.39	nr	**35.51**
80mm diameter	16.58	21.69	0.89	18.26	nr	**39.95**
100mm diameter	20.42	26.71	0.89	18.26	nr	**44.97**
125mm diameter	22.84	29.86	0.89	18.26	nr	**48.12**
150mm diameter	26.47	34.62	0.89	18.26	nr	**52.88**
200mm diameter	34.90	45.63	1.15	23.59	nr	**69.23**
250mm diameter	40.90	53.48	1.15	23.59	nr	**77.07**
300mm diameter	44.95	58.78	1.15	23.59	nr	**82.38**
350mm diameter	49.86	65.20	1.15	23.59	nr	**88.80**
400mm diameter	55.84	73.02	1.15	23.59	nr	**96.61**

T:MECHANICAL/COOLING/HEATING SYSTEMS

Item	Net Price £	Material £	Labour hours	Labour £	Unit	Total rate £
T31 : LOW TEMPERATURE HOT WATER HEATING (cont'd)						
Y50 - THERMAL INSULATION (cont'd)						
Mineral fibre PIB clad; External pipework (cont'd)						
40mm thick; Extra over for fittings external Insulation (cont'd)						
Bends						
15mm diameter	1.68	2.20	0.30	6.16	nr	8.35
20mm diameter	1.74	2.28	0.30	6.16	nr	8.43
25mm diameter	1.86	2.43	0.30	6.16	nr	8.59
32mm diameter	1.98	2.59	0.30	6.16	nr	8.74
40mm diameter	2.08	2.71	0.30	6.16	nr	8.87
50mm diameter	2.32	3.03	0.30	6.16	nr	9.18
65mm diameter	2.58	3.37	0.30	6.16	nr	9.53
80mm diameter	2.80	3.66	0.40	8.21	nr	11.86
100mm diameter	3.49	4.57	0.40	8.21	nr	12.77
125mm diameter	3.92	5.13	0.40	8.21	nr	13.34
150mm diameter	4.54	5.93	0.40	8.21	nr	14.14
200mm diameter	6.05	7.91	0.50	10.26	nr	18.17
250mm diameter	7.08	9.26	0.50	10.26	nr	19.52
300mm diameter	7.67	10.03	0.50	10.26	nr	20.29
350mm diameter	8.42	11.02	0.50	10.26	nr	21.27
400mm diameter	9.38	12.27	0.50	10.26	nr	22.53
Tees						
15mm diameter	1.68	2.20	0.30	6.16	nr	8.35
20mm diameter	1.74	2.28	0.30	6.16	nr	8.43
25mm diameter	1.86	2.43	0.30	6.16	nr	8.59
32mm diameter	1.98	2.59	0.30	6.16	nr	8.74
40mm diameter	2.08	2.71	0.30	6.16	nr	8.87
50mm diameter	2.32	3.03	0.30	6.16	nr	9.18
65mm diameter	2.58	3.37	0.30	6.16	nr	9.53
80mm diameter	2.80	3.66	0.40	8.21	nr	11.86
100mm diameter	3.49	4.57	0.40	8.21	nr	12.77
125mm diameter	3.92	5.13	0.40	8.21	nr	13.34
150mm diameter	4.54	5.93	0.40	8.21	nr	14.14
200mm diameter	6.05	7.91	0.50	10.26	nr	18.17
250mm diameter	7.08	9.26	0.50	10.26	nr	19.52
300mm diameter	7.67	10.03	0.50	10.26	nr	20.29
350mm diameter	8.42	11.02	0.50	10.26	nr	21.27
400mm diameter	9.38	12.27	0.50	10.26	nr	22.53
Valves						
15mm diameter	15.96	20.87	1.03	21.13	nr	42.00
20mm diameter	16.68	21.81	1.03	21.13	nr	42.94
25mm diameter	17.78	23.26	1.03	21.13	nr	44.39
32mm diameter	18.73	24.50	1.03	21.13	nr	45.63
40mm diameter	19.80	25.89	1.03	21.13	nr	47.02
50mm diameter	21.94	28.69	1.03	21.13	nr	49.82
65mm diameter	24.43	31.95	1.03	21.13	nr	53.08
80mm diameter	26.33	34.43	1.25	25.65	nr	60.07
100mm diameter	32.44	42.42	1.25	25.65	nr	68.06
125mm diameter	36.28	47.44	1.25	25.65	nr	73.08
150mm diameter	42.05	54.98	1.25	25.65	nr	80.63
200mm diameter	55.43	72.48	1.55	31.80	nr	104.28
250mm diameter	64.94	84.93	1.55	31.80	nr	116.73

T:MECHANICAL/COOLING/HEATING SYSTEMS

Item	Net Price £	Material £	Labour hours	Labour £	Unit	Total rate £
300mm diameter	71.40	93.37	1.55	31.80	nr	**125.17**
350mm diameter	79.19	103.55	1.55	31.80	nr	**135.35**
400mm diameter	88.67	115.95	1.55	31.80	nr	**147.75**
Expansion bellows						
15mm diameter	23.65	30.93	1.42	29.13	nr	**60.06**
20mm diameter	24.71	32.31	1.42	29.13	nr	**61.44**
25mm diameter	26.34	34.44	1.42	29.13	nr	**63.58**
32mm diameter	27.74	36.28	1.42	29.13	nr	**65.41**
40mm diameter	29.33	38.35	1.42	29.13	nr	**67.49**
50mm diameter	32.51	42.51	1.42	29.13	nr	**71.64**
65mm diameter	36.19	47.33	1.42	29.13	nr	**76.46**
80mm diameter	39.01	51.01	1.75	35.91	nr	**86.92**
100mm diameter	48.06	62.85	1.75	35.91	nr	**98.75**
125mm diameter	53.74	70.27	1.75	35.91	nr	**106.17**
150mm diameter	62.30	81.47	1.75	35.91	nr	**117.38**
200mm diameter	82.12	107.38	2.17	44.52	nr	**151.90**
250mm diameter	96.22	125.82	2.17	44.52	nr	**170.34**
300mm diameter	105.78	138.33	2.17	44.52	nr	**182.85**
350mm diameter	117.32	153.42	2.17	44.52	nr	**197.94**
400mm diameter	131.36	171.78	2.17	44.52	nr	**216.30**
50mm thick						
15mm diameter	8.87	11.60	0.30	6.16	m	**17.75**
20mm diameter	9.30	12.16	0.30	6.16	m	**18.32**
25mm diameter	9.84	12.87	0.30	6.16	m	**19.02**
32mm diameter	10.62	13.89	0.30	6.16	m	**20.04**
40mm diameter	10.79	14.11	0.30	6.16	m	**20.26**
50mm diameter	11.95	15.63	0.30	6.16	m	**21.78**
65mm diameter	13.03	17.04	0.30	6.16	m	**23.20**
80mm diameter	13.94	18.23	0.40	8.21	m	**26.44**
100mm diameter	17.26	22.57	0.40	8.21	m	**30.77**
125mm diameter	19.31	25.25	0.40	8.21	m	**33.46**
150mm diameter	22.08	28.87	0.40	8.21	m	**37.08**
200mm diameter	29.24	38.24	0.50	10.26	m	**48.50**
250mm diameter	33.77	44.16	0.50	10.26	m	**54.42**
300mm diameter	36.22	47.36	0.50	10.26	m	**57.62**
350mm diameter	39.74	51.97	0.50	10.26	m	**62.23**
400mm diameter	44.03	57.57	0.50	10.26	m	**67.83**
Extra over for fittings external insulation						
Flange/union						
15mm diameter	13.07	17.09	0.75	15.39	nr	**32.48**
20mm diameter	13.75	17.98	0.75	15.39	nr	**33.37**
25mm diameter	14.53	19.00	0.75	15.39	nr	**34.39**
32mm diameter	15.26	19.96	0.75	15.39	nr	**35.35**
40mm diameter	15.96	20.87	0.75	15.39	nr	**36.26**
50mm diameter	17.60	23.02	0.75	15.39	nr	**38.41**
65mm diameter	19.03	24.89	0.75	15.39	nr	**40.28**
80mm diameter	20.40	26.68	0.89	18.26	nr	**44.94**
100mm diameter	25.08	32.80	0.89	18.26	nr	**51.06**
125mm diameter	27.88	36.45	0.89	18.26	nr	**54.71**
150mm diameter	31.92	41.74	0.89	18.26	nr	**60.00**
200mm diameter	41.76	54.61	1.15	23.59	nr	**78.20**
250mm diameter	48.26	63.11	1.15	23.59	nr	**86.71**
300mm diameter	52.36	68.46	1.15	23.59	nr	**92.06**
350mm diameter	57.94	75.76	1.15	23.59	nr	**99.36**
400mm diameter	64.56	84.42	1.15	23.59	nr	**108.02**

T:MECHANICAL/COOLING/HEATING SYSTEMS

Item	Net Price £	Material £	Labour hours	Labour £	Unit	Total rate £
T31 : LOW TEMPERATURE HOT WATER HEATING (cont'd)						
Y50 - THERMAL INSULATION (cont'd)						
Mineral fibre PIB clad; External pipework (cont'd)						
50mm thick; Extra over for fittings external insulation (cont'd)						
Bend						
15mm diameter	2.22	2.90	0.30	6.16	nr	**9.06**
20mm diameter	2.33	3.04	0.30	6.16	nr	**9.20**
25mm diameter	2.46	3.22	0.30	6.16	nr	**9.37**
32mm diameter	2.58	3.37	0.30	6.16	nr	**9.53**
40mm diameter	2.70	3.53	0.30	6.16	nr	**9.69**
50mm diameter	2.99	3.91	0.30	6.16	nr	**10.06**
65mm diameter	3.26	4.27	0.30	6.16	nr	**10.42**
80mm diameter	3.48	4.55	0.40	8.21	nr	**12.76**
100mm diameter	4.31	5.63	0.40	8.21	nr	**13.84**
125mm diameter	4.82	6.31	0.40	8.21	nr	**14.52**
150mm diameter	5.52	7.22	0.40	8.21	nr	**15.43**
200mm diameter	7.31	9.56	0.50	10.26	nr	**19.82**
250mm diameter	8.45	11.05	0.50	10.26	nr	**21.31**
300mm diameter	9.06	11.85	0.50	10.26	nr	**22.11**
350mm diameter	9.94	12.99	0.50	10.26	nr	**23.25**
400mm diameter	11.00	14.39	0.50	10.26	nr	**24.65**
Tee						
15mm diameter	2.22	2.90	0.30	6.16	nr	**9.06**
20mm diameter	2.33	3.04	0.30	6.16	nr	**9.20**
25mm diameter	2.46	3.22	0.30	6.16	nr	**9.37**
32mm diameter	2.58	3.37	0.30	6.16	nr	**9.53**
40mm diameter	2.70	3.53	0.30	6.16	nr	**9.69**
50mm diameter	2.99	3.91	0.30	6.16	nr	**10.06**
65mm diameter	3.26	4.27	0.30	6.16	nr	**10.42**
80mm diameter	3.48	4.55	0.40	8.21	nr	**12.76**
100mm diameter	4.31	5.63	0.40	8.21	nr	**13.84**
125mm diameter	4.82	6.31	0.40	8.21	nr	**14.52**
150mm diameter	5.52	7.22	0.40	8.21	nr	**15.43**
200mm diameter	7.31	9.56	0.50	10.26	nr	**19.82**
250mm diameter	8.45	11.05	0.50	10.26	nr	**21.31**
300mm diameter	9.06	11.85	0.50	10.26	nr	**22.11**
350mm diameter	9.94	12.99	0.50	10.26	nr	**23.25**
400mm diameter	11.00	14.39	0.50	10.26	nr	**24.65**
Valves						
15mm diameter	20.76	27.15	1.03	21.13	nr	**48.28**
20mm diameter	21.84	28.56	1.03	21.13	nr	**49.69**
25mm diameter	23.09	30.19	1.03	21.13	nr	**51.32**
32mm diameter	24.24	31.70	1.03	21.13	nr	**52.83**
40mm diameter	25.36	33.16	1.03	21.13	nr	**54.29**
50mm diameter	27.96	36.56	1.03	21.13	nr	**57.70**
65mm diameter	30.24	39.54	1.03	21.13	nr	**60.68**
80mm diameter	32.40	42.37	1.25	25.65	nr	**68.02**
100mm diameter	39.84	52.10	1.25	25.65	nr	**77.74**
125mm diameter	44.28	57.90	1.25	25.65	nr	**83.55**
150mm diameter	50.69	66.28	1.25	25.65	nr	**91.93**
200mm diameter	66.32	86.73	1.55	31.80	nr	**118.53**
250mm diameter	76.64	100.23	1.55	31.80	nr	**132.03**

T:MECHANICAL/COOLING/HEATING SYSTEMS

Item	Net Price £	Material £	Labour hours	Labour £	Unit	Total rate £
300mm diameter	83.15	108.73	1.55	31.80	nr	**140.53**
350mm diameter	92.02	120.33	1.55	31.80	nr	**152.13**
400mm diameter	102.53	134.07	1.55	31.80	nr	**165.87**
Expansion bellows						
15mm diameter	30.76	40.22	1.42	29.13	nr	**69.35**
20mm diameter	32.35	42.31	1.42	29.13	nr	**71.44**
25mm diameter	34.20	44.72	1.42	29.13	nr	**73.86**
32mm diameter	35.92	46.97	1.42	29.13	nr	**76.10**
40mm diameter	37.56	49.12	1.42	29.13	nr	**78.25**
50mm diameter	41.42	54.17	1.42	29.13	nr	**83.30**
65mm diameter	44.80	58.58	1.42	29.13	nr	**87.71**
80mm diameter	48.01	62.78	1.75	35.91	nr	**98.69**
100mm diameter	59.02	77.17	1.75	35.91	nr	**113.08**
125mm diameter	65.60	85.79	1.75	35.91	nr	**121.69**
150mm diameter	75.10	98.20	1.75	35.91	nr	**134.11**
200mm diameter	98.26	128.49	2.17	44.52	nr	**173.01**
250mm diameter	113.54	148.48	2.17	44.52	nr	**193.00**
300mm diameter	123.19	161.09	2.17	44.52	nr	**205.62**
350mm diameter	136.32	178.26	2.17	44.52	nr	**222.78**
400mm diameter	151.90	198.63	2.17	44.52	nr	**243.15**

T:MECHANICAL/COOLING/HEATING SYSTEMS

Item	Net Price £	Material £	Labour hours	Labour £	Unit	Total rate £
T33 : STEAM HEATING						
Y10 - PIPELINES						
For pipework rates refer to Section T31 - Low Temperature Hot Water Heating						
Y11 - PIPELINE ANCILLARIES						
Steam traps and accessories						
Cast iron; inverted bucket type; steam trap pressure range up to 17 bar at 210°C; screwed ends						
1/2" dia.	99.13	126.97	0.85	21.57	nr	**148.53**
3/4" dia.	146.24	187.31	1.13	28.67	nr	**215.98**
1" dia.	228.69	292.90	1.35	34.25	nr	**327.16**
11/2" dia.	422.05	540.55	1.80	45.67	nr	**586.22**
2" dia.	651.72	834.71	2.18	55.31	nr	**890.02**
Cast iron; inverted bucket type; steam trap pressure range up to 17 bar at 210°C; flanged ends to BS 4504 PN16; bolted connections						
15mm dia.	239.49	306.73	1.15	29.18	nr	**335.91**
20mm dia.	277.77	355.76	1.25	31.71	nr	**387.47**
25mm dia.	424.99	544.32	1.33	33.74	nr	**578.07**
40mm dia.	657.61	842.26	1.46	37.04	nr	**879.30**
50mm dia.	801.89	1027.05	1.60	40.60	nr	**1067.64**
Steam traps and strainers						
Stainless steel; thermodynamic trap with pressure range up to 42 bar; temperature range to 400°C; screwed ends to steel						
15mm dia.	66.25	84.85	0.84	21.32	nr	**106.17**
20mm dia.	100.11	128.22	1.14	28.93	nr	**157.16**
Stainless steel; thermodynamic trap with pressure range up to 24 bar; temperature range to 288°C; flanged ends to DIN 2456 PN64; bolted connections						
15mm dia.	266.51	341.34	1.24	31.48	nr	**372.82**
20mm dia.	272.83	349.44	1.34	34.01	nr	**383.45**
25mm dia.	295.43	378.39	1.40	35.54	nr	**413.92**
Malleable iron pipeline strainer; max steam working pressure 14 bar and temperature range to 230°C; screwed ends to steel						
1/2" dia.	10.21	13.07	0.84	21.32	nr	**34.39**
3/4" dia.	13.64	17.47	1.14	28.93	nr	**46.41**
1" dia.	20.17	25.83	1.30	32.99	nr	**58.83**
11/2" dia.	33.32	42.68	1.50	38.10	nr	**80.77**
2" dia.	59.38	76.05	1.74	44.20	nr	**120.26**

T:MECHANICAL/COOLING/HEATING SYSTEMS

Item	Net Price £	Material £	Labour hours	Labour £	Unit	Total rate £
Bronze pipeline strainer; max steam working pressure 25 bar; flanged ends to BS 4504 PN25; bolted connections						
15mm dia.	120.72	154.62	1.24	31.48	nr	**186.10**
20mm dia.	147.23	188.56	1.34	34.01	nr	**222.58**
25mm dia.	168.82	216.22	1.40	35.54	nr	**251.76**
32mm dia.	262.06	335.65	1.46	37.09	nr	**372.74**
40mm dia.	297.40	380.90	1.54	39.09	nr	**419.99**
50mm dia.	457.38	585.81	1.64	41.66	nr	**627.47**
65mm dia.	506.46	648.66	2.50	63.43	nr	**712.09**
80mm dia.	630.12	807.06	2.91	73.76	nr	**880.81**
100mm dia.	1091.43	1397.89	3.51	89.02	nr	**1486.92**
Balanced pressure thermostatic steam trap and strainer; max working pressure up to 13 bar; screwed ends to steel						
1/2" dia.	45.25	57.95	1.26	31.99	nr	**89.95**
3/4" dia.	48.93	62.67	1.71	43.44	nr	**106.11**
Bimetallic thermostatic steam trap and strainer; max working pressure up to 21 bar; flanged ends						
15mm	128.58	164.68	1.24	31.48	nr	**196.16**
20mm	141.34	181.02	1.34	34.01	nr	**215.03**
Sight glasses						
Pressed brass; straight; single window; screwed ends to steel						
15mm dia.	30.43	38.97	0.84	21.32	nr	**60.29**
20mm dia.	33.76	43.24	1.14	28.93	nr	**72.18**
25mm dia.	42.20	54.06	1.30	32.99	nr	**87.05**
Gunmetal; straight; double window; screwed ends to steel						
15mm dia.	49.08	62.85	0.84	21.32	nr	**84.17**
20mm dia.	53.98	69.14	1.14	28.93	nr	**98.07**
25mm dia.	66.74	85.48	1.30	32.99	nr	**118.48**
32mm dia.	109.93	140.79	1.35	34.29	nr	**175.08**
40mm dia.	109.93	140.79	1.74	44.20	nr	**185.00**
50mm dia.	133.48	170.97	2.08	52.86	nr	**223.82**
SG Iron flanged; BS 4504, PN 25						
15mm dia.	100.11	128.22	1.00	25.37	nr	**153.60**
20mm dia.	117.78	150.85	1.25	31.71	nr	**182.57**
25mm dia.	150.17	192.34	1.50	38.10	nr	**230.43**
32mm dia.	165.87	212.45	1.70	43.15	nr	**255.60**
40mm dia.	216.91	277.82	2.00	50.74	nr	**328.56**
50mm dia.	262.06	335.65	2.30	58.46	nr	**394.10**
Check valve and sight glass; gun metal; screwed						
15mm dia.	49.57	63.48	0.84	21.32	nr	**84.80**
20mm dia.	52.51	67.25	1.14	28.93	nr	**96.19**
25mm dia.	88.34	113.14	1.30	32.99	nr	**146.13**

T:MECHANICAL/COOLING/HEATING SYSTEMS

Item	Net Price £	Material £	Labour hours	Labour £	Unit	Total rate £
T33 : STEAM HEATING (cont'd)						
Y11 - PIPELINE ANCILLARIES (cont'd)						
Pressure reducing valves						
Pressure reducing valve for steam; maximum range of 17 bar and 232°C; screwed ends to steel						
15mm dia.	374.22	479.30	0.87	22.07	nr	**501.37**
20mm dia.	404.91	518.60	0.91	23.09	nr	**541.69**
25mm dia.	436.59	559.18	1.35	34.25	nr	**593.43**
Pressure reducing valve for steam; maximum range of 17 bar and 232°C; flanged ends to BS 4504 PN 25						
25mm dia.	525.69	673.30	1.70	43.13	nr	**716.43**
32mm dia.	596.97	764.59	1.87	47.45	nr	**812.04**
40mm dia.	712.80	912.95	2.12	53.79	nr	**966.74**
50mm dia.	822.69	1053.69	2.57	65.21	nr	**1118.90**
Safety and relief valves						
Bronze safety valve; 'pop' type; side outlet; including easing lever; working pressure saturated steam up to 20.7 bar; screwed ends to steel						
15mm dia.	85.95	110.08	0.32	8.12	nr	**118.20**
20mm dia.	105.71	135.40	0.40	10.15	nr	**145.55**
Bronze safety valve; 'pop' type; side outlet; including easing lever; working pressure saturated steam up to 17.2 bar; screwed ends to steel						
25mm dia.	138.25	177.07	0.47	11.92	nr	**188.99**
32mm dia.	185.68	237.82	0.56	14.21	nr	**252.03**
Bronze safety valve; 'pop' type; side outlet; including easing lever; working pressure saturated steam up to 13.8 bar; screwed ends to steel						
40mm dia.	258.33	330.87	0.64	16.24	nr	**347.11**
50mm dia.	341.65	437.58	0.76	19.28	nr	**456.86**
65mm dia.	498.81	638.87	0.94	23.85	nr	**662.72**
80mm dia.	559.54	716.65	1.10	27.91	nr	**744.56**

T:MECHANICAL/COOLING/HEATING SYSTEMS

Item	Net Price £	Material £	Labour hours	Labour £	Unit	Total rate £
EQUIPMENT						
Y23 - CALORIFIERS						
Non-storage calorifiers; mild steel shell construction with indirect steam heating for secondary LPHW at 82°C flow and 71°C return to BS 853; maximum test on shell 11 bar, tubes 26 bar						
Horizontal/vertical, for steam at 3 bar-5.5 bar						
88 kW capacity	453.67	581.05	8.00	202.98	nr	**784.03**
176 kW capacity	660.47	845.92	12.05	305.69	nr	**1151.61**
293 kW capacity	719.92	922.07	14.08	357.35	nr	**1279.42**
586 kW capacity	1050.80	1345.86	37.04	939.70	nr	**2285.56**
879 kW capacity	1308.01	1675.29	40.00	1014.88	nr	**2690.16**
1465 kW capacity	1596.24	2044.45	45.45	1153.27	nr	**3197.71**

T:MECHANICAL/COOLING/HEATING SYSTEMS

Item	Net Price £	Material £	Labour hours	Labour £	Unit	Total rate £
EQUIPMENT (cont'd)						
T42 : LOCAL HEATING UNITS						
Warm air unit heater for connection to LTHW or steam supplies; suitable for heights upto 3m; recirculating type; mild steel casing; heating coil; adjustable discharge louvre; axial fan; horizontal or vertical discharge; normal speed; entering air temperature 15°C; complete with enclosures; includes fixing in position; includes connections to primary heating supply; electrical work elsewhere						
Low pressure hot water						
7.5 kW, 265 l/sec	315.20	403.70	6.53	165.68	nr	**569.38**
15.4 kW, 575 l/sec	381.55	488.68	7.54	191.30	nr	**679.99**
26.9 kW, 1040 l/sec	517.02	662.20	8.65	219.47	nr	**881.67**
48.0 kW, 1620 l/sec	681.52	872.89	9.35	237.23	nr	**1110.11**
Steam, 2 Bar						
9.2 kW, 265 l/sec	450.66	577.20	6.53	165.68	nr	**742.88**
18.8 kW, 575 l/sec	487.99	625.01	6.82	173.04	nr	**798.05**
34.8 kW, 1040 l/sec	561.26	718.86	6.82	173.04	nr	**891.89**
51.6 kW, 1625 l/sec	760.34	973.84	7.10	180.14	nr	**1153.98**

T:MECHANICAL/COOLING/HEATING SYSTEMS

Item	Net Price £	Material £	Labour hours	Labour £	Unit	Total rate £
T60 : CENTRAL REFRIGERATION PLANT						
CHILLERS						
Air cooled						
Selection of air cooled chillers based on chilled water flow and return temperatures 6°C and 12°C, and an outdoor temperature of 35°C						
Air cooled liquid chiller; refrigerant 407C; scroll compressors; twin circuit; integral controls; includes placing in position; electrical work elsewhere						
Cooling load						
100 kW	17413.48	22303.01	8.00	202.98	nr	**22505.98**
150 kW	19643.49	25159.18	8.00	202.98	nr	**25362.16**
200 kW	24935.19	31936.74	8.00	202.98	nr	**32139.71**
Air cooled liquid chiller; refrigerant 407C; reciprocating compressors; twin circuit; integral controls; includes placing in position; electrical work elsewhere						
Cooling Load						
250 kW	32772.54	41974.75	8.00	202.98	nr	**42177.72**
400 kW	46124.60	59075.93	8.00	202.98	nr	**59278.90**
550 kW	58761.33	75260.92	8.00	202.98	nr	**75463.90**
700 kW	73762.73	94474.57	9.00	228.35	nr	**94702.92**
Air cooled liquid chiller; refrigerant R134a; screw compressors; twin circuit; integral controls; includes placing in position; electrical work elsewhere						
Cooling load						
250 kW	37202.40	47648.46	8.00	202.98	nr	**47851.44**
400 kW	46933.65	60112.15	8.00	202.98	nr	**60315.13**
600 kW	60967.64	78086.74	8.00	202.98	nr	**78289.72**
800 kW	90654.80	116109.75	9.00	228.35	nr	**116338.10**
1000 kW	105442.89	135050.20	9.00	228.35	nr	**135278.55**
1200 kW	122054.86	156326.64	10.00	253.72	nr	**156580.36**
Air cooled liquid chiller; ductable for indoor installation; refrigerant 407C; scroll compressors; integral controls; includes placing in position; electrical work elsewhere						
Cooling load						
40 kW	12229.51	15663.43	6.00	152.23	nr	**15815.67**
80 kW	17760.37	22747.30	6.00	152.23	nr	**22899.53**

T:MECHANICAL/COOLING/HEATING SYSTEMS

Item	Net Price £	Material £	Labour hours	Labour £	Unit	Total rate £
EQUIPMENT (cont'd)						
T60 : CENTRAL REFRIGERATION PLANT (cont'd)						
CHILLERS (cont'd)						
Higher efficiency air cooled						
Selection of air cooled chillers based on chilled water flow and return temperatures of 6°C and 12°C and an outdoor temperature of 25°C						
These machines have significantly higher part load operating efficiencies than conventional air cooled machines						
Air cooled liquid chiller, refrigerant R410A; scroll compressors; complete with free cooling facility; integral controls; including placing in position; electrical work elsewhere						
Cooling load						
250 kW	34020.00	41497.60	8.00	202.98	nr	**41700.57**
300 kW	36288.00	44264.10	8.00	202.98	nr	**44467.08**
350 kW	38556.00	47030.61	8.00	202.98	nr	**47233.58**
400 kW	41958.00	51180.37	8.00	202.98	nr	**51383.34**
450 kW	45360.00	55330.13	8.00	202.98	nr	**55533.10**
500 kW	48762.00	59479.89	8.00	202.98	nr	**59682.86**
600 kW	52164.00	63629.65	8.00	202.98	nr	**63832.62**
650 kW	56700.00	69162.66	9.00	228.35	nr	**69391.01**
700 kW	68040.00	82995.19	9.00	228.35	nr	**83223.54**
750 kW	72576.00	88528.20	9.00	228.35	nr	**88756.55**
Water cooled						
Selection of water cooled chillers based on chilled water flow and return temperatures of 6°C and 12°C, and condenser entering and leaving temperatures of 27°C and 33°C						
Water cooled liquid chiller; refrigerant 407C; reciprocating compressors; twin circuit; integral controls; includes placing in position; electrical work elsewhere						
Cooling load						
200 kw	18279.63	23412.36	8.00	202.98	nr	**23615.34**
350 kW	30655.65	39263.45	8.00	202.98	nr	**39466.42**
500 kW	39238.50	50256.28	8.00	202.98	nr	**50459.25**
650 kW	50615.86	64828.29	9.00	228.35	nr	**65056.64**
750 kW	54817.33	70209.49	9.00	228.35	nr	**70437.84**

T:MECHANICAL/COOLING/HEATING SYSTEMS

Item	Net Price £	Material £	Labour hours	Labour £	Unit	Total rate £
Water cooled condenserless liquid chiller; refrigerant 407C; reciprocating compressors; twin circuit; integral controls; includes placing in position; electrical work elsewhere						
Cooling load						
200 kW	16091.63	20610.00	8.00	202.98	nr	**20812.97**
350 kW	28247.88	36179.61	8.00	202.98	nr	**36382.58**
500 kW	35701.72	45726.41	8.00	202.98	nr	**45929.38**
650 kW	43664.05	55924.47	9.00	228.35	nr	**56152.82**
750 kW	49685.08	63636.15	9.00	228.35	nr	**63864.50**
Water cooled liquid chiller; refrigerant R134a; screw compressors; twin circuit; integral controls; includes placing in position; electrical work elsewhere						
Cooling load						
300 kW	30136.39	38598.39	8.00	202.98	nr	**38801.36**
500 kW	39731.90	50888.22	8.00	202.98	nr	**51091.20**
700 kW	58145.11	74471.68	9.00	228.35	nr	**74700.03**
900 kW	66811.99	85572.13	9.00	228.35	nr	**85800.48**
1100 kW	80584.19	103211.43	10.00	253.72	nr	**103465.15**
1300 kW	89783.26	114993.50	10.00	253.72	nr	**115247.22**
Water cooled liquid chiller; refrigerant R134a; centrifugal compressors; twin circuit; integral controls; includes placing in position; electrical work elsewhere						
Cooling load						
700 kW	57312.36	73405.10	9.00	228.35	nr	**73633.44**
1000 kW	80797.50	103484.63	10.00	253.72	nr	**103738.35**
1300 kW	106275.65	136116.78	10.00	253.72	nr	**136370.50**
1600 kW	127552.32	163367.74	11.00	279.09	nr	**163646.83**
1900 kW	155346.66	198966.45	11.00	279.09	nr	**199245.54**
2200 kW	185080.14	237048.79	13.00	329.83	nr	**237378.63**
2500 kW	191220.75	244913.62	13.00	329.83	nr	**245243.46**
3000 kW	232696.80	298035.73	15.00	380.58	nr	**298416.31**
3500 kW	316726.20	405659.75	15.00	380.58	nr	**406040.33**
4000 kW	336391.23	430846.53	20.00	507.44	nr	**431353.97**
4500 kW	358893.88	459667.69	20.00	507.44	nr	**460175.13**
5000 kW	437703.76	560606.60	25.00	634.30	nr	**561240.89**

T:MECHANICAL/COOLING/HEATING SYSTEMS

Item	Net Price £	Material £	Labour hours	Labour £	Unit	Total rate £
EQUIPMENT (cont'd)						
T60 : CENTRAL REFRIGERATION PLANT (cont'd)						
CHILLERS (cont'd)						
Absorption						
Absorption chiller, for operation using low pressure steam; selection based on chilled water flow and return temperatures of 6°C and 12°C, steam at 1 bar gauge and condenser entering and leaving temperatures of 27°C and 33°C; integral controls; includes placing in position; electrical work elsewhere						
Cooling load						
400 kW	54955.23	70386.11	8.00	202.98	nr	**70589.08**
700 kW	69779.95	89373.47	9.00	228.35	nr	**89601.81**
1000 kW	79609.24	101962.72	10.00	253.72	nr	**102216.44**
1300 kW	95004.93	121681.37	12.00	304.46	nr	**121985.83**
1600 kW	112437.80	144009.21	14.00	355.21	nr	**144364.42**
2000 kW	133259.86	170677.89	15.00	380.58	nr	**171058.47**
Absorption chiller, for operation using low pressure hot water; selection based on chilled water flow and return temperatures of 6°C and 12°C, cooling water temperatures of 27°C and 33°C and hot water at 90°C; integral controls; includes placing in position; electrical work elsewhere						
Cooling load						
700 kW	79609.24	101962.72	9.00	228.35	nr	**102191.06**
1000 kW	99514.51	127457.19	10.00	253.72	nr	**127710.91**
1300 kW	115741.88	148241.04	12.00	304.46	nr	**148545.51**
1600 kW	133259.86	170677.89	14.00	355.21	nr	**171033.10**
HEAT REJECTION						
Dry air liquid coolers						
Dry air liquid cooler; selection based on fluid temperatures 45°C on, 40°C off at 32°C dry bulb ambient temperature; includes 20% ethylene glycol; includes placing in position; electrical work elsewhere						
Flat coil configuration						
500kW	12687.33	16249.81	15.00	380.58	nr	**16630.39**
Extra for inverter panels (factory wired and mounted on units)	6744.08	8637.75	15.00	380.58	nr	**9018.33**
800kW	21464.11	27491.02	15.00	380.58	nr	**27871.60**
Extra for inverter panels (factory wired and mounted on units)	12393.81	15873.87	15.00	380.58	nr	**16254.45**
1100kW	27559.90	35298.45	15.00	380.58	nr	**35679.03**
Extra for inverter panels (factory wired and mounted on units)	13487.01	17274.02	15.00	380.58	nr	**17654.60**

T:MECHANICAL/COOLING/HEATING SYSTEMS

Item	Net Price £	Material £	Labour hours	Labour £	Unit	Total rate £
1400kW	35007.75	44837.57	15.00	380.58	nr	**45218.15**
Extra over for inverter panels (factory wired and mounted on units)	18591.29	23811.54	15.00	380.58	nr	**24192.12**
1700kW	41339.28	52946.93	15.00	380.58	nr	**53327.51**
Extra for inverter panels (factory wired and mounted on units)	20225.31	25904.38	15.00	380.58	nr	**26284.95**
2000kW	48358.39	61936.95	15.00	380.58	nr	**62317.52**
Extra for inverter panels (factory wired and mounted on units)	24877.76	31863.18	15.00	380.58	nr	**32243.76**
Note : heat rejection capacities above 500kW require multiple units. Rates are therefore for total number of units.						
'Vee' type coil configuration						
500kW	12258.60	15700.70	15.00	380.58	nr	**16081.28**
Extra for inverter panels (factory wired and mounted on units)	4743.74	6075.73	15.00	380.58	nr	**6456.31**
800kW	18,803.98	23161.79	15.00	380.58	nr	**23542.36**
Extra for inverter panels (factory wired and mounted on units)	10985.42	14070.02	15.00	380.58	nr	**14450.60**
1100kW	26590.36	34056.66	15.00	380.58	nr	**34437.24**
Extra for inverter panels (factory wired and mounted on units)	13283.91	17013.90	15.00	380.58	nr	**17394.48**
1400kW	34308.32	43941.75	15.00	380.58	nr	**44322.33**
Extra for inverter panels (factory wired and mounted on units)	16517.57	21155.54	15.00	380.58	nr	**21536.12**
1700kW	39412.32	50478.90	15.00	380.58	nr	**50859.48**
Extra for inverter panels (factory wired and mounted on units)	21970.85	28140.04	15.00	380.58	nr	**28520.62**
2000kW	51462.49	65912.64	15.00	380.58	nr	**66293.21**
Extra for inverter panels (factory wired and mounted on units)	24776.35	31733.31	15.00	380.58	nr	**32113.88**
Note : Heat rejection capacities above 1100kW require multiple units. Rates are for total number of units.						
Air cooled condensers						
Air cooled condenser; refrigerant 407C; selection based on condensing temperature of 45°C at 32°C dry bulb ambient; includes placing in position; electrical work elsewhere						
Flat coil configuration						
500kW	13543.05	17345.81	15.00	380.58	nr	**17726.39**
Extra for inverter panels (factory wired and mounted on units)	6196.91	7936.93	15.00	380.58	nr	**8317.51**
800kW	22218.72	28457.52	15.00	380.58	nr	**28838.09**
Extra for inverter panels (factory wired and mounted on units)	9486.90	12150.72	15.00	380.58	nr	**12531.30**
1100kW	29745.14	38097.28	15.00	380.58	nr	**38477.86**
Extra for inverter panels (factory wired and mounted on units)	13486.72	17273.65	15.00	380.58	nr	**17654.23**

T:MECHANICAL/COOLING/HEATING SYSTEMS

Item	Net Price £	Material £	Labour hours	Labour £	Unit	Total rate £
EQUIPMENT (cont'd)						
T60 : CENTRAL REFRIGERATION PLANT (cont'd)						
HEAT REJECTION (cont'd)						
Air cooled condensers (cont'd)						
Flat coil configuration (cont'd)						
1400kW	37136.36	47563.88	15.00	380.58	nr	**47944.46**
Extra for inverter panels (factory wired and mounted on units)	16585.17	21242.12	15.00	380.58	nr	**21622.70**
1700kW	48358.39	61936.95	15.00	380.58	nr	**62317.52**
Extra for inverter panels (factory wired and mounted on units)	24877.76	31863.18	15.00	380.58	nr	**32243.76**
2000kW	55704.54	71345.82	15.00	380.58	nr	**71726.40**
Extra for inverter panels (factory wired and mounted on units)	24877.76	31863.18	15.00	380.58	nr	**32243.76**
Note : Heat rejection capacities above 500kW require multiple units. Rates are for total number of units						
'Vee' type coil configuration						
500kW	13295.18	17028.33	15.00	380.58	nr	**17408.91**
Extra for inverter panels (factory wired and mounted on units)	4743.45	6075.36	15.00	380.58	nr	**6455.94**
800kW	19999.10	25614.65	15.00	380.58	nr	**25995.23**
Extra for inverter panels (factory wired and mounted on units)	9712.24	12439.34	15.00	380.58	nr	**12819.92**
1100kW	29350.80	37592.21	15.00	380.58	nr	**37972.78**
Extra for inverter panels (factory wired and mounted on units)	16517.57	21155.54	15.00	380.58	nr	**21536.12**
1400kW	34173.11	43768.58	15.00	380.58	nr	**44149.16**
Extra for inverter panels (factory wired and mounted on units)	19424.48	24878.68	15.00	380.58	nr	**25259.26**
1700kW	44026.20	56388.32	15.00	380.58	nr	**56768.89**
Extra for inverter panels (factory wired and mounted on units)	24787.62	31747.74	15.00	380.58	nr	**32128.31**
2000kW	51259.68	65652.88	15.00	380.58	nr	**66033.46**
Extra for inverter panels (factory wired and mounted on units)	29136.72	37318.02	15.00	380.58	nr	**37698.60**
Note : Heat rejection capacities above 1100kW require multiple units. Rates are for total number of units.						

T:MECHANICAL/COOLING/HEATING SYSTEMS

Item	Net Price £	Material £	Labour hours	Labour £	Unit	Total rate £
Cooling towers						
Cooling towers; forced draught, centrifugal fan, conterflow design; based on water temperatures of 35°C on and 29°C off at 21°C wet bulb ambient temperature; includes placing in position; electrical work elsewhere						
Open circuit type						
900kW	9021.13	11554.17	20.00	507.44	nr	**12061.61**
Extra for stainless steel construction	4044.47	5180.12	-	-	nr	**5180.12**
Extra for intake and discharge sound attenuation	4352.90	5575.15	-	-	nr	**5575.15**
Extra for fan dampers for capacity control	1009.97	1293.57	-	-	nr	**1293.57**
1500kW	13996.53	17926.61	20.00	507.44	nr	**18434.05**
Extra for stainless steel construction	6478.60	8297.73	-	-	nr	**8297.73**
Extra for intake and discharge sound attenuation	6522.95	8354.52	-	-	nr	**8354.52**
Extra for fan dampers for capacity control	1025.11	1312.95	-	-	nr	**1312.95**
2100kW	19856.24	25431.68	20.00	507.44	nr	**25939.12**
Extra for stainless steel construction	9188.59	11768.65	-	-	nr	**11768.65**
Extra for intake and discharge sound attenuation	9685.52	12405.11	-	-	nr	**12405.11**
Extra for fan dampers for capacity control	1170.55	1499.22	-	-	nr	**1499.22**
2700kW	23979.92	30713.24	20.00	507.44	nr	**31220.68**
Extra for stainless steel construction	11208.96	14356.32	-	-	nr	**14356.32**
Extra for intake and discharge sound attenuation	13048.22	16712.03	-	-	nr	**16712.03**
Extra for fan dampers for capacity control	1822.14	2333.78	-	-	nr	**2333.78**
3300kW	29551.07	37848.71	20.00	507.44	nr	**38356.15**
Extra for stainless steel construction	13680.37	17521.68	-	-	nr	**17521.68**
Extra for intake and discharge sound attenuation	12305.86	15761.23	-	-	nr	**15761.23**
Extra for fan dampers for capacity control	1490.53	1909.06	-	-	nr	**1909.06**
3900kW	33307.05	42659.34	20.00	507.44	nr	**43166.78**
Extra for stainless steel construction	15419.92	19749.68	-	-	nr	**19749.68**
Extra for intake and discharge sound attenuation	12542.07	16063.76	-	-	nr	**16063.76**
Extra for fan dampers for capacity control	1490.53	1909.06	-	-	nr	**1909.06**
4500kW	37484.25	48009.46	23.00	583.55	nr	**48593.01**
Extra for stainless steel construction	17480.05	22388.28	-	-	nr	**22388.28**
Extra for intake and discharge sound attenuation	16634.33	21305.08	-	-	nr	**21305.08**
Extra for fan dampers for capacity control	2345.75	3004.41	-	-	nr	**3004.41**
5100kW	43299.75	55457.89	23.00	583.55	nr	**56041.44**
Extra for stainless steel construction	19950.22	25552.05	-	-	nr	**25552.05**
Extra for intake and discharge sound attenuation	16590.11	21248.45	-	-	nr	**21248.45**
Extra for fan dampers for capacity control	2345.75	3004.41	-	-	nr	**3004.41**
5700kW	45776.97	58630.69	30.00	761.16	nr	**59391.85**
Extra for stainless steel construction	21375.42	27377.43	-	-	nr	**27377.43**
Extra for intake and discharge sound attenuation	24567.51	31465.82	-	-	nr	**31465.82**
Extra for fan dampers for capacity control	2579.63	3303.96	-	-	nr	**3303.96**
6300kW	55194.87	70693.04	30.00	761.16	nr	**71454.19**
Extra for stainless steel construction	25552.84	32727.82	-	-	nr	**32727.82**
Extra for intake and discharge sound attenuation	24922.40	31920.37	-	-	nr	**31920.37**
Extra for fan dampers for capacity control	2579.63	3303.96	-	-	nr	**3303.96**

T:MECHANICAL/COOLING/HEATING SYSTEMS

Item	Net Price £	Material £	Labour hours	Labour £	Unit	Total rate £
EQUIPMENT (cont'd)						
T60 : CENTRAL REFRIGERATION PLANT (cont'd)						
HEAT REJECTION (cont'd)						
Cooling towers (cont'd)						
Closed circuit type (includes 20% ethylene glycol)						
900kW	25148.14	32209.48	20.00	507.44	nr	32716.92
Extra for stainless steel construction	20477.06	26226.81	-	-	nr	26226.81
Extra for intake and discharge sound attenuation	7566.67	9691.31	-	-	nr	9691.31
Extra for fan dampers for capacity control	1025.11	1312.95	-	-	nr	1312.95
1500kW	46754.37	59882.53	20.00	507.44	nr	60389.97
Extra for stainless steel construction	37276.20	47742.98	-	-	nr	47742.98
Extra for intake and discharge sound attenuation	12078.97	15470.63	-	-	nr	15470.63
Extra for fan dampers for capacity control	1490.53	1909.06	-	-	nr	1909.06
2100kW	56977.45	72976.15	20.00	507.44	nr	73483.59
Extra for stainless steel construction	50642.21	64862.04	-	-	nr	64862.04
Extra for intake and discharge sound attenuation	14539.91	18622.57	-	-	nr	18622.57
Extra for fan dampers for capacity control	1490.53	1909.06	-	-	nr	1909.06
2700kW	61431.58	78680.95	25.00	634.30	nr	79315.25
Extra for stainless steel construction	62748.27	80367.36	-	-	nr	80367.36
Extra for intake and discharge sound attenuation	21701.66	27795.26	-	-	nr	27795.26
Extra for fan dampers for capacity control	2254.99	2888.17	-	-	nr	2888.17
3300kW	94650.20	121227.03	25.00	634.30	nr	121861.33
Extra for stainless steel construction	78280.02	100260.27	-	-	nr	100260.27
Extra for intake and discharge sound attenuation	28868.05	36973.91	-	-	nr	36973.91
Extra for fan dampers for capacity control	2579.63	3303.96	-	-	nr	3303.96
3900kW	104214.71	133477.15	25.00	634.30	nr	134111.45
Extra for stainless steel construction	89643.05	114813.93	-	-	nr	114813.93
Extra for intake and discharge sound attenuation	29340.46	37578.96	-	-	nr	37578.96
Extra for fan dampers for capacity control	2579.63	3303.96	-	-	nr	3303.96
4500kW	124765.59	159798.52	40.00	1014.88	nr	160813.40
Extra for stainless steel construction	92997.04	119109.68	-	-	nr	119109.68
Extra for intake and discharge sound attenuation	41936.05	53711.28	-	-	nr	53711.28
Extra for fan dampers for capacity control	4509.98	5776.33	-	-	nr	5776.33
5100kW	139153.07	178225.85	40.00	1014.88	nr	179240.73
Extra for stainless steel construction	120178.89	153923.92	-	-	nr	153923.92
Extra for intake and discharge sound attenuation	41936.05	53711.28	-	-	nr	53711.28
Extra for fan dampers for capacity control	4509.98	5776.33	-	-	nr	5776.33
5700kW	171583.95	219763.01	40.00	1014.88	nr	220777.88
Extra for stainless steel construction	124136.46	158992.74	-	-	nr	158992.74
Extra for intake and discharge sound attenuation	46881.20	60044.98	-	-	nr	60044.98
Extra for fan dampers for capacity control	5159.25	6607.91	-	-	nr	6607.91
6300kW	183280.10	234743.33	40.00	1014.88	nr	235758.20
Extra for stainless steel construction	141991.95	181861.87	-	-	nr	181861.87
Extra for intake and discharge sound attenuation	46881.20	60044.98	-	-	nr	60044.98
Extra for fan dampers for capacity control	5159.25	6607.91	-	-	nr	6607.91

T:MECHANICAL/COOLING/HEATING SYSTEMS

Item	Net Price £	Material £	Labour hours	Labour £	Unit	Total rate £
T61 : CHILLED WATER						
SCREWED STEEL						
Y10 - PIPELINES						
For pipework rates refer to Section T31 - Low Temperature Hot Water Heating, with the exception of chilled water blocks within brackets as detailed hereafter. For minimum fixing distances, refer to the Tables and Memoranda to the rear of the book						
FIXINGS						
For steel pipes; black malleable iron						
Oversized pipe clip, to contain 30mm insulation block for vapour barrier						
15mm dia	2.38	2.90	0.10	2.54	nr	**5.44**
20mm dia	2.40	2.93	0.11	2.79	nr	**5.72**
25mm dia	2.53	3.09	0.12	3.04	nr	**6.13**
32mm dia	2.60	3.17	0.14	3.55	nr	**6.72**
40mm dia	2.75	3.35	0.15	3.81	nr	**7.16**
50mm dia	3.92	4.78	0.16	4.06	nr	**8.84**
65mm dia	4.23	5.16	0.30	7.61	nr	**12.77**
80mm dia	4.49	5.48	0.35	8.88	nr	**14.36**
100mm dia	7.12	8.68	0.40	10.15	nr	**18.83**
125mm dia	10.76	13.12	0.60	15.23	nr	**28.35**
150mm dia	12.92	15.76	0.77	19.54	nr	**35.30**
200mm dia	16.49	21.12	0.90	22.83	nr	**43.95**
250mm dia	20.36	26.08	1.10	27.91	nr	**53.99**
300mm dia	24.23	31.04	1.25	31.71	nr	**62.75**
350mm dia	28.11	36.00	1.50	38.06	nr	**74.06**
400mm dia	31.48	40.32	1.75	44.40	nr	**84.72**
Screw on backplate (Male), black malleable iron; plugged and screwed						
M12	0.57	0.73	0.10	2.54	nr	**3.27**
Screw on backplate (Female), black malleable iron; plugged and screwed						
M12	0.57	0.73	0.10	2.54	nr	**3.27**
Extra Over channel sections for fabricated hangers and brackets						
Galvanised steel; including inserts, bolts, nuts, washers; fixed to backgrounds						
41 x 21mm	2.40	3.07	0.15	3.81	m	**6.88**
41 x 41mm	3.75	4.80	0.15	3.81	m	**8.61**
Threaded rods; metric thread; including nuts, washers etc						
12mm dia x 600mm long for ring clips	1.30	1.66	0.10	2.54	nr	**4.20**

T:MECHANICAL/COOLING/HEATING SYSTEMS

Item	Net Price £	Material £	Labour hours	Labour £	Unit	Total rate £
T61 : CHILLED WATER (cont'd)						
SCREWED STEEL (cont'd)						
Y10 - PIPELINES (cont'd)						
For plastic pipework suitable for chilled water systems, refer to ABS pipework details in Section S10 - Cold Water with the exception of chilled water blocks within brackets as detailed for the aforementioned steel pipe. For minimum fixing distances, refer to the Tables and Memoranda to the rear of the book						
For copper pipework, refer to Section S10 – Cold Water with the exception of chilled water blocks within brackets as detailed hereafter. For minimum fixing distances, refer to the Tables and Memoranda to the rear of the book						
For Copper pipework						
Oversized pipe clip, to contain 30mm insulation block for vapour barrier						
15mm dia	2.38	2.90	0.10	2.54	nr	**5.44**
22mm dia	2.40	2.93	0.11	2.79	nr	**5.72**
28mm dia	2.53	3.09	0.12	3.04	nr	**6.13**
35mm dia	2.60	3.17	0.14	3.55	nr	**6.72**
42mm dia	2.75	3.35	0.15	3.81	nr	**7.16**
54mm dia	3.92	4.78	0.16	4.06	nr	**8.84**
67mm dia	4.23	5.16	0.30	7.61	nr	**12.77**
76mm dia	4.49	5.48	0.35	8.88	nr	**14.36**
108mm dia	7.12	8.69	0.40	10.15	nr	**18.83**
133mm dia	8.26	10.08	0.60	15.23	nr	**25.30**
159mm dia	10.12	12.34	0.77	19.54	nr	**31.88**
Screw on backplate, female						
All sizes 15mm to 54mm x 10mm	1.18	1.51	0.10	2.54	nr	**4.05**
Screw on backplate, male						
All sizes 15mm to 54mm x 10mm	1.04	1.33	0.10	2.54	nr	**3.87**
Extra Over channel sections for fabricated hangers and brackets						
Galvanised steel; including inserts, bolts, nuts, washers; fixed to backgrounds						
41 x 21mm	2.40	3.07	0.15	3.81	m	**6.88**
41 x 41mm	3.75	4.80	0.15	3.81	m	**8.61**
Threaded rods; metric thread; including nuts, washers etc						
10mm dia x 600mm long for ring clips	1.25	1.60	0.10	2.54	nr	**4.14**
12mm dia x 600mm long for ring clips	1.30	1.66	0.10	2.54	nr	**4.20**

T:MECHANICAL/COOLING/HEATING SYSTEMS

Item	Net Price £	Material £	Labour hours	Labour £	Unit	Total rate £
Y22 - HEAT EXCHANGERS						
Plate heat exchanger; for use in CHW systems; painted carbon steel frame, stainless steel plates, nitrile rubber gaskets, design pressure of 10 bar and operating temperature of 110/135°C						
Primary side; 13°C in, 8°C out; secondary side; 6°C in, 11°C out						
264 kW, 12.60 l/s	4326.00	5540.70	10.00	253.72	nr	5794.42
290 kW, 13.85 l/s	4560.00	5840.40	12.00	304.46	nr	6144.87
316 kW, 15.11 l/s	4747.00	6079.91	12.00	304.46	nr	6384.37
350 kW, 16.69 l/s	5105.00	6538.43	16.00	405.95	nr	6944.38
395 kW, 18.88 l/s	5463.00	6996.96	16.00	405.95	nr	7402.91
454 kW, 21.69 l/s	5934.00	7600.21	10.00	253.72	nr	7853.93
475 kW, 22.66 l/s	6049.00	7747.50	10.00	253.72	nr	8001.22
527 kW, 25.17 l/s	6555.00	8395.58	10.00	253.72	nr	8649.30
554 kW, 26.43 l/s	6729.00	8618.44	10.00	253.72	nr	8872.15
580 kW, 27.68 l/s	6965.00	8920.70	10.00	253.72	nr	9174.42
633 kW, 30.19 l/s	7378.00	9449.67	10.00	253.72	nr	9703.39
661 kW, 31.52 l/s	7556.00	9677.65	10.00	253.72	nr	9931.37
713 kW, 34.04 l/s	8066.00	10330.85	12.00	304.46	nr	10635.31
740 kW, 35.28 l/s	8242.00	10556.27	12.00	304.46	nr	10860.73
804 kW, 38.33 l/s	8776.00	11240.21	12.00	304.46	nr	11544.68
1925 kW, 91.82 l/s	14732.00	18868.60	15.00	380.58	nr	19249.18
2710 kW, 129.26 l/s	19568.00	25062.50	15.00	380.58	nr	25443.08
3100 kW, 147.87 l/s	21800.00	27921.22	15.00	380.58	nr	28301.80
Note : For temperature conditions different to those above, the cost of the units can vary significantly, and therefore the manufacturers advice should be sought.						
Y24 - TRACE HEATING						
Trace heating; for freeze protection or temperature maintenance of pipework; to BS 6351; including fixing to parent structures by plastic pull ties						
Straight laid						
15mm	18.98	24.31	0.27	6.85	m	31.16
25mm	18.98	24.31	0.27	6.85	m	31.16
28mm	18.98	24.31	0.27	6.85	m	31.16
32mm	18.98	24.31	0.30	7.61	m	31.92
35mm	18.98	24.31	0.31	7.87	m	32.17
50mm	18.98	24.31	0.34	8.63	m	32.94
100mm	18.98	24.31	0.40	10.15	m	34.46
150mm	18.98	24.31	0.40	10.15	m	34.46

T:MECHANICAL/COOLING/HEATING SYSTEMS

Item	Net Price £	Material £	Labour hours	Labour £	Unit	Total rate £
T61 : CHILLED WATER (cont'd)						
Y24 - TRACE HEATING (cont'd)						
Trace heating; for freeze protection or temperature maintenance of pipework (cont'd)						
Helically wound						
15mm	24.18	30.97	1.00	25.37	m	**56.34**
25mm	24.18	30.97	1.00	25.37	m	**56.34**
28mm	24.18	30.97	1.00	25.37	m	**56.34**
32mm	24.18	30.97	1.00	25.37	m	**56.34**
35mm	24.18	30.97	1.00	25.37	m	**56.34**
50mm	24.18	30.97	1.00	25.37	m	**56.34**
100mm	24.18	30.97	1.00	25.37	m	**56.34**
150mm	24.18	30.97	1.00	25.37	m	**56.34**
Accessories for trace heating; weatherproof; polycarbonate enclosure to IP standards; fully installed						
Connection junction box						
100 x 100 x 75mm	41.55	53.21	1.40	35.52	nr	**88.74**
Single air thermostat						
150 x 150 x 75mm	91.04	116.61	1.42	36.03	nr	**152.63**
Single capillary thermostat						
150 x 150 x 75mm	130.98	167.75	1.46	37.04	nr	**204.80**
Twin capillary thermostat						
150 x 150 x 75mm	234.98	300.96	1.46	37.04	nr	**338.00**
EQUIPMENT						
PRESSURISATION UNITS						
Chilled water packaged pressurisation unit complete with expansion vessel(s), inter-connecting pipework and necessary isolating and drain valves; includes placing in position; electrical work elsewhere						
Selection based on a final working pressure of 4 bar, a 3m static head and system operating temperatures of 6°/12°C						
System volume						
1800 litres	1328.04	1700.94	8.00	202.98	nr	**1903.92**
4500 litres	1328.04	1700.94	8.00	202.98	nr	**1903.92**
7200 litres	1392.30	1783.24	10.00	253.72	nr	**2036.96**
9900 litres	1392.30	1783.24	10.00	253.72	nr	**2036.96**
15300 litres	1445.85	1851.83	13.00	329.83	nr	**2181.66**
22500 litres	1532.60	1962.94	20.00	507.44	nr	**2470.38**
27000 litres	1532.60	1962.94	20.00	507.44	nr	**2470.38**

T:MECHANICAL/COOLING/HEATING SYSTEMS

Item	Net Price £	Material £	Labour hours	Labour £	Unit	Total rate £
CHILLED BEAMS						
Static (passive) beams; based on water at 14°C flow and 16°C return, 24°C room temperature; 600mm wide coil providing 350-400W/m output						
Static cooled beam for exposed installation with standard casing	110.25	141.21	4.00	101.49	m	**242.69**
Static cooled beam for installation above open grid or perforated ceiling	75.60	96.83	4.00	101.49	m	**198.32**
Ventilated (active) beams; based on water at 14°C flow and 16°C return, 24°C room temperature; air supply at 10l/s/linear metre; 300mm wide beam providing 250-350W/m output unless stated otherwise; all exposed beams c/w standard casing; electrical work elsewhere						
Ventilated cooled beam flush mounted within a false ceiling; closed type with integrated secondary air circulation; 600mm wide beam providing 400W/m output	171.15	219.21	4.50	114.17	m	**333.38**
Ventilated cooled beam flush mounted within a false ceiling; open type	151.20	193.66	4.00	101.49	m	**295.14**
Ventilated cooled beam for exposed mounting with standard casing	151.20	193.66	4.00	101.49	m	**295.14**
Ventilated cooled beam flush mounted within a false ceiling; open type; with recessed integrated flush mounted 28W or 35W T5 light fittings	274.05	351.00	4.00	101.49	m	**452.49**
Ventilated cooled beam for exposed mounting with recessed integrated flush mounted 28W or 35W T5 light fittings	274.05	351.00	4.00	101.49	m	**452.49**
Ventilated cooled beam for exposed mounting with recessed integrated flush mounted direct and indirect 28W or 35W T5 light fittings	299.25	383.28	4.00	101.49	m	**484.76**
LEAK DETECTION						
Leak detection system consisting of a central control module connected by a leader cable to water sensing cables						
Control Modules						
Alarm Only	309.39	396.26	4.00	101.57	nr	**497.82**
Alarm and location	2029.13	2598.90	8.00	203.13	nr	**2802.03**
Cables						
Sensing - 3m length	84.74	108.54	4.00	101.57	nr	**210.10**
Sensing - 7.5m length	124.86	159.92	4.00	101.57	nr	**261.48**
Sensing - 15m length	215.87	276.49	8.00	203.13	nr	**479.62**
Leader - 3.5m length	38.03	48.71	2.00	50.78	nr	**99.49**
End terminal						
End terminal	14.77	18.91	0.05	1.27	nr	**20.18**

T:MECHANICAL/COOLING/HEATING SYSTEMS

Item	Net Price £	Material £	Labour hours	Labour £	Unit	Total rate £
T61 : CHILLED WATER (cont'd)						
ENERGY METERS						
Ultrasonic						
Energy meter for measuring energy use in chilled water systems; includes ultrasonic flow meter (with sensor and signal converter), energy calculator, pair of temperature sensors with brass pockets, and 3m of interconnecting cable; includes fixing in position; electrical work elsewhere						
Pipe size (flanged connections to PN16); maximum flow rate						
50mm, 36m³/hr	1028.85	1317.74	1.80	45.67	nr	**1363.41**
65mm, 60m³/hr	1133.28	1451.50	2.32	58.86	nr	**1510.36**
80mm, 100m³/hr	1269.33	1625.75	2.56	64.95	nr	**1690.70**
125mm, 250m³/hr	1469.37	1881.95	3.60	91.34	nr	**1973.29**
150mm, 360m³/hr	1594.39	2042.08	4.80	121.79	nr	**2163.86**
200mm, 600m³/hr	1780.45	2280.38	6.24	158.32	nr	**2438.71**
250mm, 1000m³/hr	2053.29	2629.84	9.60	243.57	nr	**2873.41**
300mm, 1500m³/hr	2412.18	3089.49	10.80	274.02	nr	**3363.51**
350mm, 2000m³/hr	2904.91	3720.58	13.20	334.91	nr	**4055.49**
400mm, 2500m³/hr	3324.10	4257.47	15.60	395.80	nr	**4653.27**
500mm, 3000m³/hr	3771.23	4830.16	24.00	608.93	nr	**5439.08**
600mm, 3500m³/hr	4236.02	5425.45	28.00	710.41	nr	**6135.86**
Electromagnetic						
Energy meter for measuring energy use in chilled water systems; includes electro-magnetic flow meter (with sensor and signal converter), energy calculator, pair of temperature sensors with brass pockets, and 3m of interconnecting cable; includes fixing in position; electrical work elsewhere						
Pipe size (flanged connections to PN40); maximum flow rate						
25mm, 17.7m³/hr	845.73	1083.21	1.48	37.55	nr	**1120.76**
40mm, 45m³/hr	853.82	1093.57	1.55	39.33	nr	**1132.89**
Pipe size (flanged connections to PN16); maximum flow rate						
50mm, 70m³/hr	863.38	1105.81	1.80	45.67	nr	**1151.48**
65mm, 120m³/hr	867.06	1110.52	2.32	58.86	nr	**1169.38**
80mm, 180m³/hr	871.47	1116.17	2.56	64.95	nr	**1181.13**
125mm, 450m³/hr	948.69	1215.08	3.60	91.34	nr	**1306.41**
150mm, 625m³/hr	1006.79	1289.49	4.80	121.79	nr	**1411.27**
200mm, 1100m³/hr	1083.27	1387.45	6.24	158.32	nr	**1545.77**
250mm, 1750m³/hr	1215.65	1556.99	9.60	243.57	nr	**1800.56**
300mm, 2550m³/hr	1549.53	1984.62	10.80	274.02	nr	**2258.64**
350mm, 3450m³/hr	2015.79	2581.80	13.20	334.91	nr	**2916.71**
400mm, 4500m³/hr	2299.66	2945.38	15.60	395.80	nr	**3341.18**

T:MECHANICAL/COOLING/HEATING SYSTEMS

Item	Net Price £	Material £	Labour hours	Labour £	Unit	Total rate £
T70 : LOCAL COOLING UNITS						
Split system with ceiling void evaporator unit and external condensing unit						
Ceiling mounted 4 way blow cassette heat pump unit with remote fan speed and load control; refrigerant 470C; includes outdoor unit						
Cooling 3.6kW, heating 4.1kW	1336.58	1711.87	35.00	888.02	nr	**2599.89**
Cooling 4.9kW, heating 5.5kW	1472.90	1886.47	35.00	888.02	nr	**2774.49**
Cooling 7.1kW, heating 8.2kW	1782.81	2283.41	35.00	888.02	nr	**3171.42**
Cooling 10kW, heating 11.2kW	2108.70	2700.80	35.00	888.02	nr	**3588.82**
Cooling 12.20kW, heating 14.60kW	2306.79	2954.51	35.00	888.02	nr	**3842.53**
Ceiling mounted 4 way blow cooling only unit with remote fan speed and load control; refrigerant 470C; includes outdoor unit						
Cooling 3.80kW	1209.84	1549.55	35.00	888.02	nr	**2437.57**
Cooling 5.20kW	1351.48	1730.97	35.00	888.02	nr	**2618.98**
Cooling 7.10kW	1666.73	2134.72	35.00	888.02	nr	**3022.74**
Cooling 10kw	1936.17	2479.83	35.00	888.02	nr	**3367.84**
Cooling 12.2kW	2052.26	2628.51	35.00	888.02	nr	**3516.52**
In ceiling, ducted heat pump unit with remote fan speed and load control; refrigerant 407C; includes outdoor unit						
Cooling 3.60kW, heating 4.10kW	979.80	1254.92	35.00	888.02	nr	**2142.93**
Cooling 4.90kW, heating 5.50kW	1126.77	1443.16	35.00	888.02	nr	**2331.17**
Cooling 7.10kW, heating 8.20kW	1126.77	1443.16	35.00	888.02	nr	**2331.17**
Cooling 10kW, heating 11.20kW	1702.94	2181.10	35.00	888.02	nr	**3069.12**
Cooling 12.20kW, heating 14.50kW	2271.65	2909.50	35.00	888.02	nr	**3797.52**
In ceiling, ducted cooling only unit with remote fan speed and load control; refrigerant 407C; includes outdoor unit						
Cooling 3.70kW	883.95	1132.15	35.00	888.02	nr	**2020.17**
Cooling 4.90kW	1051.16	1346.31	35.00	888.02	nr	**2234.33**
Cooling 7.10kW	1586.85	2032.42	35.00	888.02	nr	**2920.44**
Cooling 10kW	1809.44	2317.51	35.00	888.02	nr	**3205.52**
Cooling 12.3kW	2017.11	2583.49	35.00	888.02	nr	**3471.51**
Room Units						
Ceiling mounted 4 way blow cassette heat pump unit with remote fan speed and load control; refrigerant 407C; excludes outdoor unit						
Cooling 3.6kW, heating 4.1kW	843.48	1080.32	17.00	431.32	nr	**1511.64**
Cooling 4.9kW, heating 5.5kW	863.72	1106.24	17.00	431.32	nr	**1537.56**
Cooling 7.1kW, heating 8.2kW	935.07	1197.63	17.00	431.32	nr	**1628.95**
Cooling 10kW, heating 11.2kW	1012.82	1297.20	17.00	431.32	nr	**1728.53**
Cooling 12.20kW, heating 14.60kW	1098.02	1406.33	17.00	431.32	nr	**1837.65**
Ceiling mounted 4 way blow cooling unit with remote fan speed and load control; refrigerant 407C; excludes outdoor unit						
Cooling 3.80kW	796.62	1020.30	17.00	431.32	nr	**1451.63**
Cooling 5.20kW	803.01	1028.49	17.00	431.32	nr	**1459.81**
Cooling 7.10Kw	935.07	1197.63	17.00	431.32	nr	**1628.95**
Cooling 10kW	1012.82	1297.20	17.00	431.32	nr	**1728.53**
Cooling 12.2kW	1098.02	1406.33	17.00	431.32	nr	**1837.65**

T:MECHANICAL/COOLING/HEATING SYSTEMS

Item	Net Price £	Material £	Labour hours	Labour £	Unit	Total rate £
T70 : LOCAL COOLING UNITS (cont'd)						
Room Units (cont'd)						
In ceiling , ducted heat pump unit with remote fan speed and load control; refrigerant 407C; excludes outdoor unit						
Cooling 3.60kW, heating 4.10kW	486.71	623.37	17.00	431.32	nr	**1054.69**
Cooling 4.90kW, heating 5.50kW	517.59	662.92	17.00	431.32	nr	**1094.25**
Cooling 7.10kW, heating 8.20kW	855.20	1095.33	17.00	431.32	nr	**1526.65**
Cooling 10kW, heating 11.20kW	886.08	1134.88	17.00	431.32	nr	**1566.20**
Cooling 12.20kW, heating 14.50kW	1062.87	1361.31	17.00	431.32	nr	**1792.64**
In ceiling , ducted cooling unit only with remote fan speed and load control; refrigerant 407C; excludes outdoor unit						
Cooling 3.70kW	470.73	602.91	17.00	431.32	nr	**1034.23**
Cooling 4.90kW	502.68	643.83	17.00	431.32	nr	**1075.15**
Cooling 7.10kW	855.20	1095.33	17.00	431.32	nr	**1526.65**
Cooling 10kW	886.08	1134.88	17.00	431.32	nr	**1566.20**
Cooling 12.3kW	1062.87	1361.31	17.00	431.32	nr	**1792.64**
External condensing units suitable for connection to multiple indoor units; inverter driven; refrigerant 407C						
Cooling only						
9kW	1841.39	2358.43	17.00	431.32	nr	**2789.75**
Heat pump						
Cooling 5.20kW, heating 6.10kW	1312.08	1680.50	17.00	431.32	nr	**2111.82**
Cooling 6.80kW, heating 2.50kW	1673.12	2142.91	17.00	431.32	nr	**2574.23**
Cooling 8kW, heating 9.60kW	1937.23	2481.19	17.00	431.32	nr	**2912.51**
Cooling 14.50kW, heating 16.50kW	3215.24	4118.04	21.00	532.81	nr	**4650.85**

U:VENTILATION/AIR CONDITIONING SYSTEMS

Item	Net Price £	Material £	Labour hours	Labour £	Unit	Total rate £
U10 : DUCTWORK : CIRCULAR						
Y30 - AIR DUCTLINES						
Galvanised sheet metal DW144 class B spirally wound circular section ductwork; including all necessary stiffeners, joints, couplers in the running length and duct supports						
Straight duct						
80mm dia.	2.69	4.58	0.87	25.77	m	**30.34**
100mm dia.	2.78	4.73	0.87	25.77	m	**30.50**
160mm dia.	3.88	6.59	0.87	25.77	m	**32.36**
200mm dia.	4.95	8.42	0.87	25.77	m	**34.19**
250mm dia.	6.07	10.33	1.21	35.84	m	**46.17**
315mm dia.	7.43	12.64	1.21	35.84	m	**48.47**
355mm dia.	10.21	17.36	1.21	35.84	m	**53.20**
400mm dia.	11.41	19.40	1.21	35.84	m	**55.23**
450mm dia.	12.57	21.37	1.21	35.84	m	**57.21**
500mm dia.	13.65	23.22	1.21	35.84	m	**59.06**
630mm dia.	24.11	40.99	1.39	41.17	m	**82.16**
710mm dia.	26.32	44.76	1.39	41.17	m	**85.92**
800mm dia.	30.57	51.99	1.44	42.65	m	**94.64**
900mm dia.	38.03	64.66	1.46	43.24	m	**107.90**
1000mm dia.	46.28	78.70	1.65	48.87	m	**127.57**
1120mm dia.	55.38	94.17	2.43	71.97	m	**166.14**
1250mm dia.	60.61	103.06	2.43	71.97	m	**175.03**
1400mm dia.	68.38	116.27	2.77	82.04	m	**198.31**
1600mm dia.	78.47	133.43	3.06	90.63	m	**224.06**
Extra over fittings; circular duct class B						
End cap						
80mm dia.	1.29	2.19	0.15	4.44	nr	**6.64**
100mm dia.	1.35	2.30	0.15	4.44	nr	**6.74**
160mm dia.	2.04	3.47	0.15	4.44	nr	**7.92**
200mm dia.	2.38	4.05	0.20	5.92	nr	**9.98**
250mm dia.	3.46	5.88	0.29	8.59	nr	**14.47**
315mm dia.	4.25	7.23	0.29	8.59	nr	**15.82**
355mm dia.	6.49	11.04	0.44	13.03	nr	**24.07**
400mm dia.	6.63	11.27	0.44	13.03	nr	**24.30**
450mm dia.	6.94	11.80	0.44	13.03	nr	**24.83**
500mm dia.	7.22	12.27	0.44	13.03	nr	**25.30**
630mm dia.	17.48	29.72	0.58	17.18	nr	**46.89**
710mm dia.	20.01	34.02	0.69	20.44	nr	**54.45**
800mm dia.	27.77	47.22	0.81	23.99	nr	**71.21**
900mm dia.	31.41	53.41	0.92	27.25	nr	**80.65**
1000mm dia.	41.94	71.31	1.04	30.80	nr	**102.11**
1120mm dia.	46.70	79.42	1.16	34.36	nr	**113.77**
1250mm dia.	51.74	87.98	1.16	34.36	nr	**122.34**
1400mm dia.	67.63	115.00	1.16	34.36	nr	**149.35**
1600mm dia.	76.40	129.91	1.16	34.36	nr	**164.27**

U:VENTILATION/AIR CONDITIONING SYSTEMS

Item	Net Price £	Material £	Labour hours	Labour £	Unit	Total rate £
U10:DUCTWORK:CIRCULAR (cont'd)						
Y30 - AIR DUCTLINES (cont'd)						
Galvanised sheet metal DW144 class B spirally wound circular section ductwork (cont'd)						
Extra over fittings; circular duct class B (cont'd)						
Reducer						
80mm dia.	6.57	11.18	0.29	8.59	nr	19.77
100mm dia.	6.75	11.48	0.29	8.59	nr	20.07
160mm dia.	8.59	14.60	0.29	8.59	nr	23.19
200mm dia.	9.64	16.39	0.44	13.03	nr	29.42
250mm dia.	11.84	20.14	0.58	17.18	nr	37.32
315mm dia.	14.99	25.50	0.58	17.18	nr	42.67
355mm dia.	18.04	30.67	0.87	25.77	nr	56.44
400mm dia.	21.14	35.94	0.87	25.77	nr	61.71
450mm dia.	22.77	38.73	0.87	25.77	nr	64.49
500mm dia.	25.35	43.10	0.87	25.77	nr	68.87
630mm dia.	68.17	115.91	0.87	25.77	nr	141.68
710mm dia.	72.12	122.64	0.96	28.43	nr	151.07
800mm dia.	95.95	163.15	1.06	31.39	nr	194.54
900mm dia.	101.07	171.86	1.16	34.36	nr	206.22
1000mm dia.	126.69	215.43	1.25	37.02	nr	252.45
1120mm dia.	140.84	239.48	3.47	102.77	nr	342.25
1250mm dia.	158.48	269.47	3.47	102.77	nr	372.24
1400mm dia.	202.09	343.64	4.05	119.95	nr	463.59
1600mm dia.	212.25	360.90	4.62	136.83	nr	497.73
90° segmented radius bend						
80mm dia.	2.17	3.69	0.29	8.59	nr	12.28
100mm dia.	4.23	7.19	0.29	8.59	nr	15.78
160mm dia.	4.23	7.19	0.29	8.59	nr	15.78
200mm dia.	5.84	9.93	0.44	13.03	nr	22.96
250mm dia.	8.96	15.23	0.58	17.18	nr	32.41
315mm dia.	9.45	16.07	0.58	17.18	nr	33.25
355mm dia.	9.86	16.76	0.87	25.77	nr	42.53
400mm dia.	11.76	20.00	0.87	25.77	nr	45.76
450mm dia.	13.57	23.07	0.87	25.77	nr	48.84
500mm dia.	14.21	24.16	0.87	25.77	nr	49.93
630mm dia.	30.79	52.36	0.87	25.77	nr	78.12
710mm dia.	44.24	75.22	0.96	28.43	nr	103.65
800mm dia.	49.93	84.91	1.06	31.39	nr	116.30
900mm dia.	55.87	94.99	1.16	34.36	nr	129.35
1000mm dia.	84.68	143.99	1.25	37.02	nr	181.01
1120mm dia.	93.93	159.71	3.47	102.77	nr	262.49
1250mm dia.	123.14	209.39	3.47	102.77	nr	312.16
1400mm dia.	256.86	436.76	4.05	119.95	nr	556.71
1600mm dia.	259.46	441.17	4.62	136.83	nr	578.00
45° radius bend						
80mm dia.	2.27	3.85	0.29	8.59	nr	12.44
100mm dia.	2.27	3.85	0.29	8.59	nr	12.44
160mm dia.	3.44	5.86	0.29	8.59	nr	14.45
200mm dia.	4.55	7.74	0.40	11.85	nr	19.58
250mm dia.	6.54	11.12	0.58	17.18	nr	28.29

U:VENTILATION/AIR CONDITIONING SYSTEMS

Item	Net Price £	Material £	Labour hours	Labour £	Unit	Total rate £
315mm dia.	8.22	13.98	0.58	17.18	nr	31.16
355mm dia.	8.78	14.94	0.87	25.77	nr	40.70
400mm dia.	10.67	18.14	0.87	25.77	nr	43.90
450mm dia.	10.64	18.10	0.87	25.77	nr	43.86
500mm dia.	11.36	19.32	0.87	25.77	nr	45.09
630mm dia.	31.02	52.74	0.87	25.77	nr	78.51
710mm dia.	39.24	66.73	0.96	28.43	nr	95.16
800mm dia.	47.61	80.96	1.06	31.39	nr	112.35
900mm dia.	52.62	89.47	1.16	34.36	nr	123.83
1000mm dia.	76.21	129.58	1.25	37.02	nr	166.60
1120mm dia.	83.03	141.19	3.47	102.77	nr	243.96
1250mm dia.	92.74	157.69	3.47	102.77	nr	260.46
1400mm dia.	130.64	222.13	4.05	119.95	nr	342.09
1600mm dia.	138.21	235.01	4.62	136.83	nr	371.85
90° equal twin bend						
80mm dia.	6.89	11.71	0.58	17.18	nr	28.89
100mm dia.	6.99	11.89	0.58	17.18	nr	29.07
160mm dia.	12.60	21.42	0.58	17.18	nr	38.60
200mm dia.	17.93	30.49	0.87	25.77	nr	56.26
250mm dia.	28.19	47.93	1.16	34.36	nr	82.28
315mm dia.	30.67	52.15	1.16	34.36	nr	86.50
355mm dia.	34.03	57.86	1.73	51.24	nr	109.09
400mm dia.	37.71	64.12	1.73	51.24	nr	115.36
450mm dia.	40.62	69.08	1.73	51.24	nr	120.31
500mm dia.	43.54	74.03	1.73	51.24	nr	125.27
630mm dia.	84.19	143.16	1.73	51.24	nr	194.40
710mm dia.	112.17	190.73	1.82	53.90	nr	244.63
800mm dia.	138.78	235.98	1.93	57.16	nr	293.14
900mm dia.	160.89	273.57	2.02	59.83	nr	333.39
1000mm dia.	221.94	377.38	2.11	62.49	nr	439.87
1120mm dia.	258.91	440.25	4.62	136.83	nr	577.08
1250mm dia.	311.62	529.87	4.62	136.83	nr	666.70
1400mm dia.	506.42	861.10	4.62	136.83	nr	997.93
1600mm dia.	542.34	922.18	4.62	136.83	nr	1059.01
Conical branch						
80mm dia.	8.73	14.85	0.58	17.18	nr	32.03
100mm dia.	8.88	15.10	0.58	17.18	nr	32.28
160mm dia.	9.50	16.15	0.58	17.18	nr	33.33
200mm dia.	10.02	17.05	0.87	25.77	nr	42.81
250mm dia.	13.15	22.36	1.16	34.36	nr	56.71
315mm dia.	14.58	24.79	1.16	34.36	nr	59.15
355mm dia.	15.78	26.84	1.73	51.24	nr	78.08
400mm dia.	16.22	27.57	1.73	51.24	nr	78.81
450mm dia.	20.14	34.25	1.73	51.24	nr	85.49
500mm dia.	20.65	35.12	1.73	51.24	nr	86.36
630mm dia.	39.71	67.52	1.73	51.24	nr	118.76
710mm dia.	43.57	74.09	1.82	53.90	nr	128.00
800mm dia.	57.62	97.98	1.93	57.16	nr	155.15
900mm dia.	60.73	103.26	2.02	59.83	nr	163.09
1000mm dia.	75.16	127.79	2.11	62.49	nr	190.29
1120mm dia.	106.96	181.88	4.62	136.83	nr	318.71
1250mm dia.	106.96	181.88	5.20	154.01	nr	335.89
1400mm dia .	129.68	220.51	5.20	154.01	nr	374.52
1600mm dia.	153.39	260.82	5.20	154.01	nr	414.83

U:VENTILATION/AIR CONDITIONING SYSTEMS

Item	Net Price £	Material £	Labour hours	Labour £	Unit	Total rate £
U10:DUCTWORK:CIRCULAR (cont''d)						
Y30 - AIR DUCTLINES (cont'd)						
Galvanised sheet metal DW144 class B spirally wound circular section ductwork (cont'd)						
Extra over fittings; circular duct class B (cont'd)						
45° branch						
80mm dia.	6.90	11.72	0.58	17.18	nr	**28.90**
100mm dia.	7.01	11.92	0.58	17.18	nr	**29.10**
160mm dia.	10.65	18.12	0.58	17.18	nr	**35.29**
200mm dia.	10.96	18.63	0.87	25.77	nr	**44.40**
250mm dia.	11.39	19.37	1.16	34.36	nr	**53.72**
315mm dia.	11.99	20.39	1.16	34.36	nr	**54.75**
355mm dia.	15.59	26.50	1.73	51.24	nr	**77.74**
400mm dia.	16.12	27.42	1.73	51.24	nr	**78.65**
450mm dia.	16.72	28.43	1.73	51.24	nr	**79.67**
500mm dia.	17.37	29.54	1.73	51.24	nr	**80.77**
630mm dia.	32.65	55.53	1.73	51.24	nr	**106.76**
710mm dia.	35.80	60.87	1.82	53.90	nr	**114.77**
800mm dia.	52.95	90.04	2.13	63.09	nr	**153.12**
900mm dia.	60.01	102.05	2.31	68.42	nr	**170.46**
1000mm dia.	72.01	122.45	2.31	68.42	nr	**190.87**
1120mm dia.	90.83	154.44	4.62	136.83	nr	**291.28**
1250mm dia.	96.92	164.80	4.62	136.83	nr	**301.64**
1400mm dia.	112.18	190.75	4.62	136.83	nr	**327.58**
1600mm dia.	138.35	235.25	4.62	136.83	nr	**372.09**
For galvanised sheet metal DW144 class C rates, refer to galvanised sheet metal DW144 class B						

U:VENTILATION/AIR CONDITIONING SYSTEMS

Item	Net Price £	Material £	Labour hours	Labour £	Unit	Total rate £
U10 : DUCTWORK : FLAT OVAL						
Y30 - AIR DUCTLINES						
Galvanised sheet metal DW144 class B spirally wound flat oval section ductwork; including all necessary stiffeners, joints, couplers in the running length and duct supports						
Straight duct						
345 x 102mm	9.98	16.96	2.71	80.26	m	**97.23**
427 x 102mm	11.58	19.69	2.99	88.56	m	**108.25**
508 x 102mm	12.90	21.93	3.14	93.00	m	**114.93**
559 x 152mm	17.91	30.45	3.43	101.59	m	**132.04**
531 x 203mm	17.62	29.96	3.43	101.59	m	**131.54**
851 x 203mm	19.54	33.23	5.72	169.41	m	**202.64**
582 x 254mm	21.53	36.61	3.62	107.22	m	**143.83**
823 x 254mm	25.26	42.95	5.80	171.78	m	**214.73**
1303 x 254mm	25.56	43.45	8.13	240.79	m	**284.24**
632 x 305mm	25.59	43.51	3.93	116.40	m	**159.90**
1275 x 305mm	25.12	42.72	8.13	240.79	m	**283.51**
765 x 356mm	26.99	45.89	5.72	169.41	m	**215.30**
1247 x 356mm	23.92	40.67	8.13	240.79	m	**281.46**
1727 x 356mm	23.56	40.05	10.41	308.32	m	**348.37**
737 x 406mm	29.84	50.74	5.72	169.41	m	**220.15**
818 x 406mm	29.84	50.74	6.21	183.92	m	**234.66**
978 x 406mm	53.49	90.96	6.92	204.95	m	**295.91**
1379 x 406mm	58.07	98.74	8.75	259.15	m	**357.89**
1699 x 406mm	57.46	97.71	10.41	308.32	m	**406.03**
709 x 457mm	62.39	106.09	5.72	169.41	m	**275.50**
1189 x 457mm	63.06	107.22	8.80	260.63	m	**367.85**
1671 x 457mm	63.17	107.42	10.31	305.36	m	**412.78**
678 x 508mm	76.82	130.63	5.72	169.41	m	**300.04**
919 x 508mm	78.11	132.81	7.30	216.21	m	**349.02**
1321 x 508mm	78.16	132.90	8.75	259.15	m	**392.06**
Extra over fittings; flat oval duct class B						
End cap						
345 x 102mm	11.32	19.25	0.20	5.92	nr	**25.17**
427 x 102mm	11.60	19.72	0.20	5.92	nr	**25.65**
508 x 102mm	11.86	20.17	0.20	5.92	nr	**26.10**
559 x 152mm	17.35	29.49	0.29	8.59	nr	**38.08**
531 x 203mm	17.27	29.36	0.29	8.59	nr	**37.95**
851 x 203mm	17.78	30.23	0.44	13.03	nr	**43.26**
582 x 254mm	17.72	30.12	0.44	13.03	nr	**43.16**
823 x 254mm	25.21	42.87	0.44	13.03	nr	**55.90**
1303 x 254mm	18.55	31.54	0.69	20.44	nr	**51.98**
632 x 305mm	19.09	32.45	0.69	20.44	nr	**52.89**
1275 x 305mm	18.52	31.50	0.69	20.44	nr	**51.94**
765 x 356mm	19.53	33.21	0.69	20.44	nr	**53.65**
1727 x 356mm	18.80	31.96	0.69	20.44	nr	**52.40**
737 x 406mm	19.11	32.50	1.04	30.80	nr	**63.30**
818 x 406mm	27.10	46.08	0.69	20.44	nr	**66.52**
978 x 406mm	20.43	34.74	0.69	20.44	nr	**55.18**

U:VENTILATION/AIR CONDITIONING SYSTEMS

Item	Net Price £	Material £	Labour hours	Labour £	Unit	Total rate £
U10 : DUCTWORK : FLAT OVAL (cont'd)						
Y30 - AIR DUCTLINES (cont'd)						
Galvanised sheet metal DW144 class B spirally wound flat oval section ductwork (cont'd)						
Extra over fittings; flat oval duct class B (cont'd)						
End cap (cont'd)						
1379 x 406mm	53.75	91.40	1.04	30.80	nr	122.20
1699 x 406mm	55.04	93.60	1.04	30.80	nr	124.40
709 x 457mm	56.68	96.38	1.04	30.80	nr	127.18
1189 x 457mm	64.22	109.19	1.04	30.80	nr	140.00
1671 x 457mm	79.93	135.92	1.04	30.80	nr	166.72
678 x 508mm	80.01	136.05	1.04	30.80	nr	166.86
919 x 508mm	82.19	139.75	1.04	30.80	nr	170.55
1321 x 508mm	81.71	138.93	1.04	30.80	nr	169.73
Reducer						
345 x 102mm	21.49	36.54	0.95	28.14	nr	64.67
427 x 102mm	21.88	37.20	1.06	31.39	nr	68.59
508 x 102mm	22.38	38.06	1.13	33.47	nr	71.53
559 x 152mm	31.64	53.79	1.26	37.32	nr	91.11
531 x 203mm	31.45	53.48	1.26	37.32	nr	90.80
851 x 203mm	33.48	56.92	1.34	39.69	nr	96.61
582 x 254mm	36.90	62.75	1.34	39.69	nr	102.43
823 x 254mm	46.83	79.64	1.34	39.69	nr	119.32
1303 x 254mm	44.21	75.17	1.34	39.69	nr	114.86
632 x 305mm	43.83	74.52	0.70	20.73	nr	95.26
1275 x 305mm	39.82	67.70	1.16	34.36	nr	102.06
765 x 356mm	44.61	75.86	1.16	34.36	nr	110.21
1247 x 356mm	42.52	72.30	1.16	34.36	nr	106.66
1727 x 356mm	41.45	70.48	1.25	37.02	nr	107.50
737 x 406mm	50.70	86.22	1.16	34.36	nr	120.57
818 x 406mm	48.90	83.15	1.27	37.61	nr	120.77
978 x 406mm	92.49	157.27	1.44	42.65	nr	199.92
1379 x 406mm	103.46	175.92	1.44	42.65	nr	218.57
1699 x 406mm	101.83	173.15	1.44	42.65	nr	215.80
709 x 457mm	100.44	170.79	1.16	34.36	nr	205.14
1189 x 457mm	104.57	177.81	1.34	39.69	nr	217.50
1671 x 457mm	106.07	180.36	1.44	42.65	nr	223.01
678 x 508mm	107.40	182.63	1.16	34.36	nr	216.98
919 x 508mm	109.11	185.53	1.26	37.32	nr	222.85
1321 x 508mm	109.88	186.84	1.44	42.65	nr	229.48
90° radius bend						
345 x 102mm	21.76	37.00	0.29	8.59	nr	45.59
427 x 102mm	22.14	37.65	0.58	17.18	nr	54.83
508 x 102mm	22.51	38.27	0.58	17.18	nr	55.45
559 x 152mm	33.99	57.80	0.58	17.18	nr	74.98
531 x 203mm	33.42	56.83	0.87	25.77	nr	82.59
851 x 203mm	37.16	63.18	0.87	25.77	nr	88.94
582 x 254mm	41.45	70.47	0.87	25.77	nr	96.24
823 x 254mm	56.33	95.79	0.87	25.77	nr	121.56
1303 x 254mm	52.73	89.66	0.96	28.43	nr	118.09
632 x 305mm	51.62	87.77	0.87	25.77	nr	113.54

U:VENTILATION/AIR CONDITIONING SYSTEMS

Item	Net Price £	Material £	Labour hours	Labour £	Unit	Total rate £
1275 x 305mm	46.75	79.50	0.96	28.43	nr	**107.93**
765 x 356mm	52.61	89.46	0.87	25.77	nr	**115.23**
1247 x 356mm	46.92	79.78	0.96	28.43	nr	**108.22**
1727 x 356mm	44.12	75.03	1.25	37.02	nr	**112.05**
737 x 406mm	62.24	105.83	0.96	28.43	nr	**134.27**
818 x 406mm	58.54	99.54	0.87	25.77	nr	**125.30**
978 x 406mm	94.44	160.58	0.96	28.43	nr	**189.02**
1379 x 406mm	105.25	178.96	1.16	34.36	nr	**213.32**
1699 x 406mm	102.35	174.03	1.25	37.02	nr	**211.05**
709 x 457mm	101.97	173.39	0.87	25.77	nr	**199.16**
1189 x 457mm	115.45	196.31	0.96	28.43	nr	**224.75**
1671 x 457mm	114.98	195.51	1.25	37.02	nr	**232.53**
678 x 508mm	134.22	228.23	0.87	25.77	nr	**254.00**
919 x 508mm	133.96	227.78	0.96	28.43	nr	**256.21**
1321 x 508mm	133.63	227.23	1.16	34.36	nr	**261.58**
45° radius bend						
345 x 102mm	22.99	39.09	0.79	23.40	nr	**62.49**
427 x 102mm	24.18	41.11	0.85	25.17	nr	**66.29**
508 x 102mm	25.49	43.34	0.95	28.14	nr	**71.48**
559 x 152mm	36.46	62.00	0.79	23.40	nr	**85.40**
531 x 203mm	36.70	62.41	0.85	25.17	nr	**87.59**
851 x 203mm	39.74	67.58	0.98	29.03	nr	**96.60**
582 x 254mm	44.17	75.10	0.76	22.51	nr	**97.61**
823 x 254mm	56.59	96.23	0.95	28.14	nr	**124.37**
1303 x 254mm	54.59	92.83	1.16	34.36	nr	**127.19**
632 x 305mm	55.14	93.75	0.58	17.18	nr	**110.93**
1275 x 305mm	50.78	86.34	1.16	34.36	nr	**120.70**
765 x 356mm	57.83	98.33	0.87	25.77	nr	**124.10**
1247 x 356mm	53.48	90.94	1.16	34.36	nr	**125.30**
1727 x 356mm	53.98	91.78	1.26	37.32	nr	**129.10**
737 x 406mm	67.58	114.90	0.69	20.44	nr	**135.34**
818 x 406mm	67.85	115.38	0.78	23.10	nr	**138.48**
978 x 406mm	109.04	185.41	0.87	25.77	nr	**211.17**
1379 x 406mm	124.61	211.88	1.16	34.36	nr	**246.24**
1699 x 406mm	124.13	211.07	1.27	37.61	nr	**248.69**
709 x 457mm	126.02	214.28	0.81	23.99	nr	**238.27**
1189 x 457mm	133.48	226.97	0.95	28.14	nr	**255.11**
1671 x 457mm	138.30	235.17	1.26	37.32	nr	**272.49**
678 x 508mm	169.58	288.36	0.92	27.25	nr	**315.60**
919 x 508mm	172.37	293.10	1.10	32.58	nr	**325.68**
1321 x 508mm	175.12	297.77	1.25	37.02	nr	**334.79**
90° hard bend with turning vanes						
345 x 102mm	29.42	50.03	0.55	16.29	nr	**66.32**
427 x 102mm	28.99	49.30	1.16	34.36	nr	**83.66**
508 x 102mm	28.59	48.61	1.16	34.36	nr	**82.96**
559 x 152mm	31.17	53.00	1.16	34.36	nr	**87.35**
531 x 203mm	30.98	52.67	1.73	51.24	nr	**103.91**
851 x 203mm	35.36	60.13	1.73	51.24	nr	**111.37**
582 x 254mm	46.28	78.70	1.73	51.24	nr	**129.94**
823 x 254mm	54.17	92.12	1.73	51.24	nr	**143.35**
1303 x 254mm	56.15	95.47	1.82	53.90	nr	**149.38**
632 x 305mm	50.77	86.33	1.73	51.24	nr	**137.57**
1275 x 305mm	46.60	79.24	1.82	53.90	nr	**133.14**
765 x 356mm	49.77	84.63	1.73	51.24	nr	**135.87**
1247 x 356mm	48.95	83.23	1.82	53.90	nr	**137.13**
1727 x 356mm	49.32	83.87	1.82	53.90	nr	**137.77**

U:VENTILATION/AIR CONDITIONING SYSTEMS

Item	Net Price £	Material £	Labour hours	Labour £	Unit	Total rate £
U10 : DUCTWORK : FLAT OVAL (cont'd)						
Y30 - AIR DUCTLINES (cont'd)						
Galvanised sheet metal DW144 class B spirally wound flat oval section ductwork (cont'd)						
Extra over fittings; flat oval duct class B (cont'd)						
90° hard bend with turning vanes (cont'd)						
737 x 406mm	61.22	104.09	1.73	51.24	nr	**155.33**
818 x 406mm	55.40	94.19	1.73	51.24	nr	**145.43**
978 x 406mm	97.60	165.96	1.73	51.24	nr	**217.20**
1379 x 406mm	91.91	156.28	1.82	53.90	nr	**210.18**
1699 x 406mm	91.40	155.42	2.11	62.49	nr	**217.91**
709 x 457mm	91.09	154.88	1.73	51.24	nr	**206.12**
1189 x 457mm	124.34	211.42	1.82	53.90	nr	**265.32**
1671 x 457mm	119.87	203.83	2.11	62.49	nr	**266.32**
678 x 508mm	158.24	269.06	1.82	53.90	nr	**322.97**
919 x 508mm	158.10	268.83	1.82	53.90	nr	**322.74**
1321 x 508mm	156.54	266.18	2.11	62.49	nr	**328.67**
90° branch						
345 x 102mm	20.08	34.14	0.58	17.18	nr	**51.32**
427 x 102mm	19.86	33.77	0.58	17.18	nr	**50.94**
508 x 102mm	19.66	33.43	1.16	34.36	nr	**67.79**
559 x 152mm	27.71	47.12	1.16	34.36	nr	**81.47**
531 x 203mm	27.59	46.92	1.16	34.36	nr	**81.27**
851 x 203mm	27.27	46.37	1.73	51.24	nr	**97.61**
582 x 254mm	29.82	50.70	1.73	51.24	nr	**101.94**
823 x 254mm	35.61	60.55	1.73	51.24	nr	**111.79**
1303 x 254mm	36.64	62.30	1.82	53.90	nr	**116.20**
632 x 305mm	32.90	55.95	1.73	51.24	nr	**107.18**
1275 x 305mm	31.35	53.30	1.82	53.90	nr	**107.20**
765 x 356mm	45.25	76.95	1.73	51.24	nr	**128.19**
1247 x 356mm	41.18	70.02	1.82	53.90	nr	**123.92**
1727 x 356mm	39.72	67.54	2.11	62.49	nr	**130.04**
737 x 406mm	42.59	72.42	1.73	51.24	nr	**123.66**
818 x 406mm	43.62	74.18	1.73	51.24	nr	**125.42**
978 x 406mm	109.61	186.38	1.82	53.90	nr	**240.29**
1379 x 406mm	138.33	235.21	1.93	57.16	nr	**292.37**
1699 x 406mm	137.93	234.54	2.11	62.49	nr	**297.03**
709 x 457mm	121.72	206.97	1.73	51.24	nr	**258.21**
1189 x 457mm	135.64	230.64	1.82	53.90	nr	**284.54**
1671 x 457mm	134.01	227.88	2.11	62.49	nr	**290.37**
678 x 508mm	162.41	276.16	1.73	51.24	nr	**327.40**
919 x 508mm	143.45	243.92	1.82	53.90	nr	**297.83**
1321 x 508mm	142.11	241.64	2.11	62.49	nr	**304.13**
45° branch						
345 x 102mm	28.90	49.13	0.58	17.18	nr	**66.31**
427 x 102mm	28.11	47.79	0.58	17.18	nr	**64.97**
508 x 102mm	27.58	46.90	1.16	34.36	nr	**81.25**
559 x 152mm	38.85	66.06	1.73	51.24	nr	**117.30**
531 x 203mm	38.57	65.59	1.73	51.24	nr	**116.83**
851 x 203mm	37.53	63.81	1.73	51.24	nr	**115.05**
582 x 254mm	36.29	61.71	1.73	51.24	nr	**112.95**
823 x 254mm	48.27	82.07	1.82	53.90	nr	**135.97**

U:VENTILATION/AIR CONDITIONING SYSTEMS

Item	Net Price £	Material £	Labour hours	Labour £	Unit	Total rate £
1303 x 254mm	50.12	85.23	1.92	56.87	nr	**142.10**
632 x 305mm	44.47	75.62	1.73	51.24	nr	**126.85**
1275 x 305mm	38.36	65.23	1.82	53.90	nr	**119.14**
765 x 356mm	62.67	106.56	1.73	51.24	nr	**157.80**
1247 x 356mm	56.77	96.53	1.82	53.90	nr	**150.44**
1727 x 356mm	54.87	93.29	1.82	53.90	nr	**147.20**
737 x 406mm	57.04	97.00	1.73	51.24	nr	**148.24**
818 x 406mm	58.65	99.73	1.73	51.24	nr	**150.96**
978 x 406mm	121.92	207.31	1.73	51.24	nr	**258.55**
1379 x 406mm	162.03	275.52	1.93	57.16	nr	**332.68**
1699 x 406mm	161.34	274.34	2.19	64.86	nr	**339.21**
709 x 457mm	132.33	225.01	1.73	51.24	nr	**276.25**
1189 x 457mm	153.94	261.76	1.82	53.90	nr	**315.66**
1671 x 457mm	150.11	255.24	2.11	62.49	nr	**317.73**
678 x 508mm	252.49	429.33	1.73	51.24	nr	**480.57**
919 x 508mm	210.69	358.25	1.82	53.90	nr	**412.16**
1321 x 508mm	208.41	354.37	1.93	57.16	nr	**411.53**

**For rates for access doors, refer to ancillaries
Within Section U10 : DUCTWORK:
RECTANGULAR: CLASS B**

U:VENTILATION/AIR CONDITIONING SYSTEMS

Item	Net Price £	Material £	Labour hours	Labour £	Unit	Total rate £
U10 : DUCTWORK : FLEXIBLE						
Y30 - AIR DUCTLINES						
Aluminium foil flexible ductwork, DW 144 class B; multiply aluminium polyester laminate fabric, with high tensile steel wire helix						
Duct						
102mm dia	1.12	1.90	0.33	9.77	m	**11.68**
152mm dia	1.64	2.79	0.33	9.77	m	**12.56**
203mm dia	2.16	3.67	0.33	9.77	m	**13.44**
254mm dia	2.71	4.61	0.33	9.77	m	**14.39**
304mm dia	3.32	5.65	0.33	9.77	m	**15.42**
355mm dia	4.53	7.70	0.33	9.77	m	**17.48**
406mm dia	5.03	8.55	0.33	9.77	m	**18.32**
Insulated aluminium foil flexible ductwork, DW144 class B; laminate construction of aluminium and polyester multiply inner core with 25mm insulation; outer layer of multiply aluminium polyester laminate, with high tensile steel wire helix						
Duct						
102mm dia	2.51	4.27	0.50	14.81	m	**19.07**
152mm dia	3.27	5.55	0.50	14.81	m	**20.36**
203mm dia	3.97	6.74	0.50	14.81	m	**21.55**
254mm dia	4.89	8.32	0.50	14.81	m	**23.13**
304mm dia	6.29	10.70	0.50	14.81	m	**25.51**
355mm dia	7.80	13.26	0.50	14.81	m	**28.07**
406mm dia	8.64	14.70	0.50	14.81	m	**29.51**

U:VENTILATION/AIR CONDITIONING SYSTEMS

Item	Net Price £	Material £	Labour hours	Labour £	Unit	Total rate £
U10 : DUCTWORK : PLASTIC						
Y30 - AIR DUCTLINES						
Rigid grey PVC DW 154 circular section duct-work; solvent welded or filler rod welded joints; excludes couplers and supports (these are detailed separately); ductwork to conform to current HSE regulations						
Straight duct (standard length 6m)						
110mm	7.26	12.34	0.17	5.04	m	**17.38**
160mm	14.01	23.82	0.25	7.40	m	**31.22**
200mm	17.51	29.77	0.33	9.77	m	**39.54**
225mm	22.70	38.61	0.42	12.44	m	**51.05**
250mm	22.02	37.44	0.50	14.81	m	**52.25**
315mm	27.71	47.11	0.58	17.18	m	**64.29**
355mm	37.28	63.39	0.67	19.84	m	**83.23**
400mm	46.72	79.44	0.75	22.21	m	**101.65**
450mm	58.59	99.62	0.83	24.58	m	**124.20**
500mm	71.73	121.97	0.92	27.25	m	**149.22**
600mm	107.26	182.38	1.00	29.62	m	**211.99**
Extra for supports (BZP finish) **Horizontal - Maximum 2.4m centres** **Vertical - Maximum 4.0m centres**						
Duct Size						
110mm	6.59	11.21	0.17	5.04	m	**16.25**
160mm	7.25	12.33	0.25	7.40	m	**19.73**
200mm	7.74	13.16	0.33	9.77	m	**22.93**
225mm	8.07	13.72	0.42	12.44	m	**26.15**
250mm	8.72	14.83	0.50	14.81	m	**29.64**
315mm	9.21	15.66	0.58	17.18	m	**32.84**
355mm	12.13	20.63	0.67	19.84	m	**40.47**
400mm	12.46	21.18	0.75	22.21	m	**43.40**
450mm	14.50	24.65	0.83	24.58	m	**49.23**
500mm	14.90	25.34	0.92	27.25	m	**52.58**
600mm	16.30	27.71	1.00	29.62	m	**57.33**
Note - These are maximum figures and may be reduced subject to local conditions (i.e. a high number of changes of direction)						
Extra over fittings; Rigid grey PVC DW 154						
90° Bend						
110mm	16.48	28.02	0.34	10.07	m	**38.09**
160mm	21.55	36.64	0.50	14.81	m	**51.45**
200mm	26.63	45.28	0.66	19.55	m	**64.83**
225mm	31.70	53.90	0.84	24.88	m	**78.78**
250mm	36.35	61.81	1.00	29.62	m	**91.43**
315mm	57.49	97.75	1.16	34.36	m	**132.11**
355mm	77.77	132.24	1.34	39.69	m	**171.93**
400mm	96.78	164.56	1.50	44.43	m	**208.99**
450mm	276.01	469.32	1.66	49.16	m	**518.49**
500mm	326.30	554.84	1.84	54.50	m	**609.33**
600mm	552.01	938.62	2.00	59.23	m	**997.86**

U:VENTILATION/AIR CONDITIONING SYSTEMS

Item	Net Price £	Material £	Labour hours	Labour £	Unit	Total rate £
U10 : DUCTWORK : PLASTIC (cont'd)						
Y30 - AIR DUCTLINES (cont'd)						
Rigid grey PVC DW 154 circular section duct-work (cont'd)						
Extra over fittings; Rigid grey PVC DW 154 (cont'd)						
45° Bend						
110mm	12.26	20.85	0.34	10.07	m	**30.92**
160mm	16.06	27.30	0.50	14.81	m	**42.11**
200mm	19.02	32.34	0.66	19.55	m	**51.89**
225mm	21.99	37.38	0.84	24.88	m	**62.26**
250mm	24.51	41.68	1.00	29.62	m	**71.30**
315mm	38.05	64.70	1.16	34.36	m	**99.06**
355mm	49.86	84.78	1.34	39.69	m	**124.46**
400mm	60.46	102.81	1.50	44.43	m	**147.23**
450mm	195.30	332.08	1.66	49.16	m	**381.24**
500mm	219.36	373.00	1.84	54.50	m	**427.50**
600mm	335.61	570.67	2.00	59.23	m	**629.90**
Tee						
110mm	16.48	28.02	0.51	15.10	m	**43.13**
160mm	21.55	36.64	0.75	22.21	m	**58.86**
200mm	26.52	45.09	0.99	29.32	m	**74.41**
225mm	31.70	53.90	1.26	37.32	m	**91.22**
250mm	36.35	61.81	1.50	44.43	m	**106.24**
315mm	57.49	97.75	1.74	51.53	m	**149.28**
355mm	77.77	132.24	2.01	59.53	m	**191.77**
400mm	96.78	164.56	2.25	66.64	m	**231.20**
450mm	276.01	469.32	2.49	73.75	m	**543.07**
500mm	326.30	554.84	2.76	81.74	m	**636.58**
Coupler						
110mm	7.70	13.09	0.34	10.07	m	**23.16**
160mm	10.70	18.20	0.50	14.81	m	**33.01**
200mm	14.44	24.56	0.66	19.55	m	**44.11**
225mm	19.14	32.55	0.84	24.88	m	**57.42**
250mm	22.14	37.64	1.00	29.62	m	**67.26**
315mm	29.72	50.54	1.16	34.36	m	**84.90**
355mm	29.82	50.71	1.34	39.69	m	**90.40**
400mm	32.46	55.19	1.50	44.43	m	**99.62**
450mm	72.22	122.81	1.66	49.16	m	**171.97**
500mm	81.98	139.40	1.84	54.50	m	**193.89**
Damper						
110mm	45.65	77.62	0.34	10.07	m	**87.69**
160mm	52.41	89.11	0.50	14.81	m	**103.92**
200mm	58.33	99.18	0.66	19.55	m	**118.72**
225mm	62.12	105.63	0.84	24.88	m	**130.50**
250mm	64.26	109.26	1.00	29.62	m	**138.88**
315mm	76.08	129.37	1.16	34.36	m	**163.72**
355mm	80.73	137.26	1.34	39.69	m	**176.95**
400mm	87.49	148.77	1.50	44.43	m	**193.20**

U:VENTILATION/AIR CONDITIONING SYSTEMS

Item	Net Price £	Material £	Labour hours	Labour £	Unit	Total rate £
Reducer						
160 x 110	17.25	29.34	0.42	12.44	m	**41.78**
200 x 110	25.89	44.02	0.50	14.81	m	**58.83**
200 x 160	23.17	39.40	0.58	17.18	m	**56.58**
225 x 200	31.51	53.58	0.75	22.21	m	**75.80**
250 x 160	31.51	53.58	0.75	22.21	m	**75.80**
250 x 200	32.46	55.19	0.83	24.58	m	**79.78**
250 x 225	33.21	56.47	0.92	27.25	m	**83.72**
315 x 200	35.08	59.64	0.92	27.25	m	**86.89**
315 x 250	38.08	64.76	1.10	32.58	m	**97.34**
355 x 200	52.15	88.67	1.10	32.58	m	**121.25**
355 x 250	41.83	71.13	1.17	34.65	m	**105.79**
355 x 315	49.56	84.27	1.25	37.02	m	**121.30**
400 x 225	58.34	99.19	1.17	34.65	m	**133.85**
400 x 315	63.22	107.50	1.33	39.39	m	**146.89**
400 x 355	61.53	104.63	1.42	42.06	m	**146.68**
450 x 315	63.22	107.50	1.45	42.95	m	**150.44**
Flange						
110mm	11.35	19.29	0.34	10.07	m	**29.36**
160mm	13.27	22.56	0.50	14.81	m	**37.36**
200mm	15.19	25.84	0.66	19.55	m	**45.38**
225mm	15.38	26.15	0.84	24.88	m	**51.03**
250mm	15.86	26.97	1.00	29.62	m	**56.58**
315mm	23.84	40.53	1.16	34.36	m	**74.89**
355mm	26.15	44.46	1.34	39.69	m	**84.15**
400mm	29.18	49.62	1.50	44.43	m	**94.04**

U:VENTILATION/AIR CONDITIONING SYSTEMS

Item	Net Price £	Material £	Labour hours	Labour £	Unit	Total rate £
U10 : DUCTWORK : PLASTIC (cont'd)						
Y30 - AIR DUCTLINES (cont'd)						
Polypropylene (PPS) DW154 circular section ductwork; filler rod welded joints; excludes couplers and supports (these are detailed separately); ductwork to conform to current HSE regulations						
Straight duct (standard length 6m)						
110mm	9.41	16.00	0.21	6.22	m	**22.21**
160mm	15.48	26.32	0.31	9.18	m	**35.50**
200mm	19.24	32.71	0.41	12.14	m	**44.86**
225mm	25.02	42.54	0.52	15.40	m	**57.94**
250mm	27.87	47.39	0.63	18.66	m	**66.05**
315mm	49.73	84.55	0.73	21.62	m	**106.17**
355mm	56.09	95.38	0.84	24.88	m	**120.26**
400mm	74.80	127.18	0.94	27.84	m	**155.02**
Extra for supports (BZP finish) **Horizontal - Maximum 2.4m centres** **Vertical - Maximum 4.0m centres**						
Duct Size						
110mm	6.59	11.21	0.21	6.22	m	**17.43**
160mm	7.25	12.33	0.31	9.18	m	**21.51**
200mm	7.74	13.16	0.41	12.14	m	**25.30**
225mm	8.07	13.72	0.52	15.40	m	**29.12**
250mm	8.72	14.83	0.63	18.66	m	**33.49**
315mm	9.21	15.66	0.73	21.62	m	**37.28**
355mm	12.13	20.63	0.84	24.88	m	**45.51**
400mm	12.46	21.18	0.94	27.84	m	**49.02**
Note - These are maximum figures and may be reduced subject to local conditions (i.e. a high number of changes of direction)						
Extra over fittings; Polypropylene (PPS DW 154)						
90° Bend						
110mm	24.70	42.00	0.43	12.74	m	**54.73**
160mm	36.35	61.81	0.63	18.66	m	**80.47**
200mm	43.08	73.25	0.83	24.58	m	**97.83**
225mm	57.49	97.75	1.05	31.10	m	**128.85**
250mm	61.70	104.92	1.25	37.02	m	**141.94**
315mm	135.26	229.99	1.45	42.95	m	**272.94**
355mm	185.98	316.23	1.68	49.76	m	**365.99**
400mm	198.65	337.78	1.88	55.68	m	**393.47**
45° Bend						
110mm	19.02	32.34	0.43	12.74	m	**45.08**
160mm	30.43	51.75	0.63	18.66	m	**70.41**
200mm	34.66	58.94	0.83	24.58	m	**83.52**
225mm	43.12	73.32	1.05	31.10	m	**104.42**
250mm	48.20	81.96	1.25	37.02	m	**118.98**
315mm	122.58	208.44	1.45	42.95	m	**251.38**
355mm	139.49	237.18	1.68	49.76	m	**286.94**
400mm	147.92	251.53	1.88	55.68	m	**307.21**

U:VENTILATION/AIR CONDITIONING SYSTEMS

Item	Net Price £	Material £	Labour hours	Labour £	Unit	Total rate £
Tee						
110mm	85.73	145.77	0.64	18.96	m	**164.73**
160mm	123.00	209.14	0.94	27.84	m	**236.98**
200mm	149.12	253.57	1.24	36.73	m	**290.29**
225mm	165.87	282.03	1.58	46.80	m	**328.83**
250mm	195.70	332.76	1.88	55.68	m	**388.44**
315mm	270.25	459.54	2.18	64.57	m	**524.10**
355mm	335.51	570.50	2.51	74.34	m	**644.84**
400mm	391.42	665.56	2.81	83.22	m	**748.79**
Coupler						
110mm	19.02	32.34	0.43	12.74	m	**45.08**
160mm	23.24	39.51	0.63	18.66	m	**58.17**
200mm	27.89	47.43	0.83	24.58	m	**72.01**
225mm	30.86	52.47	1.05	31.10	m	**83.57**
250mm	32.55	55.34	1.25	37.02	m	**92.36**
315mm	44.37	75.45	1.45	42.95	m	**118.40**
355mm	57.49	97.75	1.68	49.76	m	**147.51**
400mm	62.57	106.39	1.88	55.68	m	**162.07**
Damper						
110mm	82.19	139.75	0.43	12.74	m	**152.48**
160mm	96.77	164.55	0.63	18.66	m	**183.21**
200mm	107.63	183.01	0.83	24.58	m	**207.59**
225mm	114.54	194.76	1.05	31.10	m	**225.86**
250mm	121.31	206.27	1.25	37.02	m	**243.29**
315mm	138.64	235.74	1.45	42.95	m	**278.68**
355mm	151.75	258.03	1.68	49.76	m	**307.79**
400mm	166.12	282.46	1.88	55.68	m	**338.14**
Reducer						
160 x 110	60.87	103.49	0.53	15.70	m	**119.19**
200 x 160	66.78	113.56	0.73	21.62	m	**135.18**
225 x 200	92.99	158.11	0.94	27.84	m	**185.96**
250 x 200	88.77	150.94	1.04	30.80	m	**181.74**
250 x 225	113.17	192.44	1.15	34.06	m	**226.50**
315 x 200	144.56	245.80	1.15	34.06	m	**279.86**
315 x 250	120.76	205.34	1.38	40.87	m	**246.21**
355 x 200	166.53	283.16	1.38	40.87	m	**324.04**
355 x 250	136.11	231.44	1.46	43.24	m	**274.68**
355 x 315	155.54	264.48	1.56	46.20	m	**310.69**
400 x 315	164.84	280.29	1.66	49.16	m	**329.46**
400 x 355	195.30	332.08	1.78	52.72	m	**384.80**
Flange						
110mm	18.00	30.60	0.43	12.74	m	**43.34**
160mm	21.95	37.33	0.63	18.66	m	**55.99**
200mm	25.69	43.69	0.83	24.58	m	**68.27**
225mm	27.04	45.98	1.05	31.10	m	**77.08**
250mm	30.11	51.19	1.25	37.02	m	**88.21**
315mm	33.98	57.77	1.45	42.95	m	**100.72**
355mm	38.22	64.98	1.68	49.76	m	**114.74**
400mm	42.38	72.06	1.88	55.68	m	**127.74**

U:VENTILATION/AIR CONDITIONING SYSTEMS

Item	Net Price £	Material £	Labour hours	Labour £	Unit	Total rate £
U10 : DUCTWORK : RECTANGULAR – CLASS B						
Y30 - AIR DUCTLINES						
Galvanised sheet metal DW144 class B rectangular section ductwork; including all necessary stiffeners, joints, couplers in the running length and duct supports						
Ductwork up to 400mm longest side						
Sum of two sides 200mm	19.86	33.78	1.19	35.24	m	**69.02**
Sum of two sides 300mm	21.15	35.97	1.19	35.24	m	**71.22**
Sum of two sides 400mm	17.57	29.87	1.16	34.36	m	**64.23**
Sum of two sides 500mm	18.99	32.30	1.16	34.36	m	**66.65**
Sum of two sides 600mm	20.13	34.23	1.27	37.61	m	**71.84**
Sum of two sides 700mm	21.25	36.14	1.27	37.61	m	**73.75**
Sum of two sides 800mm	22.50	38.26	1.27	37.61	m	**75.87**
Extra Over fittings; Rectangular ductwork class B; upto 400mm longest side						
End Cap						
Sum of two sides 200mm	9.15	15.57	0.38	11.25	nr	**26.82**
Sum of two sides 300mm	10.18	17.31	0.38	11.25	nr	**28.56**
Sum of two sides 400mm	11.20	19.05	0.38	11.25	nr	**30.31**
Sum of two sides 500mm	12.23	20.80	0.38	11.25	nr	**32.06**
Sum of two sides 600mm	13.26	22.55	0.38	11.25	nr	**33.80**
Sum of two sides 700mm	14.28	24.28	0.38	11.25	nr	**35.53**
Sum of two sides 800mm	15.30	26.02	0.38	11.25	nr	**37.28**
Reducer						
Sum of two sides 200mm	9.69	16.47	1.40	41.46	nr	**57.93**
Sum of two sides 300mm	10.93	18.58	1.40	41.46	nr	**60.04**
Sum of two sides 400mm	18.01	30.63	1.42	42.06	nr	**72.69**
Sum of two sides 500mm	19.56	33.25	1.42	42.06	nr	**75.31**
Sum of two sides 600mm	21.10	35.88	1.69	50.05	nr	**85.93**
Sum of two sides 700mm	22.65	38.51	1.69	50.05	nr	**88.56**
Sum of two sides 800mm	24.17	41.10	1.92	56.87	nr	**97.97**
Offset						
Sum of two sides 200mm	11.71	19.91	1.63	48.28	nr	**68.19**
Sum of two sides 300mm	13.24	22.51	1.63	48.28	nr	**70.79**
Sum of two sides 400mm	19.70	33.49	1.65	48.87	nr	**82.36**
Sum of two sides 500mm	21.47	36.51	1.65	48.87	nr	**85.38**
Sum of two sides 600mm	22.93	38.99	1.92	56.87	nr	**95.86**
Sum of two sides 700mm	24.55	41.74	1.92	56.87	nr	**98.61**
Sum of two sides 800mm	25.98	44.18	1.92	56.87	nr	**101.04**
Square to round						
Sum of two sides 200mm	22.32	37.95	1.63	48.28	nr	**86.23**
Sum of two sides 300mm	25.07	42.64	1.63	48.28	nr	**90.91**
Sum of two sides 400mm	33.51	56.98	1.65	48.87	nr	**105.85**
Sum of two sides 500mm	36.53	62.11	1.65	48.87	nr	**110.98**
Sum of two sides 600mm	39.56	67.26	1.92	56.87	nr	**124.13**
Sum of two sides 700mm	42.55	72.35	1.92	56.87	nr	**129.22**
Sum of two sides 800mm	45.54	77.44	1.92	56.87	nr	**134.31**

U:VENTILATION/AIR CONDITIONING SYSTEMS

Item	Net Price £	Material £	Labour hours	Labour £	Unit	Total rate £
90° radius bend						
Sum of two sides 200mm	11.26	19.15	1.22	36.13	nr	**55.28**
Sum of two sides 300mm	12.14	20.65	1.22	36.13	nr	**56.78**
Sum of two sides 400mm	20.63	35.08	1.25	37.02	nr	**72.10**
Sum of two sides 500mm	22.04	37.47	1.25	37.02	nr	**74.49**
Sum of two sides 600mm	23.87	40.59	1.33	39.39	nr	**79.98**
Sum of two sides 700mm	25.48	43.33	1.33	39.39	nr	**82.72**
Sum of two sides 800mm	27.27	46.37	1.40	41.46	nr	**87.84**
45° radius bend						
Sum of two sides 200mm	11.15	18.97	0.89	26.36	nr	**45.33**
Sum of two sides 300mm	12.24	20.81	1.12	33.17	nr	**53.99**
Sum of two sides 400mm	19.80	33.66	1.10	32.58	nr	**66.24**
Sum of two sides 500mm	21.29	36.20	1.10	32.58	nr	**68.78**
Sum of two sides 600mm	22.98	39.07	1.16	34.36	nr	**73.42**
Sum of two sides 700mm	24.57	41.78	1.16	34.36	nr	**76.13**
Sum of two sides 800mm	26.24	44.62	1.22	36.13	nr	**80.75**
90° mitre bend						
Sum of two sides 200mm	15.37	26.14	1.29	38.21	nr	**64.34**
Sum of two sides 300mm	16.81	28.58	1.29	38.21	nr	**66.79**
Sum of two sides 400mm	24.56	41.75	1.29	38.21	nr	**79.96**
Sum of two sides 500mm	26.51	45.08	1.29	38.21	nr	**83.29**
Sum of two sides 600mm	28.99	49.29	1.39	41.17	nr	**90.46**
Sum of two sides 700mm	31.27	53.17	1.39	41.17	nr	**94.34**
Sum of two sides 800mm	33.77	57.43	1.46	43.24	nr	**100.67**
Branch						
Sum of two sides 200mm	19.13	32.53	0.92	27.25	nr	**59.78**
Sum of two sides 300mm	21.23	36.11	0.92	27.25	nr	**63.36**
Sum of two sides 400mm	26.64	45.29	0.95	28.14	nr	**73.43**
Sum of two sides 500mm	29.00	49.31	0.95	28.14	nr	**77.45**
Sum of two sides 600mm	31.32	53.26	1.03	30.51	nr	**83.76**
Sum of two sides 700mm	33.64	57.19	1.03	30.51	nr	**87.70**
Sum of two sides 800mm	35.95	61.13	1.03	30.51	nr	**91.64**
Grille neck						
Sum of two sides 200mm	16.72	28.42	1.10	32.58	nr	**61.00**
Sum of two sides 300mm	18.65	31.72	1.10	32.58	nr	**64.30**
Sum of two sides 400mm	20.59	35.01	1.16	34.36	nr	**69.36**
Sum of two sides 500mm	22.53	38.30	1.16	34.36	nr	**72.66**
Sum of two sides 600mm	24.46	41.60	1.18	34.95	nr	**76.55**
Sum of two sides 700mm	26.40	44.88	1.18	34.95	nr	**79.83**
Sum of two sides 800mm	28.33	48.18	1.18	34.95	nr	**83.13**
Ductwork 401 to 600mm longest side						
Sum of two sides 600mm	21.24	36.12	1.27	37.61	m	**73.73**
Sum of two sides 700mm	22.86	38.87	1.27	37.61	m	**76.48**
Sum of two sides 800mm	24.33	41.38	1.27	37.61	m	**78.99**
Sum of two sides 900mm	25.70	43.71	1.27	37.61	m	**81.32**
Sum of two sides 1000mm	27.06	46.02	1.37	40.58	m	**86.59**
Sum of two sides 1100mm	28.77	48.91	1.37	40.58	m	**89.49**
Sum of two sides 1200mm	30.14	51.25	1.37	40.58	m	**91.83**

U:VENTILATION/AIR CONDITIONING SYSTEMS

Item	Net Price £	Material £	Labour hours	Labour £	Unit	Total rate £
U10 : DUCTWORK : RECTANGULAR–CLASS B (cont'd)						
Y30 - AIR DUCTLINES (cont'd)						
Extra over fittings; Ductwork 401 to 600mm longest side						
End Cap						
Sum of two sides 600mm	13.31	22.63	0.38	11.25	nr	**33.88**
Sum of two sides 700mm	14.35	24.39	0.38	11.25	nr	**35.65**
Sum of two sides 800mm	15.38	26.16	0.38	11.25	nr	**37.41**
Sum of two sides 900mm	16.43	27.94	0.58	17.18	nr	**45.12**
Sum of two sides 1000mm	17.47	29.70	0.58	17.18	nr	**46.88**
Sum of two sides 1100mm	18.51	31.47	0.58	17.18	nr	**48.65**
Sum of two sides 1200mm	19.54	33.23	0.58	17.18	nr	**50.41**
Reducer						
Sum of two sides 600mm	19.04	32.37	1.69	50.05	nr	**82.42**
Sum of two sides 700mm	20.46	34.78	1.69	50.05	nr	**84.84**
Sum of two sides 800mm	21.86	37.18	1.92	56.87	nr	**94.04**
Sum of two sides 900mm	23.29	39.60	1.92	56.87	nr	**96.47**
Sum of two sides 1000mm	24.72	42.03	2.18	64.57	nr	**106.59**
Sum of two sides 1100mm	26.26	44.65	2.18	64.57	nr	**109.22**
Sum of two sides 1200mm	27.69	47.08	2.18	64.57	nr	**111.64**
Offset						
Sum of two sides 600mm	22.04	37.48	1.92	56.87	nr	**94.35**
Sum of two sides 700mm	23.70	40.31	1.92	56.87	nr	**97.17**
Sum of two sides 800mm	25.06	42.60	1.92	56.87	nr	**99.47**
Sum of two sides 900mm	26.41	44.91	1.92	56.87	nr	**101.78**
Sum of two sides 1000mm	27.93	47.49	2.18	64.57	nr	**112.05**
Sum of two sides 1100mm	29.33	49.88	2.18	64.57	nr	**114.44**
Sum of two sides 1200mm	30.67	52.16	2.18	64.57	nr	**116.72**
Square to round						
Sum of two sides 600mm	31.04	52.79	1.33	39.39	nr	**92.18**
Sum of two sides 700mm	33.46	56.90	1.33	39.39	nr	**96.29**
Sum of two sides 800mm	35.86	60.97	1.40	41.46	nr	**102.44**
Sum of two sides 900mm	38.28	65.10	1.40	41.46	nr	**106.56**
Sum of two sides 1000mm	40.71	69.22	1.82	53.90	nr	**123.13**
Sum of two sides 1100mm	43.18	73.42	1.82	53.90	nr	**127.33**
Sum of two sides 1200mm	45.60	77.54	1.82	53.90	nr	**131.44**
90° radius bend						
Sum of two sides 600mm	19.44	33.06	1.16	34.36	nr	**67.42**
Sum of two sides 700mm	20.58	34.99	1.16	34.36	nr	**69.35**
Sum of two sides 800mm	22.17	37.70	1.22	36.13	nr	**73.84**
Sum of two sides 900mm	23.80	40.46	1.22	36.13	nr	**76.60**
Sum of two sides 1000mm	25.19	42.83	1.40	41.46	nr	**84.29**
Sum of two sides 1100mm	26.90	45.74	1.40	41.46	nr	**87.21**
Sum of two sides 1200mm	28.53	48.50	1.40	41.46	nr	**89.97**
45° bend						
Sum of two sides 600mm	22.33	37.97	1.16	34.36	nr	**72.32**
Sum of two sides 700mm	23.77	40.41	1.39	41.17	nr	**81.58**
Sum of two sides 800mm	25.46	43.30	1.46	43.24	nr	**86.54**
Sum of two sides 900mm	27.18	46.22	1.46	43.24	nr	**89.46**
Sum of two sides 1000mm	28.76	48.90	1.88	55.68	nr	**104.58**
Sum of two sides 1100mm	30.62	52.07	1.88	55.68	nr	**107.75**
Sum of two sides 1200mm	32.34	54.99	1.88	55.68	nr	**110.67**

U:VENTILATION/AIR CONDITIONING SYSTEMS

Item	Net Price £	Material £	Labour hours	Labour £	Unit	Total rate £
90° mitre bend						
Sum of two sides 600mm	26.68	45.37	1.39	41.17	nr	86.53
Sum of two sides 700mm	28.27	48.06	2.16	63.97	nr	112.04
Sum of two sides 800mm	30.46	51.80	2.26	66.94	nr	118.73
Sum of two sides 900mm	32.69	55.59	2.26	66.94	nr	122.52
Sum of two sides 1000mm	34.69	58.99	3.01	89.15	nr	148.14
Sum of two sides 1100mm	37.03	62.96	3.01	89.15	nr	152.11
Sum of two sides 1200mm	39.30	66.82	3.01	89.15	nr	155.97
Branch						
Sum of two sides 600mm	28.20	47.95	1.03	30.51	nr	78.45
Sum of two sides 700mm	30.30	51.52	1.03	30.51	nr	82.02
Sum of two sides 800mm	32.40	55.10	1.03	30.51	nr	85.60
Sum of two sides 900mm	34.50	58.66	1.03	30.51	nr	89.17
Sum of two sides 1000mm	36.61	62.25	1.29	38.21	nr	100.46
Sum of two sides 1100mm	38.82	66.01	1.29	38.21	nr	104.22
Sum of two sides 1200mm	40.92	69.58	1.29	38.21	nr	107.79
Grille neck						
Sum of two sides 600mm	24.44	41.56	1.18	34.95	nr	76.50
Sum of two sides 700mm	26.40	44.88	1.18	34.95	nr	79.83
Sum of two sides 800mm	28.36	48.23	1.18	34.95	nr	83.18
Sum of two sides 900mm	30.33	51.57	1.18	34.95	nr	86.52
Sum of two sides 1000mm	32.18	54.72	1.44	42.65	nr	97.37
Sum of two sides 1100mm	34.25	58.24	1.44	42.65	nr	100.89
Sum of two sides 1200mm	36.22	61.58	1.44	42.65	nr	104.23
Ductwork 601 to 800mm longest side						
Sum of two sides 900mm	29.62	50.37	1.27	37.61	m	87.99
Sum of two sides 1000mm	30.83	52.42	1.37	40.58	m	92.99
Sum of two sides 1100mm	32.27	54.86	1.37	40.58	m	95.44
Sum of two sides 1200mm	33.64	57.21	1.37	40.58	m	97.78
Sum of two sides 1300mm	34.95	59.43	1.40	41.46	m	100.89
Sum of two sides 1400mm	36.23	61.60	1.40	41.46	m	103.07
Sum of two sides 1500mm	37.52	63.80	1.48	43.83	m	107.63
Sum of two sides 1600mm	38.78	65.95	1.55	45.91	m	111.86
Extra over fittings: Ductwork 601 to 800mm longest side						
End Cap						
Sum of two sides 900mm	17.11	29.09	0.58	17.18	nr	46.26
Sum of two sides 1000mm	18.14	30.84	0.58	17.18	nr	48.02
Sum of two sides 1100mm	19.15	32.56	0.58	17.18	nr	49.74
Sum of two sides 1200mm	20.16	34.28	0.58	17.18	nr	51.46
Sum of two sides 1300mm	21.15	35.97	0.58	17.18	nr	53.15
Sum of two sides 1400mm	22.17	37.70	0.58	17.18	nr	54.88
Sum of two sides 1500mm	24.46	41.59	0.58	17.18	nr	58.76
Sum of two sides 1600mm	26.84	45.64	0.58	17.18	nr	62.82
Reducer						
Sum of two sides 900mm	22.52	38.29	1.92	56.87	nr	95.16
Sum of two sides 1000mm	23.86	40.57	2.18	64.57	nr	105.13
Sum of two sides 1100mm	25.12	42.72	2.18	64.57	nr	107.29
Sum of two sides 1200mm	26.64	45.29	2.18	64.57	nr	109.86
Sum of two sides 1300mm	28.00	47.61	2.30	68.12	nr	115.73
Sum of two sides 1400mm	29.39	49.97	2.30	68.12	nr	118.09
Sum of two sides 1500mm	33.41	56.81	2.47	73.16	nr	129.96
Sum of two sides 1600mm	37.67	64.05	2.47	73.16	nr	137.20

U:VENTILATION/AIR CONDITIONING SYSTEMS

Item	Net Price £	Material £	Labour hours	Labour £	Unit	Total rate £
U10 : DUCTWORK : RECTANGULAR– CLASS B (cont'd)						
Y30 - AIR DUCTLINES (cont'd)						
Extra over fittings: Ductwork 601 to 800mm longest side (cont'd)						
Offset						
Sum of two sides 900mm	26.64	45.29	1.92	56.87	nr	**102.16**
Sum of two sides 1000mm	27.49	46.74	2.18	64.57	nr	**111.31**
Sum of two sides 1100mm	28.34	48.19	2.18	64.57	nr	**112.75**
Sum of two sides 1200mm	29.29	49.81	2.18	64.57	nr	**114.37**
Sum of two sides 1300mm	30.11	51.20	2.30	68.12	nr	**119.32**
Sum of two sides 1400mm	31.25	53.13	2.30	68.12	nr	**121.25**
Sum of two sides 1500mm	35.12	59.71	2.47	73.16	nr	**132.87**
Sum of two sides 1600mm	39.30	66.82	2.47	73.16	nr	**139.97**
Square to round						
Sum of two sides 900mm	34.29	58.31	1.40	41.46	nr	**99.77**
Sum of two sides 1000mm	36.98	62.88	1.82	53.90	nr	**116.79**
Sum of two sides 1100mm	39.66	67.44	1.82	53.90	nr	**121.34**
Sum of two sides 1200mm	42.43	72.15	1.82	53.90	nr	**126.06**
Sum of two sides 1300mm	45.24	76.92	2.15	63.68	nr	**140.59**
Sum of two sides 1400mm	48.07	81.74	2.15	63.68	nr	**145.41**
Sum of two sides 1500mm	54.80	93.18	2.38	70.49	nr	**163.67**
Sum of two sides 1600mm	61.90	105.25	2.38	70.49	nr	**175.74**
90° radius bend						
Sum of two sides 900mm	22.85	38.85	1.22	36.13	nr	**74.98**
Sum of two sides 1000mm	24.76	42.10	1.40	41.46	nr	**83.57**
Sum of two sides 1100mm	26.57	45.18	1.40	41.46	nr	**86.64**
Sum of two sides 1200mm	28.53	48.50	1.40	41.46	nr	**89.97**
Sum of two sides 1300mm	30.48	51.83	1.91	56.57	nr	**108.40**
Sum of two sides 1400mm	31.92	54.28	1.91	56.57	nr	**110.85**
Sum of two sides 1500mm	36.95	62.83	2.11	62.49	nr	**125.32**
Sum of two sides 1600mm	42.28	71.89	2.11	62.49	nr	**134.38**
45° bend						
Sum of two sides 900mm	26.54	45.13	1.22	36.13	nr	**81.27**
Sum of two sides 1000mm	28.33	48.18	1.40	41.46	nr	**89.64**
Sum of two sides 1100mm	30.03	51.06	1.88	55.68	nr	**106.75**
Sum of two sides 1200mm	32.02	54.44	1.88	55.68	nr	**110.13**
Sum of two sides 1300mm	33.89	57.63	2.26	66.94	nr	**124.56**
Sum of two sides 1400mm	35.51	60.38	2.26	66.94	nr	**127.31**
Sum of two sides 1500mm	40.44	68.76	2.49	73.75	nr	**142.51**
Sum of two sides 1600mm	45.69	77.68	2.49	73.75	nr	**151.43**
90° mitre bend						
Sum of two sides 900mm	32.48	55.23	1.22	36.13	nr	**91.37**
Sum of two sides 1000mm	34.98	59.47	1.40	41.46	nr	**100.94**
Sum of two sides 1100mm	37.40	63.60	3.01	89.15	nr	**152.75**
Sum of two sides 1200mm	39.91	67.86	3.01	89.15	nr	**157.01**
Sum of two sides 1300mm	42.45	72.17	3.67	108.70	nr	**180.87**
Sum of two sides 1400mm	44.55	75.75	3.67	108.70	nr	**184.45**
Sum of two sides 1500mm	49.91	84.87	4.07	120.54	nr	**205.42**
Sum of two sides 1600mm	55.81	94.90	4.07	120.54	nr	**215.44**

U:VENTILATION/AIR CONDITIONING SYSTEMS

Item	Net Price £	Material £	Labour hours	Labour £	Unit	Total rate £
Branch						
Sum of two sides 900mm	36.24	61.62	1.22	36.13	nr	**97.76**
Sum of two sides 1000mm	38.52	65.50	1.40	41.46	nr	**106.96**
Sum of two sides 1100mm	40.74	69.27	1.29	38.21	nr	**107.47**
Sum of two sides 1200mm	43.15	73.37	1.29	38.21	nr	**111.58**
Sum of two sides 1300mm	45.44	77.26	1.39	41.17	nr	**118.43**
Sum of two sides 1400mm	47.74	81.18	1.39	41.17	nr	**122.35**
Sum of two sides 1500mm	54.05	91.91	1.64	48.57	nr	**140.48**
Sum of two sides 1600mm	60.65	103.13	1.64	48.57	nr	**151.70**
Grille neck						
Sum of two sides 900mm	29.09	49.47	1.22	36.13	nr	**85.60**
Sum of two sides 1000mm	31.01	52.73	1.40	41.46	nr	**94.20**
Sum of two sides 1100mm	32.92	55.98	1.44	42.65	nr	**98.63**
Sum of two sides 1200mm	34.85	59.26	1.44	42.65	nr	**101.91**
Sum of two sides 1300mm	36.81	62.59	1.69	50.05	nr	**112.64**
Sum of two sides 1400mm	39.03	66.37	1.69	50.05	nr	**116.42**
Sum of two sides 1500mm	44.87	76.29	1.79	53.02	nr	**129.30**
Sum of two sides 1600mm	51.00	86.72	1.79	53.02	nr	**139.74**
Ductwork 801 to 1000mm longest side						
Sum of two sides 1100mm	40.54	68.93	1.37	40.58	m	**109.51**
Sum of two sides 1200mm	42.88	72.92	1.37	40.58	m	**113.49**
Sum of two sides 1300mm	45.23	76.91	1.40	41.46	m	**118.37**
Sum of two sides 1400mm	47.84	81.35	1.40	41.46	m	**122.81**
Sum of two sides 1500mm	50.19	85.34	1.48	43.83	m	**129.17**
Sum of two sides 1600mm	52.53	89.32	1.55	45.91	m	**135.23**
Sum of two sides 1700mm	54.87	93.30	1.55	45.91	m	**139.21**
Sum of two sides 1800mm	57.49	97.75	1.61	47.68	m	**145.44**
Sum of two sides 1900mm	59.83	101.74	1.61	47.68	m	**149.43**
Sum of two sides 2000mm	62.17	105.72	1.61	47.68	m	**153.40**
Extra over fittings; Ductwork 801 to 1000mm longest side						
End Cap						
Sum of two sides 1100mm	19.88	33.81	1.44	42.65	nr	**76.46**
Sum of two sides 1200mm	21.00	35.71	1.44	42.65	nr	**78.36**
Sum of two sides 1300mm	22.11	37.60	1.44	42.65	nr	**80.25**
Sum of two sides 1400mm	23.07	39.22	1.44	42.65	nr	**81.87**
Sum of two sides 1500mm	26.85	45.66	1.44	42.65	nr	**88.31**
Sum of two sides 1600mm	30.64	52.10	1.44	42.65	nr	**94.75**
Sum of two sides 1700mm	34.43	58.54	1.44	42.65	nr	**101.19**
Sum of two sides 1800mm	38.21	64.97	1.44	42.65	nr	**107.62**
Sum of two sides 1900mm	42.01	71.43	1.44	42.65	nr	**114.08**
Sum of two sides 2000mm	45.79	77.86	1.44	42.65	nr	**120.51**
Reducer						
Sum of two sides 1100mm	17.24	29.32	1.44	42.65	nr	**71.97**
Sum of two sides 1200mm	18.09	30.75	1.44	42.65	nr	**73.40**
Sum of two sides 1300mm	18.93	32.19	1.69	50.05	nr	**82.25**
Sum of two sides 1400mm	19.72	33.53	1.69	50.05	nr	**83.58**
Sum of two sides 1500mm	22.94	39.02	2.47	73.16	nr	**112.17**
Sum of two sides 1600mm	26.17	44.49	2.47	73.16	nr	**117.65**
Sum of two sides 1700mm	29.39	49.97	2.47	73.16	nr	**123.13**
Sum of two sides 1800mm	32.70	55.60	2.59	76.71	nr	**132.31**
Sum of two sides 1900mm	35.92	61.08	2.71	80.26	nr	**141.34**
Sum of two sides 2000mm	39.15	66.57	2.71	80.26	nr	**146.83**

U:VENTILATION/AIR CONDITIONING SYSTEMS

Item	Net Price £	Material £	Labour hours	Labour £	Unit	Total rate £
U10 : DUCTWORK : RECTANGULAR – CLASS B (cont'd)						
Y30 - AIR DUCTLINES (cont'd)						
Extra over fittings; Ductwork 801 to 1000mm longest side (cont'd)						
Offset						
Sum of two sides 1100mm	26.79	45.55	1.44	42.65	nr	**88.20**
Sum of two sides 1200mm	27.30	46.43	1.44	42.65	nr	**89.07**
Sum of two sides 1300mm	27.73	47.15	1.69	50.05	nr	**97.20**
Sum of two sides 1400mm	27.92	47.48	1.69	50.05	nr	**97.53**
Sum of two sides 1500mm	31.80	54.07	2.47	73.16	nr	**127.22**
Sum of two sides 1600mm	35.59	60.51	2.47	73.16	nr	**133.67**
Sum of two sides 1700mm	39.30	66.82	2.59	76.71	nr	**143.53**
Sum of two sides 1800mm	42.96	73.05	2.61	77.30	nr	**150.36**
Sum of two sides 1900mm	47.08	80.06	2.71	80.26	nr	**160.32**
Sum of two sides 2000mm	50.56	85.97	2.71	80.26	nr	**166.23**
Square to round						
Sum of two sides 1100mm	34.07	57.93	1.44	42.65	nr	**100.58**
Sum of two sides 1200mm	35.86	60.97	1.44	42.65	nr	**103.62**
Sum of two sides 1300mm	37.66	64.03	1.69	50.05	nr	**114.08**
Sum of two sides 1400mm	39.12	66.53	1.69	50.05	nr	**116.58**
Sum of two sides 1500mm	46.49	79.05	2.38	70.49	nr	**149.54**
Sum of two sides 1600mm	53.85	91.56	2.38	70.49	nr	**162.05**
Sum of two sides 1700mm	61.20	104.07	2.55	75.52	nr	**179.60**
Sum of two sides 1800mm	68.59	116.63	2.55	75.52	nr	**192.15**
Sum of two sides 1900mm	75.95	129.14	2.83	83.82	nr	**212.95**
Sum of two sides 2000mm	83.30	141.65	2.83	83.82	nr	**225.47**
90° radius bend						
Sum of two sides 1100mm	15.98	27.18	1.44	42.65	nr	**69.82**
Sum of two sides 1200mm	17.10	29.07	1.44	42.65	nr	**71.72**
Sum of two sides 1300mm	18.22	30.97	1.69	50.05	nr	**81.03**
Sum of two sides 1400mm	19.21	32.66	1.69	50.05	nr	**82.72**
Sum of two sides 1500mm	23.12	39.32	2.11	62.49	nr	**101.81**
Sum of two sides 1600mm	27.03	45.96	2.11	62.49	nr	**108.46**
Sum of two sides 1700mm	30.95	52.62	2.26	66.94	nr	**119.55**
Sum of two sides 1800mm	34.91	59.36	2.26	66.94	nr	**126.29**
Sum of two sides 1900mm	38.09	64.76	2.48	73.45	nr	**138.21**
Sum of two sides 2000mm	41.99	71.40	2.48	73.45	nr	**144.85**
45° bend						
Sum of two sides 1100mm	23.03	39.15	1.44	42.65	nr	**81.80**
Sum of two sides 1200mm	24.36	41.43	1.44	42.65	nr	**84.08**
Sum of two sides 1300mm	25.70	43.71	1.69	50.05	nr	**93.76**
Sum of two sides 1400mm	26.97	45.86	1.69	50.05	nr	**95.91**
Sum of two sides 1500mm	31.35	53.30	2.49	73.75	nr	**127.05**
Sum of two sides 1600mm	35.73	60.75	2.49	73.75	nr	**134.50**
Sum of two sides 1700mm	40.11	68.19	2.67	79.08	nr	**147.27**
Sum of two sides 1800mm	44.61	75.85	2.67	79.08	nr	**154.93**
Sum of two sides 1900mm	48.57	82.60	3.06	90.63	nr	**173.23**
Sum of two sides 2000mm	52.95	90.04	3.06	90.63	nr	**180.67**

U:VENTILATION/AIR CONDITIONING SYSTEMS

Item	Net Price £	Material £	Labour hours	Labour £	Unit	Total rate £
90° mitre bend						
Sum of two sides 1100mm	31.05	52.80	1.44	42.65	nr	**95.45**
Sum of two sides 1200mm	33.07	56.23	1.44	42.65	nr	**98.88**
Sum of two sides 1300mm	35.09	59.67	1.69	50.05	nr	**109.73**
Sum of two sides 1400mm	36.90	62.74	1.69	50.05	nr	**112.79**
Sum of two sides 1500mm	42.90	72.95	4.07	120.54	nr	**193.49**
Sum of two sides 1600mm	48.90	83.15	4.07	120.54	nr	**203.70**
Sum of two sides 1700mm	54.90	93.35	2.80	82.93	nr	**176.28**
Sum of two sides 1800mm	60.93	103.61	2.67	79.08	nr	**182.69**
Sum of two sides 1900mm	66.15	112.48	2.95	87.37	nr	**199.85**
Sum of two sides 2000mm	72.16	122.70	2.95	87.37	nr	**210.07**
Branch						
Sum of two sides 1100mm	35.36	60.13	1.44	42.65	nr	**102.78**
Sum of two sides 1200mm	37.27	63.37	1.44	42.65	nr	**106.02**
Sum of two sides 1300mm	39.18	66.62	1.64	48.57	nr	**115.19**
Sum of two sides 1400mm	40.92	69.58	1.64	48.57	nr	**118.15**
Sum of two sides 1500mm	46.98	79.89	1.64	48.57	nr	**128.46**
Sum of two sides 1600mm	53.04	90.20	1.64	48.57	nr	**138.77**
Sum of two sides 1700mm	59.10	100.49	1.69	50.05	nr	**150.55**
Sum of two sides 1800mm	65.25	110.96	1.69	50.05	nr	**161.01**
Sum of two sides 1900mm	71.32	121.26	1.85	54.79	nr	**176.06**
Sum of two sides 2000mm	77.37	131.56	1.85	54.79	nr	**186.35**
Grille neck						
Sum of two sides 1100mm	34.46	58.60	1.44	42.65	nr	**101.25**
Sum of two sides 1200mm	36.44	61.96	1.44	42.65	nr	**104.61**
Sum of two sides 1300mm	38.41	65.31	1.69	50.05	nr	**115.36**
Sum of two sides 1400mm	40.11	68.21	1.69	50.05	nr	**118.26**
Sum of two sides 1500mm	46.58	79.21	1.79	53.02	nr	**132.22**
Sum of two sides 1600mm	53.05	90.21	1.79	53.02	nr	**143.22**
Sum of two sides 1700mm	59.51	101.20	1.86	55.09	nr	**156.28**
Sum of two sides 1800mm	65.99	112.21	2.02	59.83	nr	**172.03**
Sum of two sides 1900mm	72.45	123.20	2.02	59.83	nr	**183.02**
Sum of two sides 2000mm	78.93	134.21	2.02	59.83	nr	**194.03**
Ductwork 1001 to 1250mm longest side						
Sum of two sides 1300mm	50.80	86.37	1.40	41.46	m	**127.84**
Sum of two sides 1400mm	52.37	89.05	1.40	41.46	m	**130.52**
Sum of two sides 1500mm	54.77	93.13	1.48	43.83	m	**136.97**
Sum of two sides 1600mm	57.41	97.62	1.55	45.91	m	**143.52**
Sum of two sides 1700mm	59.81	101.70	1.55	45.91	m	**147.61**
Sum of two sides 1800mm	62.21	105.78	1.61	47.68	m	**153.47**
Sum of two sides 1900mm	64.61	109.87	1.61	47.68	m	**157.55**
Sum of two sides 2000mm	66.06	112.32	1.61	47.68	m	**160.01**
Sum of two sides 2100mm	68.46	116.42	2.17	64.27	m	**180.68**
Sum of two sides 2200mm	70.85	120.48	2.19	64.86	m	**185.34**
Sum of two sides 2300mm	73.73	125.37	2.19	64.86	m	**190.23**
Sum of two sides 2400mm	76.13	129.45	2.38	70.49	m	**199.94**
Sum of two sides 2500mm	78.78	133.95	2.38	70.49	m	**204.44**

U:VENTILATION/AIR CONDITIONING SYSTEMS

Item	Net Price £	Material £	Labour hours	Labour £	Unit	Total rate £
U10 : DUCTWORK : RECTANGULAR– CLASS B (cont'd)						
Y30 - AIR DUCTLINES (cont'd)						
Extra over fittings; Ductwork 1001 to 1250mm longest side						
End Cap						
Sum of two sides 1300mm	20.45	34.77	1.69	50.05	nr	**84.83**
Sum of two sides 1400mm	22.12	37.61	1.69	50.05	nr	**87.66**
Sum of two sides 1500mm	25.45	43.28	1.69	50.05	nr	**93.33**
Sum of two sides 1600mm	28.78	48.93	1.69	50.05	nr	**98.99**
Sum of two sides 1700mm	32.11	54.60	1.69	50.05	nr	**104.66**
Sum of two sides 1800mm	35.44	60.26	1.69	50.05	nr	**110.31**
Sum of two sides 1900mm	38.77	65.93	1.69	50.05	nr	**115.98**
Sum of two sides 2000mm	40.44	68.76	1.69	50.05	nr	**118.82**
Sum of two sides 2100mm	43.77	74.43	1.69	50.05	nr	**124.48**
Sum of two sides 2200mm	47.10	80.09	1.69	50.05	nr	**130.14**
Sum of two sides 2300mm	50.43	85.74	1.69	50.05	nr	**135.80**
Sum of two sides 2400mm	53.76	91.41	1.69	50.05	nr	**141.47**
Sum of two sides 2500mm	57.10	97.09	1.69	50.05	nr	**147.14**
Reducer						
Sum of two sides 1300mm	15.94	27.11	1.69	50.05	nr	**77.17**
Sum of two sides 1400mm	17.28	29.39	1.69	50.05	nr	**79.44**
Sum of two sides 1500mm	19.98	33.97	2.47	73.16	nr	**107.12**
Sum of two sides 1600mm	22.74	38.66	2.47	73.16	nr	**111.81**
Sum of two sides 1700mm	25.43	43.23	2.47	73.16	nr	**116.39**
Sum of two sides 1800mm	28.11	47.80	2.59	76.71	nr	**124.51**
Sum of two sides 1900mm	30.80	52.38	2.71	80.26	nr	**132.64**
Sum of two sides 2000mm	32.22	54.78	2.59	76.71	nr	**131.49**
Sum of two sides 2100mm	34.91	59.36	2.92	86.48	nr	**145.84**
Sum of two sides 2200mm	37.60	63.93	2.92	86.48	nr	**150.42**
Sum of two sides 2300mm	40.35	68.61	2.92	86.48	nr	**155.10**
Sum of two sides 2400mm	43.04	73.19	3.12	92.41	nr	**165.60**
Sum of two sides 2500mm	45.80	77.88	3.12	92.41	nr	**170.29**
Offset						
Sum of two sides 1300mm	35.83	60.93	1.69	50.05	nr	**110.98**
Sum of two sides 1400mm	37.93	64.50	1.69	50.05	nr	**114.55**
Sum of two sides 1500mm	42.09	71.56	2.47	73.16	nr	**144.72**
Sum of two sides 1600mm	46.23	78.61	2.47	73.16	nr	**151.76**
Sum of two sides 1700mm	50.22	85.39	2.59	76.71	nr	**162.10**
Sum of two sides 1800mm	54.14	92.05	2.61	77.30	nr	**169.35**
Sum of two sides 1900mm	57.98	98.58	2.71	80.26	nr	**178.85**
Sum of two sides 2000mm	59.90	101.85	2.71	80.26	nr	**182.11**
Sum of two sides 2100mm	63.61	108.17	2.92	86.48	nr	**194.65**
Sum of two sides 2200mm	67.90	115.46	3.26	96.55	nr	**212.01**
Sum of two sides 2300mm	72.84	123.86	3.26	96.55	nr	**220.41**
Sum of two sides 2400mm	77.75	132.21	3.48	103.07	nr	**235.28**
Sum of two sides 2500mm	82.69	140.61	3.48	103.07	nr	**243.68**
Square to round						
Sum of two sides 1300mm	29.43	50.05	1.69	50.05	nr	**100.10**
Sum of two sides 1400mm	32.30	54.92	1.69	50.05	nr	**104.97**
Sum of two sides 1500mm	38.03	64.66	2.38	70.49	nr	**135.15**
Sum of two sides 1600mm	43.77	74.42	2.38	70.49	nr	**144.91**
Sum of two sides 1700mm	49.49	84.16	2.55	75.52	nr	**159.68**
Sum of two sides 1800mm	55.22	93.90	2.55	75.52	nr	**169.42**

U:VENTILATION/AIR CONDITIONING SYSTEMS

Item	Net Price £	Material £	Labour hours	Labour £	Unit	Total rate £
Sum of two sides 1900mm	60.95	103.64	2.83	83.82	nr	187.46
Sum of two sides 2000mm	63.83	108.53	2.83	83.82	nr	192.35
Sum of two sides 2100mm	69.56	118.27	3.85	114.03	nr	232.30
Sum of two sides 2200mm	75.29	128.01	4.18	123.80	nr	251.81
Sum of two sides 2300mm	81.03	137.79	4.22	124.99	nr	262.77
Sum of two sides 2400mm	86.76	147.53	4.68	138.61	nr	286.14
Sum of two sides 2500mm	92.50	157.29	4.70	139.20	nr	296.49
90° radius bend						
Sum of two sides 1300mm	17.19	29.23	1.69	50.05	nr	79.29
Sum of two sides 1400mm	19.46	33.08	1.69	50.05	nr	83.14
Sum of two sides 1500mm	23.99	40.80	2.11	62.49	nr	103.29
Sum of two sides 1600mm	28.56	48.56	2.11	62.49	nr	111.05
Sum of two sides 1700mm	33.09	56.27	2.19	64.86	nr	121.13
Sum of two sides 1800mm	37.62	63.96	2.19	64.86	nr	128.83
Sum of two sides 1900mm	42.16	71.68	2.48	73.45	nr	145.13
Sum of two sides 2000mm	44.45	75.58	2.26	66.94	nr	142.52
Sum of two sides 2100mm	48.99	83.30	2.48	73.45	nr	156.75
Sum of two sides 2200mm	53.53	91.01	2.48	73.45	nr	164.47
Sum of two sides 2300mm	55.97	95.17	2.48	73.45	nr	168.62
Sum of two sides 2400mm	60.49	102.85	3.90	115.51	nr	218.36
Sum of two sides 2500mm	65.03	110.57	3.90	115.51	nr	226.08
45° bend						
Sum of two sides 1300mm	19.64	33.39	1.69	50.05	nr	83.44
Sum of two sides 1400mm	21.39	36.37	1.69	50.05	nr	86.42
Sum of two sides 1500mm	24.91	42.35	2.49	73.75	nr	116.10
Sum of two sides 1600mm	28.49	48.45	2.49	73.75	nr	122.20
Sum of two sides 1700mm	32.01	54.43	2.67	79.08	nr	133.51
Sum of two sides 1800mm	35.53	60.41	2.67	79.08	nr	139.49
Sum of two sides 1900mm	39.04	66.38	3.06	90.63	nr	157.01
Sum of two sides 2000mm	40.88	69.51	3.06	90.63	nr	160.14
Sum of two sides 2100mm	44.39	75.48	4.05	119.95	nr	195.43
Sum of two sides 2200mm	47.90	81.45	4.05	119.95	nr	201.40
Sum of two sides 2300mm	50.67	86.15	4.39	130.02	nr	216.17
Sum of two sides 2400mm	54.17	92.12	4.85	143.64	nr	235.76
Sum of two sides 2500mm	57.76	98.21	4.85	143.64	nr	241.86
90° mitre bend						
Sum of two sides 1300mm	38.25	65.05	1.69	50.05	nr	115.10
Sum of two sides 1400mm	41.24	70.12	1.69	50.05	nr	120.17
Sum of two sides 1500mm	47.19	80.23	2.80	82.93	nr	163.16
Sum of two sides 1600mm	53.14	90.36	2.80	82.93	nr	173.29
Sum of two sides 1700mm	59.09	100.48	2.95	87.37	nr	187.85
Sum of two sides 1800mm	65.05	110.61	2.95	87.37	nr	197.98
Sum of two sides 1900mm	71.00	120.73	4.05	119.95	nr	240.68
Sum of two sides 2000mm	73.98	125.80	4.05	119.95	nr	245.75
Sum of two sides 2100mm	79.94	135.93	4.07	120.54	nr	256.47
Sum of two sides 2200mm	85.89	146.05	4.07	120.54	nr	266.59
Sum of two sides 2300mm	90.13	153.25	4.39	130.02	nr	283.27
Sum of two sides 2400mm	96.11	163.42	4.85	143.64	nr	307.06
Sum of two sides 2500mm	102.08	173.58	4.85	143.64	nr	317.22

U:VENTILATION/AIR CONDITIONING SYSTEMS

Item	Net Price £	Material £	Labour hours	Labour £	Unit	Total rate £
U10 : DUCTWORK : RECTANGULAR – **CLASS B (cont'd)**						
Y30 - AIR DUCTLINES (cont'd)						
Extra over fittings; Ductwork 1001 to 1250mm longest side (cont'd)						
Branch						
Sum of two sides 1300mm	34.31	58.34	1.44	42.65	nr	**100.99**
Sum of two sides 1400mm	36.83	62.62	1.44	42.65	nr	**105.27**
Sum of two sides 1500mm	41.86	71.18	1.64	48.57	nr	**119.75**
Sum of two sides 1600mm	46.96	79.86	1.64	48.57	nr	**128.43**
Sum of two sides 1700mm	51.99	88.40	1.64	48.57	nr	**136.97**
Sum of two sides 1800mm	57.02	96.95	1.64	48.57	nr	**145.53**
Sum of two sides 1900mm	62.05	105.51	1.69	50.05	nr	**155.56**
Sum of two sides 2000mm	64.64	109.91	1.69	50.05	nr	**159.96**
Sum of two sides 2100mm	69.67	118.47	1.85	54.79	nr	**173.26**
Sum of two sides 2200mm	74.70	127.02	1.85	54.79	nr	**181.81**
Sum of two sides 2300mm	79.80	135.70	2.61	77.30	nr	**213.00**
Sum of two sides 2400mm	84.83	144.25	2.61	77.30	nr	**221.55**
Sum of two sides 2500mm	89.94	152.93	2.61	77.30	nr	**230.23**
Grille neck						
Sum of two sides 1300mm	38.34	65.19	1.79	53.02	nr	**118.21**
Sum of two sides 1400mm	41.39	70.38	1.79	53.02	nr	**123.39**
Sum of two sides 1500mm	47.49	80.76	1.79	53.02	nr	**133.77**
Sum of two sides 1600mm	53.60	91.14	1.79	53.02	nr	**144.16**
Sum of two sides 1700mm	59.70	101.52	1.86	55.09	nr	**156.61**
Sum of two sides 1800mm	65.81	111.90	2.02	59.83	nr	**171.73**
Sum of two sides 1900mm	71.91	122.28	2.02	59.83	nr	**182.11**
Sum of two sides 2000mm	74.96	127.47	2.02	59.83	nr	**187.29**
Sum of two sides 2100mm	81.08	137.86	2.61	77.30	nr	**215.16**
Sum of two sides 2200mm	87.18	148.24	2.61	77.30	nr	**225.54**
Sum of two sides 2300mm	93.29	158.62	2.61	77.30	nr	**235.92**
Sum of two sides 2400mm	99.39	169.00	2.88	85.30	nr	**254.30**
Sum of two sides 2500mm	105.50	179.38	2.88	85.30	nr	**264.68**
Ductwork 1251 to 1600mm longest side						
Sum of two sides 1700mm	70.42	119.74	1.55	45.91	m	**165.65**
Sum of two sides 1800mm	73.10	124.30	1.61	47.68	m	**171.98**
Sum of two sides 1900mm	75.85	128.98	1.61	47.68	m	**176.66**
Sum of two sides 2000mm	78.37	133.26	1.61	47.68	m	**180.95**
Sum of two sides 2100mm	80.89	137.54	2.17	64.27	m	**201.81**
Sum of two sides 2200mm	83.41	141.83	2.19	64.86	m	**206.69**
Sum of two sides 2300mm	86.16	146.50	2.19	64.86	m	**211.36**
Sum of two sides 2400mm	88.82	151.03	2.38	70.49	m	**221.52**
Sum of two sides 2500mm	91.34	155.31	2.38	70.49	m	**225.80**
Sum of two sides 2600mm	93.85	159.59	2.64	78.19	m	**237.78**
Sum of two sides 2700mm	96.37	163.87	2.66	78.78	m	**242.65**
Sum of two sides 2800mm	99.05	168.42	2.95	87.37	m	**255.80**
Sum of two sides 2900mm	111.47	189.54	2.96	87.67	m	**277.21**
Sum of two sides 3000mm	114.00	193.85	3.15	93.29	m	**287.14**
Sum of two sides 3100mm	116.66	198.36	3.15	93.29	m	**291.65**
Sum of two sides 3200mm	119.18	202.64	3.18	94.18	m	**296.83**

U:VENTILATION/AIR CONDITIONING SYSTEMS

Item	Net Price £	Material £	Labour hours	Labour £	Unit	Total rate £
Extra over fittings; Ductwork 1251 to 1600mm longest side						
End Cap						
Sum of two sides 1700mm	40.91	69.56	0.58	17.18	nr	**86.74**
Sum of two sides 1800mm	45.38	77.17	0.58	17.18	nr	**94.35**
Sum of two sides 1900mm	49.86	84.78	0.58	17.18	nr	**101.96**
Sum of two sides 2000mm	54.33	92.39	0.58	17.18	nr	**109.57**
Sum of two sides 2100mm	58.80	99.99	0.87	25.77	nr	**125.76**
Sum of two sides 2200mm	63.28	107.60	0.87	25.77	nr	**133.37**
Sum of two sides 2300mm	67.75	115.21	0.87	25.77	nr	**140.98**
Sum of two sides 2400mm	72.23	122.82	0.87	25.77	nr	**148.59**
Sum of two sides 2500mm	76.72	130.45	0.87	25.77	nr	**156.22**
Sum of two sides 2600mm	81.19	138.05	0.87	25.77	nr	**163.82**
Sum of two sides 2700mm	85.66	145.66	0.87	25.77	nr	**171.43**
Sum of two sides 2800mm	90.14	153.27	1.16	34.36	nr	**187.62**
Sum of two sides 2900mm	94.61	160.88	1.16	34.36	nr	**195.23**
Sum of two sides 3000mm	99.09	168.49	1.73	51.24	nr	**219.73**
Sum of two sides 3100mm	102.63	174.50	1.80	53.31	nr	**227.81**
Sum of two sides 3200mm	105.84	179.96	1.80	53.31	nr	**233.27**
Reducer						
Sum of two sides 1700mm	18.28	31.09	2.47	73.16	nr	**104.25**
Sum of two sides 1800mm	21.04	35.77	2.59	76.71	nr	**112.48**
Sum of two sides 1900mm	23.83	40.53	2.71	80.26	nr	**120.79**
Sum of two sides 2000mm	26.56	45.17	2.71	80.26	nr	**125.43**
Sum of two sides 2100mm	29.30	49.82	2.92	86.48	nr	**136.30**
Sum of two sides 2200mm	32.04	54.48	2.92	86.48	nr	**140.96**
Sum of two sides 2300mm	34.83	59.22	2.92	86.48	nr	**145.70**
Sum of two sides 2400mm	37.58	63.90	3.12	92.41	nr	**156.31**
Sum of two sides 2500mm	40.32	68.55	3.12	92.41	nr	**160.96**
Sum of two sides 2600mm	43.05	73.20	3.12	92.41	nr	**165.61**
Sum of two sides 2700mm	45.78	77.85	3.12	92.41	nr	**170.26**
Sum of two sides 2800mm	48.54	82.53	3.95	116.99	nr	**199.52**
Sum of two sides 2900mm	48.14	81.86	3.97	117.58	nr	**199.44**
Sum of two sides 3000mm	50.88	86.51	4.52	133.87	nr	**220.38**
Sum of two sides 3100mm	52.92	89.99	4.52	133.87	nr	**223.86**
Sum of two sides 3200mm	54.78	93.16	4.52	133.87	nr	**227.03**
Offset						
Sum of two sides 1700mm	58.43	99.35	2.59	76.71	nr	**176.06**
Sum of two sides 1800mm	64.85	110.27	2.61	77.30	nr	**187.58**
Sum of two sides 1900mm	69.20	117.67	2.71	80.26	nr	**197.94**
Sum of two sides 2000mm	73.37	124.75	2.71	80.26	nr	**205.01**
Sum of two sides 2100mm	77.40	131.61	2.92	86.48	nr	**218.10**
Sum of two sides 2200mm	81.32	138.28	3.26	96.55	nr	**234.83**
Sum of two sides 2300mm	85.16	144.80	3.26	96.55	nr	**241.35**
Sum of two sides 2400mm	91.30	155.24	3.47	102.77	nr	**258.01**
Sum of two sides 2500mm	94.91	161.38	3.48	103.07	nr	**264.45**
Sum of two sides 2600mm	100.96	171.67	3.49	103.36	nr	**275.03**
Sum of two sides 2700mm	107.01	181.96	3.50	103.66	nr	**285.63**
Sum of two sides 2800mm	113.09	192.30	4.34	128.54	nr	**320.84**
Sum of two sides 2900mm	113.11	192.32	4.76	140.98	nr	**333.30**
Sum of two sides 3000mm	119.16	202.61	5.32	157.56	nr	**360.18**
Sum of two sides 3100mm	123.88	210.64	5.35	158.45	nr	**369.09**
Sum of two sides 3200mm	128.26	218.08	5.35	158.45	nr	**376.54**

U:VENTILATION/AIR CONDITIONING SYSTEMS

Item	Net Price £	Material £	Labour hours	Labour £	Unit	Total rate £
U10 : DUCTWORK : RECTANGULAR – CLASS B (cont'd)						
Y30 - AIR DUCTLINES (cont'd)						
Extra over fittings; Ductwork 1251 to 1600mm longest side (cont'd)						
Square to round						
Sum of two sides 1700mm	46.30	78.73	2.55	75.52	nr	**154.26**
Sum of two sides 1800mm	53.74	91.38	2.55	75.52	nr	**166.91**
Sum of two sides 1900mm	61.12	103.94	2.83	83.82	nr	**187.75**
Sum of two sides 2000mm	68.53	116.52	2.83	83.82	nr	**200.34**
Sum of two sides 2100mm	75.92	129.09	3.85	114.03	nr	**243.12**
Sum of two sides 2200mm	83.32	141.68	4.18	123.80	nr	**265.48**
Sum of two sides 2300mm	90.71	154.23	4.22	124.99	nr	**279.22**
Sum of two sides 2400mm	98.14	166.88	4.68	138.61	nr	**305.49**
Sum of two sides 2500mm	105.55	179.47	4.70	139.20	nr	**318.67**
Sum of two sides 2600mm	112.94	192.04	4.70	139.20	nr	**331.24**
Sum of two sides 2700mm	120.35	204.64	4.71	139.50	nr	**344.14**
Sum of two sides 2800mm	127.79	217.29	8.19	242.57	nr	**459.85**
Sum of two sides 2900mm	128.18	217.96	8.62	255.30	nr	**473.26**
Sum of two sides 3000mm	135.58	230.53	8.75	259.15	nr	**489.68**
Sum of two sides 3100mm	141.11	239.94	8.75	259.15	nr	**499.09**
Sum of two sides 3200mm	146.12	248.46	8.75	259.15	nr	**507.61**
90° radius bend						
Sum of two sides 1700mm	52.74	89.67	2.19	64.86	nr	**154.53**
Sum of two sides 1800mm	57.66	98.04	2.19	64.86	nr	**162.90**
Sum of two sides 1900mm	64.98	110.50	2.26	66.94	nr	**177.43**
Sum of two sides 2000mm	72.16	122.69	2.26	66.94	nr	**189.63**
Sum of two sides 2100mm	79.33	134.89	2.48	73.45	nr	**208.34**
Sum of two sides 2200mm	86.50	147.09	2.48	73.45	nr	**220.54**
Sum of two sides 2300mm	93.83	159.54	2.48	73.45	nr	**233.00**
Sum of two sides 2400mm	98.50	167.49	3.90	115.51	nr	**283.00**
Sum of two sides 2500mm	105.65	179.65	3.90	115.51	nr	**295.15**
Sum of two sides 2600mm	112.79	191.79	4.26	126.17	nr	**317.96**
Sum of two sides 2700mm	119.94	203.94	4.55	134.76	nr	**338.70**
Sum of two sides 2800mm	124.44	211.60	4.55	134.76	nr	**346.36**
Sum of two sides 2900mm	124.60	211.87	6.87	203.47	nr	**415.34**
Sum of two sides 3000mm	131.72	223.98	7.00	207.32	nr	**431.30**
Sum of two sides 3100mm	137.23	233.34	7.00	207.32	nr	**440.67**
Sum of two sides 3200mm	142.30	241.96	7.00	207.32	nr	**449.28**
45° bend						
Sum of two sides 1700mm	25.18	42.81	2.67	79.08	nr	**121.89**
Sum of two sides 1800mm	27.60	46.93	2.67	79.08	nr	**126.01**
Sum of two sides 1900mm	31.23	53.10	3.06	90.63	nr	**143.73**
Sum of two sides 2000mm	34.79	59.16	3.06	90.63	nr	**149.79**
Sum of two sides 2100mm	38.35	65.21	4.05	119.95	nr	**185.16**
Sum of two sides 2200mm	41.91	71.27	4.05	119.95	nr	**191.22**
Sum of two sides 2300mm	45.54	77.44	4.39	130.02	nr	**207.46**
Sum of two sides 2400mm	47.84	81.35	4.85	143.64	nr	**224.99**
Sum of two sides 2500mm	51.39	87.38	4.85	143.64	nr	**231.03**
Sum of two sides 2600mm	54.93	93.41	4.87	144.24	nr	**237.64**
Sum of two sides 2700mm	58.48	99.44	4.87	144.24	nr	**243.68**
Sum of two sides 2800mm	60.69	103.19	8.81	260.93	nr	**364.12**
Sum of two sides 2900mm	60.53	102.92	8.81	260.93	nr	**363.85**
Sum of two sides 3000mm	64.06	108.93	9.31	275.74	nr	**384.67**
Sum of two sides 3100mm	66.79	113.56	9.31	275.74	nr	**389.30**
Sum of two sides 3200mm	69.30	117.84	9.39	278.11	nr	**395.95**

U:VENTILATION/AIR CONDITIONING SYSTEMS

Item	Net Price £	Material £	Labour hours	Labour £	Unit	Total rate £
90° mitre bend						
Sum of two sides 1700mm	62.67	106.56	2.67	79.08	nr	**185.64**
Sum of two sides 1800mm	66.87	113.70	2.80	82.93	nr	**196.63**
Sum of two sides 1900mm	74.53	126.73	2.95	87.37	nr	**214.10**
Sum of two sides 2000mm	82.22	139.81	2.95	87.37	nr	**227.18**
Sum of two sides 2100mm	89.91	152.88	4.05	119.95	nr	**272.83**
Sum of two sides 2200mm	97.59	165.95	4.05	119.95	nr	**285.90**
Sum of two sides 2300mm	105.26	178.98	4.39	130.02	nr	**309.00**
Sum of two sides 2400mm	109.37	185.96	4.85	143.64	nr	**329.61**
Sum of two sides 2500mm	117.08	199.07	4.85	143.64	nr	**342.72**
Sum of two sides 2600mm	124.78	212.17	4.87	144.24	nr	**356.41**
Sum of two sides 2700mm	132.48	225.27	4.87	144.24	nr	**369.51**
Sum of two sides 2800mm	136.50	232.10	8.81	260.93	nr	**493.02**
Sum of two sides 2900mm	136.67	232.39	14.81	438.63	nr	**671.02**
Sum of two sides 3000mm	144.39	245.52	15.20	450.18	nr	**695.70**
Sum of two sides 3100mm	150.69	256.24	15.60	462.03	nr	**718.27**
Sum of two sides 3200mm	156.69	266.43	15.60	462.03	nr	**728.46**
Branch						
Sum of two sides 1700mm	49.50	84.17	1.69	50.05	nr	**134.22**
Sum of two sides 1800mm	54.54	92.74	1.69	50.05	nr	**142.79**
Sum of two sides 1900mm	59.64	101.42	1.85	54.79	nr	**156.21**
Sum of two sides 2000mm	64.67	109.97	1.85	54.79	nr	**164.76**
Sum of two sides 2100mm	69.71	118.54	2.61	77.30	nr	**195.84**
Sum of two sides 2200mm	74.74	127.09	2.61	77.30	nr	**204.39**
Sum of two sides 2300mm	79.85	135.77	2.61	77.30	nr	**213.07**
Sum of two sides 2400mm	84.88	144.34	2.88	85.30	nr	**229.63**
Sum of two sides 2500mm	89.92	152.89	2.88	85.30	nr	**238.19**
Sum of two sides 2600mm	94.94	161.43	2.88	85.30	nr	**246.73**
Sum of two sides 2700mm	99.97	169.99	2.88	85.30	nr	**255.29**
Sum of two sides 2800mm	105.01	178.55	3.94	116.69	nr	**295.25**
Sum of two sides 2900mm	110.11	187.23	3.94	116.69	nr	**303.93**
Sum of two sides 3000mm	115.14	195.79	4.83	143.05	nr	**338.84**
Sum of two sides 3100mm	119.16	202.61	4.83	143.05	nr	**345.66**
Sum of two sides 3200mm	122.80	208.81	4.83	143.05	nr	**351.87**
Grille neck						
Sum of two sides 1700mm	60.56	102.98	1.86	55.09	nr	**158.07**
Sum of two sides 1800mm	67.09	114.07	2.02	59.83	nr	**173.90**
Sum of two sides 1900mm	73.62	125.18	2.02	59.83	nr	**185.01**
Sum of two sides 2000mm	80.14	136.27	2.02	59.83	nr	**196.10**
Sum of two sides 2100mm	86.67	147.37	2.61	77.30	nr	**224.67**
Sum of two sides 2200mm	93.20	158.47	2.61	77.30	nr	**235.78**
Sum of two sides 2300mm	99.72	169.57	2.61	77.30	nr	**246.87**
Sum of two sides 2400mm	106.25	180.66	2.88	85.30	nr	**265.96**
Sum of two sides 2500mm	112.78	191.77	2.88	85.30	nr	**277.07**
Sum of two sides 2600mm	119.30	202.85	2.88	85.30	nr	**288.15**
Sum of two sides 2700mm	125.82	213.95	2.88	85.30	nr	**299.25**
Sum of two sides 2800mm	132.35	225.05	3.94	116.69	nr	**341.75**
Sum of two sides 2900mm	138.88	236.15	4.12	122.02	nr	**358.17**
Sum of two sides 3000mm	145.40	247.24	5.00	148.09	nr	**395.33**
Sum of two sides 3100mm	150.59	256.06	5.00	148.09	nr	**404.15**
Sum of two sides 3200mm	155.31	264.09	5.00	148.09	nr	**412.18**

U:VENTILATION/AIR CONDITIONING SYSTEMS

Item	Net Price £	Material £	Labour hours	Labour £	Unit	Total rate £
U10 : DUCTWORK : RECTANGULAR – CLASS B (cont'd)						
Y30 - AIR DUCTLINES (cont'd)						
Ductwork 1601 to 2000mm longest side						
Sum of two sides 2100mm	87.37	148.57	2.17	64.27	m	**212.84**
Sum of two sides 2200mm	90.07	153.15	2.17	64.27	m	**217.42**
Sum of two sides 2300mm	92.82	157.83	2.19	64.86	m	**222.70**
Sum of two sides 2400mm	95.43	162.27	2.38	70.49	m	**232.76**
Sum of two sides 2500mm	98.20	166.98	2.38	70.49	m	**237.47**
Sum of two sides 2600mm	100.74	171.29	2.64	78.19	m	**249.48**
Sum of two sides 2700mm	103.29	175.63	2.66	78.78	m	**254.41**
Sum of two sides 2800mm	105.82	179.94	2.95	87.37	m	**267.31**
Sum of two sides 2900mm	108.59	184.64	2.96	87.67	m	**272.31**
Sum of two sides 3000mm	111.13	188.96	2.96	87.67	m	**276.62**
Sum of two sides 3100mm	120.60	205.07	2.96	87.67	m	**292.73**
Sum of two sides 3200mm	123.14	209.38	3.15	93.29	m	**302.68**
Sum of two sides 3300mm	135.26	229.99	3.15	93.29	m	**323.28**
Sum of two sides 3400mm	137.79	234.30	3.15	93.29	m	**327.59**
Sum of two sides 3500mm	140.34	238.63	3.15	93.29	m	**331.93**
Sum of two sides 3600mm	142.88	242.95	3.18	94.18	m	**337.13**
Sum of two sides 3700mm	145.64	247.65	3.18	94.18	m	**341.83**
Sum of two sides 3800mm	148.18	251.97	3.18	94.18	m	**346.15**
Sum of two sides 3900mm	150.73	256.30	3.18	94.18	m	**350.48**
Sum of two sides 4000mm	153.27	260.61	3.18	94.18	m	**354.80**
Extra over fittings; Ductwork 1601 to 2000mm longest side						
End Cap						
Sum of two sides 2100mm	74.00	125.83	0.87	25.77	nr	**151.60**
Sum of two sides 2200mm	79.63	135.40	0.87	25.77	nr	**161.17**
Sum of two sides 2300mm	85.26	144.98	0.87	25.77	nr	**170.74**
Sum of two sides 2400mm	90.89	154.55	0.87	25.77	nr	**180.32**
Sum of two sides 2500mm	96.53	164.14	0.87	25.77	nr	**189.91**
Sum of two sides 2600mm	102.17	173.73	0.87	25.77	nr	**199.49**
Sum of two sides 2700mm	107.80	183.30	0.87	25.77	nr	**209.07**
Sum of two sides 2800mm	113.43	192.87	1.16	34.36	nr	**227.23**
Sum of two sides 2900mm	119.06	202.44	1.16	34.36	nr	**236.80**
Sum of two sides 3000mm	124.69	212.02	1.73	51.24	nr	**263.25**
Sum of two sides 3100mm	129.14	219.59	1.80	53.31	nr	**272.91**
Sum of two sides 3200mm	133.18	226.46	1.80	53.31	nr	**279.77**
Sum of two sides 3300mm	137.22	233.33	1.80	53.31	nr	**286.65**
Sum of two sides 3400mm	141.26	240.20	1.80	53.31	nr	**293.51**
Sum of two sides 3500mm	145.30	247.06	1.80	53.31	nr	**300.37**
Sum of two sides 3600mm	149.35	253.95	1.80	53.31	nr	**307.26**
Sum of two sides 3700mm	153.38	260.80	1.80	53.31	nr	**314.11**
Sum of two sides 3800mm	157.42	267.67	1.80	53.31	nr	**320.98**
Sum of two sides 3900mm	161.47	274.55	1.80	53.31	nr	**327.86**
Sum of two sides 4000mm	165.50	281.41	1.80	53.31	nr	**334.72**
Reducer						
Sum of two sides 2100mm	29.43	50.04	2.61	77.30	nr	**127.34**
Sum of two sides 2200mm	32.17	54.70	2.61	77.30	nr	**132.00**
Sum of two sides 2300mm	34.96	59.45	2.61	77.30	nr	**136.75**
Sum of two sides 2400mm	37.66	64.03	2.88	85.30	nr	**149.33**
Sum of two sides 2500mm	40.45	68.78	3.12	92.41	nr	**161.19**

U:VENTILATION/AIR CONDITIONING SYSTEMS

Item	Net Price £	Material £	Labour hours	Labour £	Unit	Total rate £
Sum of two sides 2600mm	43.19	73.44	3.12	92.41	nr	165.85
Sum of two sides 2700mm	45.93	78.10	3.12	92.41	nr	170.51
Sum of two sides 2800mm	48.67	82.76	3.95	116.99	nr	199.75
Sum of two sides 2900mm	51.48	87.53	3.97	117.58	nr	205.11
Sum of two sides 3000mm	54.22	92.19	4.52	133.87	nr	226.06
Sum of two sides 3100mm	53.85	91.56	4.52	133.87	nr	225.43
Sum of two sides 3200mm	55.70	94.71	4.52	133.87	nr	228.58
Sum of two sides 3300mm	54.30	92.33	4.52	133.87	nr	226.20
Sum of two sides 3400mm	56.16	95.49	4.52	133.87	nr	229.36
Sum of two sides 3500mm	58.01	98.63	4.52	133.87	nr	232.50
Sum of two sides 3600mm	59.86	101.78	4.52	133.87	nr	235.65
Sum of two sides 3700mm	64.67	109.97	4.52	133.87	nr	243.84
Sum of two sides 3800mm	66.53	113.12	4.52	133.87	nr	246.99
Sum of two sides 3900mm	68.38	116.27	4.52	133.87	nr	250.14
Sum of two sides 4000mm	70.23	119.42	4.52	133.87	nr	253.29
Offset						
Sum of two sides 2100mm	88.50	150.48	2.61	77.30	nr	227.78
Sum of two sides 2200mm	95.11	161.73	2.61	77.30	nr	239.03
Sum of two sides 2300mm	98.92	168.20	2.61	77.30	nr	245.51
Sum of two sides 2400mm	108.35	184.23	2.88	85.30	nr	269.53
Sum of two sides 2500mm	112.03	190.49	3.48	103.07	nr	293.56
Sum of two sides 2600mm	115.48	196.36	3.49	103.36	nr	299.72
Sum of two sides 2700mm	118.80	202.00	3.50	103.66	nr	305.66
Sum of two sides 2800mm	121.98	207.41	4.34	128.54	nr	335.95
Sum of two sides 2900mm	125.06	212.66	4.76	140.98	nr	353.64
Sum of two sides 3000mm	127.97	217.59	5.32	157.56	nr	375.15
Sum of two sides 3100mm	128.24	218.05	5.35	158.45	nr	376.50
Sum of two sides 3200mm	132.74	225.71	5.35	158.45	nr	384.17
Sum of two sides 3300mm	130.96	222.68	5.35	158.45	nr	381.13
Sum of two sides 3400mm	135.47	230.34	5.35	158.45	nr	388.80
Sum of two sides 3500mm	139.98	238.02	5.35	158.45	nr	396.47
Sum of two sides 3600mm	144.48	245.68	5.35	158.45	nr	404.13
Sum of two sides 3700mm	152.72	259.68	5.35	158.45	nr	418.13
Sum of two sides 3800mm	157.23	267.35	5.35	158.45	nr	425.81
Sum of two sides 3900mm	161.74	275.02	5.35	158.45	nr	433.47
Sum of two sides 4000mm	166.24	282.68	5.35	158.45	nr	441.13
Square to round						
Sum of two sides 2100mm	81.83	139.15	2.61	77.30	nr	216.45
Sum of two sides 2200mm	89.74	152.60	2.61	77.30	nr	229.90
Sum of two sides 2300mm	97.60	165.96	2.61	77.30	nr	243.26
Sum of two sides 2400mm	105.55	179.47	2.88	85.30	nr	264.76
Sum of two sides 2500mm	113.41	192.84	4.70	139.20	nr	332.04
Sum of two sides 2600mm	121.29	206.24	4.70	139.20	nr	345.44
Sum of two sides 2700mm	129.18	219.66	4.71	139.50	nr	359.16
Sum of two sides 2800mm	137.06	233.05	8.19	242.57	nr	475.62
Sum of two sides 2900mm	144.92	246.42	8.19	242.57	nr	488.99
Sum of two sides 3000mm	152.81	259.84	8.19	242.57	nr	502.40
Sum of two sides 3100mm	154.16	262.13	8.19	242.57	nr	504.69
Sum of two sides 3200mm	159.52	271.24	8.19	242.57	nr	513.80
Sum of two sides 3300mm	158.56	269.61	8.19	242.57	nr	512.18
Sum of two sides 3400mm	163.91	278.71	8.62	255.30	nr	534.01
Sum of two sides 3500mm	169.26	287.81	8.62	255.30	nr	543.11
Sum of two sides 3600mm	174.61	296.91	8.62	255.30	nr	552.21
Sum of two sides 3700mm	182.71	310.67	8.62	255.30	nr	565.97
Sum of two sides 3800mm	188.06	319.77	8.75	259.15	nr	578.92
Sum of two sides 3900mm	193.42	328.88	8.75	259.15	nr	588.03
Sum of two sides 4000mm	198.77	337.98	8.75	259.15	nr	597.14

U:VENTILATION/AIR CONDITIONING SYSTEMS

Item	Net Price £	Material £	Labour hours	Labour £	Unit	Total rate £
U10 : DUCTWORK : RECTANGULAR– CLASS B (cont'd)						
Y30 - AIR DUCTLINES (cont'd)						
Extra over fittings; Ductwork 1601 to 2000mm longest side (cont'd)						
90° radius bend						
Sum of two sides 2100mm	64.13	109.05	2.61	77.30	nr	**186.35**
Sum of two sides 2200mm	124.09	211.00	2.61	77.30	nr	**288.30**
Sum of two sides 2300mm	134.16	228.12	2.61	77.30	nr	**305.42**
Sum of two sides 2400mm	137.60	233.96	2.88	85.30	nr	**319.26**
Sum of two sides 2500mm	147.60	250.98	3.90	115.51	nr	**366.49**
Sum of two sides 2600mm	157.39	267.62	4.26	126.17	nr	**393.79**
Sum of two sides 2700mm	167.16	284.24	4.55	134.76	nr	**419.00**
Sum of two sides 2800mm	176.94	300.86	4.55	134.76	nr	**435.62**
Sum of two sides 2900mm	186.95	317.88	6.87	203.47	nr	**521.35**
Sum of two sides 3000mm	196.73	334.52	6.87	203.47	nr	**537.99**
Sum of two sides 3100mm	198.68	337.84	6.87	203.47	nr	**541.31**
Sum of two sides 3200mm	205.71	349.78	6.87	203.47	nr	**553.25**
Sum of two sides 3300mm	205.15	348.83	6.87	203.47	nr	**552.30**
Sum of two sides 3400mm	212.17	360.77	7.00	207.32	nr	**568.09**
Sum of two sides 3500mm	219.20	372.73	7.00	207.32	nr	**580.05**
Sum of two sides 3600mm	226.23	384.67	7.00	207.32	nr	**591.99**
Sum of two sides 3700mm	242.47	412.30	7.00	207.32	nr	**619.62**
Sum of two sides 3800mm	249.50	424.24	7.00	207.32	nr	**631.56**
Sum of two sides 3900mm	256.52	436.19	7.00	207.32	nr	**643.51**
Sum of two sides 4000mm	263.55	448.14	7.00	207.32	nr	**655.46**
45° bend						
Sum of two sides 2100mm	28.73	48.85	2.61	77.30	nr	**126.15**
Sum of two sides 2200mm	83.00	141.12	2.61	77.30	nr	**218.43**
Sum of two sides 2300mm	89.33	151.89	2.61	77.30	nr	**229.19**
Sum of two sides 2400mm	92.38	157.08	2.88	85.30	nr	**242.38**
Sum of two sides 2500mm	98.69	167.82	4.85	143.64	nr	**311.46**
Sum of two sides 2600mm	104.84	178.26	4.87	144.24	nr	**322.50**
Sum of two sides 2700mm	110.98	188.70	4.87	144.24	nr	**332.94**
Sum of two sides 2800mm	117.12	199.15	8.81	260.93	nr	**460.08**
Sum of two sides 2900mm	123.43	209.89	8.81	260.93	nr	**470.81**
Sum of two sides 3000mm	129.58	220.34	9.31	275.74	nr	**496.08**
Sum of two sides 3100mm	131.61	223.78	9.31	275.74	nr	**499.52**
Sum of two sides 3200mm	136.04	231.32	9.31	275.74	nr	**507.06**
Sum of two sides 3300mm	136.79	232.59	9.31	275.74	nr	**508.33**
Sum of two sides 3400mm	141.22	240.14	9.31	275.74	nr	**515.87**
Sum of two sides 3500mm	145.66	247.67	9.39	278.11	nr	**525.78**
Sum of two sides 3600mm	150.09	255.21	9.39	278.11	nr	**533.32**
Sum of two sides 3700mm	160.27	272.53	9.39	278.11	nr	**550.63**
Sum of two sides 3800mm	164.71	280.06	9.39	278.11	nr	**558.17**
Sum of two sides 3900mm	169.14	287.60	9.39	278.11	nr	**565.71**
Sum of two sides 4000mm	173.57	295.14	9.39	278.11	nr	**573.24**
90° mitre bend						
Sum of two sides 2100mm	151.71	257.96	2.61	77.30	nr	**335.26**
Sum of two sides 2200mm	160.21	272.42	2.61	77.30	nr	**349.72**
Sum of two sides 2300mm	172.70	293.66	2.61	77.30	nr	**370.96**
Sum of two sides 2400mm	176.80	300.63	2.88	85.30	nr	**385.92**
Sum of two sides 2500mm	189.28	321.84	4.85	143.64	nr	**465.48**

U:VENTILATION/AIR CONDITIONING SYSTEMS

Item	Net Price £	Material £	Labour hours	Labour £	Unit	Total rate £
Sum of two sides 2600mm	201.84	343.21	4.87	144.24	nr	**487.45**
Sum of two sides 2700mm	214.41	364.57	4.87	144.24	nr	**508.81**
Sum of two sides 2800mm	226.97	385.93	8.81	260.93	nr	**646.86**
Sum of two sides 2900mm	239.44	407.14	14.81	438.63	nr	**845.78**
Sum of two sides 3000mm	252.01	428.51	15.20	450.18	nr	**878.70**
Sum of two sides 3100mm	255.07	433.71	15.20	450.18	nr	**883.89**
Sum of two sides 3200mm	264.73	450.14	15.20	450.18	nr	**900.32**
Sum of two sides 3300mm	266.44	453.05	15.20	450.18	nr	**903.24**
Sum of two sides 3400mm	276.10	469.48	15.20	450.18	nr	**919.67**
Sum of two sides 3500mm	285.76	485.91	15.60	462.03	nr	**947.94**
Sum of two sides 3600mm	295.43	502.33	15.60	462.03	nr	**964.37**
Sum of two sides 3700mm	308.16	523.99	15.60	462.03	nr	**986.02**
Sum of two sides 3800mm	317.82	540.42	15.60	462.03	nr	**1002.45**
Sum of two sides 3900mm	327.48	556.84	15.60	462.03	nr	**1018.87**
Sum of two sides 4000mm	337.14	573.26	15.60	462.03	nr	**1035.29**
Branch						
Sum of two sides 2100mm	69.17	117.61	2.61	77.30	nr	**194.91**
Sum of two sides 2200mm	74.03	125.87	2.61	77.30	nr	**203.17**
Sum of two sides 2300mm	78.98	134.29	2.61	77.30	nr	**211.59**
Sum of two sides 2400mm	83.75	142.40	2.88	85.30	nr	**227.70**
Sum of two sides 2500mm	88.69	150.81	2.88	85.30	nr	**236.11**
Sum of two sides 2600mm	93.56	159.08	2.88	85.30	nr	**244.38**
Sum of two sides 2700mm	98.43	167.36	2.88	85.30	nr	**252.66**
Sum of two sides 2800mm	103.29	175.64	3.94	116.69	nr	**292.33**
Sum of two sides 2900mm	108.24	184.05	3.94	116.69	nr	**300.75**
Sum of two sides 3000mm	113.11	192.32	3.94	116.69	nr	**309.02**
Sum of two sides 3100mm	116.99	198.93	3.94	116.69	nr	**315.62**
Sum of two sides 3200mm	120.52	204.93	3.94	116.69	nr	**321.62**
Sum of two sides 3300mm	124.13	211.07	3.94	116.69	nr	**327.76**
Sum of two sides 3400mm	127.66	217.08	4.83	143.05	nr	**360.13**
Sum of two sides 3500mm	131.20	223.09	4.83	143.05	nr	**366.14**
Sum of two sides 3600mm	134.73	229.09	4.83	143.05	nr	**372.15**
Sum of two sides 3700mm	139.98	238.02	4.83	143.05	nr	**381.07**
Sum of two sides 3800mm	143.51	244.02	4.83	143.05	nr	**387.07**
Sum of two sides 3900mm	147.04	250.02	4.83	143.05	nr	**393.08**
Sum of two sides 4000mm	150.58	256.04	4.83	143.05	nr	**399.09**
Grille neck						
Sum of two sides 2100mm	84.54	143.76	2.61	77.30	nr	**221.06**
Sum of two sides 2200mm	90.91	154.58	2.61	77.30	nr	**231.88**
Sum of two sides 2300mm	97.28	165.41	2.61	77.30	nr	**242.71**
Sum of two sides 2400mm	103.64	176.23	2.88	85.30	nr	**261.53**
Sum of two sides 2500mm	110.01	187.06	2.88	85.30	nr	**272.35**
Sum of two sides 2600mm	116.37	197.88	2.88	85.30	nr	**283.18**
Sum of two sides 2700mm	122.74	208.70	2.88	85.30	nr	**294.00**
Sum of two sides 2800mm	129.11	219.53	3.94	116.69	nr	**336.22**
Sum of two sides 2900mm	135.47	230.35	4.12	122.02	nr	**352.38**
Sum of two sides 3000mm	141.84	241.19	4.12	122.02	nr	**363.21**
Sum of two sides 3100mm	146.90	249.78	4.12	122.02	nr	**371.81**
Sum of two sides 3200mm	151.50	257.61	4.12	122.02	nr	**379.64**
Sum of two sides 3300mm	156.10	265.42	4.12	122.02	nr	**387.45**
Sum of two sides 3400mm	160.70	273.25	5.00	148.09	nr	**421.34**
Sum of two sides 3500mm	165.30	281.07	5.00	148.09	nr	**429.16**
Sum of two sides 3600mm	169.90	288.89	5.00	148.09	nr	**436.98**
Sum of two sides 3700mm	174.50	296.72	5.00	148.09	nr	**444.81**
Sum of two sides 3800mm	179.10	304.54	5.00	148.09	nr	**452.63**
Sum of two sides 3900mm	183.70	312.36	5.00	148.09	nr	**460.45**
Sum of two sides 4000mm	188.30	320.18	5.00	148.09	nr	**468.27**

U:VENTILATION/AIR CONDITIONING SYSTEMS

Item	Net Price £	Material £	Labour hours	Labour £	Unit	Total rate £
U10 : DUCTWORK : RECTANGULAR – CLASS B (cont'd)						
Y30 - AIR DUCTLINES (cont'd)						
Ductwork 2001 to 2500mm longest side						
Sum of two sides 2500mm	102.24	173.85	2.38	70.49	m	**244.34**
Sum of two sides 2600mm	107.33	182.50	2.64	78.19	m	**260.69**
Sum of two sides 2700mm	114.77	195.15	2.66	78.78	m	**273.93**
Sum of two sides 2800mm	118.08	200.78	2.95	87.37	m	**288.16**
Sum of two sides 2900mm	125.65	213.65	2.96	87.67	m	**301.32**
Sum of two sides 3000mm	132.46	225.23	3.15	93.29	m	**318.53**
Sum of two sides 3100mm	142.90	242.99	3.15	93.29	m	**336.29**
Sum of two sides 3200mm	144.64	245.94	3.15	93.29	m	**339.24**
Sum of two sides 3300mm	157.46	267.74	3.15	93.29	m	**361.04**
Sum of two sides 3400mm	161.11	273.96	3.15	93.29	m	**367.25**
Sum of two sides 3500mm	164.70	280.05	2.66	78.78	m	**358.84**
Sum of two sides 3600mm	169.35	287.97	3.18	94.18	m	**382.15**
Sum of two sides 3700mm	178.96	304.30	3.18	94.18	m	**398.48**
Sum of two sides 3800mm	183.68	312.32	3.18	94.18	m	**406.50**
Sum of two sides 3900mm	188.47	320.47	3.18	94.18	m	**414.66**
Sum of two sides 4000mm	191.95	326.39	3.18	94.18	m	**420.58**
Extra over fittings; Ductwork 2001 to 2500mm longest side						
End Cap						
Sum of two sides 2500mm	100.29	170.52	0.87	25.77	nr	**196.29**
Sum of two sides 2600mm	105.90	180.08	0.87	25.77	nr	**205.84**
Sum of two sides 2700mm	111.93	190.33	0.87	25.77	nr	**216.10**
Sum of two sides 2800mm	117.48	199.77	1.16	34.36	nr	**234.12**
Sum of two sides 2900mm	123.51	210.01	1.16	34.36	nr	**244.37**
Sum of two sides 3000mm	129.53	220.25	1.73	51.24	nr	**271.48**
Sum of two sides 3100mm	134.24	228.26	1.73	51.24	nr	**279.50**
Sum of two sides 3200mm	138.27	235.11	1.73	51.24	nr	**286.35**
Sum of two sides 3300mm	142.12	241.66	1.73	51.24	nr	**292.90**
Sum of two sides 3400mm	146.53	249.16	1.80	53.31	nr	**302.47**
Sum of two sides 3500mm	150.89	256.57	1.80	53.31	nr	**309.88**
Sum of two sides 3600mm	155.19	263.89	1.80	53.31	nr	**317.20**
Sum of two sides 3700mm	160.51	272.93	1.80	53.31	nr	**326.24**
Sum of two sides 3800mm	163.69	278.33	1.80	53.31	nr	**331.64**
Sum of two sides 3900mm	167.88	285.46	1.80	53.31	nr	**338.77**
Sum of two sides 4000mm	172.02	292.50	1.80	53.31	nr	**345.81**
Reducer						
Sum of two sides 2500mm	35.12	59.71	3.12	92.41	nr	**152.12**
Sum of two sides 2600mm	38.06	64.71	3.12	92.41	nr	**157.12**
Sum of two sides 2700mm	40.93	69.60	3.12	92.41	nr	**162.01**
Sum of two sides 2800mm	43.76	74.41	3.95	116.99	nr	**191.40**
Sum of two sides 2900mm	46.57	79.19	3.97	117.58	nr	**196.78**
Sum of two sides 3000mm	49.32	83.87	3.97	117.58	nr	**201.45**
Sum of two sides 3100mm	50.33	85.58	3.97	117.58	nr	**203.16**
Sum of two sides 3200mm	53.62	91.17	3.97	117.58	nr	**208.75**
Sum of two sides 3300mm	52.41	89.12	3.97	117.58	nr	**206.71**
Sum of two sides 3400mm	54.05	91.91	4.52	133.87	nr	**225.78**
Sum of two sides 3500mm	54.98	93.48	4.52	133.87	nr	**227.35**
Sum of two sides 3600mm	54.28	92.29	4.52	133.87	nr	**226.17**
Sum of two sides 3700mm	57.41	97.63	4.52	133.87	nr	**231.50**

U:VENTILATION/AIR CONDITIONING SYSTEMS

Item	Net Price £	Material £	Labour hours	Labour £	Unit	Total rate £
Sum of two sides 3800mm	58.57	99.59	4.52	133.87	nr	**233.46**
Sum of two sides 3900mm	60.03	102.07	4.52	133.87	nr	**235.94**
Sum of two sides 4000mm	62.04	105.50	4.52	133.87	nr	**239.37**
Offset						
Sum of two sides 2500mm	99.94	169.94	3.48	103.07	nr	**273.00**
Sum of two sides 2600mm	106.59	181.24	3.49	103.36	nr	**284.61**
Sum of two sides 2700mm	112.37	191.07	3.50	103.66	nr	**294.73**
Sum of two sides 2800mm	118.17	200.93	4.34	128.54	nr	**329.47**
Sum of two sides 2900mm	117.59	199.95	4.76	140.98	nr	**340.92**
Sum of two sides 3000mm	119.94	203.94	5.32	157.56	nr	**361.51**
Sum of two sides 3100mm	119.82	203.74	5.35	158.45	nr	**362.20**
Sum of two sides 3200mm	123.84	210.58	5.35	158.45	nr	**369.03**
Sum of two sides 3300mm	116.11	197.44	5.32	157.56	nr	**355.00**
Sum of two sides 3400mm	116.64	198.34	5.32	157.56	nr	**355.90**
Sum of two sides 3500mm	117.13	199.16	5.32	157.56	nr	**356.72**
Sum of two sides 3600mm	117.61	199.99	5.35	158.45	nr	**358.44**
Sum of two sides 3700mm	126.34	214.83	5.35	158.45	nr	**373.28**
Sum of two sides 3800mm	128.50	218.50	5.35	158.45	nr	**376.96**
Sum of two sides 3900mm	132.37	225.07	5.35	158.45	nr	**383.53**
Sum of two sides 4000mm	137.40	233.64	5.35	158.45	nr	**392.09**
Square to round						
Sum of two sides 2500mm	101.50	172.59	4.70	139.20	nr	**311.79**
Sum of two sides 2600mm	109.55	186.27	4.70	139.20	nr	**325.47**
Sum of two sides 2700mm	117.03	199.00	4.71	139.50	nr	**338.50**
Sum of two sides 2800mm	124.63	211.91	8.19	242.57	nr	**454.48**
Sum of two sides 2900mm	132.27	224.92	8.19	242.57	nr	**467.48**
Sum of two sides 3000mm	139.19	236.68	8.19	242.57	nr	**479.25**
Sum of two sides 3100mm	142.77	242.77	8.19	242.57	nr	**485.34**
Sum of two sides 3200mm	149.81	254.73	8.62	255.30	nr	**510.03**
Sum of two sides 3300mm	153.02	260.19	8.62	255.30	nr	**515.50**
Sum of two sides 3400mm	158.03	268.72	8.62	255.30	nr	**524.02**
Sum of two sides 3500mm	162.08	275.59	8.62	255.30	nr	**530.89**
Sum of two sides 3600mm	165.20	280.90	8.75	259.15	nr	**540.06**
Sum of two sides 3700mm	172.03	292.51	8.75	259.15	nr	**551.66**
Sum of two sides 3800mm	175.16	297.83	8.75	259.15	nr	**556.99**
Sum of two sides 3900mm	180.18	306.38	8.75	259.15	nr	**565.53**
Sum of two sides 4000mm	185.21	314.92	8.75	259.15	nr	**574.07**
90° radius bend						
Sum of two sides 2500mm	132.70	225.64	3.90	115.51	nr	**341.15**
Sum of two sides 2600mm	138.28	235.13	4.26	126.17	nr	**361.30**
Sum of two sides 2700mm	145.10	246.72	4.55	134.76	nr	**381.48**
Sum of two sides 2800mm	154.61	262.89	4.55	134.76	nr	**397.65**
Sum of two sides 2900mm	163.78	278.49	6.87	203.47	nr	**481.96**
Sum of two sides 3000mm	172.94	294.07	6.87	203.47	nr	**497.54**
Sum of two sides 3100mm	178.05	302.76	6.87	203.47	nr	**506.23**
Sum of two sides 3200mm	189.31	321.89	6.87	203.47	nr	**525.36**
Sum of two sides 3300mm	187.75	319.25	6.87	203.47	nr	**522.72**
Sum of two sides 3400mm	204.06	346.98	6.87	203.47	nr	**550.45**
Sum of two sides 3500mm	208.51	354.55	7.00	207.32	nr	**561.87**
Sum of two sides 3600mm	211.28	359.26	7.00	207.32	nr	**566.58**
Sum of two sides 3700mm	223.97	380.84	7.00	207.32	nr	**588.16**
Sum of two sides 3800mm	231.66	393.91	7.00	207.32	nr	**601.23**
Sum of two sides 3900mm	240.20	408.43	7.00	207.32	nr	**615.76**
Sum of two sides 4000mm	246.93	419.88	7.00	207.32	nr	**627.20**

U:VENTILATION/AIR CONDITIONING SYSTEMS

Item	Net Price £	Material £	Labour hours	Labour £	Unit	Total rate £
U10 : DUCTWORK : RECTANGULAR – CLASS B (cont'd)						
Y30 - AIR DUCTLINES (cont'd)						
Extra over fittings; Ductwork 2001 to 2500mm longest side (cont'd)						
45° bend						
Sum of two sides 2500mm	89.56	152.28	4.85	143.64	nr	295.93
Sum of two sides 2600mm	93.75	159.41	4.87	144.24	nr	303.64
Sum of two sides 2700mm	98.82	168.04	4.87	144.24	nr	312.27
Sum of two sides 2800mm	103.05	175.22	8.81	260.93	nr	436.14
Sum of two sides 2900mm	108.85	185.09	8.81	260.93	nr	446.02
Sum of two sides 3000mm	114.67	194.98	9.31	275.74	nr	470.72
Sum of two sides 3100mm	118.92	202.20	9.31	275.74	nr	477.94
Sum of two sides 3200mm	124.29	211.33	9.31	275.74	nr	487.07
Sum of two sides 3300mm	125.00	212.55	9.31	275.74	nr	488.29
Sum of two sides 3400mm	129.09	219.50	9.31	275.74	nr	495.24
Sum of two sides 3500mm	132.60	225.47	9.31	275.74	nr	501.21
Sum of two sides 3600mm	136.22	231.63	9.39	278.11	nr	509.74
Sum of two sides 3700mm	145.87	248.04	9.39	278.11	nr	526.15
Sum of two sides 3800mm	149.83	254.77	9.39	278.11	nr	532.88
Sum of two sides 3900mm	154.02	261.88	9.39	278.11	nr	539.99
Sum of two sides 4000mm	158.19	268.99	9.39	278.11	nr	547.10
90° mitre bend						
Sum of two sides 2500mm	170.85	290.52	4.85	143.64	nr	434.16
Sum of two sides 2600mm	178.39	303.33	4.87	144.24	nr	447.57
Sum of two sides 2700mm	187.62	319.03	4.87	144.24	nr	463.26
Sum of two sides 2800mm	194.89	331.38	8.81	260.93	nr	592.31
Sum of two sides 2900mm	208.76	354.98	14.81	438.63	nr	793.61
Sum of two sides 3000mm	220.38	374.73	14.81	438.63	nr	813.36
Sum of two sides 3100mm	227.50	386.84	15.20	450.18	nr	837.03
Sum of two sides 3200mm	239.13	406.61	15.20	450.18	nr	856.79
Sum of two sides 3300mm	260.91	443.64	15.20	450.18	nr	893.82
Sum of two sides 3400mm	269.83	458.82	15.20	450.18	nr	909.00
Sum of two sides 3500mm	277.60	472.02	15.20	450.18	nr	922.21
Sum of two sides 3600mm	282.48	480.32	15.60	462.03	nr	942.36
Sum of two sides 3700mm	287.54	488.92	15.60	462.03	nr	950.95
Sum of two sides 3800mm	302.12	513.72	15.60	462.03	nr	975.75
Sum of two sides 3900mm	310.33	527.67	15.60	462.03	nr	989.70
Sum of two sides 4000mm	318.67	541.85	15.60	462.03	nr	1003.89
Branch						
Sum of two sides 2500mm	87.90	149.46	2.88	85.30	nr	234.76
Sum of two sides 2600mm	92.71	157.63	2.88	85.30	nr	242.93
Sum of two sides 2700mm	100.17	170.33	2.88	85.30	nr	255.63
Sum of two sides 2800mm	104.87	178.32	3.94	116.69	nr	295.02
Sum of two sides 2900mm	110.13	187.27	3.94	116.69	nr	303.96
Sum of two sides 3000mm	114.85	195.30	3.94	116.69	nr	311.99
Sum of two sides 3100mm	119.01	202.37	3.94	116.69	nr	319.06
Sum of two sides 3200mm	122.59	208.45	3.94	116.69	nr	325.14
Sum of two sides 3300mm	126.64	215.34	3.94	116.69	nr	332.04
Sum of two sides 3400mm	131.14	223.00	3.94	116.69	nr	339.69
Sum of two sides 3500mm	138.60	235.66	4.83	143.05	nr	378.72
Sum of two sides 3600mm	138.77	235.96	4.83	143.05	nr	379.01
Sum of two sides 3700mm	149.45	254.12	4.83	143.05	nr	397.17
Sum of two sides 3800mm	147.49	250.79	4.83	143.05	nr	393.84

U:VENTILATION/AIR CONDITIONING SYSTEMS

Item	Net Price £	Material £	Labour hours	Labour £	Unit	Total rate £
Sum of two sides 3900mm	155.34	264.13	4.83	143.05	nr	**407.18**
Sum of two sides 4000mm	155.00	263.56	4.83	143.05	nr	**406.62**
Grille neck						
Sum of two sides 2500mm	109.35	185.93	2.88	85.30	nr	**271.23**
Sum of two sides 2600mm	118.79	201.99	2.88	85.30	nr	**287.29**
Sum of two sides 2700mm	124.38	211.49	2.88	85.30	nr	**296.79**
Sum of two sides 2800mm	130.76	222.33	3.94	116.69	nr	**339.03**
Sum of two sides 2900mm	137.14	233.20	3.94	116.69	nr	**349.89**
Sum of two sides 3000mm	146.50	249.10	3.94	116.69	nr	**365.79**
Sum of two sides 3100mm	152.48	259.28	4.12	122.02	nr	**381.30**
Sum of two sides 3200mm	156.61	266.29	4.12	122.02	nr	**388.32**
Sum of two sides 3300mm	162.46	276.24	4.12	122.02	nr	**398.27**
Sum of two sides 3400mm	165.19	280.88	4.12	122.02	nr	**402.91**
Sum of two sides 3500mm	172.20	292.81	5.00	148.09	nr	**440.89**
Sum of two sides 3600mm	174.81	297.24	5.00	148.09	nr	**445.32**
Sum of two sides 3700mm	184.15	313.13	5.00	148.09	nr	**461.21**
Sum of two sides 3800mm	155.94	265.16	5.00	148.09	nr	**413.25**
Sum of two sides 3900mm	191.69	325.95	5.00	148.09	nr	**474.04**
Sum of two sides 4000mm	194.05	329.96	5.00	148.09	nr	**478.05**
Ductwork 2501 to 4000mm longest side						
Sum of two sides 3000mm	153.97	261.80	2.38	70.49	m	**332.29**
Sum of two sides 3100mm	157.77	268.27	2.38	70.49	m	**338.76**
Sum of two sides 3200mm	161.24	274.16	2.38	70.49	m	**344.65**
Sum of two sides 3300mm	164.71	280.07	2.38	70.49	m	**350.56**
Sum of two sides 3400mm	168.77	286.97	2.64	78.19	m	**365.16**
Sum of two sides 3500mm	172.24	292.88	2.66	78.78	m	**371.66**
Sum of two sides 3600mm	175.71	298.78	2.95	87.37	m	**386.15**
Sum of two sides 3700mm	180.00	306.07	2.96	87.67	m	**393.74**
Sum of two sides 3800mm	183.47	311.97	3.15	93.29	m	**405.27**
Sum of two sides 3900mm	193.35	328.77	3.15	93.29	m	**422.06**
Sum of two sides 4000mm	196.82	334.68	3.15	93.29	m	**427.97**
Sum of two sides 4100mm	200.52	340.96	3.35	99.22	m	**440.18**
Sum of two sides 4200mm	204.00	346.87	3.35	99.22	m	**446.09**
Sum of two sides 4300mm	207.47	352.77	3.60	106.62	m	**459.39**
Sum of two sides 4400mm	212.23	360.88	3.60	106.62	m	**467.50**
Sum of two sides 4500mm	215.70	366.77	3.60	106.62	m	**473.40**
Extra over fittings; Ductwork 2501 to 4000mm longest side						
End Cap						
Sum of two sides 3000mm	94.14	160.08	1.73	51.24	nr	**211.32**
Sum of two sides 3100mm	97.51	165.80	1.73	51.24	nr	**217.04**
Sum of two sides 3200mm	100.56	170.99	1.73	51.24	nr	**222.22**
Sum of two sides 3300mm	103.61	176.17	1.73	51.24	nr	**227.41**
Sum of two sides 3400mm	106.66	181.36	1.73	51.24	nr	**232.59**
Sum of two sides 3500mm	109.71	186.54	1.73	51.24	nr	**237.78**
Sum of two sides 3600mm	112.76	191.73	1.73	51.24	nr	**242.96**
Sum of two sides 3700mm	115.80	196.91	1.73	51.24	nr	**248.15**
Sum of two sides 3800mm	118.85	202.10	1.73	51.24	nr	**253.34**
Sum of two sides 3900mm	121.91	207.29	1.80	53.31	nr	**260.60**
Sum of two sides 4000mm	124.96	212.48	1.80	53.31	nr	**265.79**
Sum of two sides 4100mm	128.01	217.66	1.80	53.31	nr	**270.97**
Sum of two sides 4200mm	131.06	222.85	1.80	53.31	nr	**276.16**
Sum of two sides 4300mm	134.11	228.03	1.88	55.68	nr	**283.71**
Sum of two sides 4400mm	137.16	233.22	1.88	55.68	nr	**288.90**
Sum of two sides 4500mm	140.21	238.40	1.88	55.68	nr	**294.08**

U:VENTILATION/AIR CONDITIONING SYSTEMS

Item	Net Price £	Material £	Labour hours	Labour £	Unit	Total rate £
U10 : DUCTWORK : RECTANGULAR – CLASS B (cont'd)						
Y30 - AIR DUCTLINES (cont'd)						
Extra over fittings; Ductwork 2501 to 4000mm longest side (cont'd)						
Reducer						
Sum of two sides 3000mm	37.80	64.28	3.12	92.41	nr	**156.69**
Sum of two sides 3100mm	39.72	67.53	3.12	92.41	nr	**159.94**
Sum of two sides 3200mm	41.28	70.20	3.12	92.41	nr	**162.61**
Sum of two sides 3300mm	42.86	72.88	3.12	92.41	nr	**165.28**
Sum of two sides 3400mm	44.41	75.51	3.12	92.41	nr	**167.92**
Sum of two sides 3500mm	45.98	78.19	3.12	92.41	nr	**170.59**
Sum of two sides 3600mm	47.55	80.85	3.95	116.99	nr	**197.84**
Sum of two sides 3700mm	49.22	83.70	3.97	117.58	nr	**201.28**
Sum of two sides 3800mm	50.79	86.36	4.52	133.87	nr	**220.23**
Sum of two sides 3900mm	49.16	83.58	4.52	133.87	nr	**217.45**
Sum of two sides 4000mm	50.73	86.26	4.52	133.87	nr	**220.13**
Sum of two sides 4100mm	52.41	89.12	4.52	133.87	nr	**223.00**
Sum of two sides 4200mm	53.99	91.80	4.52	133.87	nr	**225.67**
Sum of two sides 4300mm	55.56	94.47	4.92	145.72	nr	**240.18**
Sum of two sides 4400mm	57.22	97.30	4.92	145.72	nr	**243.02**
Sum of two sides 4500mm	58.79	99.97	5.12	151.64	nr	**251.61**
Offset						
Sum of two sides 3000mm	164.35	279.46	3.48	103.07	nr	**382.52**
Sum of two sides 3100mm	161.27	274.23	3.48	103.07	nr	**377.30**
Sum of two sides 3200mm	157.11	267.15	3.48	103.07	nr	**370.22**
Sum of two sides 3300mm	152.54	259.38	3.48	103.07	nr	**362.44**
Sum of two sides 3400mm	167.76	285.26	3.49	103.36	nr	**388.62**
Sum of two sides 3500mm	162.78	276.79	3.50	103.66	nr	**380.45**
Sum of two sides 3600mm	157.39	267.62	3.50	103.66	nr	**371.28**
Sum of two sides 3700mm	173.15	294.41	4.76	140.98	nr	**435.39**
Sum of two sides 3800mm	167.32	284.51	5.32	157.56	nr	**442.08**
Sum of two sides 3900mm	178.06	302.77	5.35	158.45	nr	**461.22**
Sum of two sides 4000mm	171.34	291.35	5.35	158.45	nr	**449.80**
Sum of two sides 4100mm	164.31	279.38	5.85	173.26	nr	**452.64**
Sum of two sides 4200mm	156.75	266.53	5.85	173.26	nr	**439.80**
Sum of two sides 4300mm	161.26	274.20	6.15	182.15	nr	**456.34**
Sum of two sides 4400mm	178.53	303.57	6.15	182.15	nr	**485.71**
Sum of two sides 4500mm	170.34	289.64	6.30	186.59	nr	**476.23**
Square to round						
Sum of two sides 3000mm	138.60	235.66	4.70	139.20	nr	**374.87**
Sum of two sides 3100mm	144.62	245.91	4.70	139.20	nr	**385.11**
Sum of two sides 3200mm	149.97	255.00	4.70	139.20	nr	**394.20**
Sum of two sides 3300mm	155.30	264.07	4.70	139.20	nr	**403.27**
Sum of two sides 3400mm	160.64	273.16	4.71	139.50	nr	**412.66**
Sum of two sides 3500mm	165.98	282.24	4.71	139.50	nr	**421.73**
Sum of two sides 3600mm	171.33	291.33	8.19	242.57	nr	**533.89**
Sum of two sides 3700mm	176.65	300.37	8.62	255.30	nr	**555.68**
Sum of two sides 3800mm	182.00	309.46	8.62	255.30	nr	**564.77**
Sum of two sides 3900mm	224.28	381.36	8.62	255.30	nr	**636.67**
Sum of two sides 4000mm	230.68	392.25	8.75	259.15	nr	**651.40**
Sum of two sides 4100mm	237.08	403.12	11.23	332.60	nr	**735.73**
Sum of two sides 4200mm	243.47	414.00	11.23	332.60	nr	**746.60**
Sum of two sides 4300mm	249.88	424.89	11.25	333.20	nr	**758.09**
Sum of two sides 4400mm	256.28	435.77	11.25	333.20	nr	**768.96**
Sum of two sides 4500mm	262.68	446.65	11.26	333.49	nr	**780.14**

U:VENTILATION/AIR CONDITIONING SYSTEMS

Item	Net Price £	Material £	Labour hours	Labour £	Unit	Total rate £
90° radius bend						
Sum of two sides 3000mm	225.62	383.63	3.90	115.51	nr	**499.14**
Sum of two sides 3100mm	236.31	401.81	3.90	115.51	nr	**517.32**
Sum of two sides 3200mm	245.49	417.43	4.26	126.17	nr	**543.60**
Sum of two sides 3300mm	254.68	433.05	4.26	126.17	nr	**559.22**
Sum of two sides 3400mm	247.20	420.34	4.26	126.17	nr	**546.51**
Sum of two sides 3500mm	256.20	435.63	4.55	134.76	nr	**570.39**
Sum of two sides 3600mm	265.19	450.92	4.55	134.76	nr	**585.68**
Sum of two sides 3700mm	257.11	437.18	6.87	203.47	nr	**640.66**
Sum of two sides 3800mm	265.91	452.15	6.87	203.47	nr	**655.62**
Sum of two sides 3900mm	245.88	418.09	6.87	203.47	nr	**621.56**
Sum of two sides 4000mm	254.49	432.72	7.00	207.32	nr	**640.04**
Sum of two sides 4100mm	263.62	448.26	7.20	213.25	nr	**661.50**
Sum of two sides 4200mm	272.23	462.89	7.20	213.25	nr	**676.13**
Sum of two sides 4300mm	280.84	477.53	7.41	219.46	nr	**697.00**
Sum of two sides 4400mm	259.12	440.60	7.41	219.46	nr	**660.06**
Sum of two sides 4500mm	267.44	454.74	7.55	223.61	nr	**678.36**
45° bend						
Sum of two sides 3000mm	104.58	177.82	4.85	143.64	nr	**321.46**
Sum of two sides 3100mm	109.66	186.46	4.85	143.64	nr	**330.10**
Sum of two sides 3200mm	114.03	193.89	4.85	143.64	nr	**337.53**
Sum of two sides 3300mm	118.40	201.32	4.87	144.24	nr	**345.56**
Sum of two sides 3400mm	114.67	194.98	4.87	144.24	nr	**339.22**
Sum of two sides 3500mm	118.95	202.25	4.87	144.24	nr	**346.49**
Sum of two sides 3600mm	123.22	209.53	8.81	260.93	nr	**470.46**
Sum of two sides 3700mm	119.21	202.70	8.81	260.93	nr	**463.62**
Sum of two sides 3800mm	123.39	209.81	9.31	275.74	nr	**485.55**
Sum of two sides 3900mm	113.43	192.88	9.31	275.74	nr	**468.62**
Sum of two sides 4000mm	117.53	199.84	9.31	275.74	nr	**475.58**
Sum of two sides 4100mm	121.86	207.21	10.01	296.47	nr	**503.68**
Sum of two sides 4200mm	125.95	214.17	10.01	296.47	nr	**510.64**
Sum of two sides 4300mm	130.05	221.13	9.31	275.74	nr	**496.86**
Sum of two sides 4400mm	119.42	203.06	10.52	311.58	nr	**514.64**
Sum of two sides 4500mm	123.38	209.79	10.52	311.58	nr	**521.37**
90° mitre bend						
Sum of two sides 3000mm	312.95	532.13	4.85	143.64	nr	**675.78**
Sum of two sides 3100mm	327.68	557.19	4.85	143.64	nr	**700.83**
Sum of two sides 3200mm	341.56	580.77	4.87	144.24	nr	**725.01**
Sum of two sides 3300mm	355.43	604.37	4.87	144.24	nr	**748.61**
Sum of two sides 3400mm	346.11	588.52	4.87	144.24	nr	**732.76**
Sum of two sides 3500mm	359.77	611.75	8.81	260.93	nr	**872.68**
Sum of two sides 3600mm	373.43	634.98	8.81	260.93	nr	**895.91**
Sum of two sides 3700mm	365.59	621.65	14.81	438.63	nr	**1060.28**
Sum of two sides 3800mm	379.05	644.53	14.81	438.63	nr	**1083.16**
Sum of two sides 3900mm	351.35	597.42	14.81	438.63	nr	**1036.06**
Sum of two sides 4000mm	364.59	619.95	15.20	450.18	nr	**1070.13**
Sum of two sides 4100mm	377.51	641.92	16.30	482.76	nr	**1124.68**
Sum of two sides 4200mm	390.76	664.44	16.50	488.69	nr	**1153.13**
Sum of two sides 4300mm	404.01	686.97	17.01	503.79	nr	**1190.76**
Sum of two sides 4400mm	377.15	641.31	17.01	503.79	nr	**1145.10**
Sum of two sides 4500mm	390.10	663.32	17.01	503.79	nr	**1167.11**

U:VENTILATION/AIR CONDITIONING SYSTEMS

Item	Net Price £	Material £	Labour hours	Labour £	Unit	Total rate £
U10 : DUCTWORK : RECTANGULAR – CLASS B (cont'd)						
Y30 - AIR DUCTLINES (cont'd)						
Extra over fittings; Ductwork 2501 to 4000mm longest side (cont'd)						
Branch						
Sum of two sides 3000mm	115.00	195.54	2.88	85.30	nr	**280.83**
Sum of two sides 3100mm	327.68	557.19	2.88	85.30	nr	**642.49**
Sum of two sides 3200mm	122.58	208.44	2.88	85.30	nr	**293.73**
Sum of two sides 3300mm	126.15	214.50	2.88	85.30	nr	**299.80**
Sum of two sides 3400mm	129.71	220.55	2.88	85.30	nr	**305.85**
Sum of two sides 3500mm	133.27	226.62	3.94	116.69	nr	**343.31**
Sum of two sides 3600mm	136.84	232.67	3.94	116.69	nr	**349.37**
Sum of two sides 3700mm	140.51	238.92	3.94	116.69	nr	**355.61**
Sum of two sides 3800mm	144.08	244.99	4.83	143.05	nr	**388.04**
Sum of two sides 3900mm	147.79	251.30	4.83	143.05	nr	**394.36**
Sum of two sides 4000mm	151.36	257.37	4.83	143.05	nr	**400.42**
Sum of two sides 4100mm	155.04	263.63	5.44	161.12	nr	**424.75**
Sum of two sides 4200mm	158.61	269.69	5.44	161.12	nr	**430.81**
Sum of two sides 4300mm	162.18	275.76	5.85	173.26	nr	**449.02**
Sum of two sides 4400mm	165.84	281.98	5.85	173.26	nr	**455.25**
Sum of two sides 4500mm	169.40	288.05	5.85	173.26	nr	**461.31**
Grille neck						
Sum of two sides 3000mm	144.18	245.16	2.88	85.30	nr	**330.46**
Sum of two sides 3100mm	149.32	253.90	2.88	85.30	nr	**339.19**
Sum of two sides 3200mm	154.00	261.85	2.88	85.30	nr	**347.15**
Sum of two sides 3300mm	158.68	269.81	2.88	85.30	nr	**355.11**
Sum of two sides 3400mm	163.35	277.75	3.94	116.69	nr	**394.45**
Sum of two sides 3500mm	168.03	285.71	3.94	116.69	nr	**402.40**
Sum of two sides 3600mm	172.71	293.67	3.94	116.69	nr	**410.36**
Sum of two sides 3700mm	177.39	301.62	4.12	122.02	nr	**423.65**
Sum of two sides 3800mm	182.07	309.58	4.12	122.02	nr	**431.60**
Sum of two sides 3900mm	186.74	317.53	4.12	122.02	nr	**439.55**
Sum of two sides 4000mm	191.42	325.48	5.00	148.09	nr	**473.57**
Sum of two sides 4100mm	196.10	333.44	5.00	148.09	nr	**481.52**
Sum of two sides 4200mm	200.78	341.39	5.00	148.09	nr	**489.48**
Sum of two sides 4300mm	205.45	349.35	5.23	154.90	nr	**504.25**
Sum of two sides 4400mm	210.13	357.30	5.23	154.90	nr	**512.20**
Sum of two sides 4500mm	214.81	365.25	5.39	159.64	nr	**524.89**

U:VENTILATION/AIR CONDITIONING SYSTEMS

Item	Net Price £	Material £	Labour hours	Labour £	Unit	Total rate £
ANCILLARIES						
Access doors, hollow steel construction; 25mm mineral wool insulation; removable or hinged; fixed with cams; including sub-frame and integral sealing gaskets						
Rectangular duct						
150 x 150mm	15.05	25.60	1.25	37.02	nr	**62.62**
200 x 200mm	16.27	27.66	1.25	37.02	nr	**64.69**
300 x 150mm	16.72	28.43	1.25	37.02	nr	**65.45**
300 x 300mm	19.15	32.57	1.25	37.02	nr	**69.59**
400 x 400mm	21.84	37.14	1.35	40.02	nr	**77.16**
450 x 300mm	21.84	37.14	1.50	44.47	nr	**81.61**
450 x 450mm	24.08	40.95	1.50	44.47	nr	**85.42**
Access doors, hollow steel construction; 25mm mineral wool insulation; removable or hinged; fixed with cams; including sub-frame and integral sealing gaskets						
Flat oval duct						
235 x 90mm	30.58	51.99	1.25	37.02	nr	**89.01**
235 x 140mm	32.59	55.42	1.35	40.02	nr	**95.44**
335 x 235mm	37.24	63.32	1.50	44.47	nr	**107.79**
535 x 235mm	41.89	71.23	1.50	44.47	nr	**115.70**

U:VENTILATION/AIR CONDITIONING SYSTEMS

Item	Net Price £	Material £	Labour hours	Labour £	Unit	Total rate £
U10 : DUCTWORK : RECTANGULAR - CLASS C						
Y30 - AIR DUCTLINES						
Galvanised sheet metal DW144 class C rectangular section ductwork; including all necessary stiffeners, joints, couplers in the running length and duct supports						
Ductwork up to 400mm longest side						
Sum of two sides 200mm	19.69	33.48	1.19	35.24	m	**68.72**
Sum of two sides 300mm	21.22	36.08	1.16	34.36	m	**70.44**
Sum of two sides 400mm	18.17	30.89	1.17	34.65	m	**65.54**
Sum of two sides 500mm	19.86	33.76	1.17	34.65	m	**68.41**
Sum of two sides 600mm	21.26	36.15	1.19	35.24	m	**71.40**
Sum of two sides 700mm	22.68	38.56	1.19	35.24	m	**73.81**
Sum of two sides 800mm	24.19	41.14	1.19	35.24	m	**76.38**
Extra over fittings; Ductwork up to 400mm longest side						
End Cap						
Sum of two sides 200mm	9.45	16.07	0.38	11.25	nr	**27.32**
Sum of two sides 300mm	10.53	17.91	0.38	11.25	nr	**29.16**
Sum of two sides 400mm	11.60	19.73	0.38	11.25	nr	**30.98**
Sum of two sides 500mm	12.67	21.55	0.38	11.25	nr	**32.80**
Sum of two sides 600mm	13.74	23.37	0.38	11.25	nr	**34.63**
Sum of two sides 700mm	14.82	25.19	0.38	11.25	nr	**36.45**
Sum of two sides 800mm	15.89	27.01	0.38	11.25	nr	**38.27**
Reducer						
Sum of two sides 200mm	10.01	17.01	1.40	41.46	nr	**58.48**
Sum of two sides 300mm	11.30	19.21	1.40	41.46	nr	**60.68**
Sum of two sides 400mm	18.61	31.64	1.42	42.06	nr	**73.69**
Sum of two sides 500mm	20.20	34.35	1.42	42.06	nr	**76.41**
Sum of two sides 600mm	21.80	37.06	1.69	50.05	nr	**87.12**
Sum of two sides 700mm	23.39	39.78	1.69	50.05	nr	**89.83**
Sum of two sides 800mm	24.94	42.40	1.92	56.87	nr	**99.27**
Offset						
Sum of two sides 200mm	12.11	20.59	1.63	48.28	nr	**68.86**
Sum of two sides 300mm	13.69	23.28	1.63	48.28	nr	**71.56**
Sum of two sides 400mm	20.35	34.60	1.65	48.87	nr	**83.47**
Sum of two sides 500mm	22.17	37.69	1.65	48.87	nr	**86.56**
Sum of two sides 600mm	23.69	40.28	1.92	56.87	nr	**97.14**
Sum of two sides 700mm	25.35	43.10	1.92	56.87	nr	**99.96**
Sum of two sides 800mm	26.82	45.60	1.92	56.87	nr	**102.46**
Square to round						
Sum of two sides 200mm	23.08	39.24	1.22	36.13	nr	**75.38**
Sum of two sides 300mm	25.92	44.08	1.22	36.13	nr	**80.21**
Sum of two sides 400mm	34.63	58.88	1.25	37.02	nr	**95.90**
Sum of two sides 500mm	37.75	64.19	1.25	37.02	nr	**101.21**
Sum of two sides 600mm	40.86	69.47	1.33	39.39	nr	**108.86**
Sum of two sides 700mm	43.97	74.77	1.33	39.39	nr	**114.16**
Sum of two sides 800mm	47.05	80.00	1.40	41.46	nr	**121.47**

U:VENTILATION/AIR CONDITIONING SYSTEMS

Item	Net Price £	Material £	Labour hours	Labour £	Unit	Total rate £
90° radius bend						
Sum of two sides 200mm	11.64	19.80	1.10	32.58	nr	**52.38**
Sum of two sides 300mm	12.56	21.35	1.10	32.58	nr	**53.93**
Sum of two sides 400mm	21.32	36.24	1.12	33.17	nr	**69.42**
Sum of two sides 500mm	22.76	38.71	1.12	33.17	nr	**71.88**
Sum of two sides 600mm	24.65	41.92	1.16	34.36	nr	**76.28**
Sum of two sides 700mm	26.30	44.72	1.16	34.36	nr	**79.08**
Sum of two sides 800mm	28.16	47.88	1.22	36.13	nr	**84.02**
45° radius bend						
Sum of two sides 200mm	11.54	19.62	1.29	38.21	nr	**57.83**
Sum of two sides 300mm	12.65	21.51	1.29	38.21	nr	**59.72**
Sum of two sides 400mm	20.46	34.80	1.29	38.21	nr	**73.00**
Sum of two sides 500mm	22.00	37.40	1.29	38.21	nr	**75.61**
Sum of two sides 600mm	23.74	40.37	1.39	41.17	nr	**81.54**
Sum of two sides 700mm	25.38	43.15	1.39	41.17	nr	**84.32**
Sum of two sides 800mm	27.10	46.08	1.46	43.24	nr	**89.32**
90° mitre bend						
Sum of two sides 200mm	15.90	27.03	2.04	60.42	nr	**87.45**
Sum of two sides 300mm	17.38	29.55	2.04	60.42	nr	**89.97**
Sum of two sides 400mm	25.37	43.14	2.09	61.90	nr	**105.04**
Sum of two sides 500mm	27.38	46.56	2.09	61.90	nr	**108.46**
Sum of two sides 600mm	29.94	50.90	2.15	63.68	nr	**114.58**
Sum of two sides 700mm	32.30	54.92	2.15	63.68	nr	**118.60**
Sum of two sides 800mm	34.88	59.31	2.26	66.94	nr	**126.25**
Branch						
Sum of two sides 200mm	19.76	33.60	0.92	27.25	nr	**60.85**
Sum of two sides 300mm	21.95	37.31	0.92	27.25	nr	**64.56**
Sum of two sides 400mm	27.55	46.85	0.95	28.14	nr	**74.99**
Sum of two sides 500mm	30.03	51.06	0.95	28.14	nr	**79.20**
Sum of two sides 600mm	32.46	55.19	1.03	30.51	nr	**85.69**
Sum of two sides 700mm	34.87	59.29	1.03	30.51	nr	**89.80**
Sum of two sides 800mm	37.30	63.42	1.03	30.51	nr	**93.92**
Grille neck						
Sum of two sides 200mm	21.03	35.76	1.10	32.58	nr	**68.34**
Sum of two sides 300mm	23.18	39.42	1.10	32.58	nr	**72.00**
Sum of two sides 400mm	25.37	43.14	1.16	34.36	nr	**77.49**
Sum of two sides 500mm	27.52	46.80	1.16	34.36	nr	**81.15**
Sum of two sides 600mm	29.69	50.49	1.18	34.95	nr	**85.44**
Sum of two sides 700mm	31.86	54.17	1.18	34.95	nr	**89.12**
Sum of two sides 800mm	34.02	57.85	1.18	34.95	nr	**92.80**
Ductwork 401 to 600mm longest side						
Sum of two sides 600mm	22.81	38.78	1.17	34.65	m	**73.43**
Sum of two sides 700mm	25.18	42.81	1.17	34.65	m	**77.47**
Sum of two sides 800mm	27.43	46.63	1.17	34.65	m	**81.29**
Sum of two sides 900mm	29.60	50.33	1.27	37.61	m	**87.94**
Sum of two sides 1000mm	31.77	54.03	1.48	43.83	m	**97.86**
Sum of two sides 1100mm	34.21	58.17	1.49	44.13	m	**102.30**
Sum of two sides 1200mm	36.38	61.86	1.49	44.13	m	**105.99**

U:VENTILATION/AIR CONDITIONING SYSTEMS

Item	Net Price £	Material £	Labour hours	Labour £	Unit	Total rate £
U10 : DUCTWORK : RECTANGULAR – CLASS C (cont'd)						
Y30 - AIR DUCTLINES (cont'd)						
Extra over fittings: Ductwork 401 to 600mm longest side						
End Cap						
Sum of two sides 600mm	13.76	23.39	0.38	11.25	nr	**34.64**
Sum of two sides 700mm	14.83	25.21	0.38	11.25	nr	**36.46**
Sum of two sides 800mm	15.90	27.03	0.38	11.25	nr	**38.29**
Sum of two sides 900mm	16.98	28.87	0.38	11.25	nr	**40.12**
Sum of two sides 1000mm	18.06	30.71	0.38	11.25	nr	**41.96**
Sum of two sides 1100mm	19.13	32.53	0.38	11.25	nr	**43.78**
Sum of two sides 1200mm	20.20	34.35	0.38	11.25	nr	**45.61**
Reducer						
Sum of two sides 600mm	18.85	32.05	1.69	50.05	nr	**82.10**
Sum of two sides 700mm	20.10	34.17	1.69	50.05	nr	**84.23**
Sum of two sides 800mm	21.33	36.26	1.92	56.87	nr	**93.13**
Sum of two sides 900mm	22.58	38.39	1.92	56.87	nr	**95.25**
Sum of two sides 1000mm	23.82	40.51	2.18	64.57	nr	**105.08**
Sum of two sides 1100mm	25.21	42.87	2.18	64.57	nr	**107.43**
Sum of two sides 1200mm	26.46	44.99	2.18	64.57	nr	**109.56**
Offset						
Sum of two sides 600mm	21.97	37.35	1.92	56.87	nr	**94.22**
Sum of two sides 700mm	23.48	39.92	1.92	56.87	nr	**96.79**
Sum of two sides 800mm	24.31	41.33	1.92	56.87	nr	**98.20**
Sum of two sides 900mm	25.07	42.64	1.92	56.87	nr	**99.50**
Sum of two sides 1000mm	26.11	44.40	2.18	64.57	nr	**108.97**
Sum of two sides 1100mm	26.79	45.55	2.18	64.57	nr	**110.11**
Sum of two sides 1200mm	27.28	46.38	2.18	64.57	nr	**110.95**
Square to round						
Sum of two sides 600mm	31.51	53.58	1.33	39.39	nr	**92.97**
Sum of two sides 700mm	33.73	57.35	1.33	39.39	nr	**96.74**
Sum of two sides 800mm	35.92	61.08	1.40	41.46	nr	**102.54**
Sum of two sides 900mm	38.14	64.85	1.40	41.46	nr	**106.31**
Sum of two sides 1000mm	40.36	68.63	1.82	53.90	nr	**122.53**
Sum of two sides 1100mm	42.63	72.49	1.82	53.90	nr	**126.39**
Sum of two sides 1200mm	44.85	76.25	1.82	53.90	nr	**130.16**
90° radius bend						
Sum of two sides 600mm	19.34	32.89	1.16	34.36	nr	**67.24**
Sum of two sides 700mm	19.79	33.65	1.16	34.36	nr	**68.01**
Sum of two sides 800mm	21.11	35.89	1.22	36.13	nr	**72.02**
Sum of two sides 900mm	22.45	38.17	1.22	36.13	nr	**74.31**
Sum of two sides 1000mm	23.25	39.53	1.40	41.46	nr	**80.99**
Sum of two sides 1100mm	24.68	41.96	1.40	41.46	nr	**83.42**
Sum of two sides 1200mm	26.00	44.21	1.40	41.46	nr	**85.67**
45° bend						
Sum of two sides 600mm	22.65	38.51	1.39	41.17	nr	**79.68**
Sum of two sides 700mm	23.74	40.37	1.39	41.17	nr	**81.54**
Sum of two sides 800mm	25.29	43.01	1.46	43.24	nr	**86.25**
Sum of two sides 900mm	26.87	45.69	1.46	43.24	nr	**88.93**
Sum of two sides 1000mm	28.15	47.87	1.88	55.68	nr	**103.55**
Sum of two sides 1100mm	29.86	50.78	1.88	55.68	nr	**106.46**
Sum of two sides 1200mm	31.43	53.44	1.88	55.68	nr	**109.12**

U:VENTILATION/AIR CONDITIONING SYSTEMS

Item	Net Price £	Material £	Labour hours	Labour £	Unit	Total rate £
90° mitre bend						
Sum of two sides 600mm	26.48	45.03	2.15	63.68	nr	108.71
Sum of two sides 700mm	27.44	46.65	2.15	63.68	nr	110.33
Sum of two sides 800mm	29.38	49.96	2.26	66.94	nr	116.89
Sum of two sides 900mm	31.35	53.31	2.26	66.94	nr	120.25
Sum of two sides 1000mm	32.81	55.79	3.03	89.74	nr	145.53
Sum of two sides 1100mm	34.87	59.29	3.03	89.74	nr	149.03
Sum of two sides 1200mm	36.86	62.67	3.03	89.74	nr	152.41
Branch						
Sum of two sides 600mm	29.15	49.56	1.03	30.51	nr	80.07
Sum of two sides 700mm	31.31	53.24	1.03	30.51	nr	83.75
Sum of two sides 800mm	33.48	56.94	1.03	30.51	nr	87.44
Sum of two sides 900mm	35.66	60.63	1.03	30.51	nr	91.14
Sum of two sides 1000mm	37.86	64.38	1.29	38.21	nr	102.59
Sum of two sides 1100mm	40.12	68.22	1.29	38.21	nr	106.43
Sum of two sides 1200mm	42.29	71.92	1.29	38.21	nr	110.12
Grille neck						
Sum of two sides 600mm	28.66	48.74	1.18	34.95	nr	83.69
Sum of two sides 700mm	31.86	54.17	1.18	34.95	nr	89.12
Sum of two sides 800mm	34.02	57.85	1.18	34.95	nr	92.80
Sum of two sides 900mm	36.18	61.52	1.18	34.95	nr	96.47
Sum of two sides 1000mm	38.36	65.22	1.44	42.65	nr	107.87
Sum of two sides 1100mm	40.51	68.88	1.44	42.65	nr	111.53
Sum of two sides 1200mm	42.69	72.59	1.44	42.65	nr	115.24
Ductwork 601 to 800mm longest side						
Sum of two sides 900mm	32.66	55.53	1.27	37.61	m	93.14
Sum of two sides 1000mm	34.83	59.22	1.48	43.83	m	103.06
Sum of two sides 1100mm	37.01	62.94	1.49	44.13	m	107.07
Sum of two sides 1200mm	39.44	67.06	1.49	44.13	m	111.19
Sum of two sides 1300mm	41.62	70.77	1.51	44.72	m	115.50
Sum of two sides 1400mm	43.80	74.47	1.55	45.91	m	120.38
Sum of two sides 1500mm	45.97	78.16	1.61	47.68	m	125.85
Sum of two sides 1600mm	48.14	81.86	1.62	47.98	m	129.84
Extra over fittings; Ductwork 601 to 800mm longest side						
End Cap						
Sum of two sides 900mm	16.98	28.87	0.38	11.25	nr	40.12
Sum of two sides 1000mm	18.06	30.71	0.38	11.25	nr	41.96
Sum of two sides 1100mm	19.13	32.53	0.38	11.25	nr	43.78
Sum of two sides 1200mm	20.16	34.28	0.38	11.25	nr	45.54
Sum of two sides 1300mm	21.15	35.97	0.38	11.25	nr	47.23
Sum of two sides 1400mm	22.20	37.74	0.38	11.25	nr	49.00
Sum of two sides 1500mm	25.83	43.92	0.38	11.25	nr	55.18
Sum of two sides 1600mm	29.48	50.13	0.38	11.25	nr	61.39
Reducer						
Sum of two sides 900mm	22.81	38.78	1.92	56.87	nr	95.64
Sum of two sides 1000mm	24.06	40.90	2.18	64.57	nr	105.47
Sum of two sides 1100mm	25.31	43.03	2.18	64.57	nr	107.59
Sum of two sides 1200mm	26.69	45.38	2.18	64.57	nr	109.95
Sum of two sides 1300mm	27.94	47.51	2.30	68.12	nr	115.63
Sum of two sides 1400mm	28.99	49.29	2.30	68.12	nr	117.41
Sum of two sides 1500mm	33.47	56.92	2.47	73.16	nr	130.07
Sum of two sides 1600mm	37.95	64.52	2.47	73.16	nr	137.68

U:VENTILATION/AIR CONDITIONING SYSTEMS

Item	Net Price £	Material £	Labour hours	Labour £	Unit	Total rate £
U10 : DUCTWORK : RECTANGULAR – CLASS C (cont'd)						
Y30 - AIR DUCTLINES (cont'd)						
Extra over fittings; Ductwork 601 to 800mm longest side (cont'd)						
Offset						
Sum of two sides 900mm	26.58	45.19	1.92	56.87	nr	**102.05**
Sum of two sides 1000mm	27.24	46.31	2.18	64.57	nr	**110.88**
Sum of two sides 1100mm	27.81	47.30	2.18	64.57	nr	**111.86**
Sum of two sides 1200mm	28.37	48.24	2.18	64.57	nr	**112.81**
Sum of two sides 1300mm	28.76	48.90	2.47	73.16	nr	**122.06**
Sum of two sides 1400mm	29.33	49.87	2.47	73.16	nr	**123.02**
Sum of two sides 1500mm	33.56	57.06	2.47	73.16	nr	**130.22**
Sum of two sides 1600mm	37.71	64.11	2.47	73.16	nr	**137.27**
Square to round						
Sum of two sides 900mm	37.80	64.27	1.40	41.46	nr	**105.74**
Sum of two sides 1000mm	40.02	68.04	1.82	53.90	nr	**121.95**
Sum of two sides 1100mm	42.23	71.81	1.82	53.90	nr	**125.71**
Sum of two sides 1200mm	44.51	75.68	1.82	53.90	nr	**129.59**
Sum of two sides 1300mm	46.73	79.45	2.32	68.71	nr	**148.16**
Sum of two sides 1400mm	48.56	82.57	2.32	68.71	nr	**151.29**
Sum of two sides 1500mm	57.15	97.18	2.56	75.82	nr	**173.00**
Sum of two sides 1600mm	65.75	111.80	2.58	76.41	nr	**188.21**
90° radius bend						
Sum of two sides 900mm	19.94	33.90	1.22	36.13	nr	**70.04**
Sum of two sides 1000mm	21.20	36.05	1.40	41.46	nr	**77.51**
Sum of two sides 1100mm	22.45	38.17	1.40	41.46	nr	**79.64**
Sum of two sides 1200mm	23.79	40.46	1.40	41.46	nr	**81.92**
Sum of two sides 1300mm	25.04	42.58	1.91	56.57	nr	**99.15**
Sum of two sides 1400mm	25.34	43.08	1.91	56.57	nr	**99.65**
Sum of two sides 1500mm	30.10	51.19	2.11	62.49	nr	**113.68**
Sum of two sides 1600mm	34.87	59.29	2.11	62.49	nr	**121.79**
45° bend						
Sum of two sides 900mm	25.96	44.14	1.46	43.24	nr	**87.38**
Sum of two sides 1000mm	27.48	46.72	1.88	55.68	nr	**102.40**
Sum of two sides 1100mm	29.01	49.33	1.88	55.68	nr	**105.01**
Sum of two sides 1200mm	30.68	52.17	1.88	55.68	nr	**107.85**
Sum of two sides 1300mm	32.20	54.76	2.26	66.94	nr	**121.69**
Sum of two sides 1400mm	33.11	56.29	2.44	72.27	nr	**128.56**
Sum of two sides 1500mm	38.31	65.15	2.44	72.27	nr	**137.42**
Sum of two sides 1600mm	43.51	73.99	2.68	79.37	nr	**153.36**
90° mitre bend						
Sum of two sides 900mm	28.51	48.47	2.26	66.94	nr	**115.41**
Sum of two sides 1000mm	30.54	51.94	3.03	89.74	nr	**141.68**
Sum of two sides 1100mm	32.58	55.40	3.03	89.74	nr	**145.14**
Sum of two sides 1200mm	34.68	58.97	3.03	89.74	nr	**148.71**
Sum of two sides 1300mm	36.72	62.44	3.85	114.03	nr	**176.46**
Sum of two sides 1400mm	37.79	64.26	3.85	114.03	nr	**178.28**
Sum of two sides 1500mm	44.04	74.88	4.25	125.87	nr	**200.75**
Sum of two sides 1600mm	50.27	85.48	4.26	126.17	nr	**211.66**

U:VENTILATION/AIR CONDITIONING SYSTEMS

Item	Net Price £	Material £	Labour hours	Labour £	Unit	Total rate £
Branch						
Sum of two sides 900mm	36.54	62.13	1.03	30.51	nr	**92.64**
Sum of two sides 1000mm	38.77	65.92	1.29	38.21	nr	**104.12**
Sum of two sides 1100mm	40.97	69.67	1.29	38.21	nr	**107.87**
Sum of two sides 1200mm	43.29	73.61	1.29	38.21	nr	**111.82**
Sum of two sides 1300mm	45.52	77.40	1.39	41.17	nr	**118.57**
Sum of two sides 1400mm	47.43	80.65	1.39	41.17	nr	**121.81**
Sum of two sides 1500mm	54.47	92.63	1.64	48.57	nr	**141.20**
Sum of two sides 1600mm	61.51	104.59	1.64	48.57	nr	**153.16**
Grille neck						
Sum of two sides 900mm	36.17	61.51	1.18	34.95	nr	**96.46**
Sum of two sides 1000mm	38.35	65.20	1.44	42.65	nr	**107.85**
Sum of two sides 1100mm	40.52	68.90	1.44	42.65	nr	**111.55**
Sum of two sides 1200mm	42.70	72.61	1.44	42.65	nr	**115.26**
Sum of two sides 1300mm	44.89	76.33	1.69	50.05	nr	**126.38**
Sum of two sides 1400mm	47.07	80.04	1.69	50.05	nr	**130.09**
Sum of two sides 1500mm	54.67	92.97	1.79	53.02	nr	**145.98**
Sum of two sides 1600mm	61.55	104.66	1.79	53.02	nr	**157.68**
Ductwork 801 to 1000mm longest side						
Sum of two sides 1100mm	41.90	71.24	1.49	44.13	m	**115.37**
Sum of two sides 1200mm	44.32	75.36	1.49	44.13	m	**119.49**
Sum of two sides 1300mm	46.75	79.49	1.51	44.72	m	**124.21**
Sum of two sides 1400mm	49.44	84.07	1.55	45.91	m	**129.98**
Sum of two sides 1500mm	51.87	88.20	1.61	47.68	m	**135.88**
Sum of two sides 1600mm	54.28	92.31	1.62	47.98	m	**140.29**
Sum of two sides 1700mm	56.71	96.43	1.74	51.53	m	**147.96**
Sum of two sides 1800mm	59.42	101.04	1.76	52.13	m	**153.16**
Sum of two sides 1900mm	61.83	105.14	1.81	53.61	m	**158.75**
Sum of two sides 2000mm	64.26	109.27	1.82	53.90	m	**163.17**
Extra over fittings; Ductwork 801 to 1000mm longest side						
End Cap						
Sum of two sides 1100mm	20.55	34.94	0.38	11.25	nr	**46.19**
Sum of two sides 1200mm	21.70	36.90	0.38	11.25	nr	**48.16**
Sum of two sides 1300mm	22.86	38.87	0.38	11.25	nr	**50.12**
Sum of two sides 1400mm	23.85	40.55	0.38	11.25	nr	**51.80**
Sum of two sides 1500mm	27.75	47.19	0.38	11.25	nr	**58.44**
Sum of two sides 1600mm	31.67	53.85	0.38	11.25	nr	**65.10**
Sum of two sides 1700mm	35.58	60.51	0.58	17.18	nr	**77.69**
Sum of two sides 1800mm	39.49	67.15	0.58	17.18	nr	**84.33**
Sum of two sides 1900mm	43.41	73.81	0.58	17.18	nr	**90.99**
Sum of two sides 2000mm	47.32	80.47	0.58	17.18	nr	**97.65**
Reducer						
Sum of two sides 1100mm	17.82	30.30	2.18	64.57	nr	**94.86**
Sum of two sides 1200mm	18.70	31.80	2.18	64.57	nr	**96.36**
Sum of two sides 1300mm	19.57	33.28	2.30	68.12	nr	**101.40**
Sum of two sides 1400mm	20.37	34.64	2.30	68.12	nr	**102.76**
Sum of two sides 1500mm	23.71	40.31	2.47	73.16	nr	**113.47**
Sum of two sides 1600mm	27.04	45.97	2.47	73.16	nr	**119.13**
Sum of two sides 1700mm	30.38	51.65	2.59	76.71	nr	**128.36**
Sum of two sides 1800mm	33.79	57.45	2.59	76.71	nr	**134.16**
Sum of two sides 1900mm	37.13	63.13	2.71	80.26	nr	**143.39**
Sum of two sides 2000mm	40.46	68.79	2.71	80.26	nr	**149.05**

U:VENTILATION/AIR CONDITIONING SYSTEMS

Item	Net Price £	Material £	Labour hours	Labour £	Unit	Total rate £
U10 : DUCTWORK : RECTANGULAR – CLASS C (cont'd)						
Y30 - AIR DUCTLINES (cont'd)						
Extra over fittings; Ductwork 801 to 1000mm longest side (cont'd)						
Offset						
Sum of two sides 1100mm	27.69	47.08	2.18	64.57	nr	**111.65**
Sum of two sides 1200mm	28.21	47.97	2.18	64.57	nr	**112.54**
Sum of two sides 1300mm	28.66	48.74	2.47	73.16	nr	**121.90**
Sum of two sides 1400mm	28.85	49.06	2.47	73.16	nr	**122.22**
Sum of two sides 1500mm	32.87	55.88	2.47	73.16	nr	**129.04**
Sum of two sides 1600mm	36.78	62.54	2.47	73.16	nr	**135.70**
Sum of two sides 1700mm	40.61	69.06	2.61	77.30	nr	**146.36**
Sum of two sides 1800mm	44.40	75.50	2.61	77.30	nr	**152.81**
Sum of two sides 1900mm	48.66	82.74	2.71	80.26	nr	**163.00**
Sum of two sides 2000mm	52.26	88.86	2.71	80.26	nr	**169.12**
Square to round						
Sum of two sides 1100mm	35.21	59.86	1.82	53.90	nr	**113.77**
Sum of two sides 1200mm	37.07	63.02	1.82	53.90	nr	**116.93**
Sum of two sides 1300mm	38.91	66.17	2.32	68.71	nr	**134.88**
Sum of two sides 1400mm	40.44	68.76	2.32	68.71	nr	**137.47**
Sum of two sides 1500mm	48.05	81.70	2.56	75.82	nr	**157.52**
Sum of two sides 1600mm	55.65	94.63	2.58	76.41	nr	**171.04**
Sum of two sides 1700mm	63.25	107.55	2.84	84.11	nr	**191.67**
Sum of two sides 1800mm	70.89	120.53	2.84	84.11	nr	**204.65**
Sum of two sides 1900mm	78.49	133.46	3.13	92.70	nr	**226.16**
Sum of two sides 2000mm	86.10	146.40	3.13	92.70	nr	**239.11**
90° radius bend						
Sum of two sides 1100mm	16.52	28.08	1.40	41.46	nr	**69.55**
Sum of two sides 1200mm	17.67	30.05	1.40	41.46	nr	**71.51**
Sum of two sides 1300mm	18.83	32.01	1.91	56.57	nr	**88.58**
Sum of two sides 1400mm	19.86	33.76	1.91	56.57	nr	**90.33**
Sum of two sides 1500mm	23.90	40.64	2.11	62.49	nr	**103.13**
Sum of two sides 1600mm	27.94	47.51	2.11	62.49	nr	**110.00**
Sum of two sides 1700mm	31.98	54.38	2.55	75.52	nr	**129.91**
Sum of two sides 1800mm	36.08	61.35	2.55	75.52	nr	**136.87**
Sum of two sides 1900mm	39.36	66.93	2.80	82.93	nr	**149.86**
Sum of two sides 2000mm	43.40	73.79	2.80	82.93	nr	**156.72**
45° bend						
Sum of two sides 1100mm	23.79	40.46	1.88	55.68	nr	**96.14**
Sum of two sides 1200mm	25.18	42.81	1.88	55.68	nr	**98.49**
Sum of two sides 1300mm	26.57	45.17	2.26	66.94	nr	**112.11**
Sum of two sides 1400mm	27.87	47.38	2.44	72.27	nr	**119.65**
Sum of two sides 1500mm	32.39	55.08	2.68	79.37	nr	**134.45**
Sum of two sides 1600mm	36.93	62.79	2.69	79.67	nr	**142.46**
Sum of two sides 1700mm	41.45	70.49	2.96	87.67	nr	**158.16**
Sum of two sides 1800mm	46.10	78.38	2.96	87.67	nr	**166.05**
Sum of two sides 1900mm	50.20	85.36	3.26	96.55	nr	**181.91**
Sum of two sides 2000mm	54.73	93.06	3.26	96.55	nr	**189.61**

U:VENTILATION/AIR CONDITIONING SYSTEMS

Item	Net Price £	Material £	Labour hours	Labour £	Unit	Total rate £
90° mitre bend						
Sum of two sides 1100mm	32.09	54.56	3.03	89.74	nr	**144.30**
Sum of two sides 1200mm	34.18	58.11	3.03	89.74	nr	**147.86**
Sum of two sides 1300mm	36.27	61.67	3.85	114.03	nr	**175.69**
Sum of two sides 1400mm	38.14	64.85	3.85	114.03	nr	**178.87**
Sum of two sides 1500mm	44.34	75.40	4.25	125.87	nr	**201.27**
Sum of two sides 1600mm	50.54	85.93	4.26	126.17	nr	**212.10**
Sum of two sides 1700mm	56.74	96.48	4.68	138.61	nr	**235.09**
Sum of two sides 1800mm	62.98	107.09	4.68	138.61	nr	**245.70**
Sum of two sides 1900mm	68.37	116.25	4.87	144.24	nr	**260.48**
Sum of two sides 2000mm	74.58	126.82	4.87	144.24	nr	**271.05**
Branch						
Sum of two sides 1100mm	36.55	62.15	1.29	38.21	nr	**100.36**
Sum of two sides 1200mm	38.51	65.49	1.29	38.21	nr	**103.69**
Sum of two sides 1300mm	40.49	68.85	1.39	41.17	nr	**110.01**
Sum of two sides 1400mm	42.29	71.92	1.39	41.17	nr	**113.08**
Sum of two sides 1500mm	48.55	82.56	1.64	48.57	nr	**131.13**
Sum of two sides 1600mm	54.82	93.22	1.64	48.57	nr	**141.79**
Sum of two sides 1700mm	61.08	103.86	1.69	50.05	nr	**153.91**
Sum of two sides 1800mm	67.44	114.68	1.69	50.05	nr	**164.73**
Sum of two sides 1900mm	73.71	125.34	1.85	54.79	nr	**180.13**
Sum of two sides 2000mm	79.97	135.98	1.85	54.79	nr	**190.77**
Grille neck						
Sum of two sides 1100mm	39.47	67.11	1.44	42.65	nr	**109.76**
Sum of two sides 1200mm	41.70	70.90	1.44	42.65	nr	**113.55**
Sum of two sides 1300mm	43.90	74.65	1.69	50.05	nr	**124.70**
Sum of two sides 1400mm	46.14	78.45	1.69	50.05	nr	**128.50**
Sum of two sides 1500mm	50.40	85.70	1.79	53.02	nr	**138.71**
Sum of two sides 1600mm	57.30	97.43	1.79	53.02	nr	**150.44**
Sum of two sides 1700mm	67.59	114.93	1.86	55.09	nr	**170.01**
Sum of two sides 1800mm	74.49	126.66	1.86	55.09	nr	**181.74**
Sum of two sides 1900mm	81.37	138.37	2.02	59.83	nr	**198.20**
Sum of two sides 2000mm	88.27	150.10	2.02	59.83	nr	**209.93**
Ductwork 1001 to 1250mm longest side						
Sum of two sides 1300mm	55.71	94.73	1.51	44.72	m	**139.46**
Sum of two sides 1400mm	58.34	99.20	1.55	45.91	m	**145.10**
Sum of two sides 1500mm	60.80	103.37	1.61	47.68	m	**151.06**
Sum of two sides 1600mm	63.47	107.93	1.62	47.98	m	**155.91**
Sum of two sides 1700mm	65.93	112.11	1.74	51.53	m	**163.64**
Sum of two sides 1800mm	68.39	116.28	1.76	52.13	m	**168.41**
Sum of two sides 1900mm	70.85	120.48	1.81	53.61	m	**174.09**
Sum of two sides 2000mm	73.53	125.03	1.82	53.90	m	**178.94**
Sum of two sides 2100mm	75.99	129.21	2.53	74.93	m	**204.14**
Sum of two sides 2200mm	87.65	149.05	2.55	75.52	m	**224.57**
Sum of two sides 2300mm	90.54	153.96	2.56	75.82	m	**229.78**
Sum of two sides 2400mm	93.00	158.13	2.76	81.74	m	**239.88**
Sum of two sides 2500mm	95.68	162.69	2.77	82.04	m	**244.73**

U:VENTILATION/AIR CONDITIONING SYSTEMS

Item	Net Price £	Material £	Labour hours	Labour £	Unit	Total rate £
U10 : DUCTWORK : RECTANGULAR– CLASS C (cont'd)						
Y30 - AIR DUCTLINES (cont'd)						
Extra over fittings; Ductwork 1001 to 1250mm longest side						
End Cap						
Sum of two sides 1300mm	20.25	34.44	0.38	11.25	nr	**45.69**
Sum of two sides 1400mm	21.14	35.94	0.38	11.25	nr	**47.19**
Sum of two sides 1500mm	24.58	41.80	0.38	11.25	nr	**53.05**
Sum of two sides 1600mm	28.02	47.65	0.38	11.25	nr	**58.91**
Sum of two sides 1700mm	31.47	53.51	0.58	17.18	nr	**70.69**
Sum of two sides 1800mm	34.91	59.36	0.58	17.18	nr	**76.54**
Sum of two sides 1900mm	38.36	65.22	0.58	17.18	nr	**82.40**
Sum of two sides 2000mm	41.79	71.06	0.58	17.18	nr	**88.24**
Sum of two sides 2100mm	45.23	76.92	0.87	25.77	nr	**102.68**
Sum of two sides 2200mm	48.68	82.77	0.87	25.77	nr	**108.54**
Sum of two sides 2300mm	52.12	88.63	0.87	25.77	nr	**114.39**
Sum of two sides 2400mm	55.57	94.48	0.87	25.77	nr	**120.25**
Sum of two sides 2500mm	59.01	100.34	0.87	25.77	nr	**126.11**
Reducer						
Sum of two sides 1300mm	13.14	22.34	2.30	68.12	nr	**90.46**
Sum of two sides 1400mm	13.68	23.26	2.30	68.12	nr	**91.38**
Sum of two sides 1500mm	16.41	27.91	2.47	73.16	nr	**101.06**
Sum of two sides 1600mm	19.22	32.67	2.47	73.16	nr	**105.83**
Sum of two sides 1700mm	21.95	37.31	2.59	76.71	nr	**114.02**
Sum of two sides 1800mm	24.68	41.96	2.59	76.71	nr	**118.67**
Sum of two sides 1900mm	27.41	46.60	2.71	80.26	nr	**126.86**
Sum of two sides 2000mm	30.21	51.37	2.71	80.26	nr	**131.63**
Sum of two sides 2100mm	32.94	56.01	2.92	86.48	nr	**142.49**
Sum of two sides 2200mm	33.43	56.85	2.92	86.48	nr	**143.33**
Sum of two sides 2300mm	36.24	61.61	2.92	86.48	nr	**148.10**
Sum of two sides 2400mm	38.98	66.27	3.12	92.41	nr	**158.68**
Sum of two sides 2500mm	41.77	71.02	3.12	92.41	nr	**163.43**
Offset						
Sum of two sides 1300mm	35.14	59.76	2.47	73.16	nr	**132.91**
Sum of two sides 1400mm	36.24	61.61	2.47	73.16	nr	**134.77**
Sum of two sides 1500mm	40.25	68.43	2.47	73.16	nr	**141.59**
Sum of two sides 1600mm	44.23	75.20	2.47	73.16	nr	**148.36**
Sum of two sides 1700mm	48.03	81.66	2.61	77.30	nr	**158.97**
Sum of two sides 1800mm	51.72	87.95	2.61	77.30	nr	**165.25**
Sum of two sides 1900mm	55.34	94.09	2.71	80.26	nr	**174.35**
Sum of two sides 2000mm	58.86	100.09	2.71	80.26	nr	**180.35**
Sum of two sides 2100mm	62.25	105.86	2.92	86.48	nr	**192.34**
Sum of two sides 2200mm	61.46	104.50	3.26	96.55	nr	**201.05**
Sum of two sides 2300mm	66.62	113.28	3.26	96.55	nr	**209.84**
Sum of two sides 2400mm	71.76	122.01	3.47	102.77	nr	**224.79**
Sum of two sides 2500mm	76.91	130.78	3.48	103.07	nr	**233.85**
Square to round						
Sum of two sides 1300mm	25.33	43.06	2.32	68.71	nr	**111.78**
Sum of two sides 1400mm	26.42	44.92	2.32	68.71	nr	**113.63**
Sum of two sides 1500mm	32.46	55.19	2.56	75.82	nr	**131.01**
Sum of two sides 1600mm	38.52	65.51	2.58	76.41	nr	**141.92**
Sum of two sides 1700mm	44.56	75.77	2.84	84.11	nr	**159.89**

U:VENTILATION/AIR CONDITIONING SYSTEMS

Item	Net Price £	Material £	Labour hours	Labour £	Unit	Total rate £
Sum of two sides 1800mm	50.60	86.04	2.84	84.11	nr	**170.15**
Sum of two sides 1900mm	60.95	103.64	3.13	92.70	nr	**196.34**
Sum of two sides 2000mm	62.71	106.62	3.13	92.70	nr	**199.33**
Sum of two sides 2100mm	68.75	116.91	4.26	126.17	nr	**243.08**
Sum of two sides 2200mm	70.76	120.32	4.27	126.47	nr	**246.78**
Sum of two sides 2300mm	76.82	130.62	4.30	127.35	nr	**257.97**
Sum of two sides 2400mm	82.87	140.90	4.77	141.28	nr	**282.18**
Sum of two sides 2500mm	88.92	151.21	4.79	141.87	nr	**293.07**
90° radius bend						
Sum of two sides 1300mm	10.05	17.09	1.91	56.57	nr	**73.66**
Sum of two sides 1400mm	9.75	16.59	1.91	56.57	nr	**73.16**
Sum of two sides 1500mm	14.72	25.03	2.11	62.49	nr	**87.52**
Sum of two sides 1600mm	19.72	33.53	2.11	62.49	nr	**96.02**
Sum of two sides 1700mm	24.68	41.96	2.55	75.52	nr	**117.48**
Sum of two sides 1800mm	29.64	50.40	2.55	75.52	nr	**125.93**
Sum of two sides 1900mm	34.60	58.83	2.80	82.93	nr	**141.76**
Sum of two sides 2000mm	39.61	67.35	2.80	82.93	nr	**150.27**
Sum of two sides 2100mm	44.56	75.77	2.80	82.93	nr	**158.70**
Sum of two sides 2200mm	41.91	71.26	2.80	82.93	nr	**154.18**
Sum of two sides 2300mm	43.13	73.34	4.35	128.84	nr	**202.18**
Sum of two sides 2400mm	48.07	81.74	4.35	128.84	nr	**210.57**
Sum of two sides 2500mm	53.01	90.14	4.35	128.84	nr	**218.98**
45° bend						
Sum of two sides 1300mm	16.58	28.19	2.26	66.94	nr	**95.13**
Sum of two sides 1400mm	17.00	28.91	2.44	72.27	nr	**101.17**
Sum of two sides 1500mm	20.62	35.07	2.68	79.37	nr	**114.44**
Sum of two sides 1600mm	24.34	41.39	2.69	79.67	nr	**121.06**
Sum of two sides 1700mm	27.97	47.56	2.96	87.67	nr	**135.23**
Sum of two sides 1800mm	31.61	53.74	2.96	87.67	nr	**141.41**
Sum of two sides 1900mm	35.24	59.92	3.26	96.55	nr	**156.47**
Sum of two sides 2000mm	38.96	66.24	3.26	96.55	nr	**162.79**
Sum of two sides 2100mm	42.58	72.40	7.50	222.13	nr	**294.53**
Sum of two sides 2200mm	43.31	73.65	7.50	222.13	nr	**295.78**
Sum of two sides 2300mm	45.68	77.66	7.55	223.61	nr	**301.28**
Sum of two sides 2400mm	49.30	83.82	8.13	240.79	nr	**324.61**
Sum of two sides 2500mm	52.99	90.11	8.30	245.82	nr	**335.93**
90° mitre bend						
Sum of two sides 1300mm	34.01	57.83	3.85	114.03	nr	**171.86**
Sum of two sides 1400mm	35.31	60.04	3.85	114.03	nr	**174.07**
Sum of two sides 1500mm	41.56	70.67	4.25	125.87	nr	**196.54**
Sum of two sides 1600mm	47.80	81.27	4.26	126.17	nr	**207.44**
Sum of two sides 1700mm	59.09	100.48	4.68	138.61	nr	**239.09**
Sum of two sides 1800mm	65.05	110.61	4.68	138.61	nr	**249.22**
Sum of two sides 1900mm	71.00	120.73	4.87	144.24	nr	**264.97**
Sum of two sides 2000mm	72.74	123.69	4.87	144.24	nr	**267.93**
Sum of two sides 2100mm	78.98	134.30	7.50	222.13	nr	**356.43**
Sum of two sides 2200mm	79.55	135.26	7.50	222.13	nr	**357.39**
Sum of two sides 2300mm	82.91	140.98	7.55	223.61	nr	**364.59**
Sum of two sides 2400mm	89.17	151.62	8.13	240.79	nr	**392.41**
Sum of two sides 2500mm	95.40	162.22	8.30	245.82	nr	**408.05**

U:VENTILATION/AIR CONDITIONING SYSTEMS

Item	Net Price £	Material £	Labour hours	Labour £	Unit	Total rate £
U10 : DUCTWORK : RECTANGULAR– CLASS C (cont'd)						
Y30 - AIR DUCTLINES (cont'd)						
Extra over fittings; Ductwork 1001 to 1250mm longest side (cont'd)						
Branch						
Sum of two sides 1300mm	34.02	57.85	1.39	41.17	nr	99.02
Sum of two sides 1400mm	35.46	60.29	1.39	41.17	nr	101.46
Sum of two sides 1500mm	40.67	69.15	1.64	48.57	nr	117.72
Sum of two sides 1600mm	45.94	78.11	1.64	48.57	nr	126.68
Sum of two sides 1700mm	51.14	86.95	1.69	50.05	nr	137.00
Sum of two sides 1800mm	56.33	95.79	1.69	50.05	nr	145.84
Sum of two sides 1900mm	61.53	104.62	1.85	54.79	nr	159.42
Sum of two sides 2000mm	66.80	113.59	1.85	54.79	nr	168.38
Sum of two sides 2100mm	72.01	122.44	2.61	77.30	nr	199.74
Sum of two sides 2200mm	77.21	131.28	2.61	77.30	nr	208.58
Sum of two sides 2300mm	82.48	140.24	2.61	77.30	nr	217.54
Sum of two sides 2400mm	87.68	149.08	2.88	85.30	nr	234.38
Sum of two sides 2500mm	92.96	158.06	2.88	85.30	nr	243.36
Grille neck						
Sum of two sides 1300mm	40.82	69.41	1.69	50.05	nr	119.46
Sum of two sides 1400mm	43.19	73.43	1.69	50.05	nr	123.49
Sum of two sides 1500mm	45.54	77.44	1.79	53.02	nr	130.45
Sum of two sides 1600mm	52.16	88.69	1.79	53.02	nr	141.71
Sum of two sides 1700mm	59.05	100.41	1.86	55.09	nr	155.50
Sum of two sides 1800mm	65.95	112.14	1.86	55.09	nr	167.23
Sum of two sides 1900mm	72.85	123.88	2.02	59.83	nr	183.70
Sum of two sides 2000mm	79.73	135.57	2.02	59.83	nr	195.40
Sum of two sides 2100mm	86.63	147.30	2.61	77.30	nr	224.60
Sum of two sides 2200mm	93.53	159.04	2.80	82.93	nr	241.97
Sum of two sides 2300mm	100.43	170.77	2.80	82.93	nr	253.70
Sum of two sides 2400mm	107.31	182.46	3.06	90.63	nr	273.09
Sum of two sides 2500mm	114.21	194.20	3.06	90.63	nr	284.83
Ductwork 1251 to 1600mm longest side						
Sum of two sides 1700mm	73.58	125.12	1.74	51.53	m	176.66
Sum of two sides 1800mm	76.45	129.99	1.76	52.13	m	182.12
Sum of two sides 1900mm	79.37	134.96	1.81	53.61	m	188.57
Sum of two sides 2000mm	82.07	139.55	1.82	53.90	m	193.45
Sum of two sides 2100mm	84.77	144.14	2.53	74.93	m	219.07
Sum of two sides 2200mm	87.47	148.72	2.55	75.52	m	224.25
Sum of two sides 2300mm	90.39	153.71	2.56	75.82	m	229.53
Sum of two sides 2400mm	93.24	158.54	2.76	81.74	m	240.29
Sum of two sides 2500mm	95.94	163.13	2.77	82.04	m	245.17
Sum of two sides 2600mm	108.31	184.16	2.97	87.96	m	272.13
Sum of two sides 2700mm	111.01	188.75	2.99	88.56	m	277.31
Sum of two sides 2800mm	113.90	193.68	3.30	97.74	m	291.42
Sum of two sides 2900mm	116.84	198.68	3.31	98.03	m	296.71
Sum of two sides 3000mm	119.53	203.25	3.53	104.55	m	307.80
Sum of two sides 3100mm	122.38	208.09	3.55	105.14	m	313.23
Sum of two sides 3200mm	125.09	212.69	3.56	105.44	m	318.13

U:VENTILATION/AIR CONDITIONING SYSTEMS

Item	Net Price £	Material £	Labour hours	Labour £	Unit	Total rate £
Extra over fittings; Ductwork 1251 to 1600mm longest side						
End Cap						
Sum of two sides 1700mm	42.28	71.90	0.58	17.18	nr	89.08
Sum of two sides 1800mm	46.90	79.75	0.58	17.18	nr	96.93
Sum of two sides 1900mm	51.53	87.63	0.58	17.18	nr	104.81
Sum of two sides 2000mm	56.15	95.48	0.58	17.18	nr	112.66
Sum of two sides 2100mm	60.77	103.34	0.87	25.77	nr	129.11
Sum of two sides 2200mm	65.40	111.21	0.87	25.77	nr	136.98
Sum of two sides 2300mm	70.02	119.07	0.87	25.77	nr	144.84
Sum of two sides 2400mm	74.66	126.94	0.87	25.77	nr	152.71
Sum of two sides 2500mm	79.29	134.82	0.87	25.77	nr	160.58
Sum of two sides 2600mm	83.91	142.67	0.87	25.77	nr	168.44
Sum of two sides 2700mm	88.54	150.54	0.87	25.77	nr	176.31
Sum of two sides 2800mm	93.16	158.40	1.16	34.36	nr	192.76
Sum of two sides 2900mm	97.79	166.27	1.16	34.36	nr	200.63
Sum of two sides 3000mm	56.32	95.77	1.73	51.24	nr	147.01
Sum of two sides 3100mm	106.07	180.36	1.73	51.24	nr	231.60
Sum of two sides 3200mm	109.38	185.99	1.73	51.24	nr	237.22
Reducer						
Sum of two sides 1700mm	20.52	34.89	2.59	76.71	nr	111.60
Sum of two sides 1800mm	23.29	39.60	2.59	76.71	nr	116.31
Sum of two sides 1900mm	26.12	44.42	2.71	80.26	nr	124.68
Sum of two sides 2000mm	56.15	95.48	2.71	80.26	nr	175.75
Sum of two sides 2100mm	31.67	53.85	2.92	86.48	nr	140.33
Sum of two sides 2200mm	34.44	58.56	2.92	86.48	nr	145.04
Sum of two sides 2300mm	37.28	63.38	2.92	86.48	nr	149.86
Sum of two sides 2400mm	40.05	68.10	3.12	92.41	nr	160.50
Sum of two sides 2500mm	42.82	72.81	3.12	92.41	nr	165.21
Sum of two sides 2600mm	42.33	71.97	3.16	93.59	nr	165.56
Sum of two sides 2700mm	45.10	76.68	3.16	93.59	nr	170.27
Sum of two sides 2800mm	47.86	81.38	4.00	118.47	nr	199.85
Sum of two sides 2900mm	53.55	91.06	4.01	118.77	nr	209.82
Sum of two sides 3000mm	56.32	95.77	4.56	135.06	nr	230.82
Sum of two sides 3100mm	58.38	99.27	4.56	135.06	nr	234.32
Sum of two sides 3200mm	60.26	102.46	4.56	135.06	nr	237.52
Offset						
Sum of two sides 1700mm	64.06	108.93	2.61	77.30	nr	186.23
Sum of two sides 1800mm	70.68	120.18	2.61	77.30	nr	197.48
Sum of two sides 1900mm	75.08	127.66	2.71	80.26	nr	207.92
Sum of two sides 2000mm	79.24	134.74	2.71	80.26	nr	215.01
Sum of two sides 2100mm	83.28	141.60	2.92	86.48	nr	228.08
Sum of two sides 2200mm	87.18	148.24	3.26	96.55	nr	244.79
Sum of two sides 2300mm	90.98	154.70	3.26	96.55	nr	251.26
Sum of two sides 2400mm	97.27	165.40	3.47	102.77	nr	268.17
Sum of two sides 2500mm	100.82	171.43	3.48	103.07	nr	274.50
Sum of two sides 2600mm	100.88	171.54	3.49	103.36	nr	274.91
Sum of two sides 2700mm	107.09	182.09	3.50	103.66	nr	285.75
Sum of two sides 2800mm	113.28	192.63	4.33	128.24	nr	320.87
Sum of two sides 2900mm	123.15	209.41	4.74	140.39	nr	349.80
Sum of two sides 3000mm	129.36	219.96	5.31	157.27	nr	377.23
Sum of two sides 3100mm	134.22	228.23	5.34	158.16	nr	386.38
Sum of two sides 3200mm	138.75	235.92	5.35	158.45	nr	394.38

U:VENTILATION/AIR CONDITIONING SYSTEMS

Item	Net Price £	Material £	Labour hours	Labour £	Unit	Total rate £
U10 : DUCTWORK : RECTANGULAR– CLASS C (cont'd)						
Y30 - AIR DUCTLINES (cont'd)						
Extra over fittings; Ductwork 1251 to 1600mm longest side (cont'd)						
Square to round						
Sum of two sides 1700mm	48.69	82.79	2.84	84.11	nr	**166.90**
Sum of two sides 1800mm	56.26	95.66	2.84	84.11	nr	**179.78**
Sum of two sides 1900mm	63.78	108.45	3.13	92.70	nr	**201.15**
Sum of two sides 2000mm	71.33	121.28	3.13	92.70	nr	**213.98**
Sum of two sides 2100mm	78.86	134.08	4.26	126.17	nr	**260.25**
Sum of two sides 2200mm	86.40	146.92	4.27	126.47	nr	**273.39**
Sum of two sides 2300mm	93.93	159.72	4.30	127.35	nr	**287.08**
Sum of two sides 2400mm	101.49	172.58	4.77	141.28	nr	**313.85**
Sum of two sides 2500mm	109.03	185.40	4.79	141.87	nr	**327.26**
Sum of two sides 2600mm	109.40	186.02	4.95	146.61	nr	**332.63**
Sum of two sides 2700mm	116.94	198.84	4.95	146.61	nr	**345.45**
Sum of two sides 2800mm	124.47	211.64	8.49	251.45	nr	**463.09**
Sum of two sides 2900mm	135.15	229.80	8.88	263.00	nr	**492.80**
Sum of two sides 3000mm	142.68	242.62	9.02	267.15	nr	**509.77**
Sum of two sides 3100mm	148.30	252.17	9.02	267.15	nr	**519.32**
Sum of two sides 3200mm	153.37	260.79	9.09	269.22	nr	**530.02**
90° radius bend						
Sum of two sides 1700mm	58.50	99.46	2.55	75.52	nr	**174.99**
Sum of two sides 1800mm	63.32	107.66	2.55	75.52	nr	**183.18**
Sum of two sides 1900mm	70.74	120.28	2.80	82.93	nr	**203.21**
Sum of two sides 2000mm	78.02	132.66	2.80	82.93	nr	**215.58**
Sum of two sides 2100mm	85.28	145.01	2.61	77.30	nr	**222.31**
Sum of two sides 2200mm	92.56	157.38	2.62	77.60	nr	**234.98**
Sum of two sides 2300mm	99.99	170.02	2.63	77.89	nr	**247.92**
Sum of two sides 2400mm	104.50	177.68	4.34	128.54	nr	**306.22**
Sum of two sides 2500mm	111.74	190.00	4.35	128.84	nr	**318.84**
Sum of two sides 2600mm	112.12	190.64	4.53	134.17	nr	**324.81**
Sum of two sides 2700mm	119.36	202.96	4.53	134.17	nr	**337.13**
Sum of two sides 2800mm	123.24	209.55	7.13	211.17	nr	**420.72**
Sum of two sides 2900mm	137.33	233.51	7.17	212.36	nr	**445.87**
Sum of two sides 3000mm	144.53	245.76	7.26	215.02	nr	**460.78**
Sum of two sides 3100mm	150.08	255.19	7.26	215.02	nr	**470.21**
Sum of two sides 3200mm	155.18	263.86	7.31	216.50	nr	**480.37**
45° bend						
Sum of two sides 1700mm	27.96	47.55	2.96	87.67	nr	**135.21**
Sum of two sides 1800mm	30.32	51.56	2.96	87.67	nr	**139.23**
Sum of two sides 1900mm	34.00	57.81	3.26	96.55	nr	**154.36**
Sum of two sides 2000mm	37.61	63.95	3.26	96.55	nr	**160.51**
Sum of two sides 2100mm	41.21	70.08	7.50	222.13	nr	**292.21**
Sum of two sides 2200mm	44.82	76.22	7.50	222.13	nr	**298.35**
Sum of two sides 2300mm	48.51	82.49	7.55	223.61	nr	**306.10**
Sum of two sides 2400mm	50.72	86.23	8.13	240.79	nr	**327.02**
Sum of two sides 2500mm	54.31	92.34	8.30	245.82	nr	**338.17**
Sum of two sides 2600mm	54.24	92.23	8.56	253.52	nr	**345.76**
Sum of two sides 2700mm	57.84	98.36	8.62	255.30	nr	**353.66**
Sum of two sides 2800mm	59.73	101.57	9.09	269.22	nr	**370.79**
Sum of two sides 2900mm	66.73	113.46	9.09	269.22	nr	**382.68**
Sum of two sides 3000mm	70.30	119.53	9.62	284.92	nr	**404.45**
Sum of two sides 3100mm	73.04	124.19	9.62	284.92	nr	**409.11**
Sum of two sides 3200mm	75.57	128.50	9.62	284.92	nr	**413.41**

U:VENTILATION/AIR CONDITIONING SYSTEMS

Item	Net Price £	Material £	Labour hours	Labour £	Unit	Total rate £
90° mitre bend						
Sum of two sides 1700mm	68.19	115.94	4.68	138.61	nr	**254.55**
Sum of two sides 1800mm	72.35	123.01	4.68	138.61	nr	**261.62**
Sum of two sides 1900mm	80.23	136.42	4.87	144.24	nr	**280.66**
Sum of two sides 2000mm	88.13	149.85	4.87	144.24	nr	**294.09**
Sum of two sides 2100mm	96.03	163.29	7.50	222.13	nr	**385.42**
Sum of two sides 2200mm	103.93	176.72	7.50	222.13	nr	**398.85**
Sum of two sides 2300mm	111.80	190.11	7.55	223.61	nr	**413.72**
Sum of two sides 2400mm	115.80	196.91	8.13	240.79	nr	**437.70**
Sum of two sides 2500mm	123.71	210.36	8.30	245.82	nr	**456.18**
Sum of two sides 2600mm	122.54	208.36	8.56	253.52	nr	**461.88**
Sum of two sides 2700mm	130.44	221.80	8.62	255.30	nr	**477.10**
Sum of two sides 2800mm	133.71	227.35	15.20	450.18	nr	**677.54**
Sum of two sides 2900mm	147.36	250.56	15.20	450.18	nr	**700.75**
Sum of two sides 3000mm	155.27	264.02	15.60	462.03	nr	**726.06**
Sum of two sides 3100mm	161.74	275.02	16.04	475.06	nr	**750.09**
Sum of two sides 3200mm	167.91	285.50	16.04	475.06	nr	**760.57**
Branch						
Sum of two sides 1700mm	52.51	89.29	1.69	50.05	nr	**139.34**
Sum of two sides 1800mm	57.69	98.09	1.69	50.05	nr	**148.14**
Sum of two sides 1900mm	62.95	107.03	1.85	54.79	nr	**161.83**
Sum of two sides 2000mm	68.13	115.85	1.85	54.79	nr	**170.65**
Sum of two sides 2100mm	73.31	124.66	2.61	77.30	nr	**201.96**
Sum of two sides 2200mm	78.50	133.48	2.61	77.30	nr	**210.78**
Sum of two sides 2300mm	83.76	142.42	2.61	77.30	nr	**219.72**
Sum of two sides 2400mm	88.94	151.22	2.88	85.30	nr	**236.52**
Sum of two sides 2500mm	94.11	160.03	2.88	85.30	nr	**245.32**
Sum of two sides 2600mm	99.30	168.85	2.88	85.30	nr	**254.14**
Sum of two sides 2700mm	104.49	177.67	2.88	85.30	nr	**262.96**
Sum of two sides 2800mm	109.65	186.45	3.94	116.69	nr	**303.14**
Sum of two sides 2900mm	116.62	198.30	3.94	116.69	nr	**315.00**
Sum of two sides 3000mm	121.81	207.12	4.83	143.05	nr	**350.18**
Sum of two sides 3100mm	125.95	214.16	4.83	143.05	nr	**357.21**
Sum of two sides 3200mm	129.71	220.55	4.83	143.05	nr	**363.60**
Grille neck						
Sum of two sides 1700mm	67.95	115.53	1.86	55.09	nr	**170.62**
Sum of two sides 1800mm	74.85	127.28	1.86	55.09	nr	**182.37**
Sum of two sides 1900mm	81.77	139.05	2.02	59.83	nr	**198.87**
Sum of two sides 2000mm	88.67	150.78	2.02	59.83	nr	**210.60**
Sum of two sides 2100mm	95.60	162.56	2.80	82.93	nr	**245.49**
Sum of two sides 2200mm	102.51	174.31	2.80	82.93	nr	**257.24**
Sum of two sides 2300mm	109.43	186.07	2.80	82.93	nr	**269.00**
Sum of two sides 2400mm	116.34	197.82	3.06	90.63	nr	**288.45**
Sum of two sides 2500mm	123.25	209.57	3.06	90.63	nr	**300.20**
Sum of two sides 2600mm	130.17	221.34	3.08	91.22	nr	**312.56**
Sum of two sides 2700mm	137.09	233.10	3.08	91.22	nr	**324.32**
Sum of two sides 2800mm	139.97	237.99	4.13	122.32	nr	**360.31**
Sum of two sides 2900mm	157.94	268.56	4.13	122.32	nr	**390.88**
Sum of two sides 3000mm	157.82	268.35	5.02	148.68	nr	**417.02**
Sum of two sides 3100mm	163.13	277.38	5.02	148.68	nr	**426.06**
Sum of two sides 3200mm	168.45	286.43	5.02	148.68	nr	**435.11**

U:VENTILATION/AIR CONDITIONING SYSTEMS

Item	Net Price £	Material £	Labour hours	Labour £	Unit	Total rate £
U10 : DUCTWORK : RECTANGULAR – CLASS C (cont'd)						
Y30 - AIR DUCTLINES (cont'd)						
Ductwork 1601 to 2000mm longest side						
Sum of two sides 2100mm	91.91	156.28	2.53	74.93	m	**231.21**
Sum of two sides 2200mm	94.84	161.26	2.55	75.52	m	**236.78**
Sum of two sides 2300mm	97.82	166.33	2.55	75.52	m	**241.85**
Sum of two sides 2400mm	100.66	171.17	2.56	75.82	m	**246.99**
Sum of two sides 2500mm	103.65	176.24	2.76	81.74	m	**257.98**
Sum of two sides 2600mm	106.42	180.95	2.77	82.04	m	**262.99**
Sum of two sides 2700mm	109.19	185.66	2.97	87.96	m	**273.63**
Sum of two sides 2800mm	111.96	190.38	2.99	88.56	m	**278.93**
Sum of two sides 2900mm	114.94	195.45	3.30	97.74	m	**293.19**
Sum of two sides 3000mm	117.72	200.16	3.31	98.03	m	**298.19**
Sum of two sides 3100mm	136.65	232.35	3.53	104.55	m	**336.90**
Sum of two sides 3200mm	139.42	237.07	3.53	104.55	m	**341.61**
Sum of two sides 3300mm	142.40	242.14	3.53	104.55	m	**346.69**
Sum of two sides 3400mm	145.18	246.87	3.55	105.14	m	**352.01**
Sum of two sides 3500mm	147.95	251.56	3.55	105.14	m	**356.70**
Sum of two sides 3600mm	150.72	256.28	3.55	105.14	m	**361.42**
Sum of two sides 3700mm	153.70	261.35	3.56	105.44	m	**366.78**
Sum of two sides 3800mm	156.48	266.08	3.56	105.44	m	**371.52**
Sum of two sides 3900mm	159.24	270.77	3.56	105.44	m	**376.21**
Sum of two sides 4000mm	162.03	275.50	3.56	105.44	m	**380.94**
Extra over fittings; Ductwork 1601 to 2000mm longest side						
End Cap						
Sum of two sides 2100mm	76.77	130.53	0.87	25.77	nr	**156.30**
Sum of two sides 2200mm	82.58	140.42	0.87	25.77	nr	**166.19**
Sum of two sides 2300mm	88.41	150.33	0.87	25.77	nr	**176.10**
Sum of two sides 2400mm	94.21	160.19	0.87	25.77	nr	**185.95**
Sum of two sides 2500mm	100.03	170.10	0.87	25.77	nr	**195.86**
Sum of two sides 2600mm	105.85	179.99	0.87	25.77	nr	**205.75**
Sum of two sides 2700mm	111.67	189.88	0.87	25.77	nr	**215.64**
Sum of two sides 2800mm	117.50	199.79	1.16	34.36	nr	**234.14**
Sum of two sides 2900mm	123.31	209.68	1.16	34.36	nr	**244.03**
Sum of two sides 3000mm	129.13	219.57	1.73	51.24	nr	**270.81**
Sum of two sides 3100mm	133.75	227.42	1.73	51.24	nr	**278.66**
Sum of two sides 3200mm	137.93	234.53	1.73	51.24	nr	**285.77**
Sum of two sides 3300mm	142.13	241.67	1.73	51.24	nr	**292.91**
Sum of two sides 3400mm	146.30	248.76	1.73	51.24	nr	**300.00**
Sum of two sides 3500mm	150.48	255.87	1.73	51.24	nr	**307.10**
Sum of two sides 3600mm	154.67	262.99	1.73	51.24	nr	**314.23**
Sum of two sides 3700mm	158.84	270.10	1.73	51.24	nr	**321.33**
Sum of two sides 3800mm	163.02	277.20	1.73	51.24	nr	**328.44**
Sum of two sides 3900mm	167.22	284.34	1.73	51.24	nr	**335.58**
Sum of two sides 4000mm	171.40	291.45	1.73	51.24	nr	**342.69**
Reducer						
Sum of two sides 2100mm	30.29	51.51	2.92	86.48	nr	**137.99**
Sum of two sides 2200mm	33.13	56.33	2.92	86.48	nr	**142.81**
Sum of two sides 2300mm	36.02	61.24	2.92	86.48	nr	**147.72**
Sum of two sides 2400mm	38.79	65.95	3.12	92.41	nr	**158.36**
Sum of two sides 2500mm	41.67	70.86	3.12	92.41	nr	**163.27**

U:VENTILATION/AIR CONDITIONING SYSTEMS

Item	Net Price £	Material £	Labour hours	Labour £	Unit	Total rate £
Sum of two sides 2600mm	44.51	75.68	3.16	93.59	nr	**169.27**
Sum of two sides 2700mm	47.33	80.49	3.16	93.59	nr	**174.08**
Sum of two sides 2800mm	50.17	85.31	4.00	118.47	nr	**203.78**
Sum of two sides 2900mm	53.06	90.22	4.01	118.77	nr	**208.98**
Sum of two sides 3000mm	55.88	95.02	4.01	118.77	nr	**213.78**
Sum of two sides 3100mm	52.07	88.54	4.01	118.77	nr	**207.30**
Sum of two sides 3200mm	53.98	91.79	4.56	135.06	nr	**226.84**
Sum of two sides 3300mm	58.95	100.23	4.56	135.06	nr	**235.29**
Sum of two sides 3400mm	60.86	103.48	4.56	135.06	nr	**238.54**
Sum of two sides 3500mm	62.77	106.73	4.56	135.06	nr	**241.79**
Sum of two sides 3600mm	64.68	109.98	4.56	135.06	nr	**245.04**
Sum of two sides 3700mm	66.65	113.34	4.56	135.06	nr	**248.39**
Sum of two sides 3800mm	68.55	116.57	4.56	135.06	nr	**251.62**
Sum of two sides 3900mm	70.47	119.82	4.56	135.06	nr	**254.87**
Sum of two sides 4000mm	72.38	123.07	4.56	135.06	nr	**258.12**
Offset						
Sum of two sides 2100mm	91.39	155.40	2.92	86.48	nr	**241.88**
Sum of two sides 2200mm	98.24	167.04	3.26	96.55	nr	**263.59**
Sum of two sides 2300mm	102.14	173.68	3.26	96.55	nr	**270.24**
Sum of two sides 2400mm	111.92	190.31	3.47	102.77	nr	**293.08**
Sum of two sides 2500mm	115.69	196.72	3.48	103.07	nr	**299.78**
Sum of two sides 2600mm	119.23	202.73	3.49	103.36	nr	**306.10**
Sum of two sides 2700mm	122.64	208.53	3.50	103.66	nr	**312.20**
Sum of two sides 2800mm	125.90	214.07	4.33	128.24	nr	**342.31**
Sum of two sides 2900mm	129.06	219.44	4.74	140.39	nr	**359.83**
Sum of two sides 3000mm	132.03	224.50	5.31	157.27	nr	**381.76**
Sum of two sides 3100mm	125.76	213.84	5.34	158.16	nr	**371.99**
Sum of two sides 3200mm	130.41	221.75	5.35	158.45	nr	**380.20**
Sum of two sides 3300mm	138.93	236.23	5.35	158.45	nr	**394.68**
Sum of two sides 3400mm	143.58	244.14	5.35	158.45	nr	**402.59**
Sum of two sides 3500mm	148.22	252.03	5.35	158.45	nr	**410.48**
Sum of two sides 3600mm	152.88	259.95	5.35	158.45	nr	**418.41**
Sum of two sides 3700mm	157.55	267.90	5.35	158.45	nr	**426.35**
Sum of two sides 3800mm	162.20	275.81	5.35	158.45	nr	**434.26**
Sum of two sides 3900mm	166.86	283.72	5.35	158.45	nr	**442.17**
Sum of two sides 4000mm	171.51	291.63	5.35	158.45	nr	**450.08**
Square to round						
Sum of two sides 2100mm	84.44	143.58	4.26	126.17	nr	**269.75**
Sum of two sides 2200mm	92.61	157.47	4.27	126.47	nr	**283.94**
Sum of two sides 2300mm	100.74	171.29	4.30	127.35	nr	**298.65**
Sum of two sides 2400mm	172.07	292.59	4.77	141.28	nr	**433.87**
Sum of two sides 2500mm	117.06	199.05	4.79	141.87	nr	**340.92**
Sum of two sides 2600mm	125.21	212.91	4.95	146.61	nr	**359.51**
Sum of two sides 2700mm	133.35	226.75	4.95	146.61	nr	**373.35**
Sum of two sides 2800mm	141.51	240.62	8.49	251.45	nr	**492.07**
Sum of two sides 2900mm	149.63	254.42	8.88	263.00	nr	**517.42**
Sum of two sides 3000mm	157.78	268.29	9.02	267.15	nr	**535.44**
Sum of two sides 3100mm	152.65	259.56	9.02	267.15	nr	**526.71**
Sum of two sides 3200mm	158.18	268.97	9.09	269.22	nr	**538.19**
Sum of two sides 3300mm	166.53	283.16	9.09	269.22	nr	**552.39**
Sum of two sides 3400mm	172.07	292.59	9.09	269.22	nr	**561.81**
Sum of two sides 3500mm	177.60	301.98	9.09	269.22	nr	**571.20**
Sum of two sides 3600mm	183.12	311.37	9.09	269.22	nr	**580.60**
Sum of two sides 3700mm	188.63	320.75	9.09	269.22	nr	**589.97**
Sum of two sides 3800mm	194.16	330.14	9.09	269.22	nr	**599.36**
Sum of two sides 3900mm	199.69	339.55	9.09	269.22	nr	**608.77**
Sum of two sides 4000mm	205.22	348.96	9.09	269.22	nr	**618.18**

U:VENTILATION/AIR CONDITIONING SYSTEMS

Item	Net Price £	Material £	Labour hours	Labour £	Unit	Total rate £
U10 : DUCTWORK : RECTANGULAR– CLASS C (cont'd)						
Y30 - AIR DUCTLINES (cont'd)						
Extra over fittings; Ductwork 1601 to 2000mm longest side (cont'd)						
90° radius bend						
Sum of two sides 2100mm	105.82	179.93	2.61	77.30	nr	**257.23**
Sum of two sides 2200mm	128.53	218.55	2.62	77.60	nr	**296.15**
Sum of two sides 2300mm	139.09	236.51	2.63	77.89	nr	**314.41**
Sum of two sides 2400mm	142.47	242.26	4.34	128.54	nr	**370.80**
Sum of two sides 2500mm	153.01	260.17	4.35	128.84	nr	**389.00**
Sum of two sides 2600mm	163.29	277.65	4.53	134.17	nr	**411.81**
Sum of two sides 2700mm	173.57	295.13	4.53	134.17	nr	**429.29**
Sum of two sides 2800mm	183.86	312.62	7.13	211.17	nr	**523.80**
Sum of two sides 2900mm	194.38	330.51	7.17	212.36	nr	**542.87**
Sum of two sides 3000mm	204.66	347.99	7.26	215.02	nr	**563.01**
Sum of two sides 3100mm	198.80	338.03	7.26	215.02	nr	**553.05**
Sum of two sides 3200mm	206.24	350.69	7.31	216.50	nr	**567.19**
Sum of two sides 3300mm	223.20	379.52	7.31	216.50	nr	**596.03**
Sum of two sides 3400mm	230.64	392.18	7.31	216.50	nr	**608.68**
Sum of two sides 3500mm	238.09	404.84	7.31	216.50	nr	**621.34**
Sum of two sides 3600mm	245.52	417.48	7.31	216.50	nr	**633.98**
Sum of two sides 3700mm	253.22	430.57	7.31	216.50	nr	**647.07**
Sum of two sides 3800mm	260.65	443.21	7.31	216.50	nr	**659.71**
Sum of two sides 3900mm	268.10	455.87	7.31	216.50	nr	**672.37**
Sum of two sides 4000mm	275.55	468.54	7.31	216.50	nr	**685.05**
45° bend						
Sum of two sides 2100mm	82.40	140.12	7.50	222.13	nr	**362.25**
Sum of two sides 2200mm	85.95	146.15	7.50	222.13	nr	**368.28**
Sum of two sides 2300mm	92.59	157.44	7.55	223.61	nr	**381.05**
Sum of two sides 2400mm	95.64	162.63	8.13	240.79	nr	**403.42**
Sum of two sides 2500mm	102.27	173.90	8.30	245.82	nr	**419.72**
Sum of two sides 2600mm	108.71	184.84	8.56	253.52	nr	**438.37**
Sum of two sides 2700mm	115.15	195.80	8.62	255.30	nr	**451.11**
Sum of two sides 2800mm	121.59	206.75	9.09	269.22	nr	**475.97**
Sum of two sides 2900mm	128.21	218.00	9.09	269.22	nr	**487.22**
Sum of two sides 3000mm	134.64	228.94	9.62	284.92	nr	**513.86**
Sum of two sides 3100mm	132.87	225.92	9.62	284.92	nr	**510.84**
Sum of two sides 3200mm	137.54	233.87	9.62	284.92	nr	**518.79**
Sum of two sides 3300mm	148.16	251.92	9.62	284.92	nr	**536.84**
Sum of two sides 3400mm	152.84	259.88	9.62	284.92	nr	**544.80**
Sum of two sides 3500mm	157.50	267.81	9.62	284.92	nr	**552.73**
Sum of two sides 3600mm	162.18	275.77	9.62	284.92	nr	**560.69**
Sum of two sides 3700mm	167.03	284.02	9.62	284.92	nr	**568.94**
Sum of two sides 3800mm	171.71	291.97	9.62	284.92	nr	**576.89**
Sum of two sides 3900mm	176.38	299.91	9.62	284.92	nr	**584.83**
Sum of two sides 4000mm	181.05	307.86	9.62	284.92	nr	**592.78**
90° mitre bend						
Sum of two sides 2100mm	156.53	266.17	7.50	222.13	nr	**488.30**
Sum of two sides 2200mm	165.30	281.08	7.50	222.13	nr	**503.21**
Sum of two sides 2300mm	178.20	303.00	7.55	223.61	nr	**526.61**
Sum of two sides 2400mm	182.37	310.11	8.13	240.79	nr	**550.90**
Sum of two sides 2500mm	195.26	332.01	8.30	245.82	nr	**577.84**
Sum of two sides 2600mm	208.23	354.06	8.56	253.52	nr	**607.59**
Sum of two sides 2700mm	221.20	376.13	8.62	255.30	nr	**631.43**

U:VENTILATION/AIR CONDITIONING SYSTEMS

Item	Net Price £	Material £	Labour hours	Labour £	Unit	Total rate £
Sum of two sides 2800mm	234.18	398.20	15.20	450.18	nr	**848.38**
Sum of two sides 2900mm	247.07	420.10	15.20	450.18	nr	**870.29**
Sum of two sides 3000mm	260.03	442.15	15.60	462.03	nr	**904.19**
Sum of two sides 3100mm	252.90	430.03	16.04	475.06	nr	**905.09**
Sum of two sides 3200mm	262.88	446.99	16.04	475.06	nr	**922.06**
Sum of two sides 3300mm	278.18	473.01	16.04	475.06	nr	**948.07**
Sum of two sides 3400mm	288.14	489.95	16.04	475.06	nr	**965.01**
Sum of two sides 3500mm	298.12	506.91	16.04	475.06	nr	**981.97**
Sum of two sides 3600mm	308.08	523.85	16.04	475.06	nr	**998.92**
Sum of two sides 3700mm	317.97	540.67	16.04	475.06	nr	**1015.74**
Sum of two sides 3800mm	327.94	557.62	16.04	475.06	nr	**1032.68**
Sum of two sides 3900mm	337.91	574.58	16.04	475.06	nr	**1049.64**
Sum of two sides 4000mm	347.88	591.52	16.04	475.06	nr	**1066.58**
Branch						
Sum of two sides 2100mm	72.28	122.91	2.61	77.30	nr	**200.21**
Sum of two sides 2200mm	77.12	131.14	2.61	77.30	nr	**208.44**
Sum of two sides 2300mm	82.25	139.85	2.61	77.30	nr	**217.15**
Sum of two sides 2400mm	87.21	148.30	2.88	85.30	nr	**233.59**
Sum of two sides 2500mm	92.35	157.03	2.88	85.30	nr	**242.32**
Sum of two sides 2600mm	97.41	165.63	2.88	85.30	nr	**250.93**
Sum of two sides 2700mm	102.45	174.20	2.88	85.30	nr	**259.50**
Sum of two sides 2800mm	107.51	182.81	3.94	116.69	nr	**299.50**
Sum of two sides 2900mm	112.64	191.54	3.94	116.69	nr	**308.23**
Sum of two sides 3000mm	117.71	200.14	4.83	143.05	nr	**343.20**
Sum of two sides 3100mm	121.74	207.00	4.83	143.05	nr	**350.05**
Sum of two sides 3200mm	125.42	213.27	4.83	143.05	nr	**356.32**
Sum of two sides 3300mm	130.86	222.51	4.83	143.05	nr	**365.57**
Sum of two sides 3400mm	134.54	228.76	4.83	143.05	nr	**371.82**
Sum of two sides 3500mm	138.22	235.03	4.83	143.05	nr	**378.08**
Sum of two sides 3600mm	141.90	241.28	4.83	143.05	nr	**384.33**
Sum of two sides 3700mm	145.66	247.67	4.83	143.05	nr	**390.72**
Sum of two sides 3800mm	149.33	253.92	4.83	143.05	nr	**396.97**
Sum of two sides 3900mm	153.02	260.19	4.83	143.05	nr	**403.24**
Sum of two sides 4000mm	156.69	266.44	4.83	143.05	nr	**409.49**
Grille neck						
Sum of two sides 2100mm	97.17	165.22	2.80	82.93	nr	**248.15**
Sum of two sides 2200mm	104.14	177.08	2.80	82.93	nr	**260.00**
Sum of two sides 2300mm	111.13	188.97	2.80	82.93	nr	**271.90**
Sum of two sides 2400mm	118.13	200.86	3.06	90.63	nr	**291.49**
Sum of two sides 2500mm	125.11	212.73	3.06	90.63	nr	**303.36**
Sum of two sides 2600mm	132.10	224.62	3.08	91.22	nr	**315.84**
Sum of two sides 2700mm	102.45	174.20	3.08	91.22	nr	**265.42**
Sum of two sides 2800mm	107.51	182.81	4.13	122.32	nr	**305.13**
Sum of two sides 2900mm	112.64	191.54	4.13	122.32	nr	**313.86**
Sum of two sides 3000mm	160.04	272.13	5.02	148.68	nr	**420.81**
Sum of two sides 3100mm	165.43	281.29	5.02	148.68	nr	**429.97**
Sum of two sides 3200mm	170.82	290.47	5.02	148.68	nr	**439.15**
Sum of two sides 3300mm	175.98	299.23	5.02	148.68	nr	**447.91**
Sum of two sides 3400mm	181.13	307.98	5.02	148.68	nr	**456.66**
Sum of two sides 3500mm	186.29	316.77	5.02	148.68	nr	**465.44**
Sum of two sides 3600mm	191.43	325.50	5.02	148.68	nr	**474.18**
Sum of two sides 3700mm	196.58	334.26	5.02	148.68	nr	**482.94**
Sum of two sides 3800mm	201.74	343.03	5.02	148.68	nr	**491.71**
Sum of two sides 3900mm	206.89	351.80	5.02	148.68	nr	**500.47**
Sum of two sides 4000mm	212.05	360.56	5.02	148.68	nr	**509.24**

U:VENTILATION/AIR CONDITIONING SYSTEMS

Item	Net Price £	Material £	Labour hours	Labour £	Unit	Total rate £
U10 : DUCTWORK : RECTANGULAR– CLASS C (cont'd)						
Y30 - AIR DUCTLINES (cont'd)						
Ductwork 2001 to 2500mm longest side						
Sum of two sides 2500mm	187.74	309.01	2.77	82.04	m	**391.05**
Sum of two sides 2600mm	193.05	317.75	2.97	87.96	m	**405.72**
Sum of two sides 2700mm	198.21	326.24	2.99	88.56	m	**414.79**
Sum of two sides 2800mm	203.41	334.79	3.30	97.74	m	**432.53**
Sum of two sides 2900mm	208.55	343.26	3.31	98.03	m	**441.29**
Sum of two sides 3000mm	213.38	351.21	3.53	104.55	m	**455.76**
Sum of two sides 3100mm	218.35	359.38	3.55	105.14	m	**464.53**
Sum of two sides 3200mm	223.18	367.33	3.56	105.44	m	**472.77**
Sum of two sides 3300mm	228.32	375.80	3.56	105.44	m	**481.24**
Sum of two sides 3400mm	233.15	383.75	3.56	105.44	m	**489.19**
Sum of two sides 3500mm	255.78	421.00	3.56	105.44	m	**526.43**
Sum of two sides 3600mm	260.61	428.95	3.56	105.44	m	**534.38**
Sum of two sides 3700mm	265.76	437.41	3.56	105.44	m	**542.85**
Sum of two sides 3800mm	270.59	445.36	3.56	105.44	m	**550.80**
Sum of two sides 3900mm	275.42	453.31	3.56	105.44	m	**558.75**
Sum of two sides 4000mm	281.10	477.97	3.56	105.44	m	**583.41**
Extra over fittings; Ductwork 2001 to 2500mm longest side						
End Cap						
Sum of two sides 2500mm	100.03	170.10	0.87	25.77	nr	**195.86**
Sum of two sides 2600mm	105.85	179.99	0.87	25.77	nr	**205.75**
Sum of two sides 2700mm	111.68	189.90	0.87	25.77	nr	**215.66**
Sum of two sides 2800mm	117.50	199.79	1.16	34.36	nr	**234.14**
Sum of two sides 2900mm	123.31	209.68	1.16	34.36	nr	**244.03**
Sum of two sides 3000mm	129.14	219.59	1.73	51.24	nr	**270.82**
Sum of two sides 3100mm	133.75	227.42	1.73	51.24	nr	**278.66**
Sum of two sides 3200mm	137.93	234.53	1.73	51.24	nr	**285.77**
Sum of two sides 3300mm	142.12	241.65	1.73	51.24	nr	**292.89**
Sum of two sides 3400mm	146.30	248.76	1.73	51.24	nr	**300.00**
Sum of two sides 3500mm	150.49	255.88	1.73	51.24	nr	**307.12**
Sum of two sides 3600mm	154.67	262.99	1.73	51.24	nr	**314.23**
Sum of two sides 3700mm	158.84	270.10	1.73	51.24	nr	**321.33**
Sum of two sides 3800mm	163.03	277.22	1.73	51.24	nr	**328.46**
Sum of two sides 3900mm	167.22	284.34	1.73	51.24	nr	**335.58**
Sum of two sides 4000mm	171.41	291.47	1.73	51.24	nr	**342.70**
Reducer						
Sum of two sides 2500mm	21.21	36.07	3.12	92.41	nr	**128.47**
Sum of two sides 2600mm	23.48	39.92	3.16	93.59	nr	**133.51**
Sum of two sides 2700mm	25.85	43.96	3.16	93.59	nr	**137.55**
Sum of two sides 2800mm	28.01	47.63	4.00	118.47	nr	**166.10**
Sum of two sides 2900mm	30.39	51.67	4.01	118.77	nr	**170.44**
Sum of two sides 3000mm	32.67	55.54	4.56	135.06	nr	**190.60**
Sum of two sides 3100mm	34.21	58.17	4.56	135.06	nr	**193.22**
Sum of two sides 3200mm	35.56	60.47	4.56	135.06	nr	**195.53**
Sum of two sides 3300mm	37.02	62.95	4.56	135.06	nr	**198.01**
Sum of two sides 3400mm	38.38	65.26	4.56	135.06	nr	**200.31**
Sum of two sides 3500mm	33.48	56.94	4.56	135.06	nr	**191.99**
Sum of two sides 3600mm	34.84	59.24	4.56	135.06	nr	**194.29**
Sum of two sides 3700mm	39.36	66.93	4.56	135.06	nr	**201.99**

U:VENTILATION/AIR CONDITIONING SYSTEMS

Item	Net Price £	Material £	Labour hours	Labour £	Unit	Total rate £
Sum of two sides 3800mm	40.72	69.24	4.56	135.06	nr	**204.29**
Sum of two sides 3900mm	42.07	71.54	4.56	135.06	nr	**206.60**
Sum of two sides 4000mm	43.52	74.00	4.56	135.06	nr	**209.06**
Offset						
Sum of two sides 2500mm	100.36	170.65	3.48	103.07	nr	**273.72**
Sum of two sides 2600mm	106.73	181.49	3.49	103.36	nr	**284.85**
Sum of two sides 2700mm	106.30	180.75	3.50	103.66	nr	**284.41**
Sum of two sides 2800mm	119.50	203.20	4.33	128.24	nr	**331.44**
Sum of two sides 2900mm	118.73	201.89	4.74	140.39	nr	**342.28**
Sum of two sides 3000mm	117.50	199.79	5.31	157.27	nr	**357.05**
Sum of two sides 3100mm	114.54	194.77	5.34	158.16	nr	**352.93**
Sum of two sides 3200mm	110.88	188.54	5.35	158.45	nr	**346.99**
Sum of two sides 3300mm	106.94	181.84	5.35	158.45	nr	**340.30**
Sum of two sides 3400mm	102.59	174.43	5.35	158.45	nr	**332.89**
Sum of two sides 3500mm	94.10	160.01	5.35	158.45	nr	**318.46**
Sum of two sides 3600mm	97.72	166.17	5.35	158.45	nr	**324.62**
Sum of two sides 3700mm	105.34	179.11	5.35	158.45	nr	**337.56**
Sum of two sides 3800mm	108.97	185.29	5.35	158.45	nr	**343.74**
Sum of two sides 3900mm	112.60	191.47	5.35	158.45	nr	**349.92**
Sum of two sides 4000mm	116.28	197.72	5.35	158.45	nr	**356.17**
Square to round						
Sum of two sides 2500mm	71.32	121.26	4.79	141.87	nr	**263.13**
Sum of two sides 2600mm	78.21	132.99	4.95	146.61	nr	**279.60**
Sum of two sides 2700mm	85.10	144.71	4.95	146.61	nr	**291.31**
Sum of two sides 2800mm	92.01	156.45	8.49	251.45	nr	**407.91**
Sum of two sides 2900mm	98.90	168.17	8.88	263.00	nr	**431.17**
Sum of two sides 3000mm	105.80	179.90	9.02	267.15	nr	**447.05**
Sum of two sides 3100mm	110.66	188.16	9.02	267.15	nr	**455.31**
Sum of two sides 3200mm	114.94	195.45	9.09	269.22	nr	**464.67**
Sum of two sides 3300mm	119.21	202.70	9.09	269.22	nr	**471.92**
Sum of two sides 3400mm	123.49	209.98	9.09	269.22	nr	**479.20**
Sum of two sides 3500mm	115.58	196.54	9.09	269.22	nr	**465.76**
Sum of two sides 3600mm	119.86	203.80	9.09	269.22	nr	**473.03**
Sum of two sides 3700mm	127.03	216.00	9.09	269.22	nr	**485.22**
Sum of two sides 3800mm	131.31	223.28	9.09	269.22	nr	**492.50**
Sum of two sides 3900mm	135.60	230.57	9.09	269.22	nr	**499.79**
Sum of two sides 4000mm	139.84	237.78	9.09	269.22	nr	**507.00**
90° radius bend						
Sum of two sides 2500mm	103.78	176.47	4.35	128.84	nr	**305.30**
Sum of two sides 2600mm	106.34	180.83	4.53	134.17	nr	**314.99**
Sum of two sides 2700mm	115.67	196.68	4.53	134.17	nr	**330.85**
Sum of two sides 2800mm	110.65	188.15	7.13	211.17	nr	**399.32**
Sum of two sides 2900mm	119.78	203.68	7.17	212.36	nr	**416.04**
Sum of two sides 3000mm	128.55	218.59	7.26	215.02	nr	**433.61**
Sum of two sides 3100mm	135.08	229.69	7.26	215.02	nr	**444.71**
Sum of two sides 3200mm	141.01	239.78	7.31	216.50	nr	**456.28**
Sum of two sides 3300mm	147.30	250.47	7.31	216.50	nr	**466.98**
Sum of two sides 3400mm	153.24	260.56	7.31	216.50	nr	**477.06**
Sum of two sides 3500mm	142.00	241.46	7.31	216.50	nr	**457.96**
Sum of two sides 3600mm	147.95	251.56	7.31	216.50	nr	**468.07**
Sum of two sides 3700mm	163.70	278.34	7.31	216.50	nr	**494.85**
Sum of two sides 3800mm	169.63	288.43	7.31	216.50	nr	**504.93**
Sum of two sides 3900mm	175.56	298.52	7.31	216.50	nr	**515.02**
Sum of two sides 4000mm	173.12	294.38	7.31	216.50	nr	**510.88**

U:VENTILATION/AIR CONDITIONING SYSTEMS

Item	Net Price £	Material £	Labour hours	Labour £	Unit	Total rate £
U10 : DUCTWORK : RECTANGULAR – CLASS C (cont'd)						
Y30 - AIR DUCTLINES (cont'd)						
Extra over fittings; Ductwork 2001 to 2500mm longest side (cont'd)						
45° bend						
Sum of two sides 2500mm	81.04	137.80	8.30	245.82	nr	383.62
Sum of two sides 2600mm	83.87	142.62	8.56	253.52	nr	396.14
Sum of two sides 2700mm	90.03	153.08	8.62	255.30	nr	408.38
Sum of two sides 2800mm	89.18	151.63	9.09	269.22	nr	420.86
Sum of two sides 2900mm	95.24	161.94	9.09	269.22	nr	431.16
Sum of two sides 3000mm	101.02	171.77	9.62	284.92	nr	456.69
Sum of two sides 3100mm	105.44	179.29	9.62	284.92	nr	464.21
Sum of two sides 3200mm	109.45	186.11	9.62	284.92	nr	471.03
Sum of two sides 3300mm	113.75	193.41	9.62	284.92	nr	478.33
Sum of two sides 3400mm	117.77	200.25	9.62	284.92	nr	485.17
Sum of two sides 3500mm	113.54	193.06	9.62	284.92	nr	477.97
Sum of two sides 3600mm	117.55	199.88	9.62	284.92	nr	484.79
Sum of two sides 3700mm	127.73	217.19	9.62	284.92	nr	502.11
Sum of two sides 3800mm	131.75	224.03	9.62	284.92	nr	508.95
Sum of two sides 3900mm	135.78	230.87	9.62	284.92	nr	515.79
Sum of two sides 4000mm	135.95	231.17	9.62	284.92	nr	516.09
90° mitre bend						
Sum of two sides 2500mm	120.54	204.96	8.30	245.82	nr	450.79
Sum of two sides 2600mm	125.97	214.19	8.56	253.52	nr	467.72
Sum of two sides 2700mm	136.92	232.82	8.62	255.30	nr	488.12
Sum of two sides 2800mm	129.79	220.69	15.20	450.18	nr	670.88
Sum of two sides 2900mm	140.51	238.92	15.20	450.18	nr	689.11
Sum of two sides 3000mm	151.41	257.45	15.20	450.18	nr	707.64
Sum of two sides 3100mm	159.92	271.92	16.04	475.06	nr	746.98
Sum of two sides 3200mm	167.81	285.34	16.04	475.06	nr	760.41
Sum of two sides 3300mm	175.53	298.47	16.04	475.06	nr	773.53
Sum of two sides 3400mm	183.42	311.89	16.04	475.06	nr	786.95
Sum of two sides 3500mm	168.39	286.32	16.04	475.06	nr	761.39
Sum of two sides 3600mm	176.29	299.77	16.04	475.06	nr	774.83
Sum of two sides 3700mm	189.92	322.94	16.04	475.06	nr	798.01
Sum of two sides 3800mm	197.83	336.39	16.04	475.06	nr	811.45
Sum of two sides 3900mm	205.73	349.81	16.04	475.06	nr	824.88
Sum of two sides 4000mm	201.94	343.37	16.04	475.06	nr	818.43
Branch						
Sum of two sides 2500mm	96.38	163.88	2.88	85.30	nr	249.18
Sum of two sides 2600mm	101.55	172.67	2.88	85.30	nr	257.96
Sum of two sides 2700mm	106.84	181.66	2.88	85.30	nr	266.96
Sum of two sides 2800mm	111.87	190.22	3.94	116.69	nr	306.91
Sum of two sides 2900mm	117.16	199.21	3.94	116.69	nr	315.91
Sum of two sides 3000mm	122.34	208.02	4.83	143.05	nr	351.07
Sum of two sides 3100mm	126.51	215.12	4.83	143.05	nr	358.17
Sum of two sides 3200mm	130.28	221.53	4.83	143.05	nr	364.58
Sum of two sides 3300mm	134.20	228.19	4.83	143.05	nr	371.24
Sum of two sides 3400mm	137.99	234.64	4.83	143.05	nr	377.69
Sum of two sides 3500mm	141.96	241.39	4.83	143.05	nr	384.44
Sum of two sides 3600mm	145.76	247.85	4.83	143.05	nr	390.90
Sum of two sides 3700mm	151.40	257.44	4.83	143.05	nr	400.49
Sum of two sides 3800mm	155.19	263.88	4.83	143.05	nr	406.93
Sum of two sides 3900mm	158.99	270.35	4.83	143.05	nr	413.40
Sum of two sides 4000mm	162.91	277.00	4.83	143.05	nr	420.06

U:VENTILATION/AIR CONDITIONING SYSTEMS

Item	Net Price £	Material £	Labour hours	Labour £	Unit	Total rate £
Grille neck						
Sum of two sides 2500mm	130.50	221.91	3.06	90.63	nr	**312.54**
Sum of two sides 2600mm	137.79	234.30	3.08	91.22	nr	**325.52**
Sum of two sides 2700mm	145.08	246.69	3.08	91.22	nr	**337.91**
Sum of two sides 2800mm	152.37	259.08	4.13	122.32	nr	**381.40**
Sum of two sides 2900mm	159.65	271.47	4.13	122.32	nr	**393.79**
Sum of two sides 3000mm	166.94	283.86	5.02	148.68	nr	**432.54**
Sum of two sides 3100mm	172.57	293.43	5.02	148.68	nr	**442.11**
Sum of two sides 3200mm	178.19	302.98	5.02	148.68	nr	**451.66**
Sum of two sides 3300mm	183.56	312.12	5.02	148.68	nr	**460.80**
Sum of two sides 3400mm	188.93	321.25	5.02	148.68	nr	**469.93**
Sum of two sides 3500mm	194.31	330.41	5.02	148.68	nr	**479.09**
Sum of two sides 3600mm	199.68	339.53	5.02	148.68	nr	**488.21**
Sum of two sides 3700mm	205.05	348.67	5.02	148.68	nr	**497.35**
Sum of two sides 3800mm	210.43	357.81	5.02	148.68	nr	**506.49**
Sum of two sides 3900mm	215.81	366.95	5.02	148.68	nr	**515.63**
Sum of two sides 4000mm	221.18	376.09	5.02	148.68	nr	**524.77**

Y30 - DUCTWORK ANCILLARIES: ACCESS DOORS

Refer to ancillaries in U10: DUCTWORK: RECTANGULAR: CLASS B for details of access doors

U:VENTILATION/AIR CONDITIONING SYSTEMS

Item	Net Price £	Material £	Labour hours	Labour £	Unit	Total rate £
U10 : DUCTWORK : VOLUME/FIRE DAMPERS						
Y30 - DUCTWORK ANCILLARIES: VOLUME CONTROL AND FIRE DAMPERS						
Volume control damper; opposed blade; galvanised steel casing; aluminium aerofoil blades; manually operated						
Rectangular						
Sum of two sides 200mm	13.48	22.92	1.60	47.39	nr	**70.31**
Sum of two sides 300mm	14.42	24.52	1.60	47.39	nr	**71.91**
Sum of two sides 400mm	15.77	26.82	1.60	47.39	nr	**74.20**
Sum of two sides 500mm	17.26	29.35	1.60	47.39	nr	**76.74**
Sum of two sides 600mm	19.08	32.44	1.70	50.37	nr	**82.81**
Sum of two sides 700mm	20.97	35.66	2.10	62.22	nr	**97.88**
Sum of two sides 800mm	22.92	38.97	2.15	63.68	nr	**102.65**
Sum of two sides 900mm	25.15	42.76	2.30	68.12	nr	**110.88**
Sum of two sides 1000mm	27.31	46.44	2.40	71.08	nr	**117.52**
Sum of two sides 1100mm	29.74	50.57	2.60	77.01	nr	**127.57**
Sum of two sides 1200mm	33.64	57.20	2.80	82.93	nr	**140.13**
Sum of two sides 1300mm	36.27	61.67	3.10	91.81	nr	**153.49**
Sum of two sides 1400mm	39.10	66.48	3.25	96.26	nr	**162.74**
Sum of two sides 1500mm	42.34	71.99	3.40	100.70	nr	**172.69**
Sum of two sides 1600mm	45.50	77.37	3.45	102.18	nr	**179.55**
Sum of two sides 1700mm	48.54	82.54	3.60	106.62	nr	**189.16**
Sum of two sides 1800mm	52.18	88.73	3.90	115.51	nr	**204.23**
Sum of two sides 1900mm	55.55	94.46	4.20	124.44	nr	**218.90**
Sum of two sides 2000mm	59.59	101.33	4.33	128.24	nr	**229.57**
Circular						
100mm dia.	18.00	30.61	0.80	23.69	nr	**54.30**
160mm dia.	21.44	36.46	0.90	26.66	nr	**63.11**
200mm dia.	23.26	39.55	1.05	31.11	nr	**70.66**
250mm dia.	25.88	44.01	1.20	35.56	nr	**79.56**
315mm dia.	29.93	50.89	1.35	40.02	nr	**90.92**
355mm dia.	25.48	43.33	1.65	48.87	nr	**92.20**
400mm dia.	34.38	58.46	1.90	56.31	nr	**114.76**
450mm dia.	37.14	63.15	2.10	62.22	nr	**125.37**
500mm dia.	40.45	68.78	2.95	87.37	nr	**156.15**
630mm dia.	51.10	86.89	4.55	134.76	nr	**221.65**
710mm dia.	59.59	101.33	5.20	154.01	nr	**255.34**
800mm dia.	64.12	109.03	5.80	171.78	nr	**280.81**
900mm dia.	73.75	125.40	6.40	189.55	nr	**314.95**
1000mm dia.	83.93	142.71	7.00	207.32	nr	**350.03**
Flat oval						
345 x 102mm	31.42	53.43	1.20	35.54	nr	**88.97**
508 x 102mm	34.25	58.24	1.60	47.39	nr	**105.63**
559 x 152mm	38.69	65.79	1.90	56.31	nr	**122.09**
531 x 203mm	42.27	71.88	1.90	56.31	nr	**128.18**
851 x 203mm	48.61	82.66	4.55	134.76	nr	**217.41**
582 x 254mm	46.92	79.78	2.10	62.22	nr	**142.00**
823 x 254mm	52.45	89.18	4.10	121.43	nr	**210.62**
632 x 305mm	51.71	87.93	2.95	87.37	nr	**175.29**
765 x 356mm	57.85	98.37	4.55	134.76	nr	**233.13**
737 x 406mm	61.61	104.76	4.55	134.76	nr	**239.52**
818 x 406mm	62.90	106.95	5.20	154.01	nr	**260.96**
978 x 406mm	66.74	113.48	5.50	162.90	nr	**276.38**

U:VENTILATION/AIR CONDITIONING SYSTEMS

Item	Net Price £	Material £	Labour hours	Labour £	Unit	Total rate £
709 x 457mm	64.12	109.03	4.50	133.41	nr	**242.44**
678 x 508mm	67.82	115.32	4.55	134.76	nr	**250.08**
919 x 508mm	73.35	124.72	6.00	177.70	nr	**302.43**
Fire damper; galvanised steel casing; stainless steel folding shutter; fusible link with manual reset; BS 476 4 hour fire rated						
Rectangular						
Sum of two sides 200mm	38.28	65.05	1.60	47.39	nr	**112.44**
Sum of two sides 300mm	38.28	65.05	1.60	47.39	nr	**112.44**
Sum of two sides 400mm	38.28	65.05	1.60	47.39	nr	**112.44**
Sum of two sides 500mm	42.72	72.65	1.60	47.39	nr	**120.04**
Sum of two sides 600mm	47.48	80.73	1.70	50.37	nr	**131.10**
Sum of two sides 700mm	52.36	89.03	2.10	62.22	nr	**151.25**
Sum of two sides 800mm	57.55	97.86	2.15	63.69	nr	**161.55**
Sum of two sides 900mm	62.74	106.69	2.30	68.24	nr	**174.93**
Sum of two sides 1000mm	68.07	115.74	2.40	71.19	nr	**186.94**
Sum of two sides 1100mm	72.69	123.60	2.60	77.13	nr	**200.73**
Sum of two sides 1200mm	78.45	133.40	2.80	82.96	nr	**216.36**
Sum of two sides 1300mm	84.39	143.50	3.10	91.81	nr	**235.31**
Sum of two sides 1400mm	91.15	154.58	3.25	96.26	nr	**251.84**
Sum of two sides 1500mm	96.40	163.92	3.40	100.74	nr	**264.66**
Sum of two sides 1600mm	102.63	174.51	3.45	102.18	nr	**276.69**
Sum of two sides 1700mm	108.86	185.10	3.60	106.62	nr	**291.72**
Sum of two sides 1800mm	115.52	196.43	3.90	115.51	nr	**311.94**
Sum of two sides 1900mm	127.77	207.06	4.20	124.44	nr	**331.50**
Sum of two sides 2000mm	127.56	216.89	4.33	128.24	nr	**345.13**
Sum of two sides 2100mm	137.19	233.28	4.43	131.21	nr	**364.49**
Sum of two sides 2200mm	146.39	248.92	4.55	134.76	nr	**383.68**
Circular						
100mm dia.	46.57	79.19	0.80	23.69	nr	**102.88**
160mm dia.	50.58	86.00	0.90	26.66	nr	**112.66**
200mm dia.	54.14	92.06	1.05	31.10	nr	**123.16**
250mm dia.	61.27	104.18	1.20	35.56	nr	**139.74**
315mm dia.	71.63	121.80	1.35	40.02	nr	**161.82**
355mm dia.	77.57	131.90	1.65	48.87	nr	**180.77**
400mm dia.	86.92	147.80	1.90	56.31	nr	**204.11**
450mm dia.	96.12	163.44	2.10	62.22	nr	**225.66**
500mm dia.	106.35	180.83	2.95	87.37	nr	**268.20**
630mm dia.	139.26	236.79	4.55	134.76	nr	**371.55**
710mm dia.	164.03	278.92	5.20	154.01	nr	**432.93**
800mm dia.	177.08	301.10	5.80	171.78	nr	**472.88**
900mm dia.	205.13	348.80	6.40	189.55	nr	**538.35**
1000mm dia.	235.25	400.01	7.00	207.32	nr	**607.33**

U:VENTILATION/AIR CONDITIONING SYSTEMS

Item	Net Price £	Material £	Labour hours	Labour £	Unit	Total rate £
U10 : DUCTWORK : VOLUME/FIRE DAMPERS (cont'd)						
Y30 - DUCTWORK ANCILLARIES: VOLUME CONTROL AND FIRE DAMPERS (cont'd)						
Fire damper; galvanised steel casing; stainless steel folding shutter (cont'd)						
Flat oval						
345 x 102mm	57.17	97.21	1.20	35.56	nr	**132.77**
427 x 102mm	62.50	106.27	1.35	40.02	nr	**146.29**
508 x 102mm	64.67	109.97	1.60	47.39	nr	**157.36**
559 x 152mm	66.98	113.89	1.90	56.31	nr	**170.20**
531 x 203mm	76.70	130.42	1.90	56.31	nr	**186.73**
851 x 203mm	98.30	167.14	4.55	134.76	nr	**301.90**
582 x 254mm	107.64	183.03	2.10	62.22	nr	**245.25**
632 x 305mm	117.34	199.53	2.95	87.37	nr	**286.90**
765 x 356mm	131.42	223.41	4.55	134.76	nr	**358.22**
737 x 406mm	137.61	233.99	4.55	134.76	nr	**368.75**
818 x 406mm	141.73	241.00	5.20	154.01	nr	**395.01**
978 x 406mm	159.10	270.53	5.50	162.90	nr	**433.43**
709 x 457mm	138.82	236.04	4.50	133.41	nr	**369.45**
678 x 508mm	140.63	239.13	4.55	134.76	nr	**373.89**
Smoke/fire damper; galvanised steel casing; stainless steel folding shutter; fusible link and 24V d.c. electro-magnetic shutter release mechanism; spring operated; BS 476 4 hour fire rating						
Rectangular						
Sum of two sides 200mm	187.35	318.57	1.60	47.39	nr	**365.95**
Sum of two sides 300mm	187.35	318.57	1.60	47.39	nr	**365.95**
Sum of two sides 400mm	187.35	318.57	1.60	47.39	nr	**365.95**
Sum of two sides 500mm	192.27	326.93	1.60	47.39	nr	**374.32**
Sum of two sides 600mm	196.31	333.80	1.70	50.37	nr	**384.17**
Sum of two sides 700mm	200.57	341.05	2.10	62.22	nr	**403.27**
Sum of two sides 800mm	204.95	348.49	2.15	63.69	nr	**412.19**
Sum of two sides 900mm	209.53	356.28	2.30	68.24	nr	**424.52**
Sum of two sides 1000mm	214.31	364.41	2.40	71.19	nr	**435.60**
Sum of two sides 1100mm	219.24	372.79	2.60	77.13	nr	**449.92**
Sum of two sides 1200mm	224.36	381.50	2.80	82.96	nr	**464.46**
Sum of two sides 1300mm	229.69	390.56	3.10	91.81	nr	**482.37**
Sum of two sides 1400mm	235.16	399.86	3.25	96.26	nr	**496.12**
Sum of two sides 1500mm	240.81	409.47	3.40	100.74	nr	**510.21**
Sum of two sides 1600mm	246.68	419.45	3.45	102.18	nr	**521.63**
Sum of two sides 1700mm	252.68	429.65	3.60	106.62	nr	**536.27**
Sum of two sides 1800mm	258.88	440.19	3.90	115.51	nr	**555.70**
Sum of two sides 1900mm	265.21	450.96	4.20	124.44	nr	**575.40**
Sum of two sides 2000mm	271.75	462.08	4.33	128.24	nr	**590.32**
Circular						
100mm dia.	198.95	338.29	0.80	23.69	nr	**361.98**
160mm dia.	198.95	338.29	0.90	26.66	nr	**364.95**
200mm dia.	198.95	338.29	1.05	31.10	nr	**369.39**
250mm dia.	202.18	343.78	1.20	35.56	nr	**379.34**
315mm dia.	205.55	349.51	1.35	40.02	nr	**389.54**
355mm dia.	215.15	365.84	1.65	48.87	nr	**414.71**

U:VENTILATION/AIR CONDITIONING SYSTEMS

Item	Net Price £	Material £	Labour hours	Labour £	Unit	Total rate £
400mm dia.	222.95	379.10	1.90	56.31	nr	**435.41**
450mm dia.	226.52	385.17	2.10	62.22	nr	**447.39**
500mm dia.	234.54	398.81	2.95	87.37	nr	**486.18**
630mm dia.	256.93	436.88	4.55	134.76	nr	**571.64**
710mm dia.	266.43	453.03	5.20	154.01	nr	**607.04**
800mm dia.	285.04	484.68	5.80	171.78	nr	**656.46**
900mm dia.	301.76	513.11	6.40	189.55	nr	**702.66**
1000mm dia.	319.10	542.59	7.00	207.32	nr	**749.91**
Flat oval						
531 x 203mm	245.20	416.93	1.90	56.31	nr	**473.24**
851 x 203mm	254.50	432.75	4.55	134.76	nr	**567.51**
582 x 254mm	252.68	429.65	2.10	62.22	nr	**491.87**
632 x 305mm	260.16	442.37	2.95	87.37	nr	**529.74**
765 x 356mm	269.53	458.30	4.55	134.76	nr	**593.06**
737 x 406mm	275.60	468.62	4.55	134.76	nr	**603.38**
818 x 406mm	277.35	471.60	5.20	154.01	nr	**625.61**
978 x 406mm	282.62	480.56	5.50	162.90	nr	**643.46**
709 x 457mm	279.83	475.82	4.50	133.41	nr	**609.23**
678 x 508mm	283.90	482.74	4.55	134.76	nr	**617.50**

U:VENTILATION/AIR CONDITIONING SYSTEMS

Item	Net Price £	Material £	Labour hours	Labour £	Unit	Total rate £
U10 : PLANT/EQUIPMENT						
Y41 - FANS						
Axial flow fan; including ancillaries, anti vibration mountings, mounting feet, matching flanges, flexible connectors and clips; 415V, 3 phase, 50Hz motor; includes fixing in position; electrical work elsewhere						
Aerofoil blade fan unit; short duct case						
315mm dia.; 0.47 m³/s duty; 147 Pa	470.80	603.00	4.50	114.17	nr	**717.17**
500mm dia.; 1.89 m³/s duty; 500 Pa	767.19	982.61	5.00	126.86	nr	**1109.47**
560mm dia.; 2.36 m³/s duty; 147 Pa	454.75	582.44	5.50	139.55	nr	**721.98**
710mm dia.; 5.67 m³/s duty; 245 Pa	750.07	960.68	6.00	152.23	nr	**1112.91**
Aerofoil blade fan unit; long duct case						
315mm dia.; 0.47 m³/s duty; 147 Pa	524.30	671.52	4.50	114.17	nr	**785.69**
500mm dia.; 1.89 m³/s duty; 500 Pa	767.19	982.61	5.00	126.86	nr	**1109.47**
560mm dia.; 2.36 m³/s duty; 147 Pa	535.00	685.22	5.50	139.55	nr	**824.77**
710mm dia.; 5.67 m³/s duty; 245 Pa	874.19	1119.65	6.00	152.23	nr	**1271.89**
Aerofoil blade fan unit; two stage parallel fan arrangement; long duct case						
315mm; 0.47 m³/s at 500 Pa	1030.41	1319.74	4.50	114.17	nr	**1433.91**
355mm; 0.83 m³/s at 147 Pa	1189.84	1523.94	4.75	120.52	nr	**1644.45**
710mm; 3.77 m³/s at 431 Pa	2433.18	3116.39	6.00	152.23	nr	**3268.62**
710mm; 6.61 m³/s at 500 Pa	2569.07	3290.44	6.00	152.23	nr	**3442.67**
Axial flow fan; suitable for operation at 300°C for 90 minutes; including ancillaries, anti vibration mountings, mounting feet, matching flanges, flexible connectors and clips; 415V, 3 phase, 50Hz motor; includes fixing in position; electrical work elsewhere						
450mm; 2.0m³/s at 300Pa	1269.02	1625.35	5.00	126.86	nr	**1752.21**
630mm; 4.6m³/s at 200Pa	1530.10	1959.74	5.50	139.55	nr	**2099.28**
800mm; 9.0m³/s at 300Pa	2457.79	3147.91	6.50	164.92	nr	**3312.83**
1000mm; 15.0m³/s at 400Pa	3705.41	4745.85	7.50	190.29	nr	**4936.14**
Bifurcated fan; suitable for temperature up to 200°C with motor protection to IP55; including ancillaries, anti vibration mountings, mounting feet, matching flanges, flexible connectors and clips; 415V, 3 phase, 50Hz motor; includes fixing in position; electrical work elsewhere						
300mm; 0.50m³/s at 100Pa	799.29	1023.72	4.50	114.17	nr	**1137.90**
400mm; 1.97m³/s at 200Pa	1050.74	1345.78	5.00	126.86	nr	**1472.64**
630mm; 3.86m³/s at 200Pa	1542.94	1976.18	5.50	139.55	nr	**2115.73**
800mm; 6.10m³/s at 400Pa	1756.94	2250.27	6.50	164.92	nr	**2415.19**

U:VENTILATION/AIR CONDITIONING SYSTEMS

Item	Net Price £	Material £	Labour hours	Labour £	Unit	Total rate £
Duct mounted in line fan with backward curved centrifugal impellor; including ancillaries, matching flanges, flexible connectors and clips; 415V, 3phase, 50Hz motor; includes fixing in position; electrical work elsewhere						
0.5m³/s at 200Pa	1011.15	1295.07	4.50	114.17	nr	1409.24
1.0m³/s at 300Pa	1246.55	1596.57	5.00	126.86	nr	1723.43
3.0m³/s at 500Pa	2059.75	2638.11	5.50	139.55	nr	2777.65
5.0m³/s at 750Pa	2728.50	3494.64	6.50	164.92	nr	3659.55
7.0m³/s at 1000Pa	3317.00	4248.38	7.00	177.60	nr	4425.98
Twin fan extract unit; belt driven; located internally; complete with anti-vibration mounts and non return shutter; including ancillaries, matching flanges, flexible connectors and clips; 3 phase, 50Hz motor; includes fixing in position; electrical work elsewhere						
0.25m³/s at 150Pa	1395.28	1787.06	4.50	114.17	nr	1901.23
Extra for external unit	158.36	202.83	-	-	nr	202.83
0.50m³/s at 200Pa	1623.19	2078.97	5.00	126.86	nr	2205.83
Extra for external unit	185.11	237.09	-	-	nr	237.09
1.00m³/s at 200Pa	1934.56	2477.77	5.00	126.86	nr	2604.62
Extra for external unit	216.14	276.83	-	-	nr	276.83
1.50m³/s at 250Pa	2062.96	2642.22	5.50	139.55	nr	2781.76
Extra for external unit	221.49	283.68	-	-	nr	283.68
2.00m³/s at 250Pa	2284.45	2925.90	6.50	164.92	nr	3090.82
Extra for external unit	243.96	312.46	-	-	nr	312.46
Extra for auto changeover panel	184.04	235.72	2.50	63.43	nr	299.15
Roof mounted extract fan; including ancillaries, fibreglass cowling, fitted shutters and bird guard; 415V, 3 phase, 50Hz motor; includes fixing in position; electrical work elsewhere						
Flat roof installation, fixed to curb						
315mm; 900rpm	465.12	595.72	4.50	114.17	nr	709.89
315mm; 1380rpm	465.12	595.72	4.50	114.17	nr	709.89
400mm; 900rpm	654.36	838.10	5.50	139.55	nr	977.64
400mm; 1360rpm	564.30	722.75	5.50	139.55	nr	862.30
800mm; 530rpm	1881.00	2409.17	7.00	177.60	nr	2586.77
800mm; 700rpm	1767.00	2263.16	7.00	177.60	nr	2440.76
800mm; 920rpm	1567.50	2007.64	7.00	177.60	nr	2185.24
1000mm; 470rpm	2508.00	3212.22	8.00	202.98	nr	3415.20
1000mm; 570rpm	2394.00	3066.21	8.00	202.98	nr	3269.19
1000mm; 710rpm	2329.02	2982.99	8.00	202.98	nr	3185.96

U:VENTILATION/AIR CONDITIONING SYSTEMS

Item	Net Price £	Material £	Labour hours	Labour £	Unit	Total rate £
U10 : PLANT/EQUIPMENT (cont'd)						
Y41 – FANS (cont'd)						
Roof mounted extract fan; including Ancillaries (cont'd)						
Pitched roof installation; including purlin mounting box						
315mm; 900rpm	465.12	662.24	4.50	114.17	nr	**776.42**
315mm; 1380rpm	465.12	662.24	4.50	114.17	nr	**776.42**
400mm; 900rpm	654.36	918.48	5.50	139.55	nr	**1058.03**
400mm; 1360rpm	564.30	803.13	5.50	139.55	nr	**942.68**
800mm; 530rpm	1881.00	2569.29	7.00	177.60	nr	**2746.90**
800mm; 700rpm	1767.00	2423.28	7.00	177.60	nr	**2600.89**
800mm; 920rpm	1567.50	2167.76	7.00	177.60	nr	**2345.37**
1000mm; 470rpm	2508.00	3449.05	8.00	202.98	nr	**3652.02**
1000mm; 570rpm	2394.00	3303.04	8.00	202.98	nr	**3506.01**
1000mm; 710rpm	2329.02	3219.81	8.00	202.98	nr	**3422.79**
Centrifugal fan; single speed for internal domestic kitchens/utility rooms; fitted with standard overload protection; complete with housing; includes placing in position; electrical work elsewhere						
Window mounted						
245m³/hr	76.84	98.41	0.50	12.69	nr	**111.10**
500m³/hr	123.00	315.07	0.50	12.69	nr	**327.75**
Wall mounted						
245m³/hr	91.65	232.68	0.83	21.14	nr	**253.82**
500m³/hr	137.81	508.45	0.83	21.14	nr	**529.59**
Centrifugal fan; various speeds, simultaneous ventilation from separate areas fitted with standard overload protection; complete with housing; includes placing in position; ducting and electrical work elsewhere						
Fan unit						
147-300m³/hr	113.42	145.27	1.00	25.37	nr	**170.64**
175-411m³/hr	187.25	239.83	1.00	25.37	nr	**265.20**
Toilet extract units; centrifugal fan; various speeds for internal domestic bathrooms/ W.Cs, with built in filter; complete with housing; includes placing in position; electrical work elsewhere						
Fan unit; fixed to wall; including shutter						
Single speed 85m³/hr	80.50	103.11	0.75	19.03	nr	**122.13**
Two speed 60-85m³/hr	94.44	120.96	0.83	21.14	nr	**142.10**
Humidity controlled; autospeed; fixed to wall; including shutter						
30-60-85m³/hr	153.31	196.36	1.00	25.37	nr	**221.74**

U:VENTILATION/AIR CONDITIONING SYSTEMS

Item	Net Price £	Material £	Labour hours	Labour £	Unit	Total rate £
Y42 - AIR FILTRATION						
High efficiency duct mounted filters; 99.997% H13 (EU13); tested to BS 3928						
Standard; 1700m³/ hr air volume; continuous rating up to 80°C; sealed wood case, aluminium spacers, neoprene gaskets; water repellent filter media; includes placing in position						
610 x 610 x 292mm	176.40	299.94	1.00	29.62	nr	**329.56**
Side withdrawal frame	84.60	143.85	2.50	74.04	nr	**217.90**
High capacity; 3400m³/hr air volume; continuous rating up to 80°C; anti-corrosion coated mild steel frame, polyurethane sealant and neoprene gaskets; water repellent filter media; includes placing in position						
610 x 610 x 292mm	320.17	544.40	1.00	29.62	nr	**574.02**
Side withdrawal frame	84.60	143.85	2.50	74.04	nr	**217.90**
Bag Filters; 40/60% F5 (EU5); tested to BSEN 779						
Duct mounted bag filter; continuous rating up to 60°C; rigid filter assembly; sealed into one piece coated mild steel header with sealed pocket separators; includes placing in position						
6 pocket, 592 x 592 x 25mm header; pockets 380mm long; 1690m³/hr	57.98	98.59	1.00	29.62	nr	**128.21**
Side withdrawal frame	58.66	99.75	2.00	59.23	nr	**158.98**
6 pocket, 592 x 592 x 25mm header; pockets 500mm long; 2550m³/hr	61.91	105.27	1.50	44.43	nr	**149.69**
Side withdrawal frame	58.66	99.75	2.50	74.04	nr	**173.79**
6 pocket, 592 x 592 x 25mm header; pockets 635mm long; 3380m³/hr	65.44	111.27	1.50	44.43	nr	**155.70**
Side withdrawal frame	58.66	99.75	3.00	88.85	nr	**188.60**
Bag Filters; 80/90% F7, (EU7); tested to BSEN 779						
Duct mounted bag filter; continuous rating up to 60°C; rigid filter assembly; sealed into one piece coated mild steel header with sealed pocket separators; includes placing in position						
6 pocket, 592 x 592 x 25mm header; pockets 500mm long; 1688m³/hr	80.72	137.25	1.00	29.62	nr	**166.87**
Side withdrawal frame	58.66	99.75	2.00	59.23	nr	**158.98**
6 pocket, 592 x 592 x 25mm header; pockets 635mm long; 2047m³/hr	86.60	147.26	1.50	44.43	nr	**191.68**
Side withdrawal frame	58.66	99.75	2.50	74.04	nr	**173.79**
6 pocket, 592 x 592 x 25mm header; pockets 762mm long; 2729m³/hr	101.40	172.42	1.50	44.43	nr	**216.84**
Side withdrawal frame	58.66	99.75	3.00	88.85	nr	**188.60**

U:VENTILATION/AIR CONDITIONING SYSTEMS

Item	Net Price £	Material £	Labour hours	Labour £	Unit	Total rate £
U10 : PLANT/EQUIPMENT (cont'd)						
Y42 - AIR FILTRATION (cont'd)						
Bag Filters; 80/90% F7, (EU7); tested to BSEN 779 (cont'd)						
Grease filters, washable; minimum 65% Double sided extract unit; lightweight stainless steel construction; demountable composite filter media of woven metal mat and expanded metal mesh supports; for mounting on hood and extract systems (hood not included); includes placing in position						
500 x 686 x 565mm, 4080m³/hr	359.30	610.95	2.00	59.23	nr	670.19
1000 x 686 x 565mm, 8160m³/hr;	547.93	931.69	3.00	88.85	nr	1020.54
1500 x 686 x 565mm, 12240m³/hr;	751.56	1277.94	3.50	103.66	nr	1381.61
Panel filters; 82% G3 (EU3); tested to BS EN779						
Modular duct mounted filter panels; continuous rating up to 100°C; graduated density media; rigid cardboard frame; includes placing in position						
596 x 596 x 47mm, 2360m³/hr	5.04	8.57	1.00	29.62	nr	38.18
Side withdrawal frame	51.48	87.54	2.50	74.04	nr	161.58
596 x 287 x 47mm, 1140m³/hr	3.59	6.10	1.00	29.62	nr	35.72
Side withdrawal frame	51.48	87.54	2.50	74.04	nr	161.58
Panel filters; 90% G4 (EU4); tested to BS EN779						
Modular duct mounted filter panels; continuous rating up to 100°C; pleated media with wire support; rigid cardboard frame; includes placing in position						
596 x 596 x 47mm, 2560m³/hr	10.29	17.49	1.00	29.62	nr	47.11
Side withdrawal frame	64.93	110.41	3.00	88.85	nr	199.26
596 x 287 x 47mm, 1230m³/hr	7.94	13.50	1.00	29.62	nr	43.12
Side withdrawal frame	51.48	87.54	3.00	88.85	nr	176.39
Carbon filters; standard duty disposable carbon filters; steel frame with bonded carbon panels; for fixing to ductwork; including placing in position						
12 panels						
597 x 597 x 298mm, 1460m³/hr	306.22	520.69	0.33	9.77	nr	530.46
597 x 597 x 451mm, 2200m³/hr	345.21	586.99	0.33	9.77	nr	596.77
597 x 597 x 597mm, 2930m³/hr	384.97	654.59	0.33	9.77	nr	664.36
8 panels						
451 x 451 x 298mm, 740m³/hr	230.86	392.54	0.29	8.59	nr	401.13
451 x 451 x 451mm, 1105m³/hr	255.51	434.46	0.29	8.59	nr	443.05
451 x 451 x 597mm, 1460m³/hr	279.04	474.47	0.29	8.59	nr	483.06
6 panels						
298 x 298 x 298mm, 365m³/hr	163.58	278.15	0.25	7.40	nr	285.55
298 x 298 x 451mm, 550m³/hr	175.14	297.81	0.25	7.40	nr	305.21
298 x 298 x 597mm, 780m³/hr	186.20	316.61	0.25	7.40	nr	324.01

U:VENTILATION/AIR CONDITIONING SYSTEMS

Item	Net Price £	Material £	Labour hours	Labour £	Unit	Total rate £
Y45 - SILENCERS/ACOUSTIC TREATMENT						
Attenuators; DW144 galvanised construction c/w splitters; self securing; fitted to ductwork						
To suit rectangular ducts; unit length 600mm						
100 x 100mm	77.34	131.50	0.75	22.21	nr	153.72
150 x 150mm	80.73	137.27	0.75	22.21	nr	159.48
200 x 200mm	84.12	143.03	0.75	22.21	nr	165.24
300 x 300mm	94.11	160.02	0.75	22.21	nr	182.23
400 x 400mm	105.97	180.19	1.00	29.62	nr	209.81
500 x 500mm	123.29	209.65	1.25	37.02	nr	246.67
600 x 300mm	118.77	201.96	1.25	37.02	nr	238.98
600 x 600mm	166.23	282.66	1.25	37.02	nr	319.68
700 x 300mm	125.93	214.12	1.50	44.43	nr	258.55
700 x 700mm	187.51	318.84	1.50	44.43	nr	363.27
800 x 300mm	130.45	221.81	2.00	59.23	nr	281.04
800 x 800mm	216.33	367.84	2.00	59.23	nr	427.07
1000 x 1000mm	276.92	470.87	3.00	88.85	nr	559.72
To suit rectangular ducts; unit length 1200mm						
200 x 200mm	92.75	157.72	1.00	29.62	nr	187.33
300 x 300mm	114.68	194.99	1.00	29.62	nr	224.61
400 x 400mm	137.88	234.45	1.33	39.39	nr	273.84
500 x 500mm	166.42	282.98	1.66	49.16	nr	332.15
600 x 300mm	158.69	269.83	1.66	49.16	nr	319.00
600 x 600mm	232.17	394.78	1.66	49.16	nr	443.94
700 x 300mm	170.10	289.24	2.00	59.23	nr	348.48
700 x 700mm	265.69	451.77	2.00	59.23	nr	511.00
800 x 300mm	177.84	302.39	2.66	78.78	nr	381.18
800 x 800mm	306.57	521.29	2.66	78.78	nr	600.07
1000 x 1000mm	414.99	705.65	4.00	118.47	nr	824.12
1300 x 1300mm	712.38	1211.32	8.00	236.94	nr	1448.26
1500 x 1500mm	815.39	1386.47	8.00	236.94	nr	1623.41
1800 x 1800mm	1115.65	1897.03	10.66	315.72	nr	2212.75
2000 x 2000mm	1251.90	2128.70	13.33	394.80	nr	2523.50
To suit rectangular ducts; unit length 1800mm						
200 x 200mm	117.06	199.05	1.00	29.62	nr	228.67
300 x 300mm	155.56	264.51	1.00	29.62	nr	294.13
400 x 400mm	185.95	316.18	1.33	39.39	nr	355.58
500 x 500mm	226.65	385.39	1.66	49.16	nr	434.55
600 x 300mm	216.70	368.48	1.66	49.16	nr	417.64
600 x 600mm	316.70	538.52	1.66	49.16	nr	587.68
700 x 300mm	240.27	408.55	2.00	59.23	nr	467.79
700 x 700mm	378.40	643.42	2.00	59.23	nr	702.65
800 x 300mm	254.08	432.04	2.66	78.78	nr	510.82
800 x 800mm	440.28	748.64	2.66	78.78	nr	827.42
1000 x 1000mm	591.79	1006.27	4.00	118.47	nr	1124.74
1300 x 1300mm	988.63	1681.05	8.00	236.94	nr	1917.99
1500 x 1500mm	1120.37	1905.06	8.00	236.94	nr	2141.99
1800 x 1800mm	1555.44	2644.84	10.66	315.72	nr	2960.56
2000 x 2000mm	1761.30	2994.89	13.33	394.80	nr	3389.69
2300 x 2300mm	2201.60	3743.56	16.00	473.88	nr	4217.44
2500 x 2500mm	3070.96	5221.80	18.66	552.66	nr	5774.46

U:VENTILATION/AIR CONDITIONING SYSTEMS

Item	Net Price £	Material £	Labour hours	Labour £	Unit	Total rate £
U10 : PLANT/EQUIPMENT (cont'd)						
Y45 - SILENCERS/ACOUSTIC TREATMENT (cont'd)						
To suit rectangular ducts; unit length 2400mm						
500 x 500mm	280.24	476.51	2.08	61.69	nr	**538.21**
600 x 300mm	267.53	454.91	2.08	61.69	nr	**516.60**
600 x 600mm	398.47	677.56	2.08	61.69	nr	**739.25**
700 x 300mm	290.00	493.10	2.50	74.04	nr	**567.15**
700 x 700mm	470.66	800.31	2.50	74.04	nr	**874.35**
800 x 300mm	306.57	521.29	3.33	98.63	nr	**619.91**
800 x 800mm	544.70	926.19	3.33	98.63	nr	**1024.82**
1000 x 1000mm	734.89	1249.59	5.00	148.09	nr	**1397.68**
1300 x 1300mm	1190.30	2023.95	10.00	296.17	nr	**2320.13**
1500 x 1500mm	1358.50	2309.96	10.00	296.17	nr	**2606.14**
1800 x 1800mm	1938.32	3295.89	13.33	394.74	nr	**3690.63**
2000 x 2000mm	2175.13	3698.54	16.66	493.43	nr	**4191.97**
2300 x 2300mm	2691.12	4575.92	20.00	592.35	nr	**5168.27**
2500 x 2500mm	3778.71	6425.25	23.32	690.80	nr	**7116.04**
To suit circular ducts; unit length 600mm						
100mm dia.	67.17	114.21	0.75	22.21	nr	**136.42**
200mm dia.	79.03	134.38	0.75	22.21	nr	**156.60**
250mm dia.	89.59	152.33	0.75	22.21	nr	**174.54**
315mm dia.	104.28	177.31	1.00	29.62	nr	**206.93**
355mm dia.	120.66	205.17	1.00	29.62	nr	**234.79**
400mm dia.	129.70	220.54	1.00	29.62	nr	**250.16**
450mm dia.	156.63	266.33	1.00	29.62	nr	**295.95**
500mm dia.	172.45	293.23	1.26	37.26	nr	**330.49**
630mm dia.	203.33	345.74	1.50	44.43	nr	**390.17**
710mm dia.	221.41	376.48	1.50	44.43	nr	**420.91**
800mm dia.	246.27	418.76	2.00	59.23	nr	**477.99**
1000mm dia.	330.59	562.13	3.00	88.85	nr	**650.99**
To suit circular ducts; unit length 1200mm						
100mm dia.	98.81	168.01	1.00	29.62	nr	**197.63**
200mm dia.	108.41	184.34	1.00	29.62	nr	**213.96**
250mm dia.	121.79	207.09	1.00	29.62	nr	**236.71**
315mm dia.	139.31	236.87	1.33	39.39	nr	**276.26**
355mm dia.	166.99	283.95	1.33	39.39	nr	**323.34**
400mm dia.	177.16	301.24	1.33	39.39	nr	**340.63**
450mm dia.	205.22	348.95	1.33	39.39	nr	**388.34**
500mm dia.	210.30	357.60	1.66	49.16	nr	**406.76**
630mm dia.	278.48	473.52	2.00	59.23	nr	**532.75**
710mm dia.	307.86	523.47	2.00	59.23	nr	**582.71**
800mm dia.	307.29	522.51	2.66	78.78	nr	**601.30**
1000mm dia.	412.52	701.44	4.00	118.47	nr	**819.91**
1250mm dia.	777.09	1321.35	8.00	236.94	nr	**1558.29**
1400mm dia.	844.89	1436.63	8.00	236.94	nr	**1673.57**
1600mm dia.	936.67	1592.69	10.66	315.72	nr	**1908.41**
To suit circular ducts; unit length 1800mm						
200mm dia.	151.92	258.32	1.25	37.02	nr	**295.34**
250mm dia.	170.00	289.06	1.25	37.02	nr	**326.08**
315mm dia.	202.02	343.51	1.66	49.16	nr	**392.68**
355mm dia.	238.18	405.00	1.66	49.16	nr	**454.16**
400mm dia.	251.74	428.06	1.66	49.16	nr	**477.22**
450mm dia.	246.46	419.08	1.66	49.16	nr	**468.25**
500mm dia.	584.33	993.59	2.08	61.60	nr	**1055.19**

U:VENTILATION/AIR CONDITIONING SYSTEMS

Item	Net Price £	Material £	Labour hours	Labour £	Unit	Total rate £
Attenuators; DW144 galvanised construction c/w splitters (cont'd)						
To suit circular ducts; unit length 1800mm (cont'd)						
630mm dia.	306.16	520.59	2.50	74.04	nr	**594.64**
710mm dia.	358.14	608.98	2.50	74.04	nr	**683.02**
800mm dia.	370.57	630.11	3.33	98.63	nr	**728.74**
1000mm dia.	599.53	1019.43	5.00	148.09	nr	**1167.52**
1250mm dia.	983.31	1672.01	6.66	197.25	nr	**1869.26**
1400mm dia.	967.49	1645.11	6.66	197.25	nr	**1842.36**
1600mm dia.	1086.39	1847.28	10.00	296.17	nr	**2143.46**

U:VENTILATION/AIR CONDITIONING SYSTEMS

Item	Net Price £	Material £	Labour hours	Labour £	Unit	Total rate £
U10 : PLANT/EQUIPMENT (cont'd)						
Y46 - GRILLES/DIFFUSERS/LOUVRES						
Supply grilles; single deflection; extruded aluminium alloy frame and adjustable horizontal vanes; silver grey polyester powder coated; screw fixed						
Rectangular; for duct, ceiling and sidewall applications						
100 x 100mm	7.12	12.11	0.60	17.77	nr	**29.88**
150 x 150mm	9.04	15.37	0.60	17.77	nr	**33.14**
200 x 150mm	11.88	20.20	0.65	19.25	nr	**39.45**
200 x 200mm	10.86	18.46	0.72	21.32	nr	**39.79**
300 x 100mm	9.55	16.23	0.72	21.32	nr	**37.56**
300 x 150mm	10.81	18.37	0.80	23.69	nr	**42.07**
300 x 200mm	12.29	20.90	0.88	26.06	nr	**46.97**
300 x 300mm	15.25	25.92	1.04	30.80	nr	**56.73**
400 x 100mm	10.55	17.93	0.88	26.06	nr	**43.99**
400 x 150mm	12.08	20.55	0.94	27.84	nr	**48.39**
400 x 200mm	13.72	23.33	1.04	30.80	nr	**54.13**
400 x 300mm	17.00	28.91	1.12	33.17	nr	**62.08**
600 x 200mm	17.08	29.04	1.26	37.32	nr	**66.35**
600 x 300mm	21.60	36.73	1.40	41.46	nr	**78.19**
600 x 400mm	27.12	46.12	1.61	47.68	nr	**93.80**
600 x 500mm	31.15	52.96	1.76	52.13	nr	**105.09**
600 x 600mm	37.68	64.07	2.17	64.27	nr	**128.34**
800 x 300mm	26.62	45.27	1.76	52.13	nr	**97.40**
800 x 400mm	32.15	54.66	2.17	64.27	nr	**118.93**
800 x 600mm	44.20	75.16	3.00	88.85	nr	**164.01**
1000 x 300mm	29.64	50.40	2.60	77.01	nr	**127.40**
1000 x 400mm	36.67	62.35	3.00	88.85	nr	**151.21**
1000 x 600mm	51.24	87.13	3.80	112.55	nr	**199.67**
1000 x 800mm	53.59	91.13	3.80	112.55	nr	**203.67**
1200 x 600mm	59.27	100.78	4.61	136.54	nr	**237.32**
1200 x 800mm	60.33	102.58	4.61	136.54	nr	**239.12**
1200 x 1000mm	72.06	122.53	4.61	136.54	nr	**259.07**
Rectangular; for duct, ceiling and sidewall applications; including opposed blade damper volume regulator						
100 x 100mm	14.84	25.24	0.72	21.34	nr	**46.57**
150 x 150mm	17.94	30.50	0.72	21.34	nr	**51.84**
200 x 150mm	18.79	31.95	0.83	24.60	nr	**56.56**
200 x 200mm	21.48	36.52	0.90	26.66	nr	**63.18**
300 x 100mm	19.14	32.54	0.90	26.66	nr	**59.20**
300 x 150mm	21.76	36.99	0.98	29.03	nr	**66.02**
300 x 200mm	24.60	41.82	1.06	31.41	nr	**73.23**
300 x 300mm	30.72	52.24	1.20	35.56	nr	**87.79**
400 x 100mm	23.78	40.44	1.06	31.41	nr	**71.85**
400 x 150mm	26.54	45.13	1.13	33.46	nr	**78.59**
400 x 200mm	28.09	47.76	1.20	35.56	nr	**83.32**
400 x 300mm	34.54	58.73	1.34	39.70	nr	**98.43**
600 x 200mm	33.11	56.31	1.50	44.47	nr	**100.78**
600 x 300mm	40.20	68.35	1.66	49.20	nr	**117.55**
600 x 400mm	47.82	81.31	1.80	53.36	nr	**134.68**
600 x 500mm	57.20	97.26	2.00	59.23	nr	**156.50**
600 x 600mm	66.14	112.47	2.60	77.13	nr	**189.60**

U:VENTILATION/AIR CONDITIONING SYSTEMS

Item	Net Price £	Material £	Labour hours	Labour £	Unit	Total rate £
800 x 300mm	65.17	110.81	2.00	59.23	nr	**170.04**
800 x 400mm	77.18	131.23	2.60	77.13	nr	**208.36**
800 x 600mm	107.35	182.53	3.61	106.92	nr	**289.46**
1000 x 300mm	72.14	122.67	3.00	88.94	nr	**211.61**
1000 x 400mm	84.80	144.19	3.61	106.92	nr	**251.12**
1000 x 600mm	116.29	197.74	4.61	136.49	nr	**334.23**
1000 x 800mm	161.47	274.56	4.61	136.49	nr	**411.05**
1200 x 600mm	122.54	208.37	5.62	166.39	nr	**374.76**
1200 x 800mm	169.99	289.04	6.10	180.67	nr	**469.71**
1200 x 1000mm	203.58	346.16	6.50	192.51	nr	**538.68**
Supply grilles; double deflection; extruded aluminium alloy frame and adjustable horizontal and vertical vanes; white polyester powder coated; screw fixed						
Rectangular; for duct, ceiling and sidewall applications						
100 x 100mm	8.89	15.12	0.88	26.06	nr	**41.18**
150 x 150mm	14.06	23.91	0.88	26.06	nr	**49.97**
200 x 150mm	15.07	25.62	1.08	31.99	nr	**57.61**
200 x 200mm	16.08	27.34	1.25	37.02	nr	**64.36**
300 x 100mm	15.57	26.47	1.25	37.02	nr	**63.49**
300 x 150mm	49.84	84.74	1.50	44.43	nr	**129.17**
300 x 200mm	18.58	31.60	1.75	51.83	nr	**83.43**
300 x 300mm	21.32	36.25	2.15	63.68	nr	**99.93**
400 x 100mm	17.59	29.90	1.75	51.83	nr	**81.73**
400 x 150mm	19.08	32.45	1.95	57.75	nr	**90.20**
400 x 200mm	20.94	35.60	2.15	63.68	nr	**99.28**
400 x 300mm	25.27	42.97	2.55	75.52	nr	**118.50**
600 x 200mm	28.13	47.84	3.01	89.15	nr	**136.98**
600 x 300mm	33.19	56.43	3.36	99.51	nr	**155.94**
600 x 400mm	39.18	66.62	3.80	112.55	nr	**179.16**
600 x 500mm	51.24	87.13	4.20	124.39	nr	**211.52**
600 x 600mm	63.29	107.62	4.51	133.57	nr	**241.20**
800 x 300mm	46.22	78.59	4.20	124.39	nr	**202.98**
800 x 400mm	52.75	89.69	4.51	133.57	nr	**223.27**
800 x 600mm	84.90	144.35	5.10	151.05	nr	**295.40**
1000 x 300mm	53.25	90.54	4.80	142.16	nr	**232.71**
1000 x 400mm	64.29	109.32	5.10	151.05	nr	**260.37**
1000 x 600mm	103.98	176.80	5.72	169.41	nr	**346.22**
1000 x 800mm	91.51	155.60	5.72	169.41	nr	**325.01**
1200 x 600mm	122.06	207.56	6.33	187.48	nr	**395.03**
1200 x 800mm	106.36	180.85	6.33	187.48	nr	**368.33**
1200 x 1000mm	126.13	214.47	6.33	187.48	nr	**401.95**

U:VENTILATION/AIR CONDITIONING SYSTEMS

Item	Net Price £	Material £	Labour hours	Labour £	Unit	Total rate £
U10 : PLANT/EQUIPMENT (cont'd)						
Y46 - GRILLES/DIFFUSERS/LOUVRES (cont'd)						
Rectangular; for duct, ceiling and sidewall applications; including opposed blade damper volume regulator						
100 x 100mm	16.61	28.24	1.00	29.62	nr	**57.86**
150 x 150mm	20.52	34.89	1.00	29.62	nr	**64.51**
200 x 150mm	22.19	37.74	1.26	37.32	nr	**75.06**
200 x 200mm	25.03	42.57	1.43	42.35	nr	**84.92**
300 x 100mm	23.84	40.53	1.43	42.35	nr	**82.88**
300 x 150mm	26.79	45.55	1.68	49.76	nr	**95.31**
300 x 200mm	29.98	50.98	1.93	57.16	nr	**108.14**
300 x 300mm	36.78	62.55	2.31	68.42	nr	**130.96**
400 x 100mm	29.95	50.93	1.93	57.16	nr	**108.09**
400 x 150mm	33.23	56.50	2.14	63.38	nr	**119.88**
400 x 200mm	35.31	60.04	2.31	68.42	nr	**128.45**
400 x 300mm	42.82	72.80	2.77	82.04	nr	**154.85**
600 x 200mm	43.99	74.80	3.25	96.26	nr	**171.06**
600 x 300mm	52.86	89.89	3.62	107.22	nr	**197.10**
600 x 400mm	62.31	105.94	3.99	118.17	nr	**224.12**
600 x 500mm	73.47	124.92	4.44	131.50	nr	**256.42**
600 x 600mm	84.22	143.20	4.94	146.31	nr	**289.51**
800 x 300mm	82.25	139.86	4.44	131.50	nr	**271.36**
800 x 400mm	96.80	164.60	4.94	146.31	nr	**310.91**
800 x 600mm	132.06	224.55	5.71	169.12	nr	**393.67**
1000 x 300mm	93.63	159.21	5.20	154.01	nr	**313.22**
1000 x 400mm	109.57	186.32	5.71	169.12	nr	**355.43**
1000 x 600mm	147.64	251.04	6.53	193.40	nr	**444.44**
1000 x 800mm	199.38	339.02	6.53	193.40	nr	**532.42**
1200 x 600mm	160.52	272.95	7.34	217.39	nr	**490.34**
1200 x 800mm	216.02	367.31	8.80	260.63	nr	**627.95**
1200 x 1000mm	257.67	438.14	8.80	260.63	nr	**698.77**
Floor grille suitable for mounting in raised access floors; heavy duty; extruded aluminium; standard mill finish; complete with opposed blade volume control damper						
Diffuser						
600mm x 600mm	119.55	203.28	0.70	20.73	nr	**224.01**
Extra for nylon coated black finish	20.45	34.77	-	-	nr	**34.77**
Exhaust grilles; aluminium						
0° fixed blade core						
150 x 150mm	11.48	19.52	0.60	17.84	nr	**37.36**
200 x 200mm	13.51	22.97	0.72	21.34	nr	**44.31**
250 x 250mm	15.72	26.74	0.80	23.69	nr	**50.43**
300 x 300mm	18.14	30.84	1.00	29.62	nr	**60.46**
350 x 350mm	20.73	35.24	1.20	35.54	nr	**70.79**
0° fixed blade core; including opposed blade damper volume regulator						
150 x 150mm	20.51	34.87	0.62	18.37	nr	**53.24**
200 x 200mm	24.13	41.03	0.72	21.34	nr	**62.37**
250 x 250mm	28.06	47.71	0.80	23.69	nr	**71.41**
300 x 300mm	33.60	57.14	1.00	29.62	nr	**86.75**
350 x 350mm	38.18	64.92	1.20	35.54	nr	**100.46**

U:VENTILATION/AIR CONDITIONING SYSTEMS

Item	Net Price £	Material £	Labour hours	Labour £	Unit	Total rate £
45° fixed blade core						
150 x 150mm	11.48	19.52	0.62	18.37	nr	**37.89**
200 x 200mm	13.51	22.97	0.72	21.34	nr	**44.31**
250 x 250mm	15.72	26.74	0.80	23.69	nr	**50.43**
300 x 300mm	18.14	30.84	1.00	29.62	nr	**60.46**
350 x 350mm	20.73	35.24	1.20	35.54	nr	**70.79**
45° fixed blade core; including opposed blade damper volume regulator						
150 x 150mm	20.51	34.87	0.62	18.37	nr	**53.24**
200 x 200mm	24.13	41.03	0.72	21.34	nr	**62.37**
250 x 250mm	28.06	47.71	0.80	23.69	nr	**71.41**
300 x 300mm	33.60	57.14	1.00	29.62	nr	**86.75**
350 x 350mm	38.18	64.92	1.20	35.54	nr	**100.46**
Eggcrate core						
150 x 150mm	8.54	14.52	0.62	18.37	nr	**32.89**
200 x 200mm	9.55	16.23	1.00	29.62	nr	**45.85**
250 x 250mm	12.05	20.50	0.80	23.69	nr	**44.19**
300 x 300mm	15.07	25.62	1.00	29.62	nr	**55.24**
350 x 350mm	17.08	29.04	1.20	35.54	nr	**64.58**
Eggcrate core; including opposed blade damper volume regulator						
150 x 150mm	16.73	28.45	0.62	18.37	nr	**46.83**
200 x 200mm	20.12	34.22	0.72	21.34	nr	**55.56**
250 x 250mm	23.93	40.69	0.80	23.69	nr	**64.38**
300 x 300mm	30.22	51.39	1.00	29.62	nr	**81.01**
350 x 350mm	33.96	57.74	1.20	35.54	nr	**93.28**
Mesh/perforated plate core						
150 x 150mm	12.56	21.36	0.62	18.36	nr	**39.73**
200 x 200mm	13.06	22.21	0.72	21.32	nr	**43.54**
250 x 250mm	15.57	26.47	0.80	23.69	nr	**50.17**
300 x 300mm	18.58	31.60	1.00	29.62	nr	**61.22**
350 x 350mm	20.59	35.01	1.20	35.54	nr	**70.56**
Mesh/perforated plate core; including opposed blade damper volume regulator						
150 x 150mm	16.73	28.45	0.62	18.36	nr	**46.82**
200 x 200mm	20.12	34.22	0.72	21.32	nr	**55.54**
250 x 250mm	23.93	40.69	0.80	23.69	nr	**64.38**
300 x 300mm	30.22	51.39	0.80	23.69	nr	**75.08**
350 x 350mm	33.96	57.74	1.20	35.54	nr	**93.28**
Plastic air diffusion system						
Eggcrate grilles						
150 x 150mm	6.01	7.70	0.62	15.74	nr	**23.44**
200 x 200mm	8.07	10.34	0.72	18.28	nr	**28.62**
250 x 250mm	9.59	12.28	0.80	20.30	nr	**32.58**
300 x 300mm	11.58	14.83	1.00	25.37	nr	**40.20**
Single deflection grilles						
150 x 150mm	5.78	7.41	0.62	15.74	nr	**23.14**
200 x 200mm	7.45	9.54	0.72	18.28	nr	**27.82**
250 x 250mm	8.08	10.35	0.80	20.30	nr	**30.65**
300 x 300mm	9.48	12.15	1.00	25.37	nr	**37.52**

U:VENTILATION/AIR CONDITIONING SYSTEMS

Item	Net Price £	Material £	Labour hours	Labour £	Unit	Total rate £
U10 : PLANT/EQUIPMENT (cont'd)						
Y46 - GRILLES/DIFFUSERS/LOUVRES (cont'd)						
Plastic air diffusion system (cont'd)						
Double deflection grilles						
150 x 150mm	7.60	9.74	0.62	15.74	nr	**25.48**
200 x 200mm	9.16	11.74	0.72	18.28	nr	**30.02**
250 x 250mm	11.26	14.43	0.80	20.30	nr	**34.72**
300 x 300mm	15.76	20.18	1.00	25.37	nr	**45.55**
Door transfer grilles						
150 x 150mm	11.89	15.23	0.62	15.74	nr	**30.96**
200 x 200mm	16.29	20.86	0.72	18.28	nr	**39.14**
250 x 250mm	16.22	20.78	0.80	20.30	nr	**41.08**
300 x 300mm	19.55	25.04	1.00	25.37	nr	**50.41**
Opposed blade dampers						
150 x 150mm	4.23	5.42	0.62	15.74	nr	**21.16**
200 x 200mm	5.63	7.21	0.72	18.28	nr	**25.49**
250 x 250mm	6.86	8.79	0.80	20.30	nr	**29.09**
300 x 300mm	7.45	9.54	1.00	25.37	nr	**34.91**
Neck reducers						
150 x 150mm	5.63	7.21	0.62	15.74	nr	**22.94**
200 x 200mm	6.16	7.89	0.72	18.28	nr	**26.17**
250 x 250mm	7.45	9.54	0.80	20.30	nr	**29.83**
300 x 300mm	8.25	10.56	1.00	25.37	nr	**35.93**
Ceiling mounted diffusers; circular aluminium multi-core diffuser						
Circular; for ceiling mounting						
152mm dia. neck	39.37	66.95	0.80	23.69	nr	**90.65**
203mm dia. neck	57.64	98.00	1.10	32.58	nr	**130.59**
305mm dia. neck	78.07	132.75	1.40	41.48	nr	**174.24**
381mm dia. neck	66.79	113.57	1.50	44.47	nr	**158.04**
457mm dia. neck	105.30	179.05	2.00	59.23	nr	**238.28**
Circular; for ceiling mounting; including louvre damper volume control						
152mm dia. neck	54.92	93.39	1.00	29.62	nr	**123.01**
203mm dia. neck	73.82	125.52	1.20	35.56	nr	**161.08**
305mm dia. neck	98.02	166.67	1.60	47.39	nr	**214.06**
381mm dia. neck	86.74	147.48	1.90	56.31	nr	**203.79**
457mm dia. neck	127.23	216.35	2.40	71.19	nr	**287.54**
Ceiling mounted diffusers; rectangular aluminium multi-cone diffuser; four way flow						
Rectangular; for ceiling mounting						
150 x 150 mm neck	19.50	33.16	1.80	53.36	nr	**86.52**
300 x 150 mm neck	27.90	47.45	2.30	68.12	nr	**115.57**
300 x 300 mm neck	28.20	47.96	2.80	82.93	nr	**130.89**
450 x 150 mm neck	35.55	60.44	2.80	82.93	nr	**143.37**
450 x 300 mm neck	34.80	59.17	3.20	94.78	nr	**153.95**
450 x 450 mm neck	36.17	61.50	3.40	100.74	nr	**162.25**
600 x 150 mm neck	43.54	74.04	3.20	94.78	nr	**168.82**
600 x 300 mm neck	43.32	73.65	3.50	103.66	nr	**177.31**
600 x 600 mm neck	53.75	91.39	4.00	118.47	nr	**209.86**

U:VENTILATION/AIR CONDITIONING SYSTEMS

Item	Net Price £	Material £	Labour hours	Labour £	Unit	Total rate £
Rectangular; for ceiling mounting; including opposed blade damper volume regulator						
150 x 150 mm neck	27.07	46.03	1.80	53.36	nr	**99.40**
300 x 150 mm neck	38.83	66.03	2.30	68.12	nr	**134.15**
300 x 300 mm neck	41.24	70.12	2.80	82.96	nr	**153.08**
450 x 150 mm neck	50.17	85.31	2.80	82.93	nr	**168.24**
450 x 300 mm neck	52.50	89.27	3.30	97.74	nr	**187.01**
450 x 450 mm neck	57.58	97.92	3.51	103.92	nr	**201.84**
600 x 150 mm neck	60.53	102.92	3.30	97.74	nr	**200.66**
600 x 300 mm neck	63.00	107.13	4.00	118.47	nr	**225.60**
600 x 600 mm neck	79.16	134.61	5.62	166.39	nr	**301.00**
Slot diffusers; continuous aluminium slot diffuser with flanged frame						
Diffuser						
1 slot	18.46	31.39	3.76	111.34	m	**142.73**
2 slot	22.48	38.23	3.76	111.31	m	**149.55**
3 slot	28.13	47.84	3.76	111.31	m	**159.15**
4 slot	34.82	59.21	4.50	133.41	m	**192.62**
6 slot	49.57	84.28	4.50	133.41	m	**217.69**
8 slot	50.26	85.47	5.20	154.01	m	**239.48**
Diffuser; including equalizing deflector						
1 slot	18.46	31.39	5.26	155.88	m	**187.27**
2 slot	22.48	38.23	5.26	155.88	m	**194.11**
3 slot	27.66	47.04	5.26	155.88	m	**202.92**
4 slot	31.13	52.93	6.33	187.45	m	**240.38**
6 slot	41.02	69.75	6.33	187.45	m	**257.20**
8 slot	50.26	85.47	7.20	213.25	m	**298.71**
Extra over for ends						
1 slot	95.22	161.91	1.00	29.62	nr	**191.53**
2 slot	9.17	15.60	1.00	29.62	nr	**45.21**
3 slot	10.48	17.83	1.00	29.62	nr	**47.44**
4 slot	11.79	20.05	1.30	38.50	nr	**58.56**
6 slot	14.41	24.51	1.40	41.46	nr	**65.97**
8 slot	4.51	7.67	1.40	41.46	nr	**49.14**
Plenum boxes; 1.0m long; circular spigot; including cord operated flap damper						
1 slot	36.57	62.18	2.75	81.59	nr	**143.77**
2 slot	37.73	64.16	2.75	81.59	nr	**145.75**
3 slot	38.25	65.04	2.75	81.59	nr	**146.63**
4 slot	42.52	72.29	3.51	103.92	nr	**176.21**
6 slot	45.67	77.65	3.51	103.92	nr	**181.57**
8 slot	62.20	105.77	4.20	124.39	nr	**230.16**
Plenum boxes; 2.0m long; circular spigot; including cord operated flap damper						
1 slot	43.24	135.71	3.26	96.47	nr	**232.18**
2 slot	43.24	73.53	3.26	96.47	nr	**170.00**
3 slot	45.21	76.87	3.26	96.47	nr	**173.34**
4 slot	49.24	83.73	3.76	111.34	nr	**195.08**
6 slot	53.59	91.13	3.76	111.34	nr	**202.47**
8 slot	70.52	119.91	4.10	121.43	nr	**241.35**

U:VENTILATION/AIR CONDITIONING SYSTEMS

Item	Net Price £	Material £	Labour hours	Labour £	Unit	Total rate £
U10 : PLANT/EQUIPMENT (cont'd)						
Y46 - GRILLES/DIFFUSERS/LOUVRES (cont'd)						
Perforated diffusers; rectangular face aluminium perforated diffuser; quick release face plate; for integration with rectangular ceiling tiles						
Circular spigot; rectangular diffuser						
150mm dia. spigot; 300 x 300 diffuser	47.22	80.29	1.00	29.62	nr	**109.90**
300mm dia. spigot; 600 x 600 diffuser	86.90	147.77	1.40	41.48	nr	**189.25**
Circular spigot; rectangular diffuser; including louvre damper volume regulator						
150mm dia. spigot; 300 x 300 diffuser	63.06	107.22	1.00	29.62	nr	**136.84**
300mm dia. spigot; 600 x 600 diffuser	100.96	171.68	1.60	47.39	nr	**219.06**
Rectangular spigot; rectangular diffuser						
150 x 150mm dia. spigot; 300 x 300mm diffuser	30.47	51.81	1.00	25.37	nr	**77.19**
300 x 150mm dia. spigot; 600 x 300mm diffuser	64.56	109.78	1.20	30.45	nr	**140.23**
300 x 300mm dia. spigot; 600 x 600mm diffuser	64.56	109.78	1.40	35.52	nr	**145.30**
600 x 300mm dia. spigot; 1200 x 600mm diffuser	129.13	219.56	1.60	40.60	nr	**260.16**
Rectangular spigot; rectangular diffuser; including opposed blade damper volume regulator						
150 x 150mm dia. spigot; 300 x 300mm diffuser	42.94	73.02	1.20	30.45	nr	**103.46**
300 x 150mm dia. spigot; 600 x 300mm diffuser	85.74	145.79	1.40	35.54	nr	**181.32**
300 x 300mm dia. spigot; 600 x 600mm diffuser	85.74	145.79	1.60	40.60	nr	**186.38**
600 x 300mm dia. spigot; 1200 x 600mm diffuser	175.68	298.72	1.80	45.67	nr	**344.39**
Floor swirl diffuser; manual adjustment of air discharge direction; complete with damper and dirt trap						
Plastic Diffuser						
150 dia	20.43	34.73	0.50	14.81	nr	**49.54**
200 dia	27.03	45.96	0.50	14.81	nr	**60.77**
Aluminium Diffuser						
150 dia	25.23	42.90	0.50	14.81	nr	**57.71**
200 dia	31.83	54.13	0.50	14.81	nr	**68.94**
Plastic air diffusion system						
Cellular diffusers						
300 x 300mm	16.79	21.50	2.80	71.07	nr	**92.57**
600 x 600mm	40.78	52.23	4.00	101.49	nr	**153.71**
Multi-cone diffusers						
300 x 300mm	18.25	23.38	2.80	71.07	nr	**94.45**
450 x 450mm	27.96	35.80	3.40	86.30	nr	**122.10**
500 x 500mm	28.56	36.58	3.80	96.47	nr	**133.05**
600 x 600mm	35.38	45.32	4.00	101.49	nr	**146.80**
625 x 625mm	39.58	50.70	4.26	107.97	nr	**158.66**

U:VENTILATION/AIR CONDITIONING SYSTEMS

Item	Net Price £	Material £	Labour hours	Labour £	Unit	Total rate £
Opposed blade dampers						
300 x 300mm	5.79	7.42	1.20	30.46	nr	**37.88**
450 x 450mm	9.40	12.04	1.50	38.10	nr	**50.14**
600 x 600mm	16.49	21.13	2.60	66.07	nr	**87.20**
Plenum boxes						
300mm	9.08	11.63	2.80	71.07	nr	**82.70**
450mm	11.14	14.27	3.40	86.30	nr	**100.57**
600mm	17.31	22.16	4.00	101.49	nr	**123.65**
Plenum spigot reducer						
600mm	5.83	7.47	1.00	25.37	nr	**32.84**
Blanking kits for cellular diffusers						
300mm	4.97	6.37	0.88	22.33	nr	**28.70**
600mm	6.82	8.74	1.10	27.91	nr	**36.65**
Blanking kits for multi-cone diffusers						
300mm	4.97	6.37	0.88	22.33	nr	**28.70**
450mm	5.93	7.59	0.90	22.84	nr	**30.43**
600mm	6.92	8.86	1.10	27.91	nr	**36.77**
Acoustic louvres; opening mounted; 300mm deep steel louvres with blades packed with acoustic infill; 12mm galvanised mesh bird-screen; screw fixing in opening						
Louvre units; self finished galvanised steel						
900 high x 600 wide	137.85	176.56	3.00	76.12	nr	**252.68**
900 high x 900 wide	171.03	219.05	3.00	76.12	nr	**295.17**
900 high x 1200 wide	202.49	259.34	3.34	84.74	nr	**344.09**
900 high x 1500 wide	264.26	338.47	3.34	84.74	nr	**423.21**
900 high x 1800 wide	296.30	379.49	3.34	84.74	nr	**464.24**
900 high x 2100 wide	328.33	420.52	3.34	84.74	nr	**505.26**
900 high x 2400 wide	359.79	460.81	3.68	93.37	nr	**554.18**
900 high x 2700 wide	406.69	520.89	3.68	93.37	nr	**614.26**
900 high x 3000 wide	435.29	557.52	3.68	93.37	nr	**650.89**
1200 high x 600 wide	180.75	231.51	3.00	76.12	nr	**307.62**
1200 high x 900 wide	222.51	284.99	3.34	84.74	nr	**369.73**
1200 high x 1200 wide	263.69	337.73	3.34	84.74	nr	**422.48**
1200 high x 1500 wide	349.49	447.63	3.34	84.74	nr	**532.37**
1200 high x 1800 wide	390.68	500.37	3.68	93.37	nr	**593.74**
1200 high x 2100 wide	433.00	554.59	3.68	93.37	nr	**647.96**
1200 high x 2400 wide	474.19	607.34	3.68	93.37	nr	**700.70**
1500 high x 600 wide	224.22	287.18	3.00	76.12	nr	**363.30**
1500 high x 900 wide	275.13	352.39	3.34	84.74	nr	**437.13**
1500 high x 1200 wide	326.04	417.59	3.34	84.74	nr	**502.33**
1500 high x 1500 wide	434.72	556.79	3.68	93.37	nr	**650.15**
1500 high x 1800 wide	485.63	621.99	3.68	93.37	nr	**715.36**
1500 high x 2100 wide	537.11	687.92	4.00	101.49	nr	**789.41**
1800 high x 600 wide	266.55	341.40	3.34	84.74	nr	**426.14**
1800 high x 900 wide	327.18	419.05	3.34	84.74	nr	**503.80**
1800 high x 1200 wide	388.39	497.44	3.68	93.37	nr	**590.81**
1800 high x 1500 wide	520.52	666.68	3.68	93.37	nr	**760.05**

U:VENTILATION/AIR CONDITIONING SYSTEMS

Item	Net Price £	Material £	Labour hours	Labour £	Unit	Total rate £
U10 : PLANT/EQUIPMENT (cont'd)						
Y46 - GRILLES/DIFFUSERS/LOUVRES (cont'd)						
Acoustic louvres; opening mounted (cont'd)						
Louvre units; polyester powder coated steel						
900 high x 600 wide	200.20	256.41	3.00	76.12	nr	**332.53**
900 high x 900 wide	263.69	337.73	3.00	76.12	nr	**413.85**
900 high x 1200 wide	326.04	417.59	3.34	84.74	nr	**502.33**
900 high x 1500 wide	418.70	536.27	3.34	84.74	nr	**621.01**
900 high x 1800 wide	481.62	616.86	3.34	84.74	nr	**701.60**
900 high x 2100 wide	545.12	698.18	3.34	84.74	nr	**782.92**
900 high x 2400 wide	607.46	778.03	3.68	93.37	nr	**871.40**
900 high x 2700 wide	684.68	876.94	3.68	93.37	nr	**970.31**
900 high x 3000 wide	744.17	953.13	3.68	93.37	nr	**1046.50**
1200 high x 600 wide	263.12	337.00	3.00	76.12	nr	**413.12**
1200 high x 900 wide	346.06	443.23	3.34	84.74	nr	**527.97**
1200 high x 1200 wide	429.00	549.46	3.34	84.74	nr	**634.20**
1200 high x 1500 wide	555.41	711.37	3.34	84.74	nr	**796.11**
1200 high x 1800 wide	637.78	816.86	3.68	93.37	nr	**910.23**
1200 high x 2100 wide	721.29	923.82	3.68	93.37	nr	**1017.19**
1200 high x 2400 wide	803.66	1029.32	3.68	93.37	nr	**1122.69**
1500 high x 600 wide	327.18	419.05	3.00	76.12	nr	**495.17**
1500 high x 900 wide	429.57	550.19	3.34	84.74	nr	**634.93**
1500 high x 1200 wide	531.96	681.33	3.34	84.74	nr	**766.07**
1500 high x 1500 wide	692.12	886.46	3.68	93.37	nr	**979.83**
1500 high x 1800 wide	795.08	1018.33	3.68	93.37	nr	**1111.70**
1500 high x 2100 wide	897.47	1149.47	4.00	101.49	nr	**1250.96**
1800 high x 600 wide	390.10	499.64	3.34	84.74	nr	**584.38**
1800 high x 900 wide	512.51	656.42	3.34	84.74	nr	**741.16**
1800 high x 1200 wide	635.49	813.93	3.68	93.37	nr	**907.30**
1800 high x 1500 wide	829.40	1062.29	3.68	93.37	nr	**1155.66**
Weather louvres; opening mounted; 300mm deep galvanised steel louvres; screw fixing in position						
Louvre units; including 12mm galvanised mesh birdscreen						
900 x 600mm	113.80	145.75	2.25	57.15	nr	**202.89**
900 x 900mm	157.71	201.99	2.25	57.15	nr	**259.13**
900 x 1200mm	191.21	244.91	2.50	63.43	nr	**308.34**
900 x 1500mm	217.43	278.49	2.50	63.43	nr	**341.92**
900 x 1800mm	259.28	332.09	2.50	63.43	nr	**395.52**
900 x 2100mm	335.05	429.12	2.50	63.43	nr	**492.55**
900 x 2400mm	352.32	451.25	2.76	70.09	nr	**521.34**
900 x 2700mm	405.26	519.05	2.76	70.09	nr	**589.14**
900 x 3000mm	468.57	600.14	2.76	70.09	nr	**670.23**
1200 x 600mm	157.71	201.99	2.25	57.15	nr	**259.13**
1200 x 900mm	213.74	273.76	2.50	63.43	nr	**337.19**
1200 x 1200mm	270.93	347.00	2.50	63.43	nr	**410.43**
1200 x 1500mm	295.77	378.81	2.50	63.43	nr	**442.24**
1200 x 1800mm	351.80	450.58	2.76	70.09	nr	**520.67**
1200 x 2100mm	483.51	619.27	2.76	70.09	nr	**689.36**
1200 x 2400mm	483.51	619.27	2.76	70.09	nr	**689.36**
1500 x 600mm	184.28	236.02	2.25	57.15	nr	**293.17**
1500 x 900mm	241.47	309.27	2.50	63.43	nr	**372.70**
1500 x 1200mm	321.76	412.10	2.50	63.43	nr	**475.53**

U:VENTILATION/AIR CONDITIONING SYSTEMS

Item	Net Price £	Material £	Labour hours	Labour £	Unit	Total rate £
1500 x 1500mm	354.69	454.29	2.76	70.09	nr	**524.37**
1500 x 1800mm	422.28	540.85	2.76	70.09	nr	**610.94**
1500 x 2100mm	570.73	730.99	3.00	76.19	nr	**807.18**
1800 x 600mm	198.72	254.52	2.50	63.43	nr	**317.95**
1800 x 900mm	261.68	335.16	2.50	63.43	nr	**398.59**
1800 x 1200mm	344.28	440.95	2.76	70.09	nr	**511.04**
1800 x 1500mm	405.32	519.13	3.00	76.19	nr	**595.32**

U:VENTILATION/AIR CONDITIONING SYSTEMS

Item	Net Price £	Material £	Labour hours	Labour £	Unit	Total rate £
U10 : PLANT/EQUIPMENT (cont'd)						
Y50 - THERMAL INSULATION						
Concealed Ductwork						
Flexible wrap; 20kg-45kg Bright Class O aluminium foil faced; Bright Class O foil taped joints; 62mm metal pins and washers; aluminium bands						
40mm thick insulation	6.33	8.27	0.40	8.21	m²	**16.48**
Semi-rigid slab; 45kg Bright Class O aluminium foil faced mineral fibre; Bright Class O foil taped joints; 62mm metal pins and washers; aluminium bands						
40mm thick insulation	8.28	10.83	0.65	13.34	m²	**24.16**
Plantroom Ductwork						
Semi-rigid slab; 45kg Bright Class O aluminium foil faced mineral fibre; Bright Class O foil taped joints; 62mm metal pins and washers; 22 swg plain/embossed aluminium cladding; pop riveted						
50mm thick insulation	20.89	27.32	1.50	30.78	m²	**58.09**
External Ductwork						
Semi-rigid slab; 45kg Bright Class O aluminium foil faced mineral fibre; Bright Class O foil taped joints; 62mm metal pins and washers; 0.8mm polyisobutylene sheeting; welded joints						
50mm thick insulation	15.24	19.92	1.25	25.65	m²	**45.57**

U:VENTILATION/AIR CONDITIONING SYSTEMS

Item	Net Price £	Material £	Labour hours	Labour £	Unit	Total rate £
U14 : DUCTWORK : FIRE RATED						
Y30 - DUCTLINES						
The relevant BS requires that the fire rating of ductwork meets 3 criteria; stability (hours), integrity (hours) and insulation (hours). The least of the 3 periods defines the fire rating. The BS does however allow stability and integrity to be considered in isolation. Rates are therefore provided for both types of system.						
Care should to be taken when using the rates within this section to ensure that the requirements for stability, integrity and insulation are known and the appropriate rates are used						
High density single layer mineral wool fire rated ductwork slab, in accordance with BS476, Part 24 (ISO 6944: 1985), ducts "Type A" and "Type B"; 165kg class O foil faced mineral fibre; 100mm wide bright class O foil taped joints; welded pins; includes protection to all supports.						
1/2 hour stability, integrity and insulation						
25mm thick, vertical and horizontal ductwork	24.55	32.10	1.25	25.65	m²	**57.75**
1 hour stability, integrity and insulation						
30mm thick, vertical ductwork	28.84	37.72	1.50	30.78	m²	**68.49**
40mm thick, horizontal ductwork	33.91	44.35	1.50	30.78	m²	**75.13**
1½ hour stability, integrity and insulation						
50mm thick, vertical ductwork	40.51	52.97	1.75	35.91	m²	**88.88**
70mm thick, horizontal ductwork	50.67	66.25	1.75	35.91	m²	**102.16**
2 hour stability, integrity and insulation						
70mm, vertical ductwork	52.21	68.28	2.00	41.03	m²	**109.31**
90mm horizontal ductwork	62.36	81.54	2.00	41.03	m²	**122.58**
Kitchen extract, 1 hour stability, integrity and insulation						
90mm, vertical and horizontal	62.36	81.54	2.00	41.03	m²	**122.58**

U:VENTILATION/AIR CONDITIONING SYSTEMS

Item	Net Price £	Material £	Labour hours	Labour £	Unit	Total rate £
U14 : DUCTWORK : FIRE RATED (cont'd)						
Y30 - DUCTLINES (cont'd)						
Galvanised sheet metal rectangular section ductwork to BS476 Part 24 (ISO 6944:1985), ducts "Type A" and "Type B"; provides 2 hours stability and 2 hours integrity at 1100°C (no rating for insulation); including all necessary stiffeners, joints and supports in the running length						
Ductwork up to 600mm longest side						
Sum of two sides 200mm	42.13	71.63	2.91	86.19	m	**157.82**
Sum of two sides 300mm	45.74	77.77	2.99	88.56	m	**166.33**
Sum of two sides 400mm	49.34	83.89	3.17	93.89	m	**177.78**
Sum of two sides 500mm	52.95	90.03	3.37	99.81	m	**189.84**
Sum of two sides 600mm	56.55	96.15	3.54	104.85	m	**201.00**
Sum of two sides 700mm	60.16	102.29	3.72	110.18	m	**212.47**
Sum of two sides 800mm	63.77	108.43	3.90	115.51	m	**223.94**
Sum of two sides 900mm	67.37	114.55	5.04	149.27	m	**263.82**
Sum of two sides 1000mm	70.98	120.69	5.58	165.27	m	**285.96**
Sum of two sides 1100mm	74.58	126.81	5.84	172.97	m	**299.78**
Sum of two sides 1200mm	78.19	132.95	6.11	180.96	m	**313.91**
Extra over fittings; Ductwork up to 600mm longest side						
End Cap						
Sum of two sides 200mm	11.96	20.33	0.81	23.99	nr	**44.32**
Sum of two sides 300mm	12.69	21.57	0.84	24.88	nr	**46.45**
Sum of two sides 400mm	13.42	22.82	0.87	25.77	nr	**48.59**
Sum of two sides 500mm	14.14	24.04	0.90	26.66	nr	**50.70**
Sum of two sides 600mm	14.87	25.29	0.93	27.54	nr	**52.83**
Sum of two sides 700mm	15.60	26.53	0.96	28.43	nr	**54.97**
Sum of two sides 800mm	16.34	27.78	0.98	29.03	nr	**56.80**
Sum of two sides 900mm	17.07	29.02	1.17	34.65	nr	**63.67**
Sum of two sides 1000mm	17.79	30.25	1.22	36.13	nr	**66.38**
Sum of two sides 1100mm	18.52	31.49	1.25	37.02	nr	**68.51**
Sum of two sides 1200mm	19.25	32.74	1.28	37.91	nr	**70.65**
Reducer						
Sum of two sides 200mm	47.45	80.68	2.23	66.05	nr	**146.72**
Sum of two sides 300mm	49.63	84.39	2.37	70.19	nr	**154.58**
Sum of two sides 400mm	51.80	88.08	2.51	74.34	nr	**162.42**
Sum of two sides 500mm	53.99	91.80	2.65	78.49	nr	**170.28**
Sum of two sides 600mm	56.16	95.49	2.79	82.63	nr	**178.12**
Sum of two sides 700mm	58.34	99.20	2.87	85.00	nr	**184.20**
Sum of two sides 800mm	60.52	102.91	3.01	89.15	nr	**192.06**
Sum of two sides 900mm	62.70	106.61	3.06	90.63	nr	**197.24**
Sum of two sides 1000mm	64.88	110.32	3.17	93.89	nr	**204.21**
Sum of two sides 1100mm	67.05	114.01	3.22	95.37	nr	**209.38**
Sum of two sides 1200mm	69.24	117.73	3.28	97.15	nr	**214.87**

U:VENTILATION/AIR CONDITIONING SYSTEMS

Item	Net Price £	Material £	Labour hours	Labour £	Unit	Total rate £
Offset						
Sum of two sides 200mm	114.51	194.71	2.95	87.37	nr	**282.08**
Sum of two sides 300mm	117.35	199.54	3.11	92.11	nr	**291.65**
Sum of two sides 400mm	120.19	204.38	3.26	96.55	nr	**300.93**
Sum of two sides 500mm	123.04	209.21	3.42	101.29	nr	**310.50**
Sum of two sides 600mm	125.88	214.04	3.57	105.73	nr	**319.78**
Sum of two sides 700mm	128.72	218.88	3.73	110.47	nr	**329.35**
Sum of two sides 800mm	131.56	223.71	3.23	95.66	nr	**319.37**
Sum of two sides 900mm	134.41	228.54	3.45	102.18	nr	**330.72**
Sum of two sides 1000mm	137.25	233.38	3.67	108.70	nr	**342.07**
Sum of two sides 1100mm	140.09	238.21	3.78	111.95	nr	**350.16**
Sum of two sides 1200mm	142.94	243.04	3.89	115.21	nr	**358.26**
90° radius bend						
Sum of two sides 200mm	46.43	78.95	2.06	61.01	nr	**139.97**
Sum of two sides 300mm	49.95	84.93	2.21	65.45	nr	**150.38**
Sum of two sides 400mm	53.46	90.90	2.36	69.90	nr	**160.80**
Sum of two sides 500mm	56.99	96.90	2.51	74.34	nr	**171.24**
Sum of two sides 600mm	60.50	102.87	2.66	78.78	nr	**181.65**
Sum of two sides 700mm	64.01	108.85	2.81	83.22	nr	**192.07**
Sum of two sides 800mm	67.53	114.82	2.97	87.96	nr	**202.79**
Sum of two sides 900mm	71.04	120.80	2.99	88.56	nr	**209.35**
Sum of two sides 1000mm	74.57	126.79	3.04	90.04	nr	**216.83**
Sum of two sides 1100mm	78.08	132.77	3.07	90.93	nr	**223.69**
Sum of two sides 1200mm	81.59	138.74	3.09	91.52	nr	**230.26**
45° radius bend						
Sum of two sides 200mm	57.25	97.35	1.52	45.02	nr	**142.37**
Sum of two sides 300mm	58.68	99.78	1.59	47.09	nr	**146.87**
Sum of two sides 400mm	60.10	102.19	1.66	49.16	nr	**151.35**
Sum of two sides 500mm	61.52	104.62	1.73	51.24	nr	**155.85**
Sum of two sides 600mm	62.94	107.02	1.80	53.31	nr	**160.33**
Sum of two sides 700mm	64.37	109.45	1.87	55.38	nr	**164.83**
Sum of two sides 800mm	65.78	111.86	1.93	57.16	nr	**169.02**
Sum of two sides 900mm	67.21	114.28	1.99	58.94	nr	**173.22**
Sum of two sides 1000mm	68.63	116.69	2.05	60.72	nr	**177.40**
Sum of two sides 1100mm	70.05	119.12	2.11	62.49	nr	**181.61**
Sum of two sides 1200mm	71.47	121.52	2.17	64.27	nr	**185.79**
90° mitre bend						
Sum of two sides 200mm	48.13	81.84	2.47	73.16	nr	**154.99**
Sum of two sides 300mm	55.79	94.87	2.65	78.49	nr	**173.35**
Sum of two sides 400mm	63.44	107.87	2.84	84.11	nr	**191.99**
Sum of two sides 500mm	71.10	120.90	3.02	89.44	nr	**210.34**
Sum of two sides 600mm	78.75	133.91	3.20	94.78	nr	**228.68**
Sum of two sides 700mm	86.41	146.93	3.38	100.11	nr	**247.04**
Sum of two sides 800mm	94.07	159.96	3.56	105.44	nr	**265.40**
Sum of two sides 900mm	101.72	172.97	3.59	106.33	nr	**279.30**
Sum of two sides 1000mm	109.39	186.00	3.66	108.40	nr	**294.40**
Sum of two sides 1100mm	117.03	199.00	3.69	109.29	nr	**308.29**
Sum of two sides 1200mm	124.70	212.03	3.72	110.18	nr	**322.21**

U:VENTILATION/AIR CONDITIONING SYSTEMS

Item	Net Price £	Material £	Labour hours	Labour £	Unit	Total rate £
U14 : DUCTWORK : FIRE RATED (cont'd)						
Y30 - DUCTLINES (cont'd)						
Galvanised sheet metal rectangular section ductwork to BS476 Part 24 - 2 hour (cont'd)						
Extra over fittings; Ductwork up to 600mm longest side (cont'd)						
Branch (Side-on Shoe)						
Sum of two sides 200mm	17.87	30.39	0.98	29.03	nr	**59.42**
Sum of two sides 300mm	17.97	30.56	1.06	31.39	nr	**61.95**
Sum of two sides 400mm	18.07	30.72	1.14	33.76	nr	**64.49**
Sum of two sides 500mm	18.17	30.89	1.22	36.13	nr	**67.02**
Sum of two sides 600mm	18.26	31.05	1.30	38.50	nr	**69.56**
Sum of two sides 700mm	18.36	31.22	1.38	40.87	nr	**72.09**
Sum of two sides 800mm	18.45	31.37	1.46	43.24	nr	**74.61**
Sum of two sides 900mm	18.59	31.61	1.37	40.58	nr	**72.19**
Sum of two sides 1000mm	18.64	31.70	1.46	43.24	nr	**74.94**
Sum of two sides 1100mm	18.74	31.86	1.50	44.43	nr	**76.29**
Sum of two sides 1200mm	18.84	32.03	1.55	45.91	nr	**77.94**
Ductwork 601 to 800mm longest side						
Sum of two sides 900mm	67.36	114.53	5.04	149.27	m	**263.80**
Sum of two sides 1000mm	70.97	120.67	5.58	165.27	m	**285.94**
Sum of two sides 1100mm	74.58	126.81	5.84	172.97	m	**299.78**
Sum of two sides 1200mm	78.19	132.95	6.11	180.96	m	**313.91**
Sum of two sides 1300mm	81.81	139.11	6.38	188.96	m	**328.07**
Sum of two sides 1400mm	85.42	145.25	6.65	196.96	m	**342.21**
Sum of two sides 1500mm	89.04	151.39	6.91	204.66	m	**356.05**
Sum of two sides 1600mm	92.65	157.53	7.18	212.65	m	**370.19**
Extra over fittings; Ductwork 601 to 800mm longest side						
End Cap						
Sum of two sides 900mm	15.86	26.97	1.17	34.65	nr	**61.62**
Sum of two sides 1000mm	16.99	28.90	1.22	36.13	nr	**65.03**
Sum of two sides 1100mm	18.13	30.83	1.25	37.02	nr	**67.85**
Sum of two sides 1200mm	19.26	32.76	1.28	37.91	nr	**70.67**
Sum of two sides 1300mm	20.39	34.66	1.31	38.80	nr	**73.46**
Sum of two sides 1400mm	21.52	36.59	1.34	39.69	nr	**76.28**
Sum of two sides 1500mm	22.66	38.52	1.36	40.28	nr	**78.80**
Sum of two sides 1600mm	23.79	40.45	1.39	41.17	nr	**81.62**
Reducer						
Sum of two sides 900mm	59.34	100.90	3.06	90.63	nr	**191.53**
Sum of two sides 1000mm	62.65	106.52	3.17	93.89	nr	**200.41**
Sum of two sides 1100mm	65.94	112.12	3.22	95.37	nr	**207.49**
Sum of two sides 1200mm	69.25	117.75	3.28	97.15	nr	**214.89**
Sum of two sides 1300mm	72.55	123.37	3.33	98.63	nr	**221.99**
Sum of two sides 1400mm	75.86	128.99	3.38	100.11	nr	**229.10**
Sum of two sides 1500mm	79.15	134.59	3.44	101.88	nr	**236.48**
Sum of two sides 1600mm	82.46	140.21	3.49	103.36	nr	**243.58**

U:VENTILATION/AIR CONDITIONING SYSTEMS

Item	Net Price £	Material £	Labour hours	Labour £	Unit	Total rate £
Offset						
Sum of two sides 900mm	126.21	214.60	3.45	102.18	nr	**316.78**
Sum of two sides 1000mm	132.22	224.83	3.67	108.70	nr	**333.53**
Sum of two sides 1100mm	138.23	235.04	3.78	111.95	nr	**346.99**
Sum of two sides 1200mm	144.24	245.26	3.89	115.21	nr	**360.48**
Sum of two sides 1300mm	150.24	255.47	4.00	118.47	nr	**373.94**
Sum of two sides 1400mm	156.26	265.70	4.11	121.73	nr	**387.42**
Sum of two sides 1500mm	162.26	275.90	4.23	125.28	nr	**401.19**
Sum of two sides 1600mm	168.27	286.13	4.34	128.54	nr	**414.67**
90° radius bend						
Sum of two sides 900mm	63.10	107.29	2.99	88.56	nr	**195.85**
Sum of two sides 1000mm	69.30	117.83	3.04	90.04	nr	**207.87**
Sum of two sides 1100mm	75.49	128.37	3.07	90.93	nr	**219.29**
Sum of two sides 1200mm	81.69	138.91	3.09	91.52	nr	**230.42**
Sum of two sides 1300mm	87.88	149.42	3.12	92.41	nr	**241.83**
Sum of two sides 1400mm	94.07	159.96	3.15	93.29	nr	**253.26**
Sum of two sides 1500mm	100.27	170.50	3.17	93.89	nr	**264.39**
Sum of two sides 1600mm	106.47	181.04	3.20	94.78	nr	**275.81**
45° bend						
Sum of two sides 900mm	62.66	106.54	1.74	51.53	nr	**158.08**
Sum of two sides 1000mm	65.60	111.54	1.85	54.79	nr	**166.34**
Sum of two sides 1100mm	68.54	116.54	1.91	56.57	nr	**173.11**
Sum of two sides 1200mm	71.48	121.54	1.96	58.05	nr	**179.59**
Sum of two sides 1300mm	74.41	126.52	2.02	59.83	nr	**186.35**
Sum of two sides 1400mm	77.35	131.52	2.07	61.31	nr	**192.83**
Sum of two sides 1500mm	80.29	136.52	2.13	63.09	nr	**199.61**
Sum of two sides 1600mm	83.23	141.52	2.18	64.57	nr	**206.09**
90° mitre bend						
Sum of two sides 900mm	94.48	160.65	3.59	106.33	nr	**266.97**
Sum of two sides 1000mm	104.55	177.78	3.66	108.40	nr	**286.18**
Sum of two sides 1100mm	114.62	194.90	3.69	109.29	nr	**304.18**
Sum of two sides 1200mm	124.70	212.03	3.72	110.18	nr	**322.21**
Sum of two sides 1300mm	134.76	229.15	3.75	111.07	nr	**340.21**
Sum of two sides 1400mm	144.84	246.28	3.78	111.95	nr	**358.23**
Sum of two sides 1500mm	154.90	263.39	3.81	112.84	nr	**376.24**
Sum of two sides 1600mm	164.98	280.53	3.84	113.73	nr	**394.26**
Branch (Side-on Shoe)						
Sum of two sides 900mm	17.53	29.81	1.37	40.58	nr	**70.39**
Sum of two sides 1000mm	17.97	30.56	1.46	43.24	nr	**73.80**
Sum of two sides 1100mm	18.41	31.30	1.50	44.43	nr	**75.73**
Sum of two sides 1200mm	18.85	32.05	1.55	45.91	nr	**77.96**
Sum of two sides 1300mm	19.29	32.80	1.59	47.09	nr	**79.89**
Sum of two sides 1400mm	19.73	33.54	1.63	48.28	nr	**81.82**
Sum of two sides 1500mm	18.95	32.22	1.68	49.76	nr	**81.97**
Sum of two sides 1600mm	20.61	35.04	1.72	50.94	nr	**85.98**

U:VENTILATION/AIR CONDITIONING SYSTEMS

Item	Net Price £	Material £	Labour hours	Labour £	Unit	Total rate £
U14 : DUCTWORK : FIRE RATED (cont'd)						
Y30 - DUCTLINES (cont'd)						
Galvanised sheet metal rectangular section ductwork to BS476 Part 24 - 2 hour (cont'd)						
Extra over fittings; Ductwork 601 to 800mm longest side (cont'd)						
Ductwork 801 to 1000mm longest side						
Sum of two sides 1100mm	74.58	126.81	5.84	172.97	m	**299.78**
Sum of two sides 1200mm	78.19	132.95	6.11	180.96	m	**313.91**
Sum of two sides 1300mm	81.80	139.09	6.38	188.96	m	**328.05**
Sum of two sides 1400mm	85.41	145.23	6.65	196.96	m	**342.19**
Sum of two sides 1500mm	89.02	151.37	6.91	204.66	m	**356.03**
Sum of two sides 1600mm	92.63	157.51	7.18	212.65	m	**370.17**
Sum of two sides 1700mm	96.25	163.65	7.45	220.65	m	**384.30**
Sum of two sides 1800mm	99.86	169.79	7.71	228.35	m	**398.15**
Sum of two sides 1900mm	103.47	175.94	7.98	236.35	m	**412.28**
Sum of two sides 2000mm	107.08	182.08	8.25	244.34	m	**426.42**
Extra over fittings; Ductwork 801 to 1000mm longest side						
End Cap						
Sum of two sides 1100mm	18.13	30.83	1.25	37.02	nr	**67.85**
Sum of two sides 1200mm	19.26	32.76	1.28	37.91	nr	**70.67**
Sum of two sides 1300mm	20.39	34.66	1.31	38.80	nr	**73.46**
Sum of two sides 1400mm	21.52	36.59	1.34	39.69	nr	**76.28**
Sum of two sides 1500mm	22.66	38.52	1.36	40.28	nr	**78.80**
Sum of two sides 1600mm	23.78	40.43	1.39	41.17	nr	**81.60**
Sum of two sides 1700mm	24.91	42.36	1.42	42.06	nr	**84.42**
Sum of two sides 1800mm	26.05	44.29	1.45	42.95	nr	**87.24**
Sum of two sides 1900mm	27.17	46.20	1.48	43.83	nr	**90.03**
Sum of two sides 2000mm	28.30	48.13	1.51	44.72	nr	**92.85**
Reducer						
Sum of two sides 1100mm	65.94	112.12	3.22	95.37	nr	**207.49**
Sum of two sides 1200mm	69.26	117.77	3.28	97.15	nr	**214.91**
Sum of two sides 1300mm	72.58	123.41	3.33	98.63	nr	**222.04**
Sum of two sides 1400mm	75.90	129.05	3.38	100.11	nr	**229.16**
Sum of two sides 1500mm	79.21	134.69	3.44	101.88	nr	**236.58**
Sum of two sides 1600mm	82.55	140.36	3.49	103.36	nr	**243.72**
Sum of two sides 1700mm	85.86	146.00	3.54	104.85	nr	**250.85**
Sum of two sides 1800mm	89.18	151.64	3.60	106.62	nr	**258.27**
Sum of two sides 1900mm	92.50	157.29	3.65	108.10	nr	**265.39**
Sum of two sides 2000mm	97.48	165.75	3.70	109.58	nr	**275.33**
Offset						
Sum of two sides 1100mm	137.14	233.19	3.78	111.95	nr	**345.14**
Sum of two sides 1200mm	142.94	243.04	3.89	115.21	nr	**358.26**
Sum of two sides 1300mm	148.73	252.90	4.00	118.47	nr	**371.37**
Sum of two sides 1400mm	154.53	262.75	4.11	121.73	nr	**384.48**
Sum of two sides 1500mm	160.32	272.61	4.23	125.28	nr	**397.89**
Sum of two sides 1600mm	166.13	282.48	4.34	128.54	nr	**411.02**
Sum of two sides 1700mm	171.92	292.33	4.45	131.80	nr	**424.13**
Sum of two sides 1800mm	177.72	302.19	4.56	135.06	nr	**437.24**
Sum of two sides 1900mm	183.51	312.04	4.67	138.31	nr	**450.35**
Sum of two sides 2000mm	192.21	326.83	5.53	163.78	nr	**490.62**

U:VENTILATION/AIR CONDITIONING SYSTEMS

Item	Net Price £	Material £	Labour hours	Labour £	Unit	Total rate £
90° radius bend						
Sum of two sides 1100mm	75.37	128.16	3.07	90.93	nr	**219.09**
Sum of two sides 1200mm	81.58	138.72	3.09	91.52	nr	**230.24**
Sum of two sides 1300mm	87.79	149.28	3.12	92.41	nr	**241.68**
Sum of two sides 1400mm	94.00	159.84	3.15	93.29	nr	**253.13**
Sum of two sides 1500mm	100.21	170.40	3.17	93.89	nr	**264.28**
Sum of two sides 1600mm	106.41	180.93	3.20	94.78	nr	**275.71**
Sum of two sides 1700mm	112.62	191.49	3.22	95.37	nr	**286.86**
Sum of two sides 1800mm	118.83	202.05	3.25	96.26	nr	**298.31**
Sum of two sides 1900mm	125.04	212.61	3.28	97.15	nr	**309.76**
Sum of two sides 2000mm	131.25	223.17	3.30	97.74	nr	**320.91**
45° bend						
Sum of two sides 1100mm	68.99	117.31	1.91	56.57	nr	**173.88**
Sum of two sides 1200mm	71.88	122.23	1.96	58.05	nr	**180.28**
Sum of two sides 1300mm	74.76	127.12	2.02	59.83	nr	**186.95**
Sum of two sides 1400mm	77.65	132.04	2.07	61.31	nr	**193.35**
Sum of two sides 1500mm	80.53	136.94	2.13	63.09	nr	**200.02**
Sum of two sides 1600mm	83.42	141.85	2.18	64.57	nr	**206.42**
Sum of two sides 1700mm	86.30	146.75	2.24	66.34	nr	**213.09**
Sum of two sides 1800mm	89.19	151.66	2.30	68.12	nr	**219.78**
Sum of two sides 1900mm	92.07	156.56	2.35	69.60	nr	**226.16**
Sum of two sides 2000mm	94.96	161.48	2.76	81.74	nr	**243.22**
90° mitre bend						
Sum of two sides 1100mm	114.62	194.90	3.69	109.29	nr	**304.18**
Sum of two sides 1200mm	124.70	212.03	3.72	110.18	nr	**322.21**
Sum of two sides 1300mm	134.76	229.15	3.75	111.07	nr	**340.21**
Sum of two sides 1400mm	144.84	246.28	3.78	111.95	nr	**358.23**
Sum of two sides 1500mm	154.92	263.42	3.81	112.84	nr	**376.26**
Sum of two sides 1600mm	164.98	280.53	3.84	113.73	nr	**394.26**
Sum of two sides 1700mm	175.06	297.66	3.87	114.62	nr	**412.28**
Sum of two sides 1800mm	185.13	314.80	3.90	115.51	nr	**430.31**
Sum of two sides 1900mm	195.20	331.91	3.93	116.40	nr	**448.31**
Sum of two sides 2000mm	205.28	349.05	3.96	117.28	nr	**466.33**
Branch (Side-on Shoe)						
Sum of two sides 1100mm	18.39	31.26	1.50	44.43	nr	**75.69**
Sum of two sides 1200mm	18.82	32.01	1.55	45.91	nr	**77.92**
Sum of two sides 1300mm	19.28	32.78	1.59	47.09	nr	**79.87**
Sum of two sides 1400mm	19.72	33.52	1.63	48.28	nr	**81.80**
Sum of two sides 1500mm	20.17	34.29	1.68	49.76	nr	**84.05**
Sum of two sides 1600mm	20.61	35.04	1.72	50.94	nr	**85.98**
Sum of two sides 1700mm	21.06	35.81	1.77	52.42	nr	**88.23**
Sum of two sides 1800mm	21.50	36.55	1.81	53.61	nr	**90.16**
Sum of two sides 1900mm	21.95	37.32	1.86	55.09	nr	**92.41**
Sum of two sides 2000mm	22.61	38.44	1.90	56.27	nr	**94.71**

U:VENTILATION/AIR CONDITIONING SYSTEMS

Item	Net Price £	Material £	Labour hours	Labour £	Unit	Total rate £
U14 : DUCTWORK : FIRE RATED (cont'd)						
Y30 - DUCTLINES (cont'd)						
Galvanised sheet metal rectangular section ductwork to BS476 Part 24 - 2 hour (cont'd)						
Extra over fittings; Ductwork 801 to 1000mm longest side (cont'd)						
Ductwork 1001 to 1250mm longest side						
Sum of two sides 1300mm	92.71	157.64	6.38	188.96	m	**346.60**
Sum of two sides 1400mm	96.71	164.44	6.65	196.96	m	**361.40**
Sum of two sides 1500mm	100.71	171.25	6.91	204.66	m	**375.90**
Sum of two sides 1600mm	103.71	176.35	6.91	204.66	m	**381.01**
Sum of two sides 1700mm	106.70	181.43	7.45	220.65	m	**402.08**
Sum of two sides 1800mm	110.69	188.22	7.71	228.35	m	**416.57**
Sum of two sides 1900mm	114.68	195.00	7.98	236.35	m	**431.35**
Sum of two sides 2000mm	118.68	201.80	8.25	244.34	m	**446.15**
Sum of two sides 2100mm	122.67	208.59	9.62	284.92	m	**493.51**
Sum of two sides 2200mm	125.66	213.67	9.62	284.92	m	**498.59**
Sum of two sides 2300mm	128.66	218.77	10.07	298.25	m	**517.02**
Sum of two sides 2400mm	132.65	225.56	10.47	310.09	m	**535.65**
Sum of two sides 2500mm	136.64	232.34	10.87	321.94	m	**554.28**
Extra over fittings; Ductwork 1001 to 1250mm longest side						
End Cap						
Sum of two sides 1300mm	20.40	34.69	1.31	38.80	nr	**73.48**
Sum of two sides 1400mm	21.52	36.59	1.34	39.69	nr	**76.28**
Sum of two sides 1500mm	22.64	38.50	1.36	40.28	nr	**78.78**
Sum of two sides 1600mm	23.78	40.43	1.39	41.17	nr	**81.60**
Sum of two sides 1700mm	24.91	42.36	1.42	42.06	nr	**84.42**
Sum of two sides 1800mm	26.03	44.27	1.45	42.95	nr	**87.21**
Sum of two sides 1900mm	27.17	46.20	1.48	43.83	nr	**90.03**
Sum of two sides 2000mm	28.30	48.13	1.51	44.72	nr	**92.85**
Sum of two sides 2100mm	29.44	50.06	2.66	78.78	nr	**128.84**
Sum of two sides 2200mm	30.57	51.99	2.80	82.93	nr	**134.91**
Sum of two sides 2300mm	31.70	53.89	2.95	87.37	nr	**141.27**
Sum of two sides 2400mm	32.83	55.82	3.10	91.81	nr	**147.64**
Sum of two sides 2500mm	33.96	57.75	3.24	95.96	nr	**153.71**
Reducer						
Sum of two sides 1300mm	72.55	123.37	3.33	98.63	nr	**221.99**
Sum of two sides 1400mm	75.82	128.93	3.38	100.11	nr	**229.03**
Sum of two sides 1500mm	79.09	134.49	3.44	101.88	nr	**236.37**
Sum of two sides 1600mm	82.37	140.07	3.49	103.36	nr	**243.43**
Sum of two sides 1700mm	85.66	145.65	3.54	104.85	nr	**250.49**
Sum of two sides 1800mm	88.94	151.23	3.60	106.62	nr	**257.85**
Sum of two sides 1900mm	92.22	156.81	3.65	108.10	nr	**264.91**
Sum of two sides 2000mm	95.50	162.39	3.70	109.58	nr	**271.97**
Sum of two sides 2100mm	98.77	167.95	3.75	111.07	nr	**279.01**
Sum of two sides 2200mm	102.05	173.53	3.80	112.55	nr	**286.08**
Sum of two sides 2300mm	105.33	179.11	3.85	114.03	nr	**293.14**
Sum of two sides 2400mm	108.62	184.69	3.90	115.51	nr	**300.20**
Sum of two sides 2500mm	111.90	190.27	3.95	116.99	nr	**307.26**

U:VENTILATION/AIR CONDITIONING SYSTEMS

Item	Net Price £	Material £	Labour hours	Labour £	Unit	Total rate £
Offset						
Sum of two sides 1300mm	148.24	252.07	4.00	118.47	nr	370.54
Sum of two sides 1400mm	154.12	262.07	4.11	121.73	nr	383.79
Sum of two sides 1500mm	160.00	272.07	4.23	125.28	nr	397.35
Sum of two sides 1600mm	165.88	282.06	4.34	128.54	nr	410.60
Sum of two sides 1700mm	171.75	292.04	4.45	131.80	nr	423.84
Sum of two sides 1800mm	177.63	302.04	4.56	135.06	nr	437.10
Sum of two sides 1900mm	183.50	312.02	4.67	138.31	nr	450.33
Sum of two sides 2000mm	189.38	322.02	5.53	163.78	nr	485.80
Sum of two sides 2100mm	195.26	332.02	5.76	170.60	nr	502.61
Sum of two sides 2200mm	201.13	342.00	5.99	177.41	nr	519.40
Sum of two sides 2300mm	207.01	352.00	6.22	184.22	nr	536.22
Sum of two sides 2400mm	212.88	361.97	6.45	191.03	nr	553.01
Sum of two sides 2500mm	218.76	371.97	6.68	197.84	nr	569.82
90° radius bend						
Sum of two sides 1300mm	87.68	149.09	3.12	92.41	nr	241.50
Sum of two sides 1400mm	97.56	165.89	3.15	93.29	nr	259.19
Sum of two sides 1500mm	100.13	170.25	3.17	93.89	nr	264.14
Sum of two sides 1600mm	106.35	180.83	3.20	94.78	nr	275.61
Sum of two sides 1700mm	112.58	191.43	3.22	95.37	nr	286.80
Sum of two sides 1800mm	118.80	202.01	3.25	96.26	nr	298.27
Sum of two sides 1900mm	125.03	212.59	3.28	97.15	nr	309.74
Sum of two sides 2000mm	131.26	223.19	3.30	97.74	nr	320.93
Sum of two sides 2100mm	137.48	233.77	3.32	98.33	nr	332.10
Sum of two sides 2200mm	143.70	244.35	3.34	98.92	nr	343.27
Sum of two sides 2300mm	149.93	254.93	3.36	99.51	nr	354.45
Sum of two sides 2400mm	156.16	265.53	3.38	100.11	nr	365.64
Sum of two sides 2500mm	162.38	276.11	3.40	100.70	nr	376.81
45° bend						
Sum of two sides 1300mm	71.02	120.75	2.02	59.83	nr	180.58
Sum of two sides 1400mm	74.21	126.19	2.07	61.31	nr	187.50
Sum of two sides 1500mm	77.41	131.62	2.13	63.09	nr	194.71
Sum of two sides 1600mm	80.61	137.06	2.18	64.57	nr	201.63
Sum of two sides 1700mm	83.80	142.49	2.24	66.34	nr	208.84
Sum of two sides 1800mm	87.00	147.93	2.30	68.12	nr	216.05
Sum of two sides 1900mm	90.19	153.37	2.35	69.60	nr	222.97
Sum of two sides 2000mm	93.40	158.82	2.76	81.74	nr	240.56
Sum of two sides 2100mm	96.60	164.26	2.89	85.59	nr	249.85
Sum of two sides 2200mm	99.80	169.69	3.01	89.15	nr	258.84
Sum of two sides 2300mm	102.99	175.13	3.13	92.70	nr	267.83
Sum of two sides 2400mm	106.19	180.56	3.26	96.55	nr	277.11
Sum of two sides 2500mm	109.39	186.00	3.37	99.81	nr	285.81
90° mitre bend						
Sum of two sides 1300mm	134.46	228.63	3.75	111.07	nr	339.69
Sum of two sides 1400mm	144.63	245.93	3.78	111.95	nr	357.88
Sum of two sides 1500mm	154.81	263.23	3.81	112.84	nr	376.07
Sum of two sides 1600mm	164.97	280.51	3.84	113.73	nr	394.24
Sum of two sides 1700mm	175.13	297.79	3.87	114.62	nr	412.41
Sum of two sides 1800mm	185.29	315.07	3.90	115.51	nr	430.58
Sum of two sides 1900mm	195.46	332.35	3.93	116.40	nr	448.75
Sum of two sides 2000mm	205.63	349.65	3.96	117.28	nr	466.94
Sum of two sides 2100mm	215.79	366.93	3.99	118.17	nr	485.10
Sum of two sides 2200mm	225.96	384.21	4.02	119.06	nr	503.27
Sum of two sides 2300mm	236.12	401.49	4.05	119.95	nr	521.44
Sum of two sides 2400mm	246.28	418.77	4.08	120.84	nr	539.61
Sum of two sides 2500mm	256.44	436.05	4.11	121.73	nr	557.78

U:VENTILATION/AIR CONDITIONING SYSTEMS

Item	Net Price £	Material £	Labour hours	Labour £	Unit	Total rate £
U14 : DUCTWORK : FIRE RATED (cont'd)						
Y30 - DUCTLINES (cont'd)						
Galvanised sheet metal rectangular section ductwork to BS476 Part 24 - 2 hour (cont'd)						
Extra over fittings; Ductwork 1001 to 1250mm longest side (cont'd)						
Branch (Side-on Shoe)						
Sum of two sides 1300mm	19.31	32.84	1.59	47.09	nr	**79.93**
Sum of two sides 1400mm	19.75	33.59	1.63	48.28	nr	**81.86**
Sum of two sides 1500mm	20.19	34.33	1.68	49.76	nr	**84.09**
Sum of two sides 1600mm	20.63	35.08	1.72	50.94	nr	**86.02**
Sum of two sides 1700mm	21.08	35.85	1.77	52.42	nr	**88.27**
Sum of two sides 1800mm	21.52	36.59	1.81	53.61	nr	**90.20**
Sum of two sides 1900mm	21.96	37.34	1.86	55.09	nr	**92.43**
Sum of two sides 2000mm	22.41	38.11	1.90	56.27	nr	**94.38**
Sum of two sides 2100mm	22.85	38.85	2.58	76.41	nr	**115.27**
Sum of two sides 2200mm	23.29	39.60	2.61	77.30	nr	**116.90**
Sum of two sides 2300mm	23.73	40.35	2.64	78.19	nr	**118.54**
Sum of two sides 2400mm	24.18	41.12	2.88	85.30	nr	**126.41**
Sum of two sides 2500mm	24.62	41.86	2.91	86.19	nr	**128.05**
Ductwork 1251 to 2000mm longest side						
Sum of two sides 1700mm	101.67	172.89	7.45	220.65	m	**393.54**
Sum of two sides 1800mm	108.19	183.96	7.71	228.35	m	**412.31**
Sum of two sides 1900mm	114.70	195.04	7.98	236.35	m	**431.39**
Sum of two sides 2000mm	121.22	206.12	8.25	244.34	m	**450.46**
Sum of two sides 2100mm	127.73	217.20	9.62	284.92	m	**502.12**
Sum of two sides 2200mm	134.25	228.27	9.66	286.10	m	**514.38**
Sum of two sides 2300mm	140.75	239.33	10.07	298.25	m	**537.58**
Sum of two sides 2400mm	147.27	250.41	10.47	310.09	m	**560.50**
Sum of two sides 2500mm	153.78	261.49	10.87	321.94	m	**583.43**
Sum of two sides 2600mm	160.30	272.56	11.27	333.79	m	**606.35**
Sum of two sides 2700mm	166.81	283.64	11.67	345.64	m	**629.28**
Sum of two sides 2800mm	173.33	294.72	12.08	357.78	m	**652.50**
Sum of two sides 2900mm	179.84	305.80	12.48	369.63	m	**675.42**
Sum of two sides 3000mm	186.35	316.87	12.88	381.47	m	**698.35**
Sum of two sides 3100mm	192.87	327.95	13.26	392.73	m	**720.68**
Sum of two sides 3200mm	199.38	339.03	13.69	405.46	m	**744.49**
Sum of two sides 3300mm	205.90	350.11	14.09	417.31	m	**767.42**
Sum of two sides 3400mm	212.41	361.18	14.49	429.16	m	**790.34**
Sum of two sides 3500mm	218.92	372.24	14.89	441.00	m	**813.24**
Sum of two sides 3600mm	225.43	383.32	15.29	452.85	m	**836.17**
Sum of two sides 3700mm	231.95	394.40	15.69	464.70	m	**859.09**
Sum of two sides 3800mm	238.46	405.47	16.09	476.54	m	**882.02**
Sum of two sides 3900mm	244.98	416.55	16.49	488.39	m	**904.94**
Sum of two sides 4000mm	251.49	427.63	16.89	500.24	m	**927.87**

U:VENTILATION/AIR CONDITIONING SYSTEMS

Item	Net Price £	Material £	Labour hours	Labour £	Unit	Total rate £
End Cap						
Sum of two sides 1700mm	37.01	62.94	1.42	42.06	nr	**105.00**
Sum of two sides 1800mm	39.39	66.98	1.45	42.95	nr	**109.93**
Sum of two sides 1900mm	41.77	71.03	1.48	43.83	nr	**114.86**
Sum of two sides 2000mm	44.15	75.07	1.51	44.72	nr	**119.80**
Sum of two sides 2100mm	46.53	79.12	2.66	78.78	nr	**157.90**
Sum of two sides 2200mm	48.91	83.17	2.80	82.93	nr	**166.09**
Sum of two sides 2300mm	51.29	87.21	2.95	87.37	nr	**174.58**
Sum of two sides 2400mm	53.67	91.26	3.10	91.81	nr	**183.07**
Sum of two sides 2500mm	56.05	95.30	3.24	95.96	nr	**191.26**
Sum of two sides 2600mm	58.43	99.35	3.39	100.40	nr	**199.75**
Sum of two sides 2700mm	60.80	103.39	3.54	104.85	nr	**208.24**
Sum of two sides 2800mm	63.18	107.44	3.68	108.99	nr	**216.43**
Sum of two sides 2900mm	65.58	111.50	3.83	113.43	nr	**224.94**
Sum of two sides 3000mm	67.95	115.55	3.98	117.88	nr	**233.42**
Sum of two sides 3100mm	70.33	119.59	4.12	122.02	nr	**241.62**
Sum of two sides 3200mm	72.71	123.64	4.27	126.47	nr	**250.10**
Sum of two sides 3300mm	75.09	127.68	4.42	130.91	nr	**258.59**
Sum of two sides 3400mm	77.47	131.73	4.57	135.35	nr	**267.08**
Sum of two sides 3500mm	79.43	135.07	4.72	139.79	nr	**274.86**
Sum of two sides 3600mm	81.03	137.79	4.87	144.24	nr	**282.02**
Sum of two sides 3700mm	83.41	141.83	5.02	148.68	nr	**290.51**
Sum of two sides 3800mm	80.22	136.40	5.17	153.12	nr	**289.53**
Sum of two sides 3900mm	89.36	151.95	5.32	157.56	nr	**309.52**
Sum of two sides 4000mm	91.74	156.00	5.47	162.01	nr	**318.01**
Reducer						
Sum of two sides 1700mm	79.92	135.90	3.54	104.85	nr	**240.74**
Sum of two sides 1800mm	89.66	152.45	3.60	106.62	nr	**259.07**
Sum of two sides 1900mm	99.38	168.99	3.65	108.10	nr	**277.09**
Sum of two sides 2000mm	109.12	185.54	3.70	109.58	nr	**295.12**
Sum of two sides 2100mm	118.85	202.09	2.92	86.48	nr	**288.58**
Sum of two sides 2200mm	128.58	218.63	3.10	91.81	nr	**310.44**
Sum of two sides 2300mm	138.31	235.18	3.29	97.44	nr	**332.62**
Sum of two sides 2400mm	148.05	251.74	3.48	103.07	nr	**354.80**
Sum of two sides 2500mm	157.77	268.27	3.66	108.40	nr	**376.67**
Sum of two sides 2600mm	167.51	284.82	3.85	114.03	nr	**398.85**
Sum of two sides 2700mm	177.24	301.38	4.03	119.36	nr	**420.74**
Sum of two sides 2800mm	186.97	317.91	4.22	124.99	nr	**442.90**
Sum of two sides 2900mm	196.70	334.47	4.40	130.32	nr	**464.78**
Sum of two sides 3000mm	211.29	359.28	4.59	135.94	nr	**495.22**
Sum of two sides 3100mm	211.29	359.28	4.78	141.57	nr	**500.85**
Sum of two sides 3200mm	221.03	375.83	4.96	146.90	nr	**522.73**
Sum of two sides 3300mm	235.62	400.64	5.15	152.53	nr	**553.17**
Sum of two sides 3400mm	245.35	417.20	5.34	158.16	nr	**575.35**
Sum of two sides 3500mm	255.09	433.75	5.53	163.78	nr	**597.53**
Sum of two sides 3600mm	264.81	450.28	5.72	169.41	nr	**619.69**
Sum of two sides 3700mm	274.55	466.84	5.91	175.04	nr	**641.88**
Sum of two sides 3800mm	284.28	483.39	6.10	180.67	nr	**664.06**
Sum of two sides 3900mm	294.01	499.93	6.29	186.29	nr	**686.22**
Sum of two sides 4000mm	303.74	516.48	6.48	191.92	nr	**708.40**

U:VENTILATION/AIR CONDITIONING SYSTEMS

Item	Net Price £	Material £	Labour hours	Labour £	Unit	Total rate £
U14 : DUCTWORK : FIRE RATED (cont'd)						
Y30 - DUCTLINES (cont'd)						
Galvanised sheet metal rectangular section ductwork to BS476 Part 24 - 2 hour (cont'd)						
Extra over fittings; Ductwork 1251 to 2000mm longest side						
Offset						
Sum of two sides 1700mm	79.92	135.90	4.45	131.80	nr	267.70
Sum of two sides 1800mm	86.51	147.10	4.56	135.06	nr	282.16
Sum of two sides 1900mm	93.10	158.30	4.67	138.31	nr	296.62
Sum of two sides 2000mm	99.67	169.48	5.53	163.78	nr	333.27
Sum of two sides 2100mm	106.26	180.69	5.76	170.60	nr	351.28
Sum of two sides 2200mm	112.85	191.89	5.99	177.41	nr	369.30
Sum of two sides 2300mm	119.44	203.09	6.22	184.22	nr	387.31
Sum of two sides 2400mm	126.01	214.27	6.45	191.03	nr	405.30
Sum of two sides 2500mm	132.60	225.47	6.68	197.84	nr	423.32
Sum of two sides 2600mm	139.19	236.68	6.91	204.66	nr	441.33
Sum of two sides 2700mm	145.78	247.88	7.14	211.47	nr	459.35
Sum of two sides 2800mm	152.37	259.08	7.37	218.28	nr	477.36
Sum of two sides 2900mm	158.94	270.26	7.60	225.09	nr	495.35
Sum of two sides 3000mm	165.53	281.46	7.83	231.90	nr	513.37
Sum of two sides 3100mm	172.12	292.67	8.06	238.72	nr	531.38
Sum of two sides 3200mm	178.71	303.87	8.29	245.53	nr	549.40
Sum of two sides 3300mm	185.29	315.07	8.52	252.34	nr	567.41
Sum of two sides 3400mm	191.87	326.25	8.75	259.15	nr	585.40
Sum of two sides 3500mm	198.46	337.45	8.98	265.96	nr	603.42
Sum of two sides 3600mm	205.05	348.66	9.21	272.78	nr	621.43
Sum of two sides 3700mm	211.63	359.86	9.44	279.59	nr	639.45
Sum of two sides 3800mm	218.21	371.04	9.67	286.40	nr	657.44
Sum of two sides 3900mm	224.80	382.24	9.90	293.21	nr	675.45
Sum of two sides 4000mm	231.39	393.44	10.13	300.02	nr	693.47
90° radius bend						
Sum of two sides 1700mm	111.23	189.13	3.22	95.37	nr	284.50
Sum of two sides 1800mm	121.28	206.22	3.25	96.26	nr	302.48
Sum of two sides 1900mm	131.33	223.32	3.28	97.15	nr	320.46
Sum of two sides 2000mm	141.39	240.41	3.30	97.74	nr	338.15
Sum of two sides 2100mm	151.44	257.50	3.32	98.33	nr	355.83
Sum of two sides 2200mm	161.49	274.60	3.34	98.92	nr	373.52
Sum of two sides 2300mm	171.54	291.69	3.36	99.51	nr	391.20
Sum of two sides 2400mm	181.60	308.78	3.38	100.11	nr	408.89
Sum of two sides 2500mm	191.65	325.88	3.40	100.70	nr	426.58
Sum of two sides 2600mm	201.70	342.97	3.42	101.29	nr	444.26
Sum of two sides 2700mm	211.76	360.06	3.44	101.88	nr	461.95
Sum of two sides 2800mm	221.81	377.16	3.46	102.48	nr	479.63
Sum of two sides 2900mm	231.85	394.23	3.48	103.07	nr	497.30
Sum of two sides 3000mm	241.90	411.32	3.50	103.66	nr	514.99
Sum of two sides 3100mm	251.95	428.42	3.52	104.25	nr	532.67
Sum of two sides 3200mm	262.01	445.51	3.54	104.85	nr	550.36
Sum of two sides 3300mm	272.06	462.61	3.56	105.44	nr	568.04
Sum of two sides 3400mm	282.23	479.91	3.58	106.03	nr	585.94
Sum of two sides 3500mm	292.17	496.79	3.60	106.62	nr	603.42
Sum of two sides 3600mm	302.22	513.89	3.62	107.22	nr	621.10
Sum of two sides 3700mm	312.27	530.98	3.64	107.81	nr	638.79
Sum of two sides 3800mm	322.32	548.07	3.66	108.40	nr	656.47
Sum of two sides 3900mm	332.38	565.17	3.68	108.99	nr	674.16
Sum of two sides 4000mm	342.43	582.26	3.70	109.58	nr	691.84

U:VENTILATION/AIR CONDITIONING SYSTEMS

Item	Net Price £	Material £	Labour hours	Labour £	Unit	Total rate £
45° bend						
Sum of two sides 1700mm	82.46	140.21	2.24	66.34	nr	206.56
Sum of two sides 1800mm	91.57	155.71	2.30	68.12	nr	223.83
Sum of two sides 1900mm	100.67	171.18	2.35	69.60	nr	240.79
Sum of two sides 2000mm	109.79	186.68	2.76	81.74	nr	268.43
Sum of two sides 2100mm	151.44	257.50	2.89	85.59	nr	343.10
Sum of two sides 2200mm	128.00	217.65	3.01	89.15	nr	306.80
Sum of two sides 2300mm	137.12	233.15	3.13	92.70	nr	325.85
Sum of two sides 2400mm	146.22	248.62	3.26	96.55	nr	345.18
Sum of two sides 2500mm	155.33	264.12	3.37	99.81	nr	363.93
Sum of two sides 2600mm	164.43	279.60	3.49	103.36	nr	382.96
Sum of two sides 2700mm	173.55	295.09	3.62	107.22	nr	402.31
Sum of two sides 2800mm	182.66	310.59	3.74	110.77	nr	421.36
Sum of two sides 2900mm	191.76	326.06	3.86	114.32	nr	440.39
Sum of two sides 3000mm	200.87	341.56	3.98	117.88	nr	459.44
Sum of two sides 3100mm	209.99	357.06	4.10	121.43	nr	478.49
Sum of two sides 3200mm	219.09	372.53	4.22	124.99	nr	497.52
Sum of two sides 3300mm	228.20	388.03	4.34	128.54	nr	516.57
Sum of two sides 3400mm	237.30	403.50	4.46	132.09	nr	535.60
Sum of two sides 3500mm	246.42	419.00	4.58	135.65	nr	554.65
Sum of two sides 3600mm	255.53	434.50	4.70	139.20	nr	573.70
Sum of two sides 3700mm	264.63	449.97	4.82	142.76	nr	592.73
Sum of two sides 3800mm	273.74	465.47	4.94	146.31	nr	611.78
Sum of two sides 3900mm	282.84	480.94	5.06	149.86	nr	630.81
Sum of two sides 4000mm	291.96	496.44	5.18	153.42	nr	649.86
90° mitre bend						
Sum of two sides 1700mm	166.59	283.27	3.87	114.62	nr	397.89
Sum of two sides 1800mm	195.38	332.23	3.90	115.51	nr	447.73
Sum of two sides 1900mm	224.18	381.18	3.93	116.40	nr	497.58
Sum of two sides 2000mm	252.95	430.12	3.96	117.28	nr	547.40
Sum of two sides 2100mm	281.75	479.08	3.82	113.14	nr	592.22
Sum of two sides 2200mm	310.54	528.03	3.83	113.43	nr	641.47
Sum of two sides 2300mm	339.33	576.99	3.83	113.43	nr	690.43
Sum of two sides 2400mm	368.12	625.95	3.84	113.73	nr	739.68
Sum of two sides 2500mm	396.91	674.91	3.85	114.03	nr	788.93
Sum of two sides 2600mm	425.69	723.84	3.85	114.03	nr	837.87
Sum of two sides 2700mm	454.49	772.80	3.86	114.32	nr	887.12
Sum of two sides 2800mm	483.28	821.76	3.86	114.32	nr	936.08
Sum of two sides 2900mm	512.07	870.71	3.87	114.62	nr	985.33
Sum of two sides 3000mm	540.86	919.67	3.87	114.62	nr	1034.29
Sum of two sides 3100mm	569.65	968.63	3.88	114.92	nr	1083.54
Sum of two sides 3200mm	598.43	1017.57	3.88	114.92	nr	1132.48
Sum of two sides 3300mm	627.23	1066.52	3.89	115.21	nr	1181.73
Sum of two sides 3400mm	656.02	1115.48	3.90	115.51	nr	1230.99
Sum of two sides 3500mm	684.81	1164.44	3.91	115.80	nr	1280.24
Sum of two sides 3600mm	713.60	1213.40	3.92	116.10	nr	1329.50
Sum of two sides 3700mm	742.39	1262.35	3.93	116.40	nr	1378.75
Sum of two sides 3800mm	771.17	1311.29	3.94	116.69	nr	1427.98
Sum of two sides 3900mm	799.97	1360.25	3.95	116.99	nr	1477.24
Sum of two sides 4000mm	828.76	1409.20	3.96	117.28	nr	1526.49

U:VENTILATION/AIR CONDITIONING SYSTEMS

Item	Net Price £	Material £	Labour hours	Labour £	Unit	Total rate £
U14 : DUCTWORK : FIRE RATED (cont'd)						
Y30 - DUCTLINES (cont'd)						
Galvanised sheet metal rectangular section ductwork to BS476 Part 24 - 2 hour (cont'd)						
Extra over fittings; Ductwork 1251 to 2000mm longest side (cont'd)						
Branch (Side-on Shoe)						
Sum of two sides 1700mm	18.89	32.11	1.77	52.42	nr	**84.54**
Sum of two sides 1800mm	20.64	35.10	1.81	53.61	nr	**88.71**
Sum of two sides 1900mm	22.40	38.09	1.86	55.09	nr	**93.18**
Sum of two sides 2000mm	24.16	41.07	1.90	56.27	nr	**97.35**
Sum of two sides 2100mm	25.91	44.06	2.58	76.41	nr	**120.47**
Sum of two sides 2200mm	27.67	47.05	2.61	77.30	nr	**124.35**
Sum of two sides 2300mm	29.43	50.04	2.65	78.49	nr	**128.52**
Sum of two sides 2400mm	31.18	53.02	2.68	79.37	nr	**132.40**
Sum of two sides 2500mm	32.94	56.01	2.71	80.26	nr	**136.27**
Sum of two sides 2600mm	34.70	59.00	2.75	81.45	nr	**140.45**
Sum of two sides 2700mm	36.45	61.99	2.78	82.34	nr	**144.32**
Sum of two sides 2800mm	38.21	64.97	2.81	83.22	nr	**148.20**
Sum of two sides 2900mm	39.97	67.96	2.84	84.11	nr	**152.07**
Sum of two sides 3000mm	41.72	70.95	2.87	85.00	nr	**155.95**
Sum of two sides 3100mm	43.48	73.93	2.90	85.89	nr	**159.82**
Sum of two sides 3200mm	45.24	76.92	2.93	86.78	nr	**163.70**
Sum of two sides 3300mm	46.99	79.91	2.93	86.78	nr	**166.69**
Sum of two sides 3400mm	48.75	82.90	3.00	88.85	nr	**171.75**
Sum of two sides 3500mm	50.51	85.88	3.03	89.74	nr	**175.62**
Sum of two sides 3600mm	52.26	88.87	3.06	90.63	nr	**179.50**
Sum of two sides 3700mm	54.02	91.86	3.09	91.52	nr	**183.38**
Sum of two sides 3800mm	55.78	94.84	3.12	92.41	nr	**187.25**
Sum of two sides 3900mm	57.54	97.83	3.15	93.29	nr	**191.13**
Sum of two sides 4000mm	59.29	100.82	3.18	94.18	nr	**195.00**
Rectangular section ductwork to BS476 Part 24 (ISO 6944:1985), ducts "Type A" and "Type B"; manufactured from 6mm thick laminate fire board consisting of steel circular hole punched facings pressed to a fibre cement core; provides upto 4 hours stability, 4 hours integrity and 32 minutes insulation; including all necessary stiffeners, joints and supports in the running length						
Ductwork up to 600mm longest side						
Sum of two sides 200 mm	143.83	244.56	4.00	118.47	m	**363.03**
Sum of two sides 400mm	143.83	244.56	4.00	118.47	m	**363.03**
Sum of two sides 600mm	143.83	244.56	4.00	118.47	m	**363.03**
Sum of two sides 800mm	203.16	345.46	5.50	162.90	m	**508.35**
Sum of two sides 1000mm	203.16	345.46	5.50	162.90	m	**508.35**
Sum of two sides 1200mm	282.90	481.04	6.00	177.70	m	**658.74**

U:VENTILATION/AIR CONDITIONING SYSTEMS

Item	Net Price £	Material £	Labour hours	Labour £	Unit	Total rate £
Extra over fittings; Ductwork up to 600mm longest side						
End Cap						
Sum of two sides 200 mm	33.47	56.92	0.81	23.99	m	**80.91**
Sum of two sides 400mm	33.47	56.92	0.87	25.77	m	**82.69**
Sum of two sides 600mm	33.47	56.92	0.93	27.54	m	**84.46**
Sum of two sides 800mm	45.94	78.11	0.98	29.03	m	**107.14**
Sum of two sides 1000mm	45.94	78.11	1.22	36.13	m	**114.24**
Sum of two sides 1200mm	65.63	111.59	1.28	37.91	m	**149.50**
Reducer						
Sum of two sides 200 mm	38.06	64.72	2.23	66.05	m	**130.77**
Sum of two sides 400mm	38.06	64.72	2.51	74.34	m	**139.06**
Sum of two sides 600mm	38.06	64.72	2.79	82.63	m	**147.35**
Sum of two sides 800mm	66.29	112.71	3.01	89.15	m	**201.86**
Sum of two sides 1000mm	66.29	112.71	3.17	93.89	m	**206.60**
Sum of two sides 1200mm	96.47	164.04	3.28	97.15	m	**261.19**
Offset						
Sum of two sides 200 mm	38.06	64.72	2.95	87.37	m	**152.09**
Sum of two sides 400mm	38.06	64.72	3.26	96.55	m	**161.27**
Sum of two sides 600mm	38.06	64.72	3.57	105.73	m	**170.45**
Sum of two sides 800mm	66.29	112.71	3.23	95.66	m	**208.38**
Sum of two sides 1000mm	66.29	112.71	3.67	108.70	m	**221.41**
Sum of two sides 1200mm	83.35	141.72	3.89	115.21	m	**256.94**
90° radius bend						
Sum of two sides 200 mm	64.31	109.36	2.06	61.01	m	**170.37**
Sum of two sides 400mm	64.31	109.36	2.36	69.90	m	**179.25**
Sum of two sides 600mm	64.31	109.36	2.66	78.78	m	**188.14**
Sum of two sides 800mm	79.41	135.03	2.97	87.96	m	**222.99**
Sum of two sides 1000mm	79.41	135.03	3.04	90.04	m	**225.07**
Sum of two sides 1200mm	96.47	164.04	3.09	91.52	m	**255.56**
45° radius bend						
Sum of two sides 200 mm	64.31	109.36	1.52	45.02	m	**154.37**
Sum of two sides 400mm	64.31	109.36	1.66	49.16	m	**158.52**
Sum of two sides 600mm	64.31	109.36	1.80	53.31	m	**162.67**
Sum of two sides 800mm	79.41	135.03	1.93	57.16	m	**192.19**
Sum of two sides 1000mm	79.41	135.03	2.05	60.72	m	**195.75**
Sum of two sides 1200mm	96.47	164.04	2.17	64.27	m	**228.31**
90° mitre bend						
Sum of two sides 200 mm	64.31	109.36	2.47	73.16	m	**182.51**
Sum of two sides 400mm	64.31	109.36	2.84	84.11	m	**193.47**
Sum of two sides 600mm	64.31	109.36	3.20	94.78	m	**204.13**
Sum of two sides 800mm	79.41	135.03	3.56	105.44	m	**240.47**
Sum of two sides 1000mm	79.41	135.03	3.66	108.40	m	**243.43**
Sum of two sides 1200mm	96.47	164.04	3.72	110.18	m	**274.22**
Branch						
Sum of two sides 200 mm	44.63	75.88	0.98	29.03	m	**104.90**
Sum of two sides 400mm	44.63	75.88	1.14	33.76	m	**109.64**
Sum of two sides 600mm	44.63	75.88	1.30	38.50	m	**114.38**
Sum of two sides 800mm	53.16	90.39	1.46	43.24	m	**133.64**
Sum of two sides 1000mm	53.16	90.39	1.46	43.24	m	**133.64**
Sum of two sides 1200mm	70.22	119.41	1.55	45.91	m	**165.31**

U:VENTILATION/AIR CONDITIONING SYSTEMS

Item	Net Price £	Material £	Labour hours	Labour £	Unit	Total rate £
U14 : DUCTWORK : FIRE RATED (cont'd)						
Y30 - DUCTLINES (cont'd)						
Galvanised sheet metal rectangular section ductwork to BS476 Part 24 - 4 hour (cont'd)						
Ductwork 601 to 1000mm longest side						
Sum of two sides 1000mm	203.16	345.46	6.00	177.70	m	**523.16**
Sum of two sides 1100mm	282.90	481.04	6.00	177.70	m	**658.74**
Sum of two sides 1300mm	282.90	481.04	6.00	177.70	m	**658.74**
Sum of two sides 1500mm	282.90	481.04	6.00	177.70	m	**658.74**
Sum of two sides 1700mm	338.92	576.29	6.00	177.70	m	**754.00**
Sum of two sides 1900mm	338.92	576.29	6.00	177.70	m	**754.00**
Extra over fittings; Ductwork 601 to 1000mm longest side						
End Cap						
Sum of two sides 1000mm	45.94	78.11	1.25	37.02	m	**115.13**
Sum of two sides 1100mm	65.63	111.59	1.25	37.02	m	**148.61**
Sum of two sides 1300mm	65.63	111.59	1.31	38.80	m	**150.39**
Sum of two sides 1500mm	65.63	111.59	1.36	40.28	m	**151.87**
Sum of two sides 1700mm	131.24	223.16	1.42	42.06	m	**265.21**
Sum of two sides 1900mm	131.24	223.16	1.48	43.83	m	**266.99**
Reducer						
Sum of two sides 1000mm	66.29	112.71	3.22	95.37	m	**208.08**
Sum of two sides 1100mm	96.47	164.04	3.22	95.37	m	**259.41**
Sum of two sides 1300mm	96.47	164.04	3.33	98.63	m	**262.67**
Sum of two sides 1500mm	96.47	164.04	3.44	101.88	m	**265.93**
Sum of two sides 1700mm	100.52	170.92	3.54	104.85	m	**275.76**
Sum of two sides 1900mm	100.52	170.92	3.65	108.10	m	**279.02**
Offset						
Sum of two sides 1000mm	66.29	112.71	3.78	111.95	m	**224.67**
Sum of two sides 1100mm	83.35	141.72	3.78	111.95	m	**253.68**
Sum of two sides 1300mm	83.35	141.72	4.00	118.47	m	**260.19**
Sum of two sides 1500mm	83.35	141.72	4.23	125.28	m	**267.01**
Sum of two sides 1700mm	115.38	196.20	4.45	131.80	m	**327.99**
Sum of two sides 1900mm	115.38	196.20	4.67	138.31	m	**334.51**
90° radius bend						
Sum of two sides 1000mm	79.41	135.03	3.78	111.95	m	**246.98**
Sum of two sides 1100mm	96.47	164.04	3.07	90.93	m	**254.97**
Sum of two sides 1300mm	96.47	164.04	3.12	92.41	m	**256.45**
Sum of two sides 1500mm	96.47	164.04	3.17	93.89	m	**257.93**
Sum of two sides 1700mm	141.17	240.05	3.22	95.37	m	**335.41**
Sum of two sides 1900mm	141.17	240.05	3.28	97.15	m	**337.19**
45° bend						
Sum of two sides 1000mm	79.41	135.03	3.78	111.95	m	**246.98**
Sum of two sides 1100mm	96.47	164.04	1.91	56.57	m	**220.61**
Sum of two sides 1300mm	96.47	164.04	2.02	59.83	m	**223.87**
Sum of two sides 1500mm	96.47	164.04	2.13	63.09	m	**227.13**
Sum of two sides 1700mm	141.17	240.05	2.24	66.34	m	**306.39**
Sum of two sides 1900mm	141.17	240.05	2.35	69.60	m	**309.65**

U:VENTILATION/AIR CONDITIONING SYSTEMS

Item	Net Price £	Material £	Labour hours	Labour £	Unit	Total rate £
90° mitre bend						
Sum of two sides 1000mm	79.41	135.03	3.78	111.95	m	**246.98**
Sum of two sides 1100mm	96.47	164.04	3.69	109.29	m	**273.33**
Sum of two sides 1300mm	96.47	164.04	3.75	111.07	m	**275.11**
Sum of two sides 1500mm	96.47	164.04	3.81	112.84	m	**276.88**
Sum of two sides 1700mm	141.17	240.05	3.87	114.62	m	**354.67**
Sum of two sides 1900mm	141.17	240.05	3.93	116.40	m	**356.44**
Branch						
Sum of two sides 1000mm	53.16	90.39	3.78	111.95	m	**202.35**
Sum of two sides 1100mm	70.22	119.41	1.50	44.43	m	**163.83**
Sum of two sides 1300mm	70.22	119.41	1.59	47.09	m	**166.50**
Sum of two sides 1500mm	70.22	119.41	1.68	49.76	m	**169.16**
Sum of two sides 1700mm	88.54	150.54	1.77	52.42	m	**202.97**
Sum of two sides 1900mm	88.54	150.54	1.86	55.09	m	**205.63**
Ductwork 1001 to 1250mm longest side						
Sum of two sides 1300mm	282.90	481.04	6.00	177.70	m	**658.74**
Sum of two sides 1500mm	282.90	481.04	6.00	177.70	m	**658.74**
Sum of two sides 1700mm	338.92	576.29	6.00	177.70	m	**754.00**
Sum of two sides 1900mm	338.92	576.29	6.00	177.70	m	**754.00**
Sum of two sides 2100mm	473.92	805.84	6.50	192.51	m	**998.35**
Sum of two sides 2300mm	473.92	805.84	6.50	192.51	m	**998.35**
Sum of two sides 2500mm	473.92	805.84	6.50	192.51	m	**998.35**
Extra over fittings; Ductwork 1001 to 1250mm longest side						
End Cap						
Sum of two sides 1300mm	65.63	111.59	1.31	38.80	m	**150.39**
Sum of two sides 1500mm	65.63	111.59	1.36	40.28	m	**151.87**
Sum of two sides 1700mm	131.24	223.16	1.42	42.06	m	**265.21**
Sum of two sides 1900mm	131.24	223.16	1.48	43.83	m	**266.99**
Sum of two sides 2100mm	154.70	263.04	2.66	78.78	m	**341.83**
Sum of two sides 2300mm	154.70	263.04	2.95	87.37	m	**350.41**
Sum of two sides 2500mm	154.70	263.04	3.24	95.96	m	**359.00**
Reducer						
Sum of two sides 1300mm	96.47	164.04	3.33	98.63	m	**262.67**
Sum of two sides 1500mm	96.47	164.04	3.44	101.88	m	**265.93**
Sum of two sides 1700mm	100.52	170.92	3.54	104.85	m	**275.76**
Sum of two sides 1900mm	100.52	170.92	3.65	108.10	m	**279.02**
Sum of two sides 2100mm	117.47	199.75	3.75	111.07	m	**310.82**
Sum of two sides 2300mm	117.47	199.75	3.85	114.03	m	**313.78**
Sum of two sides 2500mm	117.47	199.75	3.95	116.99	m	**316.74**
Offset						
Sum of two sides 1300mm	83.35	141.72	4.00	118.47	m	**260.19**
Sum of two sides 1500mm	83.35	141.72	4.23	125.28	m	**267.01**
Sum of two sides 1700mm	115.38	196.20	4.45	131.80	m	**327.99**
Sum of two sides 1900mm	115.38	196.20	4.67	138.31	m	**334.51**
Sum of two sides 2100mm	117.47	199.75	5.76	170.60	m	**370.35**
Sum of two sides 2300mm	117.47	199.75	6.22	184.22	m	**383.97**
Sum of two sides 2500mm	117.47	199.75	6.68	197.84	m	**397.59**

U:VENTILATION/AIR CONDITIONING SYSTEMS

Item	Net Price £	Material £	Labour hours	Labour £	Unit	Total rate £
U14 : DUCTWORK : FIRE RATED (cont'd)						
Y30 - DUCTLINES (cont'd)						
Galvanised sheet metal rectangular section ductwork to BS476 Part 24 - 4 hour (cont'd)						
Extra over fittings; Ductwork 1001 to 1250mm longest side (cont'd)						
90° radius bend						
Sum of two sides 1300mm	96.47	164.04	3.12	92.41	m	**256.45**
Sum of two sides 1500mm	96.47	164.04	3.17	93.89	m	**257.93**
Sum of two sides 1700mm	141.17	240.05	3.22	95.37	m	**335.41**
Sum of two sides 1900mm	141.17	240.05	3.28	97.15	m	**337.19**
Sum of two sides 2100mm	143.72	244.39	3.32	98.33	m	**342.72**
Sum of two sides 2300mm	143.72	244.39	3.36	99.51	m	**343.90**
Sum of two sides 2500mm	143.72	244.39	3.40	100.70	m	**345.08**
45° bend						
Sum of two sides 1300mm	96.47	164.04	2.02	59.83	m	**223.87**
Sum of two sides 1500mm	96.47	164.04	2.13	63.09	m	**227.13**
Sum of two sides 1700mm	141.17	240.05	2.24	66.34	m	**306.39**
Sum of two sides 1900mm	141.17	240.05	2.35	69.60	m	**309.65**
Sum of two sides 2100mm	143.72	244.39	2.89	85.59	m	**329.98**
Sum of two sides 2300mm	143.72	244.39	3.13	92.70	m	**337.09**
Sum of two sides 2500mm	143.72	244.39	3.37	99.81	m	**344.20**
90° mitre bend						
Sum of two sides 1300mm	96.47	164.04	3.75	111.07	m	**275.11**
Sum of two sides 1500mm	96.47	164.04	3.81	112.84	m	**276.88**
Sum of two sides 1700mm	141.17	240.05	3.87	114.62	m	**354.67**
Sum of two sides 1900mm	141.17	240.05	3.93	116.40	m	**356.44**
Sum of two sides 2100mm	143.72	244.39	3.99	118.17	m	**362.56**
Sum of two sides 2300mm	143.72	244.39	4.05	119.95	m	**364.34**
Sum of two sides 2500mm	143.72	244.39	4.11	121.73	m	**366.11**
Branch						
Sum of two sides 1300mm	70.22	119.41	1.59	47.09	m	**166.50**
Sum of two sides 1500mm	70.22	119.41	1.68	49.76	m	**169.16**
Sum of two sides 1700mm	88.54	150.54	1.77	52.42	m	**202.97**
Sum of two sides 1900mm	88.54	150.54	1.86	55.09	m	**205.63**
Sum of two sides 2100mm	91.22	155.12	2.58	76.41	m	**231.53**
Sum of two sides 2300mm	91.22	155.12	2.64	78.19	m	**233.31**
Sum of two sides 2500mm	91.22	155.12	2.91	86.19	m	**241.30**
Ductwork 1251 to 2000mm longest side						
Sum of two sides 1800mm	338.92	576.29	6.00	177.70	m	**754.00**
Sum of two sides 2000mm	338.92	576.29	6.00	177.70	m	**754.00**
Sum of two sides 2200mm	473.92	805.84	6.50	192.51	m	**998.35**
Sum of two sides 2400mm	473.92	805.84	6.50	192.51	m	**998.35**
Sum of two sides 2600mm	473.92	805.84	6.66	197.25	m	**1003.09**
Sum of two sides 2800mm	561.47	954.71	6.66	197.25	m	**1151.96**
Sum of two sides 3000mm	561.47	954.71	6.66	197.25	m	**1151.96**
Sum of two sides 3200mm	701.09	1192.11	9.00	266.56	m	**1458.67**
Sum of two sides 3400mm	701.09	1192.11	9.00	266.56	m	**1458.67**
Sum of two sides 3600mm	806.98	1372.17	11.70	346.52	m	**1718.69**
Sum of two sides 3800mm	806.98	1372.17	11.70	346.52	m	**1718.69**
Sum of two sides 4000mm	806.98	1372.17	11.70	346.52	m	**1718.69**

U:VENTILATION/AIR CONDITIONING SYSTEMS

Item	Net Price £	Material £	Labour hours	Labour £	Unit	Total rate £
Extra over fittings; Ductwork 1251 to 2000mm longest sides						
End Cap						
Sum of two sides 1800mm	131.24	223.16	1.45	42.95	m	**266.10**
Sum of two sides 2000mm	131.24	223.16	1.51	44.72	m	**267.88**
Sum of two sides 2200mm	154.70	263.04	2.80	82.93	m	**345.97**
Sum of two sides 2400mm	154.70	263.04	3.10	91.81	m	**354.86**
Sum of two sides 2600mm	210.96	358.70	3.39	100.40	m	**459.11**
Sum of two sides 2800mm	210.96	358.70	3.68	108.99	m	**467.70**
Sum of two sides 3000mm	210.96	358.70	3.98	117.88	m	**476.58**
Sum of two sides 3200mm	276.58	470.29	4.27	126.47	m	**596.76**
Sum of two sides 3400mm	276.58	470.29	4.57	135.35	m	**605.64**
Sum of two sides 3600mm	348.46	592.52	4.87	144.24	m	**736.76**
Sum of two sides 3800mm	348.46	592.52	5.17	153.12	m	**745.64**
Sum of two sides 4000mm	348.46	592.52	5.47	162.01	m	**754.53**
Reducer						
Sum of two sides 1800mm	100.52	170.92	3.60	106.62	m	**277.54**
Sum of two sides 2000mm	100.52	170.92	3.70	109.58	m	**280.50**
Sum of two sides 2200mm	117.47	199.75	3.10	91.81	m	**291.56**
Sum of two sides 2400mm	117.47	199.75	3.48	103.07	m	**302.82**
Sum of two sides 2600mm	164.47	279.66	3.85	114.03	m	**393.69**
Sum of two sides 2800mm	164.47	279.66	4.22	124.99	m	**404.65**
Sum of two sides 3000mm	164.47	279.66	4.59	135.94	m	**415.61**
Sum of two sides 3200mm	181.46	308.55	4.96	146.90	m	**455.46**
Sum of two sides 3400mm	181.46	308.55	5.34	158.16	m	**466.71**
Sum of two sides 3600mm	198.43	337.40	5.72	169.41	m	**506.82**
Sum of two sides 3800mm	198.43	337.40	6.10	180.67	m	**518.07**
Sum of two sides 4000mm	198.43	337.40	6.48	191.92	m	**529.33**
Offset						
Sum of two sides 1800mm	115.38	196.20	4.56	135.06	m	**331.25**
Sum of two sides 2000mm	115.38	196.20	5.53	163.78	m	**359.98**
Sum of two sides 2200mm	117.47	199.75	5.99	177.41	m	**377.16**
Sum of two sides 2400mm	117.47	199.75	6.45	191.03	m	**390.78**
Sum of two sides 2600mm	201.40	342.46	6.91	204.66	m	**547.11**
Sum of two sides 2800mm	201.40	342.46	7.37	218.28	m	**560.74**
Sum of two sides 3000mm	201.40	342.46	7.83	231.90	m	**574.36**
Sum of two sides 3200mm	249.65	424.50	8.29	245.53	m	**670.02**
Sum of two sides 3400mm	249.65	424.50	8.75	259.15	m	**683.65**
Sum of two sides 3600mm	297.91	506.55	9.21	272.78	m	**779.33**
Sum of two sides 3800mm	297.91	506.55	9.67	286.40	m	**792.95**
Sum of two sides 4000mm	297.91	506.55	10.13	300.02	m	**806.58**
90° radius bend						
Sum of two sides 1800mm	141.17	240.05	3.25	96.26	m	**336.30**
Sum of two sides 2000mm	141.17	240.05	3.30	97.74	m	**337.78**
Sum of two sides 2200mm	143.72	244.39	3.34	98.92	m	**343.31**
Sum of two sides 2400mm	143.72	244.39	3.38	100.11	m	**344.49**
Sum of two sides 2600mm	246.40	418.98	3.42	101.29	m	**520.27**
Sum of two sides 2800mm	246.40	418.98	3.46	102.48	m	**521.46**
Sum of two sides 3000mm	246.40	418.98	3.50	103.66	m	**522.64**
Sum of two sides 3200mm	305.43	519.35	3.54	104.85	m	**624.20**
Sum of two sides 3400mm	305.43	519.35	3.58	106.03	m	**625.39**
Sum of two sides 3600mm	364.48	619.75	3.62	107.22	m	**726.96**
Sum of two sides 3800mm	364.48	619.75	3.66	108.40	m	**728.15**
Sum of two sides 4000mm	364.48	619.75	3.70	109.58	m	**729.33**

U:VENTILATION/AIR CONDITIONING SYSTEMS

Item	Net Price £	Material £	Labour hours	Labour £	Unit	Total rate £
U14 : DUCTWORK : FIRE RATED (cont'd)						
Y30 - DUCTLINES (cont'd)						
Galvanised sheet metal rectangular section ductwork to BS476 Part 24 - 4 hour (cont'd)						
Extra over fittings; Ductwork 1251 to 2000mm longest side (cont'd)						
45° bend						
Sum of two sides 1800mm	141.17	240.05	2.30	68.12	m	**308.17**
Sum of two sides 2000mm	141.17	240.05	2.76	81.74	m	**321.79**
Sum of two sides 2200mm	143.72	244.39	3.01	89.15	m	**333.53**
Sum of two sides 2400mm	143.72	244.39	3.26	96.55	m	**340.94**
Sum of two sides 2600mm	246.40	418.98	3.49	103.36	m	**522.34**
Sum of two sides 2800mm	246.40	418.98	3.74	110.77	m	**529.75**
Sum of two sides 3000mm	246.40	418.98	3.98	117.88	m	**536.86**
Sum of two sides 3200mm	305.43	519.35	4.22	124.99	m	**644.34**
Sum of two sides 3400mm	305.43	519.35	4.46	132.09	m	**651.45**
Sum of two sides 3600mm	364.48	619.75	4.70	139.20	m	**758.95**
Sum of two sides 3800mm	364.48	619.75	4.94	146.31	m	**766.06**
Sum of two sides 4000mm	364.48	619.75	5.18	153.42	m	**773.17**
90° mitre band						
Sum of two sides 1800mm	141.17	240.05	3.90	115.51	m	**355.55**
Sum of two sides 2000mm	141.17	240.05	3.96	117.28	m	**357.33**
Sum of two sides 2200mm	143.72	244.39	3.83	113.43	m	**357.82**
Sum of two sides 2400mm	143.72	244.39	3.84	113.73	m	**358.12**
Sum of two sides 2600mm	246.40	418.98	3.85	114.03	m	**533.01**
Sum of two sides 2800mm	246.40	418.98	3.86	114.32	m	**533.30**
Sum of two sides 3000mm	246.40	418.98	3.87	114.62	m	**533.60**
Sum of two sides 3200mm	305.43	519.35	3.88	114.92	m	**634.27**
Sum of two sides 3400mm	305.43	519.35	3.90	115.51	m	**634.86**
Sum of two sides 3600mm	364.48	619.75	3.92	116.10	m	**735.85**
Sum of two sides 3800mm	364.48	619.75	3.94	116.69	m	**736.44**
Sum of two sides 4000mm	364.48	619.75	3.96	117.28	m	**737.03**
Branch						
Sum of two sides 1800mm	88.54	150.54	1.81	53.61	m	**204.15**
Sum of two sides 2000mm	88.54	150.54	1.90	56.27	m	**206.82**
Sum of two sides 2200mm	91.22	155.12	2.61	77.30	m	**232.42**
Sum of two sides 2400mm	91.22	155.12	2.68	79.37	m	**234.49**
Sum of two sides 2600mm	110.01	187.06	2.75	81.45	m	**268.50**
Sum of two sides 2800mm	110.01	187.06	2.81	83.22	m	**270.28**
Sum of two sides 3000mm	110.01	187.06	2.87	85.00	m	**272.06**
Sum of two sides 3200mm	112.69	191.61	2.93	86.78	m	**278.39**
Sum of two sides 3400mm	112.69	191.61	3.00	88.85	m	**280.46**
Sum of two sides 3600mm	144.89	246.37	3.06	90.63	m	**337.00**
Sum of two sides 3800mm	144.89	246.37	3.12	92.41	m	**338.77**
Sum of two sides 4000mm	144.89	246.37	3.18	94.18	m	**340.55**

U:VENTILATION/AIR CONDITIONING SYSTEMS

Item	Net Price £	Material £	Labour hours	Labour £	Unit	Total rate £
U30 : LOW VELOCITY AIR CONDITIONING						
Y40 - AIR HANDLING UNITS						
Supply air handling unit; inlet with motorised damper, LTHW frost coil (at -5°C to +5°C), panel filter (EU4), bag filter (EU6), cooling coil (at 28°C db/20°C wb to 12°C db/11.5°C wb), LTHW heating coil (at 5°C to 21°C), supply fan, outlet plenum; includes access sections; all units located internally; includes placing in position and fitting of sections together; electrical work elsewhere.						
Volume, external pressure						
2 m³/s at 350 Pa	4325.25	5539.74	40.00	1014.88	nr	**6554.61**
2 m³/s at 700 Pa	4620.90	5918.40	40.00	1014.88	nr	**6933.28**
5 m³/s at 350 Pa	7293.25	9341.12	65.00	1649.17	nr	**10990.29**
5 m³/s at 700 Pa	7548.38	9667.89	65.00	1649.17	nr	**11317.07**
8 m³/s at 350 Pa	10923.72	13990.99	77.00	1953.64	nr	**15944.63**
8 m/³s at 700 Pa	11093.99	14209.07	77.00	1953.64	nr	**16162.71**
10 m/³ at 350 Pa	12029.12	15406.78	100.00	2537.19	nr	**17943.97**
10 m³/s at 700 Pa	12276.05	15723.04	100.00	2537.19	nr	**18260.23**
13 m³/s at 350 Pa	15032.16	19253.04	108.00	2740.17	nr	**21993.21**
13 m³/s at 700 Pa	15496.44	19847.69	108.00	2740.17	nr	**22587.85**
15 m³/s at 350 Pa	16850.41	21581.83	120.00	3044.63	nr	**24626.46**
15 m³/s at 700 Pa	17157.01	21974.52	120.00	3044.63	nr	**25019.15**
18 m³/s at 350 Pa	19120.89	24489.84	133.00	3374.46	nr	**27864.31**
18 m³/s at 700 Pa	19472.39	24940.04	133.00	3374.46	nr	**28314.50**
20 m³/s at 350 Pa	22157.87	28379.58	142.00	3602.81	nr	**31982.39**
20 m³/s at 700 Pa	22476.52	28787.70	142.00	3602.81	nr	**32390.51**
Extra for inlet and discharge attenuators at 900mm long						
2 m³/s at 350 Pa	1330.43	1704.00	5.00	126.86	nr	**1830.85**
5 m³/s at 350 Pa	2792.25	3576.29	10.00	253.72	nr	**3830.00**
10 m³/s at 700 Pa	3903.68	4999.79	13.00	329.83	nr	**5329.62**
15 m³/s at 700 Pa	5818.28	7452.00	16.00	405.95	nr	**7857.95**
20 m³/s at 700 Pa	7444.91	9535.36	20.00	507.44	nr	**10042.80**
Extra for locating units externally						
2 m³/s at 350 Pa	985.77	1262.57	-	-	nr	**1262.57**
5 m³/s at 350 Pa	1250.76	1601.96	-	-	nr	**1601.96**
10 m³/s at 700 Pa	1966.18	2518.26	-	-	nr	**2518.26**
15 m³/s at 700 Pa	2792.85	3577.05	-	-	nr	**3577.05**
20 m³/s at 700 Pa	4799.88	6147.63	-	-	nr	**6147.63**

U:VENTILATION/AIR CONDITIONING SYSTEMS

Item	Net Price £	Material £	Labour hours	Labour £	Unit	Total rate £
U30 : LOW VELOCITY AIR CONDITIONING (cont'd)						
Y40 – AIR HANDLING UNITS (cont'd)						
Modular air handling unit with supply and extract sections. Supply side; inlet with motorised damper, LTHW frost coil (at -5°C to 5°C), panel filter (EU4), bag filter (EU6), cooling coil at 28°Cdb/20°Cwb to 12°Cdb/ 11.5°C wb), LTHW heating coil (at 5°C to 21°C), supply fan, outlet plenum. Extract side; inlet with motorised damper, extract fan; includes access sections; placing in position and fitting of sections together; electrical work elsewhere.						
2 m³/s at 350 Pa	6042.21	7738.80	50.00	1268.59	nr	**9007.40**
2 m³/s at 700 Pa	6334.58	8113.26	50.00	1268.59	nr	**9381.86**
5 m³/s at 350 Pa	10413.45	13337.44	86.00	2181.98	nr	**15519.43**
5 m³/s at 700 Pa	10608.91	13587.78	86.00	2181.98	nr	**15769.77**
8 m³/s at 350 Pa	14460.57	18520.95	105.00	2664.05	nr	**21185.00**
8 m³/s at 700 Pa	14807.69	18965.53	105.00	2664.05	nr	**21629.58**
10 m³/s at 350 Pa	16697.65	21386.19	120.00	3044.63	nr	**24430.82**
10 m³/s at 700 Pa	17057.36	21846.90	120.00	3044.63	nr	**24891.53**
13 m³/s at 350 Pa	20929.28	26806.02	130.00	3298.35	nr	**30104.36**
13 m³/s at 700 Pa	21585.74	27646.79	130.00	3298.35	nr	**30945.14**
15 m³/s at 350 Pa	23387.01	29953.85	145.00	3678.93	nr	**33632.77**
15 m³/s at 700 Pa	23974.48	30706.27	145.00	3678.93	nr	**34385.20**
18 m³/s at 350 Pa	26167.76	33515.41	160.00	4059.50	nr	**37574.91**
18 m³/s at 700 Pa	26780.42	34300.09	160.00	4059.50	nr	**38359.59**
20 m³/s at 350 Pa	28600.31	36630.98	175.00	4440.08	nr	**41071.07**
20 m³/s at 700 Pa	29594.57	37904.42	175.00	4440.08	nr	**42344.51**
Extra for inlet and discharge attenuators at 900mm long						
2 m³/s at 350 Pa	2425.43	3106.46	8.00	202.98	nr	**3309.44**
5 m³/s at 350 Pa	4789.53	6134.38	10.00	253.72	nr	**6388.10**
10 m³/s at 700 Pa	6734.25	8625.16	13.00	329.83	nr	**8954.99**
15 m³/s at 700 Pa	11322.30	14501.49	16.00	405.95	nr	**14907.44**
20 m³/s at 700 Pa	13104.96	16784.70	20.00	507.44	nr	**17292.14**
Extra for locating units externally						
2 m³/s at 350 Pa	1941.45	2486.59	-	-	nr	**2486.59**
5 m³/s at 350 Pa	2756.84	3530.93	-	-	nr	**3530.93**
10 m³/s at 700 Pa	3866.78	4952.53	-	-	nr	**4952.53**
15 m³/s at 700 Pa	7216.48	9242.79	-	-	nr	**9242.79**
20 m³/s at 700 Pa	10999.94	14088.61	-	-	nr	**14088.61**
Extra for humidifier, self generating type						
2 m³/s at 350 Pa (10kg/hr)	1593.15	2040.49	5.00	126.86	nr	**2167.35**
5 m³/s at 350 Pa (18kg/hr)	1971.55	2525.14	5.00	126.86	nr	**2652.00**
10 m³/s at 700 Pa (30kg/hr)	2243.53	2873.48	6.00	152.23	nr	**3025.72**
15 m³/s at 700 Pa (60kg/hr)	4154.87	5321.52	8.00	202.98	nr	**5524.50**
20 m³/s at 700 Pa (90kg/hr)	6246.83	8000.87	10.00	253.72	nr	**8254.59**
Extra for mixing box						
2 m³/s at 350 Pa	798.26	1022.40	4.00	101.49	nr	**1123.88**
5 m³/s at 350 Pa	1105.95	1416.49	4.00	101.49	nr	**1517.98**
10 m³/s at 700 Pa	1536.29	1967.66	5.00	126.86	nr	**2094.52**
15 m³/s at 700 Pa	2050.39	2626.12	6.00	152.23	nr	**2778.35**
20 m³/s at 700 Pa	3073.67	3936.72	6.00	152.23	nr	**4088.95**

U:VENTILATION/AIR CONDITIONING SYSTEMS

Item	Net Price £	Material £	Labour hours	Labour £	Unit	Total rate £
Extra for runaround coil, including pump and associated pipework; typical outputs in brackets. (based on minimal distance between the supply and extract units)						
2 m³/s at350 Pa (26kW)	2389.73	3060.74	30.00	761.16	nr	**3821.89**
5 m³/s at 350 Pa (37kW)	4164.55	5333.91	30.00	761.16	nr	**6095.07**
10 m³/s at 700 Pa (85kW)	7149.29	9156.74	40.00	1014.88	nr	**10171.61**
15 m³/s at 700 Pa (151kW)	10059.31	12883.87	50.00	1268.59	nr	**14152.46**
20 m³/s at 700 Pa (158kW)	13483.73	17269.82	60.00	1522.31	nr	**18792.13**
Extra for thermal wheel (typical outputs in brackets)						
2 m³/s at350 Pa (37kW)	5167.52	6618.51	12.00	304.46	nr	**6922.98**
5 m³/s at 350 Pa (65kW)	7421.80	9505.77	12.00	304.46	nr	**9810.23**
10 m3/s at 700 Pa (127kW)	9897.53	12676.65	15.00	380.58	nr	**13057.23**
15 m3/s at 700 Pa (160kW)	18543.75	23750.65	17.00	431.32	nr	**24181.97**
20 m³/s at 700 Pa (262kW)	21661.25	27743.51	19.00	482.07	nr	**28225.58**
Extra for plate heat exchanger, including additional filtration in extract leg (typical outputs in brackets)						
2 m³/s at350 Pa (25kW)	2882.08	3691.33	12.00	304.46	nr	**3995.80**
5 m³/s at 350 Pa (51kW)	5634.08	7216.07	12.00	304.46	nr	**7520.53**
10 m³/s at 700 Pa (98kW)	8266.75	10587.97	15.00	380.58	nr	**10968.55**
15 m³/s at 700 Pa (160kW)	14276.00	18284.56	17.00	431.32	nr	**18715.88**
20 m3/s at 700 Pa (190kW)	18436.25	23612.96	19.00	482.07	nr	**24095.03**
Extra for electric heating in lieu of LTHW						
2 m³/s at350 Pa	1085.75	1390.62	-	-	nr	**1390.62**
5 m³/s at 350 Pa	1999.50	2560.94	-	-	nr	**2560.94**
10 m3/s at 700 Pa	2806.83	3594.95	-	-	nr	**3594.95**
15 m3/s at 700 Pa	2676.75	3428.35	-	-	nr	**3428.35**
20 m³/s at 700 Pa	3257.25	4171.85	-	-	nr	**4171.85**

U:VENTILATION/AIR CONDITIONING SYSTEMS

Item	Net Price £	Material £	Labour hours	Labour £	Unit	Total rate £
U31 : VAV AIR CONDITIONING						
VAV TERMINAL BOXES						
VAV terminal box; integral acoustic silencer; factory installed and pre wired control components (excluding electronic controller); selected at 200Pa at entry to unit; includes fixing in position; electrical work elsewhere						
80 l/s - 110 l/s	331.80	564.19	2.00	59.23	nr	623.42
Extra for secondary silencer	80.85	137.48	0.50	14.81	nr	152.28
Extra for 2 row LTHW heating coil	31.50	53.56	-	-	nr	53.56
150 l/s - 190 l/s	349.13	593.65	2.00	59.23	nr	652.88
Extra for secondary silencer	88.72	150.87	0.50	14.81	nr	165.67
Extra for 2 row LTHW heating coil	37.80	64.27	-	-	nr	64.27
250 l/s - 310 l/s	389.55	662.38	2.00	59.23	nr	721.62
Extra for secondary silencer	117.08	199.07	0.50	14.81	nr	213.88
Extra for 2 row LTHW heating coil	49.35	83.91	-	-	nr	83.91
420 l/s - 520 l/s	421.58	716.84	2.00	59.23	nr	776.07
Extra for secondary silencer	133.35	226.75	0.50	14.81	nr	241.55
Extra for 2 row LTHW heating coil	51.45	87.48	-	-	nr	87.48
650 l/s - 790 l/s	510.83	868.60	2.00	59.23	nr	927.83
Extra for secondary silencer	170.63	290.13	0.50	14.81	nr	304.94
Extra for 2 row LTHW heating coil	67.20	114.27	-	-	nr	114.27
1130 l/s - 1370 l/s	593.25	1008.75	2.00	59.23	nr	1067.99
Extra for secondary silencer	238.35	405.29	0.50	14.81	nr	420.09
Extra for 2 row LTHW heating coil	85.58	145.51	-	-	nr	145.51
Extra for electric heater & thyristor controls, 3kw/1ph (per box)	415.70	706.84	-	-	nr	706.84
Extra for fitting free issue box controller	-	-	2.13	63.09	nr	63.09
Fan assisted VAV terminal box; factory installed and pre wired control components (excluding electronic controller); selected at 40 Pa external static pressure; includes fixing in position, electrical work elsewhere						
100 l/s - 175 l/s	733.95	1247.99	3.00	88.85	nr	1336.85
Extra for secondary silencer	87.68	149.08	0.50	14.81	nr	163.89
Extra for 1 row LTHW heating coil	44.63	75.88	-	-	nr	75.88
170 l/s - 360 l/s	800.63	1361.37	3.00	88.85	nr	1450.22
Extra for secondary silencer	110.78	188.36	0.50	14.81	nr	203.17
Extra for 1 row LTHW heating coil	50.92	86.59	-	-	nr	86.59
300 l/s - 640 l/s	918.23	1561.33	3.00	88.85	nr	1650.18
Extra for secondary silencer	169.58	288.34	0.50	14.81	nr	303.15
Extra for 1 row LTHW heating coil	56.70	96.41	-	-	nr	96.41
620 l/s - 850 l/s	918.23	1561.33	3.00	88.85	nr	1650.18
Extra for secondary silencer	169.58	288.34	0.50	14.81	nr	303.15
Extra for 1 row LTHW heating coil	56.70	96.41	-	-	nr	96.41
Extra for electric heater plus thyristor controls' 3kw/lph (per box)	415.70	706.84	-	-	nr	706.84
Extra for fitting free issue controller	-	-	2.13	63.09	nr	63.09

U:VENTILATION/AIR CONDITIONING SYSTEMS

Item	Net Price £	Material £	Labour hours	Labour £	Unit	Total rate £
U41 : FAN COIL AIR CONDITIONING						
FAN COIL UNITS						
All selections based on summer return air condition of 23°C at 50% RH, CHW at 6°/12°C, LTHW at 82°/71°C (where applicable), medium speed, external resistance of 30Pa.						
All selections are based on heating and cooling units. For waterside control units there is no significant reduction in cost between 4 pipe heating and cooling and 2 pipe cooling only units (excluding controls). For airside control units, there is a marginal reduction (less than 5%) between 4 pipe heating and cooling units and 2 pipe cooling only units (excluding controls).						
Ceiling void mounted horizontal waterside control fan coil unit; cooling coil ; LTHW heating coil; multi tapped speed transformer; fine wire mesh filter; includes fixing in position; electrical work elsewhere						
Total cooling load, heating load						
2800 W, 1000 W	305.91	391.81	4.00	101.49	nr	**493.29**
4000 W, 1700 W	361.53	463.04	4.00	101.49	nr	**564.53**
4500 W, 1900 W	382.13	489.43	4.00	101.49	nr	**590.92**
6000 W, 2600 W	499.55	639.82	4.00	101.49	nr	**741.31**
Ceiling void mounted horizontal waterside control fan coil unit; cooling coil ; electric heating coil; multi tapped speed transformer; fine wire mesh filter; includes fixing in position; electrical work elsewhere						
Total cooling load, heating load						
2800W, 1500W.	417.15	534.28	4.00	101.49	nr	**635.77**
4000W, 2000W.	469.68	601.56	4.00	101.49	nr	**703.05**
4500W, 2000W	495.43	634.54	4.00	101.49	nr	**736.03**
6000W, 3000W	635.51	813.95	4.00	101.49	nr	**915.44**
Ceiling void mounted horizontal airside control fan coil unit; cooling coil ; LHTW heating coil; multi tapped speed transformer; fine wire mesh filter, damper actuator & fixing kit; includes fixing in position; electrical work elsewhere						
Total cooling load, heating load						
2600W, 2200W.	391.40	501.30	4.00	101.49	nr	**602.79**
3600W, 3200W	442.90	567.26	4.00	101.49	nr	**668.75**
4000W, 3600W	474.83	608.16	4.00	101.49	nr	**709.65**
5400W, 5000W	618.00	791.53	4.00	101.49	nr	**893.02**

U:VENTILATION/AIR CONDITIONING SYSTEMS

Item	Net Price £	Material £	Labour hours	Labour £	Unit	Total rate £
U41 : FAN COIL AIR CONDITIONING (cont'd)						
FAN COIL UNITS (cont'd)						
Ceiling void mounted horizontal airside control fan coil unit; cooling coil ; electric heating coil; multi tapped speed transformer; fine wire mesh filter, damper actuator & fixing kit; includes fixing in position; electrical work elsewhere						
Total cooling load, heating load						
2600W, 1500W.	482.04	617.39	4.00	101.49	nr	**718.88**
3600W, 2000W	547.96	701.82	4.00	101.49	nr	**803.31**
4000W, 2000W	597.40	765.14	4.00	101.49	nr	**866.63**
5400W, 3000W	721.00	923.45	4.00	101.49	nr	**1024.94**
Ceiling void mounted slimline horizontal waterside control fan coil unit, 170mm deep; cooling coil; LTHW heating coil; multi tapped speed transformer; fine wire mesh filter; includes fixing in position; electrical work elsewhere						
Total cooling load, heating load						
1100W, 1500W.	329.60	422.15	3.50	88.80	nr	**510.95**
3200W, 3700W	463.50	593.65	4.00	101.49	nr	**695.13**
4600W, 5000W	611.82	783.61	4.00	101.49	nr	**885.10**
Ceiling void mounted slimline horizontal waterside control fan coil unit, 170mm deep; cooling coil; electric heating coil; multi tapped speed transformer; fine wire mesh filter; includes fixing in position; electrical work elsewhere						
Total cooling load, heating load						
1100W, 1000W.	473.80	606.84	3.50	88.80	nr	**695.64**
3400W, 2000W	587.10	751.95	4.00	101.49	nr	**853.44**
4800W, 3000W	731.30	936.64	4.00	101.49	nr	**1038.13**
Ceiling void mounted slimline horizontal airside control fan coil unit, 170mm deep; cooling coil; LTHW heating coil; multi tapped speed transformer; fine wire mesh filter, damper actuator & fixing kit; includes fixing in position; electrical work elsewhere						
Total cooling load, heating load						
1000W, 1600W.	363.59	465.68	3.50	88.80	nr	**554.48**
3000W, 3300W	566.50	725.57	4.00	101.49	nr	**827.06**
4500W, 4500W	726.15	930.05	4.00	101.49	nr	**1031.53**
Ceiling void mounted slimline horizontal airside control fan coil unit, 170mm deep; cooling coil; electric heating coil; multi tapped speed transformer; fine wire mesh filter, damper actuator & fixing kit; includes fixing in position; electrical work elsewhere						
Total cooling load, heating load						
1000W, 1000W.	535.60	685.99	3.50	88.80	nr	**774.79**
3000W, 2000W	705.55	903.66	4.00	101.49	nr	**1005.15**
4500W, 3000W	896.10	1147.72	4.00	101.49	nr	**1249.20**

U:VENTILATION/AIR CONDITIONING SYSTEMS

Item	Net Price £	Material £	Labour hours	Labour £	Unit	Total rate £
All selections are based on heating and cooling units (cont'd)						
Low level perimeter waterside control fan coil unit; cooling coil; LTHW heating coil; multi tapped speed transformer; fine wire mesh filter; includes fixing in position; electrical work elsewhere						
Total cooling load, heating load						
1700W, 1400W	257.50	329.80	3.50	88.80	nr	**418.61**
Extra over for standard cabinet	136.99	175.46	1.00	25.37	nr	**200.83**
2200W, 1900W	298.70	382.57	3.50	88.80	nr	**471.37**
Extra over for standard cabinet	166.86	213.71	1.00	25.37	nr	**239.08**
2600W, 2200W	329.60	422.15	3.50	88.80	nr	**510.95**
Extra over for standard cabinet	170.98	218.99	1.00	25.37	nr	**244.36**
3900W, 3200W	432.60	554.07	3.50	88.80	nr	**642.87**
Extra over for standard cabinet	186.43	238.78	1.00	25.37	nr	**264.15**
4600W, 3900W	494.40	633.22	3.50	88.80	nr	**722.02**
Extra over for standard cabinet	245.14	313.97	1.00	25.37	nr	**339.34**
Low level perimeter waterside control fan coil unit; cooling coil; electric heating coil; multi tapped speed transformer; fine wire mesh filter; includes fixing in position; electrical work elsewhere						
Total cooling load, heating load						
1700W, 1500W	334.75	428.74	3.50	88.80	nr	**517.55**
Extra over for standard cabinet	136.99	175.46	1.00	25.37	nr	**200.83**
2200W, 2000W	350.20	448.53	3.50	88.80	nr	**537.33**
Extra over for standard cabinet	166.86	213.71	1.00	25.37	nr	**239.08**
2600W, 2000W	437.75	560.67	3.50	88.80	nr	**649.47**
Extra over for standard cabinet	170.98	218.99	1.00	25.37	nr	**244.36**
3800W, 3000W	520.15	666.20	3.50	88.80	nr	**755.00**
Extra over for standard cabinet	186.43	238.78	1.00	25.37	nr	**264.15**
4600W, 4000W	633.45	811.32	3.50	88.80	nr	**900.12**
Extra over for standard cabinet	245.14	313.97	1.00	25.37	nr	**339.34**
Low level perimeter airside control fan coil unit; cooling coil; LTHW heating coil; multi tapped speed transformer; fine wire mesh filter; damper actuator & fixing kit; includes fixing in position; electrical work elsewhere						
Total cooling load, heating load						
1200W, 1400W	332.69	426.11	3.50	88.80	nr	**514.91**
Extra over for standard cabinet	136.99	175.46	1.00	25.37	nr	**200.83**
1800W, 2000W	371.83	476.24	3.50	88.80	nr	**565.04**
Extra over for standard cabinet	166.86	213.71	1.00	25.37	nr	**239.08**
2200W, 2400W	400.67	513.17	3.50	88.80	nr	**601.98**
Extra over for standard cabinet	170.98	218.99	-	-	nr	**218.99**
3200W, 3600W	520.15	666.20	3.50	88.80	nr	**755.00**
Extra over for standard cabinet	186.43	238.78	1.00	25.37	nr	**264.15**

U:VENTILATION/AIR CONDITIONING SYSTEMS

Item	Net Price £	Material £	Labour hours	Labour £	Unit	Total rate £
U41 : FAN COIL AIR CONDITIONING (cont'd)						
FAN COIL UNITS (cont'd)						
Low level perimeter airside control fan coil unit; cooling coil; electric heating coil; multi tapped speed transformer; fine wire mesh filter; damper actuator & fixing kit; includes fixing in position; electrical work elsewhere						
Total cooling load, heating load						
1250W, 1500W	453.20	580.45	3.50	88.80	nr	**669.26**
Extra over for standard cabinet	136.99	175.46	1.00	25.37	nr	**200.83**
1900W, 2000W	489.25	626.63	3.50	88.80	nr	**715.43**
Extra over for standard cabinet	166.86	213.71	1.00	25.37	nr	**239.08**
2300W, 2000W	559.29	716.33	3.50	88.80	nr	**805.13**
Extra over for standard cabinet	170.98	218.99	1.00	25.37	nr	**244.36**
3300W, 3000W	638.60	817.91	3.50	88.80	nr	**906.71**
Extra over for standard cabinet	186.43	238.78	1.00	25.37	nr	**264.15**
Typical DDC control pack for airside control units; controller, return air sensor (damper and damper actuator included in above rates)						
Heating and cooling, per unit	180.25	230.86	2.00	50.74	nr	**281.61**
Cooling only, per unit	180.25	230.86	2.00	50.74	nr	**281.61**
Typical DDC control pack for waterside control units; controller, return air sensor, four port valves and actuators						
Heating and cooling, per unit	247.20	316.61	2.00	50.74	nr	**367.36**
Cooling only, per unit	216.30	277.03	2.00	50.74	nr	**327.78**
Note - Care needs to be taken when using these controls rates, as they can vary significantly, depending on the equipment manufacturer specified and the degree of control required						

U:VENTILATION/AIR CONDITIONING SYSTEMS

Item	Net Price £	Material £	Labour hours	Labour £	Unit	Total rate £
U70 - AIR CURTAINS						
The selection of air curtains requires consideration of the particular conditions involved; climatic conditions, wind influence, construction and position all influence selection; consultation with a specialist manufacturer is therefore advisable.						
Commercial grade air curtains; recessed or exposed units with rigid sheet steel casing; aluminium grilles; high quality motor/ centrifugal fan assembly; includes fixing in position; electrical work elsewhere						
Ambient temperature; 240V single phase supply; mounting height 2.40m						
1000 x 590 x 270mm	2257.76	2891.72	12.05	305.69	nr	**3197.40**
1500 x 590 x 270mm	2887.09	3697.76	12.05	305.69	nr	**4003.44**
2000 x 590 x 270mm	3496.85	4478.73	12.05	305.73	nr	**4784.46**
2500 x 590 x 270mm	3877.95	4966.84	13.00	329.83	nr	**5296.67**
Ambient temperature; 240V single phase supply; mounting height 2.80m						
1000 x 590 x 270mm	2617.23	3352.12	16.13	409.22	nr	**3761.35**
1500 x 590 x 270mm	3340.29	4278.21	16.13	409.22	nr	**4687.43**
2000 x 590 x 270mm	4104.55	5257.07	16.13	409.22	nr	**5666.29**
2500 x 590 x 270mm	4810.10	6160.73	17.10	433.86	nr	**6594.59**
Ambient temperature 240V single phase supply; mounting height 3.30m						
1000 x 774 x 370mm	3393.85	4346.81	17.24	437.45	nr	**4784.26**
1500 x 774 x 370mm	4512.43	5779.48	17.24	437.45	nr	**6216.92**
2000 x 774 x 370mm	5372.48	6881.02	17.24	437.45	nr	**7318.47**
2500 x 774 x 370mm	6463.25	8278.07	18.30	464.31	nr	**8742.37**
Ambient temperature; 240V single phase supply; mounting height 4.00m						
1000 x 774 x 370mm	3759.50	4815.13	19.10	484.60	nr	**5299.73**
1500 x 774 x 370mm	4898.68	6274.18	19.10	484.60	nr	**6758.78**
2000 x 774 x 370mm	6043.01	7739.83	19.10	484.60	nr	**8224.43**
2500 x 774 x 370mm	7084.34	9073.55	19.90	504.90	nr	**9578.45**
Water heated; 240V single phase supply; mounting height 2.40m						
1000 x 590 x 270mm; 2.30 - 9.40kW output	2595.60	3324.42	12.05	305.69	nr	**3630.10**
1500 x 590 x 270mm; 3.50 - 14.20kW output	3321.75	4254.46	12.05	305.69	nr	**4560.15**
2000 x 590 x 270mm; 4.70 - 19.00kW output	4020.09	5148.89	12.05	305.69	nr	**5454.58**
2500 x 590 x 270mm; 5.90 - 23.70kW output	4468.14	5722.75	13.00	329.83	nr	**6052.58**
Water heated; 240V single phase supply; mounting height 2.80m						
1000 x 590 x 270mm; 3.30 - 11.90kW output	2999.36	3841.55	16.13	409.22	nr	**4250.77**
1500 x 590 x 270mm; 5.00 - 17.90kW output	3829.54	4904.84	16.13	409.22	nr	**5314.06**
2000 x 590 x 270mm; 6.70 - 23.90kW output	4773.02	6113.24	16.13	409.22	nr	**6522.46**
2500 x 590 x 270mm; 8.30 - 29.80kW output	5518.74	7068.35	17.10	433.86	nr	**7502.21**

U:VENTILATION/AIR CONDITIONING SYSTEMS

Item	Net Price £	Material £	Labour hours	Labour £	Unit	Total rate £
U70 - AIR CURTAINS (cont'd)						
Commercial grade air curtains (cont'd)						
Water heated; 240V single phase supply; mounting height 3.30m						
1000 x 774 x 370mm; 6.10 - 21.80kW output	3918.12	5018.29	17.24	437.45	nr	**5455.74**
1500 x 774 x 370mm; 9.20 - 32.80kW output	5298.32	6786.04	17.24	437.45	nr	**7223.48**
2000 x 774 x 370mm; 12.30 - 43.70kW output	6551.83	8391.52	17.24	437.45	nr	**8828.97**
2500 x 774 x 370mm; 15.30 - 54.60kW output	7756.93	9935.00	18.30	464.31	nr	**10399.30**
Water heated; 240V single phase supply; mounting height 4.00m						
1000 x 774 x 370mm; 7.20 - 24.20kW output	4327.03	5542.02	19.10	484.60	nr	**6026.62**
1500 x 774 x 370mm; 10.90 - 36.30kW output	5846.28	7487.86	19.10	484.60	nr	**7972.46**
2000 x 774 x 370mm; 14.50 - 48.40kW output	7279.01	9322.88	19.10	484.60	nr	**9807.49**
2500 x 774 x 370mm; 18.10 - 60.60kW output	8530.46	10925.73	19.90	504.90	nr	**11430.63**
Electrically heated; 415V three phase supply; mounting height 2.40m						
1000 x 590 x 270mm; 2.30 - 9.40kW output	3127.08	4005.13	12.05	305.69	nr	**4310.82**
1500 x 590 x 270mm; 3.50 - 14.20kW output	3967.56	5081.61	12.05	305.69	nr	**5387.30**
2000 x 590 x 270mm; 4.70 - 19.00kW output	4724.61	6051.23	12.05	305.92	nr	**6357.15**
2500 x 590 x 270mm; 5.90 - 23.70kW output	5375.57	6884.98	13.00	330.09	nr	**7215.06**
Electrically heated; 415V three phase supply; mounting height 2.80m						
1000 x 590 x 270mm; 3.30 - 11.90kW output	3754.35	4808.53	16.13	409.54	nr	**5218.07**
1500 x 590 x 270mm; 5.00 - 17.90kW output	4616.46	5912.72	16.13	409.22	nr	**6321.94**
2000 x 590 x 270mm; 6.70 - 23.90kW output	5827.74	7464.11	16.13	409.22	nr	**7873.33**
2500 x 590 x 270mm; 8.30 - 29.80kW output	6514.75	8344.03	17.10	433.86	nr	**8777.89**
Electrically heated; 415V three phase supply; mounting height 3.30m						
1000 x 774 x 370mm; 6.10 - 21.80kW output	5703.11	7304.49	17.24	437.75	nr	**7742.23**
1500 x 774 x 370mm; 9.20 - 32.80kW output	7642.60	9788.57	17.24	437.41	nr	**10225.98**
2000 x 774 x 370mm; 12.30 - 43.70kW output	9223.65	11813.56	17.24	437.41	nr	**12250.97**
2500 x 774 x 370mm; 15.30 - 54.60kW output	10919.03	13984.98	18.30	464.31	nr	**14449.29**
Electrically heated; 415V three phase supply; mounting height 4.00m						
1000 x 774 x 370mm; 7.20 - 24.20kW output	6294.33	8061.71	19.10	484.98	nr	**8546.69**
1500 x 774 x 370mm; 10.90 - 36.30kW output	8412.01	10774.02	19.10	484.60	nr	**11258.62**
2000 x 774 x 370mm; 14.50 - 48.40kW output	10167.13	13021.96	19.10	484.60	nr	**13506.56**
2500 x 774 x 370mm; 18.10 - 60.60kW output	12051.00	15434.80	19.90	504.90	nr	**15939.70**
Industrial grade air curtains; recessed or exposed units with rigid sheet steel casing; aluminium grilles; high quality motor/ centrifugal fan assembly; includes fixing in position; electrical work elsewhere						
Ambient temperature; 415V three phase supply; including wiring between multiple units; horizontally or vertically mounted; opening maximum 6.00m						
1106 x 516 x 689mm; 1.2A supply	1908.59	2444.50	17.24	437.45	nr	**2881.95**
1661 x 516 x 689mm; 1.8A supply	2743.92	3514.39	17.24	437.45	nr	**3951.83**

U:VENTILATION/AIR CONDITIONING SYSTEMS

Item	Net Price £	Material £	Labour hours	Labour £	Unit	Total rate £
Water heated; 415V three phase supply; including wiring between multiple units; horizontally or vertically mounted; opening maximum 6.00m						
1106 x 516 x 689mm; 1.2A supply; 34.80kW output	2146.52	2749.24	17.24	437.45	nr	**3186.69**
1661 x 516 x 689mm; 1.8A supply; 50.70kW output	3100.30	3970.83	17.24	437.45	nr	**4408.28**
Water heated; 415V three phase supply; including wiring between multiple units; vertically mounted in single bank for openings maximum 6.00m wide or opposing twin banks for openings maximum 10.00m wide						
1106 x 689mm; 1.2A supply; 34.80kW output	2146.52	2749.24	17.24	437.45	nr	**3186.69**
1661 x 689mm; 1.8A supply; 50.70kW output	3100.30	3970.83	17.24	437.45	nr	**4408.28**
Remote mounted electronic controller unit; 415V three phase supply; excluding wiring to units						
Five speed, 7A	360.50	461.72	5.00	126.86	nr	**588.58**

Spon's Estimating Cost Guide to Plumbing and Heating
Unit Rates and Project Costs

Bryan Spain

All the cost data you need to keep your estimating accurate, competitive and profitable.
Do you work on jobs between £50 and £25,000? Then this book is for you.
Specially written for contractors and small businesses carrying out small works, Spon's Estimating Costs Guide to Plumbing and Heating contains accurate information on thousands of rates, each broken down into labour, material overheads and profit.
The first book to include typical project costs for:
* rainwater goods installations
* bathrooms
* external waste systems
* central heating systems
* hot and cold water systems
More than just a price book, it gives easy-to-read professional advice on setting up and running a business including:
* taxation
* book-keeping, VAT and CIS4's
* legal obligations
For the cost of approximately two hours of your charge-out rate (or less), this book will help you to:
* produce estimates faster
* keep your estimates accurate and competitive
* run your business more effectively
* help save you time
No matter how big your firm is - from one-man-band to an established business - this book contains valuable commercial and cost information that you can't afford to be without.

December 2005: 216x138 mm: 264 pages
PB: 0-415-38618-7: £27.99

To Order: Tel: +44 (0) 1264 343071 Fax: +44 (0) 1264 343005, or
Post: Taylor and Francis Customer Services, Thomson Publishing Services, Cheriton House, Andover, Hants, SP10 5BE, UK Email: book.orders@tandf.co.uk

For a complete listing of all our titles visit:
www.tandf.co.uk

Electrical Installations

Material Costs/Measured Work Prices

DIRECTIONS

The following explanations are given for each of the column headings and letter codes.

Unit	Prices for each unit are given as singular (i.e. 1 metre, 1 nr) unless stated otherwise
Net price	Industry tender prices, plus nominal allowance for fixings, waste and applicable trade discounts.
Material cost	Net price plus percentage allowance for overheads, profit and preliminaries.
Labour norms	In man-hours for each operation.
Labour cost	Labour constant multiplied by the appropriate all-in man-hour cost based on gang rate. (See also relevant Rates of Wages Section) plus percentage allowance for overheads, profit and preliminaries
Measured work Price (total rate)	Material cost plus Labour cost.

MATERIAL COSTS

The Material Costs given are based at Third Quarter 2006 but exclude any charges in respect of VAT. At the time of going to press, the raw material cost of copper was falling but had not been reflected within factory exit prices for finished goods i.e. cable etc. Users of the book are advised to register on the SPON's website www.pricebooks.co.uk/updates to receive the free quarterly updates - alerts will then be provided by e-mail as changes arise.

MEASURED WORK PRICES

These prices are intended to apply to new work in the London area. The prices are for reasonable quantities of work and the user should make suitable adjustments if the quantities are especially small or especially large. Adjustments may also be required for locality (e.g. outside London – refer to cost indices in approximate estimating section for details of adjustment factors) and for the market conditions (e.g. volume of work secured or being tendered) at the time of use.

ELECTRICAL INSTALLATIONS

The labour rate has been based on average gang rates per man hour effective from 8 January 2007 including allowances for all other emoluments and expenses. To this rate has been added 8% and 6% to cover site and head office overheads and preliminary items together with a further 3% for profit, resulting in an inclusive rate of £25.39 per man hour. The rate has been calculated on a working year of 2025 hours; a detailed build-up of the rate is given at the end of these directions.

In calculating the 'Measured Work Prices' the following assumptions have been made:
- (a)   That the work is carried out as a sub-contract under the Standard Form of Building Contract.
- (b)   That, unless otherwise stated, the work is being carried out in open areas at a height which would not require more than simple scaffolding.
- (c)   That the building in which the work is being carried out is no more than six storey's high.

Where these assumptions are not valid, as for example where work is carried out in ducts and similar confined spaces or in multi-storey structures when additional time is needed to get to and from upper floors, then an appropriate adjustment must be made to the prices. Such adjustment will normally be to the labour element only.

DIRECTIONS

LABOUR RATE - ELECTRICAL

The annual cost of a notional eleven man gang

		TECHNICIAN 1 NR	APPROVED ELECTRICIANS 4 NR	ELECTRICIANS 4 NR	LABOURERS 2 NR	SUB-TOTALS
Hourly Rate from 8 January 2007		15.86	14.09	12.99	10.42	
Working hours per annum per man		1,710.00	1,710.00	1,710.00	1,710.00	
x Hourly rate x nr of men = £ per annum		27,120.60	96,375.60	88,851.60	35,636.40	247,984.20
Overtime Rate		23.79	21.14	19.49	15.63	
Overtime hours per annum per man		315.00	315.00	315.00	315.00	
x Hourly rate x nr of men = £ per annum		7,493.85	26,630.10	24,551.10	9,846.90	68,521.95
Total		34,614.45	123,005.70	113,402.70	45,483.30	316,506.15
Incentive schemes (insert percentage)	5.00%	1,730.72	6,150.29	5,670.14	2,274.17	15,825.31
Daily Travel Time Allowance (15-20 miles each way) effective from 9/01/06		4.40	4.40	4.40	4.40	
Days per annum per man		225.00	225.00	225.00	225.00	
x nr of men = £ per annum		990.00	3,960.00	3,960.00	1,980.00	10,890.00
Daily Travel Allowance (15-20 miles each way) effective from 9/01/06		3.03	3.03	3.03	3.03	
Days per annum per man		225.00	225.00	225.00	225.00	
x nr of men = £ per annum		681.75	2,727.00	2,727.00	1,636.50	7,499.25
JIB Pension Scheme @ 2.5%		1,039.38	3,711.56	3,434.41	1,393.43	9,578.78
JIB combined benefits scheme (nr of weeks per man)		52.00	52.00	52.00	52.00	
Benefit Credit		52.75	48.00	45.00	38.18	
x nr of men = £ per annum		2,743.00	9,984.00	9,360.00	3,970.72	26,057.72
Holiday Top-up Funding		28.79	25.78	23.96	19.51	
x nr of men @ 7.5 hrs per day = £ per annum		1,497.08	5,362.24	4,983.68	2,021.04	13,872.04
National Insurance Contributions:						
Annual gross pay (subject to NI) each		38,016.92	135,842.99	125,759.84	51,100.97	
% of NI Contributions		12.80	12.80	12.80	12.80	
£ Contributions/annum		4,211.67	14,769.93	13,479.29	5,231.94	37,692.82

DIRECTIONS

LABOUR RATE - ELECTRICAL

The annual cost of a notional eleven man gang

		SUB-TOTALS
SUB-TOTAL		437,922.07
TRAINING (INCLUDING ANY TRADE REGISTRATIONS) - SAY	1.00%	4,379.22
SEVERANCE PAY AND SUNDRY COSTS - SAY	1.50%	6,634.52
EMPLOYER'S LIABILITY AND THIRD PARTY INSURANCE - SAY	2.00%	8,978.72
ANNUAL COST OF NOTIONAL GANG		457,914.53
MEN ACTUALLY WORKING = 10.5 THEREFORE ANNUAL COST PER PRODUCTIVE MAN		43,610.91
AVERAGE NR OF HOURS WORKED PER MAN = 2025 THEREFORE ALL IN MAN HOUR		21.54
PRELIMINARY ITEMS - SAY	8.00%	1.72
SITE AND HEAD OFFICE OVERHEADS - SAY	6.00%	1.40
PROFIT - SAY	3.00%	0.74
THEREFORE INCLUSIVE MAN HOUR		25.39

Notes:

(1)   Hourly wage rates are those effective from 8 January 2007.
(2)   The following assumptions have been made in the above calculations:-
     (a)   Hourly rates are based on London rate and job reporting own transport.
     (b)   The working week of 37.5 hours is made up of 7.5 hours Monday to Friday.
     (c)   Five days in the year are lost through sickness or similar reason.
     (d)   A working year of 2025 hours, including overtime worked.
(3)   The incentive scheme addition of 5% is intended to reflect bonus schemes typically in use.
(4)   National insurance contributions are those effective from 6 April 2006.
(5)   Weekly Holiday Credit/Welfare Stamp values are those effective from 26 September 2005.
(6)   Fares (New Malden to Waterloo is £6.00 plus £4.00 Zone 1 Return) current at June 2006
(7)   Does not include for major project status.
(8)   Caution should be applied when utilising the labour rate and the 'all-in' rate it applies to, as the size and complexity of the project will reflect in the gang size.
(9)   Paid Holidays with effect from 9 January 2006, for all 30 days (22 Annual and 8 Public) are to be paid at normal earnings level.

Spon's First Stage Estimating Handbook

Bryan Spain

The ultimate paradox of construction project management is that the earlier a decision has to be made, the more difficult the task is of estimating the cost implications. Yet the most important decisions are taken earliest - when the cost estimates are the most difficult to obtain.

Spon's First Stage Estimating Handbook is the only comprehensive and reliable source of first stage estimating costs. Covering the whole spectrum of building, civil engineering, mechanical and electrical work, and landscaping work, vital cost data is presented as costs per square metre, elemental cost analyses, principal rates and composite rates. Compact and clear, Spon's First Stage Estimating Handbook is ideal for meetings with clients when the inevitable question 'just an idea ... a ballpark figure ...' needs an accurate response.

With additional sections on whole life costing and general information, this is an essential reference for all construction professionals and clients making early judgements on the viability of new projects.

April 2006: 216x138 mm: 240 pages
PB: 0-415-38619-5: £40.00

To Order: Tel: +44 (0) 1264 343071 Fax: +44 (0) 1264 343005, or
Post: Taylor and Francis Customer Services, Thomson Publishing Services, Cheriton House, Andover, Hants, SP10 5BE, UK Email: book.orders@tandf.co.uk

For a complete listing of all our titles visit:
www.tandf.co.uk

V:ELECTRICAL SUPPLY/POWER/LIGHTING SYSTEMS

Item	Net Price £	Material £	Labour hours	Labour £	Unit	Total rate £
V10 : ELECTRICAL GENERATION PLANT						
STANDBY GENERATORS						
Standby diesel generating sets; supply and installation; fixing to base; all supports and fixings; all necessary connections to equipment						
Three phase, 400 Volt, four wire 50 Hz packaged standby diesel generating set, complete with radio and television suppressors, daily service fuel tank and associated piping, 4 metres of exhaust pipe and primary exhaust silencer, control panel, mains failure relay, starting battery with charger, all internal wiring, interconnections, earthing and labels. Rated for standby duty; including UK delivery, installation and commissioning.						
60 kVA	18835.40	22291.14	100.00	2539.14	nr	**24830.28**
100 kVA	19435.64	23001.49	100.00	2539.14	nr	**25540.63**
150 kVA	21541.32	25493.50	100.00	2539.14	nr	**28032.64**
315 kVA	38837.75	45963.31	120.00	3046.97	nr	**49010.28**
500 kVA	45777.73	54176.57	120.00	3046.97	nr	**57223.54**
750 kVA	77902.61	92195.40	140.00	3554.80	nr	**95750.20**
1000 kVA	103538.48	122534.69	140.00	3554.80	nr	**126089.49**
1500 kVA	121854.77	144211.46	170.00	4316.54	nr	**148528.00**
2000 kVA	187701.90	222139.56	170.00	4316.54	nr	**226456.10**
2500 kVA	256989.27	304139.10	210.00	5332.19	nr	**309471.29**
Extra for residential silencer; performance 75dBA at 1m; including connection to exhaust pipe						
60 kVA	721.36	853.71	10.00	253.91	nr	**1107.62**
100 kVA	861.96	1020.10	10.00	253.91	nr	**1274.01**
150 kVA	988.49	1169.85	10.00	253.91	nr	**1423.76**
315 kVA	1922.91	2275.70	15.00	380.87	nr	**2656.57**
500 kVA	2459.33	2910.54	15.00	380.87	nr	**3291.42**
750 kVA	3664.12	4336.38	20.00	507.83	nr	**4844.21**
1000 kVA	4934.88	5840.29	20.00	507.83	nr	**6348.12**
1500 kVA	7861.42	9303.76	20.00	507.83	nr	**9811.59**
2000 kVA	8514.65	10076.83	30.00	761.74	nr	**10838.57**
2500 kVA	9735.66	11521.87	30.00	761.74	nr	**12283.61**
Synchronisation panel for paralleling generators - not generators to mains; including interconnecting cables; commissioning and testing; fixing to backgrounds						
2 x 60 kVA	6008.81	7111.25	80.00	2031.31	nr	**9142.56**
2 x 100 kVA	6459.80	7644.98	80.00	2031.31	nr	**9676.29**
2 x 150 kVA	8126.39	9617.34	80.00	2031.31	nr	**11648.65**
2 x 315 kVA	13815.08	16349.73	80.00	2031.31	nr	**18381.05**
2 x 500 kVA	19945.02	23604.34	80.00	2031.31	nr	**25635.65**
2 x 750 kVA	22507.10	26636.47	80.00	2031.31	nr	**28667.79**
2 x 1000 kVA	24256.96	28707.39	120.00	3046.97	nr	**31754.36**
2 x 1500 kVA	28715.99	33984.51	120.00	3046.97	nr	**37031.48**
2 x 2000 kVA	31113.67	36822.10	120.00	3046.97	nr	**39869.07**
2 x 2500 kVA	36273.51	42928.61	120.00	3046.97	nr	**45975.58**

V:ELECTRICAL SUPPLY/POWER/LIGHTING SYSTEMS

Item	Net Price £	Material £	Labour hours	Labour £	Unit	Total rate £
V10 : ELECTRICAL GENERATION PLANT (cont'd)						
STANDBY GENERATORS (cont'd)						
Prefabricated drop-over acoustic housing; performance 85dBA at 1m over the range from 60kVA to 315kVA, 75dBA from 500kVA to 2500 kVA.						
60 kVA	2595.60	3071.81	4.00	101.57	nr	3173.38
100 kVA	2685.36	3178.05	7.00	177.74	nr	3355.79
150 kVA	4360.61	5160.65	15.00	380.87	nr	5541.52
315 kVA	9304.14	11011.18	25.00	634.78	nr	11645.96
500 kVA	15367.03	18186.42	40.00	1015.66	nr	19202.08
750 kVA	24216.95	28660.03	40.00	1015.66	nr	29675.69
1000 kVA	30637.81	36258.93	40.00	1015.66	nr	37274.59
1500 kVA	60772.73	71922.70	40.00	1015.66	nr	72938.36
2000 kVA	73047.75	86449.83	60.00	1523.48	nr	87973.31
2500 kVA	101596.11	120235.95	70.00	1777.40	nr	122013.35
COMBINED HEAT AND POWER (CHP) UNITS						
Gas fired engine; acoustic enclosure complete with exhaust fan and attenuators; exhaust gas attenuation; includes 6m long pipe connections; dry air cooler for secondary water circuit to reject excess heat; controls and panel; commissioning						
Electrical output; Heat output						
82 kW; 132 kW	79800.00	94440.91	-	-	nr	94440.91
100kW, 148kW	85000.00	100594.95	-	-	nr	100594.95
118kW, 181kW	89200.00	105565.52	-	-	nr	105565.52
130kW, 201kW	97750.00	115684.19	-	-	nr	115684.19
140kW, 207kW	103400.00	122370.80	-	-	nr	122370.80
150kW, 208kW	108900.00	128879.88	-	-	nr	128879.88
160kW, 216kW	104000.00	123080.88	-	-	nr	123080.88
198kW, 233kW	136050.00	161011.09	-	-	nr	161011.09
210kW, 319kW	128600.00	152194.24	-	-	nr	152194.24
237kW, 359kW	141400.00	167342.66	-	-	nr	167342.66
307kW, 435kW	191750.00	226930.37	-	-	nr	226930.37
380kW, 500kW	222000.00	262730.34	-	-	nr	262730.34
490kW, 679kW	305600.00	361668.43	-	-	nr	361668.43
501kW, 518kW	282000.00	333738.54	-	-	nr	333738.54
600kW, 873kW	335750.00	397350.05	-	-	nr	397350.05
725kW, 1019kW	454000.00	537295.38	-	-	Unit	537295.38
975kW, 1293kW	563750.00	667181.21	-	-	nr	667181.21
1160kW, 1442kW	638250.00	755349.73	-	-	nr	755349.73
1379kW, 1475kW	761950.00	901744.97	-	-	nr	901744.97
1566kW, 1647kW	833500.00	986422.24	-	-	nr	986422.24
1600kW, 1625kW	831500.00	984055.31	-	-	nr	984055.31
1760kW, 1821kW	913100.00	1080626.46	-	-	nr	1080626.46
Note: The rates detailed are based on a specialist sub-contract package and include installation						

V:ELECTRICAL SUPPLY/POWER/LIGHTING SYSTEMS

Item	Net Price £	Material £	Labour hours	Labour £	Unit	Total rate £
V11 : HV SUPPLY						
Y61 - HV CABLES						
Cable; 6350/11000 volts, 3 core, XLPE; stranded copper conductors; steel wire armoured; LSOH to BS 7835						
Laid in trench/duct including marker tape (cable tiles or clips measured elsewhere)						
95mm²	29.89	35.37	0.23	5.84	m	**41.21**
120mm²	32.89	38.92	0.23	5.84	m	**44.76**
150mm²	39.35	46.57	0.25	6.35	m	**52.92**
185mm²	42.72	50.56	0.25	6.35	m	**56.90**
240mm²	49.75	58.88	0.27	6.86	m	**65.73**
300mm²	56.80	67.22	0.29	7.36	m	**74.58**
Cable tiles; single width; laid in trench above cables on prepared sand bed. (cost of excavation excluded); reinforced concrete covers; concave/convex ends						
914 x 152 x 63/38mm	9.21	10.90	0.11	2.79	m	**13.69**
914 x 229 x 63/38mm	10.84	12.83	0.11	2.79	m	**15.62**
914 x 305 x 63/38mm	12.07	14.28	0.11	2.79	m	**17.08**
Clipped direct to backgrounds including cleats						
95mm²	37.91	44.86	0.47	11.93	m	**56.80**
120mm²	40.91	48.42	0.50	12.70	m	**61.11**
150mm²	47.14	55.79	0.53	13.46	m	**69.24**
185mm²	51.67	61.16	0.55	13.97	m	**75.12**
240mm²	59.73	70.69	0.60	15.23	m	**85.93**
300mm²	67.90	80.36	0.68	17.27	m	**97.62**
Terminations for above cables, including heat-shrink kit and glanding off						
95mm²	395.33	467.86	4.75	120.61	m	**588.47**
120mm²	393.19	465.33	5.30	134.57	m	**599.91**
150mm²	422.93	500.52	6.00	152.35	m	**652.87**
185mm²	423.95	501.73	6.90	175.20	m	**676.93**
240mm²	442.96	524.23	7.43	188.66	m	**712.89**
300mm²	477.68	565.31	8.75	222.17	m	**787.49**
Cable; 6350/11000volts, 3 core, paper insulated; lead sheathed; steel wire armoured; stranded copper conductors; to BS 6480						
Laid in trench/duct including marker tape (cable tiles measured elsewhere)						
95mm²	32.55	38.52	0.22	5.59	m	**44.11**
120mm²	35.13	41.57	0.22	5.59	m	**47.16**
150mm²	42.98	50.86	0.24	6.09	m	**56.96**
185mm²	44.86	53.09	0.26	6.60	m	**59.70**
240mm²	52.05	61.60	0.34	8.63	m	**70.24**

V:ELECTRICAL SUPPLY/POWER/LIGHTING SYSTEMS

Item	Net Price £	Material £	Labour hours	Labour £	Unit	Total rate £
V11 : HV SUPPLY (cont'd)						
Y61 - HV CABLES (cont'd)						
Cable; 6350/11000volts, 3 core, paper insulated (cont'd)						
Cable tiles; single width; laid in trench above cables on prepared sand bed. (cost of excavation excluded); reinforced concrete covers; concave/convex ends						
914 x 152 x 63/38mm	9.21	10.90	0.11	2.79	m	**13.69**
914 x 229 x 63/38mm	10.84	12.83	0.11	2.79	m	**15.62**
914 x 305 x 63/38mm	12.07	14.28	0.11	2.79	m	**17.08**
Clipped direct to backgrounds including cleats						
95mm²	41.61	49.25	0.55	13.97	m	**63.21**
120mm²	44.17	52.28	0.63	16.00	m	**68.28**
150mm²	51.75	61.25	0.66	16.76	m	**78.01**
185mm²	54.81	64.86	0.72	18.28	m	**83.15**
240mm²	63.17	74.76	0.82	20.82	m	**95.58**
Terminations for above cables, including compound joint and glanding off						
95 mm²	441.36	522.34	4.75	120.61	m	**642.95**
120 mm²	449.33	531.77	5.30	134.57	m	**666.34**
150 mm²	469.35	555.46	6.10	154.89	m	**710.35**
185 mm²	470.35	556.64	6.90	175.20	m	**731.84**
240 mm²	494.66	585.41	7.43	188.66	m	**774.07**
300 mm²	524.57	620.81	8.75	222.17	m	**842.99**

V:ELECTRICAL SUPPLY/POWER/LIGHTING SYSTEMS

Item	Net Price £	Material £	Labour hours	Labour £	Unit	Total rate £
Y70 - HV SWITCHGEAR AND TRANSFORMERS						
H.V. Circuit Breakers; installed on prepared foundations including all supports, fixings and inter panel connections where relevant. Excludes main and multi core cabling and heat shrink cable termination kits.						
Three phase 11kV, 630 Amp, Air or SF6 insulated, with fixed pattern vacuum or SF6 circuit breaker panels; hand charged spring closing operation; prospective fault level up to 25 kA for 3 seconds. Feeders include ammeter with selector switch, 3 pole IDMT, overcurrent and earth fault relays with necessary current relays with necessary current transformers; incomers include 3 phase VT, voltmeter and phase selector switch; Includes IDMT overcurrent and earth fault relays/CTs.						
Single panel with cable chamber	15437.44	18269.74	31.70	804.91	nr	**19074.65**
Three panel with one incomer and two feeders; with cable chambers	46255.20	54741.64	67.83	1722.30	nr	**56463.94**
Five panel with two incoming, two feeders and a bus section; with cable chambers	77078.52	91220.12	99.17	2518.07	nr	**93738.18**
Ring Main Unit (RMU)						
Three phase 11kV, 630 Amp, SF6 insulated RMU with vacuum or SF6 200 Amp circuit breaker tee-off, Includes IDMT overcurrent and earth fault relays/CTs for tee-off and cable boxes.						
3 way Ring Main Unit	10459.76	12378.82	67.83	1722.30	nr	**14101.12**
Extra for,						
Remote actuator to ring switches (per switch)	1579.42	1869.19	-	-	nr	**1869.19**
Remote tripping of circuit breaker	789.71	934.60	-	-	nr	**934.60**
3 - Phase Neon indicators (per circuit)	92.87	109.91	-	-	nr	**109.91**
Pressure gauge with alarm contacts (for SF6 only)	203.75	241.13	-	-	nr	**241.13**
Tripping Batteries						
Battery chargers; switchgear tripping and closing; double wound transformer and earth screen; including fixing to background, commissioning and testing						
Valve regulated lead acid battery						
30 volt; 19 Ah; 3A	1798.70	2128.71	6.50	165.04	nr	**2293.75**
30 volt; 29 Ah; 3A	2881.65	3410.34	8.50	215.83	nr	**3626.17**
110 volt; 19 Ah; 3A	2939.55	3478.87	6.50	165.04	nr	**3643.92**
110 volt; 29 Ah; 3A	3207.40	3795.86	8.50	215.83	nr	**4011.69**
110 volt; 38 Ah; 3A	3427.40	4056.23	10.00	253.91	nr	**4310.14**

V:ELECTRICAL SUPPLY/POWER/LIGHTING SYSTEMS

Item	Net Price £	Material £	Labour hours	Labour £	Unit	Total rate £
V11 : HV SUPPLY (cont'd)						
Y70 - HV SWITCHGEAR AND TRANSFORMERS (cont'd)						
Step down transformers; 11 / 0.415kV, Dyn 11, 50Hz. Complete with lifting lugs, mounting skids, provisions for wheels, undrilled gland plates to air-filled cable boxes, off load tapping facility, including UK delivery						
Oil-filled in free breathing ventilated steel tank						
500kVA	6825.48	8077.75	30.00	761.74	nr	**8839.49**
800kVA	7773.13	9199.27	30.00	761.74	nr	**9961.01**
1000kVA	8809.84	10426.18	30.00	761.74	nr	**11187.92**
1250kVA	10705.13	12669.20	35.00	888.70	nr	**13557.90**
1500kVA	12685.68	15013.12	35.00	888.70	nr	**15901.82**
2000kVA	16476.26	19499.16	40.00	1015.66	nr	**20514.82**
MIDEL - filled in gasket-sealed steel tank						
500kVA	9040.44	10699.09	30.00	761.74	nr	**11460.83**
800kVA	10303.97	12194.44	30.00	761.74	nr	**12956.18**
1000kVA	11731.74	13884.16	30.00	761.74	nr	**14645.90**
1250kVA	14179.82	16781.40	35.00	888.70	nr	**17670.10**
1500kVA	16871.12	19966.47	35.00	888.70	nr	**20855.17**
2000kVA	21846.26	25854.40	40.00	1015.66	nr	**26870.05**
Extra for,						
Fluid temperature indicator with 2 N/O contacts.	250.00	295.87	2.00	50.78	nr	**346.65**
Winding temperature indicator with 2 N/O contacts.	612.50	724.88	2.00	50.78	nr	**775.66**
Dehydrating Breather	62.50	73.97	2.00	50.78	nr	**124.75**
Plain Rollers	187.50	221.90	2.00	50.78	nr	**272.68**
Pressure relief device with 1 N/O contact	375.00	443.80	2.00	50.78	nr	**494.58**
Step down transformers; 11 / 0.415kV, Dyn 11, 50Hz. Complete with lifting lugs, mounting skids, provisions for wheels, undrilled gland plates to air-filled cable boxes, off load tapping facility, including delivery						
Cast Resin type in ventilated steel enclosure, AN - Air Natural including winding temperature indicator with 2 N/O contacts						
500kVA	9672.20	11446.76	40.00	1015.66	nr	**12462.42**
800kVA	11251.61	13315.94	40.00	1015.66	nr	**14331.60**
1000kVA	13611.24	16108.49	40.00	1015.66	nr	**17124.15**
1250kVA	14732.62	17435.61	45.00	1142.61	nr	**18578.22**
1600kVA	16713.18	19779.55	45.00	1142.61	nr	**20922.16**
2000kVA	18924.36	22396.41	50.00	1269.57	nr	**23665.98**

V:ELECTRICAL SUPPLY/POWER/LIGHTING SYSTEMS

Item	Net Price £	Material £	Labour hours	Labour £	Unit	Total rate £
Cast Resin type in ventilated steel enclosure with temperature controlled fans to achieve 40% increase to AN/AF rating. Includes winding temperature indicator with 2 N/O contacts						
500/700kVA	10619.84	12568.27	42.00	1066.44	nr	**13634.71**
800/1120kVA	12357.20	14624.38	42.00	1066.44	nr	**15690.82**
1000/1400kVA	15289.19	18094.30	42.00	1066.44	nr	**19160.73**
12501750kVA	15920.95	18841.97	47.00	1193.40	nr	**20035.36**
1600/2240kVA	17986.79	21286.82	47.00	1193.40	nr	**22480.22**
2000/2800kVA	20513.84	24277.52	52.00	1320.35	nr	**25597.87**

V:ELECTRICAL SUPPLY/POWER/LIGHTING SYSTEMS

Item	Net Price £	Material £	Labour hours	Labour £	Unit	Total rate £
V20 : LV DISTRIBUTION						
Y60 - CONDUIT AND CABLE TRUNKING						
Heavy gauge, screwed drawn steel; surface fixed on saddles to backgrounds, with standard pattern boxes and fittings including all fixings and supports. (forming holes, conduit entry, draw wires etc. and components for earth continuity are included.)						
Black enamelled						
20 mm dia.	1.75	2.07	0.49	12.44	m	**14.51**
25 mm dia.	2.57	3.05	0.56	14.22	m	**17.27**
32 mm dia.	5.95	7.04	0.64	16.25	m	**23.29**
38 mm dia.	7.33	8.67	0.73	18.54	m	**27.21**
50 mm dia.	9.17	10.86	1.04	26.41	m	**37.26**
Galvanised						
20 mm dia.	2.20	2.60	0.49	12.44	m	**15.05**
25 mm dia.	2.93	3.46	0.56	14.22	m	**17.68**
32 mm dia.	6.23	7.37	0.64	16.25	m	**23.62**
38 mm dia.	9.14	10.82	0.73	18.54	m	**29.35**
50 mm dia.	13.75	16.27	1.04	26.41	m	**42.68**
High impact PVC; surface fixed on saddles to backgrounds; with standard pattern boxes and fittings; including all fixings and supports.						
Light gauge						
16 mm dia.	0.61	0.72	0.27	6.86	m	**7.58**
20 mm dia.	0.80	0.94	0.28	7.11	m	**8.05**
25 mm dia.	1.32	1.57	0.33	8.38	m	**9.94**
32 mm dia.	1.78	2.11	0.38	9.65	m	**11.76**
38 mm dia.	2.29	2.71	0.44	11.17	m	**13.88**
50 mm dia.	2.77	3.28	0.48	12.19	m	**15.47**
Heavy gauge						
16 mm dia.	1.07	1.27	0.27	6.86	m	**8.12**
20 mm dia.	1.16	1.37	0.28	7.11	m	**8.48**
25 mm dia.	1.55	1.84	0.33	8.38	m	**10.22**
32 mm dia.	2.54	3.01	0.38	9.65	m	**12.66**
38 mm dia.	3.30	3.90	0.44	11.17	m	**15.07**
50 mm dia.	5.46	6.46	0.48	12.19	m	**18.65**
Flexible conduits; including adaptors and locknuts (for connections to equipment.)						
Metallic, PVC covered conduit; not exceeding 1m long; including zinc plated mild steel adaptors, lock nuts and earth conductor						
16 mm dia.	5.75	6.81	0.46	11.68	nr	**18.49**
20 mm dia.	6.81	8.06	0.42	10.66	nr	**18.72**
25 mm dia.	10.41	12.32	0.43	10.92	nr	**23.23**
32 mm dia.	16.06	19.01	0.51	12.95	nr	**31.96**
38 mm dia.	28.78	34.06	0.56	14.22	nr	**48.27**
50 mm dia.	47.50	56.21	0.82	20.82	nr	**77.03**

V:ELECTRICAL SUPPLY/POWER/LIGHTING SYSTEMS

Item	Net Price £	Material £	Labour hours	Labour £	Unit	Total rate £
PVC conduit; not exceeding 1m long; including nylon adaptors, lock nuts						
16 mm dia.	4.03	4.77	0.46	11.68	nr	**16.45**
20 mm dia.	4.03	4.77	0.48	12.19	nr	**16.96**
25 mm dia.	5.06	5.99	0.50	12.70	nr	**18.69**
32 mm dia.	7.33	8.67	0.58	14.73	nr	**23.40**
PVC adaptable boxes; fixed to backgrounds; including all supports and fixings (cutting and connecting conduit to boxes is included.)						
Square pattern						
75 x 75 x 53 mm	2.94	3.48	0.69	17.52	nr	**21.00**
100 x 100 x 75 mm	3.80	4.50	0.71	18.03	nr	**22.53**
150 x 150 x 75 mm	6.81	8.06	0.80	20.31	nr	**28.38**
Terminal strips to be fixed in metal or polythene adaptable boxes)						
20 Amp high density polythene						
2 way	0.89	1.06	0.23	5.84	nr	**6.90**
3 way	0.99	1.17	0.23	5.84	nr	**7.01**
4 way	1.06	1.26	0.23	5.84	nr	**7.10**
5 way	1.13	1.34	0.23	5.84	nr	**7.18**
6 way	1.22	1.44	0.25	6.35	nr	**7.79**
7 way	1.30	1.54	0.25	6.35	nr	**7.89**
8 way	1.41	1.67	0.29	7.36	nr	**9.03**
9 way	1.52	1.80	0.30	7.62	nr	**9.42**
10 way	1.74	2.06	0.34	8.63	nr	**10.70**
11 way	1.81	2.14	0.34	8.63	nr	**10.77**
12 way	1.96	2.32	0.34	8.63	nr	**10.96**
13 way	2.18	2.58	0.37	9.39	nr	**11.98**
14 way	2.45	2.90	0.37	9.39	nr	**12.29**
15 way	2.52	2.98	0.39	9.90	nr	**12.88**
16 way	2.58	3.06	0.45	11.43	nr	**14.48**
18 way	2.78	3.29	0.45	11.43	nr	**14.72**

V:ELECTRICAL SUPPLY/POWER/LIGHTING SYSTEMS

Item	Net Price £	Material £	Labour hours	Labour £	Unit	Total rate £
V20 : LV DISTRIBUTION (cont'd)						
Y60 - CONDUIT AND CABLE TRUNKING (cont'd)						
TRUNKING						
Galvanised steel trunking; fixed to backgrounds; jointed with standard connectors (including plates for air gap between trunking and background); earth continuity straps included						
Single compartment						
50 x 50 mm	6.13	7.25	0.39	9.90	m	**17.15**
75 x 50 mm	7.35	8.70	0.44	11.17	m	**19.87**
75 x 75 mm	7.27	8.61	0.47	11.93	m	**20.54**
100 x 50 mm	8.23	9.74	0.50	12.70	m	**22.43**
100 x 75 mm	9.52	11.26	0.57	14.47	m	**25.73**
100 x 100 mm	8.10	9.58	0.62	15.74	m	**25.32**
150 x 50 mm	10.03	11.87	0.78	19.81	m	**31.68**
150 x 100 mm	10.82	12.81	0.78	19.81	m	**32.62**
150 x 150 mm	10.75	12.72	0.86	21.84	m	**34.56**
225 x 75 mm	10.95	12.95	0.88	22.34	m	**35.30**
225 x 150 mm	14.21	16.82	0.84	21.33	m	**38.15**
225 x 225 mm	16.03	18.97	0.99	25.14	m	**44.11**
300 x 75 mm	12.65	14.97	0.96	24.38	m	**39.35**
300 x 100 mm	13.13	15.54	0.99	25.14	m	**40.68**
300 x 150 mm	16.15	19.11	0.99	25.14	m	**44.25**
300 x 225 mm	17.93	21.22	1.09	27.68	m	**48.90**
300 x 300 mm	19.73	23.35	1.16	29.45	m	**52.81**
Double compartment						
50 x 50 mm	7.45	8.81	0.41	10.41	m	**19.22**
75 x 50 mm	8.00	9.46	0.47	11.93	m	**21.40**
75 x 75 mm	8.58	10.15	0.50	12.70	m	**22.85**
100 x 50 mm	8.78	10.39	0.54	13.71	m	**24.10**
100 x 75 mm	8.87	10.49	0.62	15.74	m	**26.24**
100 x 100 mm	9.99	11.82	0.66	16.76	m	**28.58**
150 x 50 mm	10.01	11.85	0.70	17.77	m	**29.62**
150 x 100 mm	11.48	13.59	0.83	21.07	m	**34.67**
150 x 150 mm	14.23	16.85	0.92	23.36	m	**40.21**
Triple compartment						
75 x 50 mm	8.20	9.70	0.54	13.71	m	**23.41**
75 x 75 mm	9.29	11.00	0.58	14.73	m	**25.73**
100 x 50 mm	9.22	10.91	0.61	15.49	m	**26.40**
100 x 75 mm	10.51	12.43	0.70	17.77	m	**30.21**
100 x 100 mm	11.04	13.07	0.74	18.79	m	**31.86**
150 x 50 mm	10.64	12.59	0.79	20.06	m	**32.65**
150 x 100 mm	12.09	14.31	0.78	19.81	m	**34.11**
150 x 150 mm	16.06	19.01	1.01	25.65	m	**44.65**
Four compartment						
100 x 50 mm	10.08	11.92	0.64	16.25	m	**28.18**
100 x 75 mm	11.15	13.20	0.72	18.28	m	**31.48**
100 x 100 mm	12.45	14.74	0.77	19.55	m	**34.29**
150 x 50 mm	11.47	13.58	0.82	20.82	m	**34.40**
150 x 100 mm	12.89	15.26	0.95	24.12	m	**39.38**
150 x 150 mm	18.50	21.90	1.04	26.41	m	**48.30**

V:ELECTRICAL SUPPLY/POWER/LIGHTING SYSTEMS

Item	Net Price £	Material £	Labour hours	Labour £	Unit	Total rate £
Galvanised steel trunking fittings; cutting and jointing trunking to fittings is included						
Stop end						
50 x 50 mm	0.47	0.56	0.19	4.82	nr	5.38
75 x 50 mm	0.62	0.73	0.20	5.08	nr	5.81
75 x 75 mm	0.62	0.73	0.21	5.33	nr	6.07
100 x 50 mm	0.67	0.79	0.31	7.87	nr	8.66
100 x 75 mm	0.70	0.83	0.27	6.86	nr	7.69
100 x 100 mm	0.70	0.83	0.27	6.86	nr	7.69
150 x 50 mm	0.77	0.91	0.28	7.11	nr	8.02
150 x 100 mm	0.84	0.99	0.30	7.62	nr	8.61
150 x 150 mm	0.85	1.01	0.32	8.13	nr	9.13
225 x 75 mm	0.90	1.06	0.35	8.89	nr	9.95
225 x 150 mm	1.32	1.57	0.37	9.39	nr	10.96
225 x 225 mm	1.58	1.86	0.38	9.65	nr	11.51
300 x 75 mm	1.32	1.57	0.42	10.66	nr	12.23
300 x 100 mm	1.58	1.86	0.42	10.66	nr	12.53
300 x 150 mm	1.64	1.95	0.43	10.92	nr	12.86
300 x 225 mm	3.45	4.08	0.45	11.43	nr	15.51
300 x 300 mm	1.98	2.34	0.48	12.19	nr	14.53
Flanged connector						
50 x 50 mm	0.62	0.73	0.19	4.82	nr	5.56
75 x 50 mm	0.95	1.13	0.20	5.08	nr	6.21
75 x 75 mm	0.94	1.12	0.21	5.33	nr	6.45
100 x 50 mm	1.01	1.20	0.26	6.60	nr	7.80
100 x 75 mm	1.03	1.22	0.27	6.86	nr	8.08
100 x 100 mm	1.02	1.21	0.27	6.86	nr	8.07
150 x 50 mm	1.02	1.21	0.28	7.11	nr	8.32
150 x 100 mm	1.08	1.28	0.30	7.62	nr	8.90
150 x 150 mm	1.09	1.29	0.32	8.13	nr	9.42
225 x 75 mm	0.90	1.06	0.35	8.89	nr	9.95
225 x 150 mm	1.32	1.57	0.37	9.39	nr	10.96
225 x 225 mm	1.58	1.86	0.38	9.65	nr	11.51
300 x 75 mm	1.32	1.57	0.42	10.66	nr	12.23
300 x 100 mm	1.58	1.86	0.42	10.66	nr	12.53
300 x 150 mm	1.64	1.95	0.43	10.92	nr	12.86
300 x 225 mm	1.79	2.12	0.45	11.43	nr	13.55
300 x 300 mm	1.98	2.34	0.48	12.19	nr	14.53
Bends 90°; single compartment						
50 x 50 mm	2.65	3.13	0.42	10.66	nr	13.79
75 x 50 mm	3.32	3.93	0.45	11.43	nr	15.36
75 x 75 mm	3.23	3.82	0.48	12.19	nr	16.01
100 x 50 mm	3.84	4.55	0.53	13.46	nr	18.00
100 x 75 mm	3.63	4.30	0.56	14.22	nr	18.52
100 x 100 mm	3.52	4.16	0.58	14.73	nr	18.89
150 x 50 mm	4.34	5.13	0.64	16.25	nr	21.38
150 x 100 mm	5.89	6.97	0.91	23.11	nr	30.07
150 x 150 mm	6.95	8.22	0.89	22.60	nr	30.82
225 x 75 mm	7.88	9.32	0.76	19.30	nr	28.62
225 x 150 mm	8.61	10.19	0.82	20.82	nr	31.01
225 x 225 mm	10.29	12.18	0.83	21.07	nr	33.26
300 x 75 mm	11.04	13.07	0.85	21.58	nr	34.65
300 x 100 mm	11.29	13.36	0.90	22.85	nr	36.22
300 x 150 mm	11.94	14.13	0.96	24.38	nr	38.50
300 x 225 mm	12.86	15.22	0.98	24.88	nr	40.10
300 x 300 mm	13.80	16.33	1.06	26.91	nr	43.25

V:ELECTRICAL SUPPLY/POWER/LIGHTING SYSTEMS

Item	Net Price £	Material £	Labour hours	Labour £	Unit	Total rate £
V20 : LV DISTRIBUTION (cont'd)						
Y60 - CONDUIT AND CABLE TRUNKING (cont'd)						
TRUNKING (cont'd)						
Galvanised steel trunking fittings (cont'd)						
Bends 90°; double compartment						
50 x 50 mm	3.33	3.95	0.42	10.66	nr	**14.61**
75 x 50 mm	3.99	4.72	0.45	11.43	nr	**16.15**
75 x 75 mm	4.55	5.39	0.49	12.44	nr	**17.83**
100 x 50 mm	3.99	4.72	0.53	13.46	nr	**18.18**
100 x 75 mm	5.27	6.23	0.56	14.22	nr	**20.45**
100 x 100 mm	5.34	6.32	0.58	14.73	nr	**21.04**
150 x 50 mm	6.23	7.38	0.65	16.50	nr	**23.88**
150 x 100 mm	6.28	7.43	0.69	17.52	nr	**24.95**
150 x 150 mm	9.92	11.75	0.73	18.54	nr	**30.28**
Bends 90°; triple compartment						
75 x 50 mm	4.20	4.97	0.47	11.93	nr	**16.90**
75 x 75 mm	5.11	6.04	0.51	12.95	nr	**18.99**
100 x 50 mm	4.57	5.40	0.56	14.22	nr	**19.62**
100 x 75 mm	6.12	7.24	0.59	14.98	nr	**22.22**
100 x 100 mm	6.83	8.08	0.61	15.49	nr	**23.57**
150 x 50 mm	7.34	8.68	0.68	17.27	nr	**25.95**
150 x 100 mm	8.14	9.64	0.73	18.54	nr	**28.17**
150 x 150 mm	11.75	13.91	0.77	19.55	nr	**33.46**
Bends 90°; four compartment						
100 x 50 mm	5.31	6.29	0.56	14.22	nr	**20.51**
100 x 75 mm	7.26	8.59	0.59	14.98	nr	**23.57**
100 x 100 mm	12.17	14.40	0.61	15.49	nr	**29.89**
150 x 50 mm	12.60	14.92	0.69	17.52	nr	**32.44**
150 x 100 mm	13.22	15.65	0.73	18.54	nr	**34.19**
150 x 150 mm	14.19	16.79	0.77	19.55	nr	**36.35**
Tees; single compartment						
50 x 50 mm	3.12	3.69	0.56	14.22	nr	**17.91**
75 x 50 mm	3.70	4.38	0.57	14.47	nr	**18.86**
75 x 75 mm	3.59	4.25	0.60	15.23	nr	**19.48**
100 x 50 mm	4.34	5.13	0.65	16.50	nr	**21.64**
100 x 75 mm	4.36	5.16	0.71	18.03	nr	**23.19**
100 x 100 mm	4.22	4.99	0.72	18.28	nr	**23.28**
150 x 50 mm	6.15	7.28	0.82	20.82	nr	**28.10**
150 x 100 mm	6.21	7.35	0.84	21.33	nr	**28.68**
150 x 150 mm	6.01	7.12	0.91	23.11	nr	**30.22**
225 x 75 mm	10.63	12.58	0.94	23.87	nr	**36.44**
225 x 150 mm	14.61	17.28	1.01	25.65	nr	**42.93**
225 x 225 mm	16.86	19.95	1.02	25.90	nr	**45.85**
300 x 75 mm	15.59	18.45	1.07	27.17	nr	**45.62**
300 x 100 mm	16.31	19.30	1.07	27.17	nr	**46.47**
300 x 150 mm	16.99	20.10	1.14	28.95	nr	**49.05**
300 x 225 mm	18.47	21.86	1.19	30.22	nr	**52.07**
300 x 300mm	19.95	23.61	1.26	31.99	nr	**55.61**

V:ELECTRICAL SUPPLY/POWER/LIGHTING SYSTEMS

Item	Net Price £	Material £	Labour hours	Labour £	Unit	Total rate £
Tees; double compartment						
50 x 50 mm	4.51	5.34	0.56	14.22	nr	**19.55**
75 x 50 mm	5.38	6.37	0.57	14.47	nr	**20.84**
75 x 75 mm	6.13	7.25	0.60	15.23	nr	**22.49**
100 x 50 mm	6.67	7.89	0.65	16.50	nr	**24.40**
100 x 75 mm	7.08	8.38	0.71	18.03	nr	**26.41**
100 x 100 mm	7.75	9.17	0.72	18.28	nr	**27.45**
150 x 50 mm	8.06	9.54	0.82	20.82	nr	**30.36**
150 x 100 mm	8.43	9.98	0.85	21.58	nr	**31.56**
150 x 150 mm	9.46	11.20	0.91	23.11	nr	**34.31**
Tees; triple compartment						
75 x 50 mm	5.75	6.80	0.60	15.23	nr	**22.04**
75 x 75 mm	6.99	8.27	0.63	16.00	nr	**24.27**
100 x 50 mm	6.99	8.27	0.68	17.27	nr	**25.54**
100 x 75 mm	8.36	9.89	0.74	18.79	nr	**28.68**
100 x 100 mm	9.33	11.04	0.75	19.04	nr	**30.08**
150 x 50 mm	10.22	12.10	0.87	22.09	nr	**34.19**
150 x 100 mm	11.11	13.15	0.89	22.60	nr	**35.75**
150 x 150 mm	16.01	18.95	0.96	24.38	nr	**43.32**
Tees; four compartment						
100 x 50 mm	9.84	11.65	0.66	16.76	nr	**28.41**
100 x 75 mm	9.73	11.51	0.72	18.28	nr	**29.80**
100 x 100 mm	11.03	13.05	0.72	18.28	nr	**31.33**
150 x 50 mm	7.95	9.40	0.83	21.07	nr	**30.48**
150 x 100 mm	14.86	17.58	0.85	21.58	nr	**39.17**
150 x 150 mm	18.98	22.46	0.92	23.36	nr	**45.82**
Crossovers; single compartment						
50 x 50 mm	4.08	4.83	0.65	16.50	nr	**21.34**
75 x 50 mm	5.85	6.93	0.66	16.76	nr	**23.69**
75 x 75 mm	5.65	6.68	0.69	17.52	nr	**24.20**
100 x 50 mm	6.91	8.18	0.74	18.79	nr	**26.97**
100 x 75 mm	6.95	8.22	0.80	20.31	nr	**28.53**
100 x 100 mm	6.69	7.92	0.81	20.57	nr	**28.49**
150 x 50 mm	6.91	8.18	0.91	23.11	nr	**31.29**
150 x 100 mm	8.36	9.89	0.94	23.87	nr	**33.76**
150 x 150 mm	8.08	9.57	0.99	25.14	nr	**34.71**
225 x 75 mm	14.51	17.18	1.01	25.65	nr	**42.82**
225 x 150 mm	19.39	22.95	1.08	27.42	nr	**50.37**
225 x 225 mm	22.15	26.21	1.09	27.68	nr	**53.89**
300 x 75 mm	20.86	24.69	1.14	28.95	nr	**53.63**
300 x 100 mm	21.47	25.41	1.16	29.45	nr	**54.86**
300 x 150 mm	22.15	26.21	1.19	30.22	nr	**56.43**
300 x 225 mm	24.01	28.42	1.21	30.72	nr	**59.14**
300 x 300mm	25.93	30.69	1.29	32.75	nr	**63.45**
Crossovers; double compartment						
50 x 50 mm	5.16	6.11	0.66	16.76	nr	**22.87**
75 x 50 mm	6.14	7.27	0.66	16.76	nr	**24.03**
75 x 75 mm	6.98	8.26	0.70	17.77	nr	**26.04**
100 x 50 mm	8.13	9.62	0.74	18.79	nr	**28.41**
100 x 75 mm	8.05	9.53	0.80	20.31	nr	**29.84**
100 x 100 mm	8.81	10.43	0.81	20.57	nr	**30.99**
150 x 50 mm	9.54	11.30	0.86	21.84	nr	**33.13**
150 x 100 mm	9.57	11.32	0.94	23.87	nr	**35.19**
150 x 150 mm	14.98	17.73	1.00	25.39	nr	**43.13**

V:ELECTRICAL SUPPLY/POWER/LIGHTING SYSTEMS

Item	Net Price £	Material £	Labour hours	Labour £	Unit	Total rate £
V20 : LV DISTRIBUTION (cont'd)						
Y60 - CONDUIT AND CABLE TRUNKING (cont'd)						
TRUNKING (cont'd)						
Galvanised steel trunking fittings (cont'd)						
Crossovers; triple compartment						
75 x 50 mm	6.14	7.27	0.70	17.77	nr	**25.04**
75 x 75 mm	6.98	8.26	0.73	18.54	nr	**26.80**
100 x 50 mm	7.81	9.24	0.79	20.06	nr	**29.30**
100 x 75 mm	9.31	11.02	0.85	21.58	nr	**32.61**
100 x 100 mm	10.36	12.26	0.85	21.58	nr	**33.85**
150 x 50 mm	9.84	11.65	0.97	24.63	nr	**36.28**
150 x 100 mm	10.83	12.82	0.99	25.14	nr	**37.96**
150 x 150 mm	17.73	20.99	1.06	26.91	nr	**47.90**
Crossovers; four compartment						
100 x 50 mm	9.81	11.61	0.79	20.06	nr	**31.67**
100 x 75 mm	11.02	13.04	0.85	21.58	nr	**34.62**
100 x 100 mm	12.44	14.73	0.86	21.84	nr	**36.56**
150 x 50 mm	12.56	14.86	0.97	24.63	nr	**39.49**
150 x 100 mm	13.80	16.33	1.00	25.39	nr	**41.72**
150 x 150 mm	21.38	25.30	1.06	26.91	nr	**52.22**
Galvanised steel flush floor trunking; fixed to backgrounds; supports and fixings; standard coupling joints; earth continuity straps included.						
Triple compartment						
350 x 60mm	35.12	41.56	1.32	33.52	m	**75.08**
Four compartment						
350 x 60mm	36.57	43.28	1.32	33.52	m	**76.80**
Galvanised steel flush floor trunking; fittings (cutting and jointing trunking to fittings is included.)						
Stop end; triple compartment						
350 x 60mm	3.68	4.36	0.53	13.46	nr	**17.81**
Stop end; four compartment						
350 x 60mm	3.68	4.36	0.53	13.46	nr	**17.81**
Rising bend; standard; triple compartment						
350 x 60mm	30.38	35.96	1.30	33.01	nr	**68.97**
Rising bend; standard; four compartment						
350 x 60mm	31.82	37.66	1.30	33.01	nr	**70.67**
Rising bend; skirting; triple compartment						
350 x 60mm	59.59	70.53	1.33	33.77	nr	**104.30**
Rising bend; skirting; four compartment						
350 x 60mm	67.64	80.05	1.33	33.77	nr	**113.82**

V:ELECTRICAL SUPPLY/POWER/LIGHTING SYSTEMS

Item	Net Price £	Material £	Labour hours	Labour £	Unit	Total rate £
Junction box; triple compartment						
350 x 60mm	39.61	46.87	1.16	29.45	nr	**76.33**
Junction box; four compartment						
350 x 60mm	41.20	48.76	1.16	29.45	nr	**78.22**
Body coupler (pair)						
3 and 4 Compartment	2.05	2.42	0.16	4.06	nr	**6.49**
Service outlet module comprising flat lid with flanged carpet trim; twin 13 A outlet and drilled plate for mounting 2 telephone outlets; one blank plate; triple compartment						
3 Compartment	49.13	58.14	0.47	11.93	nr	**70.08**
Service outlet module comprising flat lid with flanged carpet trim; twin 13 A outlet and drilled plate for mounting 2 telephone outlets; two blank plates; four compartment						
4 Compartment	54.72	64.75	0.47	11.93	nr	**76.69**
Single compartment PVC trunking; grey finish; clip on lid; fixed to backgrounds; including supports and fixings (standard coupling joints)						
50 x 50mm	8.93	10.56	0.27	6.86	m	**17.42**
75 x 50mm	9.69	11.47	0.28	7.11	m	**18.58**
75 x 75mm	10.98	13.00	0.29	7.36	m	**20.36**
100 x 50mm	12.42	14.70	0.34	8.63	m	**23.33**
100 x 75mm	13.63	16.13	0.37	9.39	m	**25.52**
100 x 100mm	14.41	17.05	0.37	9.39	m	**26.44**
150 x 50mm	12.49	14.79	0.41	10.41	m	**25.20**
150 x 75mm	22.26	26.34	0.44	11.17	m	**37.52**
150 x 100mm	26.80	31.71	0.44	11.17	m	**42.88**
150 x 150mm	27.42	32.45	0.48	12.19	m	**44.63**
Single compartment PVC trunking; fittings (cutting and jointing trunking to fittings is included)						
Crossover						
50 x 50mm	14.63	17.31	0.29	7.36	nr	**24.67**
75 x 50mm	16.24	19.22	0.30	7.62	nr	**26.84**
75 x 75mm	17.62	20.85	0.31	7.87	nr	**28.72**
100 x 50mm	23.49	27.80	0.35	8.89	nr	**36.68**
100 x 75mm	29.61	35.04	0.36	9.14	nr	**44.18**
100 x 100mm	26.19	30.99	0.40	10.16	nr	**41.15**
150 x 75mm	33.56	39.71	0.45	11.43	nr	**51.14**
150 x 100mm	40.26	47.64	0.46	11.68	nr	**59.32**
150 x 150mm	62.50	73.96	0.47	11.93	nr	**85.90**

V:ELECTRICAL SUPPLY/POWER/LIGHTING SYSTEMS

Item	Net Price £	Material £	Labour hours	Labour £	Unit	Total rate £
V20 : LV DISTRIBUTION (cont'd)						
Y60 - CONDUIT AND CABLE TRUNKING (cont'd)						
TRUNKING (cont'd)						
Single compartment PVC trunking fittings (cont'd)						
Stop end						
50 x 50mm	0.61	0.72	0.12	3.05	nr	3.77
75 x 50mm	0.87	1.03	0.12	3.05	nr	4.08
75 x 75mm	1.14	1.35	0.13	3.30	nr	4.66
100 x 50mm	1.57	1.86	0.16	4.06	nr	5.93
100 x 75mm	2.44	2.88	0.16	4.06	nr	6.95
100 x 100mm	2.45	2.90	0.18	4.57	nr	7.47
150 x 75mm	6.37	7.54	0.20	5.08	nr	12.62
150 x 100mm	7.89	9.33	0.21	5.33	nr	14.66
150 x 150mm	8.04	9.52	0.22	5.59	nr	15.10
Flanged coupling						
50 x 50mm	3.60	4.26	0.32	8.13	nr	12.39
75 x 50mm	4.11	4.86	0.33	8.38	nr	13.24
75 x 75mm	4.93	5.84	0.34	8.63	nr	14.47
100 x 50mm	5.57	6.59	0.44	11.17	nr	17.76
100 x 75mm	6.30	7.46	0.45	11.43	nr	18.88
100 x 100mm	6.72	7.95	0.46	11.68	nr	19.63
150 x 75mm	7.10	8.40	0.57	14.47	nr	22.87
150 x 100mm	7.48	8.85	0.57	14.47	nr	23.32
150 x 150mm	7.85	9.29	0.59	14.98	nr	24.28
Internal coupling						
50 x 50mm	1.27	1.50	0.07	1.78	nr	3.28
75 x 50mm	1.51	1.79	0.07	1.78	nr	3.57
75 x 75mm	1.50	1.78	0.07	1.78	nr	3.55
100 x 50mm	2.02	2.39	0.08	2.03	nr	4.42
100 x 75mm	2.27	2.68	0.08	2.03	nr	4.72
100 x 100mm	2.50	2.96	0.08	2.03	nr	4.99
External coupling						
50 x 50mm	1.40	1.65	0.09	2.29	nr	3.94
75 x 50mm	1.66	1.96	0.09	2.29	nr	4.25
75 x 75mm	1.65	1.95	0.09	2.29	nr	4.24
100 x 50mm	2.22	2.62	0.10	2.54	nr	5.16
100 x 75mm	2.52	2.98	0.10	2.54	nr	5.52
100 x 100mm	2.75	3.26	0.10	2.54	nr	5.79
150 x 75mm	3.15	3.73	0.11	2.79	nr	6.52
150 x 100mm	3.28	3.88	0.11	2.79	nr	6.67
150 x 150mm	3.40	4.03	0.11	2.79	nr	6.82
Angle; flat cover						
50 x 50mm	5.09	6.03	0.18	4.57	nr	10.60
75 x 50mm	6.80	8.05	0.19	4.82	nr	12.88
75 x 75mm	7.82	9.26	0.20	5.08	nr	14.34
100 x 50mm	11.67	13.81	0.23	5.84	nr	19.65
100 x 75mm	17.77	21.03	0.26	6.60	nr	27.63
100 x 100mm	16.21	19.19	0.26	6.60	nr	25.79
150 x 75mm	22.10	26.16	0.30	7.62	nr	33.77
150 x 100mm	26.28	31.10	0.33	8.38	nr	39.48
150 x 150mm	39.83	47.13	0.34	8.63	nr	55.77

V:ELECTRICAL SUPPLY/POWER/LIGHTING SYSTEMS

Item	Net Price £	Material £	Labour hours	Labour £	Unit	Total rate £
Angle; internal or external cover						
50 x 50mm	6.20	7.33	0.18	4.57	nr	**11.90**
75 x 50mm	8.56	10.13	0.19	4.82	nr	**14.95**
75 x 75mm	10.98	13.00	0.20	5.08	nr	**18.08**
100 x 50mm	11.82	13.99	0.23	5.84	nr	**19.83**
100 x 75mm	19.03	22.52	0.26	6.60	nr	**29.12**
100 x 100mm	19.10	22.60	0.26	6.60	nr	**29.21**
150 x 75mm	23.36	27.65	0.30	7.62	nr	**35.27**
150 x 100mm	27.54	32.59	0.33	8.38	nr	**40.97**
150 x 150mm	38.67	45.77	0.34	8.63	nr	**54.40**
Tee; flat cover						
50 x 50mm	4.26	5.05	0.24	6.09	nr	**11.14**
75 x 50mm	6.51	7.70	0.25	6.35	nr	**14.05**
75 x 75mm	6.70	7.93	0.26	6.60	nr	**14.53**
100 x 50mm	14.45	17.10	0.32	8.13	nr	**25.22**
100 x 75mm	15.36	18.18	0.33	8.38	nr	**26.56**
100 x 100mm	20.04	23.72	0.34	8.63	nr	**32.36**
150 x 75mm	26.50	31.36	0.41	10.41	nr	**41.77**
150 x 100mm	34.01	40.25	0.42	10.66	nr	**50.91**
150 x 150mm	46.25	54.74	0.44	11.17	nr	**65.91**
Tee; internal or external cover						
50 x 50mm	12.09	14.30	0.24	6.09	nr	**20.40**
75 x 50mm	13.29	15.73	0.25	6.35	nr	**22.08**
75 x 75mm	14.76	17.47	0.26	6.60	nr	**24.07**
100 x 50mm	18.96	22.44	0.32	8.13	nr	**30.57**
100 x 75mm	21.67	25.65	0.33	8.38	nr	**34.03**
100 x 100mm	24.32	28.78	0.34	8.63	nr	**37.41**
150 x 75mm	31.50	37.28	0.41	10.41	nr	**47.69**
150 x 100mm	37.95	44.91	0.42	10.66	nr	**55.57**
150 x 150mm	50.67	59.97	0.44	11.17	nr	**71.14**
Division Strip (1.8m long)						
50mm	6.56	7.77	0.07	1.78	nr	**9.54**
75mm	8.40	9.94	0.07	1.78	nr	**11.72**
100mm	10.69	12.65	0.08	2.03	nr	**14.68**
PVC miniature trunking; white finish; fixed to backgrounds; including supports and fixing; standard coupling joints						
Single compartment						
16 x 16mm	1.18	1.39	0.20	5.08	m	**6.47**
25 x 16mm	1.44	1.70	0.21	5.33	m	**7.03**
38 x 16mm	1.81	2.14	0.24	6.09	m	**8.23**
38 x 25mm	2.15	2.55	0.25	6.35	m	**8.90**
Compartmented						
38 x 16mm	2.10	2.49	0.24	6.09	m	**8.58**
38 x 25mm	2.52	2.98	0.25	6.35	m	**9.33**

V:ELECTRICAL SUPPLY/POWER/LIGHTING SYSTEMS

Item	Net Price £	Material £	Labour hours	Labour £	Unit	Total rate £
V20 : LV DISTRIBUTION (cont'd)						
Y60 - CONDUIT AND CABLE TRUNKING (cont'd)						
TRUNKING (cont'd)						
PVC miniature trunking Fittings; single compartment; white finish; cutting and jointing trunking to fittings is included						
Coupling						
16 x 16mm	0.36	0.42	0.10	2.54	nr	**2.96**
25 x 16mm	0.36	0.42	0.10	2.54	nr	**2.96**
38 x 16mm	0.36	0.42	0.12	3.05	nr	**3.47**
38 x 25mm	0.87	1.03	0.14	3.55	nr	**4.59**
Stop end						
16 x 16mm	0.36	0.42	0.12	3.05	nr	**3.47**
25 x 16mm	0.36	0.42	0.12	3.05	nr	**3.47**
38 x 16mm	0.36	0.42	0.15	3.81	nr	**4.23**
38 x 25mm	0.44	0.52	0.17	4.32	nr	**4.84**
Bend; flat, internal or external						
16 x 16mm	0.36	0.42	0.18	4.57	nr	**4.99**
25 x 16mm	0.36	0.42	0.18	4.57	nr	**4.99**
38 x 16mm	0.36	0.42	0.21	5.33	nr	**5.75**
38 x 25mm	0.87	1.03	0.23	5.84	nr	**6.87**
Tee						
16 x 16mm	0.61	0.72	0.23	5.84	nr	**6.56**
25 x 16mm	0.61	0.72	0.19	4.82	nr	**5.55**
38 x 16mm	0.61	0.72	0.26	6.60	nr	**7.32**
38 x 25mm	0.86	1.02	0.29	7.36	nr	**8.38**
PVC bench trunking; white or grey finish; fixed to backgrounds; including supports and fixings; standard coupling joints Trunking						
90 x 90mm	19.87	23.51	0.33	8.38	m	**31.89**
PVC bench trunking fittings; white or grey finish; cutting and jointing trunking to fittings is included						
Stop end						
90 x 90mm	4.21	4.98	0.09	2.29	nr	**7.27**
Coupling						
90 x 90mm	2.60	3.08	0.09	2.29	nr	**5.37**
Internal or external bend						
90 x 90mm	14.28	16.90	0.28	7.11	nr	**24.01**
Socket plate						
90 x 90mm - 1 gang	0.80	0.94	0.10	2.54	nr	**3.48**
90 x 90mm - 2 gang	0.97	1.14	0.10	2.54	nr	**3.68**

V:ELECTRICAL SUPPLY/POWER/LIGHTING SYSTEMS

Item	Net Price £	Material £	Labour hours	Labour £	Unit	Total rate £
PVC underfloor trunking; single compartment; fitted in floor screed; standard coupling joints						
Trunking						
60 x 25mm	9.75	11.54	0.22	5.59	m	**17.13**
90 x 35mm	14.03	16.60	0.27	6.86	m	**23.46**
PVC underfloor trunking fittings; single compartment; fitted in floor screed; (cutting and jointing trunking to fittings is included)						
Jointing sleeve						
60 x 25mm	0.65	0.77	0.08	2.03	nr	**2.80**
90 x 35mm	1.08	1.28	0.10	2.54	nr	**3.82**
Duct connector						
90 x 35mm	0.50	0.60	0.17	4.32	nr	**4.91**
Socket reducer						
90 x 35mm	0.90	1.07	0.12	3.05	nr	**4.12**
Vertical access box; 2 compartment						
Shallow	50.40	59.65	0.37	9.39	nr	**69.04**
Duct bend; vertical						
60 x 25mm	9.83	11.63	0.27	6.86	nr	**18.49**
90 x 35mm	11.10	13.13	0.35	8.89	nr	**22.02**
Duct bend; horizontal						
60 x 25mm	11.59	13.72	0.30	7.62	nr	**21.34**
90 x 35mm	11.76	13.92	0.37	9.39	nr	**23.31**
Zinc coated steel underfloor ducting; fixed to backgrounds; standard coupling joints; earth continuity straps; (including supports and fixing, packing shims where required)						
Double compartment						
150 x 25mm	8.51	10.08	0.57	14.47	m	**24.55**
Triple compartment						
225 x 25mm	15.09	17.86	0.93	23.61	m	**41.47**
Zinc coated steel underfloor ducting fittings; (cutting and jointing to fittings is included.)						
Stop end; double compartment						
150 x 25mm	2.68	3.17	0.31	7.87	nr	**11.04**
Stop end; triple compartment						
225 x 25mm	3.06	3.62	0.37	9.39	nr	**13.02**
Rising bend; double compartment; standard trunking						
150 x 25mm	17.24	20.40	0.71	18.03	nr	**38.43**

V:ELECTRICAL SUPPLY/POWER/LIGHTING SYSTEMS

Item	Net Price £	Material £	Labour hours	Labour £	Unit	Total rate £
V20 : LV DISTRIBUTION (cont'd)						
Y60 - CONDUIT AND CABLE TRUNKING (cont'd)						
TRUNKING (cont'd)						
Zinc coated steel underfloor ducting Fittings (cont'd)						
Rising bend; triple compartment; standard trunking						
225 x 25mm	30.98	36.67	0.85	21.58	nr	**58.25**
Rising bend; double compartment; to skirting						
150 x 25	38.38	45.42	0.90	22.85	nr	**68.27**
Rising bend; triple compartment; to skirting						
225 x 25	45.25	53.55	0.95	24.12	nr	**77.68**
Horizontal bend; double compartment						
150 x 25mm	26.21	31.02	0.64	16.25	nr	**47.27**
Horizontal bend; triple compartment						
225 x 25mm	30.23	35.78	0.77	19.55	nr	**55.33**
Junction or service outlet boxes; terminal; double compartment						
150mm	27.11	32.08	0.91	23.11	nr	**55.18**
Junction or service outlet boxes; terminal; triple compartment						
225mm	31.00	36.69	1.11	28.18	nr	**64.88**
Junction or service outlet boxes; through or angle; double compartment						
150mm	36.02	42.63	0.97	24.63	nr	**67.26**
Junction or service outlet boxes; through or angle; triple compartment						
225mm	40.00	47.34	1.17	29.71	nr	**77.04**
Junction or service outlet boxes; tee; double compartment						
150mm	36.02	42.63	1.02	25.90	nr	**68.53**
Junction or service outlet boxes; tee; triple compartment						
225mm	40.00	47.34	1.22	30.98	nr	**78.31**
Junction or service outlet boxes; cross; double compartment						
up to 150mm	36.02	42.63	1.03	26.15	nr	**68.78**
Junction or service outlet boxes; cross; triple compartment						
225mm	40.00	47.34	1.23	31.23	nr	**78.57**

V:ELECTRICAL SUPPLY/POWER/LIGHTING SYSTEMS

Item	Net Price £	Material £	Labour hours	Labour £	Unit	Total rate £
Plates for junction/inspection boxes; double and triple compartment						
Blank plate	7.88	9.32	0.92	23.36	nr	**32.68**
Conduit entry plate	9.87	11.68	0.86	21.84	nr	**33.51**
Trunking entry plate	9.87	11.68	0.86	21.84	nr	**33.51**
Service outlet box comprising flat lid with flanged carpet trim; twin 13A outlet and drilled plate for mounting 2 telephone outlets and terminal blocks; terminal outlet box; double compartment						
150 x 25mm trunking	56.16	66.46	1.68	42.66	nr	**109.12**
Service outlet box comprising flat lid with flanged carpet trim; twin 13A outlet and drilled plate for mounting 2 telephone outlets and terminal blocks; terminal outlet box; triple compartment						
225 x 25mm trunking	62.52	74.00	1.93	49.01	nr	**123.00**
PVC skirting/dado modular trunking; white (cutting and jointing trunking to fittings and backplates for fixing to walls is included)						
Main carrier/backplate						
50 x 170mm	15.57	18.43	2.02	51.29	m	**69.72**
Extension carrier/backplate						
50 x 42mm	9.47	11.21	0.58	14.73	m	**25.94**
Carrier/backplate						
Including cover seal	5.52	6.54	0.53	13.46	m	**19.99**
Chamfered covers for fixing to backplates						
50 x 42mm	3.20	3.79	0.33	8.38	m	**12.17**
Square covers for fixing to backplates						
50 x 42 mm	6.43	7.61	0.33	8.38	m	**15.98**
Plain covers for fixing to backplates						
85 mm	3.20	3.79	0.34	8.63	m	**12.42**
Retainers-clip to backplates to hold cables						
For chamfered covers	0.89	1.06	0.07	1.78	m	**2.83**
For square-recessed covers	0.76	0.89	0.07	1.78	m	**2.67**
For plain covers	2.96	3.50	0.07	1.78	m	**5.28**
Prepackaged corner assemblies						
Internal ; for 170 x 50 Assy	6.22	7.36	0.51	12.95	nr	**20.31**
Internal; for 215 x 50 Assy	7.74	9.16	0.53	13.46	nr	**22.62**
Internal; for 254 x 50 Assy	9.21	10.90	0.53	13.46	nr	**24.36**
External; for 170 x 50 Assy	6.22	7.36	0.56	14.22	nr	**21.58**
External ; for 215 x 50 Assy	7.74	9.16	0.58	14.73	nr	**23.89**
External ; for 254 x 50 Assy	9.21	10.90	0.58	14.73	nr	**25.63**
Clip on end caps						
170 x 50 Assy	3.73	4.41	0.11	2.79	nr	**7.20**
215 x 50 Assy	4.41	5.22	0.11	2.79	nr	**8.01**
254 x 50 Assy	5.18	6.13	0.11	2.79	nr	**8.92**

V:ELECTRICAL SUPPLY/POWER/LIGHTING SYSTEMS

Item	Net Price £	Material £	Labour hours	Labour £	Unit	Total rate £
V20 : LV DISTRIBUTION (cont'd)						
Y60 - CONDUIT AND CABLE TRUNKING (cont'd)						
TRUNKING (cont'd)						
PVC skirting/dado modular trunking (cont'd)						
Outlet box						
1 Gang; in horizontal trunking; clip in	3.32	3.93	0.34	8.63	nr	**12.56**
2 Gang; in horizontal trunking; clip in	4.15	4.91	0.34	8.63	nr	**13.54**
1 Gang; in vertical trunking; clip in	3.32	3.93	0.34	8.63	nr	**12.56**
Sheet steel adaptable boxes; with plain or knockout sides; fixed to backgrounds; including supports and fixings (cutting and connecting conduit to boxes is included.)						
Square pattern - black						
75 x 75 x 37 mm	1.81	2.14	0.69	17.52	nr	**19.66**
75 x 75 x 50 mm	1.73	2.05	0.69	17.52	nr	**19.57**
75 x 75 x 75 mm	1.97	2.33	0.69	17.52	nr	**19.85**
100 x 100 x 50 mm	1.81	2.14	0.71	18.03	nr	**20.17**
150 x 150 x 50 mm	2.76	3.27	0.79	20.06	nr	**23.33**
150 x 150 x 75 mm	3.24	3.83	0.80	20.31	nr	**24.15**
150 x 150 x 100 mm	4.32	5.11	0.80	20.31	nr	**25.43**
225 x 225 x 50 mm	5.52	6.53	0.93	23.61	nr	**30.15**
225 x 225 x 100 mm	7.32	8.66	0.94	23.87	nr	**32.53**
300 x 300 x 100 mm	7.86	9.30	0.99	25.14	nr	**34.44**
Square pattern - galvanised						
75 x 75 x 37 mm	2.59	3.07	0.69	17.52	nr	**20.59**
75 x 75 x 50 mm	2.70	3.19	0.69	17.52	nr	**20.71**
75 x 75 x 75 mm	2.33	2.76	0.70	17.77	nr	**20.53**
100 x 100 x 50 mm	2.86	3.38	0.71	18.03	nr	**21.41**
150 x 150 x 50 mm	3.33	3.94	0.84	21.33	nr	**25.27**
150 x 150 x 75 mm	3.97	4.70	0.80	20.31	nr	**25.01**
150 x 150 x 100 mm	4.78	5.65	0.80	20.31	nr	**25.97**
225 x 225 x 50 mm	6.13	7.26	0.93	23.61	nr	**30.87**
225 x 225 x 100 mm	7.81	9.25	0.94	23.87	nr	**33.11**
300 x 300 x 100 mm	12.55	14.85	0.99	25.14	nr	**39.99**
Rectangular pattern - black						
100 x 75 x 50 mm	2.91	3.44	0.69	17.52	nr	**20.96**
150 x 75 x 50 mm	3.04	3.60	0.70	17.77	Unit	**21.38**
150 x 75 x 75 mm	3.30	3.90	0.71	18.03	nr	**21.93**
150 x 100 x 75 mm	7.35	8.70	0.71	18.03	nr	**26.73**
225 x 75 x 50 mm	6.35	7.52	0.78	19.81	nr	**27.32**
225 x 150 x 75 mm	10.14	12.00	0.81	20.57	nr	**32.57**
225 x 150 x 100 mm	19.03	22.52	0.81	20.57	nr	**43.08**
300 x 150 x 50 mm	19.03	22.52	0.93	23.61	nr	**46.13**
300 x 150 x 75 mm	19.03	22.52	0.94	23.87	nr	**46.38**
300 x 150 x 100 mm	19.03	22.52	0.96	24.38	nr	**46.89**

V:ELECTRICAL SUPPLY/POWER/LIGHTING SYSTEMS

Item	Net Price £	Material £	Labour hours	Labour £	Unit	Total rate £
Rectangular pattern - galvanised						
100 x 75 x 50 mm	4.37	5.17	0.69	17.52	nr	**22.69**
150 x 75 x 50 mm	6.48	7.67	0.70	17.77	nr	**25.44**
150 x 75 x 75 mm	7.94	9.39	0.71	18.03	nr	**27.42**
150 x 100 x 75 mm	13.23	15.66	0.71	18.03	nr	**33.69**
225 x 75 x 50 mm	11.93	14.12	0.89	22.60	nr	**36.71**
225 x 150 x 75 mm	16.76	19.83	0.81	20.57	nr	**40.40**
225 x 150 x 100 mm	31.63	37.43	0.81	20.57	nr	**58.00**
300 x 150 x 50 mm	31.63	37.43	0.93	23.61	nr	**61.04**
300 x 150 x 75 mm	31.63	37.43	0.94	23.87	nr	**61.30**
300 x 150 x 100 mm	31.63	37.43	0.96	24.38	nr	**61.80**

V:ELECTRICAL SUPPLY/POWER/LIGHTING SYSTEMS

Item	Net Price £	Material £	Labour hours	Labour £	Unit	Total rate £
V20 : LV DISTRIBUTION (cont'd)						
Y61 - LV CABLES AND WIRING						
ARMOURED CABLE						
Cable; XLPE insulated; PVC sheathed; copper stranded conductors to BS 5467; laid in trench/duct including marker tape. (Cable tiles measured elsewhere.)						
600/1000 Volt grade; single core (aluminium wire armour)						
25 mm²	3.60	4.26	0.15	3.81	m	**8.07**
35 mm²	3.80	4.50	0.15	3.81	m	**8.31**
50 mm²	4.50	5.33	0.17	4.32	m	**9.64**
70 mm²	5.56	6.58	0.18	4.57	m	**11.15**
95 mm²	7.08	8.38	0.20	5.08	m	**13.46**
120 mm²	7.20	8.52	0.22	5.59	m	**14.11**
150 mm²	8.29	9.81	0.24	6.09	m	**15.90**
185 mm²	10.66	12.62	0.26	6.60	m	**19.22**
240 mm²	13.03	15.42	0.30	7.62	m	**23.04**
300 mm²	15.73	18.62	0.31	7.87	m	**26.49**
400 mm²	19.90	23.55	0.38	9.65	m	**33.20**
500 mm²	23.98	28.38	0.44	11.17	m	**39.55**
630 mm²	32.12	38.01	0.52	13.20	m	**51.22**
800 mm²	40.80	48.29	0.62	15.74	m	**64.03**
1000 mm²	46.45	54.97	0.65	16.50	m	**71.48**
600/1000 Volt grade; two core (galvanised steel wire armour)						
1.5 mm²	0.51	0.60	0.06	1.52	m	**2.13**
2.5 mm²	0.62	0.73	0.06	1.52	m	**2.26**
4 mm²	0.79	0.93	0.08	2.03	m	**2.97**
6 mm²	0.96	1.14	0.08	2.03	m	**3.17**
10 mm²	1.52	1.80	0.10	2.54	m	**4.34**
16 mm²	2.17	2.57	0.10	2.54	m	**5.11**
25 mm²	3.26	3.86	0.15	3.81	m	**7.67**
35 mm²	4.20	4.97	0.15	3.81	m	**8.78**
50 mm²	5.24	6.20	0.17	4.32	m	**10.52**
70 mm²	7.62	9.02	0.18	4.57	m	**13.59**
95 mm²	10.94	12.95	0.20	5.08	m	**18.03**
120 mm²	12.53	14.83	0.22	5.59	m	**20.41**
150 mm²	16.96	20.07	0.24	6.09	m	**26.17**
185 mm²	23.90	28.28	0.26	6.60	m	**34.89**
240 mm²	30.93	36.60	0.30	7.62	m	**44.22**
300 mm²	36.80	43.55	0.31	7.87	m	**51.42**
400 mm²	48.90	57.87	0.35	8.89	m	**66.76**
600/1000 Volt grade; three core (galvanised steel wire armour)						
1.5 mm²	0.54	0.64	0.07	1.78	m	**2.42**
2.5 mm²	0.71	0.84	0.07	1.78	m	**2.62**
4 mm²	1.03	1.22	0.09	2.29	m	**3.50**
6 mm²	1.25	1.48	0.10	2.54	m	**4.02**
10 mm²	1.94	2.30	0.11	2.79	m	**5.09**
16 mm²	2.84	3.36	0.11	2.79	m	**6.15**
25 mm²	5.30	6.27	0.16	4.06	m	**10.34**
35 mm²	5.43	6.43	0.16	4.06	m	**10.49**
50 mm²	7.71	9.12	0.19	4.82	m	**13.95**

V:ELECTRICAL SUPPLY/POWER/LIGHTING SYSTEMS

Item	Net Price £	Material £	Labour hours	Labour £	Unit	Total rate £
70 mm²	11.46	13.56	0.21	5.33	m	**18.89**
95 mm²	16.60	19.65	0.23	5.84	m	**25.49**
120 mm²	19.60	23.20	0.24	6.09	m	**29.29**
150 mm²	24.08	28.50	0.27	6.86	m	**35.35**
185 mm²	33.20	39.29	0.30	7.62	m	**46.91**
240 mm²	45.90	54.32	0.33	8.38	m	**62.70**
300 mm²	55.90	66.16	0.35	8.89	m	**75.04**
400 mm²	67.90	80.36	0.41	10.41	m	**90.77**
600/1000 Volt grade; four core (galvanised steel wire armour)						
1.5 mm²	0.65	0.77	0.08	2.03	m	**2.80**
2.5 mm²	0.87	1.03	0.09	2.29	m	**3.31**
4 mm²	1.25	1.48	0.10	2.54	m	**4.02**
6 mm²	1.71	2.02	0.10	2.54	m	**4.56**
10 mm²	2.51	2.97	0.12	3.05	m	**6.02**
16 mm²	3.86	4.57	0.12	3.05	m	**7.62**
25 mm²	7.10	8.40	0.18	4.57	m	**12.97**
35 mm²	8.04	9.52	0.19	4.82	m	**14.34**
50 mm²	10.50	12.43	0.21	5.33	m	**17.76**
70 mm²	14.90	17.63	0.23	5.84	m	**23.47**
95 mm²	20.59	24.37	0.26	6.60	m	**30.97**
120 mm²	25.79	30.52	0.28	7.11	m	**37.63**
150 mm²	31.36	37.11	0.32	8.13	m	**45.24**
185 mm²	44.90	53.14	0.35	8.89	m	**62.02**
240 mm²	58.86	69.66	0.36	9.14	m	**78.80**
300 mm²	76.41	90.43	0.40	10.16	m	**100.59**
400 mm²	93.00	110.06	0.45	11.43	m	**121.49**
600/1000 Volt grade; seven core (galvanised steel wire armour)						
1.5 mm²	1.24	1.47	0.10	2.54	m	**4.01**
2.5 mm²	1.50	1.77	0.10	2.54	m	**4.31**
4 mm²	2.92	3.45	0.11	2.79	m	**6.24**
600/1000 Volt grade; twelve core (galvanised steel wire armour)						
1.5 mm²	2.00	2.37	0.11	2.79	m	**5.16**
2.5 mm²	2.41	2.85	0.11	2.79	m	**5.64**
600/1000 Volt grade; nineteen core (galvanised steel wire armour)						
1.5 mm²	2.60	3.07	0.13	3.30	m	**6.37**
2.5 mm²	3.65	4.32	0.14	3.55	m	**7.88**
600/1000 Volt grade; twenty seven core (galvanised steel wire armour)						
1.5 mm²	3.71	4.39	0.14	3.55	m	**7.94**
2.5 mm²	4.91	5.81	0.16	4.06	m	**9.87**
600/1000 Volt grade; thirty seven core (galvanised steel wire armour)						
1.5 mm²	4.98	5.90	0.15	3.81	m	**9.71**
2.5 mm²	7.06	8.36	0.17	4.32	m	**12.67**

V:ELECTRICAL SUPPLY/POWER/LIGHTING SYSTEMS

Item	Net Price £	Material £	Labour hours	Labour £	Unit	Total rate £
V20 : LV DISTRIBUTION (cont'd)						
Y61 - LV CABLES AND WIRING (cont'd)						
ARMOURED CABLE (cont'd)						
Cable; XLPE insulated; PVC sheathed copper stranded conductors to BS 5467; clipped direct to backgrounds including cleat						
600/1000 Volt grade; single core (aluminium wire armour)						
25 mm²	4.22	4.99	0.35	8.89	m	**13.88**
35 mm²	4.42	5.23	0.36	9.14	m	**14.37**
50 mm²	5.46	6.46	0.37	9.39	m	**15.86**
70 mm²	6.52	7.72	0.39	9.90	m	**17.62**
95 mm²	8.62	10.20	0.42	10.66	m	**20.87**
120 mm²	8.74	10.34	0.47	11.93	m	**22.28**
150 mm²	9.83	11.63	0.51	12.95	m	**24.58**
185 mm²	12.20	14.44	0.59	14.98	m	**29.42**
240 mm²	14.57	17.24	0.68	17.27	m	**34.51**
300 mm²	18.05	21.36	0.74	18.79	m	**40.15**
400 mm²	20.75	24.56	0.88	22.34	m	**46.90**
500 mm²	29.42	34.82	0.88	22.34	m	**57.16**
630 mm²	39.29	46.50	1.05	26.66	m	**73.16**
800 mm²	47.97	56.77	1.33	33.77	m	**90.54**
1000 mm²	53.62	63.46	1.40	35.55	m	**99.01**
600/1000 Volt grade; two core (galvanised steel wire armour)						
1.5 mm²	1.03	1.22	0.20	5.08	m	**6.30**
2.5 mm²	1.19	1.41	0.20	5.08	m	**6.49**
4 mm²	1.58	1.87	0.21	5.33	m	**7.20**
6 mm²	1.81	2.14	0.22	5.59	m	**7.73**
10 mm²	2.64	3.12	0.24	6.09	m	**9.22**
16 mm²	3.33	3.94	0.25	6.35	m	**10.29**
25 mm²	5.80	6.86	0.35	8.89	m	**15.75**
35 mm²	7.13	8.44	0.36	9.14	m	**17.58**
50 mm²	8.58	10.15	0.37	9.39	m	**19.55**
70 mm²	13.00	15.39	0.39	9.90	m	**25.29**
95 mm²	17.63	20.86	0.42	10.66	m	**31.53**
120 mm²	19.50	23.08	0.47	11.93	m	**35.01**
150 mm²	22.31	26.40	0.51	12.95	m	**39.35**
185 mm²	27.03	31.99	0.59	14.98	m	**46.97**
240 mm²	36.82	43.58	0.68	17.27	m	**60.84**
300 mm²	40.46	47.88	0.74	18.79	m	**66.67**
400 mm²	52.26	61.85	0.88	22.34	m	**84.19**
600/1000 Volt grade; three core (galvanised steel wire armour)						
1.5 mm²	1.06	1.25	0.20	5.08	m	**6.33**
2.5 mm²	1.29	1.53	0.21	5.33	m	**6.86**
4.0 mm²	1.85	2.19	0.22	5.59	m	**7.78**
6.0 mm²	2.13	2.52	0.22	5.59	m	**8.11**
10 mm²	3.09	3.66	0.25	6.35	m	**10.00**
16 mm²	4.03	4.77	0.26	6.60	m	**11.37**
25 mm²	7.40	8.76	0.37	9.39	m	**18.15**

V:ELECTRICAL SUPPLY/POWER/LIGHTING SYSTEMS

Item	Net Price £	Material £	Labour hours	Labour £	Unit	Total rate £
35 mm²	7.40	8.76	0.39	9.90	m	**18.66**
50 mm²	10.72	12.69	0.40	10.16	m	**22.84**
70 mm²	16.27	19.26	0.42	10.66	m	**29.92**
95 mm²	22.19	26.26	0.45	11.43	m	**37.69**
120 mm²	22.19	26.26	0.52	13.20	m	**39.46**
150 mm²	31.04	36.73	0.55	13.97	m	**50.70**
185 mm²	39.64	46.91	0.63	16.00	m	**62.91**
240 mm²	58.31	69.01	0.71	18.03	m	**87.04**
300 mm²	61.32	72.57	0.78	19.81	m	**92.38**
400 mm²	73.24	86.68	0.87	22.09	m	**108.77**
600/1000 Volt grade; four core (galvanised steel wire armour)						
1.5 mm²	1.23	1.46	0.21	5.33	m	**6.79**
2.5 mm²	1.67	1.98	0.22	5.59	m	**7.56**
4.0 mm²	2.13	2.52	0.22	5.59	m	**8.11**
6.0 mm²	2.69	3.18	0.23	5.84	m	**9.02**
10 mm²	3.68	4.36	0.26	6.60	m	**10.96**
16 mm²	5.20	6.15	0.26	6.60	m	**12.76**
25 mm²	10.22	12.10	0.39	9.90	m	**22.00**
35 mm²	12.70	15.03	0.40	10.16	m	**25.19**
50 mm²	16.42	19.43	0.41	10.41	m	**29.84**
70 mm²	20.74	24.55	0.45	11.43	m	**35.97**
95 mm²	28.00	33.14	0.50	12.70	m	**45.83**
120 mm²	33.75	39.94	0.54	13.71	m	**53.65**
150 mm²	36.38	43.05	0.60	15.23	m	**58.29**
185 mm²	49.92	59.08	0.67	17.01	m	**76.09**
240 mm²	63.88	75.60	0.75	19.04	m	**94.64**
300 mm²	85.58	101.28	0.83	21.07	m	**122.36**
400 mm²	100.17	118.55	0.91	23.11	m	**141.65**
600/1000 Volt grade; seven core (galvanised steel wire armour)						
1.5 mm²	1.50	1.77	0.20	5.08	m	**6.85**
2.5 mm²	1.75	2.07	0.20	5.08	m	**7.15**
4.0 mm²	3.17	3.75	0.23	5.84	m	**9.59**
600/1000 Volt grade; twelve core (galvanised steel wire armour)						
1.5 mm²	2.25	6.04	0.23	5.84	m	**11.88**
2.5 mm²	2.66	3.15	0.24	6.09	m	**9.24**
600/1000 Volt grade; nineteen core (galvanised steel wire armour)						
1.5 mm²	2.85	3.37	0.26	6.60	m	**9.97**
2.5 mm²	3.90	4.62	0.28	7.11	m	**11.73**
600/1000 Volt grade; twenty seven core (galvanised steel wire armour)						
1.5 mm²	3.96	4.69	0.29	7.36	m	**12.05**
2.5 mm²	5.16	6.11	0.30	7.62	m	**13.72**
600/1000 Volt grade; thirty seven core (galvanised steel wire armour)						
1.5 mm²	5.24	6.20	0.32	8.13	m	**14.32**
2.5 mm²	7.32	8.66	0.33	8.38	m	**17.04**

V:ELECTRICAL SUPPLY/POWER/LIGHTING SYSTEMS

Item	Net Price £	Material £	Labour hours	Labour £	Unit	Total rate £
V20 : LV DISTRIBUTION (cont'd)						
Y61 - LV CABLES AND WIRING (cont'd)						
ARMOURED CABLE (cont'd)						
Cable termination; brass weatherproof gland with inner and outer seal, shroud, brass locknut and earth ring (including drilling and cutting mild steel gland plate)						
600/1000 Volt grade; single core (aluminium wire armour)						
25 mm²	8.01	9.49	1.70	43.17	nr	**52.65**
35 mm²	8.01	9.49	1.79	45.45	nr	**54.94**
50 mm²	8.01	9.49	2.06	52.31	nr	**61.79**
70 mm²	8.63	10.21	2.12	53.83	nr	**64.04**
95 mm²	9.44	11.17	2.39	60.69	nr	**71.85**
120 mm²	9.56	11.31	2.47	62.72	nr	**74.03**
150 mm²	9.58	11.34	2.73	69.32	nr	**80.66**
185 mm²	15.18	17.97	3.05	77.44	nr	**95.41**
240 mm²	15.40	18.22	3.45	87.60	nr	**105.82**
300 mm²	19.86	23.51	3.84	97.50	nr	**121.01**
400 mm²	27.10	32.07	4.21	106.90	nr	**138.97**
500 mm²	28.10	33.26	5.70	144.73	m	**177.99**
630 mm²	36.13	42.75	6.20	157.43	m	**200.18**
800 mm²	51.50	60.95	7.50	190.44	m	**251.39**
1000 mm²	57.83	68.44	10.00	253.91	m	**322.35**
600/1000 Volt grade; two core (galvanised steel wire armour)						
1.5 mm²	1.92	2.27	0.58	14.73	nr	**17.00**
2.5 mm²	1.92	2.27	0.58	14.73	nr	**17.00**
4 mm²	1.92	2.27	0.58	14.73	nr	**17.00**
6 mm²	2.42	2.87	0.67	17.01	nr	**19.88**
10 mm²	2.87	3.39	1.00	25.39	nr	**28.79**
16 mm²	4.32	5.11	1.11	28.18	nr	**33.29**
25 mm²	3.27	3.87	1.70	43.17	nr	**47.03**
35 mm²	4.14	4.90	1.79	45.45	nr	**50.35**
50 mm²	4.14	4.90	2.06	52.31	nr	**57.21**
70 mm²	5.17	6.12	2.12	53.83	nr	**59.95**
95 mm²	4.60	5.44	2.39	60.69	nr	**66.12**
120 mm²	11.53	13.65	2.47	62.72	nr	**76.37**
150 mm²	9.29	10.99	2.73	69.32	nr	**80.31**
185 mm²	21.00	24.86	3.05	77.44	nr	**102.30**
240 mm²	20.50	24.26	3.45	87.60	nr	**111.86**
300 mm²	20.50	24.26	3.84	97.50	nr	**121.76**
400 mm²	34.09	40.34	4.21	106.90	nr	**147.24**
600/1000 Volt grade; three core (galvanised steel wire armour)						
1.5 mm²	1.92	2.27	0.62	15.74	nr	**18.02**
2.5 mm²	1.92	2.27	0.62	15.74	nr	**18.02**
4 mm²	1.92	2.27	0.62	15.74	nr	**18.02**
6 mm²	2.42	2.87	0.71	18.03	nr	**20.90**
10 mm²	2.87	3.39	1.06	26.91	nr	**30.31**
16 mm²	4.32	5.11	1.19	30.22	nr	**35.32**
25 mm²	3.27	3.87	1.81	45.96	nr	**49.83**

V:ELECTRICAL SUPPLY/POWER/LIGHTING SYSTEMS

Item	Net Price £	Material £	Labour hours	Labour £	Unit	Total rate £
35 mm²	4.14	4.90	1.99	50.53	nr	**55.43**
50 mm²	4.14	4.90	2.23	56.62	nr	**61.52**
70 mm²	10.80	12.78	2.40	60.94	nr	**73.72**
95 mm²	9.59	11.35	2.63	66.78	nr	**78.13**
120 mm²	25.61	30.30	2.83	71.86	nr	**102.16**
150 mm²	20.62	24.40	3.22	81.76	nr	**106.16**
185 mm²	21.18	25.06	3.44	87.35	nr	**112.41**
240 mm²	34.09	40.34	3.83	97.25	nr	**137.59**
300 mm²	34.09	40.34	4.28	108.68	nr	**149.02**
400 mm²	52.64	62.30	5.00	126.96	nr	**189.25**
600/1000 Volt grade; four core (galvanised steel wire armour)						
1.5 mm²	1.92	2.27	0.67	17.01	nr	**19.28**
2.5 mm²	1.92	2.27	0.67	17.01	nr	**19.28**
4 mm²	2.38	2.82	0.71	18.03	nr	**20.85**
6 mm²	2.78	3.29	0.76	19.30	nr	**22.58**
10 mm²	4.18	4.95	1.14	28.95	nr	**33.89**
16 mm²	4.18	4.95	1.29	32.75	nr	**37.70**
25 mm²	4.14	4.90	1.99	50.53	nr	**55.43**
35 mm²	4.14	4.90	2.16	54.85	nr	**59.74**
50 mm²	8.64	10.23	2.49	63.22	nr	**73.45**
70 mm²	8.86	10.48	2.65	67.29	nr	**77.77**
95 mm²	19.18	22.70	2.98	75.67	nr	**98.37**
120 mm²	20.11	23.80	3.15	79.98	nr	**103.79**
150 mm²	19.82	23.46	3.50	88.87	nr	**112.33**
185 mm²	33.77	39.96	3.72	94.46	nr	**134.42**
240 mm²	52.64	62.30	4.33	109.94	nr	**172.24**
300 mm²	52.64	62.30	4.86	123.40	nr	**185.70**
400 mm²	52.64	62.30	5.46	138.64	nr	**200.93**
600/1000 Volt grade; seven core (galvanised steel wire armour)						
1.5 mm²	1.92	2.27	0.81	20.57	nr	**22.84**
2.5 mm²	1.89	2.23	0.85	21.58	nr	**23.82**
4 mm²	2.20	2.60	0.93	23.61	nr	**26.22**
600/1000 Volt grade; twelve core (galvanised steel wire armour)						
1.5 mm²	2.38	2.82	1.14	28.95	nr	**31.77**
2.5 mm²	4.18	4.95	1.13	28.69	nr	**33.64**
600/1000 Volt grade; nineteen core (galvanised steel wire armour)						
1.5 mm²	4.18	4.95	1.54	39.10	nr	**44.05**
2.5 mm²	4.18	4.95	1.54	39.10	nr	**44.05**
600/1000 Volt grade; twenty seven core (galvanised steel wire armour)						
1.5 mm²	3.27	3.87	1.94	49.26	nr	**53.13**
2.5 mm²	4.45	5.27	2.31	58.65	nr	**63.92**
600/1000 Volt grade; thirty seven core (galvanised steel wire armour)						
1.5 mm²	4.45	5.27	2.53	64.24	nr	**69.51**
2.5 mm²	4.45	5.27	2.87	72.87	nr	**78.14**

V:ELECTRICAL SUPPLY/POWER/LIGHTING SYSTEMS

Item	Net Price £	Material £	Labour hours	Labour £	Unit	Total rate £
V20 : LV DISTRIBUTION (cont'd)						
Y61 - LV CABLES AND WIRING (cont'd)						
ARMOURED CABLE (cont'd)						
Cable; XLPE insulated; LSOH sheathed (LSF) ; copper stranded conductors to BS 6724; laid in trench/duct including marker tape (cable tiles measured elsewhere.)						
600/1000 Volt grade; single core (aluminium wire armour)						
50 mm²	4.70	5.56	0.17	4.32	m	**9.88**
70 mm²	5.88	6.96	0.18	4.57	m	**11.53**
95 mm²	7.68	9.09	0.20	5.08	m	**14.17**
120 mm²	7.96	9.42	0.22	5.59	m	**15.01**
150 mm²	8.54	10.11	0.24	6.09	m	**16.20**
185 mm²	11.10	13.14	0.26	6.60	m	**19.74**
240 mm²	13.60	16.10	0.30	7.62	m	**23.71**
300 mm²	16.37	19.37	0.31	7.87	m	**27.24**
400 mm²	21.20	25.09	0.35	8.89	m	**33.98**
500 mm²	25.02	29.61	0.44	11.17	m	**40.78**
630 mm²	32.63	38.62	0.52	13.20	m	**51.82**
800 mm²	42.78	50.63	0.62	15.74	m	**66.37**
1000 mm²	54.59	64.61	0.65	16.50	m	**81.11**
600/1000 Volt grade; two core (galvanised steel wire armour)						
1.5 mm²	0.59	0.70	0.06	1.52	m	**2.22**
2.5 mm²	0.73	0.86	0.06	1.52	m	**2.39**
4 mm²	0.95	1.12	0.08	2.03	m	**3.16**
6 mm²	1.20	1.42	0.08	2.03	m	**3.45**
10 mm²	1.71	2.02	0.10	2.54	m	**4.56**
16 mm²	2.60	3.08	0.10	2.54	m	**5.62**
25 mm²	3.80	4.50	0.15	3.81	m	**8.31**
35 mm²	4.94	5.85	0.15	3.81	m	**9.66**
50 mm²	6.20	7.34	0.17	4.32	m	**11.65**
70 mm²	8.87	10.50	0.18	4.57	m	**15.07**
95 mm²	13.28	15.72	0.20	5.08	m	**20.79**
120 mm²	15.20	17.99	0.22	5.59	m	**23.57**
150 mm²	18.17	21.50	0.24	6.09	m	**27.60**
185 mm²	22.53	26.66	0.26	6.60	m	**33.27**
240 mm²	28.73	34.00	0.30	7.62	m	**41.62**
300 mm²	35.87	42.45	0.31	7.87	m	**50.32**
400 mm²	45.47	53.81	0.35	8.89	m	**62.70**
600/1000 Volt grade; three core (galvanised steel wire armour)						
1.5 mm²	0.66	0.78	0.07	1.78	m	**2.56**
2.5 mm²	0.85	1.01	0.07	1.78	m	**2.78**
4 mm²	1.14	1.35	0.09	2.29	m	**3.63**
6 mm²	1.48	1.75	0.10	2.54	m	**4.29**
10 mm²	2.34	2.77	0.11	2.79	m	**5.56**
16 mm²	3.39	4.01	0.11	2.79	m	**6.81**
25 mm²	5.32	6.30	0.16	4.06	m	**10.36**
35 mm²	6.97	8.25	0.16	4.06	m	**12.31**
50 mm²	9.01	10.66	0.19	4.82	m	**15.49**

V:ELECTRICAL SUPPLY/POWER/LIGHTING SYSTEMS

Item	Net Price £	Material £	Labour hours	Labour £	Unit	Total rate £
70 mm²	13.11	15.52	0.21	5.33	m	**20.85**
95 mm²	18.24	21.59	0.23	5.84	m	**27.43**
120 mm²	21.47	25.41	0.24	6.09	m	**31.50**
150 mm²	26.86	31.79	0.27	6.86	m	**38.64**
185 mm²	32.73	38.73	0.30	7.62	m	**46.35**
240 mm²	41.15	48.70	0.33	8.38	m	**57.08**
300 mm²	51.88	61.40	0.35	8.89	m	**70.29**
400 mm²	67.86	80.31	0.41	10.41	m	**90.72**
600/1000 Volt grade; four core (galvanised steel wire armour)						
1.5 mm²	0.80	0.95	0.08	2.03	m	**2.98**
2.5 mm²	1.03	1.22	0.09	2.29	m	**3.50**
4 mm²	1.40	1.66	0.10	2.54	m	**4.20**
6 mm²	2.00	2.37	0.10	2.54	m	**4.91**
10 mm²	2.97	3.51	0.12	3.05	m	**6.56**
16 mm²	4.32	5.11	0.12	3.05	m	**8.16**
25 mm²	6.69	7.92	0.18	4.57	m	**12.49**
35 mm²	8.70	10.30	0.19	4.82	m	**15.12**
50 mm²	11.23	13.29	0.21	5.33	m	**18.62**
70 mm²	16.34	19.34	0.23	5.84	m	**25.18**
95 mm²	21.89	25.91	0.26	6.60	m	**32.51**
120 mm²	27.40	32.43	0.28	7.11	m	**39.54**
150 mm²	33.56	39.72	0.32	8.13	m	**47.84**
185 mm²	41.64	49.28	0.35	8.89	m	**58.17**
240 mm²	54.07	63.99	0.36	9.14	m	**73.13**
300 mm²	68.32	80.85	0.40	10.16	m	**91.01**
400 mm²	89.49	105.91	0.45	11.43	m	**117.33**
600/1000 Volt grade; seven core (galvanised steel wire armour)						
1.5 mm²	1.50	1.77	0.10	2.54	m	**4.31**
2.5 mm²	1.79	2.12	0.10	2.54	m	**4.66**
4 mm²	3.04	3.59	0.11	2.79	m	**6.39**
600/1000 Volt grade; twelve core (galvanised steel wire armour)						
1.5 mm²	2.28	2.69	0.11	2.79	m	**5.49**
2.5 mm²	2.78	3.29	0.11	2.79	m	**6.09**
600/1000 Volt grade; nineteen core (galvanised steel wire armour)						
1.5 mm²	3.04	3.59	0.13	3.30	m	**6.89**
2.5 mm²	3.79	4.49	0.14	3.55	m	**8.05**
600/1000 Volt grade; twenty seven core (galvanised steel wire armour)						
1.5 mm²	4.11	4.87	0.14	3.55	m	**8.42**
2.5 mm²	5.45	6.44	0.16	4.06	m	**10.51**
600/1000 Volt grade; thirty seven core (galvanised steel wire armour)						
1.5 mm²	5.31	6.29	0.15	3.81	m	**10.10**
2.5 mm²	7.22	8.54	0.17	4.32	m	**12.86**

V:ELECTRICAL SUPPLY/POWER/LIGHTING SYSTEMS

Item	Net Price £	Material £	Labour hours	Labour £	Unit	Total rate £
V20 : LV DISTRIBUTION (cont'd)						
Y61 - LV CABLES AND WIRING (cont'd)						
ARMOURED CABLE (cont'd)						
Cable; XLPE insulated; LSOH sheathed (LSF) copper stranded conductors to BS 6724; clipped direct to backgrounds including cleat.						
600/1000 Volt grade; single core (aluminium wire armour)						
50 mm²	5.63	6.66	0.37	9.39	m	**16.06**
70 mm²	6.81	8.06	0.39	9.90	m	**17.96**
95 mm²	9.27	10.97	0.42	10.66	m	**21.64**
120 mm²	9.82	11.62	0.47	11.93	m	**23.56**
150 mm²	10.40	12.31	0.51	12.95	m	**25.26**
185 mm²	12.54	14.84	0.59	14.98	m	**29.82**
240 mm²	15.34	18.15	0.68	17.27	m	**35.42**
300 mm²	18.11	21.43	0.74	18.79	m	**40.22**
400 mm²	24.86	29.42	0.81	20.57	m	**49.99**
500 mm²	28.68	33.94	0.88	22.34	m	**56.29**
630 mm²	37.29	44.13	1.05	26.66	m	**70.79**
800 mm²	47.78	56.55	1.33	33.77	m	**90.32**
1000 mm²	60.25	71.30	1.40	35.55	m	**106.85**
600/1000 Volt grade; two core (galvanised steel wire armour)						
1.5 mm²	1.10	1.30	0.20	5.08	m	**6.38**
2.5 mm²	1.29	1.53	0.20	5.08	m	**6.60**
4.0 mm²	1.73	2.05	0.21	5.33	m	**7.38**
6.0 mm²	2.05	2.43	0.22	5.59	m	**8.01**
10 mm²	2.82	3.34	0.24	6.09	m	**9.43**
16 mm²	3.75	4.44	0.25	6.35	m	**10.79**
25 mm²	6.35	7.51	0.35	8.89	m	**16.40**
35 mm²	7.86	9.30	0.36	9.14	m	**18.44**
50 mm²	9.54	11.29	0.37	9.39	m	**20.69**
70 mm²	14.34	16.97	0.39	9.90	m	**26.87**
95 mm²	19.96	23.62	0.42	10.66	m	**34.29**
120 mm²	20.16	23.86	0.47	11.93	m	**35.79**
150 mm²	23.54	27.86	0.51	12.95	m	**40.81**
185 mm²	24.27	28.72	0.59	14.98	m	**43.70**
240 mm²	30.47	36.06	0.68	17.27	m	**53.33**
300 mm²	39.53	46.78	0.74	18.79	m	**65.57**
400 mm²	50.13	59.33	0.81	20.57	m	**79.89**
600/1000 Volt grade; three core (galvanised steel wire armour)						
1.5 mm²	1.18	1.40	0.20	5.08	m	**6.47**
2.5 mm²	1.65	1.95	0.21	5.33	m	**7.28**
4.0 mm²	1.97	2.33	0.22	5.59	m	**7.92**
6.0 mm²	2.36	2.79	0.22	5.59	m	**8.38**
10 mm²	3.48	4.12	0.25	6.35	m	**10.47**
16 mm²	4.58	5.42	0.26	6.60	m	**12.02**
25 mm²	8.18	9.68	0.37	9.39	m	**19.08**
35 mm²	9.70	11.48	0.39	9.90	m	**21.38**
50 mm²	12.00	14.20	0.40	10.16	m	**24.36**
70 mm²	18.10	21.42	0.42	10.66	m	**32.09**

V:ELECTRICAL SUPPLY/POWER/LIGHTING SYSTEMS

Item	Net Price £	Material £	Labour hours	Labour £	Unit	Total rate £
95 mm²	23.79	28.15	0.45	11.43	m	**39.58**
120 mm²	28.92	34.23	0.52	13.20	m	**47.43**
150 mm²	33.82	40.02	0.55	13.97	m	**53.99**
185 mm²	34.47	40.79	0.63	16.00	m	**56.79**
240 mm²	46.17	54.64	0.71	18.03	m	**72.67**
300 mm²	57.30	67.81	0.78	19.81	m	**87.62**
400 mm²	74.28	87.91	0.87	22.09	m	**110.00**
600/1000 Volt grade; four core (galvanised steel wire armour)						
1.5 mm²	1.37	1.62	0.21	5.33	m	**6.95**
2.5 mm²	1.83	2.17	0.22	5.59	m	**7.75**
4.0 mm²	1.68	1.99	0.22	5.59	m	**7.57**
6.0 mm²	2.98	3.53	0.23	5.84	m	**9.37**
10 mm²	4.15	4.91	0.26	6.60	m	**11.51**
16 mm²	5.64	6.67	0.26	6.60	m	**13.28**
25 mm²	10.90	12.90	0.39	9.90	m	**22.80**
35 mm²	13.36	15.81	0.40	10.16	m	**25.97**
50 mm²	17.15	20.30	0.41	10.41	m	**30.71**
70 mm²	22.14	26.20	0.45	11.43	m	**37.63**
95 mm²	29.29	34.66	0.50	12.70	m	**47.36**
120 mm²	35.35	41.84	0.54	13.71	m	**55.55**
150 mm²	43.60	51.60	0.60	15.23	m	**66.83**
185 mm²	46.66	55.22	0.67	17.01	m	**72.23**
240 mm²	59.49	70.40	0.75	19.04	m	**89.45**
300 mm²	75.49	89.34	0.83	21.07	m	**110.42**
400 mm²	97.66	115.58	0.91	23.11	m	**138.68**
600/1000 Volt grade; seven core (galvanised steel wire armour)						
1.5 mm²	1.75	2.07	0.20	5.08	m	**7.15**
2.5 mm²	2.05	2.42	0.20	5.08	m	**7.50**
4.0 mm²	3.29	3.89	0.23	5.84	m	**9.73**
600/1000 Volt grade; twelve core (galvanised steel wire armour)						
1.5 mm²	2.53	2.99	0.23	5.84	m	**8.83**
2.5 mm²	3.04	3.59	0.24	6.09	m	**9.69**
600/1000 Volt grade; nineteen core (galvanised steel wire armour)						
1.5 mm²	3.29	3.89	0.26	6.60	m	**10.49**
2.5 mm²	4.05	4.79	0.28	7.11	m	**11.90**
600/1000 Volt grade; twenty seven core (galvanised steel wire armour)						
1.5 mm²	4.37	5.17	0.29	7.36	m	**12.53**
2.5 mm²	5.70	6.74	0.30	7.62	m	**14.36**
600/1000 Volt grade; thirty seven core (galvanised steel wire armour)						
1.5 mm²	5.57	6.59	0.32	8.13	m	**14.71**
2.5 mm²	7.47	8.84	0.33	8.38	m	**17.22**

V:ELECTRICAL SUPPLY/POWER/LIGHTING SYSTEMS

Item	Net Price £	Material £	Labour hours	Labour £	Unit	Total rate £
V20 : LV DISTRIBUTION (cont'd)						
Y61 - LV CABLES AND WIRING (cont'd)						
ARMOURED CABLE (cont'd)						
Cable termination; brass weatherproof gland with inner and outer seal, shroud, brass locknut and earth ring (including drilling and cutting mild steel gland plate)						
600/1000 Volt grade; single core (aluminium wire armour)						
25 mm²	8.09	9.57	1.70	43.17	nr	52.74
35 mm²	8.09	9.57	1.79	45.45	nr	55.02
50 mm²	8.40	9.94	2.06	52.31	nr	62.24
70 mm²	9.16	10.84	2.12	53.83	nr	64.67
95 mm²	9.27	10.98	2.39	60.69	nr	71.66
120 mm²	9.31	11.02	2.47	62.72	nr	73.73
150 mm²	11.96	14.15	2.73	69.32	nr	83.47
185 mm²	15.42	18.25	3.05	77.44	nr	95.69
240 mm²	20.33	24.06	3.45	87.60	nr	111.66
300 mm²	20.64	24.43	3.84	97.50	nr	121.93
400 mm²	25.72	30.44	4.21	106.90	nr	137.33
500 mm²	27.34	32.36	5.70	144.73	m	177.09
630 mm²	35.15	41.60	6.20	157.43	m	199.03
800 mm²	50.12	59.32	7.50	190.44	m	249.75
1000 mm²	56.28	66.60	10.00	253.91	m	320.52
600/1000 Volt grade; two core (galvanised steel wire armour)						
1.5 mm²	2.06	2.44	0.58	14.73	nr	17.16
2.5 mm²	2.06	2.44	0.58	14.73	nr	17.16
4 mm²	2.06	2.44	0.58	14.73	nr	17.16
6 mm²	2.06	2.44	0.67	17.01	nr	19.45
10 mm²	3.20	3.79	1.00	25.39	nr	29.18
16 mm²	4.27	5.05	1.11	28.18	nr	33.24
25 mm²	4.75	5.62	1.70	43.17	nr	48.79
35 mm²	6.79	8.04	1.79	45.45	nr	53.49
50 mm²	6.93	8.20	2.06	52.31	nr	60.51
70 mm²	11.21	13.26	2.12	53.83	nr	67.09
95 mm²	9.44	11.18	2.39	60.69	nr	71.86
120 mm²	16.95	20.06	2.47	62.72	nr	82.78
150 mm²	13.29	15.73	2.73	69.32	nr	85.05
185 mm²	20.97	24.81	3.05	77.44	nr	102.26
240 mm²	23.97	28.37	3.45	87.60	nr	115.97
300 mm²	25.65	30.35	3.84	97.50	nr	127.86
400 mm²	27.90	33.02	4.21	106.90	nr	139.92
600/1000 Volt grade; three core (galvanised steel wire armour)						
1.5 mm²	1.91	2.26	0.62	15.74	nr	18.00
2.5 mm²	1.91	2.26	0.62	15.74	nr	18.00
4 mm²	1.91	2.26	0.62	15.74	nr	18.00
6 mm²	2.26	2.67	0.71	18.03	nr	20.70
10 mm²	2.83	3.34	1.06	26.91	nr	30.26
16 mm²	3.77	4.46	1.19	30.22	nr	34.67
25 mm²	5.59	6.62	1.81	45.96	nr	52.58

V:ELECTRICAL SUPPLY/POWER/LIGHTING SYSTEMS

Item	Net Price £	Material £	Labour hours	Labour £	Unit	Total rate £
35 mm²	7.80	9.23	1.99	50.53	nr	**59.76**
50 mm²	8.13	9.62	2.23	56.62	nr	**66.24**
70 mm²	12.08	14.30	2.40	60.94	nr	**75.24**
95 mm²	13.44	15.90	2.63	66.78	nr	**82.68**
120 mm²	17.47	20.68	2.83	71.86	nr	**92.54**
150 mm²	20.85	24.68	3.22	81.76	nr	**106.44**
185 mm²	23.25	27.52	3.44	87.35	nr	**114.87**
240 mm²	36.03	42.64	3.83	97.25	nr	**139.89**
300 mm²	38.46	45.51	4.28	108.68	nr	**154.19**
400 mm²	51.64	61.12	5.00	126.96	nr	**188.07**
600/1000 Volt grade; four core (galvanised steel wire armour)						
1.5 mm²	1.91	2.26	0.67	17.01	nr	**19.27**
2.5 mm²	1.91	2.26	0.67	17.01	nr	**19.27**
4 mm²	2.21	2.61	0.71	18.03	nr	**20.64**
6 mm²	2.71	3.21	0.76	19.30	nr	**22.50**
10 mm²	3.61	4.27	1.14	28.95	nr	**33.22**
16 mm²	3.61	4.27	1.29	32.75	nr	**37.03**
25 mm²	8.38	9.92	1.99	50.53	nr	**60.45**
35 mm²	8.82	10.43	2.16	54.85	nr	**65.28**
50 mm²	9.25	10.94	2.49	63.22	nr	**74.17**
70 mm²	12.96	15.34	2.65	67.29	nr	**82.63**
95 mm²	18.02	21.33	2.98	75.67	nr	**97.00**
120 mm²	19.26	22.80	3.15	79.98	nr	**102.78**
150 mm²	21.70	25.68	3.50	88.87	nr	**114.55**
185 mm²	32.42	38.37	3.72	94.46	nr	**132.83**
240 mm²	41.10	48.64	4.33	109.94	nr	**158.59**
300 mm²	54.21	64.16	4.86	123.40	nr	**187.56**
400 mm²	58.62	69.38	5.46	138.64	nr	**208.02**
600/1000 Volt grade; seven core (galvanised steel wire armour)						
1.5 mm²	1.91	2.26	0.81	20.57	nr	**22.82**
2.5 mm²	2.21	2.61	0.85	21.58	nr	**24.19**
4 mm²	3.61	4.27	0.93	23.61	nr	**27.89**
600/1000 Volt grade; twelve core (galvanised steel wire armour)						
1.5 mm²	2.21	2.61	1.14	28.95	nr	**31.56**
2.5 mm²	3.61	4.27	1.13	28.69	nr	**32.97**
600/1000 Volt grade; nineteen core (galvanised steel wire armour)						
1.5 mm²	3.61	4.27	1.54	39.10	nr	**43.38**
2.5 mm²	3.61	4.27	1.54	39.10	nr	**43.38**
600/1000 Volt grade; twenty seven core (galvanised steel wire armour)						
1.5 mm²	5.29	6.27	1.94	49.26	nr	**55.53**
2.5 mm²	5.29	6.27	2.31	58.65	nr	**64.92**
600/1000 Volt grade; thirty seven core (galvanised steel wire armour)						
1.5 mm²	5.29	6.27	2.53	64.24	nr	**70.51**
2.5 mm²	5.29	6.27	2.87	72.87	nr	**79.14**

V:ELECTRICAL SUPPLY/POWER/LIGHTING SYSTEMS

Item	Net Price £	Material £	Labour hours	Labour £	Unit	Total rate £
V20 : LV DISTRIBUTION (cont'd)						
Y61 - LV CABLES AND WIRING (cont'd)						
UN-ARMOURED CABLE						
Cable: XLPE insulated; PVC sheathed 90c copper to CMA Code 6181e; for internal wiring; clipped to backgrounds; (Supports and fixings included)						
300/500 Volt grade; single core						
6 mm²	0.52	0.61	0.09	2.29	m	**2.90**
10 mm²	0.74	0.88	0.10	2.54	m	**3.42**
16 mm²	1.01	1.20	0.12	3.05	m	**4.25**
Cable; LSF insulated to CMA Code 6491B; non-sheathed copper; laid/drawn in trunking/conduit						
450/750 Volt grade; single core						
1.5 mm²	0.17	0.20	0.03	0.76	m	**0.96**
2.5 mm²	0.22	0.26	0.03	0.76	m	**1.02**
4 mm²	0.38	0.45	0.03	0.76	m	**1.21**
6 mm²	0.56	0.66	0.04	1.02	m	**1.68**
10 mm²	0.96	1.14	0.04	1.02	m	**2.15**
16 mm²	1.35	1.60	0.05	1.27	m	**2.87**
25 mm²	2.51	2.97	0.06	1.52	m	**4.49**
35 mm²	3.23	3.82	0.06	1.52	m	**5.35**
50 mm²	3.70	4.38	0.07	1.78	m	**6.16**
70 mm²	5.03	5.95	0.08	2.03	m	**7.98**
95 mm²	6.29	7.44	0.08	2.03	m	**9.48**
120 mm²	8.76	10.36	0.10	2.54	m	**12.90**
150 mm²	11.38	13.47	0.13	3.30	m	**16.77**
Cable; twin & earth to CMA code 6242Y; clipped to backgrounds						
300/500 Volt grade; PVC/PVC						
1.5 mm² 2C+E	0.81	0.96	0.01	0.25	m	**1.21**
1.5 mm² 3C+E	2.35	2.78	0.02	0.51	m	**3.29**
2.5 mm² 2C+E	1.05	1.25	0.02	0.51	m	**1.75**
4 mm² 2C+E	3.23	3.82	0.02	0.51	m	**4.33**
6 mm² 2C+E	3.81	4.51	0.02	0.51	m	**5.01**
10 mm² 2C+E	6.16	7.29	0.03	0.76	m	**8.05**
16 mm² 2C+E	9.73	11.52	0.03	0.76	m	**12.28**
300/500 Volt grade; LSF/LSF						
1.5 mm² 2C+E	1.08	1.28	0.01	0.25	m	**1.53**
1.5 mm² 3C+E	4.24	5.02	0.02	0.51	m	**5.52**
2.5 mm² 2C+E	1.51	1.79	0.02	0.51	m	**2.30**
4 mm² 2C+E	4.16	4.92	0.02	0.51	m	**5.43**
6 mm² 2C+E	7.18	8.50	0.02	0.51	m	**9.01**
10 mm² 2C+E	8.26	9.78	0.03	0.76	m	**10.54**
16 mm² 2C+E	12.14	14.36	0.03	0.76	m	**15.12**

V:ELECTRICAL SUPPLY/POWER/LIGHTING SYSTEMS

Item	Net Price £	Material £	Labour hours	Labour £	Unit	Total rate £
EARTH CABLE						
Cable; LSF insulated to CMA Code 6491B; non-sheathed copper; laid/drawn in trunking/conduit						
450/750 Volt grade; single core						
1.5 mm²	0.17	0.20	0.03	0.76	m	**0.96**
2.5 mm²	0.22	0.26	0.03	0.76	m	**1.02**
4 mm²	0.38	0.45	0.03	0.76	m	**1.21**
6 mm²	0.56	0.66	0.04	1.02	m	**1.68**
10 mm²	0.96	1.14	0.04	1.02	m	**2.15**
16 mm²	1.35	1.60	0.05	1.27	m	**2.87**
25 mm²	2.51	2.97	0.06	1.52	m	**4.49**
35 mm²	3.23	3.82	0.06	1.52	m	**5.35**
50 mm²	3.70	4.38	0.07	1.78	m	**6.16**
70 mm²	5.03	5.95	0.08	2.03	m	**7.98**
95 mm²	6.29	7.44	0.08	2.03	m	**9.48**
120 mm²	7.56	8.94	0.10	2.54	m	**11.48**
150 mm²	9.83	11.63	0.13	3.30	m	**14.93**
185 mm²	11.79	13.96	0.16	4.06	m	**18.02**
240 mm²	13.62	16.12	0.20	5.08	m	**21.20**
FLEXIBLE CABLE						
Flexible cord; PVC insulated; PVC sheathed; copper stranded to CMA Code 218*Y (laid loose)						
300 Volt grade; two core						
0.50 mm²	0.14	0.17	0.07	1.78	m	**1.95**
0.75 mm²	0.23	0.27	0.07	1.78	m	**2.05**
300 Volt grade; three core						
0.50 mm²	0.19	0.22	0.07	1.78	m	**2.00**
0.75 mm²	0.31	0.36	0.07	1.78	m	**2.14**
1.0 mm²	0.43	0.51	0.07	1.78	m	**2.29**
1.5 mm²	0.60	0.72	0.07	1.78	m	**2.49**
2.5 mm²	0.82	0.98	0.08	2.03	m	**3.01**
Flexible cord; PVC insulated; PVC sheathed; copper stranded to CMA Code 318Y (laid loose)						
300/500 Volt grade; two core						
0.75 mm²	0.23	0.27	0.07	1.78	m	**2.05**
1.0 mm²	0.29	0.34	0.07	1.78	m	**2.12**
1.5 mm²	0.41	0.48	0.07	1.78	m	**2.26**
2.5 mm²	0.87	1.03	0.07	1.78	m	**2.81**
300/500 Volt grade; three core						
0.75 mm²	0.22	0.26	0.07	1.78	m	**2.04**
1.0 mm²	0.26	0.31	0.07	1.78	m	**2.09**
1.5 mm²	0.36	0.43	0.07	1.78	m	**2.21**
2.5 mm²	0.75	0.89	0.08	2.03	m	**2.92**

V:ELECTRICAL SUPPLY/POWER/LIGHTING SYSTEMS

Item	Net Price £	Material £	Labour hours	Labour £	Unit	Total rate £
V20 : LV DISTRIBUTION (cont'd)						
Y61 - LV CABLES AND WIRING (cont'd)						
FLEXIBLE CABLE (cont'd)						
Flexible cord; PVC insulated (cont'd)						
300/500 Volt grade; four core						
0.75 mm²	0.60	0.72	0.08	2.03	m	**2.75**
1.0 mm²	0.70	0.83	0.08	2.03	m	**2.86**
1.5 mm²	1.00	1.18	0.08	2.03	m	**3.22**
2.5 mm²	1.54	1.82	0.09	2.29	m	**4.11**
Flexible cord; PVC insulated; PVC sheathed for use in high temperature zones; copper stranded to CMA Code 309Y (laid loose)						
300/500 Volt grade; two core						
0.50 mm²	0.31	0.36	0.07	1.78	m	**2.14**
0.75 mm²	0.36	0.43	0.07	1.78	m	**2.21**
1.0 mm²	0.59	0.70	0.07	1.78	m	**2.48**
1.5 mm²	0.84	0.99	0.07	1.78	m	**2.77**
2.5 mm²	1.25	1.48	0.07	1.78	m	**3.26**
300/500 Volt grade; three core						
0.50 mm²	0.42	0.49	0.07	1.78	m	**2.27**
0.75 mm²	0.44	0.52	0.07	1.78	m	**2.30**
1.0 mm²	0.59	0.70	0.07	1.78	m	**2.48**
1.5 mm²	0.84	0.99	0.07	1.78	m	**2.77**
2.5 mm²	1.25	1.48	0.07	1.78	m	**3.26**
Flexible cord; rubber insulated; rubber sheathed; copper stranded to CMA code 318 (laid loose)						
300/500 Volt grade; two core						
0.50 mm²	0.15	0.18	0.07	1.78	m	**1.96**
0.75 mm²	0.17	0.20	0.07	1.78	m	**1.97**
1.0 mm²	0.19	0.22	0.07	1.78	m	**2.00**
1.5 mm²	0.26	0.31	0.07	1.78	m	**2.09**
2.5 mm²	0.39	0.46	0.07	1.78	m	**2.23**
300/500 Volt grade; three core						
0.50 mm²	0.19	0.22	0.07	1.78	m	**2.00**
0.75 mm²	0.20	0.23	0.07	1.78	m	**2.01**
1.0 mm²	0.23	0.27	0.07	1.78	m	**2.05**
1.5 mm²	0.32	0.38	0.07	1.78	m	**2.15**
2.5 mm²	0.50	0.59	0.07	1.78	m	**2.36**
300/500 Volt grade; four core						
0.50 mm²	0.28	0.33	0.08	2.03	m	**2.36**
0.75 mm²	0.31	0.36	0.08	2.03	m	**2.40**
1.0 mm²	0.35	0.42	0.08	2.03	m	**2.45**
1.5 mm²	0.42	0.49	0.08	2.03	m	**2.53**
2.5 mm²	0.64	0.76	0.08	2.03	m	**2.79**

V:ELECTRICAL SUPPLY/POWER/LIGHTING SYSTEMS

Item	Net Price £	Material £	Labour hours	Labour £	Unit	Total rate £
Flexible cord; rubber insulated; rubber sheathed; for 90C operation; copper stranded to CMA Code 318 (laid loose)						
450/750 Volt grade; two core						
0.50 mm²	0.23	0.27	0.07	1.78	m	**2.05**
0.75 mm²	0.32	0.38	0.07	1.78	m	**2.15**
1.0 mm²	0.35	0.42	0.07	1.78	m	**2.19**
1.5 mm²	0.45	0.53	0.07	1.78	m	**2.31**
2.5 mm²	0.58	0.69	0.07	1.78	m	**2.47**
450/750 Volt grade; three core						
0.50 mm²	0.26	0.31	0.07	1.78	m	**2.09**
0.75 mm²	0.31	0.36	0.07	1.78	m	**2.14**
1.0 mm²	0.36	0.43	0.07	1.78	m	**2.21**
1.5 mm²	0.43	0.51	0.07	1.78	m	**2.29**
2.5 mm²	0.58	0.69	0.07	1.78	m	**2.47**
450/750 Volt grade; four core						
0.75 mm²	0.43	0.51	0.08	2.03	m	**2.54**
1.0 mm²	0.51	0.60	0.08	2.03	m	**2.63**
1.5 mm²	0.64	0.76	0.08	2.03	m	**2.79**
2.5 mm²	1.00	1.18	0.08	2.03	m	**3.22**
Heavy flexible cable; rubber insulated; rubber sheathed; copper stranded to CMA Code 638P (laid loose)						
450/750 Volt grade; two core						
1.0 mm²	0.31	0.36	0.08	2.03	m	**2.40**
1.5 mm²	0.36	0.43	0.08	2.03	m	**2.46**
2.5 mm²	0.50	0.59	0.08	2.03	m	**2.62**
450/750 Volt grade; three core						
1.0 mm²	0.36	0.43	0.08	2.03	m	**2.46**
1.5 mm²	0.43	0.51	0.08	2.03	m	**2.54**
2.5 mm²	0.58	0.69	0.08	2.03	m	**2.72**
450/750 Volt grade; four core						
1.0 mm²	0.50	0.59	0.08	2.03	m	**2.62**
1.5 mm²	0.54	0.64	0.08	2.03	m	**2.67**
2.5 mm²	0.76	0.90	0.08	2.03	m	**2.93**

V:ELECTRICAL SUPPLY/POWER/LIGHTING SYSTEMS

Item	Net Price £	Material £	Labour hours	Labour £	Unit	Total rate £
V20 : LV DISTRIBUTION (cont'd)						
Y61 - LV CABLES AND WIRING (cont'd)						
FIRE RATED CABLE						
Cable, mineral insulated; copper sheathed with copper conductors; fixed with clips to backgrounds. BASEC approval to BS 6207 Part 1 1995; complies with BS 6387 Category CWZ						
Light duty 500 Volt grade; bare						
2L 1.0	1.74	2.06	0.23	5.84	m	**7.90**
2L 1.5	1.93	2.28	0.23	5.84	m	**8.12**
2L 2.5	2.24	2.66	0.25	6.35	m	**9.00**
2L 4.0	2.94	3.48	0.25	6.35	m	**9.82**
3L 1.0	1.97	2.33	0.24	6.09	m	**8.42**
3L 1.5	2.27	2.68	0.25	6.35	m	**9.03**
3L 2.5	3.05	3.61	0.25	6.35	m	**9.95**
4L 1.0	2.19	2.59	0.25	6.35	m	**8.94**
4L 1.5	2.51	2.97	0.25	6.35	m	**9.32**
4L 2.5	3.51	4.15	0.26	6.60	m	**10.75**
7L 1.5	3.51	4.15	0.28	7.11	m	**11.26**
7L 2.5	4.37	5.17	0.27	6.86	m	**12.02**
Light duty 500 Volt grade; LSF sheathed						
2L 1.0	1.87	2.21	0.23	5.84	m	**8.05**
2L 1.5	2.02	2.40	0.23	5.84	m	**8.24**
2L 2.5	2.33	2.76	0.25	6.35	m	**9.11**
2L 4.0	3.06	3.62	0.25	6.35	m	**9.97**
3L 1.0	2.13	2.53	0.24	6.09	m	**8.62**
3L 1.5	2.42	2.86	0.25	6.35	m	**9.21**
3L 2.5	3.18	3.76	0.25	6.35	m	**10.11**
4L 1.0	2.35	2.79	0.25	6.35	m	**9.13**
4L 1.5	2.69	3.19	0.25	6.35	m	**9.54**
4L 2.5	3.62	4.28	0.26	6.60	m	**10.88**
7L 1.5	3.86	4.57	0.28	7.11	m	**11.68**
7L 2.5	4.74	5.61	0.27	6.86	m	**12.47**
Heavy duty 750 Volt grade; bare						
1H 10	3.04	3.59	0.25	6.35	m	**9.94**
1H 16	4.34	5.13	0.26	6.60	m	**11.74**
1H 25	8.57	10.14	0.27	6.86	m	**17.00**
1H 35	10.91	12.91	0.32	8.13	m	**21.03**
1H 50	13.65	16.15	0.35	8.89	m	**25.04**
1H 70	14.73	17.43	0.38	9.65	m	**27.08**
1H 95	16.30	19.29	0.41	10.41	m	**29.70**
1H 120	18.00	21.30	0.46	11.68	m	**32.98**
1H 150	21.23	25.13	0.50	12.70	m	**37.82**
1H 185	24.74	29.28	0.56	14.22	m	**43.50**
1H 240	31.34	37.09	0.69	17.52	m	**54.61**
2H 1.5	2.71	3.20	0.25	6.35	m	**9.55**
2H 2.5	3.23	3.83	0.26	6.60	m	**10.43**
2H 4.0	3.89	4.61	0.26	6.60	m	**11.21**
2H 6.0	4.83	5.71	0.29	7.36	m	**13.08**
2H 10	6.01	7.11	0.34	8.63	m	**15.74**
2H 16	9.60	11.36	0.40	10.16	m	**21.52**
2H 25	13.91	16.46	0.44	11.17	m	**27.63**

V:ELECTRICAL SUPPLY/POWER/LIGHTING SYSTEMS

Item	Net Price £	Material £	Labour hours	Labour £	Unit	Total rate £
3H 1.5	2.92	3.45	0.25	6.35	m	9.80
3H 2.5	3.62	4.28	0.25	6.35	m	10.63
3H 4.0	4.30	5.09	0.27	6.86	m	11.95
3H 6.0	5.30	6.27	0.30	7.62	m	13.89
3H 10	7.18	8.50	0.35	8.89	m	17.39
3H 16	11.21	13.27	0.41	10.41	m	23.68
3H 25	16.60	19.65	0.47	11.93	m	31.58
4H 1.5	3.60	4.26	0.24	6.09	m	10.35
4H 2.5	4.22	5.00	0.26	6.60	m	11.60
4H 4.0	5.08	6.01	0.29	7.36	m	13.38
4H 6.0	6.40	7.58	0.31	7.87	m	15.45
4H 10	8.91	10.54	0.37	9.39	m	19.94
4H 16	12.68	15.00	0.44	11.17	m	26.17
4H 25	19.77	23.39	0.52	13.20	m	36.60
7H 1.5	4.61	5.45	0.30	7.62	m	13.07
7H 2.5	5.65	6.69	0.32	8.13	m	14.82
12H 2.5	9.80	11.60	0.39	9.90	m	21.50
19H 1.5	14.34	16.98	0.42	10.66	m	27.64
Heavy duty 750 Volt grade; LSF sheathed						
1H 10	3.22	3.81	0.25	6.35	m	10.16
1H 16	4.63	5.47	0.26	6.60	m	12.08
1H 25	9.32	11.03	0.27	6.86	m	17.89
1H 35	11.77	13.93	0.32	8.13	m	22.06
1H 50	14.67	17.36	0.35	8.89	m	26.25
1H 70	14.66	17.35	0.38	9.65	m	27.00
1H 95	15.35	18.16	0.41	10.41	m	28.57
1H 120	19.28	22.82	0.46	11.68	m	34.50
1H 150	22.67	26.83	0.50	12.70	m	39.53
1H 185	26.50	31.36	0.56	14.22	m	45.58
1H 240	33.37	39.50	0.68	17.27	m	56.76
2H 1.5	2.94	3.48	0.25	6.35	m	9.82
2H 2.5	3.44	4.07	0.26	6.60	m	10.68
2H 4.0	4.15	4.91	0.26	6.60	m	11.51
2H 6.0	5.15	6.09	0.29	7.36	m	13.46
2H 10	6.40	7.58	0.34	8.63	m	16.21
2H 16	10.06	11.91	0.40	10.16	m	22.07
2H 25	15.01	17.76	0.44	11.17	m	28.93
3H 1.5	3.15	3.72	0.25	6.35	m	10.07
3H 2.5	3.87	4.58	0.25	6.35	m	10.93
3H 4.0	4.40	5.21	0.27	6.86	m	12.06
3H 6.0	5.60	6.63	0.30	7.62	m	14.24
3H 10	7.58	8.97	0.35	8.89	m	17.86
3H 16	11.91	14.10	0.41	10.41	m	24.51
3H 25	17.81	21.08	0.47	11.93	m	33.01
4H 1.5	6.75	7.99	0.24	6.09	m	14.09
4H 2.5	7.13	8.44	0.26	6.60	m	15.04
4H 4.0	8.17	9.67	0.29	7.36	m	17.04
4H 6.0	8.90	10.53	0.31	7.87	m	18.40
4H 10	10.86	12.85	0.37	9.39	m	22.24
4H 16	15.05	17.81	0.44	11.17	m	28.98
4H 25	21.29	25.20	0.52	13.20	m	38.40
7H 1.5	4.94	5.85	0.30	7.62	m	13.46
7H 2.5	6.33	7.49	0.32	8.13	m	15.61
12H 2.5	10.38	12.29	0.39	9.90	m	22.19
19H 1.5	15.03	17.78	0.42	10.66	m	28.45

V:ELECTRICAL SUPPLY/POWER/LIGHTING SYSTEMS

Item	Net Price £	Material £	Labour hours	Labour £	Unit	Total rate £
V20 : LV DISTRIBUTION (cont'd)						
Y61 - LV CABLES AND WIRING (cont'd)						
FIRE RATED CABLE (cont'd)						
Cable terminations for M.I. Cable; Polymeric one piece moulding; containing grey sealing compound; testing; phase marking and connection						
Light Duty 500 Volt grade; Brass gland; polymeric one moulding containing grey sealing compound; coloured conductor sleeving; Earth tag; plastic gland shroud						
2L 1.5	2.61	3.09	0.27	6.86	m	**9.95**
2L 2.5	2.63	4.41	0.27	6.86	m	**11.27**
3L 1.5	3.01	4.44	0.27	6.86	m	**11.29**
4L 1.5	3.05	4.48	0.27	6.86	m	**11.34**
Cable Terminations; for MI copper sheathed cable. Certified for installation in potentially explosive atmospheres; testing; phase marking and connection; BS 6207 Part 2 1995						
Light duty 500 Volt grade; brass gland; brass pot with earth tail; pot closure; sealing compound; conductor sleeving; plastic gland shroud; identification markers						
2L 1.0	2.94	6.99	0.39	9.90	nr	**16.89**
2L 1.5	2.94	6.99	0.41	10.41	nr	**17.40**
2L 2.5	2.94	7.17	0.41	10.41	nr	**17.58**
2L 4.0	2.94	7.17	0.46	11.68	nr	**18.85**
3L 1.0	2.96	7.18	0.43	10.92	nr	**18.10**
3L 1.5	2.96	7.07	0.43	10.92	nr	**17.99**
3L 2.5	2.96	7.18	0.44	11.17	nr	**18.35**
4L 1.0	2.98	7.25	0.47	11.93	nr	**19.19**
4L 1.5	2.98	7.14	0.47	11.93	nr	**19.07**
4L 2.5	2.98	7.25	0.50	12.70	nr	**19.95**
7L 1.0	2.78	3.29	0.69	17.52	nr	**20.81**
7L 1.5	5.19	6.15	0.70	17.77	nr	**23.92**
7L 2.5	5.19	6.15	0.74	18.79	nr	**24.94**
Heavy duty 750 Volt grade; brass gland; brass pot with earth tail; pot closure; sealing compound; conductor sleeving; plastic gland shroud; identification markers						
1H 10	5.11	6.05	0.37	9.39	nr	**15.44**
1H 16	5.11	6.05	0.39	9.90	nr	**15.95**
1H 25	8.33	9.86	0.56	14.22	nr	**24.08**
1H 35	8.33	9.86	0.57	14.47	nr	**24.33**
1H 50	10.67	12.63	0.60	15.23	nr	**27.86**
1H 70	13.01	15.40	0.67	17.01	nr	**32.41**
1H 95	15.35	18.17	0.75	19.04	nr	**37.21**
1H 120	17.69	20.94	0.94	23.87	nr	**44.80**
1H 150	20.03	23.70	0.99	25.14	nr	**48.84**
1H 185	22.37	26.47	1.26	31.99	nr	**58.47**
1H 240	24.71	29.24	1.37	34.79	nr	**64.03**

V:ELECTRICAL SUPPLY/POWER/LIGHTING SYSTEMS

Item	Net Price £	Material £	Labour hours	Labour £	Unit	Total rate £
2H 1.5	3.61	4.27	0.42	10.66	nr	14.94
2H 2.5	3.61	4.27	0.42	10.66	nr	14.94
2H 4.0	4.28	5.06	0.47	11.93	nr	17.00
2H 6.0	5.28	6.25	0.54	13.71	nr	19.96
2H 10	11.38	13.47	0.58	14.73	nr	28.20
2H 16	18.48	21.87	0.69	17.52	nr	39.39
2H 25	22.54	26.68	0.77	19.55	nr	46.23
3H 1.5	3.61	4.27	0.44	11.17	nr	15.44
3H 2.5	4.28	5.06	0.44	11.17	nr	16.23
3H 4.0	5.28	6.25	0.57	14.47	nr	20.72
3H 6.0	6.04	7.15	0.61	15.49	nr	22.64
3H 10	11.38	13.47	0.65	16.50	nr	29.98
3H 16	18.48	21.87	0.78	19.81	nr	41.67
3H 25	27.99	33.13	0.85	21.58	nr	54.71
4H 1.5	3.61	4.27	0.52	13.20	nr	17.48
4H 2.5	5.28	6.25	0.53	13.46	nr	19.70
4H 4.0	6.04	7.15	0.60	15.23	nr	22.39
4H 6.0	10.13	11.99	0.65	16.50	nr	28.49
4H 10	10.13	11.99	0.69	17.52	nr	29.51
4H 16	20.22	23.93	0.88	22.34	nr	46.27
4H 25	27.99	33.13	0.93	23.61	nr	56.74
7H 1.5	8.52	10.09	0.71	18.03	nr	28.11
7H 2.5	7.28	8.62	0.74	18.79	nr	27.41
12H 1.5	11.72	13.87	0.85	21.58	nr	35.45
12H 2.5	12.97	15.35	1.00	25.39	nr	40.75
19H 2.5	22.94	27.15	1.11	28.18	nr	55.34
Cable; FP100; LOSH insulated; non sheathed fire resistant to LPCB Approved to BS 6387 Category CWZ; in conduit or trunking including terminations						
450/750 volt grade; single core						
1.0 mm²	0.40	0.47	0.13	3.30	m	3.77
1.5 mm²	0.40	0.47	0.13	3.30	m	3.77
2.5 mm²	0.51	0.60	0.13	3.30	m	3.90
4 mm²	0.66	0.78	0.13	3.30	m	4.08
6 mm²	1.12	1.33	0.13	3.30	m	4.63
10 mm²	1.45	1.72	0.16	4.06	m	5.78
16 mm²	2.25	2.66	0.16	4.06	m	6.73
Cable; FP200; Insudite insulated; LSOH sheathed screened fire resistant BASEC Approved to BS 7629; fixed with clips to backgrounds						
300/500 volt grade; two core						
1.5 mm²	1.61	1.90	0.13	3.30	m	5.20
2.5 mm²	1.94	2.29	0.13	3.30	m	5.59
4.0 mm²	2.52	2.98	0.13	3.30	m	6.28
300/500 volt grade; three core						
1.5 mm²	1.96	2.32	0.13	3.30	m	5.62
2.5 mm²	2.22	2.63	0.13	3.30	m	5.93
4.0 mm²	3.09	3.66	0.13	3.30	m	6.96
300/500 volt grade; four core						
1.5 mm²	2.19	2.59	0.13	3.30	m	5.89
2.5 mm²	2.71	3.20	0.13	3.30	m	6.50
4.0 mm²	3.71	4.39	0.13	3.30	m	7.69

V:ELECTRICAL SUPPLY/POWER/LIGHTING SYSTEMS

Item	Net Price £	Material £	Labour hours	Labour £	Unit	Total rate £
V20 : LV DISTRIBUTION (cont'd)						
Y61 - LV CABLES AND WIRING (cont'd)						
FIRE RATED CABLE (cont'd)						
Cable FP200, Insudite insulated; LSOH sheathed (cont'd)						
Terminations; including glanding-off, connection to equipment						
Two core						
1.5 mm²	0.68	0.81	0.35	8.89	nr	9.69
2.5 mm²	0.68	0.81	0.35	8.89	nr	9.69
4.0 mm²	0.87	1.03	0.35	8.89	nr	9.92
Three core						
1.5 mm²	0.70	0.83	0.35	8.89	nr	9.72
2.5 mm²	0.70	0.83	0.35	8.89	nr	9.72
4.0 mm²	0.90	1.07	0.35	8.89	nr	9.95
Four core						
1.5 mm²	0.92	1.09	0.35	8.89	nr	9.98
2.5 mm²	0.92	1.09	0.35	8.89	nr	9.98
4.0 mm²	0.92	1.09	0.35	8.89	nr	9.98
Cable; FP400; polymeric insulated; LSOH sheathed fire resistant; armoured; with copper stranded copper conductors; BASEC Approved to BS 7846; fixed with clips to backgrounds						
600/1000 volt grade; two core						
1.5 mm²	2.30	2.72	0.16	4.06	m	6.78
2.5 mm²	2.66	3.15	0.16	4.06	m	7.21
4.0 mm²	2.96	3.50	0.16	4.06	m	7.56
6.0 mm²	3.64	4.31	0.16	4.06	m	8.37
10 mm²	3.93	4.65	0.20	5.08	m	9.73
16 mm²	5.97	7.07	0.20	5.08	m	12.15
25 mm²	8.27	9.78	0.20	5.08	m	14.86
600/1000 volt grade; three core						
1.5 mm²	2.55	3.02	0.16	4.06	m	7.08
2.5 mm²	2.98	3.53	0.16	4.06	m	7.59
4.0 mm²	3.52	4.17	0.16	4.06	m	8.23
6.0 mm²	3.73	4.41	0.16	4.06	m	8.48
10 mm²	4.52	5.35	0.20	5.08	m	10.43
16 mm²	7.41	8.77	0.20	5.08	m	13.85
25 mm²	9.24	10.93	0.20	5.08	m	16.01
600/1000 volt grade; four core						
1.5 mm²	2.88	3.41	0.16	4.06	m	7.47
2.5 mm²	3.49	4.13	0.16	4.06	m	8.19
4.0 mm²	3.85	4.56	0.16	4.06	m	8.62
6.0 mm²	4.84	5.73	0.16	4.06	m	9.79
10 mm²	5.75	6.81	0.20	5.08	m	11.89
16 mm²	8.55	10.12	0.20	5.08	m	15.20
25 mm²	12.41	14.69	0.20	5.08	m	19.77

V:ELECTRICAL SUPPLY/POWER/LIGHTING SYSTEMS

Item	Net Price £	Material £	Labour hours	Labour £	Unit	Total rate £
Terminations; including glanding-off, connection to equipment,						
Two core						
1.5 mm²	3.75	4.44	0.58	14.73	nr	**19.17**
2.5 mm²	3.75	4.44	0.58	14.73	nr	**19.17**
4.0 mm²	3.75	4.44	0.58	14.73	nr	**19.17**
6.0 mm²	4.71	5.57	0.67	17.01	nr	**22.58**
10 mm²	4.71	5.57	1.00	25.39	nr	**30.96**
16 mm²	6.64	7.86	1.11	28.18	nr	**36.04**
25 mm²	6.64	7.86	1.70	43.17	nr	**51.02**
Three core						
1.5 mm²	3.75	4.44	0.62	15.74	nr	**20.18**
2.5 mm²	3.75	4.44	0.62	15.74	nr	**20.18**
4.0 mm²	3.75	4.44	0.62	15.74	nr	**20.18**
6.0 mm²	4.71	5.57	0.71	18.03	nr	**23.60**
10 mm²	4.71	5.57	1.06	26.91	nr	**32.48**
16 mm²	6.64	7.86	1.19	30.22	nr	**38.07**
25 mm²	6.64	7.86	1.81	45.96	nr	**53.82**
Four core						
1.5 mm²	3.75	4.44	0.67	17.01	nr	**21.45**
2.5 mm²	3.75	4.44	0.67	17.01	nr	**21.45**
4.0 mm²	4.71	5.57	0.71	18.03	nr	**23.60**
6.0 mm²	4.71	5.57	0.76	19.30	nr	**24.87**
10 mm²	6.64	7.86	1.14	28.95	nr	**36.80**
16 mm²	6.64	7.86	1.29	32.75	nr	**40.61**
25 mm²	10.95	12.96	1.99	50.53	nr	**63.49**
Cable; Firetuff fire resistant to BS 6387; fixed with clips to backgrounds						
Two core						
1.5 mm²	1.64	1.94	0.16	4.06	m	**6.00**
2.5 mm²	1.97	2.33	0.16	4.06	m	**6.39**
4.0 mm²	2.55	3.02	0.16	4.06	m	**7.08**
Three core						
1.5 mm²	1.99	2.36	0.16	4.06	m	**6.42**
2.5 mm²	2.25	2.67	0.16	4.06	m	**6.73**
4.0 mm²	3.25	3.84	0.16	4.06	m	**7.90**
Four core						
1.5 mm²	2.22	2.63	0.16	4.06	m	**6.69**
2.5 mm²	2.73	3.23	0.16	4.06	m	**7.29**
4.0 mm²	3.72	4.40	0.16	4.06	m	**8.46**

V:ELECTRICAL SUPPLY/POWER/LIGHTING SYSTEMS

Item	Net Price £	Material £	Labour hours	Labour £	Unit	Total rate £
V20 : LV DISTRIBUTION (cont'd)						
Y61 - LV CABLES AND WIRING (cont'd)						
MODULAR WIRING						
Modular wiring systems; including commissioning						
Master distribution box; steel; fixed to backgrounds; 6 Port						
4 mm 18 core armoured home run cable	114.35	135.33	0.90	22.85	nr	**158.18**
4 mm 24 core armoured cable home run cable	114.35	135.33	0.95	24.12	nr	**159.45**
4 mm 18 core armoured home run cable & data cable	121.49	143.78	0.95	24.12	nr	**167.90**
6 mm 18 core armoured home run cable	114.35	135.33	1.00	25.39	nr	**160.72**
6 mm 24 core armoured home run cable	114.35	135.33	1.10	27.93	nr	**163.26**
6 mm 18 core armoured home run cable & data cable	121.49	143.78	1.10	27.93	nr	**171.71**
Master distribution box; steel; fixed to backgrounds; 9 Port						
4 mm 27 core armoured home run cable	142.93	169.15	1.30	33.01	nr	**202.16**
4 mm 27 core armoured home run cable & data cable	142.93	169.15	1.45	36.82	nr	**205.97**
6 mm 27 core armoured home run cable	150.08	177.62	1.45	36.82	nr	**214.43**
6 mm 27 core armoured home run cable & data cable	150.08	177.62	1.55	39.36	nr	**216.97**
Metal clad cable; BSEN 60439 Part 2 1993; BASEC approved						
4 mm 18 core	9.99	11.82	0.30	7.62	m	**19.44**
4 mm 24 core	15.36	18.18	0.32	8.13	m	**26.31**
4 mm 27 core	15.36	18.18	0.32	8.13	m	**26.31**
6 mm 18 core	13.31	15.76	0.32	8.13	m	**23.88**
6 mm 27 core	20.68	24.47	0.35	8.89	m	**33.36**
Metal clad data cable						
Single twisted pair	1.94	2.30	0.18	4.57	m	**6.87**
Twin twisted pair	3.13	3.70	0.18	4.57	m	**8.28**
Distribution cables; armoured; BSEN 60439 Part 2 1993; BASEC approved						
3 wire; 6.1 metre long	30.22	35.76	0.92	23.36	nr	**59.12**
4 wire; 6.1 metre long	36.30	42.96	0.96	24.38	nr	**67.34**

V:ELECTRICAL SUPPLY/POWER/LIGHTING SYSTEMS

Item	Net Price £	Material £	Labour hours	Labour £	Unit	Total rate £
Extender cables; armoured; BSEN 60439 Part 2 1993; BASEC approved						
3 Wire						
0.9 metre long	13.21	15.64	0.13	3.30	nr	18.94
1.5 metre long	15.49	18.33	0.23	5.84	nr	24.17
2.1 metre long	17.77	21.03	0.31	7.87	nr	28.90
2.7 metre long	20.03	23.71	0.40	10.16	nr	33.87
3.4 metre long	22.31	26.40	0.51	12.95	nr	39.35
4.6 metre long	26.87	31.80	0.69	17.52	nr	49.32
6.1 metre long	32.55	38.52	0.92	23.36	nr	61.88
7.6 metre long	38.24	45.26	1.14	28.95	nr	74.21
9.1 metre long	43.94	52.00	1.37	34.79	nr	86.79
10.7 metre long	59.90	70.89	1.61	40.88	nr	111.77
4 Wire						
0.9 metre long	14.38	17.02	0.14	3.55	nr	20.57
1.5 metre long	17.22	20.38	0.24	6.09	nr	26.47
2.1 metre long	20.07	23.76	0.32	8.13	nr	31.88
2.7 metre long	22.91	27.12	0.43	10.92	nr	38.04
3.4 metre long	25.75	30.48	0.51	12.95	nr	43.43
4.6 metre long	31.45	37.22	0.67	17.01	nr	54.23
6.1 metre long	38.57	45.64	0.92	23.36	nr	69.00
7.6 metre long	45.67	54.05	1.22	30.98	nr	85.02
9.1 metre long	52.78	62.47	1.46	37.07	nr	99.54
10.7 metre long	59.90	70.89	1.71	43.42	nr	114.31
3 Wire; including twisted pair						
0.9 metre long	15.73	18.62	0.13	3.30	nr	21.92
1.5 metre long	19.34	22.88	0.23	5.84	nr	28.72
2.1 metre long	22.94	27.15	0.31	7.87	nr	35.02
2.7 metre long	26.55	31.42	0.40	10.16	nr	41.57
3.4 metre long	30.15	35.68	0.51	12.95	nr	48.63
4.6 metre long	37.36	44.21	0.69	17.52	nr	61.73
6.1 metre long	46.38	54.89	0.92	23.36	nr	78.25
7.6 metre long	55.38	65.54	1.14	28.95	nr	94.49
9.1 metre long	64.40	76.22	1.37	34.79	nr	111.00
10.7 metre long	73.42	86.89	1.61	40.88	nr	127.77
Extender whip ended cables; armoured; BSEN 60439 Part 2 1993; BASEC approved						
3 wire; 3.0 metre long	18.34	21.71	0.30	7.62	nr	29.33
4 wire; 3.0 metre long	21.45	25.39	0.30	7.62	nr	33.01
T Connectors						
3 wire						
Snap fix	11.04	13.07	0.10	2.54	nr	15.61
0.3 metre flexible cable	12.52	14.81	0.10	2.54	nr	17.35
0.3 metre armoured cable	12.50	14.80	0.15	3.81	nr	18.60
0.3 metre armoured cable with twisted pair	13.94	16.49	0.15	3.81	nr	20.30
4 Wire						
Snap fix	11.52	13.64	0.10	2.54	nr	16.18
0.3 metre flexible cable	12.96	15.34	0.10	2.54	nr	17.87
0.3 metre armoured cable	12.96	15.34	0.18	4.57	nr	19.91

V:ELECTRICAL SUPPLY/POWER/LIGHTING SYSTEMS

Item	Net Price £	Material £	Labour hours	Labour £	Unit	Total rate £
V20 : LV DISTRIBUTION (cont'd)						
Y61 - LV CABLES AND WIRING (cont'd)						
MODULAR WIRING (cont'd)						
Extender whip ended cables (cont'd)						
Splitters						
5 wire	17.67	20.92	0.20	5.08	nr	**26.00**
5 wire converter	13.69	16.21	0.20	5.08	nr	**21.29**
Switch modules						
3 wire; 6.1 metre long armoured cable	36.86	43.63	0.75	19.04	nr	**62.67**
4 wire; 6.1 metre long armoured cable	42.18	49.92	0.80	20.31	nr	**70.24**
Distribution cables; unarmoured; IEC 998 DIN/VDE 0628						
3 wire; 6.1 metre long	12.89	15.26	0.70	17.77	nr	**33.03**
4 wire; 6.1 metre long	15.42	18.25	0.75	19.04	nr	**37.30**
Extender cables; unarmoured; IEC 998 DIN/VDE 0628						
3 Wire						
0.9 metre long	8.95	10.59	0.07	1.78	nr	**12.37**
1.5 metre long	9.67	11.45	0.12	3.05	nr	**14.50**
2.1 metre long	10.40	12.31	0.17	4.32	nr	**16.62**
2.7 metre long	12.97	15.35	0.22	5.59	nr	**20.94**
3.4 metre long	11.85	14.02	0.27	6.86	nr	**20.87**
4.6 metre long	13.41	15.87	0.37	9.39	nr	**25.27**
6.1 metre long	15.22	18.02	0.49	12.44	nr	**30.46**
7.6 metre long	17.03	20.16	0.61	15.49	nr	**35.64**
9.1 metre long	18.84	22.30	0.73	18.54	nr	**40.83**
10.7 metre long	20.77	24.58	0.86	21.84	nr	**46.42**
4 Wire						
0.9 metre long	10.24	12.12	0.08	2.03	nr	**14.15**
1.5 metre long	11.15	13.19	0.14	3.55	nr	**16.75**
2.1 metre long	12.06	14.27	0.19	4.82	nr	**19.10**
2.7 metre long	12.97	15.35	0.24	6.09	nr	**21.44**
3.4 metre long	13.88	16.43	0.31	7.87	nr	**24.30**
4.6 metre long	15.85	18.76	0.41	10.41	nr	**29.17**
6.1 metre long	18.13	21.46	0.55	13.97	nr	**35.42**
7.6 metre long	20.41	24.15	0.68	17.27	nr	**41.42**
9.1 metre long	26.71	31.61	0.82	20.82	nr	**52.43**
10.7 metre long	29.14	34.49	0.96	24.38	nr	**58.87**
5 Wire						
0.9 metre long	12.07	14.29	0.09	2.29	nr	**16.57**
1.5 metre long	13.55	16.03	0.15	3.81	nr	**19.84**
2.1 metre long	15.01	17.76	0.21	5.33	nr	**23.09**
2.7 metre long	16.47	19.49	0.27	6.86	nr	**26.35**
3.4 metre long	17.94	21.23	0.34	8.63	nr	**29.87**
4.6 metre long	21.11	24.98	0.46	11.68	nr	**36.66**
6.1 metre long	24.78	29.32	0.61	15.49	nr	**44.81**
7.6 metre long	28.43	33.65	0.76	19.30	nr	**52.95**
9.1 metre long	32.09	37.98	0.91	23.11	nr	**61.09**
10.7 metre long	36.01	42.61	1.07	27.17	nr	**69.78**

V:ELECTRICAL SUPPLY/POWER/LIGHTING SYSTEMS

Item	Net Price £	Material £	Labour hours	Labour £	Unit	Total rate £
Extender whip ended cables; armoured; IEC 998 DIN/VDE 0628						
3 wire; 2.5mm; 3.0 metre long	9.13	10.80	0.30	7.62	nr	**18.42**
4 wire; 2.5mm; 3.0 metre long	10.72	12.69	0.30	7.62	nr	**20.30**
T Connectors						
3 wire						
5 pin; direct fix	9.23	10.93	0.10	2.54	nr	**13.47**
5 pin; 1.5mm flexible cable; 0.3 metre long	11.81	13.97	0.15	3.81	nr	**17.78**
4 Wire						
5 pin; direct fix	10.73	12.70	0.20	5.08	nr	**17.78**
5 pin; 1.5mm flexible cable; 0.3 metre long	13.37	15.83	0.20	5.08	nr	**20.91**
5 Wire						
5 pin; direct fix	12.23	14.48	0.20	5.08	nr	**19.56**
Splitters						
3 way; 5 pin	7.05	8.34	0.25	6.35	nr	**14.69**
Switch Modules						
3 wire	20.07	23.76	0.20	5.08	nr	**28.83**
4 wire	20.93	24.77	0.22	5.59	nr	**30.36**

V:ELECTRICAL SUPPLY/POWER/LIGHTING SYSTEMS

Item	Net Price £	Material £	Labour hours	Labour £	Unit	Total rate £
V20 : LV DISTRIBUTION (cont'd)						
Y62 - BUSBAR TRUNKING						
MAINS BUSBAR						
Low impedance busbar trunking; fixed to backgrounds including supports, fixings and connections/jointing to equipment;						
Straight copper busbar						
1000 Amp TP&N	356.40	421.79	3.41	86.58	m	**508.37**
1350 Amp TP&N	438.07	518.45	3.58	90.90	m	**609.35**
2000 Amp TP&N	549.45	650.26	5.00	126.96	m	**777.21**
2500 Amp TP&N	830.12	982.42	5.90	149.81	m	**1132.23**
Extra for fittings mains bus bar						
IP54 protection						
1000 Amp TP&N	24.30	28.76	2.16	54.85	m	**83.60**
1350 Amp TP&N	26.12	30.92	2.61	66.27	m	**97.19**
2000 Amp TP&N	36.23	42.88	3.51	89.12	m	**132.00**
2500 Amp TP&N	43.02	50.92	3.96	100.55	m	**151.47**
End cover						
1000 Amp TP&N	27.95	33.07	0.56	14.22	nr	**47.29**
1350 Amp TP&N	29.16	34.51	0.56	14.22	nr	**48.73**
2000 Amp TP&N	46.78	55.36	0.66	16.76	nr	**72.12**
2500 Amp TP&N	47.38	56.08	0.66	16.76	nr	**72.84**
Edge elbow						
1000 Amp TP&N	400.37	473.82	2.01	51.04	nr	**524.86**
1350 Amp TP&N	450.58	533.25	2.01	51.04	nr	**584.29**
2000 Amp TP&N	698.89	827.12	2.40	60.94	nr	**888.06**
2500 Amp TP&N	917.91	1086.32	2.40	60.94	nr	**1147.26**
Flat elbow						
1000 Amp TP&N	347.36	411.08	2.01	51.04	nr	**462.12**
1350 Amp TP&N	376.65	445.75	2.01	51.04	nr	**496.79**
2000 Amp TP&N	534.28	632.31	2.40	60.94	nr	**693.25**
2500 Amp TP&N	665.41	787.50	2.40	60.94	nr	**848.44**
Offset						
1000 Amp TP&N	697.50	825.47	3.00	76.17	nr	**901.64**
1350 Amp TP&N	857.92	1015.33	3.00	76.17	nr	**1091.50**
2000 Amp TP&N	1351.76	1599.76	3.50	88.87	nr	**1688.63**
2500 Amp TP&N	1538.68	1820.99	3.50	88.87	nr	**1909.86**
Edge Z unit						
1000 Amp TP&N	1044.86	1236.55	3.00	76.17	nr	**1312.73**
1350 Amp TP&N	1337.81	1583.25	3.00	76.17	nr	**1659.43**
2000 Amp TP&N	2004.62	2372.40	3.50	88.87	nr	**2461.27**
2500 Amp TP&N	2289.20	2709.19	3.50	88.87	nr	**2798.06**
Flat Z unit						
1000 Amp TP&N	896.99	1061.55	3.00	76.17	nr	**1137.73**
1350 Amp TP&N	1129.95	1337.26	3.00	76.17	nr	**1413.44**
2000 Amp TP&N	1682.37	1991.03	3.50	88.87	nr	**2079.90**
2500 Amp TP&N	2022.75	2393.86	3.50	88.87	nr	**2482.73**

V:ELECTRICAL SUPPLY/POWER/LIGHTING SYSTEMS

Item	Net Price £	Material £	Labour hours	Labour £	Unit	Total rate £
Edge tee						
1000 Amp TP&N	1044.86	1236.55	2.20	55.86	nr	**1292.42**
1350 Amp TP&N	1337.81	1583.25	2.20	55.86	nr	**1639.11**
2000 Amp TP&N	2004.62	2372.40	2.60	66.02	nr	**2438.42**
2500 Amp TP&N	2290.59	2710.84	2.60	66.02	nr	**2776.86**
Tap off; TP&N integral contactor/breaker						
18 Amp	182.74	216.26	0.82	20.82	nr	**237.09**
Tap off; TP&N fusable with on-load switch; excludes fuses						
32 Amp	517.28	612.19	0.82	20.82	nr	**633.01**
63 Amp	527.65	624.46	0.88	22.34	nr	**646.80**
100 Amp	645.64	764.09	1.18	29.96	nr	**794.05**
160 Amp	734.35	869.09	1.41	35.80	nr	**904.89**
250 Amp	948.01	1121.94	1.76	44.69	nr	**1166.63**
315 Amp	1119.16	1324.49	2.06	52.31	nr	**1376.80**
Tap off; TP&N MCCB						
63 Amp	653.65	773.58	0.88	22.34	nr	**795.92**
125 Amp	783.42	927.15	1.18	29.96	nr	**957.11**
160 Amp	855.40	1012.34	1.41	35.80	nr	**1048.14**
250 Amp	1100.14	1301.99	1.76	44.69	nr	**1346.68**
400 Amp	1402.43	1659.73	2.06	52.31	nr	**1712.04**

RISING MAINS BUSBAR

Rising mains busbar; insulated supports, earth continuity bar; including couplers; fixed to backgrounds

Item	Net Price £	Material £	Labour hours	Labour £	Unit	Total rate £
Straight aluminium bar						
200 Amp TP&N	127.26	150.61	2.13	54.08	m	**204.69**
315 Amp TP&N	142.38	168.50	2.15	54.59	m	**223.09**
400 Amp TP&N	165.06	195.34	2.15	54.59	m	**249.94**
630 Amp TP&N	204.12	241.57	2.47	62.72	m	**304.29**
800 Amp TP&N	306.18	362.35	2.88	73.13	m	**435.48**

Extra for fittings rising busbar

Item	Net Price £	Material £	Labour hours	Labour £	Unit	Total rate £
End feed unit						
200 Amp TP&N	255.15	301.96	2.57	65.26	nr	**367.22**
315 Amp TP&N	255.15	301.96	2.76	70.08	nr	**372.04**
400 Amp TP&N	285.52	337.91	2.76	70.08	nr	**407.99**
630 Amp TP&N	285.52	337.91	3.64	92.42	nr	**430.34**
800 Amp TP&N	321.98	381.05	4.54	115.28	nr	**496.32**
Top feeder unit						
200 Amp TP&N	255.15	301.96	2.57	65.26	nr	**367.22**
315 Amp TP&N	255.15	301.96	2.76	70.08	nr	**372.04**
400 Amp TP&N	285.52	337.91	2.76	70.08	nr	**407.99**
630 Amp TP&N	285.52	337.91	3.64	92.42	nr	**430.34**
800 Amp TP&N	321.98	381.05	4.54	115.28	nr	**496.32**
End cap						
200 Amp TP&N	21.87	25.88	0.18	4.57	nr	**30.45**
315 Amp TP&N	21.87	25.88	0.27	6.86	nr	**32.74**
400 Amp TP&N	24.30	28.76	0.27	6.86	nr	**35.61**
630 Amp TP&N	24.30	28.76	0.41	10.41	nr	**39.17**
800 Amp TP&N	70.47	83.40	0.41	10.41	nr	**93.81**

V:ELECTRICAL SUPPLY/POWER/LIGHTING SYSTEMS

Item	Net Price £	Material £	Labour hours	Labour £	Unit	Total rate £
V20 : LV DISTRIBUTION (cont'd)						
Y62 – BUSBAR TRUNKING (cont'd)						
RISING MAINS BUSBAR (cont'd))						
Fittings; rising busbar (cont'd)						
Edge elbow						
200 Amp TP&N	30.38	35.95	0.55	13.97	nr	49.91
315 Amp TP&N	30.38	35.95	0.94	23.87	nr	59.82
400 Amp TP&N	228.42	270.33	0.94	23.87	nr	294.20
630 Amp TP&N	228.42	270.33	1.45	36.82	nr	307.15
800 Amp TP&N	216.27	255.95	1.45	36.82	nr	292.77
Flat elbow						
200 Amp TP&N	98.42	116.47	0.55	13.97	nr	130.44
315 Amp TP&N	98.42	116.47	0.94	23.87	nr	140.34
400 Amp TP&N	133.65	158.17	0.94	23.87	nr	182.04
630 Amp TP&N	133.65	158.17	1.45	36.82	nr	194.99
800 Amp TP&N	184.68	218.56	1.45	36.82	nr	255.38
Edge tee						
200 Amp TP&N	138.51	163.92	0.61	15.49	nr	179.41
315 Amp TP&N	138.51	163.92	1.02	25.90	nr	189.82
400 Amp TP&N	194.40	230.07	1.02	25.90	nr	255.97
630 Amp TP&N	194.40	230.07	1.57	39.86	nr	269.93
800 Amp TP&N	277.02	327.84	1.57	39.86	nr	367.71
Flat tee						
200 Amp TP&N	177.39	209.94	0.61	15.49	nr	225.42
315 Amp TP&N	138.51	163.92	1.02	25.90	nr	189.82
400 Amp TP&N	279.45	330.72	1.02	25.90	nr	356.62
630 Amp TP&N	279.45	330.72	1.57	39.86	nr	370.59
800 Amp TP&N	390.01	461.57	1.57	39.86	nr	501.44
Tap off units						
TP&N fusable with on-load switch; excludes fuses						
32 Amp	139.72	165.36	0.82	20.82	nr	186.18
63 Amp	184.68	218.56	0.88	22.34	nr	240.91
100 Amp	247.86	293.33	1.18	29.96	nr	323.30
250 Amp	371.79	440.00	1.41	35.80	nr	475.80
400 Amp	541.89	641.31	2.06	52.31	nr	693.62
TP&N MCCB						
32 Amp	142.16	168.24	0.82	20.82	nr	189.06
63 Amp	196.83	232.94	0.88	22.34	nr	255.29
100 Amp	315.90	373.86	1.18	29.96	nr	403.82
250 Amp	534.60	632.68	1.41	35.80	nr	668.49
400 Amp	937.98	1110.07	2.06	52.31	nr	1162.38

V:ELECTRICAL SUPPLY/POWER/LIGHTING SYSTEMS

Item	Net Price £	Material £	Labour hours	Labour £	Unit	Total rate £
LIGHTING BUSBAR						
Pre-wired busbar, plug-in trunking for lighting; galvanised sheet steel housing (PE); tin-plated copper conductors with tap-off units at 1m intervals.						
Straight lengths - 25 Amp						
2 Pole & PE	23.01	27.23	0.16	4.06	m	**31.29**
4 Pole & PE	24.89	29.46	0.16	4.06	m	**33.53**
Straight lengths - 40 Amp						
2 Pole & PE	22.75	26.92	0.16	4.06	m	**30.99**
4 Pole & PE	30.55	36.16	0.16	4.06	m	**40.22**
Components for pre-wired busbars, plug-in trunking for lighting.						
Plug-in tap off units						
10 Amp with phase selection, 2P & PE; 2m of cable	14.55	17.22	0.10	2.54	nr	**19.76**
10 Amp 4 Pole & PE; 3m of cable	19.45	23.02	0.10	2.54	nr	**25.56**
16 Amp 4 Pole & PE; 3m of cable	18.85	22.31	0.10	2.54	nr	**24.85**
16 Amp with phase selection, 2P & PE; no cable	16.69	19.75	0.10	2.54	nr	**22.29**
Trunking components						
End feed unit & cover; 4P & PE	24.58	29.09	0.23	5.84	nr	**34.93**
Centre feed unit	121.84	144.19	0.29	7.36	nr	**151.55**
Right hand, intermediate terminal box feed unit	25.66	30.37	0.23	5.84	nr	**36.21**
End cover (for R/hand feed)	7.10	8.40	0.06	1.52	nr	**9.93**
Flexible elbow unit	58.33	69.03	0.12	3.05	nr	**72.07**
Fixing bracket - universal	4.26	5.04	0.10	2.54	nr	**7.58**
Suspension Bracket - Flat	3.74	4.42	0.10	2.54	nr	**6.96**
UNDERFLOOR BUSBAR						
Pre-wired busbar, plug-in trunking for underfloor power distribution; galvanised sheet steel housing (PE); copper conductors with tap-off units at 300mm intervals.						
Straight lengths - 63 Amp						
2 pole & PE	15.46	18.29	0.28	7.11	m	**25.40**
3 pole & PE; Clean Earth System	19.55	23.13	0.28	7.11	m	**30.24**
Components for pre-wired busbars, plug-in trunking for underfloor power distribution.						
Plug-in tap off units						
32 Amp 2P & PE; 3m metal flexible pre-wired conduit	25.62	30.32	0.25	6.35	nr	**36.67**
32 Amp 3P & PE; clean earth; 3m metal flexible pre-wired conduit	30.84	36.49	0.28	7.11	nr	**43.60**

V:ELECTRICAL SUPPLY/POWER/LIGHTING SYSTEMS

Item	Net Price £	Material £	Labour hours	Labour £	Unit	Total rate £
V20 : LV DISTRIBUTION (cont'd)						
Y62 – BUSBAR TRUNKING (cont'd)						
UNDERFLOOR BUSBAR (cont'd)						
Components for pre-wired busbars, plug in Trunking for underfloor power distribution (cont'd)						
Trunking components						
End feed unit & cover; 2P & PE	27.60	32.66	0.35	8.89	nr	**41.55**
End feed unit & cover; 3P & PE; clean earth	30.23	35.77	0.38	9.65	nr	**45.42**
End cover; 2P & PE	8.59	10.17	0.11	2.79	nr	**12.96**
End cover; 3P & PE	9.24	10.94	0.11	2.79	nr	**13.73**
Flexible interlink/corner; 2P&PE; 1m long	47.55	56.28	0.34	8.63	nr	**64.91**
Flexible interlink/corner; 3P&PE; 1m long	53.70	63.56	0.35	8.89	nr	**72.44**
Flexible interlink/corner; 2P&PE; 2m long	58.03	68.68	0.37	9.39	nr	**78.07**
Flexible interlink/corner; 3P&PE; 2m long	63.67	75.36	0.37	9.39	nr	**84.75**

V:ELECTRICAL SUPPLY/POWER/LIGHTING SYSTEMS

Item	Net Price £	Material £	Labour hours	Labour £	Unit	Total rate £
Y63 - CABLE SUPPORTS						
LADDER RACK						
Light duty Galvanised Steel Ladder Rack; fixed to backgrounds; including supports, fixings and brackets; earth continuity straps.						
Straight lengths						
150 mm wide ladder	19.00	22.48	0.69	17.52	m	**40.00**
300 mm wide ladder	19.78	23.40	0.88	22.34	m	**45.75**
450 mm wide ladder	21.07	24.94	1.26	31.99	m	**56.93**
600 mm wide ladder	21.88	25.89	1.51	38.34	m	**64.23**
Extra over; (cutting and jointing racking to fittings is included.)						
Inside riser bend						
150 mm wide ladder	33.49	39.63	0.33	8.38	nr	**48.01**
300 mm wide ladder	34.84	41.24	0.56	14.22	nr	**55.46**
450 mm wide ladder	35.68	42.23	0.85	21.58	nr	**63.81**
600 mm wide ladder	37.15	43.96	0.99	25.14	nr	**69.10**
Outside riser bend						
300 mm wide ladder	34.84	41.24	0.43	10.92	nr	**52.16**
450 mm wide ladder	35.68	42.23	0.73	18.54	nr	**60.77**
600 mm wide ladder	37.15	43.96	0.86	21.84	nr	**65.80**
Equal tee						
300 mm wide ladder	40.51	47.95	0.62	15.74	nr	**63.69**
450 mm wide ladder	45.36	53.68	1.09	27.68	nr	**81.35**
600 mm wide ladder	47.82	56.59	1.12	28.44	nr	**85.03**
Unequal tee						
300 mm wide ladder	40.51	47.95	0.57	14.47	nr	**62.42**
450 mm wide ladder	45.36	53.68	1.17	29.71	nr	**83.39**
600 mm wide ladder	47.82	56.59	1.17	29.71	nr	**86.30**
4 way cross overs						
300 mm wide ladder	64.70	76.57	0.72	18.28	nr	**94.85**
450 mm wide ladder	70.16	83.03	1.13	28.69	nr	**111.73**
600 mm wide ladder	82.92	98.13	1.29	32.75	nr	**130.88**
Heavy duty galvanised steel ladder rack; fixed to backgrounds; including supports, fixings and brackets; earth continuity straps.						
Straight lengths						
150 mm wide ladder	24.96	29.54	0.68	17.27	m	**46.81**
300 mm wide ladder	25.94	30.70	0.79	20.06	m	**50.76**
450 mm wide ladder	27.05	32.01	1.07	27.17	m	**59.18**
600 mm wide ladder	27.85	32.96	1.24	31.49	m	**64.45**
750 mm wide ladder	29.20	34.55	1.49	37.83	m	**72.39**
900 mm wide ladder	30.20	35.75	1.67	42.40	m	**78.15**

V:ELECTRICAL SUPPLY/POWER/LIGHTING SYSTEMS

Item	Net Price £	Material £	Labour hours	Labour £	Unit	Total rate £
V20 : LV DISTRIBUTION (cont'd)						
Y63 - CABLE SUPPORTS (cont'd)						
LADDER RACK (cont'd)						
Heavy duty galvanised steel ladder rack (cont'd)						
Extra over; (cutting and jointing racking to fittings is included.)						
Flat bend						
150 mm wide ladder	34.35	40.65	0.34	8.63	nr	**49.29**
300 mm wide ladder	35.72	42.27	0.39	9.90	nr	**52.17**
450 mm wide ladder	38.92	46.06	0.43	10.92	nr	**56.97**
600 mm wide ladder	42.50	50.30	0.61	15.49	nr	**65.79**
750 mm wide ladder	46.70	55.27	0.82	20.82	nr	**76.09**
900 mm wide ladder	50.91	60.25	0.97	24.63	nr	**84.88**
Inside riser bend						
150 mm wide ladder	44.51	52.67	0.27	6.86	nr	**59.53**
300 mm wide ladder	45.87	54.29	0.45	11.43	nr	**65.72**
450 mm wide ladder	47.94	56.74	0.65	16.50	nr	**73.24**
600 mm wide ladder	51.06	60.43	0.81	20.57	nr	**81.00**
750 mm wide ladder	54.00	63.91	0.92	23.36	nr	**87.27**
900 mm wide ladder	57.05	67.52	1.06	26.91	nr	**94.43**
Outside riser bend						
150 mm wide ladder	44.51	52.67	0.27	6.86	nr	**59.53**
300 mm wide ladder	45.87	54.29	0.33	8.38	nr	**62.67**
450 mm wide ladder	47.94	56.74	0.61	15.49	nr	**72.23**
600 mm wide ladder	51.06	60.43	0.76	19.30	nr	**79.73**
750 mm wide ladder	54.00	63.91	0.94	23.87	nr	**87.78**
900 mm wide ladder	57.40	67.93	1.05	26.66	nr	**94.59**
Equal tee						
150mm wide ladder	54.58	64.59	0.37	9.39	nr	**73.99**
300 mm wide ladder	58.21	68.89	0.57	14.47	nr	**83.37**
450 mm wide ladder	61.41	72.68	0.83	21.07	nr	**93.75**
600 mm wide ladder	66.13	78.26	0.92	23.36	nr	**101.62**
750 mm wide ladder	81.10	95.98	1.13	28.69	nr	**124.67**
900 mm wide ladder	84.09	99.52	1.20	30.47	nr	**129.99**
Unequal tee						
300 mm wide ladder	57.97	68.61	0.57	14.47	nr	**83.08**
450 mm wide ladder	61.05	72.25	1.17	29.71	nr	**101.96**
600 mm wide ladder	65.89	77.98	1.17	29.71	nr	**107.69**
750 mm wide ladder	80.87	95.70	1.25	31.74	nr	**127.44**
900 mm wide ladder	83.85	99.23	1.33	33.77	nr	**133.00**
4 way cross overs						
150 mm wide ladder	78.72	93.16	0.50	12.70	nr	**105.86**
300 mm wide ladder	81.22	96.13	0.67	17.01	nr	**113.14**
450 mm wide ladder	85.02	100.62	0.92	23.36	nr	**123.98**
600 mm wide ladder	89.23	105.60	1.07	27.17	nr	**132.77**
750 mm wide ladder	107.48	127.20	1.25	31.74	nr	**158.94**
900 mm wide ladder	111.16	131.55	1.36	34.53	nr	**166.09**

V:ELECTRICAL SUPPLY/POWER/LIGHTING SYSTEMS

Item	Net Price £	Material £	Labour hours	Labour £	Unit	Total rate £
Extra heavy duty galvanised steel ladder rack; fixed to backgrounds; including supports, fixings and brackets; earth continuity straps.						
Straight lengths						
150 mm wide ladder	31.15	36.87	0.63	16.00	m	**52.86**
300 mm wide ladder	32.34	38.27	0.70	17.77	m	**56.05**
450 mm wide ladder	33.59	39.75	0.83	21.07	m	**60.83**
600 mm wide ladder	34.96	41.37	0.89	22.60	m	**63.97**
750 mm wide ladder	37.03	43.83	1.22	30.98	m	**74.80**
900 mm wide ladder	38.21	45.22	1.44	36.56	m	**81.78**
Extra over; (cutting and jointing racking to fittings is included.)						
Flat bend						
150 mm wide ladder	37.19	44.01	0.36	9.14	nr	**53.16**
300 mm wide ladder	39.07	46.23	0.39	9.90	nr	**56.14**
450 mm wide ladder	42.90	50.76	0.43	10.92	nr	**61.68**
600 mm wide ladder	47.05	55.68	0.61	15.49	nr	**71.17**
750 mm wide ladder	51.76	61.26	0.82	20.82	nr	**82.08**
900 mm wide ladder	56.49	66.85	0.97	24.63	nr	**91.48**
Inside riser bend						
150 mm wide ladder	47.35	56.03	0.36	9.14	nr	**65.17**
300 mm wide ladder	48.09	56.92	0.39	9.90	nr	**66.82**
450 mm wide ladder	51.29	60.70	0.43	10.92	nr	**71.62**
600 mm wide ladder	54.42	64.40	0.61	15.49	nr	**79.89**
750 mm wide ladder	56.05	66.33	0.82	20.82	nr	**87.16**
900 mm wide ladder	61.06	72.27	0.97	24.63	nr	**96.90**
Outside riser bend						
150 mm wide ladder	47.35	56.03	0.36	9.14	nr	**65.17**
300 mm wide ladder	48.09	56.92	0.39	9.90	nr	**66.82**
450 mm wide ladder	51.29	60.70	0.41	10.41	nr	**71.11**
600 mm wide ladder	54.42	64.40	0.57	14.47	nr	**78.88**
750 mm wide ladder	56.05	66.33	0.82	20.82	nr	**87.16**
900 mm wide ladder	61.06	72.27	0.93	23.61	nr	**95.88**
Equal tee						
150 mm wide ladder	55.71	65.93	0.37	9.39	nr	**75.32**
300 mm wide ladder	62.65	74.15	0.57	14.47	nr	**88.62**
450 mm wide ladder	67.02	79.32	0.83	21.07	nr	**100.39**
600 mm wide ladder	72.93	86.31	0.92	23.36	nr	**109.67**
750 mm wide ladder	88.45	104.67	1.13	28.69	nr	**133.37**
900 mm wide ladder	92.62	109.61	1.20	30.47	nr	**140.08**
Unequal tee						
150 mm wide ladder	55.71	65.93	0.37	9.39	nr	**75.32**
300 mm wide ladder	62.65	74.15	0.57	14.47	nr	**88.62**
450 mm wide ladder	67.02	79.32	1.17	29.71	nr	**109.03**
600 mm wide ladder	72.93	86.31	1.17	29.71	nr	**116.02**
750 mm wide ladder	88.45	104.67	1.25	31.74	nr	**136.41**
900 mm wide ladder	92.62	109.61	1.33	33.77	nr	**143.38**

V:ELECTRICAL SUPPLY/POWER/LIGHTING SYSTEMS

Item	Net Price £	Material £	Labour hours	Labour £	Unit	Total rate £
V20 : LV DISTRIBUTION (cont'd)						
Y63 - CABLE SUPPORTS (cont'd)						
LADDER RACK (cont'd)						
Extra heavy duty galvinised steel ladder rack; Fittings (cont'd)						
4 way cross overs						
150 mm wide ladder	83.77	99.13	0.50	12.70	nr	**111.83**
300 mm wide ladder	87.41	103.45	0.67	17.01	nr	**120.46**
450 mm wide ladder	104.79	124.01	0.92	23.36	nr	**147.37**
600 mm wide ladder	110.69	131.00	1.07	27.17	nr	**158.16**
750 mm wide ladder	115.41	136.59	1.25	31.74	nr	**168.33**
900 mm wide ladder	120.13	142.17	1.36	34.53	nr	**176.70**
CABLE TRAY						
Galvanised steel cable tray to BS 729; including standard coupling joints, fixings and earth continuity straps. (Supports and hangers are excluded.)						
Light duty tray						
Straight lengths						
50 mm wide	1.87	2.22	0.19	4.82	m	**7.04**
75 mm wide	2.41	2.85	0.23	5.84	m	**8.69**
100 mm wide	2.30	2.73	0.31	7.87	m	**10.60**
150 mm wide	2.83	3.35	0.33	8.38	m	**11.73**
225 mm wide	4.16	4.93	0.39	9.90	m	**14.83**
300 mm wide	6.19	7.33	0.49	12.44	m	**19.77**
450 mm wide	9.61	11.38	0.60	15.23	m	**26.61**
600 mm wide	12.98	15.37	0.79	20.06	m	**35.43**
750 mm wide	16.38	19.39	1.04	26.41	m	**45.79**
900 mm wide	20.33	24.06	1.26	31.99	m	**56.05**
Extra over; (cutting and jointing tray to fittings is included.)						
Straight reducer						
75 mm wide	5.49	6.49	0.22	5.59	nr	**12.08**
100 mm wide	5.89	6.97	0.25	6.35	nr	**13.32**
150 mm wide	7.31	8.66	0.27	6.86	nr	**15.51**
225 mm wide	8.94	10.57	0.34	8.63	nr	**19.21**
300 mm wide	11.07	13.11	0.39	9.90	nr	**23.01**
450 mm wide	15.89	18.81	0.49	12.44	nr	**31.25**
600 mm wide	18.94	22.42	0.54	13.71	nr	**36.13**
750 mm wide	23.76	28.12	0.61	15.49	nr	**43.61**
900 mm wide	27.36	32.38	0.69	17.52	nr	**49.90**

V:ELECTRICAL SUPPLY/POWER/LIGHTING SYSTEMS

Item	Net Price £	Material £	Labour hours	Labour £	Unit	Total rate £
Flat bend; 90°						
50mm wide	4.09	4.85	0.19	4.82	nr	**9.67**
75 mm wide	4.16	4.93	0.24	6.09	nr	**11.02**
100 mm wide	4.62	5.47	0.28	7.11	nr	**12.58**
150 mm wide	4.73	5.59	0.30	7.62	nr	**13.21**
225 mm wide	5.99	7.09	0.36	9.14	nr	**16.23**
300 mm wide	7.62	9.02	0.44	11.17	nr	**20.20**
450 mm wide	11.79	13.95	0.57	14.47	nr	**28.42**
600 mm wide	16.93	20.03	0.69	17.52	nr	**37.55**
750 mm wide	23.43	27.72	0.81	20.57	nr	**48.29**
900 mm wide	33.74	39.93	0.94	23.87	nr	**63.80**
Adjustable riser						
50 mm wide	6.11	7.23	0.26	6.60	nr	**13.83**
75 mm wide	6.65	7.87	0.29	7.36	nr	**15.23**
100 mm wide	7.34	8.68	0.32	8.13	nr	**16.81**
150 mm wide	9.18	10.86	0.36	9.14	nr	**20.00**
225 mm wide	11.25	13.31	0.44	11.17	nr	**24.48**
300 mm wide	14.32	16.94	0.52	13.20	nr	**30.15**
450 mm wide	18.35	21.72	0.66	16.76	nr	**38.48**
600 mm wide	23.00	27.22	0.79	20.06	nr	**47.28**
750 mm wide	29.69	35.14	1.03	26.15	nr	**61.29**
900 mm wide	34.83	41.22	1.10	27.93	nr	**69.15**
Inside riser; 90°						
50 mm wide	5.28	6.25	0.28	7.11	nr	**13.36**
75 mm wide	5.80	6.86	0.31	7.87	nr	**14.73**
100 mm wide	5.80	6.86	0.33	8.38	nr	**15.24**
150 mm wide	7.20	8.52	0.37	9.39	nr	**17.91**
225 mm wide	8.75	10.36	0.44	11.17	nr	**21.53**
300 mm wide	11.27	13.34	0.53	13.46	nr	**26.80**
450 mm wide	15.20	17.99	0.67	17.01	nr	**35.00**
600 mm wide	19.79	23.42	0.79	20.06	nr	**43.48**
750 mm wide	24.05	28.46	0.95	24.12	nr	**52.58**
900 mm wide	28.29	33.48	1.11	28.18	nr	**61.66**
Outside riser; 90°						
50 mm wide	5.28	6.25	0.28	7.11	nr	**13.36**
75 mm wide	5.49	6.49	0.31	7.87	nr	**14.36**
100 mm wide	5.80	6.86	0.33	8.38	nr	**15.24**
150 mm wide	7.20	8.52	0.37	9.39	nr	**17.91**
225 mm wide	8.75	10.36	0.44	11.17	nr	**21.53**
300 mm wide	11.27	13.34	0.53	13.46	nr	**26.80**
450 mm wide	15.20	17.99	0.67	17.01	nr	**35.00**
600 mm wide	19.79	23.42	0.79	20.06	nr	**43.48**
750 mm wide	24.05	28.46	0.95	24.12	nr	**52.58**
900 mm wide	28.29	33.48	1.11	28.18	Unit	**61.66**
Equal tee						
50 mm wide	5.99	7.09	0.30	7.62	nr	**14.71**
75 mm wide	6.13	7.25	0.31	7.87	nr	**15.13**
100 mm wide	6.49	7.68	0.35	8.89	nr	**16.56**
150 mm wide	6.96	8.23	0.36	9.14	nr	**17.37**
225 mm wide	8.79	10.40	0.74	18.79	nr	**29.19**
300 mm wide	10.42	12.33	0.54	13.71	nr	**26.04**
450 mm wide	17.10	20.24	0.71	18.03	nr	**38.27**
600 mm wide	22.66	26.81	0.92	23.36	nr	**50.17**
750 mm wide	33.01	39.06	1.19	30.22	nr	**69.28**
900 mm wide	45.78	54.18	1.44	36.56	nr	**90.74**

V:ELECTRICAL SUPPLY/POWER/LIGHTING SYSTEMS

Item	Net Price £	Material £	Labour hours	Labour £	Unit	Total rate £
V20 : LV DISTRIBUTION (cont'd)						
Y63 - CABLE SUPPORTS (cont'd)						
CABLE TRAY (cont'd)						
Light duty tray (cont'd)						
Fittings; extra over (cont'd)						
Unequal tee						
75 mm wide	6.13	7.25	0.38	9.65	nr	**16.90**
100 mm wide	6.49	7.68	0.39	9.90	nr	**17.58**
150 mm wide	6.96	8.23	0.43	10.92	nr	**19.15**
225 mm wide	8.79	10.40	0.50	12.70	nr	**23.09**
300 mm wide	10.42	12.33	0.63	16.00	nr	**28.33**
450 mm wide	17.10	20.24	0.80	20.31	nr	**40.55**
600 mm wide	22.66	26.81	1.02	25.90	nr	**52.71**
750 mm wide	33.01	39.06	1.12	28.44	nr	**67.50**
900 mm wide	45.78	54.18	1.35	34.28	nr	**88.46**
4 way crossovers						
50 mm wide	8.21	9.72	0.38	9.65	nr	**19.37**
75 mm wide	8.31	9.84	0.40	10.16	nr	**20.00**
100 mm wide	8.83	10.45	0.40	10.16	nr	**20.61**
150 mm wide	9.54	11.30	0.44	11.17	nr	**22.47**
225 mm wide	11.73	13.88	0.53	13.46	nr	**27.34**
300 mm wide	15.44	18.28	0.64	16.25	nr	**34.53**
450 mm wide	23.34	27.63	0.84	21.33	nr	**48.96**
600 mm wide	30.21	35.75	1.03	26.15	nr	**61.91**
750 mm wide	43.98	52.04	1.13	28.69	nr	**80.74**
900 mm wide	62.72	74.23	1.36	34.53	nr	**108.76**
Medium duty tray with return flange						
Straight lengths						
75 mm wide	2.81	3.32	0.33	8.38	m	**11.70**
100 mm wide	3.08	3.65	0.35	8.89	m	**12.54**
150 mm wide	3.74	4.43	0.39	9.90	m	**14.33**
225 mm wide	4.36	5.16	0.45	11.43	m	**16.58**
300 mm wide	4.36	5.16	0.57	14.47	m	**19.63**
450 mm wide	9.11	10.78	0.69	17.52	m	**28.30**
600 mm wide	12.66	14.98	0.91	23.11	m	**38.09**
Extra over; (cutting and jointing tray to fittings is included.)						
Straight reducer						
100 mm wide	6.75	7.99	0.25	6.35	nr	**14.34**
150 mm wide	7.37	8.72	0.27	6.86	nr	**15.58**
225 mm wide	8.49	10.05	0.34	8.63	nr	**18.68**
300 mm wide	9.81	11.61	0.39	9.90	nr	**21.51**
450 mm wide	12.63	14.95	0.49	12.44	nr	**27.39**
600 mm wide	15.65	18.52	0.54	13.71	nr	**32.23**

V:ELECTRICAL SUPPLY/POWER/LIGHTING SYSTEMS

Item	Net Price £	Material £	Labour hours	Labour £	Unit	Total rate £
Flat bend; 90°						
75 mm wide	10.34	12.24	0.24	6.09	nr	**18.33**
100 mm wide	11.49	13.60	0.28	7.11	nr	**20.71**
150 mm wide	12.13	14.36	0.30	7.62	nr	**21.98**
225 mm wide	14.04	16.62	0.36	9.14	nr	**25.76**
300 mm wide	17.31	20.48	0.44	11.17	nr	**31.66**
450 mm wide	25.93	30.69	0.57	14.47	nr	**45.16**
600 mm wide	31.33	37.07	0.69	17.52	nr	**54.59**
Adjustable bend						
75 mm wide	10.34	12.24	0.29	7.36	nr	**19.60**
100 mm wide	11.49	13.60	0.32	8.13	nr	**21.72**
150 mm wide	12.13	14.36	0.36	9.14	nr	**23.50**
225 mm wide	14.04	16.62	0.44	11.17	nr	**27.79**
300 mm wide	17.31	20.48	0.52	13.20	nr	**33.69**
Adjustable riser						
75 mm wide	7.90	9.35	0.29	7.36	nr	**16.71**
100 mm wide	8.04	9.51	0.32	8.13	nr	**17.64**
150 mm wide	9.31	11.02	0.36	9.14	nr	**20.16**
225 mm wide	10.34	12.24	0.44	11.17	nr	**23.41**
300 mm wide	11.28	13.35	0.52	13.20	nr	**26.55**
450 mm wide	16.25	19.23	0.66	16.76	nr	**35.99**
600 mm wide	21.10	24.97	0.79	20.06	nr	**45.03**
Inside riser; 90°						
75 mm wide	6.66	7.88	0.31	7.87	nr	**15.75**
100 mm wide	6.75	7.99	0.33	8.38	nr	**16.37**
150 mm wide	7.69	9.11	0.37	9.39	nr	**18.50**
225 mm wide	9.31	11.02	0.44	11.17	nr	**22.20**
300 mm wide	11.28	13.35	0.53	13.46	nr	**26.81**
450 mm wide	16.25	19.23	0.67	17.01	nr	**36.24**
600 mm wide	25.98	30.74	0.79	20.06	nr	**50.80**
Outside riser; 90°						
75 mm wide	6.66	7.88	0.31	7.87	nr	**15.75**
100 mm wide	6.75	7.99	0.33	8.38	nr	**16.37**
150 mm wide	7.69	9.11	0.37	9.39	nr	**18.50**
225 mm wide	9.31	11.02	0.44	11.17	nr	**22.20**
300 mm wide	11.28	13.35	0.53	13.46	nr	**26.81**
450 mm wide	16.25	19.23	0.67	17.01	nr	**36.24**
600 mm wide	25.98	30.74	0.79	20.06	nr	**50.80**
Equal tee						
75 mm wide	14.73	17.43	0.31	7.87	nr	**25.31**
100 mm wide	15.69	18.56	0.35	8.89	nr	**27.45**
150 mm wide	16.89	19.99	0.36	9.14	nr	**29.13**
225 mm wide	18.32	21.68	0.74	18.79	nr	**40.47**
300 mm wide	22.21	26.28	0.54	13.71	nr	**39.99**
450 mm wide	29.19	34.54	0.71	18.03	nr	**52.57**
600 mm wide	41.84	49.51	0.92	23.36	nr	**72.87**
Unequal tee						
100 mm wide	15.69	18.56	0.39	9.90	nr	**28.47**
150 mm wide	15.69	18.56	0.43	10.92	nr	**29.48**
225 mm wide	18.32	21.68	0.50	12.70	nr	**34.38**
300 mm wide	22.21	26.28	0.63	16.00	nr	**42.28**
450 mm wide	29.19	34.54	0.80	20.31	nr	**54.85**
600 mm wide	41.84	49.51	1.02	25.90	nr	**75.41**

V:ELECTRICAL SUPPLY/POWER/LIGHTING SYSTEMS

Item	Net Price £	Material £	Labour hours	Labour £	Unit	Total rate £
V20 : LV DISTRIBUTION (cont'd)						
Y63 - CABLE SUPPORTS (cont'd)						
CABLE TRAY (cont'd)						
Medium duty tray (cont'd)						
Fittings; extra over (cont'd)						
4 way crossovers						
75 mm wide	20.86	24.69	0.40	10.16	nr	**34.84**
100 mm wide	22.48	26.61	0.40	10.16	nr	**36.76**
150 mm wide	23.72	28.08	0.44	11.17	nr	**39.25**
225 mm wide	28.18	33.34	0.53	13.46	nr	**46.80**
300 mm wide	32.71	38.71	0.64	16.25	nr	**54.96**
450 mm wide	41.61	49.24	0.84	21.33	nr	**70.57**
600 mm wide	60.54	71.64	1.03	26.15	nr	**97.80**
Heavy duty tray with return flange						
Straight lengths						
75 mm	5.08	6.01	0.34	8.63	m	**14.64**
100 mm	5.42	6.42	0.36	9.14	m	**15.56**
150 mm	6.23	7.37	0.40	10.16	m	**17.53**
225 mm	6.98	8.27	0.46	11.68	m	**19.95**
300 mm	8.46	10.01	0.58	14.73	m	**24.74**
450 mm	12.74	15.08	0.70	17.77	m	**32.86**
600 mm	15.43	18.26	0.92	23.36	m	**41.62**
750 mm	19.40	22.96	1.01	25.65	m	**48.61**
900 mm	21.46	25.39	1.14	28.95	m	**54.34**
Extra over; (cutting and jointing tray to fittings is included.)						
Straight reducer						
100 mm wide	10.82	12.81	0.25	6.35	nr	**19.15**
150 mm wide	11.75	13.91	0.27	6.86	nr	**20.77**
225 mm wide	12.27	14.52	0.34	8.63	nr	**23.15**
300 mm wide	12.66	14.98	0.39	9.90	nr	**24.89**
450 mm wide	21.77	25.76	0.49	12.44	nr	**38.21**
600 mm wide	26.74	31.64	0.54	13.71	nr	**45.35**
750 mm wide	33.12	39.20	0.60	15.23	nr	**54.43**
900 mm wide	39.05	46.22	0.66	16.76	nr	**62.98**
Flat bend; 90°						
75 mm wide	13.96	16.52	0.24	6.09	nr	**22.62**
100 mm wide	15.89	18.81	0.28	7.11	nr	**25.92**
150 mm wide	16.94	20.05	0.30	7.62	nr	**27.66**
225 mm wide	19.17	22.69	0.36	9.14	nr	**31.83**
300 mm wide	21.86	25.87	0.44	11.17	nr	**37.04**
450 mm wide	32.09	37.97	0.57	14.47	nr	**52.44**
600 mm wide	42.91	50.78	0.69	17.52	nr	**68.30**
750 mm wide	57.53	68.09	0.83	21.07	nr	**89.17**
900 mm wide	65.24	77.21	1.01	25.65	nr	**102.85**

V:ELECTRICAL SUPPLY/POWER/LIGHTING SYSTEMS

Item	Net Price £	Material £	Labour hours	Labour £	Unit	Total rate £
Adjustable bend						
75 mm wide	13.96	16.52	0.29	7.36	nr	**23.89**
100 mm wide	15.89	18.81	0.32	8.13	nr	**26.93**
150 mm wide	16.94	20.05	0.36	9.14	nr	**29.19**
225 mm wide	19.17	22.69	0.44	11.17	nr	**33.86**
300 mm wide	21.86	25.87	0.52	13.20	nr	**39.08**
Adjustable riser						
75 mm wide	11.80	13.96	0.29	7.36	nr	**21.33**
100 mm wide	12.42	14.70	0.32	8.13	nr	**22.82**
150 mm wide	14.10	16.69	0.36	9.14	nr	**25.83**
225 mm wide	15.58	18.44	0.44	11.17	nr	**29.61**
300 mm wide	17.01	20.13	0.52	13.20	nr	**33.33**
450 mm wide	21.77	25.76	0.66	16.76	nr	**42.52**
600 mm wide	26.74	31.64	0.79	20.06	nr	**51.70**
750 mm wide	33.12	39.20	1.03	26.15	nr	**65.35**
900 mm wide	39.05	46.22	1.10	27.93	nr	**74.15**
Inside riser; 90°						
75 mm wide	10.68	12.64	0.31	7.87	nr	**20.51**
100 mm wide	10.82	12.81	0.33	8.38	nr	**21.19**
150 mm wide	11.75	13.91	0.37	9.39	nr	**23.30**
225 mm wide	12.27	14.52	0.44	11.17	nr	**25.69**
300 mm wide	12.66	14.98	0.53	13.46	nr	**28.44**
450 mm wide	21.77	25.76	0.67	17.01	nr	**42.78**
600 mm wide	26.74	31.64	0.79	20.06	nr	**51.70**
750 mm wide	33.12	39.20	0.95	24.12	nr	**63.32**
900 mm wide	39.05	46.22	1.11	28.18	nr	**74.40**
Outside riser; 90°						
75 mm wide	10.68	12.64	0.31	7.87	nr	**20.51**
100 mm wide	10.82	12.81	0.33	8.38	nr	**21.19**
150 mm wide	11.75	13.91	0.37	9.39	nr	**23.30**
225 mm wide	12.27	14.52	0.44	11.17	nr	**25.69**
300 mm wide	12.66	14.98	0.53	13.46	nr	**28.44**
450 mm wide	21.77	25.76	0.67	17.01	nr	**42.78**
600 mm wide	26.74	31.64	0.79	20.06	nr	**51.70**
750 mm wide	33.12	39.20	0.95	24.12	nr	**63.32**
900 mm wide	39.05	46.22	1.11	28.18	nr	**74.40**
Equal tee						
75 mm wide	18.65	22.08	0.31	7.87	nr	**29.95**
100 mm wide	20.84	24.66	0.35	8.89	nr	**33.55**
150 mm wide	22.83	27.02	0.36	9.14	nr	**36.16**
225 mm wide	26.29	31.11	0.74	18.79	nr	**49.90**
300 mm wide	28.45	33.67	0.54	13.71	nr	**47.38**
450 mm wide	41.37	48.95	0.71	18.03	nr	**66.98**
600 mm wide	54.65	64.67	0.92	23.36	nr	**88.03**
750 mm wide	72.13	85.36	1.19	30.22	nr	**115.58**
900 mm wide	82.39	97.50	1.45	36.82	nr	**134.32**
Unequal tee						
75 mm wide	18.65	22.08	0.38	9.65	nr	**31.72**
100 mm wide	20.84	24.66	0.39	9.90	nr	**34.56**
150 mm wide	22.83	27.02	0.43	10.92	nr	**37.93**
225 mm wide	26.29	31.11	0.50	12.70	nr	**43.81**
300 mm wide	28.45	33.67	0.63	16.00	nr	**49.67**
450 mm wide	41.37	48.95	0.80	20.31	nr	**69.27**
600 mm wide	54.65	64.67	1.02	25.90	nr	**90.57**
750 mm wide	72.13	85.36	1.12	28.44	nr	**113.80**
900 mm wide	82.39	97.50	1.35	34.28	nr	**131.78**

V:ELECTRICAL SUPPLY/POWER/LIGHTING SYSTEMS

Item	Net Price £	Material £	Labour hours	Labour £	Unit	Total rate £
V20 : LV DISTRIBUTION (cont'd)						
Y63 - CABLE SUPPORTS (cont'd)						
CABLE TRAY (cont'd)						
Heavy duty tray with return flange (cont'd)						
Fittings; extra over (cont'd)						
4 way crossovers						
75 mm wide	26.19	30.99	0.40	10.16	nr	**41.15**
100 mm wide	27.12	32.09	0.40	10.16	nr	**42.25**
150 mm wide	33.15	39.24	0.44	11.17	nr	**50.41**
225 mm wide	37.59	44.49	0.53	13.46	nr	**57.95**
300 mm wide	42.04	49.76	0.64	16.25	nr	**66.01**
450 mm wide	60.02	71.03	0.84	21.33	nr	**92.36**
600 mm wide	74.78	88.51	1.03	26.15	nr	**114.66**
750 mm wide	99.59	117.86	1.13	28.69	nr	**146.55**
900 mm wide	120.15	142.20	1.36	34.53	nr	**176.73**
GRP cable tray including standard coupling joints and fixings; (supports and hangers excluded).						
Tray						
100 mm wide	17.66	20.90	0.34	8.63	m	**29.53**
200 mm wide	22.60	26.74	0.39	9.90	m	**36.64**
400 mm wide	35.45	41.95	0.53	13.46	m	**55.41**
Cover						
100 mm wide	9.94	11.77	0.10	2.54	m	**14.31**
200 mm wide	13.18	15.60	0.11	2.79	m	**18.39**
400 mm wide	22.51	26.64	0.14	3.55	m	**30.20**
Extra for; (cutting and jointing to fittings included).						
Reducer						
200 mm wide	46.92	55.53	0.23	5.84	nr	**61.37**
400 mm wide	61.30	72.55	0.30	7.62	nr	**80.16**
Reducer cover						
200 mm wide	29.37	34.76	0.25	6.35	nr	**41.10**
400 mm wide	42.92	50.80	0.28	7.11	nr	**57.91**
Bend						
100 mm wide	39.24	46.44	0.34	8.63	nr	**55.07**
200 mm wide	45.55	53.91	0.40	10.16	nr	**64.06**
400 mm wide	58.53	69.26	0.32	8.13	nr	**77.39**
Bend cover						
100 mm wide	19.49	23.06	0.10	2.54	nr	**25.60**
200 mm wide	26.04	30.82	0.10	2.54	nr	**33.36**
400 mm wide	35.98	42.59	0.13	3.30	nr	**45.89**
Tee						
100 mm wide	49.94	59.10	0.37	9.39	nr	**68.49**
200 mm wide	55.06	65.16	0.43	10.92	nr	**76.08**
400 mm wide	68.30	80.83	0.56	14.22	nr	**95.05**

V:ELECTRICAL SUPPLY/POWER/LIGHTING SYSTEMS

Item	Net Price £	Material £	Labour hours	Labour £	Unit	Total rate £
Tee cover						
100 mm wide	25.15	29.76	0.27	6.86	nr	**36.62**
200 mm wide	30.28	35.84	0.31	7.87	nr	**43.71**
400 mm wide	42.72	50.56	0.37	9.39	nr	**59.96**
BASKET TRAY						
Mild Steel Cable Basket; Zinc Plated Including Standard Coupling Joints, Fixings and Earth Continuity Straps (supports and hangers are excluded)						
Basket 54mm deep						
100 mm wide	2.54	3.00	0.22	5.59	m	**8.59**
150 mm wide	2.83	3.35	0.25	6.35	m	**9.70**
200 mm wide	3.10	3.67	0.28	7.11	m	**10.78**
300 mm wide	3.64	4.31	0.34	8.63	m	**12.94**
450 mm wide	4.41	5.21	0.44	11.17	m	**16.39**
600 mm wide	5.39	6.38	0.70	17.77	m	**24.15**
Extra for; (cutting and jointing to fittings is included)						
Reducer						
150 mm wide	9.01	10.66	0.25	6.35	nr	**17.01**
200 mm wide	10.57	12.51	0.28	7.11	nr	**19.62**
300 mm wide	10.75	12.72	0.38	9.65	nr	**22.37**
450 mm wide	11.80	13.97	0.48	12.19	nr	**26.16**
600 mm wide	14.70	17.40	0.48	12.19	nr	**29.58**
Bend						
100 mm wide	8.61	10.19	0.23	5.84	nr	**16.03**
150 mm wide	10.09	11.94	0.26	6.60	nr	**18.54**
200 mm wide	10.26	12.14	0.30	7.62	nr	**19.76**
300 mm wide	11.23	13.29	0.35	8.89	nr	**22.18**
450 mm wide	13.99	16.55	0.50	12.70	nr	**29.25**
600 mm wide	16.47	19.49	0.58	14.73	nr	**34.22**
Tee						
100 mm wide	10.82	12.81	0.28	7.11	nr	**19.92**
150 mm wide	11.48	13.59	0.30	7.62	nr	**21.20**
200 mm wide	11.71	13.86	0.33	8.38	nr	**22.23**
300 mm wide	14.73	17.43	0.39	9.90	nr	**27.34**
450 mm wide	19.30	22.84	0.56	14.22	nr	**37.06**
600 mm wide	19.96	23.62	0.65	16.50	nr	**40.12**
Cross over						
100 mm wide	14.76	17.47	0.40	10.16	nr	**27.63**
150 mm wide	15.02	17.78	0.42	10.66	nr	**28.44**
200 mm wide	16.31	19.30	0.46	11.68	nr	**30.98**
300 mm wide	18.52	21.92	0.51	12.95	nr	**34.87**
450 mm wide	21.71	25.69	0.74	18.79	nr	**44.48**
600 mm wide	22.41	26.52	0.82	20.82	nr	**47.34**

V:ELECTRICAL SUPPLY/POWER/LIGHTING SYSTEMS

Item	Net Price £	Material £	Labour hours	Labour £	Unit	Total rate £
V20 : LV DISTRIBUTION (cont'd)						
Y63 - CABLE SUPPORTS (cont'd)						
CABLE TRAY (cont'd)						
Basket Tray (cont'd)						
Mild steel cable basket; epoxy coated including standard coupling joints, fixings and earth continuity straps (supports and hangers are excluded)						
Basket 54mm deep						
100 mm wide	6.17	7.30	0.22	5.59	m	**12.88**
150 mm wide	6.98	8.26	0.25	6.35	m	**14.60**
200 mm wide	7.77	9.19	0.28	7.11	m	**16.30**
300 mm wide	8.82	10.44	0.34	8.63	m	**19.08**
450 mm wide	10.89	12.88	0.44	11.17	m	**24.06**
600 mm wide	12.39	14.66	0.70	17.77	m	**32.43**
Extra for; (cutting and jointing to fittings is included)						
Reducer						
150 mm wide	12.90	15.26	0.28	7.11	nr	**22.37**
200 mm wide	14.46	17.11	0.28	7.11	nr	**24.22**
300 mm wide	15.41	18.24	0.38	9.65	nr	**27.89**
450 mm wide	18.03	21.33	0.48	12.19	nr	**33.52**
600 mm wide	22.47	26.60	0.48	12.19	nr	**38.79**
Bend						
100 mm wide	12.50	14.79	0.23	5.84	nr	**20.63**
150 mm wide	13.98	16.54	0.26	6.60	nr	**23.14**
200 mm wide	14.15	16.74	0.30	7.62	nr	**24.36**
300 mm wide	15.92	18.84	0.35	8.89	nr	**27.73**
450 mm wide	20.21	23.91	0.50	12.70	nr	**36.61**
600 mm wide	24.25	28.69	0.58	14.73	nr	**43.42**
Tee						
100 mm wide	14.71	17.41	0.28	7.11	nr	**24.52**
150 mm wide	15.37	18.19	0.30	7.62	nr	**25.81**
200 mm wide	15.61	18.47	0.33	8.38	nr	**26.85**
300 mm wide	19.40	22.96	0.39	9.90	nr	**32.86**
450 mm wide	24.73	29.27	0.56	14.22	nr	**43.49**
600 mm wide	27.73	32.82	0.65	16.50	nr	**49.33**
Cross over						
100 mm wide	18.65	22.07	0.40	10.16	nr	**32.23**
150 mm wide	18.86	22.32	0.42	10.66	nr	**32.98**
200 mm wide	20.21	23.91	0.46	11.68	nr	**35.59**
300 mm wide	23.19	27.44	0.51	12.95	nr	**40.39**
450 mm wide	27.93	33.05	0.74	18.79	nr	**51.84**
600 mm wide	30.19	35.72	0.82	20.82	nr	**56.55**

V:ELECTRICAL SUPPLY/POWER/LIGHTING SYSTEMS

Item	Net Price £	Material £	Labour hours	Labour £	Unit	Total rate £
Y71 - LV SWITCHGEAR & DISTRIBUTION BOARDS						
LV switchboard components, factory-assembled modular construction to IP41; form 4, type 5; 2400mm high, with front and rear access; top cable entry/exit; includes delivery, offloading, positioning and commissioning (hence separate labour rates are not detailed below); excludes cabling and cable terminations						
AIR CIRCUIT BREAKERS						
Air circuit breakers (ACBs) to BSEN 60947-2, withdrawable type, fitted with adjustable instantaneous and overload protection. Includes enclosure and copper links, assembled into LV switchboard.						
ACB-100 kA fault rated.						
4 pole, 6300 A (1600mm wide)	26933.44	31874.92	-	-	nr	**31874.92**
4 pole, 5000 A (1600mm wide)	20352.03	24086.02	-	-	nr	**24086.02**
4 pole, 4000 A (1600mm wide)	17073.80	20206.33	-	-	nr	**20206.33**
4 pole, 3200 A (1600mm wide)	11526.53	13641.30	-	-	nr	**13641.30**
4 pole, 2500 A (1600mm wide)	8958.45	10602.06	-	-	nr	**10602.06**
4 pole, 2000 A (1600mm wide)	6983.87	8265.20	-	-	nr	**8265.20**
4 pole, 1600 A (1600mm wide)	5717.92	6766.99	-	-	nr	**6766.99**
4 pole, 1250 A (1600mm wide)	5060.76	5989.25	-	-	nr	**5989.25**
4 pole, 1000 A (1600mm wide)	4999.62	5916.90	-	-	nr	**5916.90**
4 pole, 800 A (1600mm wide)	4915.05	5816.81	-	-	nr	**5816.81**
3 pole, 6300 A (1600mm wide)	24404.09	28881.51	-	-	nr	**28881.51**
3 pole, 5000 A (1600mm wide)	17114.56	20254.56	-	-	nr	**20254.56**
3 pole, 4000 A (1600mm wide)	14014.62	16585.88	-	-	nr	**16585.88**
3 pole, 3200 A (1600mm wide)	9623.79	11389.47	-	-	nr	**11389.47**
3 pole, 2500 A (1600mm wide)	7705.24	9118.92	-	-	nr	**9118.92**
3 pole, 2000 A (1600mm wide)	5801.98	6866.47	-	-	nr	**6866.47**
3 pole, 1600 A (1600mm wide)	4716.88	5582.29	-	-	nr	**5582.29**
3 pole, 1250 A (1600mm wide)	4385.74	5190.39	-	-	nr	**5190.39**
3 pole, 1000 A (1600mm wide)	4291.50	5078.86	-	-	nr	**5078.86**
3 pole, 800 A (1600mm wide)	4209.48	4981.79	-	-	nr	**4981.79**
ACB-65 kA fault rated.						
4 pole, 4000 A (1600mm wide)	14982.56	17731.41	-	-	nr	**17731.41**
4 pole, 3200 A (1600mm wide)	10135.77	11995.38	-	-	nr	**11995.38**
4 pole, 2500 A (1600mm wide)	7967.60	9429.42	-	-	nr	**9429.42**
4 pole, 2000 A (1600mm wide)	6336.88	7499.51	-	-	nr	**7499.51**
4 pole, 1600 A (1600mm wide)	5228.86	6188.20	-	-	nr	**6188.20**
4 pole, 1250 A (1600mm wide)	4668.48	5525.01	-	-	nr	**5525.01**
4 pole, 1000 A (1600mm wide)	4579.33	5419.50	-	-	nr	**5419.50**
4 pole, 800 A (1600mm wide)	4497.31	5322.43	-	-	nr	**5322.43**
3 pole, 4000 A (1600mm wide)	12692.63	15021.34	-	-	nr	**15021.34**
3 pole, 3200 A (1600mm wide)	8607.46	10186.67	-	-	nr	**10186.67**
3 pole, 2500 A (1600mm wide)	6694.01	7922.16	-	-	nr	**7922.16**
3 pole, 2000 A (1600mm wide)	5320.56	6296.73	-	-	nr	**6296.73**
3 pole, 1600 A (1600mm wide)	4352.63	5151.21	-	-	nr	**5151.21**
3 pole, 1250 A (1600mm wide)	4052.07	4795.50	-	-	nr	**4795.50**
3 pole, 1000 A (1600mm wide)	3960.37	4686.98	-	-	nr	**4686.98**
3 pole, 800 A (1600mm wide)	3878.34	4589.90	-	-	nr	**4589.90**

V:ELECTRICAL SUPPLY/POWER/LIGHTING SYSTEMS

Item	Net Price £	Material £	Labour hours	Labour £	Unit	Total rate £
V20 : LV DISTRIBUTION (cont'd)						
Y71 - LV SWITCHGEAR & DISTRIBUTION BOARDS (cont'd)						
AIR CIRCUIT BREAKERS (cont'd)						
Extra for						
Cable box (one per ACB for form 4, types 6 & 7)	251.63	297.79	-	-	nr	**297.79**
Opening Coil	75.08	88.85	-	-	nr	**88.85**
Closing Coil	75.08	88.85	-	-	nr	**88.85**
Undervoltage Release	147.47	174.52	-	-	nr	**174.52**
Motor Operator	455.81	539.44	-	-	nr	**539.44**
Mechanical Interlock (per ACB)	429.00	507.71	-	-	nr	**507.71**
ACB Fortress/Castell Adaptor Kit (one per ACB)	107.25	126.93	-	-	nr	**126.93**
Fortress/Castell ACB Lock (one per ACB)	214.50	253.85	-	-	nr	**253.85**
Fortress/Castell Key	53.63	63.46	-	-	nr	**63.46**
MOULDED CASE CIRCUIT BREAKERS						
Moulded case circuit breakers (MCCBs) to BS EN 60947-2; plug-in type, fitted with electronic trip unit. Includes metalwork section and copper links, assembled into LV switchboard						
MCCB-150 kA fault rated						
4 Pole, 630 A (800mm wide, 600mm high)	2539.44	3005.35	-	-	nr	**3005.35**
4 Pole, 400 A (800mm wide, 400mm high)	1816.78	2150.10	-	-	nr	**2150.10**
4 Pole, 250 A (800mm wide, 400mm high)	1624.74	1922.83	-	-	nr	**1922.83**
4 Pole, 160 A (800mm wide, 300mm high)	1104.21	1306.80	-	-	nr	**1306.80**
4 Pole, 100 A (800mm wide, 200mm high)	919.76	1088.51	-	-	nr	**1088.51**
3 Pole, 630 A (800mm wide, 600mm high)	2266.55	2682.39	-	-	nr	**2682.39**
3 Pole, 400 A (800mm wide, 400mm high)	1730.86	2048.42	-	-	nr	**2048.42**
3 Pole, 250 A (800mm wide, 400mm high)	1394.80	1650.70	-	-	nr	**1650.70**
3 Pole, 160 A (800mm wide, 300mm high)	1020.83	1208.12	-	-	nr	**1208.12**
3 Pole, 100 A (800mm wide, 200mm high)	816.15	965.89	-	-	nr	**965.89**
MCCB-70kA fault rated						
4 Pole, 630 A (800mm wide, 600mm high)	2175.58	2574.73	-	-	nr	**2574.73**
4 Pole, 400 A (800mm wide, 400mm high)	1495.87	1770.32	-	-	nr	**1770.32**
4 Pole, 250 A (800mm wide, 400mm high)	1286.14	1522.11	-	-	nr	**1522.11**
4 Pole, 160 A (800mm wide, 300mm high)	965.23	1142.33	-	-	nr	**1142.33**
4 Pole, 100 A (800mm wide, 200mm high)	727.72	861.24	-	-	nr	**861.24**
3 Pole, 630 A (800mm wide, 600mm high)	1905.22	2254.77	-	-	nr	**2254.77**
3 Pole, 400 A (800mm wide, 400mm high)	1291.20	1528.10	-	-	nr	**1528.10**
3 Pole, 250 A (800mm wide, 400mm high)	1139.59	1348.68	-	-	nr	**1348.68**
3 Pole, 160 A (800mm wide, 300mm high)	841.42	995.80	-	-	nr	**995.80**
3 Pole, 100 A (800mm wide, 200mm high)	619.07	732.64	-	-	nr	**732.64**
MCCB-45kA fault rated						
4 Pole, 630 A (800mm wide, 600mm high)	2109.89	2496.99	-	-	nr	**2496.99**
4 Pole, 400 A (800mm wide, 400mm high)	1450.40	1716.50	-	-	nr	**1716.50**
3 Pole, 630 A (800mm wide, 600mm high)	1824.36	2159.08	-	-	nr	**2159.08**
3 Pole, 400 A (800mm wide, 400mm high)	1243.19	1471.28	-	-	nr	**1471.28**

V:ELECTRICAL SUPPLY/POWER/LIGHTING SYSTEMS

Item	Net Price £	Material £	Labour hours	Labour £	Unit	Total rate £
MCCB-36kA fault rated						
4 Pole, 250 A (800mm wide, 400mm high)	1230.56	1456.33	-	-	nr	1456.33
4 Pole, 160 A (800mm wide, 300mm high)	932.39	1103.45	-	-	nr	1103.45
3 Pole, 250 A (800mm wide, 400mm high)	698.94	827.18	-	-	nr	827.18
3 Pole, 160 A (800mm wide, 300mm high)	615.50	728.43	-	-	nr	728.43
Extra for						
Cable box (one per MCCB for form 4, types 6 & 7)	106.39	125.91	-	-	nr	125.91
Shunt trip (for ratings 100A to 630A)	37.24	44.07	-	-	nr	44.07
Undervoltage release (for ratings 100A to 630A)	53.20	62.96	-	-	nr	62.96
Motor operator for 630A MCCB	505.36	598.08	-	-	nr	598.08
Motor operator for 400A MCCB	505.36	598.08	-	-	nr	598.08
Motor operator for 250A MCCB	420.25	497.35	-	-	nr	497.35
Motor operator for 160A/100A MCCB	268.64	317.93	-	-	nr	317.93
Door handle for 630/400A MCCB	66.50	78.69	-	-	nr	78.69
Door handle for 250/160/100A MCCB	53.20	62.96	-	-	nr	62.96
MCCB earth fault protection	398.97	472.17	-	-	nr	472.17
LV SWITCHBOARD BUSBAR						
Copper busbar assembled into LV switchboard, ASTA type tested to appropriate fault level. Busbar Length may be estimated by adding the widths of the ACB sections to the width of the MCCB sections. ACB's up to 2000A rating may be stacked two high; larger ratings are one per section. To determine the number of MCCB sections, add together all the MCCB heights and divide by 1800mm, rounding up as necessary						
LV switchboard busbar						
6000 A (6 x 10mm x 100mm)	2062.05	2440.38	-	-	nr	2440.38
5000 A (4 x 10mm x 100mm)	1695.47	2006.53	-	-	nr	2006.53
4000 A (4 x 10mm x 100mm)	1695.47	2006.53	-	-	nr	2006.53
3200 A (3 x 10mm x 100mm)	1177.09	1393.05	-	-	nr	1393.05
2500 A (2 x 10mm x 100mm)	993.80	1176.13	-	-	nr	1176.13
2000 A (2 x 10mm x 80mm)	730.30	864.29	-	-	nr	864.29
1600 A (2 x 10mm x 50mm)	529.83	627.04	-	-	nr	627.04
1250 A (2 x 10mm x 40mm)	415.27	491.46	-	-	nr	491.46
1000 A (2 x 10mm x 30mm)	415.27	491.46	-	-	nr	491.46
800 A (2 x 10mm x 20mm)	323.62	383.00	-	-	nr	383.00
630 A (2 x 10mm x 20mm)	323.62	383.00	-	-	nr	383.00
400 A (2 x 10mm x 10mm)	277.58	328.51	-	-	nr	328.51
AUTOMATIC POWER FACTOR CORRECTION						
Automatic power factor correction (PFC); floor standing steel enclosure to IP 42, complete with microprocessor based relay and status indication; includes delivery, offloading, positioning and commissioning; excludes cabling and cable terminations						
Standard PFC (no de-tuning)						
100 kVAr	4042.73	4784.45	-	-	nr	4784.45
200 kVAr	5654.18	6691.55	-	-	nr	6691.55
400 kVAr	9652.01	11422.86	-	-	nr	11422.86
600 kVAr	13221.32	15647.03	-	-	nr	15647.03

V:ELECTRICAL SUPPLY/POWER/LIGHTING SYSTEMS

Item	Net Price £	Material £	Labour hours	Labour £	Unit	Total rate £
V20 : LV DISTRIBUTION (cont'd)						
Y71 - LV SWITCHGEAR & DISTRIBUTION BOARDS (cont'd)						
AUTOMATIC POWER FACTOR CORRECTION (cont'd)						
PFC with de-tuning reactors						
100 kVAr	6577.94	7784.80	-	-	nr	**7784.80**
200 kVAr	9192.69	10879.28	-	-	nr	**10879.28**
400 kVAr	15842.48	18749.10	-	-	nr	**18749.10**
600 kVAr	23298.00	27572.48	-	-	nr	**27572.48**
AUTOMATIC TRANSFER SWITCHES						
Automatic transfer switches; steel enclosure; solenoid operating; programmable controller, keypad and LCD display; fixed to backgrounds; including commissioning and testing						
Panel Mounting type 3 pole or 4 pole; overlapping neutral						
100 amp	2667.08	3156.41	2.60	66.02	nr	**3222.43**
260amp	3719.56	4401.99	3.30	83.79	nr	**4485.78**
400amp	4532.84	5364.48	4.30	109.18	nr	**5473.66**
600amp	5824.52	6893.14	5.30	134.57	nr	**7027.72**
800amp	7367.36	8719.05	5.50	139.65	nr	**8858.70**
1000amp	11720.80	13871.22	5.83	148.03	nr	**14019.25**
1600amp	12558.00	14862.02	6.20	157.43	nr	**15019.44**
2000amp	15428.40	18259.05	6.90	175.20	nr	**18434.25**
Enclosed type 3 pole or 4 pole; over lapping neutral						
100 amp	3769.79	4461.44	2.60	66.02	nr	**4527.45**
260amp	4822.27	5707.01	3.30	83.79	nr	**5790.81**
400amp	5635.55	6669.51	4.30	109.18	nr	**6778.69**
600amp	7305.17	8645.45	4.84	122.89	nr	**8768.34**
800amp	8848.01	10471.35	5.12	130.00	nr	**10601.36**
1000amp	13201.45	15623.52	5.50	139.65	nr	**15763.17**
1600amp	15430.79	18261.88	6.20	157.43	nr	**18419.31**
2000amp	18301.19	21658.91	6.90	175.20	nr	**21834.11**
3000amp	22008.79	26046.75	6.20	157.43	nr	**26204.17**
4000amp	26939.90	31882.56	6.90	175.20	nr	**32057.76**
BOARDS AND PANELS						
MCCB panelboards; IP4X construction, 50kA busbars and fully-rated neutral; fitted with doorlock, removable glandplate; form 3b Type2; BSEN 60439-1; including fixing to backgrounds Panelboards cubicle with MCCB incomer						
Up to 250A						
4 Way TPN	747.50	884.65	1.00	25.39	nr	**910.04**
Extra over for integral incomer metering	765.60	906.06	1.50	38.09	nr	**944.15**
Up to 630A						
6 way TPN	1344.67	1591.38	2.00	50.78	nr	**1642.16**
12 way TPN	1518.67	1797.30	2.50	63.48	nr	**1860.78**
18 Way TPN	1767.84	2092.19	3.00	76.17	nr	**2168.36**
Extra over for integral incomer metering	904.80	1070.80	1.50	38.09	nr	**1108.89**

V:ELECTRICAL SUPPLY/POWER/LIGHTING SYSTEMS

Item	Net Price £	Material £	Labour hours	Labour £	Unit	Total rate £
Up to 800A						
6 way TPN	2300.98	2723.14	2.00	50.78	nr	**2773.92**
12 way TPN	2760.34	3266.77	2.50	63.48	nr	**3330.25**
18 Way TPN	2938.51	3477.64	3.00	76.17	nr	**3553.82**
Extra over for integral incomer metering	904.80	1070.80	1.50	38.09	nr	**1108.89**
Up to 1200A						
20 Way TPN	6156.82	7286.41	3.50	88.87	nr	**7375.28**
Up to 1600A						
28 Way TPN	7973.38	9436.25	3.50	88.87	nr	**9525.12**
Up to 2000A						
28Way TPN	8679.12	10271.48	4.00	101.57	nr	**10373.04**
Feeder MCCBs						
Single Pole						
32A	61.25	72.49	0.75	19.04	nr	**91.53**
63A	62.64	74.13	0.75	19.04	nr	**93.18**
100A	64.03	75.78	0.75	19.04	nr	**94.82**
160A	68.21	80.72	1.00	25.39	nr	**106.11**
Double pole						
32A	91.87	108.73	0.75	19.04	nr	**127.77**
63A	93.26	110.38	0.75	19.04	nr	**129.42**
100A	136.42	161.44	0.75	19.04	nr	**180.49**
160A	169.82	200.98	1.00	25.39	nr	**226.37**
Triple pole						
32A	122.50	144.97	0.75	19.04	nr	**164.01**
63A	125.28	148.27	0.75	19.04	nr	**167.31**
100A	162.86	192.74	0.75	19.04	nr	**211.79**
160A	210.19	248.76	1.00	25.39	nr	**274.15**
250A	315.98	373.96	1.00	25.39	nr	**399.35**
400A	427.34	505.75	1.25	31.74	nr	**537.49**
630A	701.57	830.28	1.50	38.09	nr	**868.37**
MCB distribution boards; IP3X external protection enclosure; removable earth and neutral bars and DIN rail; 125/250amp incomers; including fixing to backgrounds						
SP & N						
6 way	51.56	61.02	2.00	50.78	nr	**111.81**
8 way	61.15	72.36	2.50	63.48	nr	**135.84**
12 way	70.17	83.04	3.00	76.17	nr	**159.22**
16 way	83.42	98.73	4.00	101.57	nr	**200.30**
24 way	175.72	207.96	5.00	126.96	nr	**334.91**
TP & N						
4 way	373.52	442.05	3.00	76.17	nr	**518.23**
6 way	386.65	457.59	3.50	88.87	nr	**546.46**
8 way	405.05	479.37	4.00	101.57	nr	**580.93**
12 way	431.66	510.85	4.00	101.57	nr	**612.42**
16 way	493.93	584.55	5.00	126.96	nr	**711.50**
24 way	622.04	736.16	6.40	162.50	nr	**898.67**

V:ELECTRICAL SUPPLY/POWER/LIGHTING SYSTEMS

Item	Net Price £	Material £	Labour hours	Labour £	Unit	Total rate £
V20 : LV DISTRIBUTION (cont'd)						
Y71 - LV SWITCHGEAR & DISTRIBUTION BOARDS (cont'd)						
BOARDS AND PANELS (cont'd)						
Miniature circuit breakers for distribution boards; BS EN 60 898; DIN rail mounting; including connecting to circuit						
SP&N; including connecting of wiring						
6 Amp	8.86	10.49	0.10	2.54	nr	**13.03**
10 - 40 Amp	9.21	10.90	0.10	2.54	nr	**13.44**
50 - 63 Amp	9.65	11.42	0.14	3.55	nr	**14.97**
TP&N; including connecting of wiring						
6 Amp	37.53	44.41	0.30	7.62	nr	**52.03**
10 - 40 Amp	39.01	46.17	0.45	11.43	nr	**57.59**
50 - 63 Amp	40.87	48.37	0.45	11.43	nr	**59.80**
Residual current circuit breakers for distribution boards; DIN rail mounting; including connecting to circuit						
SP&N						
10mA						
6 Amp	58.12	68.79	0.21	5.33	nr	**74.12**
10 - 32 Amp	57.02	67.48	0.26	6.60	nr	**74.08**
45 Amp	57.02	67.48	0.26	6.60	nr	**74.08**
30mA						
6 Amp	58.12	68.79	0.21	5.33	nr	**74.12**
10 - 40 Amp	57.02	67.48	0.21	5.33	nr	**72.82**
50 -63 Amp	57.02	67.48	0.26	6.60	nr	**74.08**
100mA						
6 Amp	107.47	127.19	0.21	5.33	nr	**132.52**
10 - 40 Amp	107.47	127.19	0.21	5.33	nr	**132.52**
50 -63 Amp	107.47	127.19	0.26	6.60	nr	**133.79**
HRC fused distribution boards; IP4X external protection enclosure; including earth and neutral bars; fixing to backgrounds						
SP&N						
20 Amp incomer						
4 way	117.90	139.53	1.00	25.39	nr	**164.92**
6 way	142.33	168.44	1.20	30.47	nr	**198.91**
8 way	166.86	197.47	1.40	35.55	nr	**233.02**
12 way	215.92	255.53	1.80	45.70	nr	**301.24**
32 Amp incomer						
4 way	141.95	168.00	1.00	25.39	nr	**193.39**
6 way	186.67	220.92	1.20	30.47	nr	**251.39**
8 way	219.62	259.91	1.40	35.55	nr	**295.46**
12 way	282.76	334.64	1.80	45.70	nr	**380.35**

V:ELECTRICAL SUPPLY/POWER/LIGHTING SYSTEMS

Item	Net Price £	Material £	Labour hours	Labour £	Unit	Total rate £
TP&N						
20 Amp incomer						
4 way	211.32	250.09	1.50	38.09	nr	**288.18**
6 way	267.28	316.32	2.10	53.32	nr	**369.64**
8 way	316.01	373.99	2.70	68.56	nr	**442.54**
12 way	443.35	524.69	3.90	99.03	nr	**623.72**
32 Amp incomer						
4 way	252.81	299.19	1.50	38.09	nr	**337.28**
6 way	339.41	401.68	2.10	53.32	nr	**455.00**
8 way	414.62	490.69	2.70	68.56	nr	**559.24**
12 way	575.06	680.57	3.90	99.03	nr	**779.60**
63 Amp incomer						
4 way	537.39	635.98	2.17	55.10	nr	**691.08**
6 way	689.20	815.64	2.83	71.86	nr	**887.50**
8 way	829.98	982.26	2.57	65.26	nr	**1047.51**
100 Amp incomer						
4 way	849.89	1005.81	2.40	60.94	nr	**1066.75**
6 way	1110.88	1314.70	2.73	69.32	nr	**1384.01**
8 way	1358.32	1607.53	3.87	98.26	nr	**1705.80**
200 Amp incomer						
4 way	2105.09	2491.31	5.36	136.10	nr	**2627.41**
6 way	2782.31	3292.78	6.17	156.66	nr	**3449.44**
HRC fuse; includes fixing to fuse holder						
2-30 Amp	2.41	2.85	0.10	2.54	nr	**5.39**
35 - 63 Amp	5.21	6.17	0.12	3.05	nr	**9.21**
80 Amp	7.69	9.11	0.15	3.81	nr	**12.91**
100 Amp	7.69	9.11	0.15	3.81	nr	**12.91**
125 Amp	13.98	16.55	0.15	3.81	nr	**20.35**
160 Amp	14.67	17.36	0.15	3.81	nr	**21.17**
200 Amp	15.19	17.98	0.15	3.81	nr	**21.79**
Consumer units; fixed to backgrounds; including supports, fixings, connections/jointing to equipment.						
Switched and insulated; moulded plastic case, 63 Amp 230 Volt SP&N; earth and neutral bars; 30mA RCCB protection; fitted MCB's						
2 way	97.36	115.22	1.67	42.40	nr	**157.62**
4 way	109.25	129.29	1.59	40.37	nr	**169.66**
6 way	118.75	140.54	2.50	63.48	nr	**204.02**
8 way	128.14	151.65	3.00	76.17	nr	**227.83**
12 way	148.86	176.17	4.00	101.57	nr	**277.73**
16 way	179.87	212.88	5.50	139.65	nr	**352.53**
Switched and insulated; moulded plastic case, 100 Amp 230 Volt SP&N; earth and neutral bars; 30mA RCCB protection; fitted MCB's						
2 way	97.36	115.22	1.67	42.40	nr	**157.62**
4 way	109.25	129.29	1.59	40.37	nr	**169.66**
6 way	118.75	140.54	2.50	63.48	nr	**204.02**
8 way	128.14	151.65	3.00	76.17	nr	**227.83**
12 way	148.86	176.17	4.00	101.57	nr	**277.73**
16 way	179.87	212.88	5.50	139.65	nr	**352.53**

V:ELECTRICAL SUPPLY/POWER/LIGHTING SYSTEMS

Item	Net Price £	Material £	Labour hours	Labour £	Unit	Total rate £
V20 : LV DISTRIBUTION (cont'd)						
Y71 - LV SWITCHGEAR & DISTRIBUTION BOARDS (cont'd)						
BOARDS AND PANELS (cont'd)						
Consumer units (cont'd)						
Extra for						
Residual current device; double pole; 230 volt/30mA tripping current						
16 Amp	54.08	64.00	0.22	5.59	nr	69.58
30 Amp	53.23	62.99	0.22	5.59	nr	68.58
40 Amp	54.91	64.99	0.22	5.59	nr	70.57
63 Amp	65.41	77.41	0.22	5.59	nr	83.00
80 Amp	75.63	89.50	0.22	5.59	nr	95.09
100 Amp	93.09	110.17	0.25	6.35	nr	116.51
Residual current device; double pole; 230 volt/100mA tripping current						
63 Amp	62.15	73.55	0.22	5.59	nr	79.14
80 Amp	71.90	85.09	0.22	5.59	nr	90.68
100 Amp	93.10	110.18	0.25	6.35	nr	116.53
Heavy duty fuse switches; with HRC fuses BS 5419; short circuit rating 65kA, 500 volt; including retractable operating switches						
SP&N						
63 Amp	210.12	248.67	1.30	33.01	nr	281.68
100 Amp	307.15	363.50	1.95	49.51	nr	413.02
TP&N						
63 Amp	264.71	313.27	1.83	46.47	nr	359.74
100 Amp	371.98	440.23	2.48	62.97	nr	503.20
200 Amp	573.81	679.08	3.13	79.48	nr	758.56
300 Amp	996.87	1179.76	4.45	112.99	nr	1292.75
400 Amp	1094.37	1295.15	4.45	112.99	nr	1408.14
600 Amp	1651.90	1954.97	5.72	145.24	nr	2100.21
800 Amp	2580.45	3053.88	7.88	200.08	nr	3253.97
Switch disconnectors to BSEN 60947-3; in sheet steel case; IP41 with door interlock fixed to backgrounds						
Double pole						
20 Amp	38.56	45.63	1.02	25.90	nr	71.53
32 Amp	46.48	55.00	1.02	25.90	nr	80.90
63 Amp	169.60	200.71	1.21	30.72	nr	231.44
100 Amp	155.63	184.18	1.86	47.23	nr	231.41
TP&N						
20 Amp	48.41	57.29	1.29	32.75	nr	90.05
32 Amp	56.32	66.65	1.83	46.47	nr	113.11
63 Amp	192.71	228.07	2.48	62.97	nr	291.04
100 Amp	191.14	226.21	2.48	62.97	nr	289.18
125 Amp	201.01	237.89	2.48	62.97	nr	300.86
160 Amp	462.50	547.35	2.48	62.97	nr	610.33

V:ELECTRICAL SUPPLY/POWER/LIGHTING SYSTEMS

Item	Net Price £	Material £	Labour hours	Labour £	Unit	Total rate £
Enclosed switch disconnector to BSEN 60947-3; enclosure minimum IP55 rating; complete with earth connection bar; fixed to backgrounds.						
TP						
20 Amp	38.56	45.63	1.02	25.90	nr	**71.53**
32 Amp	46.48	55.00	1.02	25.90	nr	**80.90**
63 Amp	169.60	200.71	1.21	30.72	nr	**231.44**
TP&N						
20 Amp	48.41	57.29	1.29	32.75	nr	**90.05**
32 Amp	56.32	66.65	1.83	46.47	nr	**113.11**
63 Amp	192.71	228.07	2.48	62.97	nr	**291.04**
Busbar chambers; fixed to background including all supports, fixings, connections/jointing to equipment.						
Sheet steel case enclosing 4 pole 550 Volt copper bars, detachable metal end plates						
600mm long						
200 Amp	394.30	466.65	2.62	66.53	nr	**533.17**
300 Amp	506.67	599.63	3.03	76.94	nr	**676.57**
500 Amp	868.16	1027.44	4.48	113.75	nr	**1141.19**
900mm long						
200 Amp	567.94	672.14	3.04	77.19	nr	**749.33**
300 Amp	669.34	792.15	3.59	91.16	nr	**883.30**
500 Amp	990.08	1171.73	4.42	112.23	nr	**1283.96**
1350mm long						
200 Amp	775.76	918.09	3.38	85.82	nr	**1003.91**
300 Amp	913.06	1080.57	3.94	100.04	nr	**1180.62**
500 Amp	1461.54	1729.68	4.82	122.39	nr	**1852.07**
Contactor relays; pressed steel enclosure; fixed to backgrounds including supports, fixings, connections/jointing to equipment						
Relays						
6 Amp, 415/240 Volt, 4 pole N/O	45.67	54.05	0.52	13.20	nr	**67.25**
6 Amp, 415/240 Volt, 8 pole N/O	55.85	66.09	0.85	21.58	nr	**87.68**
Push button stations; heavy gauge pressed steel enclosure; polycarbonate cover; IP65; fixed to backgrounds including supports, fixings, connections/joining to equipment						
Standard units						
One button (start or stop)	57.78	68.38	0.39	9.90	nr	**78.29**
Two button (start or stop)	61.34	72.60	0.47	11.93	nr	**84.53**
Three button (forward-reverse-stop)	86.93	102.88	0.57	14.47	nr	**117.36**

V:ELECTRICAL SUPPLY/POWER/LIGHTING SYSTEMS

Item	Net Price £	Material £	Labour hours	Labour £	Unit	Total rate £
V20 : LV DISTRIBUTION (cont'd)						
Y71 - LV SWITCHGEAR & DISTRIBUTION BOARDS (cont'd)						
BOARDS AND PANELS (cont'd)						
Weatherproof junction boxes; enclosures with rail mounted terminal blocks; side hung door to receive padlock; fixed to backgrounds, including all supports and fixings (Suitable for cable up to 2.5mm²; including glandplates and gaskets.)						
Sheet steel with zinc spray finish enclosure						
Overall Size 229 x 152; suitable to receive 3 x 20(A) glands per gland plate	60.38	71.46	1.43	36.31	nr	**107.77**
Overall Size 306 x 306; suitable to receive 14 x 20(A) glands per gland plate	81.00	95.86	2.17	55.10	nr	**150.96**
Overall Size 458 x 382; suitable to receive 18 x 20(A) glands per gland plate	117.94	139.57	3.51	89.12	nr	**228.70**
Overall Size 762 x508; suitable to receive 26 x 20(A) glands per gland plate	124.73	147.62	4.85	123.15	nr	**270.77**
Overall Size 914 x 610; suitable to receive 45 x 20(A) glands per gland plate	137.87	163.16	7.01	177.99	nr	**341.15**
Weatherproof junction boxes; enclosures with rail mounted terminal blocks; screw fixed lid; fixed to backgrounds, including all supports and fixings (suitable for cable up to 2.5mm²; including glandplates and gaskets)						
Glassfibre reinforced polycarbonate enclosure						
Overall Size 190 x 190 x 130	82.46	97.59	1.43	36.31	nr	**133.90**
Overall Size 190 x 190 x 180	120.72	142.87	1.53	38.85	nr	**181.72**
Overall Size 280 x 190 x 130	136.35	161.37	2.17	55.10	nr	**216.47**
Overall Size 280 x 190 x 180	152.76	180.79	2.37	60.18	nr	**240.97**
Overall Size 380 x 190 x 130	170.44	201.71	3.33	84.55	nr	**286.26**
Overall Size 380 x 190 x 180	183.06	216.65	3.30	83.79	nr	**300.44**
Overall Size 380 x 280 x 130	195.69	231.59	4.66	118.32	nr	**349.91**
Overall Size 380 x 280 x 180	210.84	249.52	5.36	136.10	nr	**385.62**
Overall Size 560 x 280 x 130	253.76	300.32	7.01	177.99	nr	**478.31**
Overall Size 560 x 380 x 180	261.34	309.29	7.67	194.75	nr	**504.04**

V:ELECTRICAL SUPPLY/POWER/LIGHTING SYSTEMS

Item	Net Price £	Material £	Labour hours	Labour £	Unit	Total rate £
V21 : GENERAL LIGHTING						
Y73 - LUMINAIRES (GENERAL)						
LUMINAIRES						
Fluorescent Luminaires; surface fixed to backgrounds						
Batten type; surface mounted						
600 mm Single - 18 W	7.99	9.46	0.58	14.73	nr	**24.19**
600 mm Twin - 18 W	14.15	16.75	0.59	14.98	nr	**31.73**
1200 mm Single - 36 W	10.57	12.51	0.76	19.30	nr	**31.81**
1200 mm Twin - 36 W	20.26	23.97	0.77	19.55	nr	**43.53**
1500 mm Single - 58 W	11.96	14.15	0.84	21.33	nr	**35.48**
1500 mm Twin - 58 W	24.07	28.48	0.85	21.58	nr	**50.06**
1800 mm Single - 70 W	14.43	17.08	1.05	26.66	nr	**43.74**
1800 mm Twin - 70 W	26.39	31.23	1.06	26.91	nr	**58.15**
2400 mm Single - 100 W	19.75	23.38	1.25	31.74	nr	**55.12**
2400 mm Twin - 100 W	34.57	40.91	1.27	32.25	nr	**73.15**
Surface mounted, opal diffuser						
600 mm Twin - 18 W	21.95	25.97	0.62	15.74	nr	**41.71**
1200 mm Single - 36 W	18.74	22.18	0.79	20.06	nr	**42.24**
1200 mm Twin - 36 W	29.19	34.55	0.80	20.31	nr	**54.86**
1500 mm Single - 58 W	21.55	25.50	0.88	22.34	nr	**47.85**
1500 mm Twin - 58 W	34.13	40.40	0.90	22.85	nr	**63.25**
1800 mm Single - 70 W	26.89	31.83	1.09	27.68	nr	**59.51**
1800 mm Twin - 70 W	36.92	43.69	1.10	27.93	nr	**71.62**
2400 mm Single - 100 W	35.33	41.81	1.30	33.01	nr	**74.82**
2400 mm Twin - 100 W	51.00	60.35	1.31	33.26	nr	**93.61**
Surface mounted linear fluorescent; T8 lamp; high frequency control gear; low brightness; 65° cut-off; including wedge style louvre.						
1200mm, 1 x 36 watt	49.59	58.69	1.09	27.68	nr	**86.37**
1200mm 2 x 36 watt	53.03	62.76	1.09	27.68	nr	**90.44**
Extra for emergency pack	44.73	52.94	0.25	6.35	nr	**59.29**
1500mm, 1 x 58 watt	57.80	68.40	0.90	22.85	nr	**91.25**
1500mm 2 x 58 watt	61.78	73.12	0.90	22.85	nr	**95.97**
Extra for emergency pack	44.73	52.94	0.25	6.35	nr	**59.29**
1800mm, 1 x 70 watt	87.05	103.02	0.90	22.85	nr	**125.87**
1800mm, 2 x 70 watt	99.91	118.24	0.90	22.85	nr	**141.09**
Extra for emergency pack	73.90	87.46	0.25	6.35	nr	**93.81**
Modular recessed linear fluorescent; high frequency control gear; low brightness; 65° cut off; including wedge style louvre; fitted to exposed T grid ceiling.						
600 x 600 mm, 3 x 18 watt T8	39.00	46.15	0.84	21.33	nr	**67.48**
600 x 600 mm, 4 x 18 watt T8	40.33	47.73	0.87	22.09	nr	**69.82**
Extra for emergency pack	45.10	53.37	0.25	6.35	nr	**59.72**
300 x 1200 mm, 2 x 36 watt T8	86.14	101.95	0.87	22.09	nr	**124.04**
Extra for emergency pack	51.60	61.07	0.25	6.35	nr	**67.42**
600 x 1200 mm, 3 x 36 watt T8	87.44	103.48	0.89	22.60	nr	**126.08**

V:ELECTRICAL SUPPLY/POWER/LIGHTING SYSTEMS

Item	Net Price £	Material £	Labour hours	Labour £	Unit	Total rate £
V21 : GENERAL LIGHTING (cont'd)						
Y73 - LUMINAIRES (GENERAL) (cont'd)						
LUMINAIRES (cont'd)						
Modular recessed linear fluorescent (cont'd)						
600 x 1200 mm, 4 x 36 watt T8	89.47	105.89	0.91	23.11	nr	**129.00**
Extra for emergency pack	54.43	64.41	0.25	6.35	nr	**70.76**
600 x 600, 3 x 14 watt T5	57.06	67.53	0.84	21.33	nr	**88.85**
600 x 600, 4 x 14 watt T5	59.39	70.28	0.87	22.09	nr	**92.37**
Extra for emergency pack	68.31	80.85	0.25	6.35	nr	**87.19**
Modular recessed; T8 lamp; high frequency control gear; cross-blade louvre; fitted to exposed T grid ceiling						
600 x 600 mm, 3 x 18 watt	42.59	50.40	0.84	21.33	nr	**71.73**
600 x 600 mm, 4 x 18 watt	57.63	68.21	0.87	22.09	nr	**90.30**
Extra for emergency pack	51.27	60.67	0.25	6.35	nr	**67.02**
Modular recessed compact fluorescent; TCL lamp; high frequency control gear; low brightness; 65° cut-off; including wedge style louvre; fitted to exposed T grid ceiling						
300 x 300 mm, 2 x 18 watt	91.59	108.39	0.75	19.04	nr	**127.43**
Extra for emergency pack	72.14	85.37	0.25	6.35	nr	**91.72**
500 x 500 mm, 2 x 36 watt	84.57	100.08	0.82	20.82	nr	**120.90**
600 x 600 mm, 2 x 36 watt	85.65	101.36	0.82	20.82	nr	**122.18**
600 x 600 mm, 2 x 40 watt	89.61	106.06	0.82	20.82	nr	**126.88**
Extra for emergency pack	51.69	61.18	0.25	6.35	nr	**67.53**
Ceiling recessed asymmetric compact fluorescent downlighter; high frequency control gear; TCD lamp in 200 mm diameter luminaire; for wall-washing application.						
1 x 18 watt	142.11	168.18	0.75	19.04	nr	**187.23**
1 x 26 watt	142.11	168.18	0.75	19.04	nr	**187.23**
2 x 18 watt	158.31	187.36	0.75	19.04	nr	**206.40**
2 x 26 watt	158.31	187.36	0.75	19.04	nr	**206.40**
Ceiling recessed asymmetric compact fluorescent downlights; high frequency control gear; linear 200mm x 600mm luminaire with low glare louvre; for wall washing applications						
1 x 55 watt TCL	57.40	67.94	0.75	19.04	nr	**86.98**
Wall mounted compact fluorescent uplighter; high frequency control gear; TCL lamp in 300mm x 600mm luminaire						
2 x 36 watt	225.06	266.35	0.84	21.33	nr	**287.68**
2 x 40 watt	242.03	286.44	0.84	21.33	nr	**307.77**
2 x 55 watt	242.03	286.44	0.84	21.33	nr	**307.77**

V:ELECTRICAL SUPPLY/POWER/LIGHTING SYSTEMS

Item	Net Price £	Material £	Labour hours	Labour £	Unit	Total rate £
Suspended linear fluorescent; T5 lamp; high frequency control gear; low brightness; 65° cut-off; 30% uplight, 70% downlight; including wedge style louvre						
1 x 49 watt	159.21	188.42	0.75	19.04	nr	**207.46**
Extra for emergency pack	87.95	104.09	0.25	6.35	nr	**110.43**
Semi-recessed 'architectural' linear fluorescent; T5 lamp; high frequency control gear; low brightness, delivers direct, ceiling and graduated wall washing illumination						
600 x 600 mm, 2 x 24 watt	130.43	154.36	0.87	22.09	nr	**176.45**
600 x 600 mm, 4 x 14 watt	140.36	166.11	0.87	22.09	nr	**188.20**
500 x 500 mm,2 x 24 watt	128.24	151.77	0.87	22.09	nr	**173.86**
Extra for emergency pack	62.46	73.92	0.25	6.35	nr	**80.27**
Downlighter, recessed; low voltage; mirror reflector with white/chrome bezel; dimmable transformer; for dichroic lamps						
85mm dia x 20/50 watt	14.35	16.98	0.66	16.76	nr	**33.74**
118mm dia x 50 watt	18.96	22.44	0.66	16.76	nr	**39.20**
165mm dia x 100 watt	86.63	102.52	0.66	16.76	nr	**119.28**
High/Low Bay luminaires						
Compact discharge; aluminium reflector						
150 watt	50.42	59.67	1.50	38.09	nr	**97.76**
250 watt	50.42	59.67	1.50	38.09	nr	**97.76**
400 watt	53.52	63.34	1.50	38.09	nr	**101.42**
Sealed discharge; aluminium reflector						
150 watt	163.61	193.63	1.50	38.09	nr	**231.71**
250 watt	176.59	208.99	1.50	38.09	nr	**247.08**
400 watt	230.89	273.25	1.50	38.09	nr	**311.34**
Corrosion resistant GRP body; gasket sealed; acrylic diffuser						
600 mm Single - 18 W	28.22	33.39	0.49	12.44	nr	**45.83**
600 mm Twin - 18 W	36.74	43.48	0.49	12.44	nr	**55.93**
1200 mm Single - 36 W	31.76	37.59	0.64	16.25	nr	**53.84**
1200 mm Twin - 36 W	40.72	48.19	0.64	16.25	nr	**64.44**
1500 mm Single - 58 W	35.40	41.90	0.72	18.28	nr	**60.18**
1500 mm Twin - 58 W	44.02	52.10	0.72	18.28	nr	**70.38**
1800 mm Single - 70 W	52.39	62.00	0.94	23.87	nr	**85.87**
1800 mm Twin - 70 W	65.03	76.96	0.94	23.87	nr	**100.83**
Flameproof to IIA/IIB,I.P. 64; Aluminium Body; BS 229 and 899						
600 mm Single - 18 W	292.26	345.88	1.04	26.41	nr	**372.29**
600 mm Twin - 18 W	364.09	430.89	1.04	26.41	nr	**457.30**
1200 mm Single - 36 W	320.13	378.87	1.31	33.26	nr	**412.13**
1200 mm Twin - 36 W	395.80	468.41	1.18	29.96	nr	**498.38**
1500 mm Single - 58 W	342.73	405.61	1.64	41.64	nr	**447.25**
1500 mm Twin - 58 W	413.85	489.78	1.64	41.64	nr	**531.43**
1800 mm Single - 70 W	375.71	444.64	1.97	50.02	nr	**494.66**
1800 mm Twin - 70 W	433.45	512.98	1.97	50.02	nr	**563.00**

V:ELECTRICAL SUPPLY/POWER/LIGHTING SYSTEMS

Item	Net Price £	Material £	Labour hours	Labour £	Unit	Total rate £
V21 : GENERAL LIGHTING (cont'd)						
Y73 - LUMINAIRES (GENERAL) (cont'd)						
LUMINAIRES (cont'd)						
External Lighting						
Ground mounted 50 watt	354.89	420.01	2.25	57.13	nr	**477.14**
Ceiling mounted 50 watt	159.45	188.70	2.25	57.13	nr	**245.83**
Bulkhead; aluminium body and polycarbonate bowl; vandal resistant; IP65						
60 watt	30.62	36.24	0.75	19.04	nr	**55.29**
Extra for						
Emergency version	82.29	97.39	0.25	6.35	nr	**103.73**
2D 2 pin 16 watt	25.86	30.60	0.66	16.76	nr	**47.36**
2D 2 pin 28 watt	51.65	61.13	0.66	16.76	nr	**77.89**
Extra for						
Emergency version	60.86	72.03	0.25	6.35	nr	**78.38**
Photocell	18.15	21.48	0.75	19.04	nr	**40.52**
1500 mm high circular bollard; polycarbonate visor; vandal resistant; IP54						
50 watt	182.74	216.27	1.75	44.44	nr	**260.71**
70 watt	185.66	219.72	1.75	44.44	nr	**264.16**
80 watt	222.21	262.98	1.75	44.44	nr	**307.42**
Floodlight; enclosed high performance discharge light; integral control gear; reflector; toughened glass; IP65						
70 watt	96.24	113.90	1.25	31.74	nr	**145.64**
100 watt	134.10	158.70	1.25	31.74	nr	**190.44**
150 watt	106.80	126.39	1.25	31.74	nr	**158.13**
250 watt	186.93	221.23	1.25	31.74	nr	**252.97**
400 watt	193.65	229.18	1.25	31.74	nr	**260.92**
Extra for						
Photocell	18.50	21.89	0.75	19.04	nr	**40.94**
Lighting Track						
Single circuit; extruded aluminium white finish; low voltage with copper conductors; including couplers and supports; fixed to backgrounds						
Straight track	16.06	19.01	0.50	12.70	m	**31.70**
Live end	5.44	6.44	0.33	8.38	nr	**14.82**
Dead end	1.79	2.12	0.25	6.35	nr	**8.46**
Elbow	10.84	12.83	0.33	8.38	nr	**21.21**
Tee	16.31	19.30	0.33	8.38	nr	**27.68**
Cross	21.71	25.69	0.50	12.70	nr	**38.39**
Flexible couplers	26.33	31.16	0.33	8.38	nr	**39.54**

V:ELECTRICAL SUPPLY/POWER/LIGHTING SYSTEMS

Item	Net Price £	Material £	Labour hours	Labour £	Unit	Total rate £
Three circuit; extruded aluminium white finish; low voltage with copper conductors; including couplers and supports; fixed to backgrounds						
Straight track	25.98	30.75	0.75	19.04	m	**49.79**
Live end	11.77	13.93	0.50	12.70	nr	**26.63**
Dead end	1.29	1.53	0.45	11.43	nr	**12.95**
Elbow	16.22	19.20	0.40	10.16	nr	**29.35**
Tee	20.69	24.49	0.55	13.97	nr	**38.45**
Cross	24.49	28.99	0.88	22.34	nr	**51.33**
Flexible couplers	30.21	35.75	0.45	11.43	nr	**47.17**

V:ELECTRICAL SUPPLY/POWER/LIGHTING SYSTEMS

Item	Net Price £	Material £	Labour hours	Labour £	Unit	Total rate £
V21: GENERAL LIGHTING (cont'd)						
Y74 - LIGHTING ACCESSORIES						
SWITCHES						
6 Amp metal clad surface mounted switch, gridswitch; one way						
1 Gang	4.33	5.13	0.43	10.92	nr	**16.05**
2 Gang	5.72	6.77	0.55	13.97	nr	**20.73**
3 Gang	9.32	11.03	0.77	19.55	nr	**30.58**
4 Gang	10.69	12.65	0.88	22.34	nr	**35.00**
6 Gang	19.24	22.77	1.10	27.93	nr	**50.70**
8 Gang	21.75	25.74	1.28	32.50	nr	**58.24**
12 Gang	32.57	38.55	1.67	42.40	nr	**80.95**
Extra for						
6 Amp - Two way switch	2.05	2.42	0.03	0.76	nr	**3.18**
20 Amp - Two way switch	2.75	3.25	0.04	1.02	nr	**4.27**
20 Amp - Intermediate	5.17	6.12	0.08	2.03	nr	**8.15**
20 Amp - One way SP switch	2.12	2.51	0.08	2.03	nr	**4.54**
Steel blank plate; 1 Gang	1.04	1.24	0.07	1.78	nr	**3.01**
Steel blank plate; 2 Gang	1.91	2.27	0.08	2.03	nr	**4.30**
6 Amp modular type switch; galvanised steel box, bronze or satin chrome coverplate; metalclad switches; flush mounting; one way						
1 Gang	11.46	13.56	0.43	10.92	nr	**24.48**
2 Gang	15.83	18.73	0.55	13.97	nr	**32.70**
3 Gang	22.85	27.04	0.77	19.55	nr	**46.59**
4 Gang	27.21	32.21	0.88	22.34	nr	**54.55**
6 Gang	45.88	54.30	1.18	29.96	nr	**84.26**
8 Gang	54.74	64.78	1.63	41.39	nr	**106.17**
9 Gang	68.65	81.25	1.83	46.47	nr	**127.71**
12 Gang	81.75	96.75	2.29	58.15	nr	**154.90**
6 Amp modular type switch; galvanised steel box; bronze or satin chrome coverplate; flush mounting; two way						
1 Gang	11.88	14.06	0.43	10.92	nr	**24.98**
2 Gang	16.68	19.74	0.55	13.97	nr	**33.70**
3 Gang	24.13	28.56	0.77	19.55	nr	**48.11**
4 Gang	28.92	34.22	0.88	22.34	nr	**56.57**
6 Gang	48.56	57.48	1.18	29.96	nr	**87.44**
8 Gang	58.15	68.81	1.63	41.39	nr	**110.20**
9 Gang	72.48	85.78	1.83	46.47	nr	**132.24**
12 Gang	86.86	102.79	2.22	56.37	nr	**159.16**
Plate switches; 10 Amp flush mounted, white plastic fronted; 16mm metal box; fitted brass earth terminal						
1 Gang 1 Way, Single Pole	1.71	2.02	0.28	7.11	nr	**9.13**
1 Gang 2 Way, Single Pole	1.93	2.28	0.33	8.38	nr	**10.66**
2 Gang 2 Way, Single Pole	2.79	3.31	0.44	11.17	nr	**14.48**
3 Gang 2 Way, Single Pole	5.55	6.57	0.56	14.22	nr	**20.79**
1 Gang Intermediate	6.42	7.60	0.43	10.92	nr	**18.52**
1 Gang 1 Way, Double Pole	5.73	6.78	0.33	8.38	nr	**15.16**
1 Gang Single Pole with bell symbol	4.63	5.48	0.23	5.84	nr	**11.32**
1 Gang Single Pole marked "PRESS"	3.81	4.50	0.23	5.84	nr	**10.34**
Time delay switch, suppressed	35.75	42.31	0.49	12.44	nr	**54.75**

V:ELECTRICAL SUPPLY/POWER/LIGHTING SYSTEMS

Item	Net Price £	Material £	Labour hours	Labour £	Unit	Total rate £
Plate switches; 6 Amp flush mounted white plastic fronted; 25mm metal box; fitted brass earth terminal						
4 Gang 2 Way, Single Pole	11.57	13.70	0.42	10.66	nr	**24.36**
6 Gang 2 Way, Single Way	22.03	26.08	0.47	11.93	nr	**38.01**
Architrave plate switches; 6 Amp flush mounted, white plastic fronted; 27mm metal box; brass earth terminal						
1 Gang 2 Way, Single Pole	2.97	3.51	0.30	7.62	nr	**11.13**
Ceiling switches, white moulded plastic, pull cord; standard unit						
6 Amp, 1 Way, Single Pole	3.26	3.85	0.32	8.13	nr	**11.98**
6 Amp, 2 Way, Single Pole	3.92	4.63	0.34	8.63	nr	**13.27**
16 Amp, 1 Way, Double Pole	5.87	6.95	0.37	9.39	nr	**16.35**
45 Amp, 1 Way, Double Pole with neon indicator	9.11	10.78	0.47	11.93	nr	**22.71**
10 Amp splash proof moulded switch with plain, threaded or PVC entry						
1 Gang,2 Way Single Pole	14.01	16.59	0.34	8.63	nr	**25.22**
2 Gang, 1 Way Single Pole	15.83	18.73	0.36	9.14	nr	**27.87**
2 Gang, 2 Way Single Pole	20.05	23.73	0.40	10.16	nr	**33.89**
6 Amp watertight switch; metalclad; BS 3676; ingress protected to IP65 surface mounted						
1 Gang, 2 Way; terminal entry	13.52	16.00	0.41	10.41	nr	**26.41**
1 Gang, 2 Way; through entry	13.52	16.00	0.42	10.66	nr	**26.66**
2 Gang, 2 Way; terminal entry	39.84	47.15	0.54	13.71	nr	**60.86**
2 Gang, 2 Way; through entry	39.84	47.15	0.53	13.46	nr	**60.61**
2 Way replacement switch	10.74	12.71	0.10	2.54	nr	**15.24**
15 Amp watertight switch; metalclad; BS 3676; ingress protected to IP65; surface mounted						
1 Gang 2 Way, terminal entry	18.09	21.41	0.42	10.66	nr	**32.08**
1 Gang 2 Way, through entry	18.09	21.41	0.43	10.92	nr	**32.33**
2 Gang 2 Way, terminal entry	42.48	50.28	0.55	13.97	nr	**64.24**
2 Gang 2 Way, through entry	42.48	50.28	0.54	13.71	nr	**63.99**
Intermediate interior only	10.74	12.71	0.11	2.79	nr	**15.50**
2 way interior only	10.74	12.71	0.11	2.79	nr	**15.50**
Double pole interior only	10.74	12.71	0.11	2.79	nr	**15.50**
Electrical accessories; fixed to backgrounds (including fixings)						
Dimmer switches; rotary action; for individual lights; moulded plastic case; metal backbox; flush mounted						
1 Gang, 1 Way; 250 Watt	11.26	13.33	0.28	7.11	nr	**20.44**
1 Gang, 1 Way; 400 Watt	14.83	17.55	0.28	7.11	nr	**24.66**
Dimmer switches; push on/off action; for individual lights; moulded plastic case; metal backbox; flush mounted						
1 Gang, 2 Way; 250 Watt	19.01	22.50	0.34	8.63	nr	**31.13**
3 Gang, 2 Way; 250 Watt	100.74	119.22	0.48	12.19	nr	**131.41**
4 Gang, 2 Way; 250 Watt	123.84	146.56	0.57	14.47	nr	**161.03**

V:ELECTRICAL SUPPLY/POWER/LIGHTING SYSTEMS

Item	Net Price £	Material £	Labour hours	Labour £	Unit	Total rate £
V21: GENERAL LIGHTING (cont'd)						
Y74 - LIGHTING ACCESSORIES (cont'd)						
SWITCHES (cont'd)						
Electrical accessories (cont'd)						
Dimmer switches; rotary action; metal clad; metal backbox; BS 5518 and BS 800; flush mounted						
1 Gang, 1 Way; 400 Watt	30.96	36.65	0.33	8.38	nr	**45.03**
Ceiling Roses						
Ceiling rose: white moulded plastic; flush fixed to conduit box						
Plug in type; ceiling socket with 2 terminals , loop-in and ceiling plug with 3 terminals and cover	6.08	7.20	0.34	8.63	nr	**15.83**
BC lampholder; white moulded plastic; heat resistant PVC insulated and sheathed cable; flush fixed						
2 Core; 0.75mm²	1.46	1.73	0.33	8.38	nr	**10.11**
Batten holder: white moulded plastic; 3 terminals; BS 5042; fixed to conduit						
Straight pattern; 2 terminals with loop-in and Earth	4.34	5.14	0.29	7.36	nr	**12.51**
Angled pattern; looped in terminal	4.34	5.14	0.29	7.36	nr	**12.51**
LIGHTING CONTROLS						
Lighting control system; including software, commissioning and testing. Typical component parts indicated. System requirements dependant on final lighting design						
CABLES						
Cable; Twin twisted bus; LSF sheathed; aluminium conductors	1.16	1.38	0.08	2.03	m	**3.41**
Cable; ELV 4 core 7/0.2; LSF sheathed; aluminium screened; copper conductor	1.52	1.80	0.15	3.81	m	**5.61**
EQUIPMENT						
Central supervisor controller including software	4992.00	5907.88	12.00	304.70	nr	**6212.58**
Area control unit	832.00	984.65	4.00	101.57	nr	**1086.21**
Lighting control module; plug in; 9 output, 9 channel switching						
Base and lid assembly	144.56	171.08	2.05	52.05	nr	**223.13**
Lighting control module; plug in; 9 output 9 channel dimming (DSI)						
Base and lid assembly	166.40	196.93	2.05	52.05	nr	**248.98**

V:ELECTRICAL SUPPLY/POWER/LIGHTING SYSTEMS

Item	Net Price £	Material £	Labour hours	Labour £	Unit	Total rate £
Lighting control module; plug in; 9 output 9 channel dimming (DALI)						
Base and lid assembly	197.60	233.85	2.05	52.05	nr	**285.91**
Lighting control module; hard wired; 4 circuit switching						
Base and lid assembly	156.00	184.62	1.85	46.97	nr	**231.60**
Presence detectors						
Flush mounted	52.00	61.54	0.60	15.23	nr	**76.78**
Universal presence detectors with photo cell; flush mounted	57.20	67.69	0.60	15.23	nr	**82.93**
Scene switch plate; anodised aluminium finish						
4 way	62.40	73.85	1.20	30.47	nr	**104.32**

V:ELECTRICAL SUPPLY/POWER/LIGHTING SYSTEMS

Item	Net Price £	Material £	Labour hours	Labour £	Unit	Total rate £
V22 : GENERAL LV POWER						
Y74 - ACCESSORIES						
OUTLETS						
Socket outlet: unswitched; 13 Amp metal clad; BS 1363; galvanised steel box and coverplate with white plastic inserts; fixed surface mounted						
1 Gang	5.30	6.27	0.41	10.41	nr	**16.69**
2 Gang	9.76	11.55	0.41	10.41	nr	**21.96**
Socket outlet: switched; 13 Amp metal clad; BS 1363; galvanised steel box and coverplate with white plastic inserts; fixed surface mounted						
1 Gang	6.02	7.12	0.43	10.92	nr	**18.04**
2 Gang	9.85	11.65	0.45	11.43	nr	**23.08**
Socket outlet: switched with neon indicator; 13 Amp metal clad; BS 1363; galvanised steel box and coverplate with white plastic inserts; fixed surface mounted						
1 Gang	10.97	12.98	0.43	10.92	nr	**23.90**
2 Gang	19.95	23.61	0.45	11.43	nr	**35.04**
Socket outlet: unswitched; 13 Amp; BS 1363; white moulded plastic box and coverplate; fixed surface mounted						
1 Gang	3.79	4.49	0.41	10.41	nr	**14.90**
2 Gang	7.49	8.87	0.41	10.41	nr	**19.28**
Socket outlet; switched; 13 Amp; BS 1363; white moulded plastic box and coverplate; fixed surface mounted						
1 Gang	4.57	5.40	0.43	10.92	nr	**16.32**
2 Gang	7.71	9.13	0.45	11.43	nr	**20.55**
Socket outlet: switched with neon indicator; 13 Amp; BS 1363; white moulded plastic box and coverplate; fixed surface mounted						
1 Gang	9.36	11.08	0.43	10.92	nr	**22.00**
2 Gang	12.63	14.94	0.45	11.43	nr	**26.37**
Socket outlet: switched; 13 Amp; BS 1363; galvanised steel box, white moulded coverplate; flush fitted						
1 Gang	4.57	5.40	0.43	10.92	nr	**16.32**
2 Gang	8.86	10.48	0.45	11.43	nr	**21.91**
Socket outlet: switched with neon indicator; 13 Amp; BS 1363; galvanised steel box, white moulded coverplate; flush fixed						
1 Gang	9.36	11.08	0.43	10.92	nr	**22.00**
2 Gang	16.20	19.18	0.45	11.43	nr	**30.60**
Socket outlet: switched; 13 Amp; BS 1363; galvanised steel box, satin chrome coverplate; BS 4662; flush fixed						
1 Gang	15.29	18.10	0.43	10.92	nr	**29.01**
2 Gang	21.69	25.67	0.45	11.43	nr	**37.10**

V:ELECTRICAL SUPPLY/POWER/LIGHTING SYSTEMS

Item	Net Price £	Material £	Labour hours	Labour £	Unit	Total rate £
Socket outlet: switched with neon indicator; 13 Amp; BS 1363; steel backbox, satin chrome coverplate; BS 4662; flush fixed						
1 Gang	12.02	14.23	0.43	10.92	nr	**25.15**
2 Gang	21.69	25.67	0.45	11.43	nr	**37.10**
RCD protected socket outlets, 13 amp, to BS 1363; galvanised steel box, white moulded cover plate; flush fitted						
2 gang, 10 mA tripping (active control)	55.86	66.11	0.45	11.43	nr	**77.53**
2 gang, 30 mA tripping (active control)	49.75	58.88	0.45	11.43	nr	**70.30**
2 gang, 30 mA tripping (passive control)	49.75	58.88	0.45	11.43	nr	**70.30**
Filtered socket outlets, 13 amp, to BS 1363, with separate 'clean earth' terminal; galvanised steel box, white moulded cover plate; flush fitted						
2 gang (spike protected)	42.07	49.79	0.50	12.70	nr	**62.48**
2 gang (spike and RFI protected)	53.25	63.02	0.55	13.97	nr	**76.99**
Replacement filter cassette	14.75	17.46	0.15	3.81	nr	**21.26**
Non-standard socket outlets, 13 amp, to BS 1363, with separate 'clean earth' terminal; for plugs with T-shaped earth pin; galvanised steel box, white moulded cover plate; flush fitted						
1 gang	7.81	9.24	0.43	10.92	nr	**20.16**
2 gang	14.05	16.63	0.43	10.92	nr	**27.55**
2 gang coloured RED	19.51	23.09	0.43	10.92	nr	**34.01**
Weatherproof socket outlet: 40 Amp; switched; single gang; RCD protected; water and dust protected to I.P.66; surface mounted						
40A 30mA tripping current protecting 1 socket	63.56	75.22	0.52	13.20	nr	**88.42**
40A 30mA tripping current protecting 2 sockets	67.79	80.23	0.64	16.25	nr	**96.48**
Plug for weatherproof socket outlet: protected to I.P.66						
13Amp plug	2.78	3.29	0.21	5.33	nr	**8.63**
Floor service outlet box; comprising flat lid with flanged carpet trim; twin 13A switched socket outlets; punched plate for mounting 2 telephone outlets; one blank plate; triple compartment						
3 compartment	26.62	31.50	0.88	22.34	nr	**53.85**
Floor service outlet box; comprising flat lid with flanged carpet trim; twin 13A switched socket outlets; punched plate for mounting 2 telephone outlets; two blank plates; four compartment						
4 compartment	36.30	42.96	0.88	22.34	nr	**65.30**
Floor service outlet box; comprising flat lid with flanged carpet trim; single 13A unswitched socket outlet; single compartment; circular						
1 compartment	38.72	45.82	0.79	20.06	nr	**65.88**
Floor service grommet, comprising flat lid with flanged carpet trim; circular						
Floor Grommet	19.36	22.91	0.49	12.44	nr	**35.35**

V:ELECTRICAL SUPPLY/POWER/LIGHTING SYSTEMS

Item	Net Price £	Material £	Labour hours	Labour £	Unit	Total rate £
V22 : GENERAL LV POWER (cont'd)						
Y74 – ACCESSORIES (cont'd)						
POWER POSTS/POLES/PILLARS						
Power Post						
Power post; aluminium painted body; PVC-U cover; 5 nr outlets	235.97	279.27	4.00	101.57	nr	380.83
Power Pole						
Power pole; 3.6 metres high; aluminium painted body; PVC-U cover; 6 nr outlets	302.57	358.08	4.00	101.57	nr	459.64
Extra for						
Power pole extension bar; 900mm long	33.07	39.13	1.50	38.09	nr	77.22
Vertical multi compartment pillar; PVC-U; BS 4678 Part4 EN60529; excludes accessories						
Single						
630mm long	105.39	124.73	2.00	50.78	nr	175.51
3000mm long	303.71	359.43	2.00	50.78	nr	410.21
Double						
630mm long	105.39	124.73	3.00	76.17	nr	200.90
3000mm long	322.34	381.48	3.00	76.17	nr	457.66
CONNECTION UNITS						
Connection units: moulded pattern; BS 5733; moulded plastic box; white coverplate; knockout for flex outlet; surface mounted - standard fused						
DP Switched	5.84	6.91	0.49	12.44	nr	19.35
Unswitched	5.36	6.34	0.49	12.44	nr	18.78
DP Switched with neon indicator	7.41	8.77	0.49	12.44	nr	21.22
Connection units: moulded pattern; BS 5733; galvanised steel box; white coverplate; knockout for flex outlet; surface mounted						
DP Switched	7.36	15.62	0.49	12.44	nr	28.06
DP Unswitched	6.88	8.14	0.49	12.44	nr	20.58
DP Switched with neon indicator	8.93	10.57	0.49	12.44	nr	23.01
Connection units: galvanised pressed steel pattern; galvanised steel box; satin chrome or satin brass finish; white moulded plastic inserts; flush mounted - standard fused						
DP Switched	10.77	12.74	0.49	12.44	nr	25.19
Unswitched	10.13	11.99	0.49	12.44	nr	24.43
DP Switched with neon indicator	14.53	17.20	0.49	12.44	nr	29.64
Connection units: galvanised steel box; satin chrome or satin brass finish; white moulded plastic inserts; flex outlet; flush mounted - standard fused						
Switched	10.33	12.22	0.49	12.44	nr	24.67
Unswitched	9.81	11.61	0.49	12.44	nr	24.05
Switched with neon indicator	13.30	15.74	0.49	12.44	nr	28.18

V:ELECTRICAL SUPPLY/POWER/LIGHTING SYSTEMS

Item	Net Price £	Material £	Labour hours	Labour £	Unit	Total rate £
SHAVER SOCKETS						
Shaver unit: self setting overload device; 200/250 voltage supply; white moulded plastic faceplate; unswitched						
Surface type with moulded plastic box	19.43	22.99	0.55	13.97	nr	**36.96**
Flush type with galvanised steel box	20.33	24.06	0.57	14.47	nr	**38.53**
Shaver unit: dual voltage supply unit; white moulded plastic faceplate; unswitched						
Surface type with moulded plastic box	23.42	27.72	0.62	15.74	nr	**43.46**
Flush type with galvanised steel box	24.35	28.82	0.64	16.25	nr	**45.07**
COOKER CONTROL UNITS						
Cooker control unit: BS 4177; 45 amp D.P. main switch; 13 Amp switched socket outlet; metal coverplate; plastic inserts; neon indicators						
Surface mounted with mounting box	27.50	32.55	0.61	15.49	nr	**48.03**
Flush mounted with galvanised steel box	26.29	31.11	0.61	15.49	nr	**46.60**
Cooker control unit: BS 4177; 45 Amp D.P. main switch; 13 Amp switched socket outlet; moulded plastic box and coverplate; surface mounted						
Standard	19.09	22.59	0.61	15.49	nr	**38.08**
With neon indicators	22.39	26.49	0.61	15.49	nr	**41.98**
CONTROL COMPONENTS						
Connector unit : moulded white plastic cover and block; galvanised steel back box; to immersion heaters						
3Kw up to 915mm long; fitted to thermostat	20.46	24.21	0.75	19.05	nr	**43.26**
Water heater switch : 20 Amp; switched with neon indicator						
DP Switched with neon indicator	10.15	25.84	0.45	11.43	nr	**37.27**
SWITCH DISCONNECTORS						
Switch disconnectors; moulded plastic enclosure; fixed to backgrounds						
3 pole; IP54; Grey						
16 Amp	16.43	19.45	0.80	20.31	nr	**39.76**
25 Amp	19.44	23.00	0.80	20.31	nr	**43.32**
40 Amp	31.66	37.47	0.80	20.31	nr	**57.78**
63 Amp	49.29	58.33	1.00	25.39	nr	**83.73**
80 Amp	85.44	101.11	1.25	31.74	nr	**132.85**
6 pole; IP54; Grey						
25 Amp	27.31	32.32	1.00	25.39	nr	**57.72**
63 Amp	46.13	54.60	1.25	31.74	nr	**86.34**
80 Amp	87.45	103.49	1.80	45.70	nr	**149.20**
3 pole; IP54; Yellow						
16 Amp	18.06	21.38	0.80	20.31	nr	**41.69**
25 Amp	21.34	25.26	0.80	20.31	nr	**45.57**
40 Amp	34.63	40.98	0.80	20.31	nr	**61.29**
63 Amp	53.87	63.75	1.00	25.39	nr	**89.14**

V:ELECTRICAL SUPPLY/POWER/LIGHTING SYSTEMS

Item	Net Price £	Material £	Labour hours	Labour £	Unit	Total rate £
V22 : GENERAL LV POWER (cont'd)						
Y74 – ACCESSORIES (cont'd)						
SWITCH DISCONNECTORS (cont'd)						
6 pole; IP54; Yellow						
25 Amp	27.31	32.32	1.00	25.39	nr	**57.72**
INDUSTRIAL SOCKETS/PLUGS						
Plugs; Splashproof; 100-130 volts, 50-60 Hz; IP 44 (Yellow)						
2 pole and earth						
16 Amp	2.71	3.20	0.55	13.97	nr	**17.17**
32 Amp	8.55	10.12	0.60	15.23	nr	**25.35**
3 pole and earth						
16 Amp	9.37	11.09	0.65	16.50	nr	**27.60**
32 Amp	12.57	14.88	0.72	18.28	nr	**33.16**
3 pole; neutral and earth						
16 Amp	10.05	11.90	0.72	18.28	nr	**30.18**
32 Amp	14.99	17.74	0.78	19.81	nr	**37.55**
Connectors; Splashproof; 100-130 volts, 50-60 Hz; IP 44 (Yellow)						
2 pole and earth						
16 Amp	3.90	4.62	0.42	10.66	nr	**15.29**
32 Amp	11.19	13.24	0.50	12.70	nr	**25.94**
3 pole and earth						
16 Amp	11.61	13.73	0.48	12.19	nr	**25.92**
32 Amp	16.11	19.07	0.58	14.73	nr	**33.80**
3 pole; neutral and earth						
16 Amp	14.51	17.17	0.52	13.20	nr	**30.37**
32 Amp	18.28	21.64	0.73	18.54	nr	**40.17**
Angled sockets; surface mounted; Splashproof; 100-130 volts, 50-60 Hz; IP 44 (Yellow)						
2 pole and earth						
16 Amp	5.20	6.16	0.55	13.97	nr	**20.12**
32 Amp	6.20	7.34	0.60	15.23	nr	**22.58**
3 pole and earth						
16 Amp	11.96	14.15	0.65	16.50	nr	**30.66**
32 Amp	22.32	26.41	0.72	18.28	nr	**44.70**
3 pole; neutral and earth						
16 Amp	16.52	19.55	0.72	18.28	nr	**37.84**
32 Amp	20.43	24.17	0.78	19.81	nr	**43.98**

V:ELECTRICAL SUPPLY/POWER/LIGHTING SYSTEMS

Item	Net Price £	Material £	Labour hours	Labour £	Unit	Total rate £
Plugs; Watertight; 100-130 volts, 50-60 Hz; IP67 (Yellow)						
2 pole and earth						
16 Amp	9.78	11.57	0.55	13.97	nr	**25.54**
32 Amp	18.02	21.32	0.60	15.23	nr	**36.56**
63 Amp	39.41	46.64	0.75	19.04	nr	**65.69**
Connectors; Watertight; 100-130 volts, 50-60 Hz; IP 67 (Yellow)						
2 pole and earth						
16 Amp	14.53	17.20	0.42	10.66	nr	**27.86**
32 Amp	24.65	29.17	0.50	12.70	nr	**41.87**
63 Amp	59.65	70.60	0.67	17.01	nr	**87.61**
Angled sockets; surface mounted; Watertight; 100-130 volts, 50-60 Hz; IP 67 (Yellow)						
2 pole and earth						
16 Amp	25.77	30.50	0.55	13.97	nr	**44.47**
32 Amp	37.76	44.69	0.60	15.23	nr	**59.93**
Plugs; Splashproof; 200-250 volts, 50-60 Hz; IP 44 (Blue)						
2 pole and earth						
16 Amp	2.71	3.20	0.55	13.97	nr	**17.17**
32 Amp	6.36	7.52	0.60	15.23	nr	**22.76**
63 Amp	30.68	36.31	0.75	19.04	nr	**55.35**
3 pole and earth						
16 Amp	8.03	9.50	0.65	16.50	nr	**26.01**
32 Amp	10.93	12.94	0.72	18.28	nr	**31.22**
63 Amp	30.80	36.45	0.83	21.07	nr	**57.53**
3 pole; neutral and earth						
16 Amp	8.76	10.36	0.72	18.28	nr	**28.64**
32 Amp	12.65	14.97	0.78	19.81	nr	**34.78**
Connectors; Splashproof; 200-250 volts, 50-60 Hz; IP 44 (Blue)						
2 pole and earth						
16 Amp	4.24	5.01	0.42	10.66	nr	**15.68**
32 Amp	10.53	12.46	0.50	12.70	nr	**25.15**
63 Amp	37.17	43.99	0.67	17.01	nr	**61.00**
3 pole and earth						
16 Amp	11.22	13.28	0.48	12.19	nr	**25.47**
32 Amp	14.83	17.55	0.58	14.73	nr	**32.28**
63 Amp	30.58	36.19	0.75	19.04	nr	**55.23**
3 pole; neutral and earth						
16 Amp	13.04	15.43	0.52	13.20	nr	**28.63**
32 Amp	44.66	52.85	0.73	18.54	nr	**71.39**

V:ELECTRICAL SUPPLY/POWER/LIGHTING SYSTEMS

Item	Net Price £	Material £	Labour hours	Labour £	Unit	Total rate £
V22 : GENERAL LV POWER (cont'd)						
Y74 – ACCESSORIES (cont'd)						
INDUSTRIAL SOCKETS / PLUGS (cont'd)						
Angled sockets; surface mounted; Splashproof; 200-250 volts, 50-60 Hz; IP 44 (Blue)						
2 pole and earth						
16 Amp	5.46	6.46	0.55	13.97	nr	**20.42**
32 Amp	8.04	9.52	0.60	15.23	nr	**24.75**
63 Amp	41.55	49.17	0.75	19.04	nr	**68.21**
3 pole and earth						
16 Amp	10.82	12.81	0.65	16.50	nr	**29.31**
32 Amp	18.74	22.18	0.72	18.28	nr	**40.46**
63 Amp	41.29	48.87	0.83	21.07	nr	**69.95**
3 pole; neutral and earth						
16 Amp	9.43	11.16	0.72	18.28	nr	**29.44**
32 Amp	16.50	19.53	0.78	19.81	nr	**39.33**
Plugs; Watertight; 200-250 volts, 50-60 Hz; IP67 (Blue)						
2 pole and earth						
16 Amp	9.80	11.60	0.41	10.41	nr	**22.01**
32 Amp	16.87	19.97	0.50	12.70	nr	**32.67**
63 Amp	45.64	54.01	0.66	16.76	nr	**70.77**
125 Amp	116.38	137.73	0.86	21.84	nr	**159.57**
Connectors; Watertight; 200-250 volts, 50-60 Hz; IP 67 (Blue)						
2 pole and earth						
16 Amp	14.34	16.98	0.42	10.66	nr	**27.64**
32 Amp	22.86	27.05	0.50	12.70	nr	**39.75**
63 Amp	48.02	56.82	0.67	17.01	nr	**73.84**
125 Amp	142.76	168.95	0.87	22.09	nr	**191.04**
Angled sockets; surface mounted; Watertight; 200-250 volts, 50-60 Hz; IP 67 (Blue)						
2 pole and earth						
16 Amp	15.62	18.49	0.55	13.97	nr	**32.45**
32 Amp	31.63	37.43	0.60	15.23	nr	**52.66**
125 Amp	234.77	277.85	1.00	25.39	nr	**303.24**

V:ELECTRICAL SUPPLY/POWER/LIGHTING SYSTEMS

Item	Net Price £	Material £	Labour hours	Labour £	Unit	Total rate £
V32 : UNINTERRUPTIBLE POWER SUPPLY						
Uninterruptible power supply; sheet steel enclosure; self contained battery pack; including installation, testing and commissioning.						
Single phase input and output; 5 year battery life; standard 13A socket outlet connection						
1.0kVA (10 minute supply)	818.89	969.14	0.30	7.62	nr	**976.76**
1.0kVA (30 minute supply)	1433.29	1696.26	0.50	12.70	nr	**1708.95**
2.0kVA (10 minute supply)	1546.55	1830.29	0.50	12.70	nr	**1842.99**
2.0kVA (60 minute supply)	2596.44	3072.81	0.50	12.70	nr	**3085.50**
3.0kVA (10 minute supply)	2020.10	2390.72	0.50	12.70	nr	**2403.42**
3.0kVA (40 minute supply)	3069.99	3633.24	1.00	25.39	nr	**3658.63**
5.0kVA (30 minute supply)	4877.56	5772.45	1.00	25.39	nr	**5797.84**
8.0kVA (10 minute supply)	6226.60	7369.00	2.00	50.78	nr	**7419.78**
8.0kVA (30 minute supply)	7299.60	8638.86	2.00	50.78	nr	**8689.64**
Uninterruptible power supply; including final connections and testing and commissioning						
Medium size static; single phase input and output; 10 year battery life; in cubicle						
10.0 kVA (10 minutes supply)	6650.45	7870.61	10.00	253.91	nr	**8124.52**
10.0 kVA (30 minutes supply)	7886.70	9333.67	15.00	380.87	nr	**9714.54**
15.0 kVA (10 minutes supply)	8596.25	10173.40	10.00	253.91	nr	**10427.32**
15.0 kVA (30 minutes supply)	10161.40	12025.71	15.00	380.87	nr	**12406.58**
20.0 kVA (10 minutes supply)	10908.90	12910.36	10.00	253.91	nr	**13164.27**
20.0 kVA (30 minutes supply)	11677.10	13819.50	15.00	380.87	nr	**14200.37**
Medium size static; three phase input and output; 10 year battery life; in cubicle						
10.0 kVA (10 minutes supply)	8221.35	9729.72	10.00	253.91	nr	**9983.64**
10.0 kVA (30 minutes supply)	9494.40	11236.34	15.00	380.87	nr	**11617.21**
15.0 kVA (10 minutes supply)	9155.15	10834.85	15.00	380.87	nr	**11215.72**
15.0 kVA (30 minutes supply)	10882.45	12879.05	20.00	507.83	nr	**13386.88**
20.0 kVA (10 minutes supply)	11274.60	13343.15	20.00	507.83	nr	**13850.98**
20.0 kVA (30 minutes supply)	12022.10	14227.79	25.00	634.78	nr	**14862.58**
30.0 kVA (10 minutes supply)	12229.10	14472.77	25.00	634.78	nr	**15107.56**
30.0 kVA (30 minutes supply)	13091.60	15493.52	30.00	761.74	nr	**16255.26**
Large size static; three phase input and output; 10 year battery life; in cubicle						
40 kVA (10 minutes supply)	12114.10	14336.67	30.00	761.74	nr	**15098.42**
40 kVA (30 minutes supply)	15893.00	18808.89	30.00	761.74	nr	**19570.63**
60 kVA (10 minutes supply)	17730.70	20983.75	35.00	888.70	nr	**21872.45**
60 kVA (30 minutes supply)	22496.30	26623.70	35.00	888.70	nr	**27512.40**
100 kVA (10 minutes supply)	20411.35	24156.22	40.00	1015.66	nr	**25171.88**
200 kVA (10 minutes supply)	35034.75	41462.58	40.00	1015.66	nr	**42478.23**
300 kVA (10 minutes supply)	51909.85	61433.75	40.00	1015.66	nr	**62449.41**
400 kVA (10 minutes supply)	68602.10	81188.53	50.00	1269.57	nr	**82458.10**
500 kVA (10 minutes supply)	83697.00	99052.89	60.00	1523.48	nr	**100576.37**
600 kVA (10 minutes supply)	96410.25	114098.64	70.00	1777.40	nr	**115876.04**
800 kVA (10 minutes supply)	125761.70	148835.20	80.00	2031.31	nr	**150866.51**

V:ELECTRICAL SUPPLY/POWER/LIGHTING SYSTEMS

Item	Net Price £	Material £	Labour hours	Labour £	Unit	Total rate £
V32 : UNINTERRUPTIBLE POWER SUPPLY (cont'd)						
Integral diesel rotary; three phase input and output; no break supply; including ventilation and acoustic attenuation, oil day tank and interconnecting pipework						
100 kVA	117341.40	138870.03	100.00	2539.14	nr	**141409.17**
125 kVA	133220.60	157662.58	100.00	2539.14	nr	**160201.72**
150 kVA	143238.25	169518.17	100.00	2539.14	nr	**172057.31**
180 kVA	153487.05	181647.32	100.00	2539.14	nr	**184186.46**
200 kVA	226138.30	267627.89	100.00	2539.14	nr	**270167.03**
250 kVA	233110.75	275879.58	100.00	2539.14	nr	**278418.72**
300 kVA	239855.50	283861.79	120.00	3046.97	nr	**286908.76**
400 kVA	282844.80	334738.34	120.00	3046.97	nr	**337785.30**
500 kVA	310608.10	367595.37	120.00	3046.97	nr	**370642.34**
630 kVA	375761.35	444702.28	140.00	3554.80	nr	**448257.08**
800 kVA	453684.20	536921.64	140.00	3554.80	nr	**540476.44**
1000 kVA	509834.10	603373.36	140.00	3554.80	nr	**606928.16**
1125 kVA	574921.80	680402.70	160.00	4062.62	nr	**684465.33**
1250 kVA	605174.85	716206.28	160.00	4062.62	nr	**720268.90**
1500 kVA	664205.50	786067.28	170.00	4316.54	nr	**790383.82**
1750 kVA	753651.35	891923.76	170.00	4316.54	nr	**896240.30**

V:ELECTRICAL SUPPLY/POWER/LIGHTING SYSTEMS

Item	Net Price £	Material £	Labour hours	Labour £	Unit	Total rate £
V40 : EMERGENCY LIGHTING						
Y73 - LUMINAIRES						
24 Volt/50 Volt/110 volt fluorescent slave luminaires						
For use with DC central battery systems						
Indoor, 8 Watt	32.02	37.90	0.80	20.31	nr	**58.21**
Indoor, exit sign box	39.90	47.22	0.80	20.31	nr	**67.53**
Outdoor, 8 Watt weatherproof	35.70	42.25	0.80	20.31	nr	**62.56**
Conversion module AC/DC	39.90	47.22	0.25	6.35	nr	**53.57**
Self contained; polycarbonate base and diffuser; LED charging light to European sign directive; 3 hour standby						
Non maintained						
Indoor, 8 Watt	35.36	41.85	1.00	25.39	nr	**67.24**
Outdoor, 8 Watt weatherproof, vandal resistant IP65	50.72	60.02	1.00	25.39	nr	**85.41**
Maintained						
Indoor, 8 Watt	51.58	61.04	1.00	25.39	nr	**86.43**
Outdoor, 8 Watt weatherproof, vandal resistant IP65	74.97	88.72	1.00	25.39	nr	**114.12**
Exit signage						
Exit sign; gold effect, pendular including brackets						
Non maintained, 8 Watt	96.25	113.91	1.00	25.39	nr	**139.30**
Maintained, 8 Watt	103.60	122.61	1.00	25.39	nr	**148.00**
Modification kit						
Module & battery for 58W fluorescent modification from mains fitting to emergency; 3 hour standby	33.18	39.27	0.50	12.70	nr	**51.96**
Extra for remote box (when fitting is too small for modification)	16.48	19.51	0.50	12.70	nr	**32.21**
12 Volt low voltage lighting; non maintained; 3 hour standby						
2 x 20 Watt lamp load	114.66	135.70	1.20	30.47	nr	**166.17**
1 x 50 Watt lamp load	125.16	148.12	1.00	25.39	nr	**173.51**
Maintained 3 hour standby						
2 x 20 Watt lamp load	108.93	128.91	1.20	30.47	nr	**159.38**
1 x 50 Watt lamp load	118.90	140.72	1.00	25.39	nr	**166.11**

V:ELECTRICAL SUPPLY/POWER/LIGHTING SYSTEMS

Item	Net Price £	Material £	Labour hours	Labour £	Unit	Total rate £
V40 : EMERGENCY LIGHTING (cont'd)						
Y71 – LV SWITCHGEAR						
DC central battery systems BS5266 compliant 24/50/110 Volt						
DC supply to luminaires on mains failure; metal cubicle with battery charger, changeover device and battery as integral unit; ICEL 1001 compliant; 10 year design life valve regulated lead acid battery; 24 hour recharge; LCD display & LED indication; ICEL alarm pack; Includes on-site commissioning on 110 Volt systems only						
24 Volt, wall mounted						
300 W maintained, 1 hour	1877.40	2221.85	4.00	101.57	nr	**2323.41**
635 W maintained, 3 hour	2277.45	2695.29	6.00	152.35	nr	**2847.64**
470 W non maintained, 1 hour	1673.70	1980.77	4.00	101.57	nr	**2082.34**
780 W non maintained, 3 hour	1978.20	2341.14	6.00	152.35	nr	**2493.49**
50 Volt						
935 W maintained, 3 hour	2035.95	2409.49	8.00	203.13	nr	**2612.62**
1965 W maintained, 3 hour	2503.20	2962.46	8.00	203.13	nr	**3165.59**
1311 W non maintained, 3 hour	2450.70	2900.33	8.00	203.13	nr	**3103.46**
2510 W non maintained 3, hour	3742.20	4428.78	8.00	203.13	nr	**4631.91**
110 Volt						
1603 W maintained, 3 hour	3526.95	4174.04	8.00	203.13	nr	**4377.17**
4446 W maintained, 3 hour	4027.80	4766.78	10.00	253.91	nr	**5020.69**
2492 W non maintained, 3 hour	3665.55	4338.07	10.00	253.91	nr	**4591.98**
5429 W non maintained, 3 hour	6106.80	7227.21	12.00	304.70	nr	**7531.91**
DC central battery systems; BS EN 50171 compliant; 24/50/110 Volt Central power systems						
DC supply to luminaires on mains failure; metal cubicle with battery charger, changeover device and battery as integral unit; 10 year design life valve regulated lead acid battery; 12 hour recharge to 80% of specified duty; low volts discount; LCD display & LED indication; includes on-site commissioning for CPS systems only; battery sized for 'end of life' @ 20°C test pushbutton						
24 Volt, floor standing						
400 W non maintained, 1 hour	1762.95	2086.40	4.00	101.57	nr	**2187.96**
600 W maintained, 3 hour	2266.95	2682.87	6.00	152.35	nr	**2835.22**
50 Volt						
2133 W non maintained, 3 hour	3962.70	4689.74	8.00	203.13	nr	**4892.87**
1900 W maintained, 3 hour	3787.35	4482.22	8.00	203.13	nr	**4685.35**
110 Volt						
2200 W non maintained, 3 hour	3665.55	4338.07	8.00	203.13	nr	**4541.20**
4000 W maintained, 3 hour	4090.80	4841.34	12.00	304.70	nr	**5146.04**

V:ELECTRICAL SUPPLY/POWER/LIGHTING SYSTEMS

Item	Net Price £	Material £	Labour hours	Labour £	Unit	Total rate £
Low Power Systems						
DC supply to luminaires on mains failure; metal cubicle with battery charger, changeover device and battery as integral unit; 5 Year design life valve regulated lead acid battery; low volts discount; LED display & LED indication; battery sized for 'end of life' @ 20°C test pushbutton						
24 Volt, floor standing						
300 W non maintained, 1 hour	1877.40	2221.85	4.00	101.57	nr	**2323.41**
600 W maintained, 3 hour	2077.95	2459.19	6.00	152.35	nr	**2611.54**
AC static inverter system; BS5266 compliant; one hour standby						
Central system supplying AC power on mains failure to mains luminaires; ICEL 1001 compliant metal cubicle(s) with changeover device, battery charger, battery & static inverter; 10 year design Life valve regulated lead acid battery; 24 hour recharge; LED indication and LCD display; pure sinewave output						
One hour						
750 VA, 600 W single phase I/P & O/P	2285.85	2705.23	6.00	152.35	nr	**2857.58**
3 KVA, 2.55 KW single phase I/P & O/P	3750.60	4438.72	8.00	203.13	nr	**4641.85**
5 KVA, 4.25 KW single phase I/P & O/P	4951.80	5860.31	10.00	253.91	nr	**6114.22**
8 KVA, 6.80 KW single phase I/P & O/P	5833.80	6904.13	12.00	304.70	nr	**7208.82**
10 KVA, 8.5 KW single phase I/P & O/P	7530.60	8912.24	14.00	355.48	nr	**9267.72**
13 KVA, 11.05 KW single phase I/P & O/P	10074.75	11923.16	16.00	406.26	nr	**12329.43**
15 KVA, 12.75 KW single phase I/P & O/P	10441.20	12356.85	30.00	761.74	nr	**13118.59**
20 KVA, 17.0 KW 3 phase I/P & single phase O/P	14420.70	17066.47	40.00	1015.66	nr	**18082.12**
30 KVA, 25.5 KW 3 phase I/P & O/P	19488.00	23063.46	60.00	1523.48	nr	**24586.95**
40 KVA, 34.0 KW 3 phase I/P & O/P	24294.90	28752.29	80.00	2031.31	nr	**30783.60**
50 KVA, 42.5 KW 3 phase I/P & O/P	31685.85	37499.25	90.00	2285.23	nr	**39784.48**
65 KVA, 55.25 KW 3 phase I/P & O/P	38869.95	46001.42	100.00	2539.14	nr	**48540.56**
90 KVA, 68.85 KW 3 phase I/P & O/P	51364.95	60788.88	120.00	3046.97	nr	**63835.85**
120 KVA, 102 KW 3 phase I/P & O/P	57336.30	67855.79	150.00	3808.71	nr	**71664.50**
Three hour						
750 VA, 600 W single phase I/P & O/P	2675.40	3166.26	6.00	152.35	nr	**3318.60**
3 KVA, 2.55 KW single phase I/P & O/P	4501.35	5327.21	8.00	203.13	nr	**5530.34**
5 KVA, 4.25 KW single phase I/P & O/P	7153.65	8466.13	10.00	253.91	nr	**8720.04**
8 KVA, 6.80 KW single phase I/P & O/P	8877.75	10506.55	12.00	304.70	nr	**10811.25**
10 KVA, 8.5 KW single phase I/P & O/P	11710.65	13859.20	14.00	355.48	nr	**14214.68**
13 KVA, 11.05 KW single phase I/P & O/P	14439.60	17088.83	16.00	406.26	nr	**17495.10**
15 KVA, 12.75 KW single phase I/P & O/P	14675.85	17368.43	30.00	761.74	nr	**18130.17**
20 KVA, 17.0 KW 3 phase I/P & single phase O/P	21367.50	25287.80	40.00	1015.66	nr	**26303.45**
30 KVA, 25.5 KW 3 phase I/P & O/P	31934.70	37793.76	60.00	1523.48	nr	**39317.24**
40 KVA, 34.0 KW 3 phase I/P & O/P	38181.15	45186.25	80.00	2031.31	nr	**47217.56**
50 KVA, 42.5 KW 3 phase I/P & O/P	51497.25	60945.45	90.00	2285.23	nr	**63230.68**
65 KVA, 55.25 KW 3 phase I/P & O/P	62265.00	73688.76	100.00	2539.14	nr	**76227.90**
90 KVA, 68.85 KW 3 phase I/P & O/P	73690.05	87209.96	120.00	3046.97	nr	**90256.93**
120 KVA, 102 KW 3 phase I/P & O/P	90825.00	107488.66	150.00	3808.71	nr	**111297.37**

V:ELECTRICAL SUPPLY/POWER/LIGHTING SYSTEMS

Item	Net Price £	Material £	Labour hours	Labour £	Unit	Total rate £
V40 : EMERGENCY LIGHTING (cont'd)						
Y71 – LV SWITCHGEAR (cont'd)						
AC static inverter system; BS EN 50171 compliant; one hour standby; Low power system (typically wall mounted)						
Central system supplying AC power on mains failure to mains luminaires; metal cubicle(s) with changeover device, battery charger, battery & static inverter; 5 year design life valve regulated lead acid battery; LED indication and LCD display; 12 hour recharge to 80% duty; inverter rated for 120% of load for 100% of duty; battery sized for 'end of life' @ 20°C test pushbutton						
One hour						
300 VA, 240 W single phase I/P & O/P	1029.00	1217.79	3.00	76.17	nr	1293.96
600 VA, 480 W single phase I/P & O/P	1239.00	1466.32	4.00	101.57	nr	1567.88
750 VA, 600 W single phase I/P & O/P	2285.85	2705.23	6.00	152.35	nr	2857.58
Three hour						
150 VA, 120 W single phase I/P & O/P	1134.00	1342.06	3.00	76.17	nr	1418.23
450 VA, 360 W single phase I/P & O/P	1344.00	1590.58	4.00	101.57	nr	1692.15
750 VA, 600 W single phase I/P & O/P	2675.40	3166.26	6.00	152.35	nr	3318.60
AC static inverter system central power system; CPS BS EN 50171 compliant; one hour standby						
Central system supplying AC power on mains failure to mains luminaires; metal cubicle(s) with changeover device, battery charger, battery & static inverter; LED indication and LCD display; pure sinewave output; 10 year design life valve regulated lead acid battery; 12 hour recharge to 80% duty specified; inverter rated for 120% of load for 100% of duty; battery sized for 'end of life' @ 20°C test push button; includes on-site commissioning						
One hour						
750 VA, 600 W single phase I/P & O/P	2558.85	3028.32	6.00	152.35	nr	3180.67
3 KVA, 2.55 KW single phase I/P & O/P	4076.10	4823.94	8.00	203.13	nr	5027.07
5 KVA, 4.25 KW single phase I/P & O/P	5277.30	6245.53	10.00	253.91	nr	6499.44
8 KVA, 6.80 KW single phase I/P & O/P	6159.30	7289.35	12.00	304.70	nr	7594.04
10 KVA, 8.5 KW single phase I/P & O/P	7908.60	9359.59	14.00	355.48	nr	9715.07
13 KVA, 11.05 KW single phase I/P & O/P	10452.75	12370.52	16.00	406.26	nr	12776.78
15 KVA, 12.75 KW single phase I/P & O/P	10819.20	12804.20	30.00	761.74	nr	13565.94
20 KVA, 17.0 KW 3 phase I/P & single phase O/P	14903.70	17638.08	40.00	1015.66	nr	18653.74
30 KVA, 25.5 KW 3 phase I/P & O/P	20076.00	23759.34	60.00	1523.48	nr	25282.83
40 KVA, 34.0 KW 3 phase I/P & O/P	24882.90	29448.17	80.00	2031.31	nr	31479.48
50 KVA, 42.5 KW 3 phase I/P & O/P	32405.10	38350.46	90.00	2285.23	nr	40635.69
65 KVA, 55.25 KW 3 phase I/P & O/P	40024.95	47368.33	100.00	2539.14	nr	49907.47
90 KVA, 68.85 KW 3 phase I/P & O/P	52519.95	62155.79	120.00	3046.97	nr	65202.75
120 KVA, 102 KW 3 phase I/P & O/P	58491.30	69222.70	150.00	3808.71	nr	73031.41

V:ELECTRICAL SUPPLY/POWER/LIGHTING SYSTEMS

Item	Net Price £	Material £	Labour hours	Labour £	Unit	Total rate £
Three hour						
750 VA, 600 W single phase I/P & O/P	2948.40	3489.34	6.00	152.35	nr	**3641.69**
3 KVA, 2.55 KW single phase I/P & O/P	4826.85	5712.43	8.00	203.13	nr	**5915.56**
5 KVA, 4.25 KW single phase I/P & O/P	7479.15	8851.35	10.00	253.91	nr	**9105.26**
8 KVA, 6.80 KW single phase I/P & O/P	9203.25	10891.77	12.00	304.70	nr	**11196.47**
10 KVA, 8.5 KW single phase I/P & O/P	12088.65	14306.55	14.00	355.48	nr	**14662.03**
13 KVA, 11.05 KW single phase I/P & O/P	14817.60	17536.19	16.00	406.26	nr	**17942.45**
15 KVA, 12.75 KW single phase I/P & O/P	15053.85	17815.78	30.00	761.74	nr	**18577.52**
20 KVA, 17.0 KW 3 phase I/P & single phase O/P	21850.50	25859.41	40.00	1015.66	nr	**26875.07**
30 KVA, 25.5 KW 3 phase I/P & O/P	32522.70	38489.64	60.00	1523.48	nr	**40013.12**
40 KVA, 34.0 KW 3 phase I/P & O/P	38769.15	45882.13	80.00	2031.31	nr	**47913.44**
50 KVA, 42.5 KW 3 phase I/P & O/P	52216.50	61796.66	90.00	2285.23	nr	**64081.89**
65 KVA, 55.25 KW 3 phase I/P & O/P	63420.00	75055.67	100.00	2539.14	nr	**77594.81**
90 KVA, 68.85 KW 3 phase I/P & O/P	74845.05	88576.87	120.00	3046.97	nr	**91623.84**
120 KVA, 102 KW 3 phase I/P & O/P	91980.00	108855.57	150.00	3808.71	nr	**112664.28**

W:COMMUNICATIONS/SECURITY/CONTROL

Item	Net Price £	Material £	Labour hours	Labour £	Unit	Total rate £
W10 : TELECOMMUNICATIONS						
Y61 - CABLES						
Multipair internal telephone cable; BS 6746; loose laid on tray / basket						
0.5 millimetre diameter conductor LSZH insulated and sheathed multipair cables; BT specification CW 1308						
3 pair	0.10	0.12	0.03	0.76	m	**0.88**
4 pair	0.12	0.14	0.03	0.76	m	**0.90**
6 pair	0.18	0.21	0.03	0.76	m	**0.97**
10 pair	0.36	0.43	0.05	1.27	m	**1.70**
15 pair	0.45	0.53	0.06	1.52	m	**2.06**
20 pair + 1 wire	0.60	0.71	0.06	1.52	m	**2.23**
25 pair	0.74	0.88	0.08	2.03	m	**2.91**
40 pair + earth	1.15	1.36	0.08	2.03	m	**3.39**
50 pair + earth	1.50	1.78	0.10	2.54	m	**4.31**
80 pair + earth	2.34	2.77	0.10	2.54	m	**5.31**
100 pair + earth	2.55	3.02	0.13	3.30	m	**6.32**
Multipair internal telephone cable; BS 6746; installed in conduit / trunking						
0.5 millimetre diameter conductor LSZH insulated and sheathed multipair cables; BT specification CW 1308						
3 pair	0.10	0.12	0.05	1.27	m	**1.39**
4 pair	0.12	0.14	0.06	1.52	m	**1.67**
6 pair	0.18	0.21	0.06	1.52	m	**1.74**
10 pair	0.36	0.43	0.07	1.78	m	**2.20**
15 pair	0.45	0.53	0.07	1.78	m	**2.31**
20 pair + 1 wire	0.60	0.71	0.09	2.29	m	**3.00**
25 pair	0.74	0.88	0.10	2.54	m	**3.41**
40 pair + earth	1.15	1.36	0.13	3.30	m	**4.66**
50 pair + earth	1.50	1.78	0.15	3.81	m	**5.58**
80 pair + earth	2.34	2.77	0.20	5.08	m	**7.85**
100 pair + earth	2.55	3.02	0.24	6.09	m	**9.11**
Telephone undercarpet cable; low profile; laid loose						
0.5 millimetre; PVC insulated; PVC sheathed multicore cable; BT CW 1316						
6 Core	0.62	0.73	0.05	1.27	m	**2.00**
Telephone drop wire cable; drawn in conduit or trunking						
0.5 millimetre conductor; PVC insulate twisted pair; Polyethylene sheathed; BT CW 1378						
Drop Wire 10	0.57	0.67	0.06	1.52	m	**2.20**

W:COMMUNICATIONS/SECURITY/CONTROL

Item	Net Price £	Material £	Labour hours	Labour £	Unit	Total rate £
Y74 - ACCESSORIES						
Telephone outlet: moulded plastic plate with box; fitted and connected; flush or surface mounted						
Single master outlet	6.47	7.63	0.35	8.89	nr	**16.52**
Single secondary outlet	4.78	5.64	0.35	8.89	nr	**14.53**
Telephone outlet: bronze or satin chromeplate; with box; fitted and connected; flush or surface mounted						
Single master outlet	10.53	12.42	0.35	8.89	nr	**21.31**
Single secondary outlet	11.51	13.57	0.35	8.89	nr	**22.46**

W:COMMUNICATIONS/SECURITY/CONTROL

Item	Net Price £	Material £	Labour hours	Labour £	Unit	Total rate £
W20 : RADIO/TELEVISION						
RADIO						
Y61 - CABLES						
Radio Frequency Cable; BS 2316 ; PVC sheathed; laid loose						
7/0.41mm tinned copper inner conductor; solid polyethylene dielectric insulation; bare copper wire braid; PVC sheath; 75 ohm impedance						
Cable	0.98	1.16	0.05	1.27	m	**2.43**
Twin 1/0.58mm copper covered steel solid core wire conductor; solid polyethylene dielectric insulation; barecopper wire braid; PVC sheath; 75 ohm impedance						
Cable	1.40	1.66	0.05	1.27	m	**2.93**
TELEVISION						
Y61 - CABLES						
Television aerial cable; coaxial; PVC sheathed; fixed to backgrounds						
General purpose TV aerial downlead; copper stranded inner conductor; cellular polythene insulation; copper braid outer conductor; 75 ohm impedance						
7/0.25mm	0.26	0.31	0.06	1.52	m	**1.84**
Low loss TV aerial downlead; solid copper inner conductor; cellular polythene insulation; copper braid outer; conductor; 75 ohm impedance						
1/1.12mm	0.42	0.50	0.06	1.52	m	**2.02**
Low loss air spaced; solid copper inner conductor; air spaced polythene insulation; copper braid outer conductor; 75 ohm impedance						
1/1.00mm	0.25	0.30	0.06	1.52	m	**1.82**
Satellite aerial downlead; solid copper inner conductor; air spaced polythene insulation; copper tape and braid outer conductor; 75 ohm impedance						
1/1.00mm	0.52	0.61	0.06	1.52	m	**2.13**
Satellite TV coaxial; solid copper inner conductor; semi air spaced polyethylene dielectric insulation; plain annealed copper foil and copper braid screen in outer conductor; PVC sheath; 75 ohm impedance						
1/1.25mm	0.70	0.82	0.08	2.03	m	**2.85**

W:COMMUNICATIONS/SECURITY/CONTROL

Item	Net Price £	Material £	Labour hours	Labour £	Unit	Total rate £
Satellite TV coaxial; solid copper inner conductor; air spaced polyethylene dielectric insulation; plain annealed copper foil and copper braid screen in outer conductor; PVC sheath; 75 ohm impedance						
1/1.67mm	1.15	1.36	0.09	2.29	m	**3.65**
Video cable; PVC flame retardant sheath; laid loose						
7/0.1mm silver coated copper covered annealed steel wire conductor; polyethylene dielectric insulation with tin coated copper wire braid; 75 ohm impedance						
Cable	0.66	0.78	0.05	1.27	m	**2.05**
Y74 - ACCESSORIES						
TV co-axial socket outlet: moulded plastic box; flush or surface mounted						
One way Direct Connection	6.70	7.93	0.35	8.89	nr	**16.82**
Two way Direct Connection	9.34	11.05	0.35	8.89	nr	**19.94**
One way Isolated UHF/VHF	11.80	13.96	0.35	8.89	nr	**22.85**
Two way Isolated UHF/VHF	16.08	19.03	0.35	8.89	nr	**27.91**

W:COMMUNICATIONS/SECURITY/CONTROL

Item	Net Price £	Material £	Labour hours	Labour £	Unit	Total rate £
W23 : CLOCKS						
Clock timing systems; master and slave units; fixed to background; excluding supports and fixings						
Quartz master clock with solid state digital readout for parallel loop operation; one minute, half minute and one second pulse; maximum of 80 clocks						
Over two loops only	595.00	704.16	4.40	111.72	nr	**815.89**
Power supplies for above, giving 24 hours power reserve						
2 6 Amp hour batteries	250.00	295.87	3.00	76.17	nr	**372.04**
2 15 Amp hour batteries	285.00	337.29	5.00	126.96	nr	**464.25**
Radio receiver to accept BBC Rugby Transmitter MSF signal						
To synchronise time of above Quartz master clock	130.00	153.85	7.04	178.81	nr	**332.66**
Wall clocks for slave (impulse) systems; 24V DC, white dial with black numerals fitted with axispolycarbonate disc; BS 467.7 Class O						
305mm diameter 1 minute impulse	52.54	62.18	1.37	34.79	nr	**96.97**
305mm diameter 1/2 minute impulse	52.54	62.18	1.37	34.79	nr	**96.97**
227mm diameter 1 second impulse	75.00	88.76	1.37	34.79	nr	**123.55**
305mm diameter 1 second impulse	75.00	88.76	1.37	34.79	nr	**123.55**
Quartz battery movement; BS 467.7 Class O; white dial with black numerals and sweep second hand; fitted with axispolycarbonate disc; stove enamel case						
305mm diameter	27.50	32.43	0.77	19.55	nr	**51.98**
Internal wall mounted electric clock; white dial with black numerals; 240v, 50 Hz						
305mm diameter	32.51	38.47	1.00	25.39	nr	**63.87**
458mm diameter	150.00	177.52	1.00	25.39	nr	**202.91**
Matching clock; BS 467.7 Class O; 240V AC, 50/60 Hz mains supply; 12 hour duration; dial with 1-12; IP 66; axispolycarbonate disc; spun metal movement cover; semi flush mount on 6 point fixing bezel						
227mm diameter	205.33	243.00	0.62	15.74	nr	**258.74**
Digital clocks; 240V, 50 hz supply; with/without synchronisation from masterclock;12/24 hour display; stand alone operation; 50mm digits						
Flush - hours/minutes/seconds or minutes/seconds/10th seconds	200.00	236.69	0.57	14.47	nr	**251.17**
Surface - hours/minutes/seconds or minutes/seconds/10th seconds	175.00	443.05	0.57	14.47	nr	**457.53**
Flush - hours/minutes or minutes/seconds	170.00	637.62	0.57	14.47	nr	**652.09**
Surface - hours/minutes or minutes/seconds	145.00	407.68	0.57	14.47	nr	**422.15**

W:COMMUNICATIONS/SECURITY/CONTROL

Item	Net Price £	Material £	Labour hours	Labour £	Unit	Total rate £
W30 : DATA TRANSMISSION						
Cabinets						
Floor standing; suitable for 19" patch panels with glass lockable doors, metal rear doors, side panels, vertical cable management, 2 x 4 way PDU's, 4 way fan, earth bonding kit; installed on raised floor						
600 wide x 800 deep - 18U	630.00	803.46	3.00	80.41	nr	**883.87**
600 wide x 800 deep - 24U	656.25	836.94	3.00	80.41	nr	**917.35**
600 wide x 800 deep - 33U	703.50	897.19	4.00	107.22	nr	**1004.41**
600 wide x 800 deep - 42U	756.00	964.15	4.00	107.22	nr	**1071.37**
600 wide x 800 deep - 47U	798.00	1017.71	4.00	107.22	nr	**1124.93**
800 wide x 800 deep - 42U	855.75	1091.36	4.00	107.22	nr	**1198.58**
800 wide x 800 deep - 47U	892.50	1138.23	4.00	107.22	nr	**1245.45**
Label cabinet	2.00	2.54	0.25	6.70	nr	**9.25**
Wall mounted; suitable for 19" patch panels with glass lockable doors, side panels, vertical cable management, 2 x 4 way PDU's, 4 way fan, earth bonding kit; fixed to wall						
19 wide x 500 deep - 9U	420.00	535.64	3.00	80.41	nr	**616.05**
19 wide x 500 deep - 12U	430.50	549.03	3.00	80.41	nr	**629.44**
19 wide x 500 deep - 15U	446.25	569.12	3.00	80.41	nr	**649.53**
19 wide x 500 deep - 18U	525.00	669.55	3.00	80.41	nr	**749.96**
19 wide x 500 deep - 21U	546.00	696.33	3.00	80.41	nr	**776.74**
Label cabinet	2.00	2.54	0.25	6.70	nr	**9.25**
Frames						
Floor standing; suitable for 19" patch panels with supports, vertical cable management, earth bonding kit; installed on raised floor						
19 wide x 500 deep - 25U	472.50	602.59	2.50	67.01	nr	**669.60**
19 wide x 500 deep - 39U	577.50	736.50	2.50	67.01	nr	**803.51**
19 wide x 500 deep - 42U	630.00	803.46	2.50	67.01	nr	**870.47**
19 wide x 500 deep - 47U	682.50	870.41	2.50	67.01	nr	**937.42**
Label frame	2.00	2.54	0.25	6.70	nr	**9.25**
Patch panels						
Category 5e; 19" wide fully loaded, finished in black including termination and forming of cables						
16 port - RJ45 UTP - Krone / 110	53.87	68.70	3.25	87.11	nr	**155.81**
24 port - RJ45 UTP - Krone / 110	71.82	91.59	4.75	127.32	nr	**218.91**
32 port - RJ45 UTP - Krone / 110	105.84	134.98	6.30	168.87	nr	**303.85**
48 port - RJ45 UTP - Krone / 110	138.91	177.16	9.35	250.62	nr	**427.78**
Patch panel labelling per port	0.21	0.27	0.02	0.54	nr	**0.80**
Category 6; 19" wide fully loaded, finished in black including termination and forming of cables						
16 port - RJ45 UTP - Krone / 110	80.80	103.04	3.40	91.13	nr	**194.18**
24 port - RJ45 UTP - Krone / 110	103.74	132.30	5.00	134.02	nr	**266.32**
32 port - RJ45 UTP - Krone / 110	143.64	183.19	6.60	176.91	nr	**360.10**
48 port - RJ45 UTP - Krone / 110	193.51	246.80	9.80	262.68	nr	**509.48**
Patch panel labelling per port	0.21	0.27	0.02	0.54	nr	**0.80**

W:COMMUNICATIONS/SECURITY/CONTROL

Item	Net Price £	Material £	Labour hours	Labour £	Unit	Total rate £
W30 : DATA TRANSMISSION (cont'd)						
Patch panels (cont'd)						
Cat 3/voice; 19" wide fully loaded, finished in black including termination and forming of cables (assuming 2 pairs per port)						
16 port - RJ45 UTP - Krone	68.04	86.77	2.00	53.61	nr	140.38
24 port - RJ45 UTP - Krone	81.27	103.65	2.60	69.69	nr	173.34
32 port - RJ45 UTP - Krone	117.18	149.44	3.25	87.11	nr	236.56
48 port - RJ45 UTP - Krone	125.69	160.29	4.65	124.64	nr	284.93
900 pair fully loaded PB type frame including forming and termination of 9 x 100 pair cables	514.50	656.16	25.00	670.10	nr	1326.26
Installation and termination of Krone strip (10 pair block - 237A) including designation label	4.25	5.42	0.50	13.40	nr	18.83
Patch panel labelling per port	0.21	0.27	0.02	0.54	nr	0.80
Fibre; 19" wide fully loaded, labelled, aluminium alloy c/w couplers, fibre management and glands (excludes termination of fibre cores)						
8 way ST; fixed drawer	63.00	80.35	0.50	13.40	nr	93.75
16 way ST; fixed drawer	94.50	120.52	0.50	13.40	nr	133.92
24 way ST; fixed drawer	131.25	167.39	0.50	13.40	nr	180.79
8 way ST; sliding drawer	84.00	107.13	0.50	13.40	nr	120.53
16 way ST; sliding drawer	115.50	147.30	0.50	13.40	nr	160.70
24 way ST; sliding drawer	152.25	194.17	0.50	12.70	nr	206.86
8 way (4 duplex) SC; fixed drawer	78.75	100.43	0.50	13.40	nr	113.83
16 way (8 duplex) SC; fixed drawer	115.50	147.30	0.50	13.40	nr	160.70
24 way (12 duplex) SC; fixed drawer	141.75	180.78	0.50	13.40	nr	194.18
8 way (4 duplex) SC; sliding drawer	99.75	127.21	0.50	13.40	nr	140.62
16 way (8 duplex) SC; sliding drawer	136.50	174.08	0.50	13.40	nr	187.48
24 way (12 duplex) SC; sliding drawer	162.75	207.56	0.50	13.40	nr	220.96
8 way (4 duplex) MTRJ; fixed drawer	89.25	113.82	0.50	13.40	nr	127.23
16 way (8 duplex) MTRJ; fixed drawer	136.50	174.08	0.50	13.40	nr	187.48
24 way (12 duplex) MTRJ; fixed drawer	162.75	207.56	0.50	13.40	nr	220.96
8 way (4 duplex) FC/PC; fixed drawer	89.25	113.82	0.50	13.40	nr	127.23
16 way (8 duplex) FC/PC; fixed drawer	136.50	174.08	0.50	13.40	nr	187.48
24 way (12 duplex) FC/PC; fixed drawer	162.75	207.56	0.50	13.40	nr	220.96
Patch panel label per way	0.21	0.27	0.02	0.54	nr	0.80
Patch leads						
Category 5e; straight through booted RJ45 UTP - RJ45 UTP						
Patch lead 1m length	2.41	3.08	0.09	2.41	nr	5.49
Patch lead 3m length	3.59	4.58	0.09	2.41	nr	6.99
Patch lead 5m length	5.40	6.89	0.10	2.68	nr	9.57
Patch lead 7m length	7.22	9.21	0.10	2.68	nr	11.89
Category 6; straight through booted RJ45 UTP - RJ45 UTP						
Patch lead 1 m length	4.81	6.14	0.09	2.41	nr	8.55
Patch lead 3 m length	6.15	7.84	0.09	2.41	nr	10.26
Patch lead 5 m length	7.75	9.88	0.10	2.68	nr	12.56
Patch lead 7 m length	9.09	11.59	0.10	2.68	nr	14.27
Simplex 62.5/125 ST- ST						
Fibre patch lead 1 m length	10.16	12.96	0.08	2.14	nr	15.10
Fibre patch lead 3 m length	11.57	14.76	0.08	2.14	nr	16.91
Fibre patch lead 5 m length	12.94	16.50	0.10	2.68	nr	19.18

W:COMMUNICATIONS/SECURITY/CONTROL

Item	Net Price £	Material £	Labour hours	Labour £	Unit	Total rate £
Simplex 62.5/125 ST- SC						
Fibre patch lead 1m length	12.72	16.23	0.08	2.14	nr	18.37
Fibre patch lead 3m length	14.11	18.00	0.08	2.14	nr	20.14
Fibre patch lead 5m length	15.50	19.77	0.10	2.68	nr	22.45
Duplex 62.5/125 MTRJ - MTRJ						
Fibre patch lead 1 m length	24.05	30.67	0.08	2.14	nr	32.82
Fibre patch lead 3 m length	26.19	33.40	0.08	2.14	nr	35.54
Fibre patch lead 5 m length	28.32	36.12	0.10	2.68	nr	38.80
Duplex 62.5/125 MTRJ - ST						
Fibre patch lead 1 m length	33.46	42.68	0.08	2.14	nr	44.82
Fibre patch lead 3 m length	35.60	45.40	0.08	2.14	nr	47.55
Fibre patch lead 5 m length	37.74	48.13	0.10	2.68	nr	50.81
Duplex 62.5/125 ST- ST						
Fibre patch lead 1 m length	13.90	17.73	0.08	2.14	nr	19.87
Fibre patch lead 3 m length	17.10	21.81	0.08	2.14	nr	23.95
Fibre patch lead 5 m length	18.18	23.18	0.10	2.68	nr	25.86
Duplex 62.5/125 ST - SC						
Fibre patch lead 1 m length	15.50	19.77	0.08	2.14	nr	21.91
Fibre patch lead 3 m length	17.10	21.81	0.08	2.14	nr	23.95
Fibre patch lead 5 m length	18.18	23.18	0.10	2.68	nr	25.86
Duplex 62.5/125 SC - SC						
Fibre patch lead 1 m length	17.10	21.81	0.08	2.14	nr	23.95
Fibre patch lead 3 m length	19.24	24.53	0.08	2.14	nr	26.68
Fibre patch lead 5 m length	21.38	27.26	0.10	2.68	nr	29.94
Duplex 50/125 OM3 MTRJ - MTRJ						
Fibre patch lead 1m length	30.31	38.66	0.08	2.14	nr	40.80
Fibre patch lead 3m length	33.39	42.58	0.08	2.03	nr	44.61
Fibre patch lead 5m length	36.44	46.47	0.10	2.68	nr	49.15
Duplex 50/125 OM3 MTRJ - ST						
Fibre patch lead 1m length	37.17	47.41	0.08	2.14	nr	49.55
Fibre patch lead 3m length	39.55	50.44	0.08	2.14	nr	52.58
Fibre patch lead 5m length	41.92	53.47	0.10	2.68	nr	56.15
Duplex 50/125 OM3 ST - ST						
Fibre patch lead 1m length	30.31	38.66	0.08	2.14	nr	40.80
Fibre patch lead 3m length	33.39	42.58	0.08	2.14	nr	44.72
Fibre patch lead 5m length	36.44	46.47	0.10	2.68	nr.	49.15
Duplex 50/125 OM3 ST - SC						
Fibre patch lead 1m length	37.17	47.41	0.08	2.14	nr	49.55
Fibre patch lead 3m length	39.55	50.44	0.08	2.14	nr	52.58
Fibre patch lead 5m length	41.92	53.47	0.10	2.68	nr	56.15
Duplex 50/125 OM3 SC - SC						
Fibre patch lead 1m length	37.17	47.41	0.08	2.14	nr	49.55
Fibre patch lead 3m length	39.55	50.44	0.08	2.14	nr	52.58
Fibre patch lead 5m length	41.92	53.47	0.10	2.68	nr	56.15

W:COMMUNICATIONS/SECURITY/CONTROL

Item	Net Price £	Material £	Labour hours	Labour £	Unit	Total rate £
W30 : DATA TRANSMISSION (cont'd)						
Data cabling						
Unshielded twisted pair; solid copper conductors; PVC insulation; nominal impedance 100 Ohm; Cat 5e to ISO 11801, EIA/TIA 568B and EN 50173/50174 standards to the current revisions						
4 pair 24AWG; nominal outside diameter 5.6mm; installed above ceiling	0.13	0.16	0.02	0.54	m	**0.70**
4 pair 24AWG; nominal outside diameter 5.6mm; installed in riser	0.13	0.16	0.02	0.54	m	**0.70**
4 pair 24AWG; nominal outside diameter 5.6mm; installed below floor	0.13	0.16	0.01	0.27	m	**0.43**
4 pair 24AWG; nominal outside diameter 5.6mm; installed in trunking	0.13	0.16	0.02	0.54	m	0.70
Category 5e cable test	-	-	0.15	4.02	nr	**4.02**
Unshielded twisted pair; solid copper conductors; LSOH sheathed; nominal impedance 100 Ohm; Cat 5e to ISO 11801, EIA/TIA 568B and EN 50173/50174 standards to the current revisions						
4 pair 24AWG; nominal outside diameter 5.6mm; installed above ceiling	0.19	0.24	0.02	0.54	m	**0.78**
4 pair 24AWG; nominal outside diameter 5.6mm; installed in riser	0.19	0.24	0.02	0.54	m	**0.78**
4 pair 24AWG; nominal outside diameter 5.6mm; installed below floor	0.19	0.24	0.01	0.27	m	**0.51**
4 pair 24AWG; nominal outside diameter 5.6mm; installed in trunking	0.19	0.24	0.02	0.54	m	**0.78**
Category 5e cable test	-	-	0.15	4.02	nr	**4.02**
Unshielded twisted pair; solid copper conductors; PVC insulation; nominal impedance 100 Ohm; Cat 6 to ISO 11801, EIA/TIA 568B and EN 50173/50174 standards to the current revisions						
4 pair 24AWG; nominal outside diameter 5.6mm; installed above ceiling	0.20	0.26	0.02	0.40	m	**0.66**
4 pair 24AWG; nominal outside diameter 5.6mm; installed in riser	0.20	0.26	0.02	0.40	m	**0.66**
4 pair 24AWG; nominal outside diameter 5.6mm; installed below floor	0.20	0.26	0.01	0.21	m	**0.47**
4 pair 24AWG; nominal outside diameter 5.6mm; installed in trunking	0.20	0.26	0.02	0.54	m	0.79
Category 6 cable test	-	-	0.20	5.36	nr	**5.36**
Unshielded twisted pair; solid copper conductors; LSOH sheathed; nominal impedance 100 Ohm; Cat 6 to ISO 11801, EIA/TIA 568B and EN 50173/50174 standards to the current revisions						
4 pair 24AWG; nominal outside diameter 5.6mm; installed above ceiling	0.24	0.30	0.02	0.59	m	**0.89**
4 pair 24AWG; nominal outside diameter 5.6mm; installed in riser	0.24	0.30	0.02	0.59	m	**0.89**
4 pair 24AWG; nominal outside diameter 5.6mm; installed below floor	0.24	0.30	0.01	0.29	m	**0.60**
4 pair 24AWG; nominal outside diameter 5.6mm; installed in trunking	0.24	0.30	0.02	0.59	m	**0.89**
Category 6 cable test	-	-	0.22	5.90	nr	**5.90**

W:COMMUNICATIONS/SECURITY/CONTROL

Item	Net Price £	Material £	Labour hours	Labour £	Unit	Total rate £
Fibre optic cable, tight buffered, internal/external application, single mode, LSOH sheathed						
4 core fibre optic cable	1.22	1.56	0.10	2.68	m	**4.24**
8 core fibre optic cable	2.03	2.59	0.10	2.68	m	**5.27**
12 core fibre optic cable	2.52	3.21	0.10	2.68	m	**5.89**
16 core fibre optic cable	3.06	3.90	0.10	2.68	m	**6.58**
24 core fibre optic cable	3.69	4.71	0.10	2.68	m	**7.39**
Singlemode core test per core	-	-	0.20	5.36	nr	**5.36**
Fibre optic cable OM1 and OM2, tight buffered, internal/external application, 62.5/125 multimode fibre, LSOH sheathed						
4 core fibre optic cable	1.31	1.67	0.10	2.68	m	**4.35**
8 core fibre optic cable	2.16	2.75	0.10	2.68	m	**5.44**
12 core fibre optic cable	2.75	3.51	0.10	2.68	m	**6.19**
16 core fibre optic cable	3.42	4.36	0.10	2.68	m	**7.04**
24 core fibre optic cable	3.96	5.05	0.10	2.68	m	**7.73**
Multimode core test per core	-	-	0.20	5.36	nr	**5.36**
Fibre optic cable OM3, tight buffered, internal only application, 50/125 multimode fibre, LSOH sheathed						
4 core fibre optic cable	1.62	2.07	0.10	2.68	nr	**4.75**
8 core fibre optic cable	2.52	3.21	0.10	2.68	nr	**5.89**
12 core fibre optic cable	3.65	4.66	0.10	2.68	nr	**7.34**
24 core fibre cable	6.98	8.90	0.10	2.68	nr	**11.58**
Multimode core test per core	-	-	0.20	5.36	nr	**5.36**
Fibre optic single and multimode connectors and couplers including termination						
ST singlemode booted connector	6.14	7.83	0.25	6.70	nr	**14.53**
ST multimode booted connector	2.97	3.79	0.25	6.70	nr	**10.49**
SC simplex singlemode booted connector	8.71	11.11	0.25	6.70	nr	**17.81**
SC simplex multimode booted connector	3.17	4.04	0.25	6.70	nr	**10.74**
SC duplex multimode booted connector	6.34	8.08	0.25	6.70	nr	**14.78**
ST - SC duplex adaptor	15.40	19.64	0.01	0.27	nr	**19.91**
ST inline bulkhead coupler	3.30	4.21	0.01	0.27	nr	**4.48**
SC duplex coupler	6.16	7.86	0.01	0.27	nr	**8.12**
MTRJ small form factor duplex connector	6.43	16.41	0.25	6.70	nr	**23.11**
LC simplex multimode booted connector	3.17	4.04	0.25	6.70	nr	**10.74**
Voice cabling						
Low speed data; unshielded twisted pair; solid copper conductors; PVC insulation; nominal impedance 100 Ohm; Category 3 to ISO IS 11801/EIA /TIA 568B and EN50173/50174 standards to current revisions						
Installed in riser						
25 pair 24AWG	1.13	1.43	0.03	0.80	m	**2.24**
50 pair 24AWG	2.25	2.87	0.06	1.61	m	**4.48**
100 pair 24AWG	3.94	5.02	0.10	2.68	m	**7.70**
Installed below floor						
25 pair 24AWG	1.13	1.43	0.02	0.54	m	**1.97**
50 pair 24AWG	2.25	2.87	0.05	1.34	m	**4.21**
100 pair 24AWG	3.94	5.02	0.08	2.14	m	**7.17**
Cat 3 cable circuit test per pair	-	-	0.08	2.14	nr	**2.14**

W:COMMUNICATIONS/SECURITY/CONTROL

Item	Net Price £	Material £	Labour hours	Labour £	Unit	Total rate £
W30 : DATA TRANSMISSION (cont'd)						
Voice cabling (cont'd)						
Low speed data; unshielded twisted pair; solid copper conductors; LSOH sheath; nominal impedance 100 Ohm; Category 3 to ISO IS 11801/EIA/TIA 568B and EN50173/50174 standards to current revisions						
Installed in riser						
25 pair 24AWG	1.41	1.80	0.03	0.80	m	**2.61**
50 pair 24AWG	2.81	3.59	0.06	1.61	m	**5.20**
100 pair 24AWG	4.72	6.03	0.10	2.68	m	**8.71**
Installed below floor						
25 pair 24AWG	1.41	1.80	0.02	0.54	m	**2.34**
50 pair 24AWG	2.81	3.59	0.05	1.34	m	**4.93**
100 pair 24AWG	4.72	6.03	0.08	2.14	m	**8.17**
Cat 3 cable circuit test per pair	-	-	0.08	2.14	nr	**2.14**
Accessories						
Category 5e RJ45 data outlet plate and multiway outlet boxes for wall, ceiling and below floor installations including label to ISO 11801 standards						
Wall mounted; fully loaded						
One gang LSOH PVC plate	6.38	8.14	0.15	4.02	nr	**12.16**
Two gang LSOH PVC plate	9.13	11.64	0.20	5.36	nr	**17.00**
Four gang LSOH PVC plate	15.62	19.92	0.40	10.72	nr	**30.64**
One gang satin brass plate	12.10	15.43	0.25	6.70	nr	**22.13**
Two gang satin brass plate	14.85	18.94	0.33	8.85	nr	**27.78**
Ceiling mounted; fully loaded						
One gang metal clad plate	7.70	9.82	0.33	8.85	nr	**18.67**
Two gang metal clad plate	10.45	13.33	0.45	12.06	nr	**25.39**
Below floor; fully loaded						
Four way outlet box, 5 m length 20mm flexible conduit with glands and strain relief bracket	26.02	33.18	0.80	21.44	nr	**54.62**
Six way outlet box, 5 m length 25mm flexible conduit with glands and strain relief bracket	33.64	42.90	1.20	32.16	nr	**75.06**
Eight way outlet box, 5 m length 25mm flexible conduit with glands and strain relief bracket	42.35	54.01	1.40	37.53	nr	**91.54**
Installation of outlet boxes to desks	-	-	0.40	10.72	nr	**10.72**
Category 6 RJ45 data outlet plate and multiway outlet boxes for wall, ceiling and below floor installations including label to ISO 11801 standards						
Wall mounted; fully loaded						
One gang LSOH PVC plate	8.14	10.38	0.17	4.42	nr	**14.80**
Two gang LSOH PVC plate	11.94	15.22	0.22	5.90	nr	**21.12**
Four gang LSOH PVC plate	21.56	27.50	0.44	11.79	nr	**39.29**
One gang satin brass plate	13.64	17.40	0.28	7.37	nr	**24.77**
Two gang satin brass plate	17.43	22.24	0.36	9.73	nr	**31.97**

W:COMMUNICATIONS/SECURITY/CONTROL

Item	Net Price £	Material £	Labour hours	Labour £	Unit	Total rate £
Ceiling mounted; fully loaded						
One gang metal clad plate	9.24	11.78	0.36	9.73	nr	**21.51**
Two gang metal clad plate	13.04	16.62	0.50	13.27	nr	**29.89**
Below floor; fully loaded						
Four way outlet box, 5 m length 20mm flexible conduit with glands and strain relief bracket	31.89	40.67	0.88	23.59	nr	**64.26**
Six way outlet box, 5 m length 25mm flexible conduit with glands and strain relief bracket	42.42	54.09	1.32	35.38	nr	**89.48**
Eight way outlet box, 5 m length 25mm flexible conduit with glands and strain relief bracket	54.03	68.91	1.54	41.28	nr	**110.19**
Installation of outlet boxes to desks	-	-	0.40	10.72	nr	**10.72**

W:COMMUNICATIONS/SECURITY/CONTROL

Item	Net Price £	Material £	Labour hours	Labour £	Unit	Total rate £
W40 : ACCESS CONTROL						
ACCESS CONTROL EQUIPMENT						
Equipment to control the movement of personnel into defined spaces; includes fixing to backgrounds, termination of power and data cables; excludes cable containment and cable installation						
Access control						
Magnetic swipe reader	155.00	201.81	1.50	63.61	nr	**265.42**
Proximity reader	110.34	143.66	1.50	63.61	nr	**207.27**
Exit button	9.20	11.98	1.50	63.61	nr	**75.59**
Exit PIR	33.92	44.16	2.00	84.81	nr	**128.98**
Emergency break glass double pole	20.18	26.27	1.00	42.41	nr	**68.68**
Alarm contact flush	1.15	1.50	1.00	42.41	nr	**43.90**
Alarm contact surface	1.15	1.50	1.00	42.41	nr	**43.90**
Reader controller 16 door	3030.70	3945.97	4.00	169.62	nr	**4115.60**
Reader controller 8 door	1566.70	2039.84	4.00	169.62	nr	**2209.47**
Reader controller 2 door	648.60	844.48	4.00	169.62	nr	**1014.10**
Reader interface	149.00	194.00	1.50	63.61	nr	**257.61**
Lock power supply 12 volt 3 AMP	39.10	50.91	2.00	84.81	nr	**135.72**
Lock power supply 24 volt 3 AMP	97.75	127.27	2.00	84.81	nr	**212.08**
Rechargeable battery 12 volt 7ah	9.00	11.72	0.25	10.60	nr	**22.32**
Lock Equipment						
Single slimline magnetic lock, monitored	36.80	47.91	2.00	84.81	nr	**132.73**
Single slimline magnetic lock, unmonitored	26.45	34.44	1.75	74.21	nr	**108.65**
Double slimline magnetic lock, monitored	73.60	95.83	4.00	169.62	nr	**265.45**
Double slimline magnetic lock, unmonitored	58.65	76.36	4.50	190.83	nr	**267.19**
Standard single magnetic lock, monitored	50.94	66.32	1.25	53.01	nr	**119.33**
Standard single magnetic lock, unmonitored	42.20	54.94	1.00	42.41	nr	**97.35**
Standard single magnetic lock, double monitored	95.95	124.93	1.50	63.61	nr	**188.54**
Standard double magnetic lock, monitored	100.40	130.72	1.50	63.61	nr	**194.33**
Standard double magnetic lock, unmonitored	90.35	117.64	1.25	53.01	nr	**170.64**
12V electric release fail safe, monitored	41.70	54.29	1.25	53.01	nr	**107.30**
12V electric release fail secure, monitored	41.70	54.29	1.00	42.41	nr	**96.70**
Solenoid bolt	100.86	131.32	1.50	63.61	nr	**194.93**
Electric mortise lock	124.70	162.36	1.00	42.41	nr	**204.77**

W:COMMUNICATIONS/SECURITY/CONTROL

Item	Net Price £	Material £	Labour hours	Labour £	Unit	Total rate £
W41 : SECURITY DETECTION & ALARM						
SECURITY DETECTION & ALARM EQUIPMENT						
Detection and alarm systems for the protection of property and persons; includes fixing of equipment to backgrounds and termination of power and data cabling; excludes cable containment and cable installation						
Detection, alarm equipment						
Alarm contact flush	1.15	1.50	1.00	42.41	nr	**43.90**
Alarm contact surface	1.15	1.50	1.00	42.41	nr	**43.90**
Roller shutter contact	6.90	8.98	1.00	42.41	nr	**51.39**
Personal attack button	7.50	9.77	1.00	42.41	nr	**52.17**
Acoustic break glass detectors	23.00	29.95	2.00	84.81	nr	**114.76**
Vibration detectors	25.30	32.94	2.00	84.81	nr	**117.75**
12 metre PIR detector	17.25	22.46	1.50	63.61	nr	**86.07**
15 metre dual detector	23.00	29.95	1.50	63.61	nr	**93.56**
8 zone alarm panel	110.00	143.22	3.00	127.22	nr	**270.44**
8-24 zone end station	159.85	208.12	4.00	169.62	nr	**377.75**
Remote keypad	41.50	54.03	2.00	84.81	nr	**138.85**
8 zone expansion	51.75	67.38	1.50	63.61	nr	**130.99**
Final exit set button	-	-	1.00	42.41	nr	**42.41**
Self contained external sounder	46.00	59.89	2.00	84.81	nr	**144.70**
Internal loudspeaker	5.75	7.49	2.00	84.81	nr	**92.30**
Rechargeable battery 12 volt 7ah (ampere hrs)	9.00	11.72	0.25	10.60	nr	**22.32**
Surveillance Equipment						
Vandal resistant camera, colour	218.00	283.84	3.00	127.22	nr	**411.05**
External camera, colour	448.00	583.30	5.00	212.03	nr	**795.33**
Auto dome external, colour	1146.00	1492.09	5.00	212.03	nr	**1704.12**
Auto dome external, colour/monochrome	1335.00	1738.17	5.00	212.03	nr	**1950.20**
Auto dome internal, colour	1064.00	1385.33	4.00	169.62	nr	**1554.95**
Auto dome internal colour/monochrome	1290.00	1679.58	4.00	169.62	nr	**1849.20**
Mini internal domes	172.50	224.59	3.00	127.22	nr	**351.81**
Camera switcher, 32 inputs	1811.25	2358.25	4.00	169.62	nr	**2527.87**
Full function keyboard	425.00	553.35	1.00	42.41	nr	**595.76**
16 CH multiplexers, simplex	724.50	943.30	2.00	84.81	nr	**1028.11**
16 CH multiplexers, duplex	902.75	1175.38	2.00	84.81	nr	**1260.19**
16 way DVR, 250 GB hard drive	2081.50	2710.11	3.00	127.22	nr	**2837.33**
Additional 250 GB hard drive	391.00	509.08	1.00	42.41	nr	**551.49**
10" colour monitor	221.95	288.98	2.00	84.81	nr	**373.79**
15" colour monitor	195.50	254.54	2.00	84.81	nr	**339.35**
17" colour monitor, high resolution	402.50	524.05	2.00	84.81	nr	**608.87**
21" colour monitor, high resolution	483.00	628.87	2.00	84.81	nr	**713.68**

W:COMMUNICATIONS/SECURITY/CONTROL

Item	Net Price £	Material £	Labour hours	Labour £	Unit	Total rate £
W50 : FIRE DETECTION AND ALARM						
STANDARD FIRE DETECTION CONTROL PANEL						
Zone control panel; 2 x 12 volt batteries/charge up to 48 hours standby; mild steel case; flush or surface mounting						
1 zone	161.37	190.98	3.00	76.17	nr	**267.15**
2 zone	173.33	205.13	3.51	89.09	nr	**294.22**
4 zone	262.98	311.23	4.00	101.57	nr	**412.80**
8 zone	418.38	495.14	5.00	126.96	nr	**622.10**
12 zone	546.29	646.51	6.00	152.35	nr	**798.86**
16 zone	771.02	912.48	6.00	152.35	nr	**1064.82**
24 zone	1037.59	1227.96	6.00	152.35	nr	**1380.31**
Repeater panels						
8 zone	283.31	335.28	4.00	101.57	nr	**436.85**
EQUIPMENT						
Manual call point units: plastic covered						
Surface mounted						
Call point	9.09	10.76	0.50	12.70	nr	**23.45**
Call point; Weatherproof	65.48	77.49	0.80	20.31	nr	**97.81**
Flush mounted						
Call point	7.93	9.38	0.56	14.22	nr	**23.60**
Call point; Weatherproof	61.57	72.87	0.86	21.85	nr	**94.72**
Detectors						
Smoke, ionisation type with mounting base	31.77	37.60	0.75	19.05	nr	**56.65**
Smoke, optical type with mounting base	31.77	37.60	0.75	19.05	nr	**56.65**
Fixed temperature heat detector with mounting base (60°C)	25.57	30.27	0.75	19.05	nr	**49.32**
Rate of Rise heat detector with mounting base (90°C)	23.96	28.36	0.75	19.05	nr	**47.41**
Duct detector including optical smoke detector and base	203.12	240.39	2.00	50.78	nr	**291.17**
Remote smoke detector LED indicator with base	8.18	9.68	0.50	12.70	nr	**22.38**
Sounders						
6" bell, conduit box	17.69	20.93	0.75	19.05	nr	**39.98**
6" bell, conduit box; weatherproof	24.22	28.66	0.75	19.04	nr	**47.70**
Siren; 230V	48.61	57.53	1.25	31.74	nr	**89.27**
Magnetic Door Holder; 230V ; surface fixed	49.28	58.33	1.50	38.13	nr	**96.45**
ADDRESSABLE FIRE DETECTION CONTROL PANEL						
Analogue addressable panel; BS EN54 Part 2 and 4 1998; incorporating 120 addresses per loop (maximum 1-2km length); sounders wired on loop; sealed lead acid integral battery standby providing 48 hour standby; 24 volt DC; mild steel case; surface fixed						
1 loop; 4 x 12 volt batteries	922.07	1091.24	6.00	152.35	nr	**1243.59**

W:COMMUNICATIONS/SECURITY/CONTROL

Item	Net Price £	Material £	Labour hours	Labour £	Unit	Total rate £
Extra for 1 loop panel						
Loop card	184.62	218.49	1.00	25.39	nr	**243.88**
Repeater panel	469.06	555.12	6.00	152.35	nr	**707.47**
Network nodes	984.19	1164.76	6.00	152.35	nr	**1317.11**
Interface unit; for other systems						
Mains powered	165.68	196.08	1.50	38.09	nr	**234.16**
Loop powered	121.36	143.62	1.00	25.39	nr	**169.01**
Single channel I/O	42.60	50.42	1.00	25.39	nr	**75.81**
Zone module	60.25	71.31	1.50	38.09	nr	**109.39**
4 loop; 4 x 12 volt batteries; 24 hour standby; 30 minute alarm	1475.62	1746.36	8.00	203.13	nr	**1949.49**
Extra for 4 loop panel						
Loop card	184.62	218.49	1.00	25.39	nr	**243.88**
Repeater panel	984.19	1164.76	6.00	152.35	nr	**1317.11**
Mimic panel	1484.67	1757.06	5.00	126.96	nr	**1884.02**
Network nodes	984.04	1164.58	6.00	152.35	nr	**1316.93**
Interface unit; for other systems						
Mains powered	165.68	196.08	1.50	38.09	nr	**234.16**
Loop powered	121.36	143.62	1.00	25.39	nr	**169.01**
Single channel I/O	42.60	50.42	1.00	25.39	nr	**75.81**
Zone module	60.24	71.29	1.50	38.09	nr	**109.38**
Line modules	6.16	7.29	1.00	25.39	nr	**32.68**
8 loop; 4 x 12 volt batteries; 24 hour standby; 30 minute alarm	2951.26	3492.72	12.00	304.70	nr	**3797.42**
Extra for 8 loop panel						
Loop card	184.62	218.49	1.00	25.39	nr	**243.88**
Repeater panel	984.19	1164.76	6.00	152.35	nr	**1317.11**
Mimic panel	1484.67	1757.06	5.00	126.96	nr	**1884.02**
Network nodes	984.04	1164.58	6.00	152.35	nr	**1316.93**
Interface unit; for other systems						
Mains powered	165.68	196.08	1.50	38.09	nr	**234.16**
Loop powered	121.36	143.62	1.00	25.39	nr	**169.01**
Single channel I/O	42.60	50.42	1.00	25.39	nr	**75.81**
Zone module	60.24	71.29	1.50	38.09	nr	**109.38**
Line modules	6.16	7.29	1.00	25.39	nr	**32.68**
EQUIPMENT						
Manual Call Point						
Surface mounted						
Call point	38.15	45.15	1.00	25.39	nr	**70.54**
Call point; Weather proof	80.03	94.71	1.25	31.74	nr	**126.45**
Flush mounted						
Call point	37.34	44.19	1.00	25.39	nr	**69.58**
Call point; Weather proof	57.04	67.51	1.25	31.74	nr	**99.25**
Detectors						
Smoke, ionisation type with mounting base	53.71	63.57	0.75	19.05	nr	**82.62**
Smoke, optical type with mounting base	46.10	54.56	0.75	19.05	nr	**73.60**
Fixed temperature heat detector with mounting base (60°C)	49.41	58.47	0.75	19.05	nr	**77.52**
Rate of Rise heat detector with mounting base (90°C)	49.41	58.47	0.75	19.05	nr	**77.52**

W:COMMUNICATIONS/SECURITY/CONTROL

Item	Net Price £	Material £	Labour hours	Labour £	Unit	Total rate £
W50 : FIRE DETECTION AND ALARM (cont'd)						
EQUIPMENT (cont'd)						
Detectors (cont'd)						
Duct Detector including optical smoke detector and addressable base	215.58	255.14	2.00	50.78	nr	**305.92**
Beam smoke detector with transmitter and receiver unit	450.64	533.32	2.00	50.78	nr	**584.10**
Zone short circuit isolator	37.53	44.41	0.75	19.04	nr	**63.46**
Plant interface unit	27.65	32.72	0.50	12.70	nr	**45.42**
Sounders						
Xenon flasher, 24 volt, conduit box	23.04	27.27	0.50	12.70	nr	**39.97**
Xenon flasher, 24 volt, conduit box; weatherproof	39.18	46.37	0.50	12.70	nr	**59.07**
6" bell, conduit box	17.69	20.93	0.75	19.05	nr	**39.98**
6" bell, conduit box; weatherproof	24.22	28.66	0.75	19.04	nr	**47.70**
Siren; 24V polarised	17.14	20.28	1.00	25.39	nr	**45.67**
Siren; 240V	48.61	57.53	1.25	31.74	nr	**89.27**
Magnetic Door Holder; 240V ; surface fixed	49.28	58.33	1.50	38.13	nr	**96.45**

W:COMMUNICATIONS/SECURITY/CONTROL

Item	Net Price £	Material £	Labour hours	Labour £	Unit	Total rate £
W51 : EARTHING AND BONDING						
EARTH BAR						
Earth bar; polymer insulators and base mounting; including connections						
Non disconnect link						
6 way	149.25	176.63	0.81	20.57	nr	**197.20**
8 way	164.16	194.28	0.81	20.57	nr	**214.85**
10 way	180.60	213.73	0.81	20.57	nr	**234.30**
Disconnect link						
6 way	167.65	198.41	1.01	25.65	nr	**224.05**
8 way	184.41	218.25	1.01	25.65	nr	**243.89**
10 way	202.85	240.07	1.01	25.65	nr	**265.71**
Solid earth bar; including connections						
150 x 50 x 6 mm	49.49	58.57	1.01	25.65	nr	**84.21**
Extra for earthing						
Disconnecting link						
300 x 50 x 6mm	47.01	55.64	1.16	29.45	nr	**85.09**
500 x 50 x 6mm	51.71	61.20	1.16	29.45	nr	**90.65**
Crimp lugs; including screws and connections to cable						
25 mm	0.56	0.67	0.31	7.87	nr	**8.54**
35 mm	0.81	0.96	0.31	7.87	nr	**8.83**
50 mm	0.95	1.12	0.32	8.13	nr	**9.25**
70 mm	1.56	1.85	0.32	8.13	nr	**9.97**
95 mm	1.88	2.22	0.46	11.68	nr	**13.90**
120 mm	2.00	2.37	1.25	31.74	nr	**34.11**
Earth clamps; connection to pipework						
15mm to 32mm dia	1.06	1.26	0.15	3.81	nr	**5.07**
32mm to 50mm dia	1.31	1.55	0.18	4.57	nr	**6.12**
50mm to 75mm dia	1.56	1.85	0.20	5.08	nr	**6.93**

W:COMMUNICATIONS/SECURITY/CONTROL

Item	Net Price £	Material £	Labour hours	Labour £	Unit	Total rate £
W52 : LIGHTNING PROTECTION						
CONDUCTOR TAPE						
PVC sheathed copper tape						
25 x 3 mm	8.72	10.84	0.30	7.62	m	**18.46**
25 x 6 mm	15.30	19.01	0.30	7.62	m	**26.63**
50 x 6 mm	32.57	40.47	0.30	7.62	m	**48.09**
PVC sheathed copper solid circular conductor						
8mm	5.42	6.42	0.50	12.70	m	**19.11**
Bare copper tape						
20 x 3 mm	6.38	7.54	0.30	7.62	m	**15.16**
25 x 3 mm	6.51	8.09	0.30	7.62	m	**15.71**
25 x 6 mm	13.03	16.19	0.40	10.16	m	**26.34**
50 x 6 mm	26.05	30.83	0.50	12.70	m	**43.53**
Bare copper solid circular conductor						
8mm	4.33	5.12	0.50	12.70	m	**17.81**
Tape fixings; flat; metallic PVC sheathed copper						
25 x 3 mm	5.46	6.78	0.33	8.38	nr	**15.16**
25 x 6 mm	5.65	7.02	0.33	8.38	nr	**15.40**
50 x 6 mm	9.31	11.57	0.33	8.38	nr	**19.95**
8mm	6.00	7.46	0.50	12.70	nr	**20.15**
Bare copper						
20 x 3 mm	6.38	7.92	0.30	7.62	nr	**15.54**
25 x 3 mm	6.51	8.09	0.30	7.62	nr	**15.71**
25 x 6 mm	13.03	16.19	0.40	10.16	nr	**26.34**
50 x 6 mm	26.05	30.83	0.50	12.70	nr	**43.53**
8mm	4.33	5.37	0.50	12.70	nr	**18.07**
Tape fixings; flat; non-metallic; PVC sheathed copper						
25 x 3 mm	0.71	0.88	0.30	7.62	nr	**8.50**
Tape fixings; flat; non-metallic; Bare copper						
20 x 3 mm	0.69	0.85	0.30	7.62	nr	**8.47**
25 x 3 mm	0.69	0.85	0.30	7.62	nr	**8.47**
50 x 6 mm	1.76	2.09	0.30	7.62	nr	**9.70**
Puddle flanges; copper						
600 mm long	66.41	78.60	0.93	23.61	nr	**102.21**
AIR RODS						
Pointed air rod fixed to structure; copper 10 mm diameter						
500 mm long	12.29	14.54	1.00	25.39	nr	**39.93**
1000mm long	18.64	22.06	1.50	38.09	nr	**60.14**
Extra for						
Air terminal base	15.91	18.83	0.35	8.89	nr	**27.72**
Strike Pad	22.59	26.73	0.35	8.89	nr	**35.62**

W:COMMUNICATIONS/SECURITY/CONTROL

Item	Net Price £	Material £	Labour hours	Labour £	Unit	Total rate £
Pointed air rod fixed to structure; copper						
16 mm diameter						
500 mm long	17.52	20.74	0.91	23.11	nr	**43.85**
1000mm long	32.01	37.89	1.75	44.44	nr	**82.32**
2000mm long	58.46	69.19	2.50	63.48	nr	**132.67**
Extra for						
Multiple point	33.98	40.21	0.35	8.89	nr	**49.10**
Air terminal base	17.34	20.52	0.35	8.89	nr	**29.41**
Ridge saddle	30.38	35.95	0.35	8.89	nr	**44.83**
Side mounting bracket	30.23	35.77	0.50	12.70	nr	**48.47**
Rod to tape coupling	14.07	16.66	0.50	12.70	nr	**29.35**
Strike Pad	22.59	26.73	0.35	8.89	nr	**35.62**
AIR TERMINALS						
16 mm diameter						
500 mm long	17.52	20.74	0.65	16.50	nr	**37.24**
1000mm long	32.01	37.89	0.78	19.81	nr	**57.69**
2000mm long	58.46	69.19	1.50	38.09	nr	**107.28**
Extra for						
Multiple point	33.98	40.21	0.35	8.89	nr	**49.10**
Flat saddle	17.34	20.52	0.35	8.89	nr	**29.41**
Side bracket	30.23	35.77	0.50	12.70	nr	**48.47**
Rod to cable coupling	14.07	16.66	0.50	12.70	nr	**29.35**
BONDS AND CLAMPS						
Bond to flat surface; copper						
26 mm	3.35	3.96	0.45	11.43	nr	**15.39**
8 mm diameter	13.54	16.02	0.33	8.38	nr	**24.40**
Pipe bond						
26 mm	6.51	7.71	0.45	11.43	nr	**19.13**
8 mm diameter	28.86	34.16	0.33	8.38	nr	**42.54**
Rod to tape clamp						
26 mm	5.44	6.44	0.45	11.43	nr	**17.86**
Square clamp; copper						
25 x 3 mm	5.86	6.94	0.33	8.38	nr	**15.32**
50 x 6 mm	25.99	30.76	0.50	12.70	nr	**43.45**
8 mm diameter	6.91	8.18	0.33	8.38	nr	**16.56**
Test clamp; copper						
26 x 8 mm; oblong	9.14	10.81	0.50	12.70	nr	**23.51**
26 x 8 mm; plate type	25.04	29.63	0.50	12.70	nr	**42.33**
26 x 8 mm; screw down	22.50	26.63	0.50	12.70	nr	**39.32**
Cast in earth points						
2 hole	17.94	21.23	0.75	19.04	nr	**40.27**
4 hole	28.51	33.74	1.00	25.39	nr	**59.14**
Extra for cast in earth points						
Cover plate; 25 x 3 mm	20.14	23.83	0.25	6.35	nr	**30.18**
Cover plate; 8 mm	20.14	23.83	0.25	6.35	nr	**30.18**
Rebar clamp; 8 mm	43.19	51.11	0.25	6.35	nr	**57.46**
Static earth receptacle	98.03	116.01	0.50	12.70	nr	**128.71**

W:COMMUNICATIONS/SECURITY/CONTROL

Item	Net Price £	Material £	Labour hours	Labour £	Unit	Total rate £
W52 : LIGHTNING PROTECTION (cont'd)						
BONDS AND CLAMPS (cont'd)						
Copper braided bonds						
25 x 3 mm						
200 mm hole centres	12.38	14.65	0.33	8.38	nr	**23.02**
400 mm holes centres	18.19	21.52	0.40	10.16	nr	**31.68**
U bolt clamps						
16 mm	7.24	8.57	0.33	8.38	nr	**16.94**
20 mm	8.20	9.70	0.33	8.38	nr	**18.08**
25 mm	10.05	11.89	0.33	8.38	nr	**20.27**
EARTH PITS/MATS						
Earth inspection pit; hand to others for fixing						
Concrete	36.13	42.76	1.00	25.39	nr	**68.16**
Polypropylene	35.86	42.44	1.00	25.39	nr	**67.83**
Extra for						
5 hole copper earth bar; concrete pit	29.16	34.51	0.35	8.89	nr	**43.40**
5 hole earth bar; polypropylene	24.90	29.47	0.35	8.89	nr	**38.36**
Water proof electrode seal						
Single flange	232.36	275.00	0.93	23.61	nr	**298.61**
Double flange	390.98	462.72	0.93	23.61	nr	**486.33**
Earth electrode mat; laid in ground and connected						
Copper tape lattice						
600 x 600 x 3 mm	74.63	88.32	0.93	23.61	nr	**111.93**
900 x 900 x 3 mm	133.69	158.22	0.93	23.61	nr	**181.83**
Copper tape plate						
600 x 600 x 1.5 mm	70.97	84.00	0.93	23.61	nr	**107.61**
600 x 600 x 3 mm	141.95	167.99	0.93	23.61	nr	**191.61**
900 x 900 x 1.5 mm	159.13	188.32	0.93	23.61	nr	**211.93**
900 x 900 x 3 mm	308.74	365.38	0.93	23.61	nr	**389.00**
EARTH RODS						
Solid cored copper earth electrodes driven into ground and connected						
15 mm diameter						
1200 mm long	21.48	25.41	0.93	23.61	nr	**49.03**
Extra for						
Coupling	1.20	1.42	0.06	1.52	nr	**2.94**
Driving stud	1.46	1.73	0.06	1.52	nr	**3.25**
Spike	1.38	1.63	0.06	1.52	nr	**3.15**
Rod Clamp; flat tape	5.44	6.44	0.25	6.35	nr	**12.78**
Rod Clamp; solid conductor	2.65	3.14	0.25	6.35	nr	**9.48**
20 mm diameter						
1200 mm long	39.01	46.17	0.98	24.88	nr	**71.05**

W:COMMUNICATIONS/SECURITY/CONTROL

Item	Net Price £	Material £	Labour hours	Labour £	Unit	Total rate £
Extra for						
Coupling	1.20	1.42	0.06	1.52	nr	**2.94**
Driving stud	2.42	2.87	0.06	1.52	nr	**4.39**
Spike	2.24	2.65	0.06	1.52	nr	**4.17**
Rod Clamp; flat tape	5.44	6.44	0.25	6.35	nr	**12.78**
Rod Clamp; solid conductor	2.98	3.52	0.25	6.35	nr	**9.87**
Stainless steel earth electrodes driven into ground and connected						
16 mm diameter						
1200 mm long	49.83	58.97	0.93	23.61	nr	**82.58**
Extra for						
Coupling	1.56	1.85	0.06	1.52	nr	**3.37**
Driving head	1.46	1.73	0.06	1.52	nr	**3.25**
Spike	1.38	1.63	0.06	1.52	nr	**3.15**
Rod Clamp; flat tape	5.44	6.44	0.25	6.35	nr	**12.78**
Rod Clamp; solid conductor	2.65	3.14	0.25	6.35	nr	**9.48**
SURGE PROTECTION						
Single Phase; including connection to equipment						
90 - 150v	247.25	292.61	5.00	126.96	nr	**419.57**
200 - 280v	247.25	292.61	5.00	126.96	nr	**419.57**
Three Phase; including connection to equipment						
156 - 260v	488.75	578.42	10.00	253.91	nr	**832.34**
346 - 484v	488.75	578.42	10.00	253.91	nr	**832.34**
349 - 484v; remote display	546.25	646.47	10.00	253.91	nr	**900.38**
346 - 484v; 60kA	948.75	1122.82	10.00	253.91	nr	**1376.73**
346 - 484v; 120kA	1811.25	2143.56	10.00	253.91	nr	**2397.47**

W:COMMUNICATIONS/SECURITY/CONTROL

Item	Net Price £	Material £	Labour hours	Labour £	Unit	Total rate £
W60 :CENTRAL CONTROL/BUILDING MANAGEMENT						
Equipment						
Switches/sensors; includes fixing in position; electrical work elsewhere. **Note - these are normally free issued to the mechanical contractor for fitting.** **The labour times applied assume the installation has been prepared for the fitting of the component.**						
Pressure devices						
Liquid differential pressure sensor	124.40	147.22	0.50	12.70	nr	159.92
Liquid differential pressure switch	68.22	80.74	0.50	12.70	nr	93.44
Air differential pressure transmitter	93.90	111.13	0.50	12.70	nr	123.83
Air differential pressure switch	13.24	15.66	0.50	12.70	nr	28.36
Liquid level switch	47.71	56.46	0.50	12.70	nr	69.16
Static pressure sensor	322.59	381.78	0.50	12.70	nr	394.48
High pressure switch	68.21	80.73	0.50	12.70	nr	93.42
Low pressure switch	68.21	80.73	0.50	12.70	nr	93.42
Water pressure switch	68.21	80.73	0.50	12.70	nr	93.42
Duct averaging temperature sensor	112.35	132.96	1.00	25.39	nr	158.35
Temperature devices						
Return air sensor (fan coils)	5.63	6.66	1.00	25.39	nr	32.05
Frost thermostat	23.27	27.54	0.50	12.70	nr	40.24
Immersion thermostat	48.10	56.92	0.50	12.70	nr	69.62
Temperature high limit	44.98	53.24	0.50	12.70	nr	65.93
Temperature sensor with averaging element	112.35	132.96	0.50	12.70	nr	145.66
Immersion temperature sensor	48.95	57.93	0.50	12.70	nr	70.63
Space temperature sensor	4.82	5.70	1.00	25.39	nr	31.09
Combined space temperature & humidity sensor	115.56	136.76	1.00	25.39	nr	162.15
Outside air temperature sensor	9.63	11.40	2.00	50.78	nr	62.18
Outside air temperature & humidity sensor	133.22	157.66	2.00	50.78	nr	208.44
Duct humidity sensor	124.39	147.21	0.50	12.70	nr	159.90
Space humidity sensor	115.56	136.76	1.00	25.39	nr	162.15
Immersion water flow sensor	115.56	136.76	0.50	12.70	nr	149.46
Rain sensor	164.51	194.70	2.00	50.78	nr	245.48
Wind speed and direction sensor	845.03	1000.07	2.00	50.78	nr	1050.85
Controllers; includes fixing in position; electrical work elsewhere						
Zone						
Fan coil controller	219.35	259.59	2.00	50.78	nr	310.38
VAV controller	219.35	259.59	2.00	50.78	nr	310.38
Plant						
Controller, 96 I/O points (exact configuration is dependent upon the number of I/O boards added)	4245.23	5024.10	0.50	12.70	nr	5036.79
Controller, 48 I/O points (exact configuration is dependent upon the number of I/O boards added)	2271.07	2687.75	0.50	12.70	nr	2700.44
Controller, 32 I/O points (exact configuration is dependent upon the number of I/O boards added)	1613.03	1908.97	0.50	12.70	nr	1921.66

W:COMMUNICATIONS/SECURITY/CONTROL

Item	Net Price £	Material £	Labour hours	Labour £	Unit	Total rate £
Additional Digital Input Boards (12 DI)	505.57	598.33	0.20	5.08	nr	603.41
Additional Digital Output Boards (6 DO)	329.02	389.39	0.20	5.08	nr	394.47
Additional analogue Input Boards (8 AI)	329.02	389.39	0.20	5.08	nr	394.47
Additional analogue Output Boards (8 AO)	329.02	389.39	0.20	5.08	nr	394.47
Outstation Enclosure (fitted in riser with space allowance for controller and network device)	227.73	269.51	5.00	126.96	nr	396.47
Damper actuator; electrical work elsewhere						
Damper actuator 0-10v	68.05	80.54	-	-	nr	80.54
Damper actuator with auxiliary switches	89.88	106.37	-	-	nr	106.37
Frequency inverters: not mounted within MCC; includes fixing in position; electrical work elsewhere						
2.2kW	506.39	599.30	2.00	50.78	nr	650.08
3kW	558.54	661.02	2.00	50.78	nr	711.80
7.5kW	706.20	835.77	2.00	50.78	nr	886.55
11kW	957.65	1133.35	2.00	50.78	nr	1184.13
15kW	1237.45	1464.49	2.00	50.78	nr	1515.27
18.5kW	1372.28	1624.05	2.50	63.48	nr	1687.52
20kW	1649.94	1952.65	2.50	63.48	nr	2016.13
30kW	1862.61	2204.35	2.50	63.48	nr	2267.83
55kW	3881.70	4593.88	3.00	76.17	nr	4670.05
Miscellaneous; includes fixing in position; electrical work elsewhere						
1kW Thyristor	77.04	91.17	2.00	50.78	nr	141.96
10kW Thyristor	205.44	243.13	2.00	50.78	nr	293.91
Front end and networking; electrical work elsewhere						
PC/monitor	2697.29	3192.16	2.00	50.78	nr	3242.94
Dot matrix printer	489.06	578.79	2.00	50.78	nr	629.58
PC Software	2411.48	2853.91	-	-	nr	2853.91
Lonmaker software	858.13	1015.57	-	-	nr	1015.57
Lonmaker credits	6.36	7.52	-	-	nr	7.52
Network server software	1209.26	1431.12	-	-	nr	1431.12
Router (allows connection to a network)	679.45	804.11	-	-	nr	804.11

DAVIS LANGDON

EUROPE & MIDDLE EAST
office locations

EUROPE & MIDDLE EAST

ENGLAND

**DAVIS LANGDON
LONDON**
MidCity Place
71 High Holborn
London WC1V 6QS
Tel: (020) 7061 7000
Fax: (020) 7061 7061
Email: simon.johnson@davislangdon.com

BIRMINGHAM
75-77 Colmore Row
Birmingham B3 2HD
Tel: (0121) 710 1100
Fax: (0121) 710 1399
Email: david.daly@davislangdon.com

BRISTOL
St Lawrence House
29/31 Broad Street
Bristol BS1 2HF
Tel: (0117) 927 7832
Fax: (0117) 925 1350
Email: alan.francis@davislangdon.com

CAMBRIDGE
36 Storey's Way
Cambridge CB3 0DT
Tel: (01223) 351 258
Fax: (01223) 321 002
Email: laurence.brett@davislangdon.com

LEEDS
No 4 The Embankment
Victoria Wharf
Sovereign Street
Leeds LS1 4BA
Tel: (0113) 243 2481
Fax: (0113) 242 4601
Email: duncan.sissons@davislangdon.com

LIVERPOOL
Cunard Building
Water Street
Liverpool
L3 1JR
Tel: (0151) 236 1992
Fax: (0151) 227 5401
Email: andrew.stevenson@davislangdon.com

MAIDSTONE
11 Tower View
Kings Hill
West Malling
Kent ME19 4UY
Tel: (01732) 840 429
Fax: (01732) 842 305
Email: nick.leggett@davislangdon.com

MANCHESTER
Cloister House
Riverside
New Bailey Street
Manchester
M3 5AG
Tel: (0161) 819 7600
Fax: (0161) 819 1818
Email: paul.stanion@davislangdon.com

MILTON KEYNES
Everest House
Rockingham Drive
Linford Wood
Milton Keynes
MK14 6LY
Tel: (01908) 304 700
Fax: (01908) 660 059
Email: kevin.sims@davislangdon.com

NORWICH
63 Thorpe Road
Norwich NR1 1UD
Tel: (01603) 628 194
Fax: (01603) 615 928
Email: michael.ladbrook@davislangdon.com

OXFORD
Avalon House
Marcham Road
Abingdon
Oxford OX14 1TZ
Tel: (01235) 555 025
Fax: (01235) 554 909
Email: paul.coomber@davislangdon.com

PETERBOROUGH
Clarence House
Minerva Business Park
Lynchwood
Peterborough PE2 6FT
Tel: (01733) 362 000
Fax: (01733) 230 875
Email: stuart.bremner@davislangdon.com

PLYMOUTH
1 Ensign House
Parkway Court
Longbridge Road
Plymouth PL6 8LR
Tel: (01752) 827 444
Fax: (01752) 221 219
Email: gareth.steventon@davislangdon.com

SOUTHAMPTON
Brunswick House
Brunswick Place
Southampton SO15 2AP
Tel: (023) 8033 3438
Fax: (023) 8022 6099
Email: chris.tremellen@davislangdon.com

**LEGAL SUPPORT
LONDON**
MidCity Place
71 High Holborn
London WC1V 6QS
Tel: (020) 7061 7000
Fax: (020) 7061 7061
Email: mark.hackett@davislangdon.com

**MANAGEMENT CONSULTING
LONDON**
MidCity Place
71 High Holborn
London WC1V 6QS
Tel: (020) 7061 7000
Fax: (020) 7061 7005
Email: john.connaughton@davislangdon.com

SPECIFICATION CONSULTING
Davis Langdon Schumann Smith
STEVENAGE
Southgate House
Southgate
Stevenage
SG1 1HG
Tel: (01438) 742 642
Fax: (01438) 742 632
Email: nick.schumann@schumannsmith.com

MANCHESTER
Cloister House
Riverside, New Bailey Street
Manchester M3 5AG
Tel: (0161) 819 7600
Fax: (0161) 819 1818
Email: richard.jackson@davislangdon.com

ENGINEERING SERVICES
Davis Langdon Mott Green Wall
MidCity Place
71 High Holborn
London WC1V 6QS
Tel: (020) 7061 7777
Fax: (020) 7061 7009
Email: barry.nugent@mottgreenwall.co.uk

**PROPERTY TAX & FINANCE
LONDON**
Davis Langdon Crosher & James
MidCity Place
71 High Holborn
London WC1V 6QS
Tel: (020) 7061 7077
Fax: (020) 7061 7078
Email: tony.llewellyn@crosherjames.com

BIRMINGHAM
102 New Street
Birmingham
B2 4HQ
Tel: (0121) 632 3600
Fax: (0121) 632 3601
Email: clive.searle@crosherjames.com

SOUTHAMPTON
Brunswick House
Brunswick Place
Southampton SO15 2AP
Tel: (023) 8068 2800
Fax: (0870) 048 8141
Email: david.rees@crosherjames.com

SCOTLAND

**DAVIS LANGDON
EDINBURGH**
39 Melville Street
Edinburgh EH3 7JF
Tel: (0131) 240 1350
Fax: (0131) 240 1399
Email: sam.mackenzie@davislangdon.com

GLASGOW
Monteith House
11 George Square
Glasgow G2 1DY
Tel: (0141) 248 0300
Fax: (0141) 248 0303
Email: sam.mackenzie@davislangdon.com

PROPERTY TAX & FINANCE
Davis Langdon Crosher & James
EDINBURGH
39 Melville Street
Edinburgh EH3 7JF
Tel: (0131) 220 4225
Fax: (0131) 220 4226
Email: ian.mcfarlane@crosherjames.com

GLASGOW
Monteith House
11 George Square
Glasgow G2 1DY
Tel: (0141) 248 0333
Fax: (0141) 248 0313
Email: ken.fraser@crosherjames.com

WALES

**DAVIS LANGDON
CARDIFF**
4 Pierhead Street
Capital Waterside
Cardiff CF10 4QP
Tel: (029) 2049 7497
Fax: (029) 2049 7111
Email: paul.edwards@davislangdon.com

PROPERTY TAX & FINANCE
Davis Langdon Crosher & James
CARDIFF
4 Pierhead Street
Capital Waterside
Cardiff CF10 4QP
Tel: (029) 2049 7497
Fax: (029) 2049 7111
Email: michael.murray@crosherjames.com

IRELAND

**DAVIS LANGDON PKS
CORK**
Hibernian House
80A South Mall
Cork. Ireland
Tel: (00 353 21) 4222 800
Fax: (00 353 21) 4222 801
Email: alangmaid@dlpks.ie

DUBLIN
24 Lower Hatch Street
Dublin 2, Ireland
Tel: (00 353 1) 676 3671
Fax: (00 353 1) 676 3672
Email: mwebb@dlpks.ie

GALWAY
Heritage Hall
Kirwan's Lane
Galway
Ireland
Tel: (00 353 91) 530 199
Fax: (00 353 91) 530 198
Email: joregan@dlpks.ie

LIMERICK
8 The Crescent
Limerick, Ireland
Tel: (00 353 61) 318 870
Fax: (00 353 61) 318 871
Email: cbarry@dlpks.ie

**SPECIFICATION CONSULTING
DUBLIN**
24 Lower Hatch Street
Dublin 2, Ireland
Tel: (00 353 1) 676 3671
Fax: (00 353 1) 676 3672
Email: jhartnett@dlpks.ie

SPAIN

**DAVIS LANGDON EDETCO
BARCELONA**
C/Muntaner, 479, 1-2
Barcelona 08021
Spain
Tel: (00 34 93) 418 6899
Fax: (00 34 93) 211 0003
Contact: Francesc Monells
Email: barcelona@edetco.com

GIRONA
C/Salt 10 2on
Girona 17005
Spain
Tel: (00 34 97) 223 8000
Fax: (00 34 97) 224 2661
Contact: Francesc Monells
Email: girona@edetco.com

RUSSIA

**RUPERTI PROJECT SERVICES
INTERNATIONAL
MOSCOW**
Office 5, 15 Myasnitskaya ul
Moscow 101000
Russia
Tel: (00 7 495) 933 7810
Fax: (00 7 495) 933 7811
Email: anthony.ruperti@davislangdon.co

LEBANON

**DAVIS LANGDON
BEIRUT**
PO Box 13-5422-Shouran
Beirut
Lebanon
Tel: (00 9611) 780 111
Fax: (00 9611) 809 045
Contact: Muhyidden Itani
Email: DLL.MI@cyberia.net.lb

BAHRAIN

**DAVIS LANGDON
MANAMA**
3rd Floor Building 256
Road No 3605
Area No 336
PO Box 640
Manama
Bahrain
Tel: (00 973) 1782 7567
Fax: (00 973) 1772 7210
Email: steven.coates@davislangdon-bahrain.com

UNITED ARAB EMIRATES

**DAVIS LANGDON
DUBAI**
PO Box 7856
No. 410
Oud Metha Office Building
Dubai
UAE
Tel: (00 9714) 32 42 919
Fax: (00 9714) 32 42 838
Email: neil.taylor@davislangdon-dubai.com

QATAR

**DAVIS LANGDON
DOHA**
PO Box 3206
Doha
State of Qatar
Tel: (00 974) 4580 150
Fax: (00 974) 4697 905
Email:steven.humphrey@davislangdon-qatar.com

EGYPT

**DAVIS LANGDON
CAIRO**
35 Misr Helwan Road
Maadi 11431
Cairo
Egypt
Tel: (00 20 2) 526 2319
Fax: (00 20 2) 527 1338
Email: bob.ames@dlegypt.com

Specialist Service Lines
Project Management | Cost Management | Management Consulting | Legal Support | Specification Consulting | Engineering Services | Property Tax & Finance

Specialist Sectors
Arts | Commercial Offices | Distribution | Education | Food Processing | Health | Heritage | Hotels & Leisure | Industrial | Infrastructure | Public Buildings | Regeneration | Residential | Retail | Sports | Transportation

Davis Langdon LLP is a member firm of Davis Langdon & Seah International, with offices throughout Europe and the Middle East, Asia, Australasia, Africa and the USA

Rates of Wages

4th Edition
Design of Electrical Services for Buildings

B Rigby

Electrical services are a vital component in any building, so it is necessary for construction professionals to understand the basic principle of services design. *The Design of Electrical Services for Buildings* aims to provide a basic grounding for students and graduates in the field. It covers methods of wiring, schemes of distribution and protection for lighting and power installations. Systems such as alarms and standby supplies are also covered. Each method is described in detail and examples of calculations are given.

For this fourth edition, the coverage of wiring and electrical regulations have been brought fully up-to-date, and the practical information has been revised.

December 2004: 234x156mm: 352 pages
120 line figures, 55 b+w photos, 25 tables
HB: 0-415-31082-2: £65.00
PB: 0-415-31083-0: £22.99

To Order: Tel: +44 (0) 1264 343071 Fax: +44 (0) 1264 343005, or
Post: Taylor and Francis Customer Services, Thomson Publishing Services, Cheriton House, Andover, Hants, SP10 5BE, UK Email: book.orders@tandf.co.uk

For a complete listing of all our titles visit:
www.tandf.co.uk

Taylor & Francis
Taylor & Francis Group plc

Mechanical Installations

Rates of Wages

HEATING, VENTILATING, AIR CONDITIONING, PIPING AND DOMESTIC ENGINEERING INDUSTRY

At the time of going to print the Joint Conciliation Committee of the Heating, Ventilating and Domestic Engineering Industry has not announced a new wage agreement. Therefore, having taken account of the previous three years trend and inflation all non promulgated rates have been increased by 5% this being in line with the agreements for the last three years. Users of the book are advised to register on the SPON's website www.pricebooks.co.uk/updates to receive the free quarterly updates - alerts will then be provided by e-mail as changes arise.

For full details of the existing and forthcoming wage agreement and the Heating Ventilating Air Conditioning Piping and Domestic Engineering Industry's National Working Rule Agreement, contact:

Heating and Ventilating Contractor's Association
ESCA House,
34 Palace Court,
Bayswater,
London W2 4JG
Telephone: 020 7313 4900
Internet: www.hvca.org.uk

WAGE RATES, ALLOWANCES AND OTHER PROVISIONS

Hourly rates of wages
All districts of the United Kingdom

Main Grades	From 3 October 2005 +5% p/hr
Foreman	13.86
Senior Craftsman (+2nd welding skill)	11.92
Senior Craftsman	11.45
Craftsman (+2nd welding skill)	10.97
Craftsman	10.50
Operative	9.52
Adult Trainee	8.02
Mate (over 18)	8.02
Mate (17-18)	5.16
Mate (up to 17)	3.52
Modern Apprentices	
Junior	5.22
Intermediate	7.38
Senior	9.52

Note: Ductwork Erection Operatives are entitled to the same rates and allowances as the parallel Fitter grades shown.

Trainee Rates of Pay

Junior Ductwork Trainees (Probationary)

	From 3 October2005 +5% p/hr
Age at entry	
17	4.29
18	5.17
19	6.60
20	8.03

Junior Ductwork Erectors (Year of Training)

	From 3 October 2005 +5%		
	1 yr *p/h*	*2 yr* *p/hr*	*3 yr* *p/hr*
Age at entry			
17	5.17	6.60	8.03
18	6.60	8.03	8.57
19	8.03	8.10	8.95
20	8.03	8.51	8.96

Responsibility Allowance (Craftsmen)

	From 3 October2005 +5% p/hr
Second welding skill or supervisory responsibility (one unit)	0.47
Second welding skill and supervisory responsibility (two units)	0.95

Responsibility Allowance (Senior Craftsmen)

	From 3 October 2005 +5% p/hr
Second welding skill	0.47
Supervising responsibility	0.95
Second welding skill and supervisory responsibility	1.42

Daily travelling allowance – Scale 2

C: Craftsmen including Installers
M&A: Mates, Apprentices and Adult Trainees

Direct distance from centre to job in miles

		From 3 October 2005 +5%	
		C	M&A
Over	*Not exceeding*	*p/hr*	*p/hr*
15	20	2.30	1.97
20	30	5.92	5.12
30	40	8.53	7.37
40	50	11.22	9.61

HEATING, VENTILATING, AIR CONDITIONING, PIPING & DOMESTIC ENGINEERING INDUSTRY

Daily travelling allowance – Scale 1

C: Craftsmen including Installers
M&A: Mates, Apprentices and Adult Trainees

Direct distance from centre to job in miles

		From 3 October 2005 +5%	
		C	M&A
Over	*Not exceeding*	*p/hr*	*p/hr*
15	20	8.60	8.27
20	30	12.22	11.42
30	40	14.83	13.67
40	50	17.52	15.91

Weekly Holiday Credit and Welfare Contributions

	From						
	3 October 2005 +5%						
	£	£	£	£	£	£	£
	a	b	c	d	e	f	g
Weekly Holiday Credit	66.29	61.67	59.38	57.10	54.81	52.53	50.24
Combined Weekly/Welfare Holiday Credit and Contribution	72.04	67.42	65.13	62.85	60.75	58.29	56.00

	From					
	6 October 2005					
	£	£	£	£	£	£
	h	i	j	k	l	m
Weekly Holiday Credit	45.57	38.38	35.30	25.02	24.68	17.82
Combined Weekly/Welfare Holiday Credit and Contribution	51.32	44.13	41.06	30.78	30.43	23.57

The grades of H&V Operatives entitled to the different rates of Weekly Holiday Credit and Welfare Contribution are as follows:

a Foreman	*b* Senior Craftsman (RAS & RAW)	*c* Senior Craftsman (RAS)	*d* Senior Craftsman (RAW)
e Senior Craftsman Craftsman Apprentice (+2 RA)	*f* Craftsman (+ 1RA)	*g* Craftsman	*h* Installer Senior Modern
i Adult Trainee Mate (over 18)	*j* Intermediate Modern Apprentice	*k* Junior Modern Apprentice	*l* Mate (aged 17-18)
m Mate (under 17)			

HEATING, VENTILATING, AIR CONDITIONING, PIPING & DOMESTIC ENGINEERING INDUSTRY

	From *3 October 2005*
Daily abnormal conditions money Per day	2.99 (Assume no change)

	From *3 November 2003*
Lodging allowance Per night	26.50 (Assume no change)

Explanatory Notes

1.   Working Hours

The normal working week (Monday to Friday) shall be 38 hours.

2.   Overtime

Time worked in excess of 38 hours during the normal working week shall be paid at time and a half until 12 hours have been worked since the actual starting time. Thereafter double time shall be paid until normal starting time the following morning. Weekend overtime shall be paid at time and a half for the first 5 hours worked on a Saturday and at double time thereafter until normal starting time on Monday morning.

PLUMBING MECHANICAL ENGINEERING SERVICES INDUSTRY

PLUMBING MECHANICAL ENGINEERING SERVICES INDUSTRY

The Joint Industry Board for Plumbing Mechanical Engineering Services has agreed a two year wage agreement for 2006 – 2007 with effect from 2 January 2006. Current rates of pay for 2007 will be effective from 1 January 2007.

For full details of this wage agreement and the JIB PMES National Working Rules, contact:

The Joint Industry Board for Plumbing Mechanical Engineering Services in England and Wales
Brook House,
Brook Street,
St Neots,
Huntingdon,
Cambridge PE19 2HW
Telephone: 01480 476925
E-mail: info@jib-pmes.org.uk

WAGE RATES, ALLOWANCES AND OTHER PROVISIONS

EFFECTIVE FROM 1 JANUARY 2007

Basic Rates of Hourly Pay

Applicable in England and Wales

	Hourly rate £
Operatives	
Technical plumber and gas service technician	12.88
Advanced plumber and gas service engineer	11.60
Trained plumber and gas service fitter	9.94
Apprentices	
4th year of training with NVQ level 3	9.62
4th year of training with NVQ level 2	8.71
4th year of training	7.68
3rd year of training with NVQ level 2	7.58
3rd year of training	6.24
2nd year of training	5.52
1st year of training	4.82
Adult Trainees	
3rd 6 months of employment	8.68
2nd 6 months of employment	8.32
1st 6 months of employment	7.76

PLUMBING MECHANICAL ENGINEERING SERVICES INDUSTRY

Major Projects Agreement

Where a job is designated as being a Major Project then the following Major Project Performance Payment hourly rate supplement shall be payable:

Employee Category

	National Payment £	London* Payment £
Technical Plumber and Gas Service Technician	2.20	3.57
Advanced Plumber and Gas Service Engineer	2.20	3.57
Trained Plumber and Gas Service Fitter	2.20	3.57
All 4th year apprentices	1.76	2.86
All 3rd year apprentices	1.32	2.68
2nd year apprentice	1.21	1.96
1st year apprentice	0.88	1.43
All adult trainees	1.76	2.86

* The London Payment Supplement applies only to designated Major Projects that are within the M25 London orbital motorway and are effective from 1 February 2007.

* National payment hourly rates are unchanged for 2007 and will continue to be at the rates shown in Promulgation 138A issued 14 October 2003.

Allowances

Daily travel time allowance plus return fares

All daily travel allowances are to be paid at the daily rate as follows :

Over	Not exceeding	All Operatives	3rd & 4th Year Apprentices	1st & 2nd Year Apprentices
20	30	£3.65	£2.35	£1.45
30	40	£8.50	£5.45	£3.50
40	50	£9.70	£5.80	£3.65

Responsibility/Incentive Pay Allowance

As from Monday 3rd September 2003, Employers may, in consultation with the employees concerned, enhance the basic graded rates of pay by the payment of an additional amount, as per the bands shown below, where it is agreed that their work involves extra responsibility, productivity or flexibility.

Band 1 - an additional rate of £ 0.24 per hour
Band 2 - an additional rate of £ 0.44 per hour
Band 3 - an additional rate of £ 0.64 per hour
Band 4 - an additional rate of £ 0.84 per hour

This allowance forms part of an operative's basic rate of pay and shall be used to calculate premium payments.

Mileage allowance ... . £0.40 per mile

Lodging allowance ... . £21.06 per night

Subsistence Allowance (London Only) ... ... £4.18 per night

PLUMBING MECHANICAL ENGINEERING SERVICES INDUSTRY

Plumbers welding supplement

Possession of Gas or Arc Certificate.. £0.27 per hour
Possession of Gas and Arc Certificate.. £0.46 per hour

Weekly Holiday Credit Contributions (34th Issue Stamps Option)

	Public & Annual Gross Value, £	Combined Holiday Credit*, £
Technical Plumber and Gas Service Technician	62.50	64.50
Advance Plumber and Gas Service Engineer	56.20	58.10
Trained Plumber and Gas Service Fitter	48.20	49.70
Adult Trainee	36.20	37.30
Apprentice in last year of training	36.20	37.30
Apprentice 3rd year	26.00	26.80
Apprentice 2nd year	23.10	23.90
Apprentice 1st year	20.10	20.80
Working Principal	26.20	26.90
Ancillary Employee	31.70	60.50

* Public and Annual gross value is the Combined Holiday credit value less JIB administration value.

Explanatory Notes

1. Working Hours

    The normal working week (Monday to Friday) shall be 37½ hours, with 45 hours to be worked in the same period before overtime rates become applicable.

2. Overtime

    Overtime shall be paid at time and a half up to 8.00pm (Monday to Friday) and up to 1.00pm (Saturday). Overtime worked after these times shall be paid at double time.

3. Major Projects Agreement

    Under the Major Projects Agreement the normal working week shall be 38 hours (Monday to Friday) with overtime rates payable for all hours worked in excess of 38 hours in accordance with 2 above. However, it should be noted that the hourly rate supplement shall be paid for each hour worked but does not attract premium time enhancement.

4. Pension

    In addition to their hourly rates of pay, plumbing employees are entitled to inclusion within the Industry Pension Scheme (or one providing equivalent benefits). The current levels of industry scheme contributions are 6½% (employers) and 3¼% (employees).

Electrical Installations

Rates of Wages

ELECTRICAL CONTRACTING INDUSTRY

The Joint Industry Board for the Electrical Contracting Industry has agreed a two year wage agreement for 2006/07 & 2007/08 with effect from 9 January 2006. The 2007/08 increase takes effect from 8 January 2007.

For full details of this wage agreement and the Joint Industry Board for the Electrical Contracting Industry's National Working Rules, contact:

The Joint Industry Board for the Electrical Contracting Industry
Kingswood House
47/51 Sidcup Hill,
Sidcup,
Kent DA14 6HP
Telephone : 020 8302 0031
Internet : www.jib.org.uk

WAGES (Graded Operatives)

Rates

Since **7 January 2002** two different wage rates have applied to JIB Graded Operatives working on site, depending on whether the Employer transports them to site or whether they provide their own transport. The two categories are:

Job Employed (Transport Provided)

Payable to an Operative who is transported to and from the job by his Employer. The Operative shall also be entitled to payment for Travel Time, when travelling in his own time, as detailed in the appropriate scale.

Job Employed (Own Transport)

Payable to an Operative who travels by his own means to and from the job. The Operative shall be entitled to payment for Travel Allowance and also Travel Time, when travelling in his own time, as detailed in the appropriate scale.

The JIB rates of wages are set out below:

From and including 8 January 2007, the JIB hourly rates of wages shall be as set out below:

(i)   National Standard Rate:

Grade	Transport Provided	Own Transport
Technician (or equivalent specialist grade)	£ 13.48	£ 14.16
Approved Electrician (or equivalent specialist grade)	£ 11.90	£ 12.58
Electrician (or equivalent specialist grade)	£ 10.91	£ 11.60
Senior Graded Electrical Trainee	£ 9.82	£ 10.44
Electrical Improver	£ 9.82	£ 10.44
Labourer	£ 8.66	£ 9.30
Adult Trainee	£ 8.66	£ 9.30
Adult Trainee (under 21)	£ 6.50	£ 6.98

ELECTRICAL CONTRACTING INDUSTRY

(ii)   London Rate:

Grade	Transport Provided	Own Transport
Technician (or equivalent specialist grade)	£ 15.10	£ 15.86
Approved Electrician (or equivalent specialist grade)	£ 13.33	£ 14.09
Electrician (or equivalent specialist grade)	£ 12.22	£ 12.99
Senior Graded Electrical Trainee	£ 11.00	£ 11.69
Electrical Improver	£ 11.00	£ 11.69
Labourer	£ 9.70	£ 10.42
Adult Trainee	£ 9.70	£ 10.42
Adult Trainee (under 21)	£ 7.28	£ 7.28

1999 Joint Industry Board Apprentice Training Scheme

From and including 8 January 2007, the JIB hourly rates for Job Employed apprentices shall be:

(i)   National Standard Rates

	Transport Provided	Own Transport
Stage 1	£ 3.81	£ 4.46
Stage 2	£ 5.62	£ 6.28
Stage 3	£ 8.13	£ 8.81
Stage 4	£ 8.61	£ 9.29

(ii)   London Rate

	Transport Provided	Own Transport
Stage 1	£ 4.27	£ 5.00
Stage 2	£ 6.29	£ 7.06
Stage 3	£ 9.11	£ 9.87
Stage 4	£ 9.64	£ 10.40

ELECTRICAL CONTRACTING INDUSTRY

Travelling Time and Travel Allowances

From and including 8 January 2007

Operatives required to start/finish at the normal starting and finishing time on jobs which are 15 miles and over from the shop - in a straight line - receive payment for Travelling Time and where transport is not provided by the Employer, Travel Allowance, as follows:

Distance	Total Daily Travel Allowance	Total Daily Travelling Time
(a) National Standard Rate		
Up to 15 miles	Nil	Nil
Over 15 & up to 20 miles each way	£ 3.01	£ 4.11
Over 20 & up to 25 miles each way	£ 3.99	£ 5.20
Over 25 & up to 35 miles each way	£ 5.26	£ 6.35
Over 35 & up to 55 miles each way	£ 8.38	£ 8.38
Over 55 & up to 75 miles each way	£ 10.26	£ 10.26

For each additional 10 mile band over 75 miles, additional payment of £ 1.81 for Daily Travel Allowance and £1.81 for Daily Travel Time will be made.

Note: Special arrangements may apply for work in the Merseyside area.

Distance	Total Daily Travel Allowance	Total Daily Travelling Time
(b) London Rate		
Up to 15 miles	Nil	Nil
Over 15 & up to 20 miles each way	£ 3.03	£ 4.40
Over 20 & up to 25 miles each way	£ 4.02	£ 5.76
Over 25 & up to 35 miles each way	£ 5.30	£ 6.93
Over 35 & up to 55 miles each way	£ 8.45	£ 9.36
Over 55 & up to 75 miles each way	£ 10.34	£ 10.95

For each additional 10 mile band over 75 miles, additional payments of £ 1.82 for Daily Travel Allowance and £1.82 for Daily Travel Time will be made.

Travelling time and travel allowance - Section 8 (Employed permanently at the shop)

Operatives required to start/finish at the normal starting and finishing time on jobs which are 15 miles and over from the shop - in a straight line - receive payment for Travelling Time and where transport is not provided by the Employer, Travel Allowance, as follows:

From and including 8 January 2007

Distance	Total Daily Travel Allowance	Total Daily Travelling Time
(a) National Standard Rate		
Up to 15 miles	Nil	Nil
Over 15 & up to 20 miles each way	£ 3.01	£ 2.41
Over 20 & up to 25 miles each way	£ 3.99	£ 3.62
Over 25 & up to 35 miles each way	£ 5.26	£ 4.83
Over 35 & up to 55 miles each way	£ 8.38	£ 6.04
Over 55 & up to 75 miles each way	£ 10.26	£ 7.25

For each additional 10 mile band over 75 miles, additional payments of £ 1.81 for Daily Travel Allowance and £1.21 for Daily Travelling Time will be made.

Note: Special arrangements may apply for work in the Merseyside area.

ELECTRICAL CONTRACTING INDUSTRY

Distance	Total Daily Travel Allowance	Total Daily Travelling Time
(b) **London Rate**		
Up to 15 miles	Nil	Nil
Over 15 & up to 20 miles each way	£ 3.03	£ 2.67
Over 20 & up to 25 miles each way	£ 4.02	£ 4.02
Over 25 & up to 35 miles each way	£ 5.30	£ 5.36
Over 35 & up to 55 miles each way	£ 8.45	£ 6.68
Over 55 & up to 75 miles each way	£ 10.34	£ 8.03

For each additional 10 mile band over 75 miles, additional payments of £ 1.82 for Daily Travel Allowance and £1.22 for Daily Travelling Time will be made.

Lodging Allowances
£28.35 from and including 5 January 2004

Lodgings weekend retention fee, maximum reimbursement
£28.35 from and including 5 January 2004

Annual Holiday Lodging Allowance Retention
£5.85 per night (£ 40.95 per week) from and including 5 January 2004

Responsibility money

From and including 30 March 1998 the minimum payment increased to 10p per hour and the maximum to £1.00 per hour (no change)

From and including 4 January 1992 responsibility payments are enhanced by overtime and shift premiums where appropriate (no change)

Combined JIB Benefits Stamp Value (from week commencing 26 September 2005)

JIB Grade	Weekly JIB combined credit value	Holiday Value
Technician	£ 52.75	£ 40.75
Approved Electrician	£ 48.00	£ 36.00
Electrician	£ 45.00	£ 33.00
Senior Graded Electrical Trainee and Electrical Improver	£ 41.67	£ 29.67
Labourer & Adult Trainee	£ 38.18	£ 26.18
Adult Trainee (Under 21)	£ 38.18	£ 26.18

ELECTRICAL CONTRACTING INDUSTRY

Explanatory Notes

1. Working Hours

    The normal working week (Monday to Friday) shall be 37½ hours, with 38 hours to be worked in the same period before overtime rates become applicable.

2. Overtime

    Overtime shall be paid at time and a half for all weekday overtime. Saturday overtime shall be paid at time and a half for the first 6 hours, or up to 3.00pm (whichever comes first). Thereafter double time shall be paid until normal starting time on Monday.

Daywork

When work is carried out in connection with a contract that cannot be valued in any other way, it is usual to assess the value on a cost basis with suitable allowances to cover overheads and profit. The basis of costing is a matter for agreement between the parties concerned but definitions of prime cost for the Heating and Ventilating and Electrical Industries have been published jointly by the Royal Institution of Chartered Surveyors and the appropriate bodies of the industries concerned, for those who wish to use them.

These, together with a schedule of basic plant hire charges are reproduced on the following pages, with the kind permission of the Royal Institution of Chartered Surveyors, who own the copyright.

Spon's Estimating Costs Guide to Finishings
Painting and Decorating, Plastering and Tiling

Bryan Spain

Especially written for contractors and small businesses carrying out small works, Spon's Estimating Costs Guide to Finishings contains accurate information on thousands of rates, each broken down to labour, material overheads and profit.

This is the first book to include typical project costs for painting and wallpapering, plasterwork, floor and wall tiles, rendering, and external work.

More than just a price book, it gives easy-to-read, professional advice on setting up and running a business including help on producing estimates faster, and keeping estimates accurate and competitive.

Suitable for any size firm from one-man-band to established business, this book contains valuable commercial and cost information that contractors can't afford to be without.

March 2005: 216x138 mm: 288 pages
PB: 0-415-34411-5: £27.99

To Order: Tel: +44 (0) 1264 343071 Fax: +44 (0) 1264 343005, or
Post: Taylor and Francis Customer Services, Thomson Publishing Services, Cheriton House, Andover, Hants, SP10 5BE, UK Email: book.orders@tandf.co.uk

For a complete listing of all our titles visit:
www.tandf.co.uk

Taylor & Francis
Taylor & Francis Group plc

HEATING AND VENTILATING INDUSTRY

DEFINITION OF PRIME COST OF DAYWORK CARRIED OUT UNDER A HEATING, VENTILATING, AIR CONDITIONING, REFRIGERATION, PIPEWORK AND/OR DOMESTIC ENGINEERING CONTRACT (JULY 1980 EDITION)

This Definition of Prime Cost is published by the Royal Institution of Chartered Surveyors and the Heating and Ventilating Contractors Association for convenience, and for use by people who choose to use it. Members of the Heating and Ventilating Contractors Association are not in any way debarred from defining Prime Cost and rendering accounts for work carried out on that basis in any way they choose. Building owners are advised to reach agreement with contractors on the Definition of Prime Cost to be used prior to entering into a contract or sub-contract.

SECTION 1: APPLICATION

1.1   This Definition provides a basis for the valuation of daywork executed under such heating, ventilating, air conditioning, refrigeration, pipework and or domestic engineering contracts as provide for its use.
1.2   It is not applicable in any other circumstances, such as jobbing or other work carried out as a separate or main contract nor in the case of daywork executed after a date of practical completion.
1.3   The terms 'contract' and 'contractor' herein shall be read as 'sub-contract' and 'sub-contractor' as applicable.

SECTION 2: COMPOSITION OF TOTAL CHARGES

2.1   The Prime Cost of daywork comprises the sum of the following costs:
        (a)   Labour as defined in Section 3.
        (b)   Materials and goods as defined in Section 4.
        (c)   Plant as defined in Section 5.
2.2   Incidental costs, overheads and profit as defined in Section 6, as provided in the contract and expressed therein as percentage adjustments, are applicable to each of 2.1 (a)-(c).

SECTION 3: LABOUR

3.1   The standard wage rates, emoluments and expenses referred to below and the standard working hours referred to in 3.2 are those laid down for the time being in the rules or decisions or agreements of the Joint Conciliation Committee of the Heating, Ventilating and Domestic Engineering Industry applicable to the works (or those of such other body as may be appropriate) and to the grade of operative concerned at the time when and the area where the daywork is executed.
3.2   Hourly base rates for labour are computed by dividing the annual prime cost of labour, based upon the standard working hours and as defined in 3.4, by the number of standard working hours per annum. See example.
3.3   The hourly rates computed in accordance with 3.2 shall be applied in respect of the time spent by operatives directly engaged on daywork, including those operating mechanical plant and transport and erecting and dismantling other plant (unless otherwise expressly provided in the contract) and handling and distributing the materials and goods used in the daywork.
3.4   The annual prime cost of labour comprises the following:
        (a)   Standard weekly earnings (i.e. the standard working week as determined at the appropriate rate for the operative concerned).
        (b)   Any supplemental payments.
        (c)   Any guaranteed minimum payments (unless included in Section 6.1(a)-(p)).
        (d)   Merit money.
        (e)   Differentials or extra payments in respect of skill, responsibility, discomfort, inconvenience or risk (excluding those in respect of supervisory responsibility - see 3.5)
        (f)   Payments in respect of public holidays.
        (g)   Any amounts which may become payable by the contractor to or in respect of operatives arising from the rules etc. referred to in 3.1 which are not provided for in 3.4 (a)-(f) nor in Section 6.1 (a)-(p).
        (h)   Employers contributions to the WELPLAN, the HVACR Welfare and Holiday Scheme or payments in lieu thereof.
        (i)   Employers National Insurance contributions as applicable to 3.4 (a)-(h).
        (j)   Any contribution, levy or tax imposed by Statute, payable by the contractor in his capacity as an employer.

HEATING AND VENTILATING INDUSTRY

3.5   Differentials or extra payments in respect of supervisory responsibility are excluded from the annual prime cost (see Section 6). The time of principals, staff, foremen, chargehands and the like when working manually is admissible under this Section at the rates for the appropriate grades.

SECTION 4: MATERIALS AND GOODS

4.1   The prime cost of materials and goods obtained specifically for the daywork is the invoice cost after deducting all trade discounts and any portion of cash discounts in excess of 5%.
4.2   The prime cost of all other materials and goods used in the daywork is based upon the current market prices plus any appropriate handling charges.
4.3   The prime cost referred to in 4.1 and 4.2 includes the cost of delivery to site.
4.4   Any Value Added Tax which is treated, or is capable of being treated, as input tax (as defined by the Finance Act 1972, or any re-enactment or amendment thereof or substitution therefore) by the contractor is excluded.

SECTION 5: PLANT

5.1   Unless otherwise stated in the contract, the prime cost of plant comprises the cost of the following:
        (a)   use or hire of mechanically-operated plant and transport for the time employed on and/or provided or retained for the daywork;
        (b)   use of non-mechanical plant (excluding non-mechanical hand tools) for the time employed on and/or provided or retained for the daywork;
        (c)   transport to and from the site and erection and dismantling where applicable.
5.2   The use of non-mechanical hand tools and of erected scaffolding, staging, trestles or the like is excluded (see Section 6), unless specifically retained for the daywork.

SECTION 6: INCIDENTAL COSTS, OVERHEADS AND PROFIT

6.1   The percentage adjustments provided in the contract which are applicable to each of the totals of Sections 3, 4 and 5 comprise the following:
        (a)   Head office charges.
        (b)   Site staff including site supervision.
        (c)   The additional cost of overtime (other than that referred to in 6.2).
        (d)   Time lost due to inclement weather.
        (e)   The additional cost of bonuses and all other incentive payments in excess of any included in 3.4.
        (f)   Apprentices' study time.
        (g)   Fares and travelling allowances.
        (h)   Country, lodging and periodic allowances.
        (i)   Sick pay or insurances in respect thereof, other than as included in 3.4.
        (j)   Third party and employers' liability insurance.
        (k)   Liability in respect of redundancy payments to employees.
        (l)   Employer's National Insurance contributions not included in 3.4.
        (m)   Use and maintenance of non-mechanical hand tools.
        (n)   Use of erected scaffolding, staging, trestles or the like (but see 5.2).
        (o)   Use of tarpaulins, protective clothing, artificial lighting, safety and welfare facilities, storage and the like that may be available on site.
        (p)   Any variation to basic rates required by the contractor in cases where the contract provides for the use of a specified schedule of basic plant charges (to the extent that no other provision is made for such variation - see 5.1).
        (q)   In the case of a sub-contract which provides that the sub-contractor shall allow a cash discount, such provision as is necessary for the allowance of the prescribed rate of discount.
        (r)   All other liabilities and obligations whatsoever not specifically referred to in this Section nor chargeable under any other Section.
        (s)   Profit.
6.2   The additional cost of overtime where specifically ordered by the Architect/Supervising Officer shall only be chargeable in the terms of a prior written agreement between the parties.

HEATING AND VENTILATING INDUSTRY

MECHANICAL INSTALLATIONS

Calculation of Hourly Base Rate of Labour for Typical Main Grades applicable from 3 October 2005 plus 5% Provisional Allowance, refer to notes within Section Three – Rates of Wages.

	FOREMAN	SENIOR CRAFTSMAN (+ 2nd Welding Skill)	SENIOR CRAFTSMAN	CRAFTSMAN	INSTALLER	MATE OVER 18
Hourly Rate from 3 October 2005 Plus 5% (Provisional)	13.86	11.92	11.45	10.50	9.52	8.02
Annual standard earnings excluding all holidays, 45.8 weeks x 38 hours	24,121.94	20,745.57	19,927.58	18,274.20	16,568.61	13,958.01
Employers national insurance contributions from 6 April 2005	2,491.58	2,059.40	1,954.70	1,743.07	1,524.75	1,190.60
Weekly holiday credit and welfare contributions (52 weeks) from 3 October 2005	3,746.11	3,268.36	3,149.33	2,911.82	2,668.85	2,294.84
Annual prime cost of labour	30,178.15	25,912.13	24,877.69	22,790.25	20,636.37	17,337.88
Hourly base rate	17.34	14.89	14.29	13.09	11.86	9.96

Notes:

(1) Annual industry holiday (4.6 weeks x 38 hours) and public holidays (1.6 weeks x 38 hours) are paid through weekly holiday credit and welfare stamp scheme.
(2) Where applicable, Merit money and other variables (e.g. daily abnormal conditions money), which attract Employer's National Insurance contribution, should be included.
(3) Contractors in Northern Ireland should add the appropriate amount of CITB Levy to the annual prime cost of labour prior to calculating the hourly base rate.
(4) Hourly rate based on 1,740.40 hours per annum and calculated as follows

52 Weeks @ 38 hrs/wk =		1,976.00
Less		
Public Holiday = 8/5 = 1.6 weeks @ 38 hrs/wk =	60.80	
Annual holidays = 4.6 weeks @ 38 hrs/wk =	174.80	235.60

Hours =		1,740.40
		-------------

(5) For calculation of Holiday Credits and ENI refer to detailed labour rate evaluation
(6) National Insurance contributions are those effective from 6 April 2006.
(7) Weekly holiday credit/welfare stamp values are those effective from 3 October 2005 + 5% (Provisional) increase for October 2006.
(8) Hourly rates of wages are those effective from 3 October 2005 + 5% (assumed) increase for October 2006.

ELECTRICAL INDUSTRY

DEFINITION OF PRIME COST OF DAYWORK CARRIED OUT UNDER AN ELECTRICAL CONTRACT (MARCH 1981 EDITION)

This Definition of Prime Cost is published by The Royal Institution of Chartered Surveyors and The Electrical Contractors' Associations for convenience and for use by people who choose to use it. Members of The Electrical Contractors' Association are not in any way debarred from defining Prime Cost and rendering accounts for work carried out on that basis in any way they choose. Building owners are advised to reach agreement with contractors on the Definition of Prime Cost to be used prior to entering into a contract or sub-contract.

SECTION 1: APPLICATION

    1.1   This Definition provides a basis for the valuation of daywork executed under such electrical contracts as provide for its use.

    1.2   It is not applicable in any other circumstances, such as jobbing, or other work carried out as a separate or main contract, nor in the case of daywork executed after the date of practical completion.

    1.3   The terms 'contract' and 'contractor' herein shall be read as 'sub-contract' and 'sub-contractor' as the context may require.

SECTION 2: COMPOSITION OF TOTAL CHARGES

    2.1   The Prime Cost of daywork comprises the sum of the following costs:
        (a)   Labour as defined in Section 3.
        (b)   Materials and goods as defined in Section 4.
        (c)   Plant as defined in Section 5.

    2.2   Incidental costs, overheads and profit as defined in Section 6, as provided in the contract and expressed therein as percentage adjustments, are applicable to each of 2.1 (a)-(c).

SECTION 3: LABOUR

    3.1   The standard wage rates, emoluments and expenses referred to below and the standard working hours referred to in 3.2 are those laid down for the time being in the rules and determinations or decisions of the Joint Industry Board or the Scottish Joint Industry Board for the Electrical Contracting Industry (or those of such other body as may be appropriate) applicable to the works and relating to the grade of operative concerned at the time when and in the area where daywork is executed.

    3.2   Hourly base rates for labour are computed by dividing the annual prime cost of labour, based upon the standard working hours and as defined in 3.4 by the number of standard working hours per annum. See examples.

    3.3   The hourly rates computed in accordance with 3.2 shall be applied in respect of the time spent by operatives directly engaged on daywork, including those operating mechanical plant and transport and erecting and dismantling other plant (unless otherwise expressly provided in the contract) and handling and distributing the materials and goods used in the daywork.

    3.4   The annual prime cost of labour comprises the following:
        (a)   Standard weekly earnings (i.e. the standard working week as determined at the appropriate rate for the operative concerned).
        (b)   Payments in respect of public holidays.
        (c)   Any amounts which may become payable by the Contractor to or in respect of operatives arising from operation of the rules etc. referred to in 3.1 which are not provided for in 3.4(a) and (b) nor in Section 6.
        (d)   Employer's National Insurance Contributions as applicable to 3.4 (a)-(c).
        (e)   Employer's contributions to the Joint Industry Board Combined Benefits Scheme or Scottish Joint Industry Board Holiday and Welfare Stamp Scheme, and holiday payments made to apprentices in compliance with the Joint Industry Board National Working Rules and Industrial Determinations as an employer.
        (f)   Any contribution, levy or tax imposed by Statute, payable by the Contractor in his capacity as an employer.

    3.5   Differentials or extra payments in respect of supervisory responsibility are excluded from the annual prime cost (see Section 6). The time of principals and similar categories, when working manually, is admissible under this Section at the rates for the appropriate grades.

ELECTRICAL INDUSTRY

SECTION 4: MATERIALS AND GOODS

4.1 The prime cost of materials and goods obtained specifically for the daywork is the invoice cost after deducting all trade discounts and any portion of cash discounts in excess of 5%.

4.2 The prime cost of all other materials and goods used in the daywork is based upon the current market prices plus any appropriate handling charges.

4.3 The prime cost referred to in 4.1 and 4.2 includes the cost of delivery to site.

4.4 Any Value Added Tax which is treated, or is capable of being treated, as input tax (as defined by the Finance Act 1972, or any re-enactment or amendment thereof or substitution therefore) by the Contractor is excluded.

SECTION 5: PLANT

5.1 Unless otherwise stated in the contract, the prime cost of plant comprises the cost of the following:
   (a) Use or hire of mechanically-operated plant and transport for the time employed on and/or provided or retained for the daywork;
   (b) Use of non-mechanical plant (excluding non-mechanical hand tools) for the time employed on and/or provided or retained for the daywork;
   (c) Transport to and from the site and erection and dismantling where applicable.

5.2 The use of non-mechanical hand tools and of erected scaffolding, staging, trestles or the likes is excluded (see Section 6), unless specifically retained for daywork.

5.3 Note: Where hired or other plant is operated by the Electrical Contractor's operatives, such time is to be included under Section 3 unless otherwise provided in the contract.

SECTION 6: INCIDENTAL COSTS, OVERHEADS AND PROFIT

6.1 The percentage adjustments provided in the contract which are applicable to each of the totals of Sections 3, 4 and 5, compromise the following:
   (a) Head Office charges.
   (b) Site staff including site supervision.
   (c) The additional cost of overtime (other than that referred to in 6.2).
   (d) Time lost due to inclement weather.
   (e) The additional cost of bonuses and other incentive payments.
   (f) Apprentices' study time.
   (g) Travelling time and fares.
   (h) Country and lodging allowances.
   (i) Sick pay or insurance in lieu thereof, in respect of apprentices.
   (j) Third party and employers' liability insurance.
   (k) Liability in respect of redundancy payments to employees.
   (l) Employers' National Insurance Contributions not included in 3.4.
   (m) Use and maintenance of non-mechanical hand tools.
   (n) Use of erected scaffolding, staging, trestles or the like (but see 5.2.).
   (o) Use of tarpaulins, protective clothing, artificial lighting, safety and welfare facilities, storage and the like that may be available on site.
   (p) Any variation to basic rates required by the Contractor in cases where the contract provides for the use of a specified schedule of basic plant charges (to the extent that no other provision is made for such variation - see 5.1).
   (q) All other liabilities and obligations whatsoever not specifically referred to in this Section nor chargeable under any other Section.
   (r) Profit.
   (s) In the case of a sub-contract which provides that the sub-contractor shall allow a cash discount, such provision as is necessary for the allowance of the prescribed rate of discount.

6.2 The additional cost of overtime where specifically ordered by the Architect/Supervising Officer shall only be chargeable in the terms of a prior written agreement between the parties.

Daywork

ELECTRICAL INDUSTRY

ELECTRICAL INSTALLATIONS

Calculation of Hourly Base Rate of Labour for Typical Main Grades applicable from 8 January 2007

	TECHNICIAN	APPROVED ELECTRICIAN	ELECTRICIAN	LABOURER
Hourly Rate from 8 January 2007 (London Rates)	15.86	14.09	12.99	10.42
Annual standard earnings excluding all holidays, 46 weeks x 37.5 hours	27,358.50	24,305.25	22,407.75	17,974.50
Employers national insurance contributions from 6 April 2006	2,905.79	2,515.04	2,272.16	1,704.71
JIB Combined benefits from 26 September 2005	2,743.00	2,496.00	2,340.00	1,985.36
Holiday top up funding	1,449.76	1,298.44	1,206.92	983.32
Annual prime cost of labour	34,457.05	30,614.73	28,226.83	22,647.89
Hourly base rate	19.98	17.75	16.36	13.13

Notes:

(1) Annual industry holiday (4.4 weeks x 37.5 hours) and public holidays (1.6 weeks x 37.5 hours)
(2) It should be noted that all labour costs incurred by the Contractor in his capacity as an Employer, other than those contained in the hourly rate above, must be taken into account under Section 6.
(3) Public Holidays are paid through weekly holiday credit and welfare stamp scheme.
(4) Contractors in Northern Ireland should add the appropriate amount of CITB Levy to the annual prime cost of labour prior to calculating the hourly base rate.
(5) Hourly rate based on 1,725 hours per annum and calculated as follows:

52 Weeks @ 37.5 hrs/wk	=		1,950.00
Less			
Public Holiday = 8/5 = 1.6 weeks @ 37.5 hrs/wk	=	60.00	
Annual holidays = 4.4 weeks @ 37.5 hrs/wk	=	165.00	225.00

Hours	=		1,725.00
			-------------

(6) For calculation of holiday credits and ENI refer to detailed labour rate evaluation.
(7) Hourly wage rates are those effective from 8 January 2007.
(8) National Insurance contributions are those effective from 6 April 2006.
(9) JIB Combined Benefits Values are those effective from 26 September 2005.

BUILDING INDUSTRY PLANT HIRE COSTS

SCHEDULE OF BASIC PLANT CHARGES (MAY 2001)

This Schedule is published by the Royal Institution of Chartered Surveyors and is for use in connection with Dayworks under a Building Contract.

EXPLANATORY NOTES

1   The rates in the Schedule are intended to apply solely to daywork carried out under and incidental to a Building Contract. They are NOT intended to apply to:
    (i)   jobbing or any other work carried out as a main or separate contract; or
    (ii) work carried out after the date of commencement of the Defects Liability Period.

2   The rates apply to plant and machinery already on site, whether hired or owned by the Contractor.

3   The rates, unless otherwise stated, include the cost of fuel and power of every description, lubricating oils, grease, maintenance, sharpening of tools, replacement of spare parts, all consumable stores and for licences and insurances applicable to items of plant.

4   The rates, unless otherwise stated, do not include the costs of drivers and attendants (unless otherwise stated).

5   The rates in the Schedule are base costs and may be subject to an overall adjustment for price movement, overheads and profit, quoted by the Contractor prior to the placing of the Contract.

6   The rates should be applied to the time during which the plant is actually engaged in daywork.

7   Whether or not plant is chargeable on daywork depends on the daywork agreement in use and the inclusion of an item of plant in this schedule does not necessarily indicate that item is chargeable.

8   Rates for plant not included in the Schedule or which is not already on site and is specifically provided or hired for daywork shall be settled at prices which are reasonably related to the rates in the Schedule having regard to any overall adjustment quoted by the Contractor in the Conditions of Contract.

NOTE: All rates in the schedule were calculated during the first quarter of 2001.

BUILDING INDUSTRY PLANT HIRE COSTS

MECHANICAL PLANT AND TOOLS

Item of Plant	Size/Rating	Unit	Rate/hr
PUMPS			
Mobile Pumps			
Including pump hoses, valves and strainers etc.			
Diaphragm	50mm dia.	Each	0.87
Diaphragm	76mm dia.	Each	1.29
Submersible	50mm dia.	Each	1.18
Induced flow	50mm dia.	Each	1.54
Induced flow	76mm dia.	Each	2.05
Centrifugal, self priming	50mm dia.	Each	1.96
Centrifugal, self priming	102mm dia.	Each	2.52
Centrifugal, self priming	152mm dia.	Each	3.87
SCAFFOLDING, SHORING, FENCING			
Complete Scaffolding			
Mobile working towers, single width	1.8m x 0.8m base x 7m high	Each	2.00
Mobile working towers, single width	1.8m x 0.8m base x 9m high	Each	2.80
Mobile working towers, double width	1.8m x 1.4m base x 7m high	Each	2.15
Mobile working towers, double width	1.8m x 1.4m base x 15m high	Each	5.10
Chimney scaffold, single unit		Each	1.79
Chimney scaffold, twin unit		Each	2.05
Chimney scaffold, four unit		Each	3.59
Trestles			
Trestle, adjustable	Any height	Pair	0.10
Trestle, painters	1.8m high	Pair	0.21
Trestle, Painters	2.4m high	Pair	0.26
Shoring, Planking and Struting			
'Acrow' adjustable prop	Sizes up to 4.9m (open)	Each	0.10
'Strong boy' support attachment		Each	0.15
Adjustable trench struts	Sizes up to 1.67m (open)	Each	0.10
Trench sheet		Metre	0.01
Backhoe trench box		Each	1.00
Temporary Fencing			
Including block and coupler			
Site fencing steel grid panel	3.5m x 2.0m	Each	0.08
Anti-climb site steel grid fence panel	3.5m x 2.0m	Each	0.08
LIFTING APPLIANCES AND CONVEYORS			
Cranes			
<u>Mobile Cranes</u>			
Rates are inclusive of drivers			
Lorry mounted, telescopic jib			
Two wheel drive	6 tonnes	Each	24.40
Two wheel drive	7 tonnes	Each	25.00
Two wheel drive	8 tonnes	Each	25.62
Two wheel drive	10 tonnes	Each	26.90
Two wheel drive	12 tonnes	Each	28.25
Two wheel drive	15 tonnes	Each	29.66
Two wheel drive	18 tonnes	Each	31.14
Two wheel drive	20 tonnes	Each	32.70
Two wheel drive	25 tonnes	Each	34.33

BUILDING INDUSTRY PLANT HIRE COSTS

MECHANICAL PLANT AND TOOLS

Item of Plant	Size/Rating	Unit	Rate/hr
Four wheel drive	10 tonnes	Each	27.44
Four wheel drive	12 tonnes	Each	28.81
Four wheel drive	15 tonnes	Each	30.25
Four wheel drive	20 tonnes	Each	33.35
Four wheel drive	25 tonnes	Each	35.19
Four wheel drive	30 tonnes	Each	37.12
Four wheel drive	45 tonnes	Each	39.16
Four wheel drive	50 tonnes	Each	41.32

Track-mounted tower crane
Rates inclusive of driver
Note : Capacity equals maximum lift in tonnes times
maximum radius at which it can be lifted

	Capacity (metre/tonnes) up to	Height under hook above ground (m) up to	Unit	Rate/hr
Tower crane	10	17	Each	7.99
Tower crane	15	18	Each	8.59
Tower crane	20	20	Each	9.18
Tower crane	25	22	Each	11.56
Tower crane	30	22	Each	13.78
Tower crane	40	22	Each	18.09
Tower crane	50	22	Each	22.20
Tower crane	60	22	Each	24.32
Tower crane	70	22	Each	23.00
Tower crane	80	22	Each	25.91
Tower crane	110	22	Each	26.45
Tower crane	125	30	Each	29.38
Tower crane	150	30	Each	32.35

Static tower cranes
Rates inclusive of driver
To be charged at 90% of the above rates for track
mounted tower cranes

	Size	Unit	Rate/hr
Crane Equipment	Up to 0.25m³	Each	0.56
Muck tipping skip	0.5m³	Each	0.67
Muck tipping skip	0.75m³	Each	0.82
Muck tipping skip	1.0m³	Each	1.03
Muck tipping skip	1.5m³	Each	1.18
Muck tipping skip	2.0m³	Each	1.38
Muck tipping skip			
Mortar skips	up to 0.38m³	Each	0.41
Boat skips	1.0m³	Each	1.08
Boat skips	1.5m³	Each	1.33
Boat skips	2.0m³	Each	1.59
Concrete skips, hand levered	0.5m³	Each	1.00
Concrete skips, hand levered	0.75m³	Each	1.10
Concrete skips, hand levered	1.0m³	Each	1.25
Concrete skips, hand levered	1.5m³	Each	1.50
Concrete skips, hand levered	2.0m³	Each	1.65

BUILDING INDUSTRY PLANT HIRE COSTS

MECHANICAL PLANT AND TOOLS

Item of Plant	Size/Rating		Unit	Rate/hr
Crane Equipment (cont'd)				
Concrete skips, geared	0.5m³		Each	1.30
Concrete skips, geared	0.75m³		Each	1.40
Concrete skips, geared	1.0m³		Each	1.55
Concrete skips, geared	1.5m³		Each	1.80
Concrete skips, geared	2.0m³		Each	2.05
Hoists				
Scaffold hoists	200kg		Each	1.92
Rack and pinion (goods only)	500kg		Each	3.31
Rack and pinion (goods only)	1100kg		Each	4.28
Rack and pinion goods and passenger	15 person, 1200kg		Each	5.62
Wheelbarrow chain sling				0.31
Conveyors				
Belt conveyors				
Conveyor	7.5m long x 400mm wide		Each	6.41
Miniveyor, control box and loading hopper	3m unit		Each	3.59
Other Conveying Equipment				
Wheelbarrow			Each	0.21
Hydraulic superlift			Each	2.95
Pavac slab lifter			Each	1.03
Hand pad and hose attachment			Each	0.26
Lifting Trucks				
Fork lift, two wheel drive	Payload	Max Lift		
Fork lift, two wheel drive	1100kg	up to 3.0m	Each	4.87
Fork lift, two wheel drive	2540kg	up to 3.7m	Each	5.12
Fork lift, four wheel drive	1524kg	up to 6.0m	Each	6.04
Fork lift, four wheel drive	2600kg	up to 5.4m	Each	7.69
Lifting Platforms				
Hydraulic platform (Cherry picker)	7.5m		Each	4.23
Hydraulic platform (Cherry picker)	13m		Each	9.23
Scissors lift	7.8m		Each	7.56
Telescopic handlers	7m, 2 tonne		Each	7.18
Telescopic handlers	13m, 3 tonne		Each	8.72
Lifting and Jacking Gear				
Pipe winch including gantry	1 tonne		Sets	1.92
Pipe winch including gantry	3 tonnes		Sets	3.21
Chain block	1 tonne		Each	0.45
Chain block	2 tonnes		Each	0.71
Chain block	5 tonnes		Each	1.22
Pull lift (Tirfor winch)	1 tonne		Each	0.64
Pull lift (Tirfor winch)	1.6 tonnes		Each	0.90
Pull lift (Tirfor winch)	3.2 tonnes		Each	1.15
Brother or chain slings, two legs	not exceeding 4.2 tonnes		Set	0.35
Brother or chain slings, two legs	not exceeding 7.5 tonnes		Set	0.45
Brother or chain slings, four legs	not exceeding 3.1 tonnes		Set	0.41
Brother or chain slings, four legs	not exceeding 11.2 tonnes		Set	1.28

BUILDING INDUSTRY PLANT HIRE COSTS

MECHANICAL PLANT AND TOOLS

Item of Plant	Size/Rating	Unit	Rate/hr
CONSTRUCTION VEHICLES			
Lorries			
Plated lorries			
Rates are inclusive of driver			
Platform lorries	7.5 tonnes	Each	19.00
Platform lorries	17 tonnes	Each	21.00
Platform lorries	24 tonnes	Each	26.00
Platform lorries with winch and skids	7.5 tonnes	Each	21.40
Platform lorries with crane	17 tonnes	Each	27.50
Platform lorries with crane	24 tonnes	Each	32.10
Tipper Lorries			
Rates are inclusive of driver			
Tipper lorries	15/17 tonnes	Each	19.50
Tipper lorries	24 tonnes	Each	21.40
Tipper lorries	30 tonnes	Each	27.10
Dumpers			
Site use only (excluding tax, insurance and extra			
Cost of DERV etc. when operating on highway)			
	Makers capacity		
Two wheel drive	0.8 tonnes	Each	1.20
Two wheel drive	1 tonne	Each	1.30
Two wheel drive	1.2 tonnes	Each	1.60
Four wheel drive	2 tonnes	Each	2.50
Four wheel drive	3 tonnes	Each	3.00
Four wheel drive	4 tonnes	Each	3.50
Four wheel drive	5 tonnes	Each	4.00
Four wheel drive	6 tonnes	Each	4.50
Dumper Trucks			
Rates are inclusive of drivers			
Dumper trucks	10/13 tonnes	Each	20.00
Dumper trucks	18/20 tonnes	Each	20.40
Dumper trucks	22/25 tonnes	Each	26.30
Dumper trucks	35/40 tonnes	Each	36.60
Tractors			
<u>Agricultural Type</u>			
Wheeled, rubber-clad tyred			
Light	48 h.p.	Each	4.65
Heavy	65 h.p.	Each	5.15
<u>Crawler Tractors</u>			
With bull or angle dozer	80/90 h.p.	Each	21.40
With bull or angle dozer	115/130 h.p.	Each	25.10
With bull or angle dozer	130/150 h.p.	Each	26.00
With bull or angle dozer	155/175 h.p.	Each	27.74
With bull or angle dozer	210/230 h.p.	Each	28.00
With bull or angle dozer	300/340 h.p.	Each	31.10
With bull or angle dozer	400/440 h.p.	Each	46.90
With loading shovel	0.8m³	Each	25.00
With loading shovel	1.0m³	Each	28.00
With loading shovel	1.2m³	Each	32.00
With loading shovel	1.4m³	Each	36.00
With loading shovel	1.8m³	Each	45.00

BUILDING INDUSTRY PLANT HIRE COSTS

MECHANICAL PLANT AND TOOLS

Item of Plant	Size/Rating	Unit	Rate/hr
Light Vans			
Ford Escort or the like		Each	4.74
Ford Transit or the like	1.0 tonnes	Each	6.79
Luton Box Van or the like	1.8 tonnes	Each	8.33
Water/Fuel Storage			
Mobile water container	110 litres	Each	0.28
Water bowser	1100 litres	Each	0.55
Water bowser	3000 litres	Each	0.74
Mobile fuel container	110 litres	Each	0.28
Fuel bowser	1100 litres	Each	0.65
Fuel bowser	3000 litres	Each	1.02
EXCAVATORS AND LOADERS			
Excavators			
Wheeled, hydraulic	7/10 tonnes	Each	12.00
Wheeled, hydraulic	11/13 tonnes	Each	12.70
Wheeled, hydraulic	15/16 tonnes	Each	14.80
Wheeled, hydraulic	17/18 tonnes	Each	16.70
Wheeled, hydraulic	20/23 tonnes	Each	16.70
Crawler, hydraulic	12/14 tonnes	Each	12.00
Crawler, hydraulic	15/17.5 tonnes	Each	14.00
Crawler, hydraulic	20/23 tonnes	Each	16.00
Crawler, hydraulic	25/30 tonnes	Each	21.00
Crawler, hydraulic	30/35 tonnes	Each	30.00
Mini excavators	1000/1500kg	Each	4.50
Mini excavators	2150/2400kg	Each	5.50
Mini excavators	2700/3500kg	Each	6.50
Mini excavators	3500/4500kg	Each	8.50
Mini excavators	4500/6000kg	Each	9.50
Loaders			
Wheeled skip loader		Each	4.50
Shovel loaders, four wheel drive	1.6m³	Each	12.00
Shovel loaders, four wheel drive	2.4m³	Each	19.00
Shovel loaders, four wheel drive	3.6m³	Each	22.00
Shovel loaders, four wheel drive	4.4m³	Each	23.00
Shovel loaders, crawlers	0.8m³	Each	11.00
Shovel loaders, crawlers	1.2m³	Each	14.00
Shovel loaders, crawlers	1.6m³	Each	16.00
Shovel loaders, crawlers	2m³	Each	17.00
Skid steer loaders wheeled	300/400kg payload	Each	6.00
Excavator Loaders			
Wheeled tractor type with back-hoe excavator			
Four wheel drive	2.5/3.5 tonnes	Each	7.00
Four wheel drive, 2 wheel steer	7/8 tonnes	Each	9.00
Four wheel drive, 4 wheel steer	7/8 tonnes	Each	10.00
Crawler, hydraulic	12 tonnes	Each	20.00
Crawler, hydraulic	20 tonnes	Each	16.00
Crawler, hydraulic	30 tonnes	Each	35.00
Crawler, hydraulic	40 tonnes	Each	38.00

BUILDING INDUSTRY PLANT HIRE COSTS

MECHANICAL PLANT AND TOOLS

Item of Plant	Size/Rating	Unit	Rate/hr
COMPACTION EQUIPMENT			
Rollers			
Vibrating roller	368kg - 420kg	Each	1.68
Single roller	533kg	Each	1.92
Single roller	750kg	Each	2.41
Vibrating roller	368kg - 420kg	Each	1.68
Single roller	533kg	Each	1.92
Single roller	750kg	Each	2.41
Twin roller	698kg	Each	1.93
Twin roller	851kg	Each	2.41
Twin roller with seat end steering wheel	1067kg	Each	3.03
Twin roller with seat end steering wheel	1397kg	Each	3.17
Pavement rollers	3 - 4 tonnes dead weight	Each	3.18
Pavement rollers	4 - 6 tonnes	Each	4.13
Pavement rollers	6 - 10 tonnes	Each	4.84
Rammers			
Tamper rammer 2 stroke-petrol	225mm - 275mm	Each	1.59
Soil Compactors			
Plate compactor	375mm - 400mm	Each	1.20
Plate compactor rubber pad	375mm - 1400mm	Each	0.33
Plate compactor reversible plate - petrol	400mm	Each	2.20
CONCRETE EQUIPMENT			
Concrete/Mortar Mixers			
Open drum without hopper	0.09/0/06m³	Each	0.62
Open drum without hopper	0.12/0.09m³	Each	0.68
Open drum without hopper	0.15/0.10m³	Each	0.72
Open drum with hopper	0.20/0.15m³	Each	0.80
Concrete/Mortar Transport Equipment			
Concrete pump including hose, valve and couplers			
Lorry mounted concrete pump	23m max. distance	Each	36.00
Lorry mounted concrete pump	50m max. distance	Each	46.00
Concrete Equipment			
Vibrator, poker, petrol type	up to 75mm dia.	Each	1.62
Air vibrator (excluding compressor and hose)	up to 75mm dia.	Each	0.79
Extra poker heads	5m	Each	0.77
Vibrating screed unit with beam	3m - 5m	Each	1.77
Vibrating screed unit with adjustable beam	725mm - 900mm	Each	2.18
Power float		Each	1.72
Power grouter		Each	0.92
TESTING EQUIPMENT			
Pipe Testing Equipment			
Pressure testing pump, electric		Sets	1.87
Pipe pressure testing equipment, hydraulic		Sets	2.46
Pressure test pump		Sets	0.64

BUILDING INDUSTRY PLANT HIRE COSTS

MECHANICAL PLANT AND TOOLS

Item of Plant	Size/Rating	Unit	Rate/hr
SITE ACCOMMODATION AND TEMPORARY SERVICES			
Heating Equipment			
Space heaters - propane		Each	0.77
Space heaters - propane/electric	80,000Btu/hr	Each	1.56
Space heaters - propane/electric	125,000Btu/hr	Each	1.79
Space heaters, propane	250,000Btu/hr	Each	1.33
Space heaters, propane	125,000Btu/hr	Each	1.64
Cabinet heaters	260,000Btu/hr	Each	0.41
Cabinet heater catalytic		Each	0.46
Electric halogen heaters		Each	1.28
Ceramic heaters		Each	0.79
Fan heaters	3kW	Each	0.41
Cooling Fan	3kW	Each	1.15
Mobile cooling unit - small		Each	1.38
Mobile cooling unit - large		Each	1.54
Air conditioning unit		Each	2.62
Site Lighting and Equipment			
Tripod floodlight	500W		
Tripod floodlight	1000W	Each	0.36
Towable floodlight	4 x 1000W	Each	0.34
Hand held floodlight	500W	Each	2.00
		Each	0.22
Rechargeable light		Each	0.62
Inspection light		Each	0.15
Plasterers light		Each	0.56
Lighting mast		Each	0.92
Festoon light string	33m	Each	0.31
Site Electrical Equipment			
Extension leads	240V/14m	Each	0.20
Extension leafs	110V/14m	Each	0.20
Cable reel	25m 110V/240V	Each	0.28
Cable reel	50m 110V/240V	Each	0.33
4 way junction box	110V	Each	0.17
Power Generating Units			
Generator - petrol	2kVA	Each	1.08
Generator - silenced petrol	2kVA	Each	1.54
Generator - petrol	3VA	Each	1.38
Generator - diesel	5kVA	Each	1.92
Generator - silenced diesel	8kVA	Each	3.59
Generator - silenced diesel	1.5kVA	Each	7.69
Tail adaptor	240V	Each	0.20
Transformers			
Transformer	3kVA	Each	0.36
Transformer	5kVA	Each	0.51
Transformer	7.5kVA	Each	0.82
Transformer	10kVA	Each	0.87
Rubbish Collection and Disposal Equipment			
Rubbish Chutes			
Standard plastic module	1m section	Each	0.18
Steel liner insert		Each	0.26
Steel top hopper		Each	0.20
Plastic side entry hopper		Each	0.20
Plastic side entry hopper liner		Each	0.20

BUILDING INDUSTRY PLANT HIRE COSTS

MECHANICAL PLANT AND TOOLS

Item of Plant	Size/Rating	Unit	Rate/hr
SITE EQUIPMENT			
Welding Equipment			
Arc-(Electric) Complete with Leads			
Welder generator - petrol	200 amp	Each	2.26
Welder generator - diesel	300/350 amp	Each	3.33
Welder generator - diesel	400 amp	Each	4.74
Extra welding lead sets		Each	0.29
Gas-Oxy Welder			
Welding and cutting set (including oxygen and acetylene,			
excluding underwater equipment and thermic boring)			
Small		Each	1.41
Large		Each	2.00
Mig welder		Each	1.00
Fume extractor		Each	0.92
Road Works Equipment			
Traffic lights, mains/generator	2-way	Set	4.01
Traffic lights, mains/generator	3-way	Set	7.92
Traffic lights, mains/generator	4-way	Set	9.81
Traffic lights, mains/generator - trailer mounted	2-way	Set	3.98
Flashing lights		Each	0.20
Road safety cone	450mm	10	0.26
Safety cone	750mm	10	0.38
Safety barrier plank	1.25m	Each	0.03
Safety barrier plank	2m	Each	0.04
Road sign		Each	0.26
DPC Equipment			
Damp proofing injection machine		Each	1.49
Cleaning Equipment			
Vacuum cleaner (industrial wet) single motor		Each	0.62
Vacuum cleaner (industrial wet) twin motor		Each	1.23
Vacuum cleaner (industrial wet) triple motor		Each	1.44
Vacuum cleaner (industrial wet) back pack		Each	0.97
Pressure washer, light duty, electric	1450 PSI	Each	0.97
Pressure washer, heavy duty, diesel	2500 PSI	Each	2.69
Cold pressure washer, electric		Each	1.79
Hot pressure washer, petrol		Each	2.92
Cold pressure washer, petrol		Each	2.00
Sandblast attachment to last washer		Each	0.54
Drain cleaning attachment to last washer		Each	0.31
Surface Preparation Equipment			
Rotavators	5 h.p.	Each	1.67
Scabbler, up to three heads		Each	1.15
Scabbler, pole		Each	1.50
Scabbler, multi-headed floor		Each	4.00
Floor preparation machine		Each	2.82

BUILDING INDUSTRY PLANT HIRE COSTS

MECHANICAL PLANT AND TOOLS

Item of Plant	Size/Rating	Unit	Rate/hr
Compressors and Equipment			
Portable Compressors			
Compressors - electric	0.23m³/min	Each	1.59
Compressors - petrol	0.28m³/min	Each	1.74
Compressors - petrol	0.71m³/min	Each	2.00
Compressors - diesel	up to 2.83m³/min	Each	1.24
Compressors - diesel	up to 3.68m³/min	Each	1.49
Compressors - diesel	up to 4.25m³/min	Each	1.60
Compressors - diesel	up to 4.81m³/min	Each	1.92
Compressors - diesel	up to 7.64m³/min	Each	3.08
Compressors - diesel	up to 11.32m³/min	Each	4.23
Compressors - diesel	up to 18.40m³/min	Each	5.73
Mobile Compressors			
Lorry mounted compressors	2.86-4.24m³/min	Each	12.50
(machine plus lorry only)			
Tractor mounted compressors	2.86-3.40m³/min	Each	13.50
(machine plus rubber tyred tractor)			
Accessories (Pneumatic Tools)			
(with and including up to 15m of air hose)			
Demolition pick		Each	1.03
Breakers (with six steels) light	up to 150kg	Each	0.79
Breakers (with six steels) medium	295kg	Each	1.08
Breakers (with six steels) heavy	386kg	Each	1.44
Rock drill (for use with compressor) hand held		Each	0.90
Additional hoses	15m	Each	0.16
Muffler, tool silencer		Each	0.14
Breakers			
Demolition hammer drill, heavy duty, electric		Each	1.00
Road breaker, electric		Each	1.65
Road breaker, 2 stroke, petrol		Each	2.05
Hydraulic breaker unit, light duty, petrol		Each	2.05
Hydraulic breaker unit, heavy duty, petrol		Each	2.60
Hydraulic breaker unit, heavy duty, diesel		Each	2.95
Quarrying and Tooling Equipment			
Block and stone splitter, hydraulic	600mm x 600mm	Each	1.35
Block and slab splitter, manual		Each	1.10
Steel Reinforcement Equipment			
Bar bending machine - manual	up to 13mm dia. rods	Each	0.90
Bar bending machine - manual	up to 20mm dia. rods	Each	1.28
Bar shearing machine - electric	up to 38mm dia. rods	Each	2.82
Bar shearing machine - electric	up to 40mm dia. rods	Each	3.85
Bar cropper machine - electric	up to 13mm dia. rods	Each	1.54
Bar cropper machine - electric	up to 20mm dia. rods	Each	2.05
Bar cropper machine - electric	up to 40mm dia. rods	Each	2.82
Bar cropper machine - 3 phase	up to 40mm dia. rods	Each	3.85
Dehumidifiers			
110/240v Water	68 litres extraction per 24 hrs	Each	1.28
110/240v Water	90 litres extraction per 24 hrs	Each	1.85

BUILDING INDUSTRY PLANT HIRE COSTS

MECHANICAL PLANT AND TOOLS

Item of Plant	Size/Rating	Unit	Rate/hr
Compressors and Equipment (cont'd)			
SMALL TOOLS			
Saws			
Masonry bench saw	350mm - 500mm dia.	Each	2.80
Floor saw	350mm dia., 125mm max. cut	Each	1.90
Floor saw	450mm dia., 150mm max. cut	Each	2.60
Floor saw, reversible	Max. cut 300mm	Each	13.00
Chop/cut off saw, electric	350mm dia.	Each	1.33
Circular saw, electric	230mm dia.	Each	0.60
Tyrannosaw		Each	1.20
Reciprocating saw		Each	0.60
Door trimmer		Each	0.90
Chainsaw, petrol	500mm	Each	2.13
Full chainsaw safety kit		Each	0.50
Worktop jig		Each	0.60
Pipework Equipment			
Pipe bender	15mm - 22mm	Each	0.33
Pipe bender, hydraulic	50mm	Each	0.60
Pipe bender, electric	50mm - 150mm dia.	Each	1.35
Pipe cutter, hydraulic		Each	1.84
Tripod pipe vice		Set	0.40
Ratchet threader	12mm - 32mm	Each	0.55
Pipe threading machine, electric	12mm - 75mm	Each	2.40
Pipe threading machine, electric	12mm - 100mm	Each	3.00
Impact wrench, electric		Each	0.54
Impact wrench, two stroke, petrol		Each	4.49
Impact wrench, heavy duty, electric		Each	1.13
Plumber's furnace, calor gas or similar		Each	2.16
Hand-held Drills and Equipment			
Impact or hammer drill	up to 25mm dia.	Each	0.50
Impact or hammer drill	35mm dia.	Each	0.90
Angle head drills		Each	0.70
Stirrer, mixer drills		Each	0.70
Paint, Insulation Application Equipment			
Airless spray unit		Each	4.20
Portaspray unit		Each	1.65
HVLP turbine spray unit		Each	1.65
Compressor and spray gun		Each	2.20
Other Handtools			
Screwing machine	13mm - 50mm dia.	Each	0.77
Screwing machine	25mm - 100mm dia.	Each	1.57
Staple gun		Each	0.33
Air nail gun	110V	Each	3.33
Cartridge hammer		Each	1.00
Tongue and groove nailer complete with mallet		Each	0.93
Chasing machine	152mm	Each	1.72
Chasing machine	76mm - 203mm	Each	5.99
Floor grinder		Each	3.00
Floor plane		Each	3.67
Diamond concrete planer		Each	2.05
Autofeed screwdriver, electric		Each	1.13
Laminate trimmer		Each	0.64

BUILDING INDUSTRY PLANT HIRE COSTS

MECHANICAL PLANT AND TOOLS

Item of Plant	Size/Rating	Unit	Rate/hr
Biscuit jointer		Each	0.87
Random orbital sander		Each	0.72
Floor sander		Each	1.33
Palm, delta, flap or belt sander	300mm	Each	0.38
Saw cutter, 2 stroke, petrol	up to 225mm	Each	1.26
Grinder, angle or cutter	300mm	Each	0.60
Grinder, angle or cutter		Each	1.10
Mortar raking tool attachment	325mm	Each	0.15
Floor/polisher scrubber		Each	1.03
Floor tile stripper		Each	1.74
Wallpaper stripper, electric		Each	0.56
Electric scraper		Each	0.51
Hot air paint stripper	All sizes	Each	0.38
Electric diamond tile cutter		Each	1.38
Hand tile cutter		Each	0.36
Electric needle gun		Each	1.08
Needle chipping gun	1.2m wide	Each	0.72
Pedestrian floor sweeper		Each	0.87

Tables and Memoranda

CONVERSION TABLES

LENGTH

Millimetre	(mm)	1 in	=	25.4	mm	: 1 mm	=	0.0394	in
Metre	(m)	1 ft	=	0.3048	m	: 1 m	=	3.2808	ft
Kilometre	(km)	1 yd	=	0.9144	m	: 1 m	=	1.0936	yd
Kilometre	(km)	1 mile	=	1.6093	km	: 1 km	=	0.6214	mile
NOTE :		1 cm	=	10	mm	1 ft	=	12	in
		1 m	=	100	cm	1 yd	=	3	ft
		1 km	=	1000	m	1 mile	=	1760	yd

AREA

Square Millimetre	(mm^2)	$1\ in^2$	=	645.2	mm^2	: $1\ mm^2$	=	0.0016	in^2
Square Centimetre	(cm^2)	$1\ in^2$	=	6.4516	cm^2	: $1\ cm^2$	=	0.1550	in^2
Square Metre	(m^2)	$1\ ft^2$	=	0.0929	m^2	: $1\ m^2$	=	10.764	ft^2
Square Metre	(m^2)	$1\ yd^2$	=	0.8361	m^2	: $1\ m^2$	=	1.1960	yd^2
Square Kilometre	(km^2)	$1\ mile^2$	=	2.590	km^2	: $1\ km^2$	=	0.3861	$mile^2$
Hectare	(ha)	1 acre	=	0.405	ha	: 1 ha		2.471	acre
NOTE :		$1\ cm^2$	=	100	mm^2	$1\ ft^2$	=	144	in^2
		$1\ m^2$	=	10000	cm^2	$1\ yd^2$	=	9	ft^2
		$1\ km^2$	=	100	ha	$1\ mile^2$	=	640	acre
						1 acre	=	4840	yd^2

VOLUME

Cubic Centimetre	(cm^3)	$1\ cm^3$	=	0.0610	in^3	: $1\ in^3$	=	16.387	cm^3
Cubic Decimetre	(dm^3)	$1\ dm^3$	=	0.0353	ft^3	: $1\ ft^3$	=	28.329	dm^3
Cubic Metre	(m^3)	$1\ m^3$	=	35.315	ft^3	: $1\ ft^3$	=	0.0283	m^3
Cubic Metre	(m^3)	$1\ m^3$	=	1.3080	yd^3	: $1\ yd^3$	=	0.7646	m^3
Litre	(L)	1 L	=	1.76	pint	: 1 pint	=	0.5683	L
Litre	(L)	1 L	=	2.113	US pt	: 1 pint	=	0.4733	US L
Litre	(L)	1L	=	0.220	gal	1 gal	=	4.546	L
NOTE :		$1\ dm^3$	=	1000	cm^3	$1\ ft^3$	=	1728	in^3
		$1\ m^3$	=	1000	dm^3	$1\ yd^3$	=	27	ft^3
		1 L	=	1	dm^3	1 pint	=	20	fl oz
		1 HL	=	100	L	1 gal	=	8	pints

MASS

Milligram	(mg)	1 mg	=	0.0154	grain	: 1 grain	=	64.935	mg
Gram	(g)	1 g	=	0.0353	oz	: 1 oz	=	28.35	g
Kilogram	(kg)	1 kg	=	2.2046	lb	: 1 lb	=	0.4536	kg
Kilogram	(kg)	1 kg	=	0.020	cwt	: 1 cwt	=	50.802	kg
Tonne	(t)	1 t	=	0.9842	ton	: 1 ton	=	1.016	t
NOTE :		1 g	=	1000	mg	1 oz	=	437.5	grains
		1 kg	=	1000	g	1 lb	=	16	oz
		1 t	=	1000	kg	1 stone	=	14	lb
						1 cwt	=	112	lb
						1 ton	=	20	cwt

CONVERSION TABLES

FORCE

Newton	(N)	1 lb f	=	4.448	N	: 1 kg f	=	9.807 N
Kilonewton	(kN)	1 lb f	=	0.004448	kN	: 1 ton f	=	9.964 kN
Meganewton	(MN)	100 ton f	=	0.9964	MN			

POWER

| Kilowatt | (kW) | 1 kW | = | 1.310 | HP | : 1 HP | = | 0.746 kW |

PRESSURE AND STRESS

Kilonewton					
per square metre	(kN/m^2)	1 lb f/in^2	=	6.895	kN/m^2
		1 bar	=	100	kN/m^2
		1 ton f/ft^2	=	107.3	kN/m^2
		1 kg f/cm^2	=	98.07	kN/m^2
		1 lb f/ft^2	=	0.0479	kN/m^2

TEMPERATURE

Degrees $\quad\quad\quad\quad °C = 5/9 \ (°F - 32°) \quad\quad\quad\quad °F = 9/5 \ (°C + 32°)$

FORMULAE

Two dimensional figures

Figure	Area

Triangle — 0.5 x base x height,
or $N(s(s - a)(s - b)(s - c))$ where s = 0.5 x the sum of the
three sides and a, b and c are the lengths of the three sides,

or $a^2 = b^2 + c^2 - 2 \times bc \times COS\ A$ where A is the angle opposite side a

Hexagon — $2.6 \times (side)^2$

Octagon — $4.83 \times (side)^2$

Trapezoid — height x 0.5 (base + top)

Circle — $3.142 \times radius^2$ or $0.7854 \times diameter^2$
(circumference = 2 x 3.142 x radius or 3.142 x diameter)

Sector of a circle — 0.5 x length of arc x radius

Segment of a circle — area of sector - area of triangle

Ellipse of a circle — 3.142 x AB
(where A = 0.5 x height and B = 0.5 x length)

Spandrel — $3/14 \times radius^2$

Three dimensional figure

Figure	Volume Surface	Area
Prism x height	Area of base x height	circumference of base
Cube	(length of side) cubed	$6 \times (length\ of\ side)^2$
Cylinder	$3.142 \times radius^2 \times$ height	2 x 3.142 x radius x (height - radius)
Sphere	$4/3 \times 3.142 \times radius^3$	$4 \times 3.142 \times radius^2$
Segment of a sphere	$[(3.142 \times h)/6] \times$ $(3 \times r^2 + h^2)$	$(2 \times 3.142 \times r \times h)$
Pyramid	1/3 x (area of base x height)	0.5 x circumference of base x slant height

FRACTIONS, DECIMALS AND MILLIMETRE EQUIVALENTS

Fractions	Decimals	mm	Fractions	Decimals	mm
1/64	0.015625	0.396875	33/64	0.515625	13.096875
1/32	0.03125	0.79375	17/32	0.53125	13.49375
3/64	0.046875	1.190625	35/64	0.546875	13.890625
1/16	0.0625	1.5875	9/16	0.5625	14.2875
5/64	0.078125	1.984375	37/64	0.578125	14.684375
3/32	0.09375	2.38125	19/32	0.59375	15.08125
7/64	0.109375	2.778125	39/64	0.609375	15.478125
1/8	0.125	3.175	5/8	0.625	15.875
9/64	0.140625	3.571875	41/64	0.640625	16.271875
5/32	0.15625	3.96875	21/32	0.65625	16.66875
11/64	0.171875	4.365625	43/64	0.671875	17.065625
3/16	0.1875	4.7625	11/16	0.6875	17.4625
13/64	0.203125	5.159375	45/64	0.703125	17.859375
7/32	0.21875	5.55625	23/32	0.71875	18.25625
15/64	0.234375	5.953125	47/64	0.734375	18.653125
1/4	0.25	6.35	3/4	0.75	19.05
17/64	0.265625	6.746875	49/64	0.765625	19.446875
9/32	0.28125	7.14375	25/32	0.78125	19.84375
19/64	0.296875	7.540625	51/64	0.796875	20.240625
5/16	0.3125	7.9375	13/16	0.8125	20.6375
21/64	0.328125	8.334375	53/64	0.828125	21.034375
11/32	0.34375	8.73125	27/32	0.84375	21.43125
23/64	0.359375	9.128125	55/64	0.859375	21.828125
3/8	0.375	9.525	7/8	0.875	22.225
25/64	0.390625	9.921875	57/64	0.890625	22.621875
13/32	0.40625	10.31875	29/32	0.90625	23.01875
27/64	0.421875	10.71563	59/64	0.921875	23.415625
7/16	0.4375	11.1125	15/16	0.9375	23.8125
29/64	0.453125	11.50938	61/64	0.953125	24.209375
15/32	0.46875	11.90625	31/32	0.96875	24.60625
31/64	0.484375	12.30313	63/64	0.984375	25.003125
1/2	0.5	12.7	1.0	1	25.4

IMPERIAL STANDARD WIRE GAUGE (SWG)

SWG No	Diameter inches	mm	SWG No	Diameter inches	mm
7/0	0.5	12.7	23	0.024	0.61
6/0	0.464	11.79	24	0.022	0.559
5/0	0.432	10.97	25	0.02	0.508
4/0	0.4	10.16	26	0.018	0.457
3/0	0.372	9.45	27	0.0164	0.417
2/0	0.348	8.84	28	0.0148	0.376
1/0	0.324	8.23	29	0.0136	0.345
1	0.3	7.62	30	0.0124	0.315
2	0.276	7.01	31	0.0116	0.295
3	0.252	6.4	32	0.0108	0.274
4	0.232	5.89	33	0.01	0.254
5	0.212	5.38	34	0.009	0.234
6	0.192	4.88	35	0.008	0.213
7	0.176	4.47	36	0.008	0.193
8	0.16	4.06	37	0.007	0.173
9	0.144	3.66	38	0.006	0.152
10	0.128	3.25	39	0.005	0.132
11	0.116	2.95	40	0.005	0.122
12	0.104	2.64	41	0.004	0.112
13	0.092	2.34	42	0.004	0.102
14	0.08	2.03	43	0.004	0.091
15	0.072	1.83	44	0.003	0.081
16	0.064	1.63	45	0.003	0.071
17	0.056	1.42	46	0.002	0.061
18	0.048	1.22	47	0.002	0.051
19	0.04	1.016	48	0.002	0.041
20	0.036	0.914	49	0.001	0.031
21	0.032	0.813	50	0.001	0.025
22	0.028	0.711			

WATER PRESSURE DUE TO HEIGHT

Imperial

Head Feet	Pressure lb/in^2		Head Feet	Pressure lb/in^2
1	0.43		70	30.35
5	2.17		75	32.51
10	4.34		80	34.68
15	6.5		85	36.85
20	8.67		90	39.02
25	10.84		95	41.18
30	13.01		100	43.35
35	15.17		105	45.52
40	17.34		110	47.69
45	19.51		120	52.02
50	21.68		130	56.36
55	23.84		140	60.69
60	26.01		150	65.03
65	28.18			

Metric

Head m	Pressure bar		Head m	Pressure bar
0.5	0.049		18.0	1.766
1.0	0.098		19.0	1.864
1.5	0.147		20.0	1.962
2.0	0.196		21.0	2.06
3.0	0.294		22.0	2.158
4.0	0.392		23.0	2.256
5.0	0.491		24.0	2.354
6.0	0.589		25.0	2.453
7.0	0.687		26.0	2.551
8.0	0.785		27.0	2.649
9.0	0.883		28.0	2.747
10.0	0.981		29.0	2.845
11.0	1.079		30.0	2.943
12.0	1.177		32.5	3.188
13.0	1.275		35.0	3.434
14.0	1.373		37.5	3.679
15.0	1.472		40.0	3.924
16.0	1.57		42.5	4.169
17.0	1.668		45.0	4.415

1 bar	=	14.5038 lbf/in^2
1 lbf/in^2	=	0.06895 bar
1 metre	=	3.2808 ft or 39.3701 in
1 foot	=	0.3048 metres
1 in wg	=	2.5 mbar (249.1 N/m^2)

TABLE OF WEIGHTS FOR STEELWORK

Mild Steel Bar

Diameter (mm)	Weight (kg/m)	Diameter (mm)	Weight (kg/m)
6	0.22	20	2.47
10	0.62	25	3.85
12	0.89	30	5.55
16	1.58	32	6.31

Mild Steel Flat

Size (mm)	Weight (kg/m)	Size (mm)	Weight (kg/m)
15 x 3	0.36	15 x 5	0.59
20 x 3	0.47	20 x 5	0.79
25 x 3	0.59	25 x 5	0.98
30 x 3	0.71	30 x 5	1.18
40 x 3	0.94	40 x 5	1.57
45 x 3	1.06	45 x 5	1.77
50 x 3	1.18	50 x 5	1.96
20 x 6	0.94	20 x 8	1.26
25 x 6	1.18	25 x 8	1.57
30 x 6	1.41	30 x 8	1.88
40 x 6	1.88	40 x 8	2.51
45 x 6	2.12	45 x 8	2.83
50 x 6	2.36	50 x 8	3.14
55 x 6	2.60	55 x 8	3.45
60 x 6	2.83	60 x 8	3.77
65 x 6	3.06	65 x 8	4.08
70 x 6	3.30	70 x 8	4.40
75 x 6	3.53	75 x 8	4.71
100 x 6	4.71	100 x 8	6.28
20 x 10	1.57	20 x 12	1.88
25 x 10	1.96	25 x 12	2.36
30 x 10	2.36	30 x 12	2.83
40 x 10	3.14	40 x 12	3.77
45 x 10	3.53	45 x 12	4.24
50 x 10	3.93	50 x 12	4.71
55 x 10	4.32	55 x 12	5.12
60 x 10	4.71	60 x 12	5.65
65 x 10	5.10	65 x 12	6.12
70 x 10	5.50	70 x 12	6.59
75 x 10	5.89	75 x 12	7.07
100 x 10	7.85	100 x 12	9.42

TABLE OF WEIGHTS FOR STEELWORK

Mild Steel Equal Angle

Size (mm)	Weight (kg/m)	Size (mm)	Weight (kg/m)
13 x 13 x 3	0.56	60 x 60 x 10	8.69
20 x 20 x 3	0.88	70 x 70 x 10	10.30
25 x 25 x 3	1.11	75 x 75 x 10	11.05
30 x 30 x 3	1.36	80 x 80 x 10	11.90
40 x 40 x 3	1.82	90 x 90 x 10	13.40
45 x 45 x 3	2.06	100 x 100 x 10	15.00
50 x 50 x 3	2.30	120 x 120 x 10	18.20
		150 x 156 x 10	23.00
30 x 30 x 6	2.56	75 x 75 x 12	13.07
40 x 40 x 6	3.52	80 x 80 x 12	14.00
45 x 45 x 6	4.00	90 x 90 x 12	15.90
50 x 50 x 6	4.47	100 x 120 x 12	21.60
60 x 60 x 6	5.42	120 x 120 x 12	21.90
70 x 70 x 6	6.38	150 x 150 x 12	27.30
75 x 75 x 6	6.82	200 x 200 x 12	36.74
80 x 80 x 6	7.34		
90 x 90 x 6	8.30		
40 x 40 x 8	4.55		
50 x 50 x 8	5.82		
60 x 60 x 8	7.09		
70 x 70 x 8	8.36		
75 x 75 x 8	8.96		
80 x 80 x 8	9.63		
90 x 90 x 8	10.90		
100 x 100 x 8	12.20		
120 x 120 x 8	14.70		

Mild Steel Unequal Angle

Size (mm)	Weight (kg/m)	Size (mm)	Weight (kg/m)
40 x 25 x 6	2.79	100 x 65 x 10	12.30
50 x 40 x 6	4.24	100 x 75 x 10	13.00
60 x 30 x 6	3.99	125 x 75 x 10	15.00
65 x 50 x 6	5.16	150 x 75 x 10	17.00
75 x 50 x 6	5.65	150 x 90 x 10	18.20
80 x 60 x 6	6.37	200 x 100 x 10	23.00
125 x 75 x 6	9.18		
75 x 50 x 8	7.39	100 x 75 x 12	15.40
80 x 60 x 8	8.34	125 x 75 x 12	17.80
100 x 65 x 8	9.94	150 x 75 x 12	20.20
100 x 75 x 8	10.60	150 x 90 x 12	21.60
125 x 75 x 8	12.20	200 x 100 x 12	27.30
137 x 102 x 8	14.88	200 x 150 x 12	32.00

TABLE OF WEIGHTS FOR STEELWORK

Rolled Steel Channels

Size (mm)	Weight (kg/m)	Size (mm)	Weight (kg/m)
32 x 27	2.80	178 x 76	20.84
38 x 19	2.49	178 x 79	26.81
51 x 25	4.46	203 x 76	23.82
51 x 38	5.80	203 x 89	29.78
64 x 25	6.70	229 x 76	26.06
76 x 38	7.46	229 x 89	32.76
76 x 51	9.45	254 x 76	28.29
102 x 51	10.42	254 x 89	35.74
127 x 64	14.90	305 x 89	41.67
152 x 76	17.88	305 x 102	46.18
152 x 89	23.84	381 x 102	55.10

Rolled Steel Joists

Size (mm)	Weight (kg/m)	Size (mm)	Weight (kg/m)
76 x 38	6.25	152 x 76	17.86
76 x 76	12.65	152 x 89	17.09
102 x 44	7.44	152 x 127	37.20
102 x 64	9.65	178 x 102	21.54
102 x 102	23.06	203 x 102	25.33
127 x 76	13.36	203 x 152	52.03
127 x 114	26.78	254 x 114	37.20
127 x 114	29.76	254 x 203	81.84
		305 x 203	96.72

Universal Columns

Size (mm)	Weight (kg/m)	Size (mm)	Weight (kg/m)
152 x 152	23.00	254 x 254	89.00
152 x 152	30.00	254 x 254	107.00
152 x 152	37.00	254 x 254	132.00
203 x 203	46.00	254 x 254	167.00
203 x 203	52.00	305 x 305	97.00
203 x 203	60.00	305 x 305	118.00
203 x 203	71.00	305 x 305	137.00
203 x 203	86.00	305 x 305	158.00
254 x 254	73.00	305 x 305	198.00

TABLE OF WEIGHTS FOR STEELWORK

Universal Beams

Size (mm)	Weight (kg/m)	Size (mm)	Weight (kg/m)
203 x 133	25.00	305 x 127	48.00
203 x 133	30.00	305 x 165	40.00
254 x 102	22.00	305 x 165	46.00
254 x 102	25.00	305 x 165	54.00
254 x 102	28.00	356 x 127	33.00
254 x 146	31.00	356 x 127	39.00
254 x 146	37.00	356 x 171	45.00
254 x 146	43.00	356 x 171	51.00
305 x 102	25.00	356 x 171	57.00
305 x 102	28.00	356 x 171	67.00
305 x 102	33.00	381 x 152	52.00
305 x 127	37.00	381 x 152	60.00
305 x 127	42.00	381 x 152	67.00

Circular Hollow Sections

Size (mm)	Weight (kg/m)	Size (mm)	Weight (kg/m)
21.3 x 3.2	1.43	76.1 x 3.2	5.75
26.9 x 3.2	1.87	76.1 x 4.0	7.11
33.7 x 2.6	1.99	76.1 x 5.0	8.77
33.7 x 3.2	2.41	88.9 x 3.2	6.76
33.7 x 4.0	2.93	88.9 x 4.0	8.36
42.4 x 2.6	2.55	88.9 x 5.0	10.30
42.4 x 3.2	3.09	114.3 x 3.6	9.83
42.4 x 4.0	3.79	114.3 x 5.0	13.50
48.3 x 3.2	3.56	114.3 x 6.3	16.80
48.3 x 4.0	4.37	139.7 x 5.0	16.60
48.3 x 5.0	5.34	139.7 x 6.3	20.70
60.3 x 3.2	4.51	139.7 x 8.0	26.00
60.3 x 4.0	5.55	139.7 x 10.0	32.00
60.3 x 5.0	6.82	168.3 x 5.0	20.10

TABLE OF WEIGHTS FOR STEELWORK

Square Hollow Sections

Size (mm)	Weight (kg/m)	Size (mm)	Weight (kg/m)
20 x 20 x 2.0	1.12	90 x 90 x 3.6	9.72
20 x 20 x 2.6	1.39	90 x 90 x 5.0	13.30
30 x 30 x 2.6	2.21	90 x 90 x 6.3	16.40
30 x 30 x 3.2	2.65	100 x 100 x 4.0	12.00
40 x 40 x 2.6	3.03	100 x 100 x 5.0	14.80
40 x 40 x 3.2	3.66	100 x 100 x 6.3	18.40
40 x 40 x 4.0	4.46	100 x 100 x 8.0	22.90
50 x 50 x 3.2	4.66	100 x 100 x 10.0	27.90
50 x 50 x 4.0	5.72	120 x 120 x 5.0	18.00
50 x 50 x 5.0	6.97	120 x 120 x 6.3	22.30
60 x 60 x 3.2	5.67	120 x 120 x 8.0	27.90
60 x 60 x 4.0	6.97	120 x 120 x 10.0	34.20
60 x 60 x 5.0	8.54	150 x 150 x 5.0	22.70
70 x 70 x 3.2	7.46	150 x 150 x 6.3	28.30
70 x 70 x 5.0	10.10	150 x 150 x 8.0	35.40
80 x 80 x 3.6	8.59	150 x 150 x 10.0	43.60
80 x 80 x 5.0	11.70		
80 x 80 x 6.3	14.40		

Rectangular Hollow Sections

Size (mm)	Weight (kg/m)	Size (mm)	Weight (kg/m)
50 x 30 x 2.6	3.03	120 x 80 x 5.0	14.80
50 x 30 x 3.2	3.66	120 x 80 x 6.3	18.40
60 x 40 x 3.2	4.66	120 x 80 x 8.0	22.90
60 x 40 x 4.0	5.72	120 x 80 x 10.0	27.90
80 x 40 x 3.2	5.67	150 x 100 x 5.0	18.70
80 x 40 x 4.0	6.97	150 x 100 x 6.3	23.30
90 x 50 x 3.6	7.46	150 x 100 x 8.0	29.10
90 x 50 x 5.0	10.10	150 x 100 x 10.0	35.70
100 x 50 x 3.2	7.18	160 x 80 x 5.0	18.00
100 x 50 x 4.0	8.86	160 x 80 x 6.3	22.30
100 x 50 x 5.0	10.90	160 x 80 x 8.0	27.90
100 x 60 x 3.6	8.59	160 x 80 x 10.0	34.20
100 x 60 x 5.0	11.70	200 x 100 x 5.0	22.70
100 x 60 x 6.3	14.40	200 x 100 x 6.3	28.30
120 x 60 x 3.6	9.72	200 x 100 x 8.0	35.40
120 x 60 x 5.0	13.30	200 x 100 x 10.0	43.60
120 x 60 x 6.3	16.40		

DIMENSIONS AND WEIGHTS OF COPPER PIPES TO
BSEN 1057, BSEN 12499, BSEN 14251

Outside Diameter (mm)	Internal Diameter (mm)	Weight per Metre (kg)	Internal Diameter (mm)	Weight per Metre (kg)	Internal Diameter (mm)	Weight per Metre (kg)
	Formerly Table X		Formerly Table Y		Formerly Table Z	
6	4.80	0.0911	4.40	0.1170	5.00	0.0774
8	6.80	0.1246	6.40	0.1617	7.00	0.1054
10	8.80	0.1580	8.40	0.2064	9.00	0.1334
12	10.80	0.1914	10.40	0.2511	11.00	0.1612
15	13.60	0.2796	13.00	0.3923	14.00	0.2031
18	16.40	0.3852	16.00	0.4760	16.80	0.2918
22	20.22	0.5308	19.62	0.6974	20.82	0.3589
28	26.22	0.6814	25.62	0.8985	26.82	0.4594
35	32.63	1.1334	32.03	1.4085	33.63	0.6701
42	39.63	1.3675	39.03	1.6996	40.43	0.9216
54	51.63	1.7691	50.03	2.9052	52.23	1.3343
76.1	73.22	3.1287	72.22	4.1437	73.82	2.5131
108	105.12	4.4666	103.12	7.3745	105.72	3.5834
133	130.38	5.5151	--	--	130.38	5.5151
159	155.38	8.7795	--	--	156.38	6.6056

DIMENSIONS OF STAINLESS STEEL PIPES TO BS 4127

Outside Diameter (mm)	Maximum Outside Diameter (mm)	Minimum Outside Diameter (mm)	Wall Thickness (mm)	Working Pressure (bar)
6	6.045	5.940	0.6	330
8	8.045	7.940	0.6	260
10	10.045	9.940	0.6	210
12	12.045	11.940	0.6	170
15	15.045	14.940	0.6	140
18	18.045	17.940	0.7	135
22	22.055	21.950	0.7	110
28	28.055	27.950	0.8	121
35	35.070	34.965	1.0	100
42	42.070	41.965	1.1	91
54	54.090	53.940	1.2	77

DIMENSIONS OF STEEL PIPES TO BS 1387

Nominal Size	Approx. Outside Diameter	Outside Diameter				Thickness		
		Light		Medium & Heavy		Light	Medium	Heavy
		Max.	Min.	Max.	Min.			
mm	mm	mm	mm	mm	mm	mm	mm	mm
6	10.20	10.10	9.70	10.40	9.80	1.80	2.00	2.65
8	13.50	13.60	13.20	13.90	13.30	1.80	2.35	2.90
10	17.20	17.10	16.70	17.40	16.80	1.80	2.35	2.90
15	21.30	21.40	21.00	21.70	21.10	2.00	2.65	3.25
20	26.90	26.90	26.40	27.20	26.60	2.35	2.65	3.25
25	33.70	33.80	33.20	34.20	33.40	2.65	3.25	4.05
32	42.40	42.50	41.90	42.90	42.10	2.65	3.25	4.05
40	48.30	48.40	47.80	48.80	48.00	2.90	3.25	4.05
50	60.30	60.20	59.60	60.80	59.80	2.90	3.65	4.50
65	76.10	76.00	75.20	76.60	75.40	3.25	3.65	4.50
80	88.90	88.70	87.90	89.50	88.10	3.25	4.05	4.85
100	114.30	113.90	113.00	114.90	113.30	3.65	4.50	5.40
125	139.70	--	--	140.60	138.70	--	4.85	5.40
150	165.1*	--	--	166.10	164.10	--	4.85	5.40

* 165.1mm (6.5in) outside diameter is not generally recommended except where screwing to BS 21 is necessary.

All dimensions are in accordance with ISO R65 except approximate outside diameters which are in accordance with ISO R64.

Light quality is equivalent to ISO R65 Light Series II.

APPROXIMATE METRES PER TONNE OF TUBES TO BS 1387

Nom.	BLACK						GALVANISED					
	Plain/screwed ends			Screwed & socketed			Plain/screwed ends			Screwed & socketed		
Size	L	M	H	L	M	H	L	M	H	L	M	H
mm	m	m	m	m	m	m	m	m	m	m	m	m
6	2765	2461	2030	2743	2443	2018	2604	2333	1948	2584	2317	1937
8	1936	1538	1300	1920	1527	1292	1826	1467	1254	1811	1458	1247
10	1483	1173	979	1471	1165	974	1400	1120	944	1386	1113	939
15	1050	817	688	1040	811	684	996	785	665	987	779	661
20	712	634	529	704	628	525	679	609	512	673	603	508
25	498	410	336	494	407	334	478	396	327	474	394	325
32	388	319	260	384	316	259	373	308	254	369	305	252
40	307	277	226	303	273	223	296	268	220	292	264	217
50	244	196	162	239	194	160	235	191	158	231	188	157
65	172	153	127	169	151	125	167	149	124	163	146	122
80	147	118	99	143	116	98	142	115	97	139	113	96
100	101	82	69	98	81	68	98	81	68	95	79	67
125	--	62	56	--	60	55	--	60	55	--	59	54
150	--	52	47	--	50	46	--	51	46	--	49	45

The figures for `plain or screwed ends' apply also to tubes to BS 1775 of equivalent size and thickness.

Key

L – Light
M – Medium
H – Heavy

FLANGE DIMENSION CHART TO BS 4504 & BS 10

Normal Pressure Rating (PN 6) 6 Bar

Nom. Size	Flange Outside Diam.	Table 6/2 Forged Welding Neck	Table 6/3 Plate Slip on	Table 6/4 Forged Bossed Screwed	Table 6/5 Forged Bossed Slip on	Table 6/8 Plate Blank	Raised Face		Nr. Bolt Hole	Size of Bolt
							Diam.	T'ness		
15	80	12	12	12	12	12	40	2	4	M10 x 40
20	90	14	14	14	14	14	50	2	4	M10 x 45
25	100	14	14	14	14	14	60	2	4	M10 x 45
32	120	14	16	14	14	14	70	2	4	M12 x 45
40	130	14	16	14	14	14	80	3	4	M12 x 45
50	140	14	16	14	14	14	90	3	4	M12 x 45
65	160	14	16	14	14	14	110	3	4	M12 x 45
80	190	16	18	16	16	16	128	3	4	M16 x 55
100	210	16	18	16	16	16	148	3	4	M16 x 55
125	240	18	20	18	18	18	178	3	8	M16 x 60
150	265	18	20	18	18	18	202	3	8	M16 x 60
200	320	20	22	--	20	20	258	3	8	M16 x 60
250	375	22	24	--	22	22	312	3	12	M16 x 65
300	440	22	24	--	22	22	365	4	12	M20 x 70

FLANGE DIMENSION CHART TO BS 4504 & BS 10

Normal Pressure Rating (PN 16) 16 Bar

Nom. Size	Flange Outside Diam.	Table 6/2 Forged Welding Neck	Table 6/3 Plate Slip on	Table 6/4 Forged Bossed Screwed	Table 6/5 Forged Bossed Slip on	Table 6/8 Plate Blank	Raised Face Diam.	Raised Face T'ness	Nr. Bolt Hole	Size of Bolt
15	95	14	14	14	14	14	45	2	4	M12 x 45
20	105	16	16	16	16	16	58	2	4	M12 x 50
25	115	16	16	16	16	16	68	2	4	M12 x 50
32	140	16	16	16	16	16	78	2	4	M16 x 55
40	150	16	16	16	16	16	88	3	4	M16 x 55
50	165	18	18	18	18	18	102	3	4	M16 x 60
65	185	18	18	18	18	18	122	3	4	M16 x 60
80	200	20	20	20	20	20	138	3	8	M16 x 60
100	220	20	20	20	20	20	158	3	8	M16 x 65
125	250	22	22	22	22	22	188	3	8	M16 x 70
150	285	22	22	22	22	22	212	3	8	M20 x 70
200	340	24	24	--	24	24	268	3	12	M20 x 75
250	405	26	26	--	26	26	320	3	12	M24 x 90
300	460	28	28	--	28	28	378	4	12	M24 x 90

MINIMUM DISTANCES BETWEEN SUPPORTS/FIXINGS

Material	BS Nominal Pipe Size		Pipes - Vertical	Pipes - Horizontal on to low gradients
	inch	mm	support distance in metres	support distance in metres
Copper	0.50	15.00	1.90	1.30
	0.75	22.00	2.50	1.90
	1.00	28.00	2.50	1.90
	1.25	35.00	2.80	2.50
	1.50	42.00	2.80	2.50
	2.00	54.00	3.90	2.50
	2.50	67.00	3.90	2.80
	3.00	76.10	3.90	2.80
	4.00	108.00	3.90	2.80
	5.00	133.00	3.90	2.80
	6.00	159.00	3.90	2.80
muPVC	1.25	32.00	1.20	0.50
	1.50	40.00	1.20	0.50
	2.00	50.00	1.20	0.60
Polypropylene	1.25	32.00	1.20	0.50
	1.50	40.00	1.20	0.50
uPVC	--	82.40	1.20	0.50
	--	110.00	1.80	0.90
	--	160.00	1.80	1.20
Steel	0.50	15.00	2.40	1.80
	0.75	20.00	3.00	2.40
	1.00	25.00	3.00	2.40
	1.25	32.00	3.00	2.40
	1.50	40.00	3.70	2.40
	2.00	50.00	3.70	2.40
	2.50	65.00	4.60	3.00
	3.00	80.40	4.60	3.00
	4.00	100.00	4.60	3.00
	5.00	125.00	5.50	3.70
	6.00	150.00	5.50	4.50
	8.00	200.00	8.50	6.00
	10.00	250.00	9.00	6.50
	12.00	300.00	10.00	7.00
	16.00	400.00	10.00	8.25

LITRES OF WATER STORAGE REQUIRED PER PERSON PER BUILDING TYPE

Type of Building	Storage lltres
Houses and flats (up to 4 bedrooms	120/bedroom
Houses and flats (more than 4 bedrooms)	100/bedroom
Hostels	90/bed
Hotels	200/bed
Nurses homes and medical quarters	120/bed
Offices with canteen	45/person
Offices without canteen	40/person
Restaurants	7/meal
Boarding schools	90/person
Day schools - Primary	15/person
Day schools - Secondary	20/person

RECOMMENDED AIR CONDITIONING DESIGN LOADS

Building Type	Design Loading
Computer rooms	500 W/m² of floor area
Restaurants	150 W/m² of floor area
Banks (main area)	100 W/m² of floor area
Supermarkets	25 W/m² of floor area
Large Office Block (exterior zone)	100 W/m² of floor area
Large Office Block (interior zone)	80 W/m² of floor area
Small Office Block (interior zone)	80 W/m² of floor area

CAPACITY AND DIMENSIONS OF GALVANISED MILD STEEL CISTERNS – BS 417

Capacity	BS type	Dimensions		
(litres)	(SCM)	Length (mm)	Width (mm)	Depth (mm)
18	45	457	305	305
36	70	610	305	371
54	90	610	406	371
68	110	610	432	432
86	135	610	457	482
114	180	686	508	508
159	230	736	559	559
191	270	762	584	610
227	320	914	610	584
264	360	914	660	610
327	450/1	1220	610	610
336	450/2	965	686	686
423	570	965	762	787
491	680	1090	864	736
709	910	1070	889	889

CAPACITY OF COLD WATER POLYPROPYLENE STORAGE CISTERNS - BS 4213

Capacity	BS type	Maximum Height
(litres)	(PC)	mm
18	4	310
36	8	380
68	15	430
91	20	510
114	25	530
182	40	610
227	50	660
273	60	660
318	70	660
455	100	760

MINIMUM INSULATION THICKNESS TO PROTECT AGAINST FREEZING FOR DOMESTIC COLD WATER SYSTEMS (8 Hour Evaluation Period)

Pipe size (mm)	Insulation thickness (mm)					
	Condition 1			Condition 2		
	$\lambda = 0.020$	$\lambda = 0.030$	$\lambda = 0.040$	$\lambda = 0.020$	$\lambda = 0.030$	$\lambda = 0.040$
Copper pipes						
15	11	20	34	12	23	41
22	6	9	13	6	10	15
28	4	6	9	4	7	10
35	3	5	7	4	5	7
42	3	4	5	8	4	6
54	2	3	4	2	3	4
76	2	2	3	2	2	3
Steel pipes						
15	9	15	24	10	18	29
20	6	9	13	6	10	15
25	4	7	9	5	7	10
32	3	5	6	3	5	7
40	3	4	5	3	4	6
50	2	3	4	2	3	4
65	2	2	3	2	3	3

Condition 1 : water temperature 7°C; ambient temperature –6°C; evaluation period 8 h; permitted ice formation 50%; normal installation i.e. inside the building and inside the envelope of the structural insulation

Condition 2 : water temperature 2°C; ambient temperature –6°C; evaluation period 8 h; permitted ice formation 50%; extreme installation, i.e. inside the building but outside the envelope of the structural insulation

λ = thermal conductivity [W/(mK)]

INSULATION THICKNESS FOR CHILLED AND COLD WATER SUPPLIES TO PREVENT CONDENSATION On A Low Emissivity Outer Surface (0.05, i.e. Bright Reinforced Aluminium Foil) With An Ambient Temperature Of +25°C And A Relative Humidity Of 80%

Steel pipe size (mm)	$t = +10$			$t = +5$			$t = 0$		
	Insulation thickness (mm)			Insulation thickness (mm)			Insulation thickness (mm)		
	$\lambda = 0.030$	$\lambda = 0.040$	$\lambda = 0.050$	$\lambda = 0.030$	$\lambda = 0.040$	$\lambda = 0.050$	$\lambda = 0.030$	$\lambda = 0.040$	$\lambda = 0.050$
15	16	20	25	22	28	34	28	36	43
25	18	24	29	25	32	39	32	41	50
50	22	28	34	30	39	47	38	49	60
100	26	34	41	36	47	57	46	60	73
150	29	38	46	40	52	64	51	67	82
250	33	43	53	46	60	74	59	77	94
Flat surfaces	39	52	65	56	75	93	73	97	122

t = temperature of contents (°C)
λ = thermal conductivity at mean temperature of insulation [W/(mK)]

INSULATION THICKNESS FOR NON-DOMESTIC HEATING INSTALLATIONS TO CONTROL HEAT LOSS

Steel pipe size (mm)	$t = 75$ Insulation thickness (mm)			$t = 100$ Insulation thickness (mm)			$t = 150$ Insulation thickness (mm)		
	$\lambda = 0.030$	$\lambda = 0.040$	$\lambda = 0.050$	$\lambda = 0.030$	$\lambda = 0.040$	$\lambda = 0.050$	$\lambda = 0.030$	$\lambda = 0.040$	$\lambda = 0.050$
10	18	32	55	20	36	62	23	44	77
15	19	34	56	21	38	64	26	47	80
20	21	36	57	23	40	65	28	50	83
25	23	38	58	26	43	68	31	53	85
32	24	39	59	28	45	69	33	55	87
40	25	40	60	29	47	70	35	57	88
50	27	42	61	31	49	72	37	59	90
65	29	43	62	33	51	74	40	63	92
80	30	44	62	35	52	75	42	65	94
100	31	46	63	37	54	76	45	68	96
150	33	48	64	40	57	77	50	73	100
200	35	49	65	42	59	79	53	76	103
250	36	50	66	43	61	80	55	78	105

t = hot face temperature (°C)
λ = thermal conductivity at mean temperature of insulation [W/(mK)]

Outdoor Lighting Guide

Institution of Lighting Engineers

An all-inclusive guide to exterior lighting from The Institution of Lighting Engineers, recognised as the pre-eminent professional source in the UK for authoritative guidance on exterior lighting. As concern grows over environmental issues and light pollution, this book fills a need for a straightforward and accessible guide to the use, design and installation of outdoor lighting.

This book provides a comprehensive source of information and advice on all forms of exterior lighting, from floodlighting, buildings and road lighting to elaborate Christmas decorations, and will be useful to practitioners and non-experts alike. Specialists will value the dependable detail on standards and related design, installation and maintenance problems, and any user can find extensive practical guidance on safety issues, the lighting of hazardous areas, and avoiding potential difficulties.

August 2005: 234x156 mm: 320 pages
HB: 0-415-37007-8: £80.00

To Order: Tel: +44 (0) 1264 343071 Fax: +44 (0) 1264 343005, or
Post: Taylor and Francis Customer Services, Thomson Publishing Services, Cheriton House, Andover, Hants, SP10 5BE, UK Email: book.orders@tandf.co.uk

For a complete listing of all our titles visit:
www.tandf.co.uk

Index

CD-Rom Single-User Licence Agreement

We welcome you as a user of this Taylor & Francis CD-ROM and hope that you find it a useful and valuable tool. Please read this document carefully. **This is a legal agreement** between you (hereinafter referred to as the "Licensee") and Taylor and Francis Books Ltd. (the "Publisher"), which defines the terms under which you may use the Product. **By breaking the seal and opening the package containing the CD-ROM you agree to these terms and conditions outlined herein. If you do not agree to these terms you must return the Product to your supplier intact, with the seal on the CD case unbroken.**

1. Definition of the Product

The product which is the subject of this Agreement, *Spon's Mechanical and Electrical Services Price Book 2007 on CD-ROM* (the "Product") consists of:

1.1 Underlying data comprised in the product (the "Data")

1.2 A compilation of the Data (the "Database")

1.3 Software (the "Software") for accessing and using the Database

1.4 A CD-ROM disk (the "CD-ROM")

2. Commencement and licence

2.1 This Agreement commences upon the breaking open of the package containing the CD-ROM by the Licensee (the "Commencement Date").

2.2 This is a licence agreement (the "Agreement") for the use of the Product by the Licensee, and not an agreement for sale.

2.3 The Publisher licenses the Licensee on a non-exclusive and non-transferable basis to use the Product on condition that the Licensee complies with this Agreement. The Licensee acknowledges that it is only permitted to use the Product in accordance with this Agreement.

3. Multiple use

For more than one user or for a wide area network or consortium, use is only permissible with the purchase from the Publisher of a multiple-user licence and adherence to the terms and conditions of that licence.

4. Installation and Use

4.1 The Licensee may provide access to the Product for individual study in the following manner: The Licensee may install the Product on a secure local area network on a single site for use by one user.

4.2 The Licensee shall be responsible for installing the Product and for the effectiveness of such installation.

4.3 Text from the Product may be incorporated in a coursepack. Such use is only permissible with the express permission of the Publisher in writing and requires the payment of the appropriate fee as specified by the Publisher and signature of a separate licence agreement.

4.4 The CD-ROM is a free addition to the book and no technical support will be provided.

5. Permitted Activities

5.1 The Licensee shall be entitled:

5.1.1 to use the Product for its own internal purposes;

5.1.2 to download onto electronic, magnetic, optical or similar storage medium reasonable portions of the Database provided that the purpose of the Licensee is to undertake internal research or study and provided that such storage is temporary;

5.1.3 to make a copy of the Database and/or the Software for back-up/archival/disaster recovery purposes.

5.2 The Licensee acknowledges that its rights to use the Product are strictly set out in this Agreement, and all other uses (whether expressly mentioned in Clause 6 below or not) are prohibited.

6. Prohibited Activities

The following are prohibited without the express permission of the Publisher:

6.1 The commercial exploitation of any part of the Product.

6.2 The rental, loan, (free or for money or money's worth) or hire purchase of this product, save with the express consent of the Publisher.

6.3 Any activity which raises the reasonable prospect of impeding the Publisher's ability or opportunities to market the Product.

6.4 Any networking, physical or electronic distribution or dissemination of the product save as expressly permitted by this Agreement.

6.5 Any reverse engineering, decompilation, disassembly or other alteration of the Product save in accordance with applicable national laws.

6.6 The right to create any derivative product or service from the Product save as expressly provided for in this Agreement.

6.7 Any alteration, amendment, modification or deletion from the Product, whether for the purposes of error correction or otherwise.

7. General Responsibilities of the License

7.1 The Licensee will take all reasonable steps to ensure that the Product is used in accordance with the terms and conditions of this Agreement.

7.2 The Licensee acknowledges that damages may not be a sufficient remedy for the Publisher in the event of breach of this Agreement by the Licensee, and that an injunction may be appropriate.

7.3 The Licensee undertakes to keep the Product safe and to use its best endeavours to ensure that the product does not fall into the hands of third parties, whether as a result of theft or otherwise.

7.4 Where information of a confidential nature relating to the product of the business affairs of the Publisher comes into the possession of the Licensee pursuant to this Agreement (or otherwise), the Licensee agrees to use such information solely for the purposes of this Agreement, and under no circumstances to disclose any element of the information to any third party save strictly as permitted under this Agreement. For the avoidance of doubt, the Licensee's obligations under this sub-clause 7.4 shall survive the termination of this Agreement.

8. Warrant and Liability

8.1 The Publisher warrants that it has the authority to enter into this agreement and that it has secured all rights and permissions necessary to enable the Licensee to use the Product in accordance with this Agreement.

8.2 The Publisher warrants that the CD-ROM as supplied on the Commencement Date shall be free of defects in materials and workmanship, and undertakes to replace any defective CD-ROM within 28 days of notice of such defect being received provided such notice is received within 30 days of such supply. As an alternative to replacement, the Publisher agrees fully to refund the Licensee in such circumstances, if the Licensee so requests, provided that the Licensee returns the Product to the Publisher. The provisions of this sub-clause 8.2 do not apply where the defect results from an accident or from misuse of the product by the Licensee.

8.3 Sub-clause 8.2 sets out the sole and exclusive remedy of the Licensee in relation to defects in the CD-ROM.

8.4 The Publisher and the Licensee acknowledge that the Publisher supplies the Product on an "as is" basis. The Publisher gives no warranties:

    8.4.1 that the Product satisfies the individual requirements of the Licensee; or

    8.4.2 that the Product is otherwise fit for the Licensee's purpose; or

    8.4.3 that the Data are accurate or complete of free of errors or omissions; or

    8.4.4 that the Product is compatible with the Licensee's hardware equipment and software operating environment.

8.5 The Publisher hereby disclaims all warranties and conditions, express or implied, which are not stated above.

8.6 Nothing in this Clause 8 limits the Publisher's liability to the Licensee in the event of death or personal injury resulting from the Publisher's negligence.

8.7 The Publisher hereby excludes liability for loss of revenue, reputation, business, profits, or for indirect or consequential losses, irrespective of whether the Publisher was advised by the Licensee of the potential of such losses.

8.8 The Licensee acknowledges the merit of independently verifying Data prior to taking any decisions of material significance (commercial or otherwise) based on such data. It is agreed that the Publisher shall not be liable for any losses which result from the Licensee placing reliance on the Data or on the Database, under any circumstances.

8.9 Subject to sub-clause 8.6 above, the Publisher's liability under this Agreement shall be limited to the purchase price.

9. Intellectual Property Rights

9.1 Nothing in this Agreement affects the ownership of copyright or other intellectual property rights in the Data, the Database of the Software.

9.2 The Licensee agrees to display the Publishers' copyright notice in the manner described in the Product.

9.3 The Licensee hereby agrees to abide by copyright and similar notice requirements required by the Publisher, details of which are as follows:

"© 2007 Taylor & Francis. All rights reserved. All materials in *Spon's Mechanical and Electrical Services Price Book 2007* are copyright protected. © 2006 Adobe Systems Incorporated. All rights reserved. No such materials may be used, displayed, modified, adapted, distributed, transmitted, transferred, published or otherwise reproduced in any form or by any means now or hereafter developed other than strictly in accordance with the terms of the licence agreement enclosed with the CD-ROM. However, text and images may be printed and copied for research and private study within the preset program limitations. Please note the copyright notice above, and that any text or images printed or copied must credit the source."

9.4 This Product contains material proprietary to and copyedited by the Publisher and others. Except for the licence granted herein, all rights, title and interest in the Product, in all languages, formats and media throughout the world, including copyrights therein, are and remain the property of the Publisher or other copyright holders identified in the Product.

10. Non-assignment

This Agreement and the licence contained within it may not be assigned to any other person or entity without the written consent of the Publisher.

11. Termination and Consequences of Termination.

11.1 The Publisher shall have the right to terminate this Agreement if:

    11.1.1 the Licensee is in material breach of this Agreement and fails to remedy such breach (where capable of remedy) within 14 days of a written notice from the Publisher requiring it to do so; or

    11.1.2 the Licensee becomes insolvent, becomes subject to receivership, liquidation or similar external administration; or

    11.1.3 the Licensee ceases to operate in business.

11.2 The Licensee shall have the right to terminate this Agreement for any reason upon two month's written notice. The Licensee shall not be entitled to any refund for payments made under this Agreement prior to termination under this sub-clause 11.2.

11.3 Termination by either of the parties is without prejudice to any other rights or remedies under the general law to which they may be entitled, or which survive such termination (including rights of the Publisher under sub-clause 7.4 above).

11.4 Upon termination of this Agreement, or expiry of its terms, the Licensee must:

    11.4.1 destroy all back up copies of the product; and

    11.4.2 return the Product to the Publisher.

12. General

12.1 **Compliance with export provisions**

The Publisher hereby agrees to comply fully with all relevant export laws and regulations of the United Kingdom to ensure that the Product is not exported, directly or indirectly, in violation of English law.

12.2 **Force majeure**

The parties accept no responsibility for breaches of this Agreement occurring as a result of circumstances beyond their control.

12.3 **No waiver**

Any failure or delay by either party to exercise or enforce any right conferred by this Agreement shall not be deemed to be a waiver of such right.

12.4 **Entire agreement**

This Agreement represents the entire agreement between the Publisher and the Licensee concerning the Product. The terms of this Agreement supersede all prior purchase orders, written terms and conditions, written or verbal representations, advertising or statements relating in any way to the Product.

12.5 **Severability**

If any provision of this Agreement is found to be invalid or unenforceable by a court of law of competent jurisdiction, such a finding shall not affect the other provisions of this Agreement and all provisions of this Agreement unaffected by such a finding shall remain in full force and effect.

12.6 **Variations**

This agreement may only be varied in writing by means of variation signed in writing by both parties.

12.7 **Notices**

All notices to be delivered to: Spon's Price Books, Taylor & Francis Books Ltd., 2 Park Square, Milton Park, Abingdon, Oxfordshire, OX14 4RN, UK.

12.8 **Governing law**

This Agreement is governed by English law and the parties hereby agree that any dispute arising under this Agreement shall be subject to the jurisdiction of the English courts.

If you have any queries about the terms of this licence, please contact:

Spon's Price Books
Taylor & Francis Books Ltd.
2 Park Square, Milton Park, Abingdon, Oxfordshire, OX14 4RN
Tel: +44 (0) 20 7017 6672
Fax: +44 (0) 20 7017 6702
www.sponpress.com

Taylor & Francis
Taylor & Francis Group

CD-ROM Installation Instructions

System requirements

Minimum

- Pentium processor
- 32 MB of RAM
- 10 MB available hard disk space
- CD-ROM drive
- Microsoft Windows 95/98/2000/NT/ME/XP
- SVGA screen
- Internet connection

Recommended

- Pentium 266 MHz processor
- 256 MB of RAM
- 100 MB available hard disk space
- CD-ROM drive
- Microsoft Windows 2000/NT/XP
- XVGA screen
- Internet connection

Microsoft ® is a registered trademark and Windows™ is a trademark of the Microsoft Corporation.

How to install *Spon's Mechanical and Electrical Services Price Book 2007 CD-ROM*

Windows 95/98/2000/NT

Spon's Mechanical and Electrical Services Price Book 2007 CD-ROM should run automatically when inserted into the CD-ROM drive. If it fails to run, follow the instructions below.

- Click the **Start** button and choose **Run.**
- Click the **Browse** button.
- Select your CD-ROM drive.
- Select the Setup file (setup.exe) then click **Open.**
- Click the OK button.
- Follow the instructions on screen.
- The installation process will create a folder containing an icon for *Spon's Mechanical and Electrical Services Price Book 2007 CD-ROM* and also an icon on your desktop.

How to run the *Spon's Mechanical and Electrical Services Price Book 2007 CD-ROM*

- Double click the icon (from the folder or desktop) installed by the Setup program.
- Follow the instructions on screen.

Installation

The CD-ROM is a free addition to the book and no technical support will be provided. For help with the use of the CD-ROM please visit www.pricebooks.co.uk

Multiple-user use of the Spon Press CD–ROM

To buy a licence to install your Spon Press Price
Book CD–ROM on a secure local area network or a
wide area network, and for the supply of network key
files, for an agreed number of users please contact:

Spon's Price Books
Taylor & Francis Books Ltd.
2 Park Square, Milton Park, Abingdon, Oxfordshire, OX14 4RN
Tel: +44 (0) 207 7017 6672
Fax: +44 (0) 207 7017 6072
www.sponpress.com

Number of users	Licence cost
2–5	£390
6–10	£780
11–20	£1170
21–30	£1750
31–50	£3500
51–75	£5000
76–100	£6000
Over 100	Please contact Spon for details